D1752015

Phytochemical Dictionary

A Handbook of Bioactive Compounds from Plants

Phytochemical Dictionary

A Handbook of Bioactive Compounds from Plants

edited by

Jeffrey B. Harborne, FRS

Professor of Botany, University of Reading

and

Herbert Baxter

formerly of the Laboratory of the Government Chemist, London

with the editorial assistance of

Dr Gerard P. Moss, *Department of Chemistry, Queen Mary and Westfield College, University of London*

and the research assistance of

Dr Renee Grayer, *Department of Botany, University of Reading*

Dr Harry Ruijgrok, *formerly of the Plant Systematics Department, University of Leiden, The Netherlands*

Dr Karin Valant-Vetschera, *Institute of Pharmacognosy, University of Vienna, Austria*

Dr Elizabeth M. Williamson, *formerly of the School of Pharmacy, University of London*

Taylor & Francis
Publishers since 1798

UK	Taylor & Francis Ltd, 4 John St., London WC1N 2ET
USA	Taylor & Francis Inc., 1900 Frost Road, Suite 101, Bristol, PA 19007

© Taylor & Francis Ltd 1993

Reprinted 1993 and 1995

Revised reprint 1995

All rights reserved. No part of this publication may be reproduced, stored in a retrieval system, or transmitted, in any form or by any means, electronic, electrostatic, magnetic tape, mechanical, photocopying, recording or otherwise, without the prior permission of the copyright owner and the publisher.

British Library Cataloguing in Publication Data
A catalogue record for this book is available from the British Library.

ISBN 0-85066-736-4

Library of Congress Cataloging-in-Publication Data are available

Cover design by John Leath.

Phototypesetting by Euroset, 2 Dover Close, Alresford, Hampshire SO24 9PG.

Printed in Great Britain by Burgess Science Press, Basingstoke, on paper which has a specified pH value on final paper manufacture of not less than 7.5 and is therefore 'acid free'.

Contents

General Introduction		vii
Key to Dictionary Entries		ix
PART I	**Carbohydrates and Lipids**	**1**
	General Introduction	2
	Chapter 1. Monosaccharides	3
	Chapter 2. Oligosaccharides	8
	Chapter 3. Polysaccharides	14
	Chapter 4. Sugar alcohols and cyclitols	20
	Chapter 5. Organic acids	24
	Chapter 6. Fatty acids and lipids	34
	Chapter 7. Hydrocarbons and derivatives	41
	Chapter 8. Acetylenes and thiophenes	46
	Chapter 9. Miscellaneous aliphatics	53
PART II	**Nitrogen-containing Compounds (excluding Alkaloids)**	**57**
	General Introduction	59
	Chapter 10. Amino acids	61
	Chapter 11. Amines	75
	Chapter 12. Cyanogenic glycosides	84
	Chapter 13. Glucosinolates	92
	Chapter 14. Purines and pyrimidines	100
	Chapter 15. Proteins and peptides	107
	Chapter 16. Miscellaneous nitrogen compounds	116
PART III	**Alkaloids**	**125**
	General Introduction	127
	Chapter 17. Amaryllidaceae alkaloids	129
	Chapter 18. Betalain alkaloids	136
	Chapter 19. Diterpenoid alkaloids	143
	Chapter 20. Indole alkaloids	158
	Chapter 21. Isoquinoline alkaloids	188
	Chapter 22. Lycopodium alkaloids	223
	Chapter 23. Monoterpene and sesquiterpene alkaloids	229
	Chapter 24. Peptide alkaloids	236
	Chapter 25. Pyrrolidine and piperidine alkaloids	243
	Chapter 26. Pyrrolizidine alkaloids	255
	Chapter 27. Quinoline alkaloids	267
	Chapter 28. Quinolizidine alkaloids	282
	Chapter 29. Steroidal alkaloids	290
	Chapter 30. Tropane alkaloids	300
	Chapter 31. Miscellaneous alkaloids	309

PART IV	**Phenolics**	**323**
	General Introduction	324
	Chapter 32. Anthocyanins and anthochlors	327
	Chapter 33. Benzofurans	338
	Chapter 34. Chromones and chromenes	343
	Chapter 35. Coumarins	351
	Chapter 36. Minor flavonoids	367
	Chapter 37. Flavones and flavonols	388
	Chapter 38. Isoflavonoids and neoflavonoids	415
	Chapter 39. Lignans	435
	Chapter 40. Phenols and phenolic acids	453
	Chapter 41. Phenolic ketones	466
	Chapter 42. Phenylpropanoids	472
	Chapter 43. Quinones	489
	Chapter 44. Stilbenoids	509
	Chapter 45. Tannins	519
	Chapter 46. Xanthones	532
	Chapter 47. Miscellaneous phenolics	540
PART V	**Terpenoids**	**550**
	General Information	552
	Chapter 48. Monoterpenoids	555
	Chapter 49. Iridoids	569
	Chapter 50. Sesquiterpenoids	579
	Chapter 51. Sesquiterpene lactones	599
	Chapter 52. Diterpenoids	645
	Chapter 53. Triterpenoid saponins	670
	Chapter 54. Steroid saponins	689
	Chapter 55. Cardenolides and bufadienolides	700
	Chapter 56. Phytosterols	712
	Chapter 57. Cucurbitacins	722
	Chapter 58. Nortriterpenoids	728
	Chapter 59. Miscellaneous triterpenoids	739
	Chapter 60. Carotenoids	745
Index		**755**

General Introduction

A vast array of natural organic compounds, the products of primary and secondary metabolism, occur in plants, but there is no simple index available for them. Most of these substances have trivial names which are widely used, but it is often difficult to find out the class of compound and its chemical structure from existing literature sources. The purpose of this Dictionary is to provide this information on the most widely encountered plant constituents, with emphasis on those which are biologically active and/or of economic value.

The continued use of plants as food, as a source of beverages and for their medicinal properties depends on knowledge of the chemical constituents that are present. Secondary substances in the diet can sometimes have adverse effects, and plant and microbial toxins can occasionally enter the food chain. There is also a resurgence of interest in many countries in medicinal plants and their curative properties. The active principles again are secondary compounds of more than one type. Many medicinally useful substances show toxic side-effects, and it is vital to know what biological activities a particular compound may have in these circumstances.

This information and much else is provided in this Dictionary, so that it should be useful to a wide audience of pharmacists, pharmacognosists, food scientists, nutritionists, chemical ecologists and biologists generally. It is also intended to serve the natural product chemists and phytochemists, since it covers a wide range of substances from phenolics and alkaloids through carbohydrates and plant glycosides to essential oils and triterpenoids. It also provides an entry into the phytochemical literature, when more detailed information is required.

Each compound is listed by trivial name and synonyms, and there is a clear indication of the class and subclass to which it belongs. The natural occurrence is covered by a short list of major sources (plant name and family) with, where possible, an indication of frequency of occurrence. The chemical structure, molecular weight and formula, biological activity and use by humans are also provided. The Dictionary is generally restricted to products of higher plants, although some lower plant substances are also included. For example, there are entries on many of the mycotoxins, since these are usually found as contaminants in higher plant tissues. Many of the higher plant constituents which are listed can be categorized as flavour substances, plant volatiles, natural sweeteners, plant poisons, carcinogens, colouring matters, oestrogenic agents, natural hallucinogens, phytoalexins, or as allelochemicals.

The entries are arranged biogenetically and there are five parts, dealing in turn with carbohydrates and lipids, nitrogen-containing compounds, alkaloids, phenolics and terpenoids. Within each part, compounds are grouped together, in alphabetical order, according to class and subclass. Points of overlap are indicated in the introductions to each chapter.

Chemical structures have been provided with almost all entries. These have been drawn to indicate the stereochemistry, using the latest data available as regards both the correctness of the basic structures as well as of the stereochemical assignments. It should be emphasized that some chemical structures shown in the earlier literature (e.g., that of the insect antifeedant azadirachtin from the neem tree) have been revised as a result of further chemical research.

In preparing this Dictionary for publication, many literature sources and compilations have been consulted, including the Chapman & Hall *Dictionary of Organic Compounds*.

The Merck Index (Budavari, 1989) provides useful information on a number of the more common plant products. A comprehensive dictionary of plant substances (excluding alkaloids) was published in German by Birkhäuser Verlag (Karrer, 1958) but it has only been updated as far as the 1966 primary literature (see Cherbuliez and Eugster, 1966; Hurlimann and Cherbuliez, 1981, 1985). The best and most up-to-date source of information on the natural distribution of plant substances is also in German. This is the monumental series *Chemotaxonomie der Pflanzen*, written by R. Hegnauer (1962–1994) in eleven volumes. Other more specialized chemical dictionaries that have been consulted are listed in the appropriate chapters or sections. For botanical nomenclature and plant classification, we have followed Mabberley (1987). Where alternative names are available for plant families, we have chosen the older conserved names (e.g., Leguminosae rather than Fabaceae, etc.).

We are most grateful to many colleagues for advice and information on particular entries. We thank Messrs Tom Faneran and Eric Hammond for their assistance in compiling some sections and Miss Valerie Norris for her secretarial help. Finally, we welcome suggestions from our readers of new bioactive substances which are appropriate for inclusion in the next edition.

J.B. Harborne
H. Baxter

References

Budavari, S. (ed.) (1989). *The Merck Index*, 11th Edition. Rahway, New Jersey: Merck & Co., Inc.

Cherbuliez, E. and Eugster, C.H. (1966). *Konstitution und Vorkommen der organischen Pflanzenstoffe*, Supplement 1. Basel: Birkhäuser-Verlag.

Hegnauer, R. (1962–1994). *Chemotaxonomie der Pflanzen*, in 11 Volumes. Basel: Birkhäuser-Verlag.

Hurlimann, H. and Cherbuliez, E. (1981, 1985). *Konstitution und Vorkommen der organischen Pflanzenstoffe*, Supplement 2, Parts 1 and 2. Basel: Birkhäuser-Verlag.

Karrer, W. (1958). *Konstitution und Vorkommen der organischen Pflanzenstoffe*. Basel: Birkhäuser-Verlag.

Mabberley, D. (1987). *The Plant Book*. Cambridge University Press.

Key to Dictionary Entries

Dictionary registry number (used for indexing) Generally accepted trivial or systematic name

2039
Camphor; Boran-2-one; 2-Camphanone

Common synonyms

Chemical structure (with stereochemistry)

Molecular formula

$C_{10}H_{16}O$ (−) form   Mol. wt 152.24

Molecular weight (calculated to 3 decimal places and rounded to 2)

Natural occurrence, plant genus and species (family in parentheses)

The (+) form occurs in rectified or Japanese oil of camphor, from the camphor tree, *Cinnamomum camphora* (Lauraceae). The (−) form, Matricaria camphor occurs in feverfew, *Chrysanthemum (Matricaria) parthenium* (Compositae), and in species of *Artemisia* spp.(Compositae), and *Lavendula* spp. (Labiatae).

Biological activity (where known)

Irritant; it affects the central nervous system and is toxic to humans.
It is used commercially as a moth repellent and as a preservative in pharmaceuticals and cosmetics. Other uses are as a rubefacient and mild analgesic and as a topical antipruritic.

Commercial or other use (where known)

PART I
Carbohydrates and Lipids

Carbohydrates and Lipids

General Introduction

The phytochemicals described in this chapter are the intermediates or products of primary plant metabolism. Some, like sucrose, are universally present in all green plants, and others, such as starch, have a very wide distribution. Sugars are produced as a result of photosynthesis within the chloroplast, and these are mainly monosaccharides. Many such sugars are formed in trace amounts as phosphate esters, and are not always easy to isolate because of their lability and fugitive nature. Other components of primary metabolism, such as the organic acids of the Krebs tricarboxylic acid cycle, accumulate within the plant at certain stages in the life cycle, e.g., in ripening fruits.

Carbohydrates and lipids provide the plant with a general storage form of energy, and they can be recovered in quality from organs such as the fruits, tubers, rhizomes and seeds. Sugars are stored primarily as oligosaccharides or polysaccharides, but sugar alcohols and cyclitols may also be regarded in part as storage forms of energy. Lipids, which are fatty acid based, represent an alternative form of stored energy in nuts and seeds. There is considerable chemical variation in the fats and oils of seeds, and some 300 fatty acids have been described from plant sources. Lipids also contribute to cell structure, e.g., as membrane components. Fatty acids, on decarboxylation, give rise to long-chain hydrocarbons, which provide the waxy coating of plant surfaces.

A number of other products of primary metabolism — the organic esters, aliphatic lactones and short-chain hydrocarbons — contribute to the volatile odours of plants. Likewise, the acetylenes and thiophenes, which are fatty acid derived, provide the plant with allelochemicals which ward off insect and microbial attack.

The literature of plant biochemistry, in which the products of primary metabolism largely figure, is considerable. There are several introductory texts (Bonner and Varner, 1976; Goodwin and Mercer, 1983) and there is a major series of books (Stumpf and Conn, 1980–) which now extends to 16 Volumes. The *Encyclopedia of Plant Physiology* (Pirson and Zimmermann, 1975–1986) also reviews some aspects of plant metabolism: Volumes 5, 6 and 19 deal with photosynthesis and the pathway of carbon metabolism, while Volumes 13A and 13B deal with carbohydrates. The methodology of plant biochemistry is reviewed in two modern series by Dey and Harborne (1989–) and by Linskens and Jackson (1985–).

References

Bonner, J. and Varner, J.E. (1976). *Plant Biochemistry*, 3rd Edition. New York: Academic Press.
Dey, P. and Harborne, J.B. (1989–). *Methods in Plant Biochemistry*, in 10 Volumes. London: Academic Press.
Goodwin, T.W. and Mercer, E.I. (1983). *Introduction to Plant Biochemistry*, 2nd Edition. Oxford: Pergamon Press.
Linskens, H.F. and Jackson, J.F. (1985–). *Modern Methods of Plant Analysis, New Series*, in several Volumes. Berlin: Springer-Verlag.
Pirson, A. and Zimmermann, M.H. (1975–1986). *Encyclopedia of Plant Physiology, New Series*, in 19 Volumes. Berlin: Springer-Verlag;
Stumpf, P.K. and Conn, E.E. (1980–). *The Biochemistry of Plants: A Comprehensive Treatise*, in at least 16 Volumes. New York: Academic Press.

Chapter 1
Monosaccharides

The major free sugars in plants are the monosaccharides glucose (**10**) and fructose (**6**) (and the disaccharide sucrose), together with traces of xylose, rhamnose and galactose. Other sugars present in trace amounts are the sugar phosphates, which are involved in metabolism; one example is ribulose (**19**), present as the 2,6-bisphosphate in the chloroplast of all green plants. The bulk of carbohydrates occurs in plants in bound form, as oligo- or polysaccharide, or attached to a range of different aglycones, as plant glycosides. Since the free sugar pool is relatively uniform in higher plants, most variation in monosaccharides is in the way they are linked in plant glycosides, oligosaccharides or polysaccharides.

Five sugars are commonly found as components of glycosides and polysaccharides: two are hexoses, glucose and galactose; two are pentoses, xylose and arabinose; and one is a methylpentose, rhamnose. Of fairly common occurrence are the uronic acids, glucuronic acid (**11**) and galacturonic acid; a third hexose, mannose (**14**), is not uncommon in polysaccharides. The pentose sugars ribose (**18**) and deoxyribose (**4**), components of RNA and DNA, respectively, should be mentioned, but these sugars are rarely encountered in plants in any other association. The only common keto sugar is fructose, a frequent component of oligosaccharides (e.g., sucrose) and of the polysaccharides known as fructans. A range of rarer sugars occurs in plant glycosides, one example being the five-carbon branched sugar apiose (**1**), present as the flavone glycoside apiin in parsley seed.

Each monosaccharide can exist as more than one optically active isomer; however, in plants, only one form is normally encountered. Thus, glucose is usually the β-D-isomer, rhamnose the α-L form and so on. Again, each sugar can theoretically exist in both a pyrano (six membered) and furano (five membered) ring form, although one or other is usually favoured. Thus, glucose normally takes up the pyrano configuration, whereas fructose is usually the furano form.

Sugars are colourless substances, normally detected and separated by paper chromatography, followed by staining with aniline hydrogen phthalate or resorcinol+sulfuric acid. Monosaccharides are normally quantitatively analysed by gas chromatography of the trimethylsilyl ethers.

More detailed accounts of the simple plant sugars can be found in most organic chemistry or biochemistry textbooks. For accounts of their comparative distribution in the plant kingdom, see Bell (1962) and Loewus and Tanner (1982).

References

Bell, D.J. (1962). Carbohydrates. In *Comparative Biochemistry* (Florkin, M. and Mason, H.S., eds), Vol. III, pp. 288–355.
Loewus, F.A. and Tanner, W. (eds) (1982). *Plant Carbohydrates I, Intracellular Carbohydrates*. Berlin: Springer-Verlag.

1
Apiose; β-D-Apiose; 3-C-(Hydroxymethyl)-D-glycero-tetrose; Tetrahydroxyisovaleraldehyde.

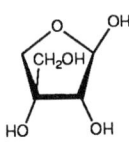

$C_5H_{10}O_5$ Mol. wt 150.14

First isolated from parsley, *Petroselinum crispum* (Umbelliferae), seed, where it occurs combined as the flavone glycoside apiin (q.v.). It also occurs in glycosidic attachment to a number of other flavonoids and phenolic compounds. It is present in polysaccharides found in several aquatic plants, notably *Lemna minor, Posidonia australis, Wolffia arrhiza* and *Zostera nana*.

2
Arabinose; β-L-Arabinopyranose; Pectin sugar; L-Arabinose; Pectinose.

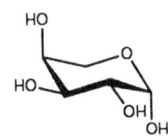

$C_5H_{10}O_5$ Mol. wt 150.14

Common component of plant glycosides, and almost universally present as a polysaccharide sugar in hemicelluloses (arabans) and gums. L-Arabinose can occur in both the pyrano form (shown) and the corresponding furano form. D-Arabinose is also known, e.g., as a sugar component of anthraquinone glycosides of *Aloe* (Liliaceae).

3
Cymarose; β-D-Cymaropyranose; D-Cymarose; 2,6-Dideoxy-3-O-methyl-*ribo*-hexose; 3-O-Methyldigitoxose.

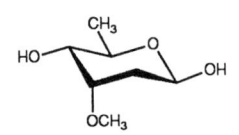

$C_7H_{14}O_4$ Mol. wt 162.19

Constituent sugar of a number of cardiac glycosides from Apocynaceae, e.g., sugar of the glycoside cymarin (q.v.) from *Strophanthus kombe*.

Constituent of cytovaricin which is active against phytopathogenic fungi.

4
Deoxyribose; 2-Deoxy-α-D-ribopyranose; D-2-Deoxyarabinose; 2-Deoxy-D-*erythro*-pentose; D-2-Deoxyribose; 2-Deoxy-D-ribose; Desoxyribose; D-*erythro*-2-Deoxypentose; D-2-Ribodesose; Thyminose.

$C_5H_{10}O_4$ Mol. wt 134.14

Universal sugar component of deoxyribonucleic acids (DNA) and the associated nucleotides and nucleosides. Readily obtainable from the acid hydrolysis of DNA.

5
Digitalose; β-D-Digitalopyranose; 6-Deoxy-3-O-methylgalactose; 3-O-Methyl-D-fucose.

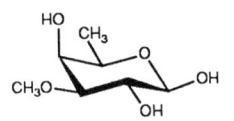

$C_7H_{14}O_5$ Mol. wt 178.19

One of the sugar constituents of digitalin (q.v.), a cardiac glycoside from the dried leaves of *Digitalis purpurea* (Scrophulariaceae), and also present in other cardiac glycosides, e.g., musaroside (q.v.).

6
Fructose; β-D-Fructopyranose; *arabino*-2-hexulose; D-Fructose; Fruit sugar; Laevulose; Levulose.

$C_6H_{12}O_6$ Mol. wt 180.16

A universal plant sugar occurring free (e.g., in nectar and phloem), in phosphorylated form and combined as oligosaccharide (most commonly as sucrose). Occurs usually in the furanose form (as in sucrose, q.v.) but crystallizes out from solutions in the pyranose form (as shown).

Sweetener: twice as sweet as glucose; helps prevent sandiness in ice cream. It is used as a substitute for glucose in parenteral nutrition (note risk of lactic acidosis) and in diabetes, and in the treatment both of severe alcohol poisoning, and of vomiting in pregnancy.

7
Fucose; α-D-Fucopyranose; L-Fucose; 6-Deoxy-L-galactose; L-Galactomethylose; Rhodeose.

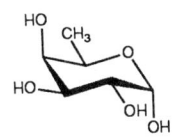

$C_6H_{12}O_5$ Mol. wt 164.16

Most abundant occurrence of L-fucose is in polysaccharide combination in seaweeds, e.g., *Fucus* spp. (Fucaceae), but also present in higher plant polysaccharides (e.g., in gum tragacanth, q.v.). The D-fucose isomer is uncommon, occurring in saponin form (q.v.) in *Convolvulus* spp. (Convolvulaceae).

8
Galactose; α-D-Galactopyranose; Brain sugar; Cerebrose; D-Galactose; L-Galactose.

$C_6H_{12}O_6$ Mol. wt 180.16

Fairly widely distributed in nature. A constituent of raffinose and stachyose (q.v.), of hemicelluloses, of pectin, of gums and mucilages and of some glycosides. Constituent sugar of agar and other seaweed polysaccharides, e.g., *Porphyra umbilicalis*.

Diagnostic aid, hepatic function.

9
Glucosamine; β-D-Glucosamine; 2-Amino-2-deoxyglucose; Chitosamine.

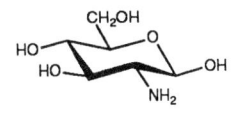

$C_6H_{13}NO_5$ Mol. wt 179.18

Best known as a sugar of animal chitin and mucoprotein. Occurs as a cell wall constituent in bacteria and fungi, e.g., *Aspergillus niger*. Also present regularly in glycoproteins of plant seeds, e.g., in *Phaseolus vulgaris* (Leguminosae).

Clinical trials have been performed in arthrosis; it is anti-arthritic.

10
Glucose; α-D-Glucopyranose; Blood sugar; Corn sugar; Dextrose; D-Glucose; Grape sugar.

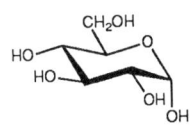

$C_6H_{12}O_6$ Mol. wt 180.16

Present in all plants, in the α- or β-form, free and combined in starch, cellulose, sucrose and many glycosides.

Sweetener: about 70% as sweet as sucrose. It is a main source of energy for living organisms. It is used in hypoglycaemia, in ketosis to counteract hepatotoxins, to reduce cerebrospinal pressure and cerebral oedema, and as a sclerosing agent in the treatment of varicose veins.

11
Glucuronic acid; β-D-Glucopyranuronic acid; β-D-Glucuronic acid.

$C_6H_{10}O_7$ Mol. wt 194.15

A component sugar of many plant gums and mucilages. Occurs in glycosidic form, e.g., as quercetin 3-glucuronide from leaves of *Gaultheria miqueliana* (Ericaceae). Can be present in polysaccharides both as the free acid and as the corresponding lactone, glucuronolactone, in which the 6-carboxyl is linked to the 3-hydroxyl.

Detoxicant in medicine.

12
Hamamelose; β-D-Hamamelopyranose; 2-*C*-Hydroxymethyl-D-ribose

$C_6H_{12}O_6$ Mol. wt 180.16

Present, combined with tannin, in the bark of witch hazel, *Hamamelis virginiana* (Hamamelidaceae). Also occurs in the free state in a variety of higher plants, especially in the leaves of *Primula* spp. (Primulaceae).

13
Mannoheptulose; α-D-Mannoheptulopyranose; Manno-2-heptulose; D-Mannoketoheptose

$C_7H_{14}O_7$ Mol. wt 210.19

From the avocado, *Persea americana* (Lauraceae).

14
Mannose; α-D-Mannopyranose; Carubinose; D-Mannose; Seminose

$C_6H_{12}O_6$ Mol. wt 180.16

Obtainable from the hydrolysis of mannans, e.g., that of vegetable ivory nut. Occurs as a sugar of many plant polysaccharides and also occasionally of plant glycosides.

Sweet taste with bitter after-taste. It is used in biochemical research.

15
Oleandrose; β-L-Oleandropyranose; 2,6-Dideoxy-3-*O*-methyl-β-L-*arabino*-hexose; L-Oleandrose

$C_7H_{14}O_4$ Mol. wt 162.19

A sugar component of cardiac glycosides, e.g., of oleandrin (q.v.) from *Nerium oleander* (Apocynaceae).

16
Quinovose; α-D-Quinovopyranose; Chinovose; 6-Deoxy-D-glucose; Epifucose; D-Epirhamnose; D-Glucomethylose; D-Isorhamnose; Isorhodeose

$C_6H_{12}O_5$ Mol. wt 164.16

From cinchona bark, *Cinchona* spp. (Rubiaceae). A constituent sugar of a number of plant saponins (q.v.), e.g., in those of seeds of *Ipomoea parasitica* (Convolvulaceae). Quinovose, substituted in the 6-position by sulfonic acid, is the sugar of a sulfolipid diglyceride widely present in algae and green plants.

17
Rhamnose; α-L-Rhamnopyranose; 6-Deoxy-D-mannose; 6-Deoxy-L-mannose; Isodulcite; L-Mannomethylose; D-Rhamnose; L-Rhamnose

$C_6H_{12}O_5$ Mol. wt 164.16

Widely distributed in plants as a trace sugar. Occurs as a constituent sugar of many plant glycosides (e.g., quercitrin, naringin, q.v.) and of many plant polysaccharides, especially pectins and gums.

Very sweet taste. It is used to assess intestinal permeability.

18
Ribose; β-L-Ribopyranose; D-Ribose

$C_5H_{10}O_5$ Mol. wt 150.14

Universal sugar component of ribonucleic acids (RNA) and of associated nucleotides and nucleosides. Also, it is the sugar of metabolic intermediates such as ATP and UDPG.

19
Ribulose; α-D-Ribulose; D-Adonose; Arabinoketose; D-Arabinulose; D-*erythro*-2-Ketopentose; D-*erythro*-2-Pentulose; D-Riboketose

$C_5H_{10}O_5$ Mol. wt 150.14

Ribulose, as the 2,6-bisphosphate, is an early product of the photosynthetic carbon cycle in all green plants.

Sweet tasting syrup.

20
Sedoheptulose; β-D-Sedoheptulopyranose; D-*altro*-2-Heptulose; *altro*-Ketoheptose; Volemulose

$C_7H_{14}O_7$ Mol. wt 210.19

May accumulate in leaves of *Sedum spectabile* (Crassulaceae) and in other succulents. It is universally present in trace amounts as an intermediate of the photosynthetic carbon cycle.

21
Sorbose; α-L-Sorbopyranose; Sorbin; Sorbinose; L-Sorbose

$C_6H_{12}O_6$ Mol. wt 180.16

Found in fruits of mountain ash, *Sorbus aucuparia* (Rosaceae), and, in combined form, in the pectins of ripe *Passiflora edulis* fruits (Passifloraceae).

An intermediate in the manufacture of ascorbic acid (vitamin C).

22
Xylose; α-D-Xylopyranose; Wood sugar; D-Xylose

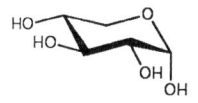

$C_5H_{10}O_5$ Mol. wt 150.14

Present free as a trace component in most plants. Otherwise, present in bound form as glycosides, or as polysaccharides known as xylans.

Very sweet. It is used as a diabetic food, and as a diagnostic aid for testing intestinal function.

Chapter 2
Oligosaccharides

Most of the common plant oligosaccharides contain from two (e.g., sucrose, **40**) to six (e.g., ajugose, **23**) monosaccharide units. Even when there are only two units, these can be joined together by ether links in a number of different ways (viz., through different hydroxyls and by α or β links) so that one of the main problems with oligosaccharides is distinguishing different isomers. In the case of disaccharides containing glucose, eight isomeric structures are possible and all are known. Four of these are recognized primarily because they represent characteristic building blocks of well known polysaccharides (q.v.), and they are obtainable by the partial hydrolysis of such polymers. Thus cellobiose (**24**) is a disaccharide with the typical $\beta\ 1\rightarrow 4$ linkage present in cellulose, laminaribiose (**28**) with the $\beta\ 1\rightarrow 3$ linkage present in laminarin, while maltose (**29**) and isomaltose (glucosyl $\alpha\ 1\rightarrow 6$ glucose) are component units of the polysaccharide, starch.

The number of oligosaccharides that accumulate as such in plants is relatively small. Sucrose (2-α-glucosylfructose) is the only one that is of universal occurrence. Fairly common are α,α-trehalose **41**, (α-glucosyl-α-glucose), raffinose (**35**, 6^G-α-galactosylsucrose), stachyose (**39**, 6^G-α-digalactosylsucrose) and verbascose (**43**, 6^G-α-trigalactosylsucrose). Raffinose and stachyose are present, for example, in many legume seeds. One of the less common storage oligosaccharides is umbelliferose (**42**, 2^G-α-galactosylsucrose), which is mainly restricted in its distribution to members of the Umbelliferae and the Araliaceae.

Considerable numbers of the known plant oligosaccharides are not present in the free state, but occur combined with other organic molecules as plant glycosides. Among flavonoid pigments, for example (q.v.), the following oligosaccharides are frequently found: rutinose (6-α-rhamnosylglucose), neohesperidose (2-α-rhamnosylglucose), sophorose (2-β-glucosylglucose), and robinobiose (2-α-rhamnosylgalactose). Other groups of plant glycoside containing different oligosaccharides are the saponins, steroidal alkaloids and cardiac glycosides (q.v.).

Oligosaccharides resemble monosaccharides in their chemical properties, and methods used for their separation are largely similar. Oligosaccharides are similar to certain monosaccharides, such as glucose and fructose, in their sweet taste. Sucrose for example is used as a standard for sweetness. Not all oligosaccharides are necessarily sweet tasting and some in fact are bitter-sweet (gentiobiose) or tasteless (cellobiose).

References

Miller, L.P. (1973). Mono- and oligosaccharides. In *Phytochemistry*, Volume 1 (Miller, L.P. (ed.), pp. 145–175. New York: Van Nostrand Reinhold.

Preiss, J. (ed.) (1980). *The Biochemistry of Plants*, Volume 3, *Carbohydrates, Structure and Function*. San Diego: Academic Press.

23
Ajugose; α-Galactosyl-(1→6)-α-galactosyl-(1→6)-α-galactosyl-(1→6)-α-galactosyl-(1→6)-α-glucosyl-(1→2)-β-fructose

$C_{36}H_{62}O_{31}$ Mol. wt 990.89

Found in the roots of *Ajuga nipponensis* (Labiatae) and *Verbascum thapsiforme* (Scrophulariaceae), and in the seeds of *Vicia sativa* (Leguminosae).

24
Cellobiose; 4-*O*-(β-D-Glucopyranosyl)-D-glucopyranose; Cellose

$C_{12}H_{22}O_{11}$ Mol. wt 342.31

This is the disaccharide that forms the repeating unit of cellulose, and is obtainable from the hydrolysis of cellulose. Also, it is reported as the sugar component of the glycoside hemocorin, from *Haemodorum corymbosum* (Haemodoraceae).

25
Gentianose; β-D-Fructofuranosyl *O*-β-D-glucopyranosyl-(1→6)-α-D-glucopyranoside

$C_{18}H_{32}O_{16}$ Mol. wt 504.45

Found in the rhizomes of many *Gentiana* spp., e.g., *Gentiana lutea* (Gentianaceae).

Reserve carbohydrate.

26
Gentiobiose; 6-*O*-β-D-Glucopyranosyl-D-glucose; Amygdalose; 6-(β-D-Glucosido)-D-glucose

$C_{12}H_{22}O_{11}$ Mol. wt 342.31

The sugar component of the cyanogenic glycoside amygdalin (q.v.) and a variety of other plant glycosides.

Bitter taste.

27
Lactose; 4-*O*-β-D-Galactopyranosyl-D-glucose; 4-(β-D-Galactosido)-D-glucose; Glucose-4-β-galactoside; Milk sugar

$C_{12}H_{22}O_{11}$ Mol. wt 342.31

Reported in the anthers of *Forsythia* flowers (Oleaceae) and the fruit of *Achras sapota* (Sapotaceae), but is present in these sources in bound form. The 3-lactoside of the flavonol isorhamnetin occurs in the seeds of *Cassia multijuga* (Leguminosae).

Less sweet than cane sugar. It is used as a nutrient in modified milk and as a food for infants and convalescents. It is a pharmaceutical diluent for tablets and capsules and is used in the treatment of hepatic encephalopathy.

28
Laminaribiose; 3-*O*-β-D-Glucopyranosyl-D-glucopyranose

$C_{12}H_{22}O_{11}$ Mol. wt 342.31

Produced during the partial hydrolysis of the seaweed polysaccharide, laminarin. Occurs as a component sugar of some plant glycosides, e.g., luteolin 7-laminaribioside from *Colchicum speciosum* petals (Liliaceae).

29
Maltose; 4-*O*-(α-D-Glucopyranosyl)-D-glucopyranose; Malt sugar; Maltobiose

$C_{12}H_{22}O_{11}$ Mol. wt 342.31

Present free in small quantities in many plants, probably as a breakdown product of starch. Also reported free in some pollens and nectars.

Used in brewing, and soft drinks and foods. Less sweet than cane sugar. It is used as a nutrient, and as a parenteral supplement of sugar for diabetics.

30
Melezitose; *O*-α-D-Glucopyranosyl-(1→3)-β-D-fructofuranosyl α-D-glucopyranoside

$C_{18}H_{32}O_{16}$ Mol. wt 504.45

Occurs in the manna of Douglas fir, *Pseudotsuga menziesii*, jack pine, *Pinus virginiana* (both Pinaceae), and other trees. It occurs as a honeydew excretion from aphids and is probably formed from sucrose by the transglucosylase action of insect sucrase.

31
Melibiose; 6-*O*-(β-D-Galactopyranosyl)-D-glucopyranose; 6-*O*-β-D-Galactopyranosyl-D-glucose; Glucose-β-galactoside

$C_{12}H_{22}O_{11}$ Mol. wt 342.31

Found in many plant exudates, and probably arises from the hydrolysis of raffinose (q.v.). Also, it is a constituent of plant nectaries.

Sweetener: less than one-third as sweet as sucrose.

32
Neohesperidose; 2-*O*-α-L-Rhamnopyranosyl-D-glucopyranose; Sophorabiose

$C_{12}H_{22}O_{10}$ Mol. wt 326.31

It is the component sugar of the *Citrus* bitter principles naringin, neohesperidin and poncirin. It is present in other plant glycosides.

33
Planteose; O-α-D-Galactopyranosyl-(1→6)-β-D-fructofuranosyl α-D-glucopyranoside

$C_{18}H_{32}O_{16}$ Mol. wt 504.45

Found in the seeds and storage organs of the genus *Plantago* (Plantaginaceae), as well as some members of *Teucrium* (Labiatae) and *Fraxinus* (Oleaceae).

Reserve carbohydrate.

34
Primeverose; 6-O-β-D-Xylopyranosyl-D-glucose

$C_{11}H_{20}O_{10}$ Mol. wt 312.28

A constituent of many phenolic glycosides such as primulaverin, gaultherin and gentiacaulin.

35
Raffinose; α-D-Galactosyl-(1→6)-α-D-glucosyl-(1→2)-β-D-fructose; Gossypose; Melitose; Melitriose

$C_{18}H_{32}O_{16}$ Mol. wt 504.45

Occurs widely in plant leaves in low concentrations, and accumulates in storage organs, e.g., of broad bean, *Vicia faba*, seeds (Leguminosae). Can be isolated in quantity from sugar beet and cotton seed.

Carbohydrate reserve.

36
Robinobiose; 6-O-(6-Deoxy-α-L-mannopyranosyl)-D-galactose; 6-O-α-L-Rhamnopyranosyl-D-galactose

$C_{12}H_{22}O_{10}$ Mol. wt 326.31

Occurs in the flavonol glycoside robinin (q.v.), a constituent of the flowers of *Robinia pseudacacia* (Leguminosae).

37
Rutinose; 6-O-α-Rhamnosyl-D-glucose

$C_{12}H_{22}O_{11}$ Mol. wt 342.31

The constituent sugar of the flavonol glycoside rutin, from *Ruta graveolens* (Rutaceae). Rutin is widely present in plants, and many other flavonoid rutinosides have also been characterized.

38
Sophorose; 2-*O*-β-D-Glucopyranosyl-β-D-glucopyranose

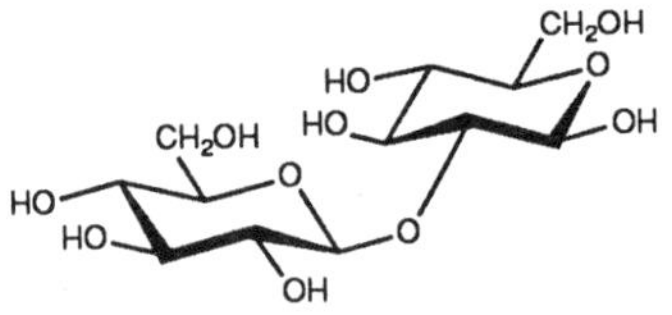

$C_{12}H_{22}O_{11}$ Mol. wt 342.31

The constituent sugar of quercetin 3-sophoroside from the pods of *Sophora japonica* (Leguminosae). Other sophorosides are known.

39
Stachyose; α-Galactosyl-(1→6)-α-galactosyl-(1→6)-α-glucosyl-(1→2)-β-fructose

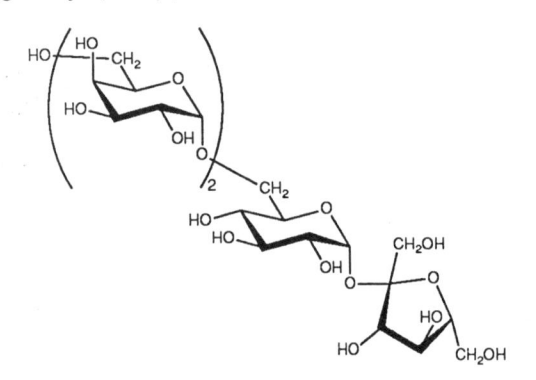

$C_{24}H_{42}O_{21}$ Mol. wt 666.60

Storage sugar in the roots of *Stachys* spp. (Labiatae), the twigs of white jasmine, the seeds of yellow lupin, *Lupinus luteus*, soybeans, *Soja max*, lentils, *Ervum lens* (Leguminosae), and ash manna, *Fraxinus ornus* (Oleaceae).

40
Sucrose; β-D-Fructofuranosyl α-D-glucopyranoside; 1-α-Glucosido-2-β-fructofuranose; Beet sugar; Cane sugar; Saccharabiose; Saccharose

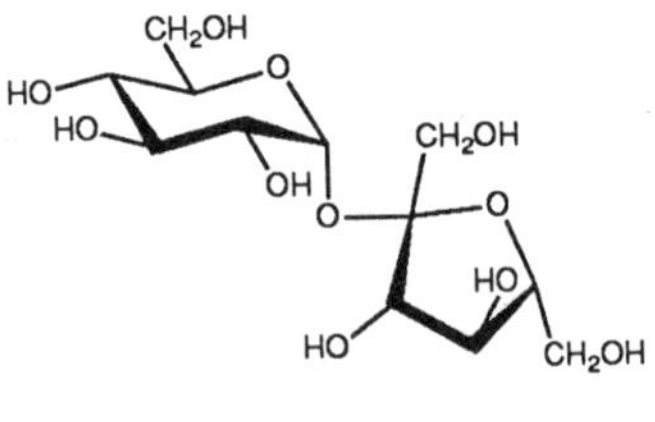

$C_{12}H_{22}O_{11}$ Mol. wt 342.31

A universal sugar of green plants, used for storage and transport. It is a major component of most plant nectars. It is produced in large amounts in sugarcane, *Saccharum officinarum* (Gramineae), sugar beet, *Beta vulgaris* (Chenopodiaceae), sugar maple, *Acer saccharum* (Aceraceae) and sweet sorghum, *Sorghum bicolor* (Gramineae).

Sweetener, preservative and antioxidant, demulcent, granulating agent, and coating and/or excipient for tablets. It is used in eyedrops as a hypertonic agent in the reduction of corneal oedema, and in wound healing, and also in the treatment of hiccups.

41
Trehalose; α-D-Glucosido-(1→1)-α-D-glucoside; Mushroom sugar; Mycose; α,α-Trehalose

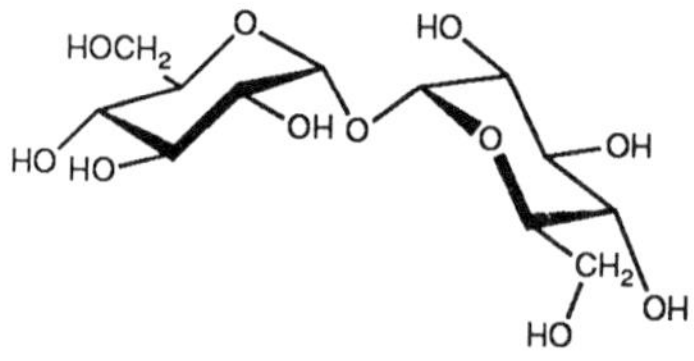

$C_{12}H_{22}O_{11}$ Mol. wt 342.31

Occurs in fungi, e.g., *Amanita muscaria*; generally, it replaces sucrose in plants lacking chlorophyll and starch. Also, it occurs in *Selaginella lepidophylla* (Selaginaceae).

42
Umbelliferose; β-D-Fructofuranosyl *O*-α-D-galactopyranosyl-(1→2)-α-D-glucopyranoside

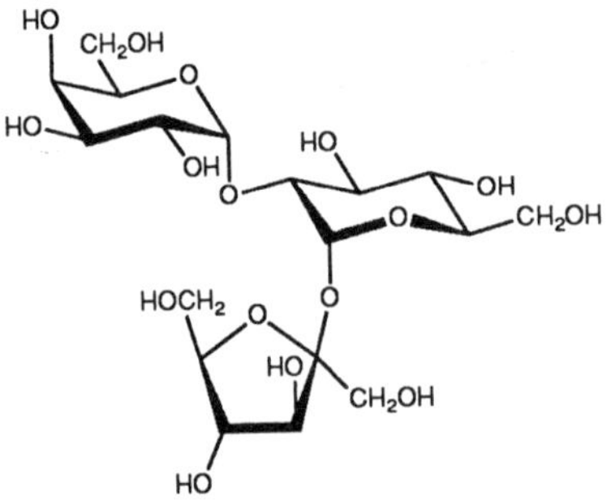

$C_{18}H_{32}O_{16}$ Mol. wt 504.45

Storage sugar in the roots of many species of Umbelliferae (e.g., *Angelica archangelica*) and Araliaceae.

A reserve carbohydrate.

43
Verbascose; β-D-Fructofuranosyl O-α-D-galactopyranosyl-(1→6)-[O-α-D-galactopyranosyl-(1→6)]$_2$-α-D-glucopyranoside

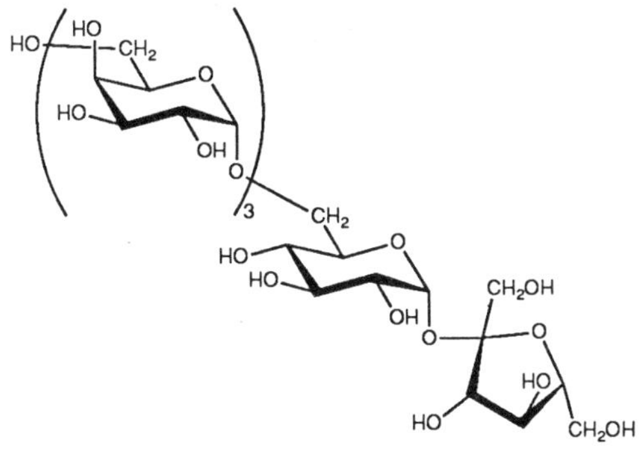

$C_{30}H_{52}O_{26}$ Mol. wt 828.75

Found in the storage organs, tubers, rhizomes and seeds of many plants. A common source is the mullein root, *Verbascum thapsus* (Scrophulariaceae).

44
Vicianose; O-α-L-Arabinopyranosyl-(1→6)-β-D-glucopyranose

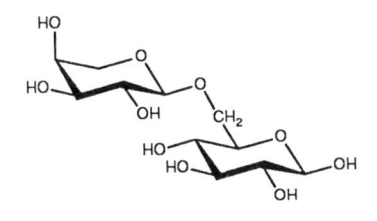

$C_{11}H_{20}O_{10}$ Mol. wt 312.28

Occurs in the cyanogenic glycoside vicianin (q.v.) from *Vicia angustifolia* (Leguminosae); other plant glycosides containing vicianose are known.

Chapter 3
Polysaccharides

The chemistry of polysaccharides is, in a sense, simpler than that of the other plant macromolecules such as the proteins (q.v.), since these polymers contain only a few simple sugars in their structures. Indeed, the best known polysaccharides, cellulose and starch, are polymers of a single sugar, glucose. The structural complexity of polysaccharides is due to the fact that two sugar units can be linked together, through an ether linkage, in a number of different ways. The reducing end of one sugar (C1) can condense with any hydroxyl of a second sugar (at C2, C3, C4 or C6) so that during polymerization some sugars may be substituted in two positions, giving rise to branched chain structures. Furthermore, the ether linkage can have either an α or a β configuration, due to the stereochemistry of simple sugars, and both types of linkage can co-exist in the same molecule.

Although a few polysaccharides (e.g., cellulose) are, in fact, simple straight-chain polymers, the majority have partly branched structures. It is difficult to determine complete sequences in such branched polysaccharides and, often, it is only possible to define their structures in terms of a "repeating unit" of an oligosaccharide, large numbers of which are linked together to produce the complete macromolecule. The structures of the individual monosaccharides from which plant polysaccharides are composed have been given earlier in this volume.

The most familiar plant polysaccharides are cellulose and starch. Cellulose (**48**) represents a very large percentage of the combined carbon in plants, and is the most abundant organic compound of all. It is the fibrous material of the cell wall and is responsible, with lignin, for the structural rigidity of plants. Cellulose occurs in almost pure form (98%) in cotton fibres; the wood of trees is a less abundant source (40–50%), but the most important commercially for cellulose production. Chemically, cellulose is a β-glucan and consists of long chains of β 1→4 linked glucose units, the molecular weight varying from 100 000 to 200 000. Cellulose occurs in the plant cell as a crystalline lattice, in which long straight chains of polymer lie side by side linked by hydrogen bonding.

Starch (**61**) differs from cellulose in having the linkage between the glucose units α 1→4 and not β 1→4 and also in having some branching in the chain. Starch, in fact, comprises two separable components, amylose and amylopectin. Amylose (about 20% of the total starch) contains approximately 300 glucose units linked in a simple chain α 1→4, which exists *in vivo* in the form of an α-helix. Amylopectin (about 80%) contains α 1→4 chains, with regular branching of the main chain by secondary α 1→6 linkages. Its structure is thus of a randomly multibranched type. Starch is the essential storage form of energy in the plant, and starch granules are frequently located within the chloroplast close to the site of photosynthesis.

The different classes of plant polysaccharide fall into two groups according to whether they easily dissolve in water or not. Those that are soluble include starch, inulin (**55**), pectin (**58**) and the various gums and mucilages. The gums (e.g., **57**) which are exuded by plants, sometimes in response to injury or infection, are almost pure polysaccharide. Their function in the plant is not entirely certain, although it may be a protective one.

The less soluble polysaccharides usually comprise the structural cell wall material and occur in close association with lignin. As well as cellulose, there are various hemicelluloses in this fraction. The hemicelluloses have a variety of sugar components, and there are three main types: the xylans, glucomannans and arabinogalactans. They are structurally complex, and other polysaccharide types may also be found with them.

From the comparative viewpoint, polysaccharides are interesting macromolecules since they do vary in type in different groups of plant. The storage polysaccharides, for example, of a number of Compositae are based on fructose rather than glucose. Thus, fructans like inulin (**55**) are isolated from tubers of chicory, *Cichorium intybus*, in place of the more usual starches.

Variation is also marked in the cell wall polysaccharides of different groups of organisms. Differences in cell wall composition distinguish the major classes of fungi and can be used, with other characters, as phylogenetic markers in the group. The cell wall polysaccharides of ferns and gymnosperms are distinguishable (from the angiosperms) by their frequent high mannose content.

Finally, the marked variation in polysaccharide composition between different algae (seaweed) must be mentioned. Three of the main classes — brown, red and green — are clearly distinct in the types of polysaccharide they have. Typical algal polysaccharides included here are agar (**45**), fucoidan (**50**) and laminaran (**56**).

Although there are numerous general accounts of polysaccharides, not many deal specifically with those from plants. An excellent comparative survey of plant polysaccharides is provided in the review by Percival (1966). The same author has also published a detailed account of algal polysaccharides (Percival and McDowell, 1967). Recent accounts of storage polysaccharides in plants are contained in Loewus and Tanner (1982), while reviews of cell wall polysaccharides in different plant groups are available in Tanner and Loewus (1981). Methods for the analysis of plant polysaccharides are reviewed in Dey (1990).

References

Dey, P.M. (ed.) (1990). *Methods in Plant Biochemistry*, Volume 2, *Carbohydrates*. London: Academic Press.
Loewus, F.A. and Tanner, W. (eds) (1982). *Plant Carbohydrates*, I, *Intracellular Carbohydrates*. Berlin: Springer-Verlag.
Percival, E. (1966). The natural distribution of plant polysaccharides. In *Comparative Phytochemistry* (Swain, T., ed.), pp. 139–158. London: Academic Press.
Percival, E. and McDowell, R.H, (1967). *Chemistry and Enzymology of Marine Algal Polysaccharides*. London: Academic Press.
Stephen, A.M. (1980). Plant carbohydrates. In *Secondary Plant Products*. (Bell, E.A. and Charlwood, B.W., eds), pp. 555–584. Berlin: Springer-Verlag.
Tanner, W. and Loewus, F.A. (eds) (1981). *Plant Carbohydrates*, II, *Extracellular Carbohydrates*. Berlin: Springer-Verlag.

45
Agar; Agar-agar; Bengal isinglass; Ceylon isinglass; Chinese isinglass; Gelose; Japan agar; Japan isinglass; Layor carang

The chemical structure is a mixture of two related polysaccharides: agarose, containing D- and L-galactose, 6-methylgalactose, xylose and 3,6-anhydro-L-galactose; and agaropectin, based on D-galactose, 3,6-anhydro-L-galactose and glucuronic acid, which is sulfated.

Obtained from the agarocytes of red algae (Rhodophyceae) (notably *Acanthopeltis, Ceranium, Gelidium, Gracilaria* and *Pterocladia*), and also from species of Gelidiaceae and Sphaerococcaceae.

It is a food gelling and thickening agent and stabilizer and a substitute for gelatin. It is used medically in making capsules and ointments. Industrial uses are corrosion inhibition, dyeing and printing of fabrics, and sizing of paper. Veterinary use is as a laxative for cats and dogs.

46
Alginic acid; Norgine; Polymannuronic acid

It is a hydrophilic colloidal polysaccharide (molecular weight about 240 000) and contains varying proportions of D-mannuronic acid and L-guluronic acid linked through the 1- and 4-positions.

Occurs in brown seaweeds, e.g., the fronds of *Laminaria digitata* and *Macrocystis pyrifera*.

It is a food emulsifier and stabilizer. Other uses include sizing paper and textiles, and as an emulsifer for mineral oils.

47
Carrageenan; Carrageen; Carrageenin; Galactan

Chemically, it is a complex mixture of hydrocolloidal polysaccharides of molecular weights from 1×10^6 to 8×10^6, dependent on the source of supply. When processed, it contains alkaline metal, calcium or magnesium salts of the sulfate esters of galactose and 3,6-anhydrogalactose copolymer.

Found in about 22 spp. of red seaweeds in the families Furcellariaceae, Gigartinaceae, Hypnaceae and Solieriaceae; it is obtained on a commercial scale from *Chondrus crispus* and *Gigartina stellata*.

Carrageenan shows blood anticoagulant activity. It is a food gelling agent, stabilizer, suspending agent and thickener, and an excipient for cosmetics and drugs.

48
Cellulose; α-Cellulose; Microcrystalline cellulose

The fibrous component of plant cell walls, with cotton its purest form (over 90%); this is the most abundant carbon compound of the plant kingdom.

Prevents caking in foods. The microcrystalline form acts as a binder–disintegrator in tablets and as a stationary phase in chromatography. Cellulose is acetylated and compounded for thermoplastics, is regenerated as rayon, and is nitrated for explosives.

49
Cholla gum

It is an arabinogalactan, containing some rhamnose, xylose, fucose and glucuronic acid residues.

Occurs in *Opuntia* spp. (Cactaceae), e.g., the stems of *Opuntia fulgida*.

50
Fucoidan; Fucoidin

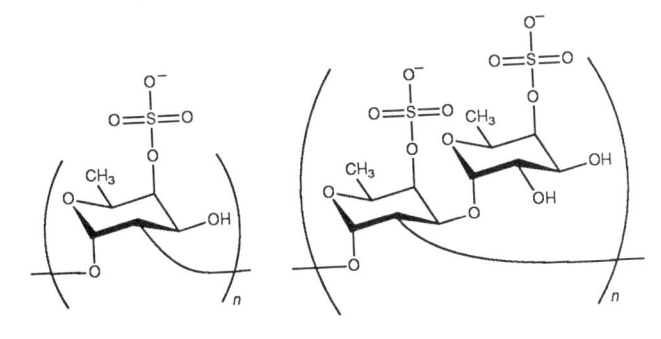

It comprises polysaccharides of molecular weight $133\,000 \pm 20\,000$ containing sulfated L-fucose. Small amounts of galactose and xylose also present.

It is found in brown algae, e.g., *Laminaria digitata* or *Fucus vesiculosus*, and is isolated from *Macrocytis pyritera* and *Pelvetia canaliculata* (23%).

Fucoidan prepared from *Fucus vesiculosus* shows anticoagulant activity. It is used to produce colloids of high viscosity. Sweetener: sweetness 65 to 70% of sucrose.

51
Gum arabic; Acacia; Gum hashab; Kordofan gum; Sudan gum

Made up of high molecular weight (240 000 to 580 000) polysaccharides and their salts, and comprising L-arabinose, D-galactose, L-rhamnose and glucuronic acid as structural units. The product from *Acacia verek* is laevo-rotatory, whereas that from other *Acacia* species is dextro-rotatory.

Occurs as a dried gummy exudation from the bark, branches and stems of *Acacia senegal* and other *Acacia* spp. (Leguminosae).

Used as a demulcent, soothes irritations of the mucous membranes. It lowers cholesterol levels in the blood of rats. It is a food emulsifier, stabilizer and thickener, may be a component of corrosion inhibitors and drilling fluids, and is a drug excipient.

52
Gum ghatti; Ghatti gum; Indian gum

Structurally, it is a calcium-magnesium salt of a water-soluble complex polysaccharide containing L-arabinose, D-galactose, D-glucuronic acid, D-mannose and D-xylose.

Occurs as a gummy exudation from the stems of *Anogeissus latifolia* (Combretaceae).

Produces a viscous aqueous mucilage. It is used as a substitute for acacia gum but is more viscous and less adhesive. It is an ingredient of pharmaceutical emulsions, oils and waxes, and table syrup, and is used to prevent fluid loss from oil-well drilling muds.

53
Gum karaya; Indian tragacanth; Karaya gum; Kadaya; Katilo; Kullo; Kuteera; Mucara; Sterculia

This is a partially acetylated polysaccharide, a galacturonorhamnan, containing about 8% acetyl groups and 37% uronic acid residues.

Occurs as an exudation from the trunk and stems of *Sterculia urens* and other *Sterculia* species (Sterculiaceae), or from *Cochlospermum gossypium* (Cochlospermaceae).

As a viscous mucilage, it is used as a binder in paper manufacture, as a denture adhesive, as a meringue stabilizer, as a substitute for gum tragacanth (q.v.), and as a thickening agent for dyes on textiles. It has laxative properties.

54
Gum tragacanth; Gum dragon; Tragacanth

This is a complex mixture of galacturonorhamnan polysaccharides (molecular weight about 840 000) containing other sugars such as arabinose, galactose and fucose.

Found as a gummy exudation of *Astragalus gummifer* and other *Astragalus* spp. (Leguminosae).

It is used as a food emulsifier and stabilizer, and used pharmaceutically as an excipient, a pastille component or stabilizer for suspensions. In textiles it is used for printing and sizing purposes.

55
Inulin; Alant starch; Alantin; Dahlin

Found in the tubers of *Dahlia variabilis* and *Helianthus tuberosus*, Jerusalem artichoke (Compositae); it is widespread in the family and also in the Campanulaceae.

It is a diagnostic aid for the glomerular filtration rate of the kidney.

56
Laminaran; Laminarin

It is a polysaccharide mainly of β-glucose joined through 1,3-linkages as drawn, allowing for varied degrees of branching.

It is found in seaweeds mainly of *Laminaria* spp., and occurs in two forms: water soluble and water insoluble, isolated from *Laminaria digitata* and *Laminaria clouston*, respectively.

Formerly used for the treatment of hysteria and other nervous conditions. Highly sulfated types produce anticoagulants comparable with heparin, while those less sulfated are antilipaemic only.

57
Mesquite gum; Prosopis gum; Sonora

It consists of arabinogalactan, with several other sugar components: rhamnose, xylose, fucose, glucuronic and galacturonic acids.

Occurs in *Prosopis* spp., notable *P. glandulosa*, *P. juliflora*, *P. torreyana* and *P. velutina* (Leguminosae).

The clear, light coloured exudate from the tree or shrub is used as an alternative to gum arabic (q.v.) or as an adhesive in gum drops. The residual pods and shoots are used as fodder.

58
Pectin

A polymer of molecular weight 20 000 to 400 000 consisting of methyl esters of araban, galactan and galacturonan in proportions that vary with source. The galacturonan molecules are linked chemically to some araban and galactan molecules.

Present in the cell walls of plant tissue as protopectin, which is enzymatically degraded to pectin. Commercial sources are found in apple residues in cider making, in orange pith and in lemon rind.

It is a food emulsifier, gelling agent and thickener for use in desserts, frozen foods, ice cream, jams and jellies; it may be amidated to provide low-sugar gels.

59
Phytoglycogen

It is a high molecular weight homopolysaccharide of branched chain structure with α-(1→4) linkages and glucopyranose residues.

Distributed in the cell protoplasm of lower plants, including fungi and yeasts. It is isolated from the fungus *Phymatotrichium omnivorum*, the causal agent of cotton root rot.

60
Porphyran

This is a low molecular weight polysaccharide mainly based on α-D-galactose; it also contains 6-*O*-methyl-D-galactose, D-galactose-6-sulfate and 3,6-anhydro-L-galactose sulfate ester.

Found notably in *Porphyra* and *Laurentia* spp., also as the sulfate in *Bangia fuscopurpurea*, and in *Gloipeltis furcata* in a highly branched form.

61
Starch; Amylum

This is a polysaccharide consisting of about 73% amylopectin branched polymer and about 27% amylose linear polymer in closely associated granular form.

Obtained from the storage granules of plants, e.g., *Zea mays*, sweet corn, *Triticum aestivum*, wheat, *Solanum tuberosum*, potato tubers and *Oryza sativa*, rice, with each characteristic of source.

Used as an antidote for iodine poisoning. In pharmacy it is used as a binder and disintegrator for tablets, and as a constituent of dusting powders and emollients.

Chapter 4
Sugar Alcohols and Cyclitols

The aldehyde or keto group of the common monosaccharides (q.v.) is readily reduced to an alcohol; reduction at the anomeric carbon atom, however, alters the possibilities for isomerism, and the same sugar alcohol may be formed from several different reducing sugars. Sorbitol (**75**), for example, is obtainable from both glucose and fructose. Such sugar alcohols, produced in this way in the laboratory by the use of sodium borohydride, also occur fairly widely in plants. They have similar solubility and R_f properties to the common monosaccharides, but do not react with some of the common sugar spray reagents and may thus be overlooked during routine plant surveys.

Glycerol is undoubtedly the best known sugar alcohol, since it is a building block of plant lipids and is universally present in plants. Of the other sugar alcohols, mannitol (**68**) (by reduction of mannose) is very common in algae, fungi and lichens as well as in higher plants (Lewis and Smith, 1967). Two that occur relatively frequently are sorbitol (**75**) (from glucose) and dulcitol (**64**) (from galactose). Sorbitol is widely distributed in the Rosaceae (e.g., in rose hips) and dulcitol in the Celastraceae. The major function of sugar alcohols is in the storage of energy; other possible functions include osmoregulation, and protection of plants from desiccation and frost damage.

A group of plant alcohols related to the alicyclic sugar alcohols are the carbocyclic inositols. These are alcohols based on cyclohexane and which usually have six hydroxyl groups, one or more of which may be methylated. Optical isomerism is again a structural feature; for example, four hexahydroxy inositols are known, *myo*-inositol (**67**), L-inositol, D-inositol and scyllitol (or *scyllo*-inositol). Other common cyclitols are pinitol (**70**, 3-methyl ether of D-inositol), present in at least 13 angiosperm families, and quebrachitol (**71**, 2-methyl ether of L-inositol), present in at least eleven angiosperm families. A rarer inositol is sequoyitol (**74**, 5-methyl ether of *myo*-inositol), which is only found in gymnosperms. The natural distribution of cyclitols is reviewed by Plouvier (1963), but see also Bieleski (1982).

References

Chemistry and Natural Distribution

Bieleski, R.L. (1982). Sugar alcohols. In *Plant Carbohydrates*, I, *Intracellular Carbohydrates*. (Loewus, F.A. and Tanner, W., eds), pp. 158–192. Berlin: Springer-Verlag.

Lewis, D.H. and Smith, D.C. (1967). Sugar alcohols in fungi and green plants. I. Distribution. *New Phytologist*, **66**, 143–184.

Plouvier, V. (1963). The distribution of aliphatic polyols and cyclitols. In *Chemical Plant Taxonomy*. (Swain, T., ed.), pp. 313–336. London: Academic Press.

Methods of Analysis

Beck, E. and Hopf, H. (1990). Branched-chain sugars and sugar alcohols. In *Methods in Plant Biochemistry*, Volume 3, *Carbohydrates*. (Dey, P.M., ed.), pp. 235–290. London: Academic Press.

Loewus, F.A. (1990). Cyclitols. In *Methods in Plant Biochemistry*, Volume 3, *Carbohydrates*. (Dey, P.M., ed.), pp. 219–234. London: Academic Press.

62
D-**Arabitol**; Arabinitol; Arabite; Lyxitol;
1,2,3,4,5-Pentanepentol

$C_5H_{12}O_5$ Mol. wt 152.15

Found in many lichens and fungi. In higher plants, it occurs in avocado seed, *Persea americana* (Lauraceae) and in *Fabiana imbricata* (Solanaceae).

Sweetener.

63
L-**Bornesitol**; (−)-Bornesitol;
L-1-*O*-Methyl-*myo*-inositol

$C_7H_{14}O_6$ Mol. wt 194.19

Occurs in *Banksia integrifolia* (Proteaceae). Also, it occurs in several members of four other families: Leguminosae (*Lathyrus*), Rhamnaceae (*Rhamnus*), Apocynaceae (*Apocynum*) and Boraginaceae (*Lithospermum ruderale*). The D form has also been found: in *Sarcocephalus diderrichii* (Rubiaceae) and in *Vinca major* (Apocynaceae).

64
Dulcitol; Dulcite; Dulcose; Euonymit; Galactitol;
meso-Galactitol; Malampyrin; Malampyrite;
Melampyrum

$C_6H_{14}O_6$ Mol. wt 182.18

Found in red algae and fungi. In higher plants, it occurs in *Melampyrum nemorosum* (Scrophulariaceae), *Gymnosporia deflexa* and *Euonymus atropurpureus* (both Celastraceae).

Mild sweetener. Its hexanitrate salt has explosive properties.

65
Erythritol; Antierythrite; *meso*-Erythritol; Erythrite;
Erythroglucin; Phycite; Phycitol

$C_4H_{10}O_4$ Mol. wt 122.12

Isolated from green algae, fungi and lichens. In higher plants, it occurs in *Primula* (Primulaceae) and in some grass species (Gramineae).

Sweetener: about twice as sweet as sucrose. It is used as a coronary vasodilator. Its tetrakis(4-hydroxybenzoyl) derivative is claimed to be antibacterial, but also cytotoxic.

66
Glycerol; Glycerin(e); Incorporation factor IFP;
Ophthalgon; 1,2,3-Propanetriol

$C_3H_8O_3$ Mol. wt 92.10

Found in all plant cells as a component of fats, and may exist as an incorporation factor for nucleic acid.

The trinitrate salt (explosive) is used as a vasodilator. It is used as a humectant, a solvent for oils, and a sweetener. It is an ingredient of antifreeze mixtures, copying inks, liqueurs, lubricants, pharmaceutical preparations, plasticizers and shock absorbing fluids.

67
myo-Inositol; Inositol; *i*-Inositol; *meso*-Inositol; Cyclohexanehexol; Cyclohexitol; Hexahydroxycyclohexane; Dambose; Inosite; Meat sugar; Nucite; Phaseomannite

$C_6H_{12}O_6$ Mol. wt 180.16

Wide occurrence in plants; it has been isolated, for example, from the tulip tree *Liriodendron tulipifera* (Magnoliaceae). It occurs linked to phosphate as phytic acid. Four isomers of *myo*-inositol have also been described from plants, but they are relatively rare.

Used as a growth factor in animals and microorganisms. Inositol is lipotropic and associated with vitamin B complexes. It is a sweetener. The hexaphosphate salt is hypocalcaemic, and the niacinate salt is a peripheral vasodilator.

68
D-Mannitol; Mannitol; Diosmol; Manicol; Manna sugar; Mannite; Mannidex; Osmitrol; Osmosal

$C_6H_{14}O_6$ Mol. wt 182.18

Widespread occurrence in higher plants, notably in *Fraxinus ornus* (Oleaceae) and *Tamarix gallica* (Tamaricaceae). Also, it occurs widely in lower plants, especially in basidiomycete fungi, many algae and lichens.

Sweetener. Its hexanitrate salt (explosive) is a vasodilator. It is used for the manufacture of artificial resins, pharmaceutical excipients and plasticizers, and as a diagnostic agent for kidney function.

69
Perseitol; D-*glycero*-D-*galacto*-Heptitol; Persitol

$C_7H_{16}O_7$ Mol. wt 212.21

Found in the leaves, fruit and seeds of the avocado, *Persea americana* (Lauraceae).

70
Pinitol; 3-*O*-Methyl-D-*chiro*-inositol

$C_7H_{14}O_6$ Mol. wt 194.19

The most widely distributed inositol monomethyl ether, having been found in six gymnosperm families (e.g., in the resin of *Pinus lambertiana* (Pinaceae)) and 13 angiosperm families (e.g., in *Bougainvillea glabra* (Nyctaginaceae)).

71
(−)-Quebrachitol; L-(−)-Quebrachitol; 2-*O*-Methyl-*chiro*-inositol

$C_7H_{14}O_6$ Mol. wt 194.19

Occurs in some 11 angiosperm families. Commercially, it is obtained from *Hevea brasiliensis* (Euphorbiaceae) by isolation from the latex, which contains 1–5% quebrachitol.

Used as a synthetic precursor.

72
(+)-Quercitol; *d*-Quercitol; "Acorn sugar"; 2-Deoxy-D-*chiro*-inositol; D-1-Deoxy-*muco*-inositol; (+)-Protoquercitol

$C_6H_{12}O_5$ Mol. wt 164.16

Found in the seeds of *Mimusops elengii* (Sapotaceae), in the acorns of *Quercus robur* (Fagaceae), and in the leaves of *Chamaerops humilis* (Palmae). There is a single record of the L-form in *Eucalyptus populnea* (Myrtaceae).

Possible use as a sweetener.

73
Ribitol; *meso*-Ribitol; Adonite; Adonitol

$C_5H_{12}O_5$ Mol. wt 152.15

Occurs in green algae, some lichens and fungi. In higher plants, it has been found in *Adonis vernalis* (Ranunculaceae), *Bupleurum falcatum* (Umbelliferae) and *Eugenia lehmannii* (Myrtaceae).

74
Sequoyitol; 5-*O*-Methyl-*myo*-inositol

$C_7H_{14}O_6$ Mol. wt 194.19

Wide occurrence in gymnosperms, e.g., the wood of *Sequoia sempervirens* (Taxodiaceae) and *Taxus baccata* (Taxaceae).

75
D-Sorbitol; Sorbitol; D-Glucitol; L-Gulitol; Sorbit

$C_6H_{14}O_6$ Mol. wt 182.18

Found in the berries of *Sorbus aucuparia* (mountain ash), and a characteristic constituent of members of the Rosaceae. Occasionally, it occurs in other plant families, e.g., in coconut milk from *Cocos nucifera* (Palmae).

Sweetener for diabetics, as it is mainly converted to carbon dioxide without appearing as glucose. Industrial uses include the manufacture of ascorbic acid, humectants, pharmaceutical excipients, plasticizers and toothpastes.

76
(−)-**Viburnitol;** 1-Deoxy-*chiro*-inositol; 1-Deoxy-*myo*-inositol; *vibo*-Quercitol

$C_6H_{12}O_5$ Mol. wt 164.16

Found in the leaves of *Viburnum tinus* (Caprifoliaceae), *Menispermum canadense* (Menispermaceae), *Gymnema sylvestre* (Asclepiadaceae) and *Achillea millefolium* (Compositae).

77
Volemitol; D-*glycero*-D-*manno*-Heptitol; D-*glycero*-D-*talo*-Heptitol; α-Sedoheptitol

$C_7H_{16}O_7$ Mol. wt 212.21

Occurs in a few fungi, lichens, and one brown alga, *Pelvetia canaliculata*. In higher plants, the chief source is the root of *Primula elatior* (Primulaceae).

Possible use as a sweetener.

Chapter 5
Organic Acids

A unique feature of plant metabolism, when compared with that of animals and microorganisms, is the ability of plants to accumulate organic acids in the cell vacuole, sometimes in considerable amounts. For example, the expressed sap of lemon fruits has a pH of 2.5, due to the presence of as much as 58 mg ml^{-1} of citric acid. Indeed, the acidity of practically all edible fruits is due to such accumulation, but the phenomenon is not confined to fruit tissue. Leaves of many plants also have this ability and members of one plant family, the Crassulaceae, are notable for the diurnal variation in the amounts of the leaf acids, mainly citric, malic and isocitric acids (Kluge and Ting, 1978). These plants carry on a modified form of photosynthesis, called Crassulacean acid metabolism (CAM), in which atmospheric carbon dioxide is converted to organic acids during the day and then the stored acids are converted to sugar during the subsequent night.

The simple organic acids that accumulate in plants fall conveniently into two groups: the tricarboxylic (Krebs) cycle acids, which include the CAM acids, and other acids. The nine tricarboxylic acid cycle acids occur, of course, in catalytic amounts in all plant tissues, but only two of these, citric and malic, regularly accumulate in such tissues. Citric acid is a major fruit constituent in the orange, lemon, strawberry, blackcurrant and gooseberry, whereas malic acid is the dominant acid in the grape, apple, plum and cherry (Ulrich, 1970). The other tricarboxylic acid cycle acids are apparently less common, although there are plants known in which they may be an abundant or major component.

Of the other plant acids, acetic acid may be considered the most important, since it serves as a universal precursor of fatty acids, lipids and many other organic plant products. Acetyl coenzyme A, together with malonyl coenzyme A, are the active forms involved in the biosynthesis of fatty acids from acetate. Acetic acid also occurs in trace amounts both free and combined (e.g., as ethyl acetate) in the essential oils of many plants. The presence of an acetic acid analogue, monofluoroacetic acid, in *Dichapetalum cymosum* (Dichapetalaceae) and a number of other plants should also be mentioned. This substance inhibits the tricarboxylic acid cycle at very low concentrations and, hence, is toxic to all living organisms. The fatal dose in man is 2–5 mg per kg body weight. Another toxic plant acid is oxalic, present in rhubarb petioles.

Three other plant acids must be mentioned because of their widespread occurrence, namely, ascorbic, shikimic and quinic. L-Ascorbic acid, or vitamin C, an organic acid in lactone form, is universal in plants. It is an essential dietary requirement in man and is, fortunately, widely distributed in food plants. Shikimic and quinic acids are two cyclohexane carboxylic acids of interest because they are precursors of aromatic compounds in plants.

Organic acids are classified chemically according to the number of carboxylic acid groups or according to whether other functional groups are present (e.g., as hydroxy acids, keto acids, etc.). The simplest monocarboxylic acid is formic, HCO_2H, the next homologue being acetic, CH_3CO_2H. The simplest dicarboxylic acid is oxalic acid, with succinic being the next higher homologue. Unsaturated derivatives of succinic acid are fumaric and maleic acids, two geometrical isomers, with fumaric being the *trans* and maleic the *cis* form. Monohydroxysuccinic acid is known as malic acid, while dihydroxysuccinic acid is tartaric acid. The keto acid corresponding to succinic is oxaloacetic, the next higher homologue of which is α-ketoglutaric, a key compound in amino acid biosynthesis.

The best known tricarboxylic acid is citric. Together with its isomer isocitric acid and its dehydration product aconitic acid, it participates in the tricarboxylic acid cycle. While citric acid has no centre of asymmetry, isocitric has two and can thus exist in four optical isomeric forms; only one of these is known naturally.

Organic acids are water-soluble, colourless liquids or relatively low melting solids. They are generally chemically stable, although α-keto acids readily undergo decarboxylation and may have to be isolated as derivatives. Acids are easily recognized by their taste in solution and by the low pH of crude aqueous plant extracts, when they occur in quantity. They are universally detected by their effect on acid–base indicators such as bromcresol green or bromothymol blue.

References

Kluge, M. and Ting, I.P. (1978). *Crassulacean Acid Metabolism.* Berlin: Springer-Verlag.
Ranson, S.L. (1965). The plant acids. In *Plant Biochemistry* (Bonner, J. and Varner, J. E., eds), pp. 493–525. New York: Academic Press.
Ulrich, R. (1970). Organic acids. In *The Biochemistry of Fruits and Their Products*, Volume 1 (Hulme, A.C., ed.), pp. 89–118. London: Academic Press.
Wang, D. (1973). Nonvolatile organic acids. In *Phytochemistry*, Volume 3 (Miller, L.P., ed.), pp. 74–111. New York: Van Nostrand.

78
Acetic acid; Ethanoic acid; Methanecarboxylic acid; Vinegar acid

$C_2H_4O_2$ Mol. wt 60.05

Wide occurrence in plant volatiles, both free (in trace amounts) and in ester form. The amyl ester, for example, is one of the flavour principles of the banana. It is universally present in all cells as the coenzyme A ester. Commercially, it is obtained by the action of *Acetobacter* on ethanol.

Glacial acetic acid is the pure compound; aqueous solutions are called acetic acid. This is a strong acid which, when ingested, causes a collapse of the circulatory system. Bacteriocidal activity begins above 5% concentration, and it is used in the bread industry to inhibit mould growth. It is an approved acidulant, flavouring agent and food stabilizer, and is widely used as a precursor in syntheses of organic compounds.

79
Aconitic acid; *cis*-Aconitic acid; Citridic acid; Achilleic acid; Equisetic acid

$C_6H_6O_6$ Mol. wt 174.11

Found in the leaves and tubers of *Aconitum napellus* (Ranunculaceae), in *Achillea* spp. (Compositae), and is the major organic acid of sugar cane juice from *Saccharum officinale* (Gramineae). It occurs universally in plants in trace amounts as an intermediate of the tricarboxylic acid cycle. The *trans* acid has been reported occasionally, e.g., from tomato, *Lycopersicon esculentum* (Solanaceae) but, since isomerization occurs readily, this may be artefactual.

Used in the manufacture of itaconic acid, and as a plasticizer for buna rubber and plastics.

80
Adipic acid; 1,4-Butanedicarboxylic acid; Hexanedioic acid

$C_6H_{10}O_4$ Mol. wt 146.15

Found in the juice of beetroot, *Beta vulgaris* (Chenopodiaceae). It is a trace constituent of many plants.

Irritant to eyes, nose and throat. It is used as an acidulant for foods. Also, it is used in the manufacture of plastics (notably nylon-6,6), resins and urethane foams.

81
Ascorbic acid; L-Ascorbic acid; Antiscorbutic vitamin; Vitamin C

$C_6H_8O_6$ Mol. wt 176.13

Wide occurrence in plants. In fruits, the concentration varies from 0.01% (in apples) and 0.2% (in blackcurrants) to as much as 1% fresh weight in rose hips and in *Citrus*. Normally, it occurs as the lactone, in equilibrium with small amounts of dehydroascorbate, the oxidized form.

Essential dietary requirement in humans, and used to treat scurvy. It is employed as an antimicrobial and antioxidant in foodstuffs.

82
Azelaic acid; Anchoic acid; 1,7-Heptanedicarboxylic acid; Lepargylic acid; Nonanedioic acid

$C_9H_{16}O_4$ Mol. wt 188.23

Occurs in rancid oleic acid; it is a constituent of olive oil.

Functions as an alternative to tetracycline in the treatment of acne and other pigmentary disorders. It is used to modify nylon moulding resins. The esters are used in hydraulic fluids, lubricants, plasticizers and in plastics (notably poly(vinyl chloride) and poly(styrene)).

83
Butyric acid; *n*-Butyric acid; Butanoic acid; Ethylacetic acid; Propylformic acid

$C_4H_8O_2$ Mol. wt 88.11

Found in grapes, *Vitis vinifera* (Vitaceae). It occurs in ester form in many plant volatiles, e.g., ethyl butyrate in apple odour. Also, it is a product of nocturnal fermentation of carbohydrates.

Esters are used as artificial flavours in confectionery and beverages. It is used in cellulose acetate butyrate manufacture, as a decalcifier for hides, and as a component of varnishes.

84
Citramalic acid; 2-Methylmalic acid; α-Hydroxypyrotartaric acid; *trans*-Methylbutanedioic acid

$C_5H_8O_5$ Mol. wt 148.12

The *S* form is found in apple peel and *Citrus* fruits. Also, it is found in lucerne, *Medicago sativa* (Leguminosae).

85
Citric acid; α-Hydroxytricarballylic acid

$C_6H_8O_7$ Mol. wt 192.13

Major acid of lemon juice (about 60 mg ml^{-1}) and of many other fruits (e.g., blackcurrant, gooseberry, raspberry). It is universally present in trace amounts in plants as the key intermediate in the tricarboxylic acid cycle, often known as the citric acid cycle.

A strong acid, responsible for the sour taste of many unripe fruits, and widely used in the food industry to add tartness to beverages. It is employed in the dyeing industry as a mordant to brighten colours.

86
Formic acid; Formylic acid; Methanoic acid

CH_2O_2 Mol. wt 46.03

Found in low concentrations in many fruits, leaves and roots of plants; it is found in stinging nettles, *Urtica dioica* (Urticaceae). Formic acid esters are found in fruit volatiles.

A strong acid with acute toxicity. Chronic absorption may cause albuminuria and haematuria. Silage treated with formic acid resists the growth of mycotoxins. Other uses are the coagulation of rubber latex, and in organic synthesis.

87
Fumaric acid; Allomaleic acid; Boletic acid

$C_4H_4O_4$ Mol. wt 116.08

Occurs in most plants in small amounts. It is found in quantity in the leaves of *Pisum sativum* (Leguminosae), *Helianthus annuus* (Compositae), and *Glaucium flavum* (Papaveraceae), and in green apples, e.g., *Malus domestica* (Rosaceae).

Substitute for tartaric acid in beverages and baking powders. It is used in the manufacture of polyhydric alcohols and synthetic resins.

88
Glutaric acid; Pentanedioic acid; 1,3-Propanedicarboxylic acid

$C_5H_8O_4$ Mol. wt 132.12

Occurs in green sugar beet, *Beta vulgaris* (Chenopodiaceae).

Moderate toxicity. It is used in the manufacture of polyamides, and polyesters.

89
D-Glyceric acid; 2,3-Dihydroxypropanoic acid; α,β-Dihydroxypropionic acid

$C_3H_6O_4$ Mol. wt 106.08

Phosphoglyceric acid is one of the first products of photosynthesis, and is therefore universally present in plants. The free acid has also been detected in a number of plants and is a biosynthetic intermediate in primary metabolism.

90
Glycolic acid; Hydroxyacetic acid; Hydroxyethanoic acid

$C_2H_4O_3$ Mol. wt 76.05

Common in low concentration in unripe grapes, apples and pears. It is a constituent of sugar cane juice. Also, it can accumulate in leaf tissue during active metabolism.

The acid is a mild irritant to the skin and mucous membranes. Its major use is pH control in cleaning metals for electroplating, copper brightening, milling or pickling; other uses are in the adhesive, leather and textile industries.

91
Glyoxylic acid; Aldehydoformic acid; Formylformic acid; Glyoxalic acid; Oxoacetic acid; Oxoethanoic acid

$C_2H_2O_3$ Mol. wt 74.04

Found in unripe fruit and young green leaves, and located in very young sugar beets. It is an intermediate in the conversion of lipid to carbohydrate in the germinating seeds of castor bean and other plant species with oil-rich seeds. Commercial production is by the action of *Aspergillus niger* on malonic or citric acid.

Irritant, and corrosive to skin.

92
Isocitric acid

$C_6H_8O_7$ Mol. wt 192.06

Accumulates in blackberries, *Rubus* spp. (Rosaceae) and in the leaves of *Bryophyllum calycinum* (Crassulaceae). It occurs in trace amounts in all plant tissues, since it is an intermediate in the tricarboxylic acid cycle. Of the four possible stereoisomers, the natural acid is the *threo*-D_s isomer.

93
Isovaleric acid; Delphinic acid; Isopropylacetic acid; 3-Methylbutanoic acid

$C_5H_{10}O_2$ Mol. wt 102.14

Found in hop oil, *Humulus lupulus* (Cannabaceae), tobacco, *Nicotiana tabacum* (Solanaceae), and *Valeriana* spp. (Valerianaceae).

Disagreeable, rancid-cheese odour. It is used in the manufacture of sedatives.

94
α-Ketoglutaric acid; 2-Oxoglutaric acid

$C_5H_6O_5$ Mol. wt 146.10

Isolated from barley grain, *Hordeum vulgare* (Gramineae). It is present in trace amounts in all plant tissues, since it is an intermediate in the tricarboxylic acid respiratory cycle.

95
L-Lactic acid; 2-Hydroxypropanoic acid; α-Hydroxypropionic acid; Tonsillosan

$C_3H_6O_3$ Mol. wt 90.08

Occurs in *Digitalis purpurea* (Scrophulariaceae), and also in apple, pear, grape and banana fruits.

An approved acidulant and preservative for foods, also with a synergistic effect on antioxidants present. It is widely used in brewing, and in the food industry. Usage is cautioned for very young; otherwise, no toxicological problems are known. Also, it is used as a solvent and in the casting of phenolaldehyde resins.

96
L-Malic acid; Hydroxybutanedioic acid; Hydroxysuccinic acid

$C_4H_6O_5$ Mol. wt 134.09

A major acid in fruits such as apple, apricot, banana, grape and peach. It accumulates at night time in the leaves of plants with Crassulacean acid metabolism, e.g., *Kalanchoe daigremontiana* (Crassulaceae). It occurs as a trace component in all plant cells, since it is an intermediate in the tricarboxylic acid cycle.

97
Malonic acid; Methanedicarboxylic acid; Propanedioic acid

$C_3H_4O_4$ Mol. wt 104.07

Occasionally accumulates in plant tissues in leaf, e.g., *Apium graveolens* (Umbelliferae), stem, e.g., *Phaseolus coccineus* (Leguminosae), root, e.g., *Beta vulgaris* (Chenopodiaceae) or seed, e.g., *Hordeum vulgare* (Gramineae). It is universally present in plants in bound form as malonyl coenzyme A, a key precursor of fatty acid synthesis. Conjugated forms of, for example, flavonoids are known in which malonic acid residues are attached.

This is a strong acid, but less toxic than oxalic acid, and a strong irritant, causing damage to mucous membranes and skin. It is used in the manufacture of barbiturates. Diesters are widely used as intermediates in organic syntheses.

98
Mevalonic acid

$C_6H_{12}O_4$ Mol. wt 148.16

Universally present in plants in trace amounts as an intermediate of terpenoid (q.v.) synthesis. It readily ring closes to form the δ-lactone mevalonolactone.

99
Monofluoroacetic acid; Fluoroacetic acid

$C_2H_3O_2F$ Mol. wt 78.04

Occurs in the South African plant "Gifblaar", *Dichapetalum cymosum* (Dichapetalaceae). It is also found in some 34 spp. of *Gastrolobium* and *Oxylobium* (Leguminosae), native to Western Australia.

It is highly poisonous to humans and to farm animals. The fatal dose in humans is between 2 and 5 mg per kg body weight. It stops respiration by blocking the tricarboxylic acid cycle.

100
Oxalacetic acid; Ketosuccinic acid; Oxaloacetic acid; Oxobutanedioic acid; Oxosuccinic acid

$C_4H_4O_5$ Mol. wt 132.08

Found in barley grain, *Hordeum vulgare* (Gramineae). It is universally present in plants in trace amounts, since it is an intermediate in the tricarboxylic acid respiratory cycle.

101
Oxalic acid; Ethanedioic acid

$C_2H_2O_4$ Mol. wt 90.04

Wide occurrence in plants as crystalline occlusions in the cell vacuoles, known as "raphides", as the insoluble calcium salt. Also, it occurs as the free acid, or as the soluble potassium salt, in spinach, *Spinacia oleracea* (Chenopodiaceae), rhubarb, *Rheum rhaponticum* (Polygonaceae) and *Oxalis* spp. (Oxalidaceae).

Poisonous, when taken in soluble form, causing paralysis of the nervous system. It is corrosive to the skin, and precipitates blood calcium. In veterinary therapy it is used as a haemostatic agent. Commercially, it is used as a condensing agent in syntheses of polyamides and intermediate for dyes, for bleaching of leather and straw, for cleaning metals and wood, and for textile finishing.

102
Pimelic acid; Heptanedioic acid

$C_7H_{12}O_4$ Mol. wt 160.17

Occurs in castor oil, *Ricinus communis* (Euphorbiaceae).

Used in the preparation of biotin by biosynthesis with fungi and bacteria. It is incorporated into polyamide and polyester fibres.

103
Propionic acid; Ethylformic acid; Methylacetic acid; Propanoic acid; Pseudoacetic acid

$C_3H_6O_2$ Mol. wt 74.08

Chiefly occurs in plants in ester form (e.g., as the methyl propionate) in fruit volatiles, e.g., in apple.

It is used as an antifungal agent, a mould inhibitor, a preservative, or in fruit flavouring. It is an esterifying agent for ester solvents, alkyl resins and perfume bases, and also used in the production of cellulose propionate and of methacrylic acid.

104
Pyruvic acid; Acetylformic acid; α-Ketopropionic acid; 2-Oxopropanoic acid; Pyroracemic acid

$C_3H_4O_3$ Mol. wt 88.07

A trace component of all plant tissues. It is present as the phosphoenol derivative as an intermediate of glycolysis.

105
Quinic acid; Chinic acid; Kinic acid

$C_7H_{12}O_6$ Mol. wt 192.17

A common plant acid, contributing to the acidity in many fruits, e.g., apple, apricot, peach, banana and pear. Also widely present in vegetative tissues.

Contributes to the sourness of unripe apples.

106
Sebacic acid; Decanedioic acid

$C_{10}H_{18}O_4$ Mol. wt 202.25

Occurs in castor oil, *Ricinus communis* (Euphorbiaceae).

Esters have low toxicity and are therefore used in food packaging. It is used for the manufacture of alkyd and polyester resins, nonmigrating plasticizers and synthetic polyamide fibres, and as a cross-linkage agent for epoxy resins.

107
Shikimic acid

$C_7H_{10}O_5$ Mol. wt 174.16

Occurs in the fruit of *Illicium religiosum* (Magnoliaceae) and recognized as a minor acid of several other plant fruits, providing 20% of the acidity of gooseberry; it is also detectable in cherry and strawberry. It is universally present in plants in trace amounts, as an intermediate in the "shikimate pathway" from photosynthetic sugars to aromatic compounds.

Is a powerful mutagen, and is thought to be responsible in part for the carcinogenic action of bracken.

108
Suberic acid; Octanedioic acid

$C_8H_{14}O_4$ Mol. wt 174.20

Occurs in castor oil, *Ricinus communis* (Euphorbiaceae).

It is an intermediate in the synthesis of drugs, dyes and high polymers.

109
Succinic acid; Amber acid; Butanedioic acid

$C_4H_6O_4$ Mol. wt 118.09

A minor acid of plant fruits, e.g., in redcurrant, strawberry, grape and pear. It occurs in leaf tissue, e.g., *Medicago sativa* (Leguminosae), and is present in all plants in trace amount as an intermediate of the tricarboxylic acid cycle. Succinyl coenzyme A, which is the metabolically active form, participates in a number of biosynthetic reactions.

It is used in foods as neutralizing agent, sequestrant and buffer. Industrially, it is used as a modifier for plastic fibre, and for colour stability in poly(vinyl chloride) and acrylic resins.

110
L-(+)-**Tartaric acid**; 2,3-Dihydroxybutanedioic acid

$C_4H_6O_6$ Mol. wt 150.09

Found in the fruits of the grape, *Vitis vinifera* (Vitaceae), of *Morus indica* (Moraceae) and of *Tamarindus indica* (Leguminosae). It is present in the leaves of *Pelargonium* spp. (Geraniaceae) as the major acid, in amounts from 0.1 to 1.0 μmol per mg dry weight and also widely present in the leaves of other angiosperms as a minor component.

Strong acid taste, and contributes to the acidity of grapes and of wines. It is used in the food industry as an acidulant.

111
Tiglic acid; *trans*-2.3-Dimethylacrylic acid; *trans*-2-Methylcrotonic acid

$C_5H_8O_2$ Mol. wt 100.12

Found as the glyceride in the seed oil (croton oil) from *Croton tiglium* (Euphorbiaceae), as the butyl ester in *Anthemis nobilis* (Compositae), as the geranyl tiglate in oil of geranium, and as a constituent of *Angelica archangelica* (Umbelliferae).

The free acid is used to break emulsions. Its esters provide bases for flavours and perfumes. *Angelica* (taste of Benedictine) is used for cake decoration.

112

Valeric acid; *n*-Valeric acid; Pentanoic acid; Propylacetic acid; Valerianic acid

$$CH_3-[CH_2]_3-C(=O)OH$$

$C_5H_{10}O_2$ Mol. wt 102.14

Occurs in numerous essential oils, e.g., valerian oil, *Valeriana officinalis* (Valerianaceae).

Low toxicity (LD_{50} intravenously in mice 1290 mg kg^{-1}). It is an intermediate in perfumery and flavours.

Chapter 6
Fatty Acids and Lipids

Fatty acids occur mainly in plants in bound form, esterified to glycerol, as fats or lipids. These lipids comprise up to 7% of the dry weight in leaves in higher plants, and are important as membrane constituents in the chloroplasts and mitochondria. Lipids also occur in considerable amounts in the seeds or fruits of many plants, and these provide a storage form of energy to use during germination. Seed oils from plants such as the olive, palm, coconut, soya bean, sunflower, rape and peanut are exploited commercially and used as food fats, for soap manufacture and in the paint industry. Plant fats, unlike animal fats, are rich in unsaturated fatty acids, and there is evidence that some of these may be essential as a dietary requirement in man.

Lipids are defined by their special solubility properties, and are extractable with alcohol or ether from living plant tissues. Such extraction removes certain other classes of lipid, such as leaf alkanes (q.v.) and steroids (q.v.), but leaves behind all the water-soluble components of the plant tissue.

There are three classes of lipid: the neutral triglycerides, the polar phospholipids and the sugar-containing glycolipids. Structural variation within each class is due to the different fatty acid residues that may be present. Triglycerides are "simple" if the same fatty acid is present in all three positions in the molecule; one example is triolein, a triglyceride with three oleic acid residues. Much more common are "mixed" triglycerides, in which different fatty acids are present in all three positions.

Phospholipids are complex in structure; all contain not only a phosphate group but also at least one other, usually basic, substituent. This basic residue may be choline, ethanolamine or serine. In addition, there are phospholipids with two or more glycerol residues or with inositol attached, instead of an organic base. Phosphatidylglycerol, for example, is a major fraction (about 20%) of the phospholipids in leaf tissue, and is located particularly in the chloroplasts and mitochondria.

By contrast with the numerous types of phospholipids in plants, there are only a few glycolipids. Most important are monogalactosyl and digalactosyl diglycerides, highly surfactant molecules which play a role in chloroplast metabolism. Finally, there is one sulfolipid; this is a diglyceride with the sugar quinovose attached. Quinovose is a 6-deoxyglucose with a sulfonic acid residue in the 6-position. First discovered in green algae, this sulfolipid is universal in plants as an essential component of the chloroplast.

Identification of lipids mainly requires the determination of their fatty acid components. Although several hundred fatty acids are now known in plants, most lipids have the same few fatty acid residues, which makes their identification that much easier. The common fatty acids are either saturated or simple unsaturated compounds of C_{16} or C_{18} chain length. Palmitic acid (**133**), a C_{16} acid, is the major saturated acid in leaf lipids and also occurs in quantity in some seed oils, e.g. peanut oil. Stearic acid, C_{18} (**138**) is less prominent in leaf lipids, but is a major saturated acid in seed fats in a number of plant families.

Unsaturated acids based on C_{16} and C_{18} are widespread in both leaf and seed oils. Oleic acid (**132**) comprises 80% of the fatty acid content of olive oil, 59% in peanut oil and is often accompanied by the di-unsaturated linoleic acid (**125**). The tri-unsaturated linolenic acid (**126**) is common, occurring in linseed oil to the extent of 52% of the total acid, with linoleic (15%) and oleic (15%). Unsaturated acids can exist in both Z and E forms, but most natural acids in fact have the Z configuration.

Some rarer fatty acids are found as lipid components, many occurring characteristically in seed oils of just a few related plants. Petroselinic acid (**135**), an isomer of the more common palmitoleic acid (**134**), occurs to the extent of 76% of the total acids in parsley seed, *Petroselinum crispum*; it appears to be present in many other Umbelliferae and also in the related Araliaceae. Erucic acid (**120**) is another unusual acid, found especially in the Cruciferae and the Tropaeoliaceae. It is present in high concentration in rape (*Brassica rapa*) seed oil and, because of its alleged toxic properties, successful efforts have been made to breed rape varieties with low erucic acid content. A third unusual acid is sterculic (**139**) with a unique cyclopropene ring in the middle of the molecule, which is found in *Sterculia* species (Sterculiaceae).

Fatty acids, obtained by acid or alkaline hydrolysis of fats or oils, are converted to their methyl esters by ethereal diazomethane and then analysed by gas chromatography. Alternatively, fatty acid methyl esters can be obtained directly by refluxing the lipids in the presence of methanol–benzene–sulfuric acid. For the identification of the more unusual fatty acids, gas chromatography linked to mass spectrometry is the recommended procedure (Gunstone, 1975).

There are many excellent up-to-date reviews and books on the chemistry and biochemistry of lipids. The best general references to plant lipids are the book of Hitchcock and Nichols (1971) and the review chapter of Harwood (1980). The chemistry of the fatty acids is covered by Gunstone (1967, 1975). The fatty acid composition of plant seed oils is reviewed by Shorland (1963), but see also Harwood (1980). For a general book on both plant and animal lipids, see Gunstone *et al.* (1986).

References

Gunstone, F.D. (1967). *An Introduction to the Chemistry and Biochemistry of Fatty Acids and their Glycerides*, 2nd Edition. London: Chapman and Hall.

Gunstone, F.D. (1975). Determination of the structure of fatty acids. In *Recent Advances in the Chemistry and Biochemistry of Plant Lipids*. (T. Galliard and E.I. Mezier, eds), pp. 21–43. London: Academic Press.

Gunstone, F.D., Harwood, J.L. and Padley, F.B. (1986). *The Lipid Handbook*. London: Chapman and Hall.

Harwood, J.L. (1980). Plant acyl lipids. In *The Biochemistry of Plants*, Volume 4, *Lipids* (P.K. Stumpf, ed.), pp. 2–56. New York: Academic Press.

Hitchcock, C. and Nichols, B.W. (1971). *Plant Lipid Biochemistry*. London: Academic Press.

Shorland, F.B. (1963). The distribution of fatty acids in plant lipids. In *Chemical Plant Taxonomy*. (Swain, T., ed.), pp. 253–312. London: Academic Press.

113
Arachidic acid; Arachic acid, Eicosanoic acid; Icosanoic acid

$CH_3-[CH_2]_{18}-COOH$

$C_{20}H_{40}O_2$ Mol. wt 312.54

Found in seed oils, notably from the peanut, *Arachis hypogaea* (Leguminosae), about 2.4%.

Used in organic syntheses, and is a component in some lubricating greases, waxes and plastics.

114
Arachidonic acid;
cis-5,*cis*-6,*cis*-11,*cis*-14-Eicosatetraenoic acid; 5,8,11,14-Icosatetraenoic acid

$CH_3-[CH_2]_4-CH_2-CH_2-CH_2-[CH_2]_3-COOH$ (with four cis double bonds)

$C_{20}H_{32}O_2$ Mol. wt 304.48

Has not been found in higher plants, but has been detected in algae, mosses and ferns. In the lipids of some mosses (e.g., *Mnium* spp.) it makes up 34% of the total fatty acids; also it occurs in lipids (26%) of the fern *Scolopendrium vulgare* (Aspleniaceae).

A precursor in the biosynthesis of thromboxanes, leukotrienes and prostaglandins. It is used in combination with linoleic and linolenic acids to treat a fat deficiency associated with vitamin F; also it is used in the treatment of eczema in dogs and swine.

115
Behenic acid; Docosanoic acid

$CH_3-[CH_2]_{20}-COOH$

$C_{22}H_{44}O_2$ Mol. wt 340.59

Jamba oil contains about 50% behenic and hydrogenated behenic acids. Rice-bran wax contains about 16% as free acid and glycerides. It occurs in most seed oils, especially from Cruciferae, and in carnauba wax.

Used in cosmetics, waxes, plasticizers, stabilizers, and in chemical syntheses.

116
Chaulmoogric acid;
(*S*)-13-(Cyclopent-2-enyl)tridecanoic acid; Hydnocarpylacetic acid

$C_{18}H_{32}O_2$ Mol. wt 280.45

Present as the glyceride ester in chaulmoogra oil (27%) extracted from seeds of *Hydnocarpus wightiana* (Flacourtiaceae).

The ethyl ester has been used in the treatment of leprosy.

117
Decanoic acid; Capric acid; Decoic acid; Decylic acid

$CH_3-[CH_2]_8-COOH$

$C_{10}H_{20}O_2$ Mol. wt 172.27

Occurs in the seed oils of *Cuphea* (Lythraceae) and as a minor constituent (7%) of coconut oil, *Cocos nucifera* (Palmae).

Used in the manufacture of esters for artificial fruit flavours, and in corrosion inhibitors, and in some surfactants.

118
(9*S*,10*R*)-9,10-Dihydroxystearic acid;
9,10-Dihydroxyoctadecanoic acid

$C_{18}H_{36}O_4$ Mol. wt 316.49

A minor ingredient (<1%) of castor oil, *Ricinus communis* (Euphorbiaceae). The 9*R*,10*R* form occurs in sclerotial lipids of the fungus *Claviceps sulcata*, and the 9*S*,10*R* form is prepared from *Shorea robusta* (Dipterocarpaceae) seed oil.

Used in the manufacture of cosmetic and toilet preparations. It yields azelaic acid on alkali cleavage.

119
α-Eleostearic acid;
cis-9, trans-11, trans-13-Octadecatrienoic acid

$C_{18}H_{30}O_2$ Mol. wt 278.44

This occurs as a major component (about 80%) of seed oil of *Aleurites cordata* (Euphorbiaceae), also known as tung oil. It is also found (50%) in the seed oil of *Kentranthus macrosiphon* (Valerianaceae).

Used in manufacture of varnishes.

120
Erucic acid; cis-13-Docosenoic acid

$C_{22}H_{42}O_2$ Mol. wt 338.58

Found in the seed fats of Cruciferae and Tropaeolaceae, constituting 40–50% of the total fatty acid content of rape, mustard and wallflower seed fats, and up to 80% of the seed fat of *Tropaeolum majus*, nasturtium. Also, it is found in small amounts in oils of unrelated plants, e.g., *Luzuriaga parriflora* (Philesiaceae).

Edible oil, subject to legal control, which causes adverse physiological effects when ingested in large amounts by animals (e.g., rats). Modern rape cultivars produce seed oils which are low in erucic acid. It is used as a lubricant and in organic syntheses.

121
Hexanoic acid; *n*-Hexanoic acid; Caproic acid; Hexoic acid; Hexylic acid

$C_6H_{12}O_2$ Mol. wt 116.16

Occurs as a minor constituent (1%) of coconut, *Cocos nucifera* (Palmae), and other palm oils.

Used as an acidulating agent in foods, and also in the manufacture of hexyl esters and phenols.

122
Lauric acid; Dodecanoic acid; Dodecoic acid; Laurostearic acid

$C_{12}H_{24}O_2$ Mol. wt 200.32

Coconut oil contains about 60% lauric acid in the form of trilaurin. Other sources of supply are *Elaeis guineenis* (Palmae), palm oil, and also *Holoptelia integrifolia* (Ulmaceae), and members of the Lauraceae.

Used in the production of alkyd resins, cocoa butter substitutes, flavourings, margarine and soaps.

123
α-Licanic acid;
4-Oxo-cis-9,trans-11,trans-13-octadecatrienoic acid

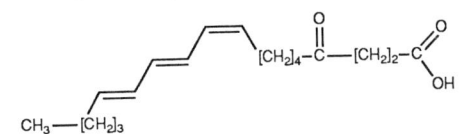

$C_{18}H_{28}O_3$ Mol. wt 292.42

Major fatty acid (40%) in Oiticica oil derived from *Licania rigidia* (Chrysobalanaceae). It also occurs (10%) in the seed oil of *Chrysobalanus icaco* (Chrysobalanaceae).

Used as a drying oil for varnishes, and as a component in the manufacture of alkyd resins.

124
Lignoceric acid; *n*-Tetracosanoic acid

$$CH_3-[CH_2]_{22}-C(=O)OH$$

$C_{24}H_{48}O_2$ Mol. wt 368.65

Major sources are rice-bran wax (about 40% as free acid and esters), and the seed fat of *Adenanthera pavonina* (Leguminosae) (about 25%). Other sources are beechwood and rotten-oakwood distillates. Small amounts (0.2–1%) occur in most natural fats.

Some uses in biochemical research.

125
Linoleic acid; 9,12-Linoleic acid; Linolic acid; *cis*-9,*cis*-12-Octadecenoic acid

$C_{18}H_{32}O_2$ Mol. wt 280.45

Widespread occurrence in plant lipids. Rich sources include sunflower oil (about 65%), peanut oil (about 25%), cottonseed, linseed and soybean oils.

It is a nutrient and an essential fatty acid component of vitamin E. Also, it is an ingredient of semi-drying oils as used in paints and coatings. It is regarded as a beneficial dietary component for men who may be prone to coronary heart disease.

126
α-Linolenic acid; *cis*-9,*cis*-12,*cis*-15-Octadecatrienoic acid

$C_{18}H_{30}O_2$ Mol. wt 278.44

Wide occurrence in vegetable oils. The major source is linseed oil extract from the dried ripe seeds of *Linum usitatissimum* (Linaceae).

Used as a nutrient, and also in the paint industry as a drying oil.

127
γ-Linolenic acid; Gamolenic acid; *cis*-6,*cis*-9,*cis*-12-Octadecatrienoic acid

$C_{18}H_{30}O_2$ Mol. wt 278.44

Widespread occurrence in vegetable oils, notably in linseed and other drying oils. The contents are about 10% in evening primrose oil, 20% in borage and blackcurrant oils.

Used as an antihypercholesterolaemic drug. It is a nutrient, and also an essential ingredient of cat food diets. A major use is as a drying oil for paints and polishes.

128
Malvalic acid; 8,9-Methylene-8-heptadecenoic acid; 7-(2-Octylcyclopropenyl)heptanoic acid

$C_{18}H_{32}O_2$ Mol. wt 280.45

Present in Malvaceae (e.g., in the seed oil of *Hibiscus syriacus*); also it is a minor constituent of cottonseed oil.

129
Myristic acid; Tetradecanoic acid

$$CH_3-[CH_2]_{12}-C(=O)OH$$

$C_{14}H_{28}O_2$ Mol. wt 228.38

Found in the fat of *Myristica fragrans* (Myristicaceae), nutmeg, 70–80%, and in coconut and cottonseed oils; it is also found in *Iris florentina*, orris root (Iridaceae).

Used in the synthesis of esters for flavours and perfumes. It is a constituent of lubricants, soaps and shaving creams, and is also used for waterproofing leather.

130
Myristoleic acid; *cis*-Tetradec-9-enoic acid

$$CH_3-[CH_2]_7 \overset{}{=} [CH_2]_7-C\overset{O}{\underset{OH}{}}$$

$C_{18}H_{34}O_2$ Mol. wt 282.47

Occurs in the seed oil of *Elaeis guineensis* (Palmae). It occurs widely in small amounts in vegetable oils and fats.

131
Octanoic acid; Caprylic acid; Octoic acid; Octylic acid

$$CH_3-[CH_2]_6-C\overset{O}{\underset{OH}{}}$$

$C_8H_{16}O_2$ Mol. wt 144.22

A minor constituent (8%) of coconut, *Cocos nucifera* (Palmae), and other palm oils.

Antifungal activity against dermatophytes and *Candida* spp. It is an intermediate used in the manufacture of dyes, and of esters used in perfumery or as flavours.

132
Oleic acid; *cis*-9-Octadecenoic acid; Red oil

$$CH_3-[CH_2]_7 \overset{}{=} [CH_2]_7-C\overset{O}{\underset{OH}{}}$$

$C_{18}H_{34}O_2$ Mol. wt 282.47

Widespread occurrence in vegetable fats and oils. Major sources are olive oil (65–86%), peanut oil (about 55%) and sunflower oil (about 20%).

Used to assist the absorption of drugs through the skin, but it is mildly irritating and not used in eye ointments. Other uses are as a soft-soap drier, as a thickening agent for lubricating oils, and for waterproofing textiles.

133
Palmitic acid; Cetylic acid; Hexadecanoic acid; Hexadecylic acid

$$CH_3-[CH_2]_{14}-C\overset{O}{\underset{OH}{}}$$

$C_{16}H_{32}O_2$ Mol. wt 256.43

A ubiquitous constituent of plant lipids, and a major component of most fats, oils and waxes, notably palm oil (about 40%), coconut oil, cottonseed oil, Japan wax, olive oil (about 9%), and peanut oil (about 8%).

Used as lubricant and as an emulsifying agent. Also, it is used in alkyd resin manufacture and as an ingredient of drying oils and soaps.

134
Palmitoleic acid; *cis*-9-Hexadecenoic acid

$$CH_3-[CH_2]_5 \overset{}{=} [CH_2]_7-C\overset{O}{\underset{OH}{}}$$

$C_{16}H_{30}O_2$ Mol. wt 254.42

Occurs in the seed oil of *Elaeis guineensis* (Palmae), palm oils, and is widespread in vegetable oils.

Used in the manufacture of soaps, and is employed in organic syntheses.

135
Petroselinic acid; 5-Heptadecylene-1-carboxylic acid; *cis*-6-Octadecenoic acid; 6-Octadecylenic acid

$$CH_3-[CH_2]_{10} \overset{}{=} [CH_2]_4-C\overset{O}{\underset{OH}{}}$$

$C_{18}H_{34}O_2$ Mol. wt 282.47

Found in parsley seed, *Petroselinum crispum* (Umbelliferae); it is the major seed oil constituent (50–90%) in plants of the Umbelliferae and closely related Araliaceae.

136
Punicic acid; *cis*-9,*trans*-11,*cis*-13-Octadecatrienoic acid; Trichosanic acid

$C_{18}H_{30}O_2$ Mol. wt 278.44

Found in the seed oil (71%) of *Punica granatum* (Punicaceae), pomegranate, and of *Trichosanthes cucumeroides* (Cucurbitaceae).

137
Ricinoleic acid; 12-Hydroxy-*cis*-9-octadecenoic acid; α-12-Hydroxyoleic acid

$C_{18}H_{34}O_3$ Mol. wt 298.47

The major source is castor oil (85–90%) from the seeds of *Ricinus* spp. (Euphorbiaceae), and it is also found in ergot oil (about 40%), and as a minor constituent of linseed oil.

In combination with nonoxinols it can be used as a spermicide. The methyl ester is used to produce nylon-11-polyamide. Safety laminations use ricinolate plasticizers. It is used as an additive to dry cleaning soaps for textile finishing, and to Turkey Red oil.

138
Stearic acid; *n*-Octadecanoic acid

$C_{18}H_{36}O_2$ Mol. wt 284.49

The most common fatty acid occurring in vegetable oils and fats. It is chiefly present as glyceryl esters.

Used in the manufacture of alkyd resins, aluminium and zinc stearates, and candles, used as a lubricant for cosmetics, employed in pharmaceuticals, notably enteric pills, ointments and suppositories, used as a thickener for greases and also as a drying oil ingredient.

139
Sterculic acid; 8-(2-Octylcyclopropenyl)octanoic acid

$C_{19}H_{34}O_2$ Mol. wt 294.48

A major fatty acid (53%) in the seed oil of *Sterculia foetida* (Sterculiaceae). It occurs in seed oils in species of Malvaceae, Sterculiaceae, Tiliaceae and Bombacaceae, and is a minor constituent of cottonseed oil, *Gossypium hirsutum*.

The sterculic acid component of cottonseed oil has been shown to be deleterious to certain organisms, and blocks the action of Δ^9-desaturase.

140
Vaccenic acid; *trans*-11-Octadecenoic acid

$C_{18}H_{34}O_2$ Mol. wt 282.47

Present at the 15% level in the seed oil of *Macfadyena unguis-cati* (Bignoniaceae) and *Asclepias syriaca* (Asclepiadaceae). It also occurs in the lipids of bacteria.

It is a growth promoting factor in rats.

141
Vernolic acid: *cis*-12,13-Epoxy-*cis*-9-octadecenoic acid; 12,13-Epoxyoleic acid; 11-(3-Pentyloxiranyl)-9-undecenoic acid

$C_{18}H_{32}O_3$ Mol. wt 296.45

Main acid (72%) of the seed oil of *Vernonia anthelmintica* (Compositae), and a constituent (up to 68%) of some *Crepis* oils (Compositae). It also occurs in the seed oils of species of Euphorbiaceae, Onagraceae, Valerianaceae and Dipsacaceae.

Chapter 7
Hydrocarbons and Derivatives

Two groups of hydrocarbon are considered in this section: short chain and long chain. Short chain hydrocarbons are widely present in volatile emanations of plant tissues, especially in flower scents or fruit odours (Maarse and Visscher, 1989). Pentadecane (**160**) is one example, from *Magnolia* flowers. Oxygenated derivatives are also found, such as octan-1-ol (**159**) in orchid flowers (Borg-Karlson *et al.*, 1985). While such hydrocarbons generally contribute pleasant odours to plants, hydrocarbons substituted by sulfur are noxious smelling. Methyl mercaptan (**154**) has the odour of rotten cabbage, while dimethyl disulfide (**145**) is equally unpleasant. Two series of sulfur compounds are found widely in species of *Allium*, with one (e.g., diallyl sulfide, **144**) or two (e.g., diallyl disulfide, **143**) sulfur atoms (Bernhard, 1970).

Long chain hydrocarbons are universally distributed in the waxy coatings of leaves and other plant organs. Simple hydrocarbons are known as alkanes, with the general formula $CH_3(CH_2)_nCH_3$, they are usually present in the range C_{25} to C_{35}, and hentriacontane (**147**) with 31 carbon atoms is a typical member. Biosynthetically, these hydrocarbons are related to fatty acids (q.v.), and are formed from them by chain elongation and decarboxylation. The function of alkanes in cuticular waxes is a protective one, providing water repellency and also a degree of disease resistance.

Besides the simple alkanes, waxes contain unsaturated hydrocarbons and substances with oxygen substitution. Hentriacontan-1-ol (**149**) is a typical alcohol from leaf waxes, while hentriacontanone (**150**) is a typical ketone. β-Diketones, such as hentriacontane-14,16-dione (**148**) are also present, but these are relatively uncommon.

Any given plant wax may have up to thirty or more major hydrocarbon constituents, and these are usually separated and identified by gas chromatography. General reviews of plant wax constituents include those of Eglinton and Hamilton (1962), Martin and Juniper (1970) and Baker (1982).

References

Baker, E.A. (1982). Chemistry and morphology of plant epicuticular waxes. In *The Plant Cuticle* (Cutler, D.F., Alvin K.L. and Price, C.E., eds), pp. 139–166. London: Academic Press.

Bernard, R.A. (1970). Chemotaxonomy: distribution of sulfur compounds in *Allium*. *Phytochemistry* **9**, 2019–2027.

Berg-Karlson, A.K., Bergstrom, G. and Groth, I. (1985). Chemical basis for the relationship between *Ophrys* orchids and their pollinators. *Chemica Scripta* **25**, 283–294.

Eglinton, G. and Hamilton, R.J. (1962). The distribution of alkanes., In *Chemical Plant Taxonomy (Swain, T., ed.), pp. 187–218.* London: Academic Press.

Maarse, H. and Visscher, C.A. (eds) (1989). *Volatile Compounds in Food*, in 3 Volumes. Zeist, The Netherlands: TNO-CIVO Food Analysis Institute.

142
Allicin; *S*-Oxodiallyl disulfide

$$CH_2=CH-CH_2-S(=O)-S-CH_2-CH=CH_2$$

$C_6H_{10}OS_2$ Mol. wt 162.28

Found in onion and garlic bulbs. It is produced by the action of allianase on the amino acid alliin (q.v.) when garlic is crushed.

Major odour principle of garlic. It has antidiabetic, antihypertensive, antibiotic and antithrombotic activities.

143
Diallyl disulfide: Allyl disulfide

$$CH_2=CH-CH_2-S-S-CH_2-CH=CH_2$$

$C_6H_{10}S_2$ Mol. wt 146.28

Principal constituent of oil of garlic from *Allium sativum* (Alliaceae), and occurs in many other *Allium* spp. including the onion, *A. cepa*.

Insecticidal. When released from onion roots into the soil, it stimulates germination of the onion fungal pathogen *Sclerotium cepivorum*. Diallyl disulfide has been used to treat soil, prior to planting onions, to reduce the chance of fungal infection.

144
Diallyl sulfide; Allyl sulfide; Thioallyl ether; Oil garlic

$$CH_2=CH-CH_2-S-CH_2-CH=CH_2$$

$C_6H_{10}S$ Mol. wt 114.21

Found in garlic, *Allium sativum* (Alliaceae).

Garlic odour. It is used in the manufacture of flavours. It is a strong irritant to eyes and skin.

145
Dimethyl disulfide: Methyl disulfide

$$CH_3-S-S-CH_3$$

$C_2H_6S_2$ Mol. wt 94.20

A volatile released from the tissue of many *Allium* spp. (Alliaceae).

Highly unpleasant odour, and used as a gas odorant. It is used in chemical syntheses.

146
Dipropyl disulfide; Propyl disulfide

$$CH_3-CH_2-CH_2-S-S-CH_2-CH_2-CH_3$$

$C_6H_{14}S_2$ Mol. wt 150.31

Found in onions, and in other *Allium* spp. (Alliaceae).
Used as a flavour compound.

147
Hentriacontane

$$CH_3-[CH_2]_{29}-CH_3$$

$C_{31}H_{64}$ Mol. wt 436.85

Occurs in many plant waxes, e.g., the epicuticular wax of *Solandra grandiflora*, and tomato, *Lycopersicon esculentum* (both Solanaceae), and in the roots of *Oenanthe crocata* (Umbelliferae).

Used in lipsticks and other cosmetics, in chewing gum, and as a protective coating for certain citrus fruits.

148
Hentriacontane-14,16-dione; Grass wax

$$CH_3-[CH_2]_{12}-C(=O)-CH_2-C(=O)-[CH_2]_{14}-CH_3$$

$C_{31}H_{60}O_2$ Mol. wt 464.82

Wide occurrence in the epicuticular waxes of cereals, e.g., wheat, *Triticum aestivum*, and grasses, e.g., *Andropogon* spp. (Gramineae).

149
Hentriacontan-1-ol; Hentriacontanol

$$CH_3-[CH_2]_{29}-CH_2OH$$

$C_{31}H_{64}O$ Mol. wt 452.85

A leaf wax constituent, isolated from, e.g., *Agave sisalana* (Agavaceae).

150
Hentriacontan-16-one; Hentriacontanone

$$CH_3-[CH_2]_{14}-\underset{\underset{O}{\|}}{C}-[CH_2]_{14}-CH_3$$

$C_{31}H_{60}O$ Mol. wt 448.82

A characteristic plant wax ketone; it is found in *Adiantum caudatum* (Adiantaceae), *Platanus orientalis* (Platanaceae), *Piper nigrum* (Piperaceae) and *Santalum album* (Santalaceae).

151
Hexacosan-1-ol; Hexacosanol; Ceryl alcohol

$$CH_3-[CH_2]_{24}-CH_2OH$$

$C_{26}H_{54}O$ Mol. wt 382.72

Occurs in many plant epicuticular waxes, e.g., in carnauba wax, *Copernicia prunifera* (Palmae), in the leaf wax of *Dactylis glomerata* (Gramineae) and of *Brassica oleracea* (Cruciferae).

152
Lenthionine; Pentathiepane

$C_2H_4S_5$ Mol. wt 188.38

Found in edible mushrooms, e.g., *Agaricus brunnescens* (Agaricaceae)

Flavour compound.

153
Methyl allyl disulfide

$$CH_3-S-S-CH_2-CH=CH_2$$

$C_4H_8S_2$ Mol. wt 120.24

Occurs in *Allium* spp. (Alliaceae).

154
Methyl mercaptan; Methanethiol; Methyl thioalcohol

$$CH_3-SH$$

CH_4S Mol. wt 48.11

Occurs in the roots of *Raphanus sativus* (Cruciferae), but also occurs as a trace component of many plant volatiles.

Odour of rotten cabbage. It is used as a pesticide and a fungicide, and also as a jet fuel additive.

155
Nonacosane

$$CH_3-[CH_2]_{27}-CH_3$$

$C_{29}H_{60}$ Mol. wt 408.80

A common leaf wax constituent, e.g., in the leaves of brussels sprouts, *Brassica oleracea* (Cruciferae).

156
Nonacosan-10-ol; Celidoniol; Ginnol

$$CH_3-[CH_2]_8-\overset{H}{\underset{[CH_2]_{18}-CH_3}{C}}-OH$$

$C_{29}H_{60}O$ Mol. wt 424.80

A common leaf wax hydrocarbon alcohol, e.g., from *Ginkgo biloba* (Ginkgoaceae), *Chelidonium majus* (Papaveraceae) and *Juniperus chinensis* (Cupressaceae).

157
Nonacosan-10-one; Celidonione; Ginnone

$$CH_3-[CH_2]_8-\underset{\underset{O}{\|}}{C}-[CH_2]_{18}-CH_3$$

$C_{29}H_{58}O$ Mol. wt 422.78

A common leaf wax ketone, e.g., in the epicuticular waxes of *Brassica oleracea* var. *gemmifera* (Cruciferae).

158
Octacosan-1-ol; *n*-Octacosyl alcohol; Montanyl alcohol

$$CH_3-[CH_2]_{26}-CH_2OH$$

$C_{28}H_{58}O$ Mol. wt 410.77

A common wax constituent of plant leaves, e.g., in wheat, *Triticum aestivum* (Gramineae) and in beetroot leaves, *Beta vulgaris* (Chenopodiaceae).

159
Octan-1-ol; *n*-Octanol; *n*-Octyl alcohol; Octoic alcohol; Octylic alcohol; Caprylic alcohol; Heptylcarbinol

$$CH_3-[CH_2]_6-CH_2OH$$

$C_8H_{18}O$ Mol. wt 130.23

Major component in the flower scent of *Ophrys lutea* and *O. fusca* (Orchidaceae). It occurs, with similar alcohols, in other flower scents.

Responsible for attracting male bees of the genus *Andrena* to pollinate orchid flowers. It is used as an antifoaming agent, as a solvent, a perfume ingredient and a flavouring agent.

160
Pentadecane; *n*-Pentadecane

$$CH_3-[CH_2]_{13}-CH_3$$

$C_{15}H_{32}$ Mol. wt 212.42

Major volatile (84% of total) in the flower scent of *Magnolia acuminata* (Magnoliaceae). It occurs in other flower scents in lower amounts, together with similar hydrocarbons.

Used in organic syntheses.

161
Propanethial *S*-oxide; Thiopropanal *S*-oxide

$$CH_3-CH_2-\underset{H}{\overset{\overset{O}{\|}}{C}}=S\to O$$

C_3H_6OS Mol. wt .90.15

Occurs in the onion, *Allium cepa* (Alliaceae).

Lachrymator.

162
Propane-1-thiol; 1-Mercaptopropane; Propyl mercaptan

$$CH_3-CH_2-CH_2SH$$

C_3H_8S Mol. wt 76.16

Occurs in onions, *Allium cepa*, leeks, *Allium ampeloprasum* (Alliaceae), potatoes, *Solanum tuberosum* (Solanaceae), peas, *Pisum sativum* and beans, *Phaseolus vulgaris* (Leguminosae).

Flavour compound.

163
Propane-2-thiol; Isopropyl mercaptan; 2-Mercaptopropane

$$CH_3-\underset{SH}{\overset{}{CH}}-CH_3$$

C_3H_8S Mol. wt 76.16

Found in potatoes, *Solanum tuberosum* (Solanaceae).

Flavour compound.

164
Triacontan-1-ol; 1-Hydroxytriacontane; Myricyl alcohol

$$CH_3-[CH_2]_{28}-CH_2OH$$

$C_{30}H_{62}O$ Mol. wt 438.83

Occurs in many plant cuticle waxes, e.g., *Trifolium repens* (Leguminosae), *Brassica oleracea* (Cruciferae), and *Melicope broadbentiana* (Rutaceae).

Plant growth regulator.

165
Tritriacontane

$$CH_3-[CH_2]_{31}-CH_3$$

$C_{33}H_{68}$ Mol. wt 464.91

A major constituent of the epicuticular waxes of members of the Crassulaceae, e.g., *Aeonium urbicum*; it is found in many other leaf waxes in minor amounts.

166
Tritriacontane-16,18-dione; Eucalyptus wax

$$CH_3-[CH_2]_{14}-\underset{O}{\overset{\parallel}{C}}-CH_2-\underset{O}{\overset{\parallel}{C}}-[CH_2]_{14}-CH_3$$

$C_{33}H_{64}O_2$ Mol. wt 492.88

Found in the plant waxes of *Eucalyptus* spp. (Myrtaceae) and *Acacia* spp. (Leguminosae).

Used in lung and bronchial medicines.

Chapter 8
Acetylenes and Thiophenes

The acetylenes (or polyacetylenes) are an unusual group of naturally occurring hydrocarbons which have one or more acetylenic groups in their structures. It is remarkable that although acetylene, $CH \equiv CH$, itself is a highly reactive, even dangerously explosive, gas, the long-chain hydrocarbon derivatives are sufficiently stable to be isolated and characterized by standard phytochemical techniques. Indeed, over 900 polyacetylenes are now known as plant products. Only a few, such as aethusin (**167**), are simple hydrocarbon derivatives; most have additional functional groups and are either alcohols, ketones, acids, esters, aromatics or furans. Thus, cicutoxin (**172**) is a typical acetylenic alcohol, while falcarinone (**179**) is a typical ketone. Capillin (**170**) is an example of an aromatic acetylene, while carlina oxide (**171**) is an acetylene substituted with both a furan and an aromatic group. Characteristic of the more complex acetylenes is mycosinol (**183**), which is an acetylenic spiroketal enol ether. In their biosynthesis, the polyacetylenes are probably formed from the corresponding fatty acid, via an olefinic intermediate, by successive dehydrogenations, followed by other modifications. Thus, falcarinone is probably produced in plants from linoleic acid via crepenynic acid (**173**), followed by dehydrogenation, oxidation and decarboxylation.

The thiophenes differ from the acetylenes in containing sulfur and characteristically have one or more thiophene substituents, as in α-terthienyl (**188**), which is a simple thiophene trimer. Many thiophenes have acetylenic substituents, as in 5-(3-buten-1-ynyl)-2,2′-bithienyl (**169**). Some thiophenes have two sulfur atoms in a 6-membered ring, as in thiarubrine A (**189**). Both biosynthetically and taxonomically, thiophenes are otherwise closely related to the polyacetylenes. Like the polyacetylenes, too, they tend to be found more regularly in root tissues than in aerial parts of plants.

Polyacetylenes have a taxonomically interesting distribution pattern in higher plant families; they occur regularly in only five families, namely the Campanulaceae, Compositae, Araliaceae, Pittosporaceae and Umbelliferae. The former two and latter three families are especially closely linked in other ways (Sorensen, 1968). Acetylenic acids (e.g., stearolic acid) have a rather different distribution from the other polyacetylenes, and are found in *Santalum* (Santalaceae) and other families in the Santalales and also in certain Malvales, where they occur in association with cyclopropene fatty acids. Acetylenics are also found in the higher fungi, in two families of the Basidiomycetes, the Agaricaceae and Polyporaceae. The fungal compounds are slightly different in having a chain length mainly between C_8 and C_{14}, whereas the higher plant acetylenes are mostly C_{14} to C_{18} compounds.

If acetylenes and thiophenes have an overall function, it is most likely as toxins in either plant–animal or plant–plant interactions. Thus, some are highly poisonous, e.g., those found in the roots of the water dropwort, *Oenanthe crocata* (Umbelliferae), and of fool's parsley, *Aethusa cynapium* (Umbelliferae), while others in fungi have antibiotic activity. Also, several acetylenes, e.g., wyerone acid in broad bean, *Vicia faba* (Leguminosae), and safynol in safflower, *Carthamus tinctorius* (Compositae), have been implicated as natural phytoalexins and are toxic to microorganisms which attack these plants. Thiophenes are noted especially for their nematocidal activities. Both polyacetylenes and thiophenes are classified as photosensitizers, their toxic effects being enhanced by exposure to light (Towers, 1980).

The most characteristic property of polyacetylenes is their UV spectrum. Most compounds show a series of three or more very intense peaks in the region 200–350 nm, and UV spectroscopy may be employed for preliminary detection. Mass spectrometry and NMR measurements are normally required for more detailed characterization.

A comprehensive review of the chemistry of the acetylenes is available (Bohlmann *et al.*, 1973). More recent investigations of their biological activities are reviewed by Towers (1980) and Lam *et al.* (1989).

References

Bohlmann, F., Burkhardt, T. and Zdero, C. (1973). *Naturally Occurring Acetylenes*. London: Academic Press.

Lam, J., Breteler, H., Arnason, T. and Hansen, L. (eds) (1988). *Chemistry and Biology of Naturally Occurring Acetylenes and Related Compounds*. Amsterdam: Elsevier.

Sorensen, N.A. (1968). The taxonomic significance of acetylenic compounds. *Recent Adv. Phytochem.* **1,** 187–228.

Towers, G.H.N. (1980). Photosensitisers from plants and their photodynamic action. *Progress in Phytochemistry* **6,** 183–202.

167
Aethusin

$C_{13}H_{14}$ Mol. wt 170.26

Occurs in *Peucedanum carvifola*, *P. austriacum*, *P. rablense* and *P. verticillare* rhizomes and/or roots (Umbelliferae).

168
Anacyclin

$C_{18}H_{25}NO$ Mol. wt 271.40

Occurs in the roots of *Achillea ptarmica*, *A. nana*, *A. millefolium*, *A. wilhelmsii* and *Anacyclus pyrethrum* (Compositae).

169
5-(3-Buten-1-ynyl)-2,2′-bithienyl; BBT

$C_{12}H_8S_2$ Mol. wt 216.33

Found in *Tagetes erecta*, *T. minuta*, *Flaveria linearis* and *F. trinervia* (Compositae).

Nematocide and photo-enhanced fungicide.

170
Capillin

$C_{12}H_8O$ Mol. wt 168.20

Found in the roots of *Chrysanthemum frutescens* and *Artemisia capillaris* (Compositae).

Antifungal activity.

171
Carlina oxide; 2-(3-Phenyl-1-propynyl)furan

$C_{13}H_{10}O$ Mol. wt 182.22

Occurs in the Carline thistle, *Carlina vulgaris* (Compositae).

Antifungal activity.

172
Cicutoxin

$C_{17}H_{22}O_2$ Mol. wt 258.36

The poisonous principle of cowbane, *Cicuta virosa* (Umbelliferae), where it is mainly present in the roots.

Very toxic. It shows antileukaemic activity, and is a possible anticancer agent.

173
Crepenynic acid

CH₃—[CH₂]₄—C≡C—CH₂—CH=CH—[CH₂]₇—COOH

$C_{18}H_{30}O_2$ Mol. wt 278.44

Occurs in the seed oil of stinking hawk's beard, *Crepis foetida* (Compositae), and *Afzelia cuanzensis* (Leguminosae).

174
Dehydrofalcarinone

$C_{17}H_{20}O$ Mol. wt 240.35

Widespread occurrence in the Compositae, e.g., in the roots of *Artemisia campestris*.

175
Dehydromatricaria ester

$C_{11}H_8O_2$ Mol. wt 172.19

Found in the roots of *Matricaria inodora, M. oreades* and *Artemisia vulgaria*, mugwort (all Compositae).

176
Dehydrosafynol

$C_{13}H_{10}O_2$ Mol. wt 198.22

Obtained from diseased safflower, *Carthamus tinctorius* (Compositae).

Phytoalexin.

177
Falcarindiol

$C_{17}H_{24}O_2$ Mol. wt 260.38

Occurs in the roots of the common carrot, *Daucus carota* (Umbelliferae), and the leaves of the tomato plant (Solanaceae) inoculated with *Cladosporium fulvum*.

Antifungal and analgesic activities. It is a phytoalexin and is toxic.

178
Falcarinol; Carotatoxin; Panaxynol

$C_{17}H_{24}O$ Mol. wt 244.38

Toxic constituent of *Falcaria vulgaris* and water hemlock, *Oenanthe crocata* roots (Umbelliferae). It occurs in the leaves of *Hedera helix* and *Schefflera arboricola* (Araliaceae), and is the agent responsible for the contact dermatitis caused when these plants are handled.

Can cause allergic reactions and is poisonous.

179
Falcarinone

$C_{17}H_{22}O$ Mol. wt 242.36

Occurs in the roots of *Falcaria vulgaris* (Umbelliferae).

Phytoalexin.

180
Ichthyotherol; Ichthyothereol; Cunaniol

$C_{14}H_{14}O_2$ Mol. wt 214.27

Occurs in the roots of *Dahlia coccinea* and in the leaves of *Ichthyothere terminalis* and *Clibadium sylvestre* (Compositae).

Extremely toxic, and used as a fish poison.

181
Lachnophyllum ester

$C_{11}H_{12}O_2$ Mol. wt 176.22

Occurs in daisy, *Bellis perennis*, *Chrysothamnus nauseusus*, *C. parryi*, Canadian fleabane, *Conyza canadensis*, *Heterotheca grandifloris* and *Baccharis* spp.; also it occurs in *Lachnophyllum gossypinium*, *Aster spinosus* and *A. sibiricus* (all Compositae).

182
(*EEE*)-*N*-(2-Methylpropyl)hexadeca-2,6,8-trien-10-ynamide

$C_{20}H_{30}NO$ Mol. wt 300.47

Found in the roots of *Achillea ageratifolia* (Compositae).

Insecticide.

183
Mycosinol

$C_{13}H_{10}O_3$ Mol. wt 214.22

Occurs in the roots of *Santolina oblongifolia* and other members of the Anthemideae tribe; it is also obtained from the stem or leaves of *Coleostephus myconis* (Compositae) infected with *Botrytis cinerea*.

Antifungal activity (a phytoalexin).

184
1-Phenylhepta-1,3,5-triyne; Phenylheptatriyne; PHT

$C_{13}H_8$ Mol. wt 164.21

Found in *Coreopsis grandiflora*, *Dahlia* spp., and *Bidens* spp. (Compositae).

Antibiotic activity; it is phototoxic, a possible anticancer agent, and a fungicide.

185
1-Phenyl-5-heptene-1,3-diyne

$C_{13}H_{10}$ Mol. wt 166.22

Occurs in *Coreopsis* spp., and *Dahlia* tubers (Compositae).

186
Safynol

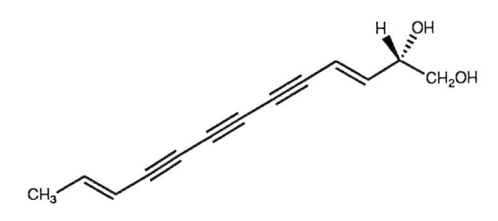

$C_{13}H_{12}O_2$　　　　　　　　　　Mol. wt 200.24

Obtained from diseased safflower, *Carthamus tinctorius*, and constitutively from *Centaurea* spp. (Compositae).

Phytoalexin.

187
Stearolic acid; 9-Octadecynoic acid

$C_{18}H_{32}O_2$　　　　　　　　　　Mol. wt 280.45

Occurs in the seed oil of the sandalwood family (Santalaceae).

188
α-Terthienyl; 2,2′:5′,2″-Terthiophene; α-T

$C_{12}H_8S_3$　　　　　　　　　　Mol. wt 248.39

Found in the roots and leaves of marigold, *Tagetes erecta* (Compositae).

Nematocide, herbicide, and antimicrobial. It is phototoxic and induces photodermatitis in humans.

189
Thiarubrine A; TR-A

$C_{13}H_8S_2$　　　　　　　　　　Mol. wt 228.34

Found in some Compositae, including various spp. of *Aspilia* and also in the hairy roots of *Chaenactis douglasii* infected with *Agrobacterium rhizogenes*.

Strongly antibiotic and also phototoxic. It is a nematocide and a potent antifungal agent.

190
Thiarubrine B; TR-B

$C_{13}H_8S_2$　　　　　　　　　　Mol. wt 228.34

Found in the hairy roots of *Chaenactis douglasii* (Compositae) infected with *Agrobacterium rhizogenes*.

Strong antifungal activity.

191
Triacetylene; 1,3,5-Hexatriyne

C_6H_2　　　　　　　　　　Mol. wt 74.08

Found in the headspace of the fungus *Fomes annosus*.

Can detonate violently at room temperature.

192
Tridec-1-ene-3,5,7,9,11-pentayne; Tridecapentaynene

$C_{13}H_6$　　　　　　　　　　Mol. wt 162.19

Occurs widely in roots of the many species of Compositae, e.g., in *Xanthium canadense* and *Zinnia elegans*.

193
Wyerone

$C_{15}H_{14}O_4$ Mol. wt 258.28

Found in broad beans, *Vicia faba* (Leguminosae) infected with *Botrytis* spp. Also, it is formed as a phytoalexin in other *Vicia* spp. and in *Lens culinaris*.

Antifungal agent (phytoalexin).

194
Wyerone acid

$C_{14}H_{12}O_4$ Mol. wt 244.25

Occurs in broad beans, *Vicia faba* (Leguminosae), infected with *Botrytis* spp.

Antifungal agent (phytoalexin).

Chapter 9
Miscellaneous Aliphatics

This section contains a number of plant substances of simple structure. These do not fit in naturally elsewhere but are related biogenetically to either organic acid or fatty acid (q.v.) metabolism. Some are simple lactones, which give rise to irritating toxins on hydrolysis, such as ranunculin (**205**) and tuliposide A (**206**). Others are odour compounds, which are found as the volatile principles in various fruits and flowers. γ-Nonalactone (**203**) is the flavour principle of coconut, while γ-undecalactone (**208**) is that of peaches. The much more pervasive "green leaf" odour is due to a mixture of leaf alcohol (**200**) and leaf aldehyde (**201**), simple hydrocarbons formed in the damaged green tissue from the breakdown of the fatty acid linoleic acid.

References

Natural Distribution

Ruijgrok, H.W.L. (1960). The distribution of ranunculin and cyanogenetic compounds in the Ranunculaceae. In *Comparative Phytochemistry* (T. Swain, ed.), pp. 175–186. London: Academic Press.

Slob, A. Jekel, B., de Jong, B. and Schlatman, E. (1975). On the occurrence of tuliposide in the Liliiflorae. *Phytochemistry* **14,** 1997–2005.

Analysis

Kameoka, H. (1986). GC-MS method for volatile flavour components of foods. *Modern Methods of Plant Analysis. New Series* **3,** 254–276.

195
Chelidonic acid

$C_7H_4O_6$ Mol. wt 184.11

Occurs in *Chelidonium majus* (Papaveraceae), and is widespread in plants of the Amaryllidaceae, Liliaceae and Papaveraceae.

196
Cucurbic acid; (1*R*,2*R*,3*S*)-Cucurbic acid

$C_{12}H_{20}O_3$ Mol. wt 212.29

Found in the pumpkin, *Cucurbita pepo* (Cucurbitaceae).

Inhibits leaf development in rice. It is a plant growth regulator similar to jasmonic acid (q.v.).

197
Ethyl 2-methylbut-2-enoate

$C_7H_{12}O_2$ Mol. wt 128.17

Occurs in the fruit of quince, *Cydonia oblonga* (Rosaceae).

Major contributor to quince flavour.

198
Jasmone

$C_{11}H_{16}O$ Mol. wt 164.25

Found in the flowers of jasmine, *Jasminum officinale* (Oleaceae).

Jasmine odour, attracting pollinators. It is used in perfumery.

199
(−)-Jasmonic acid

$C_{12}H_{18}O_3$ Mol. wt 210.28

Found as the methyl ester in oil of jasmine. The free acid occurs in *Vicia faba* seeds (Leguminosae), and is widely distributed in green plants.

Probably an endogenous growth hormone of plants. For example, it promotes leaf senescence by inhibiting chlorophyll synthesis.

200
Leaf alcohol; *cis*-Hex-3-en-1-ol

$C_6H_{12}O$ Mol. wt 100.16

A major volatile component of plants, responsible in part for the characteristic "green leaf odour" when tissue is disrupted. *Robinia pseudacacia* (Leguminosae) and mulberry leaf oil contain up to 50% leaf alcohol, green tea up to 30%.

Feeding attractant to many phytophagous insects.

201
Leaf aldehyde; *trans*-Hex-2-enal

$C_6H_{10}O$ Mol. wt 98.15

A major volatile component of plants, responsible in part for the characteristic "green leaf odour" when tissue is disrupted.

Feeding attractant to many phytophagous insects.

202
Nona-2,6-dienal

$C_9H_{14}O$ Mol. wt 138.21

Occurs in the fruit of cucumber, *Cucumis sativus* (Cucurbitaceae).

Principle of cucumber odour: odour threshold is 0.0001 ppm.

203
γ-Nonalactone

$C_9H_{16}O_2$ (−) form Mol. wt 156.23

Found in the coconut, *Cocos nucifera* (Palmae).

Coconut flavour principle. It is used in the food industry.

204
Parasorbic acid; Hexenolactone; 5-Hydroxy-2-hexenoic acid lactone; Sorbic oil

$C_6H_8O_2$ Mol. wt 112.13

Occurs in the fruits of *Sorbus aucuparia* (Rosaceae) and other *Sorbus* spp.

Controls the dormancy of *Sorbus* fruit. It is mildly toxic (LD_{50} intraperitoneally in mice 750 mg kg^{-1}).

205
Ranunculin

$C_{11}H_{15}O_8$ Mol. wt 275.24

Found in buttercups, *Ranunculus* spp., *Anemone* and *Clematis*: widespread in the Ranunculaceae.

Bitter-tasting. It is converted, when fresh plants are bruised, to protoanemonin, a vesicant oil with an acrid taste. This causes subepidermal blistering of human skin.

206
Tuliposide A

$C_{11}H_{17}O_8$ Mol. wt 277.26

Occurs in tulip bulbs, e.g., *Tulipa hybrida*: widespread in the Liliaceae.

Rearranges, in damaged tulip bulbs, to a lactone, tulipalin A, which is fungitoxic and allergenic (see tuliposide B). The lactone is responsible for the skin disease caused by excessive handling of tulip bulbs.

207
Tuliposide B

$C_{11}H_{17}O_9$ Mol. wt 293.26

Found in tulip bulbs, e.g., *Tulipa hybrida*: widespread in the Liliaceae.

Rearranges, in damaged tulip bulbs, to a lactone, tulipalin B, which is fungitoxic and a human allergen (see tuliposide A).

208
γ-Undecalactone; 4-Hydroxyundecanoic acid lactone

$C_{11}H_{20}O_2$ (−) form Mol. wt 184.28

Occurs in the fruit of peach, *Prunus persica* (Rosaceae), as the flavour principle.

Attractive flavour.

PART II
Nitrogen-containing Compounds (excluding Alkaloids)

Nitrogen-containing Compounds (excluding Alkaloids)

General Introduction

Although only 2% of the dry weight of plants is composed of the element nitrogen, compared with 40% for carbon, there is still a very large number of different nitrogen-containing organic substances known in plants. Fifteen thousand structures are a conservative estimate, since there are ten thousand of one class alone, the alkaloids. There are such diverse low molecular weight compounds as indoleacetic acid and zeatin, spermine and putrescine, linamarin and lotaustralin, morphine and strychnine, and adenine and thymine. There are also many plant proteins from abrin and arachin via ferredoxin and gliadin to urease and vicilin.

Nitrogen first appears in plants in organic form as glutamine, through the transfer of ammonium ion, produced by either nitrogen fixation or nitrate reduction, to glutamate. The other protein amino acids are formed by a variety of metabolic routes, for example from an α-keto acid through the transfer of an amino group from glutamate catalysed by an aminotransferase. The twenty protein amino acids so produced are involved in the synthesis of practically all the other nitrogenous plant substances, from the amines, alkaloids, cyanogenic glycosides, glucosinolates, porphyrins, purines, pyrimidines, cytokinins and auxins to the peptides and proteins.

All these major classes of nitrogen compound are covered in this chapter except the alkaloids, which are the subject of the next chapter. In addition to the twenty protein amino acids, there are over 300 other rarer plant amino acids and a selection of them is included in the appropriate section. Many of these nonprotein amino acids are toxic to life, and such toxicity is also associated with two further classes, the cyanogenic glycosides and the glucosinolates (or mustard oil glycosides). These two classes differ from amino acids in occurring in bound, glycosidic form; the free toxins are released following enzymatic hydrolysis of glycosidic bonds.

A further section deals with the purines and pyrimidines, the building blocks of the nucleic acids. Included under this heading are a group of plant hormones, the cytokinins, which control cell division in plants. The section on plant amines also includes compounds of physiological importance, since polyamines such as spermine and spermidine appear to have a role in the control of plant growth processes. Most plant amines are relatively volatile, and are better known for their contribution to the offensive floral odours of plants such as hogweed and cuckoo pint. Aromatic amines are distinctive for their physiological activities in humans, and especially their hallucinogenic effects.

The section on proteins and peptides only mentions a fraction of the many hundred such macromolecules present in the average higher plant. Proteins are important for the storage of nitrogen (in seeds), as the catalysts of metabolic processes and, for structural purposes, in membranes and cell walls.

Some important nitrogenous compounds are described in the miscellaneous section. For example, the two porphyrin pigments, chlorophylls a and b, have a vital function as photosynthetic catalysts. Although complex in structure, these green pigments are synthesized in the first instance from a simple amino acid precursor, glutamate. The B group vitamins are also listed in this final section, as are a number of plant toxins with nitro groups.

For general accounts of nitrogen metabolism in plants, see McKee (1962), Beevers (1976), Miflin (1981) and Miflin and Lea (1990). References to individual classes of nitrogenous compound will be found in the respective sections.

References

Beevers, L. (1976). *Nitrogen Metabolism in Plants*. London: Edward Arnold.
McKee, K.S. (1962). *Nitrogen Metabolism in Plants*. Oxford University Press.
Miflin, B.J. (ed.) (1981). *Amino Acids and Derivatives*, Volume 5, *Biochemistry of Plants* (Stumpf, P.K. and Conn, E.E, eds). New York: Academic Press.
Miflin, B.J. and Lea, P.J. (eds) (1990). *Intermediary Nitrogen Metabolism*, Volume 16, *Biochemistry of Plants* (Stumpf, P.K. and Conn, E.E., eds). New York: Academic Press.

Chapter 10
Amino acids

The plant amino acids are conveniently divided into two groups, the "protein" and "non-protein" acids, although the division between the two groups is not entirely sharp, and methods of identifying and separating acids in either group are essentially the same. The "protein" amino acids are generally recognized to be twenty in number and are those found in acid hydrolysates of plant (and animal) proteins. They also occur together in the free amino acid pool of plant tissues at concentrations varying between 20 and 200 µg per g fresh weight; there are considerable quantitative variations from tissue to tissue, depending on the metabolic status of the plant in question. In general, glutamic and aspartic acids, and their acid amides glutamine and asparagine, tend to be present in larger amounts than the others, since they represent a storage form of nitrogen. On the other hand, histidine, tryptophan, cysteine and methionine are often present in such low amounts in plant tissues that they cannot be detected readily.

Only one of the "non-protein" amino acids is regularly present in plants: the more or less ubiquitous γ-aminobutyric acid. The remainder, of which about 300 are known (Fowden, 1981), are of more restricted occurrence. Their presence (often in high concentration) in seeds, and their subsequent metabolism during germination suggest they are important as nitrogen storage materials. Most are structural analogues of one or other of the twenty "protein" amino acids. For example, two analogues of proline are pipecolic acid (**258**) which has one more methylene group than proline, and azetidine 2-carboxylic acid (**220**), which has one less. Pipecolic acid is found mainly in certain legume seeds, while azetidine 2-carboxylic acid occurs characteristically in many members of the Liliaceae. These three compounds are, strictly speaking, imino acids rather than amino acids, because the amino group present forms part of a 4, 5 or 6 membered heterocyclic ring.

Non-protein amino acids are probably formed along similar pathways to those used for the related protein amino acids, although biosynthetic details are lacking presently for most compounds. These rarer amino acids have a wide but scattered distribution in higher plants. Besides a major presence in the Leguminosae, they have been recorded, mainly as seed constituents, in the Cucurbitaceae, Euphorbiaceae, Iridaceae, Liliaceae, Resedaceae, Sapindaceae and Cycadaceae.

In addition to their value to the plant as a storage form of nitrogen, non-protein amino acids are also protective in seeds to animal herbivory. They can be very toxic to both insects and mammals. Seeds of *Lathyrus*, for example, contain compounds such as α-amino-β-oxalylaminopropionic acid (**216**) which are responsible for a neurological disease in man called lathyrism. Another toxic amino acid is hypoglycin (**244**), which is responsible for hypoglycaemia in humans, while the toxic mimosine (**255**) causes loss of hair in domestic animals (Bell, 1972).

Amino acids are colourless ionic compounds, their solubility properties and high melting points being due to the fact that they are zwitterions. They are all water soluble, although the degree of solubility varies, the aromatic amino acids (e.g., phenylalanine, tyrosine) being rather sparingly soluble. Since they are basic, they form hydrochlorides with concentrated hydrochloric acid and, being acids, they can be esterified. They are optically active, and the protein amino acids all belong to the L-series.

"Neutral" amino acids (e.g., glycine, alanine) are those in which the amino groups are balanced by an equal number of acidic groups. Basic amino acids (e.g., lysine) have an additional free amino group, and acidic amino acids (e.g., glutamic acid) have an additional carboxylic acid group. Because of their different charge properties, amino acid mixtures can be divided into neutral, basic and acidic fractions by using either electrophoresis or ion exchange chromatography.

Non-protein amino acids are detected in the first instance by the different positions they occupy on a two-dimensional chromatogram from the twenty protein amino acids, and by the different colours (e.g., red or green instead of purple) they produce when stained with ninhydrin. Characterization is then based on electrophoretic behaviour, chemical analysis, NMR spectral measurements and, in difficult cases, x-ray crystallography.

References

Barrett, G.C. (ed.) (1985). *Chemistry and Biochemistry of the Amino Acids*. London: Chapman and Hall.
Rosenthal, G.A. (1982). *Plant Non-protein Amino and Imino Acids*. New York: Academic Press.
Fowden, L. (1981). Non-protein amino acids. In *Biochemistry of Plants: Secondary Plant Products* (Conn, E.E., ed.), Volume 7, pp. 215–248. New York: Academic Press.

Ecological Aspects

Bell, E.A. (1972). Toxic amino acids in the Leguminosae. In *Phytochemical Ecology* (Harborne, J.B., ed.), pp. 163–178. London: Academic Press.
Bell, E.A. (1978). Toxins in seeds. In *Biochemical Aspects of Plant and Animal Coevolution* (Harborne, J.B., ed.), pp. 143–162. London: Academic Press.

209
L-Abrine; *N*-Methyl-L-tryptophan

$C_{12}H_{14}N_2O_2$ Mol. wt 218.26

Found in the seeds of jequirity or Indian liquorice, *Abrus precatorius* (Leguminosae), which are also known as crab's eyes or prayer beads.

Not to be confused with abrin (q.v.), a protein, which is the major toxic principle of these seeds.

210
L-Alanine; L-α-Alanine; (*S*)-2-Aminopropanoic acid; L-α-Aminopropionic acid

$C_3H_7NO_2$ Mol. wt 89.10

This protein amino acid occurs in many foods. Sources of free alanine include carob pods and flour, also known as St.John's bread or locust beans, *Ceratonia siliqua* (Leguminosae).

Dietary supplement. It reverses hypoglycaemia and ketosis caused by starvation. Also, it stimulates glucagon secretion in patients with pancreatitis.

211
β-Alanine; 3-Alanine; 3-Aminopropanoic acid; β-Aminopropionic acid

$C_3H_7NO_2$ Mol. wt 89.10

Found in the seeds of *Iris tingitana* (Iridaceae), honesty, *Lunaria* spp. (Cruciferae), and other plants.

Known to be neurotoxic in avian species. It is a constituent of homoglutathione.

212
L-Albizziine; Albizziin

$C_4H_9N_3O_3$ Mol. wt 147.14

Occurs in the seeds of many *Acacia*, *Albizia* and *Dislium* spp. (Leguminosae), including *Albizia julibrissin* and *A. lophantha*.

Insecticide, with antimetabolic activity due to blocking nitrogen transfer from glutamine to aspartic acid. It is essentially nontoxic in other systems.

213
Alliin

$C_6H_{11}NO_3S$ Mol. wt 177.23

Found in onion and garlic bulbs, *Allium cepa* and *A. sativa* (Liliaceae).

Platelet aggregation inhibitor, and antithrombotic. When garlic is crushed, alliin is converted enzymatically to allicin (q.v.)

214
γ-Aminobutyric acid; 4-Aminobutyric acid; GABA; Piperidic acid

$C_4H_9NO_2$ Mol. wt 103.12

Very widespread occurrence in seeds, including peas, beans, etc., *Pisum*, *Vicia* and *Phaseolus* spp. (Leguminosae).

Inhibitory transmitter at the neuromuscular junction in the central nervous system. It has been used to treat cerebral disorders, including coma, and is antihypertensive. It is neurotoxic in avians.

215
L-α-Amino-γ-oxalylaminobutyric acid

$C_6H_{10}N_2O_5$ Mol. wt 190.16

Occurs in the seeds of *Acacia* spp. and of the everlasting pea, *Lathyrus latifolius* (Leguminosae).

Causes neurolathyrism in a similar way to α-amino-β-oxalylaminopropionic acid (q.v.).

216
L-α-Amino-β-oxalylaminopropionic acid

$C_5H_8N_2O_5$ Mol. wt 176.13

Found in all parts, especially the seeds, of wild peas and vetchlings, *Lathyrus* spp. (Leguminosae). The most frequently eaten is *L. sativus*, the chickling vetch, sometimes known as the chick-pea, Indian pea, mutter pea and grass pea (not to be confused with *Cicer arietinum* (Leguminosae) which is called the chick-pea).

Causes neurolathyrism, a neurotoxic syndrome, in humans and domestic animals (particularly horses and cattle), characterized by paralysis of the legs and, occasionally, the arms, bladder and bowel. This may be permanent and death may occur. Neurolathyrism is common in times of famine in India: the plant is able to withstand adverse climatic conditions and is eaten in greater quantities; the susceptibility of individuals is thus increased.

217
L-Arginine; L-α-Amino-δ-guanidinovaleric acid

$C_6H_{14}N_4O_2$ Mol. wt 174.21

As a protein amino acid it is found in all plants. As the free amino acid it is found in alfalfa (lucerne; *Medicago sativa* (Leguminosae)), asparagus, *Asparagus officinale* (Liliaceae), and fenugreek, *Trigonella foenum-graecum* (Leguminosae), among others.

Thought to be essential for the growth of infants. It has been used to treat hyperammonaemia. Arginine stimulates the release of growth hormone from the pituitary gland.

218
L-Asparagine; L-β-Asparagine; (S)-α-Aminosuccinamic acid; L-Aminosuccinic acid amide; Asparamide; L-Aspartic acid amide

$C_4H_8N_2O_3$ Mol. wt 132.12

As a protein amino acid it is found in all plants. As the free amino acid it is found in coffee beans, *Coffea arabica* (Rubiaceae), and other species and varieties, asparagus, *Asparagus officinale* (Liliaceae), and legumes such as soya beans, *Glycine* spp., and vetch, *Vicia* spp.

Dietary amino acid. Asparagine is synthesized by healthy cells for their metabolism, but not by malignant cells; this property is exploited in the chemotherapy of certain cancers by the use of the enzyme asparaginase, which breaks down asparagine to aspartic acid and ammonia.

219
L-Aspartic acid; (S)-Aminosuccinic acid; L-Asparagic acid; L-Asparaginic acid

$C_4H_7NO_4$ Mol. wt 133.11

As a protein amino acid it is found in all plants. As the free amino acid it is found in coffee, *Coffea* spp. (Rubiaceae), liquorice, *Glycyrrhiza glabra* (Leguminosae), and many foods, including sugar cane and sugar beet.

Dietary amino acid; it is neuro-excitatory in large doses. It is used in the preparation of the sweetening agent aspartame (1-methyl-*N*-L-aspartyl-L-phenylalanine).

220
L-Azetidine 2-carboxylic acid; L-2-Azetidine carboxylic acid

$C_4H_7NO_2$ Mol. wt 101.11

Occurs in the rhizome and fresh foliage of many Liliaceous plants, including Solomon's seal, *Polygonatum multiflorum*, lily-of-the-valley, *Convallaria majalis*, and squill, *Drimia maritima*; it occurs in sugar beet, *Beta vulgaris* (Chenopodiaceae), *Delonix regia* (Leguminosae), and other plants.

Larvicide, microbial growth retardant, e.g., in *Escherichia coli*, and causes development aberrations in chick embryos. These effects are thought to be due to competitive inhibition of proline uptake and incorporation, with particular reference to collagen synthesis.

221
L-Baikiain; Baikaine; (S)-4,5-Dehydrohomoproline; (S)-4,5-Dehydropipecolic acid

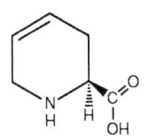

$C_6H_9NO_2$ Mol. wt 127.15

Occurs in *Baikiaea plurijuga*, *Coronilla officinalis* and *Caesalpinia tinctoria* (Leguminosae).

Phytotoxic; it causes inhibition of hypocotyl or radical growth in germinating lettuce seedlings.

222
L-Canaline; (S)-2-Amino-4-(aminoxy)butyric acid

$C_4H_{10}N_2O_3$ Mol. wt 134.14

Widespread occurrence in the seeds of the Leguminosae, particularly those containing canavanine (from which it is easily formed), including the jack bean, *Canavalia ensiformis*.

Antimetabolic activity in a variety of systems; however, the exact mode of action is unknown. Canaline affects the central nervous system in insects, and exhibits phytotoxicity which can be prevented by large excesses of ornithine, citrulline or arginine. It is a powerful inhibitor of pyridoxal phosphate-containing enzymes.

223
L-Canavanine; 2-Amino-4-(guanidinoxy)butyric acid

$C_5H_{12}N_4O_3$ Mol. wt 176.18

Widespread occurrence in the seeds of the Leguminosae, including the jack bean, *Canavalia ensiformis*, the first known source.

Numerous phytotoxic effects, including inhibition of growth and germination; it is cytotoxic in assorted cultures of human and animal cells, and inhibits placental alkaline phosphatase in humans. Most of these effects appear to be caused by canavanine acting as an antimetabolite, and thereby blocking arginine uptake.

224
L-Citrulline; (S)-α-Amino-δ-ureidovaleric acid; δ-Ureido-L-norvaline

$C_6H_{13}N_3O_3$ Mol. wt 175.19

Found in watermelon, *Citrullus vulgaris* (Cucurbitaceae), and sesame, *Sesamum indicum* (Pedaliaceae).

Protects *Lemna minor*, an aquatic microphyte, and possibly other organisms, from the toxic effects of canaline (q.v.).

225
Coprine

$C_8H_{14}N_2O_4$ Mol. wt 202.21

Occurs in the inky cap mushroom, *Coprinus atramentarius* (Coprinaceae).

Causes nausea, flushing and breathing difficulties in some individuals if alcohol is consumed after eating the mushroom, similar to the disulfiram reaction used to discourage alcoholics from drinking. This is caused by interference with alcohol metabolism, causing accumulation of acetaldehyde in the blood.

226
L-β-Cyanoalanine; 3-Cyano-L-alanine; (S)-α-Amino-β-cyanopropionic acid

$C_4H_6N_2O_2$ Mol. wt 114.11

Occurs in the seeds of common vetch, *Vicia sativa*, and other *Vicia* spp. (Leguminosae).

Neurotoxic: produces neurolathyrism in avians and other animals. Symptoms include hyperactivity, tremors, convulsions, and may lead to death.

227
L-Cysteine; (R)-α-Amino-β-mercaptopropionic acid; β-Mercapto-L-alanine

$C_3H_7NO_2S$ Mol. wt 121.16

As a protein amino acid it is found in all plants, and it is a constituent of glutathione.

Dietary amino acid. It is used in eye drops for healing alkali burns, and in some skin antibiotic preparations.

228
L-Cystine; Dicysteine; β,β'-Dithiodi-L-alanine; Gelucystine

$C_6H_{12}N_2O_4S_2$ Mol. wt 240.31

A protein amino acid found in all plants.

Dietary amino acid. It is used to treat the congenital condition homocystinuria.

229
L-α,γ-Diaminobutyric acid; DABA; (S)-2,4-Diaminobutyric acid

$C_4H_{10}N_2O_2$ Mol. wt 118.14

Found in the seeds of everlasting peas, *Lathyrus sylvestris*, and *L. latifolius*, in *Acacia* spp. (Leguminosae); in Solomon's seal, *Polygonatum multiflorum* (Liliaceae), and some Compositae spp.

Toxic. It inhibits ornithine carbamoyltransferase, causes chronic ammonia toxicity in rats, impairs liver function, and causes tremors, convulsions and weakness of the hind limbs.

230
L-Djenkolic acid

$C_7H_{14}N_2O_4S_2$ Mol. wt 254.34

Found in the djenkol bean, *Pithecolobium lobatum*, which is eaten in Java, and also *P. bubalinum*, *Albizia lophanta*, and some *Mimosa* and *Acacia* spp. (Leguminosae).

Crystallizes out of the urine if sufficient of the bean is eaten; it causes acute kidney malfunction and possible blocking of urine flow. In young children it has been reported to produce painful swelling of the genitalia.

231
L-Dopa; Dopa; Levodopa; Laevo-dopa; 3,4-Dihydroxy-L-phenylalanine; 3-Hydroxy-L-tyrosine

$C_9H_{11}NO_4$ Mol. wt 197.19

Found in broad or horse beans, also called fava beans, *Vicia faba*, and some *Lupinus* and *Mucuna* spp. (all Leguminosae).

L-Dopa is used to treat Parkinson's disease, a neurological disorder characterized by tremors, rigidity and hypokinesis. It is the biological precursor of the catecholamines. L-Dopa is thought to be responsible for favism, a haemolytic anaemia associated with individuals deficient in glucose-6-phosphate dehydrogenase and who have consumed broad beans.

232
L-Glutamic acid; L-Glutaminic acid; (+)-α-Amino-L-glutaric acid

$C_5H_9NO_4$ Mol. wt 147.13

As a protein amino acid it is found in all plants. It is a constituent of vegetable proteins such as gluten and soya protein.

A component of intermediate metabolism. Excessive doses are neurotoxic. The flavour enhancer monosodium glutamate (MSG), when given to neonates, can cause stunted growth with obesity, and atropy of the pituitary, ovaries and uteri. It is banned from baby foods but is often used in adult convenience foods, where it can cause flushing and palpitations in susceptible individuals.

233
L-Glutamine; (+)-α-Amino-L-glutaramic acid; Levoglutamide; Levoglutamine

$C_5H_{10}N_2O_3$ Mol. wt 146.15

Occurs as a protein amino acid, and is found free in all plants and bacteria.

A component of intermediate metabolism, and a dietary amino acid. It has been used to treat alcoholism and mental deficiency.

234
L-γ-Glutamyl-L-hypoglycin; Hypoglycin B

$C_{12}H_{18}N_2O_5$ Mol. wt 270.29

Occurs in the arillus and seed of the ackee, *Blighia sapida* (Sapindaceae).

Very toxic. It possesses similar properties, including hypoglycaemia, to those of hypoglycin (q.v.).

235
Glycine; Aminoacetic acid; Aminoethanoic acid; Glycocol

$C_2H_5NO_2$ Mol. wt 75.07

As a protein amino acid it is found in all plants.

A component of intermediate metabolism. It may be used to counteract gastric hyperacidity combined with antacids, and to reduce gastric irritation caused by aspirin.

236
L-Histidine

$C_6H_9N_3O_2$ Mol. wt 155.16

A dietary amino acid present in many plants, e.g., fenugreek seeds, *Trigonella foenum-graecum* (Leguminosae).

An essential amino acid for children. It is used in the investigation of folate deficiency, and occasionally in uraemia and chronic renal failure.

237
L-Homoarginine

C₇H₁₆N₄O₂   Mol. wt 188.23

Occurs in the seeds of at least 40 *Lathyrus* spp., including *L. cicera* and *L. sativus* (Leguminosae).

Toxic in rats and some insects, with considerable variation in organism susceptibility. It probably acts as a competitive inhibitor and antimetabolite of arginine. Also, it is a microbial growth inhibitor of *Streptococcus aureus*, *Escherichia coli* and *Candida albicans*, and a germination inhibitor.

238
L-Homocysteine

C₄H₉NO₂S   Mol. wt 135.19

Found in spinach, *Spinacia oleracea* (Chenopodiaceae).

The sodium salt may be used for seasoning, like monosodium glutamate (see glutamic acid).

239
L-Homoserine

C₄H₉NO₃   Mol. wt 119.12

Very wide occurrence. It is the principal amino acid in germinating peas, *Pisum sativum* (Leguminosae).

Plays a key role in the biosynthesis of threonine, isoleucine and methionine (q.v.).

240
γ-Hydroxy-L-arginine; 4-Hydroxy-L-arginine

C₆H₁₄N₄O₃   Mol. wt 190.21

Found in at least 17 spp. of vetches, *Vicia* (Leguminosae).

Acts as a substrate for arginase, arginine decarboxylase and L-amino acid oxidase. Phytotoxic; it is a germination inhibitor.

241
(+)-γ-Hydroxy-L-homoarginine; 4-Hydroxy-L-homoarginine

C₇H₁₆N₄O₃   Mol. wt 204.23

Widespread occurrence in the seeds of legumes such as *Vicia* and *Lathyrus*, and also lentils, *Lens culinaris*, and peas, *Pisum sativum*.

Involved in the intermediate metabolism of plants, e.g., in the biosynthesis of lathyrine (q.v.).

242
trans-4-Hydroxy-L-proline

C₅H₉NO₃   Mol. wt 131.13

As a protein amino acid, it is found in many plants, and it is a major constituent of plant cell wall glycoproteins.

Dietary amino acid. It is also a germination inhibitor.

243
5-Hydroxy-L-tryptophan; 5-HTP

$C_{11}H_{12}N_2O_3$ Mol. wt 220.23

Found in the seeds of the legumes *Mucuna pruriens* and *Griffonia simplicifolia*, as well as banana and cotton.

Precursor of serotonin (q.v.), and has been used clinically to treat mental disorders associated with this since, unlike serotonin, it crosses the blood–brain barrier. It is toxic to some insects.

244
L-Hypoglycin; Hypoglycin A; Hypoglycine

$C_7H_{11}NO_2$ Mol. wt 141.17

Found in the arillus of the fruits of the ackee, *Blighia sapida* (Sapindaceae) and *Billia hippocastanum* (Hippocastanaceae).

Very toxic, and also hypoglycaemic. It is responsible for the syndrome known as "vomiting sickness", characterized by violent retching, vomiting, convulsions and coma (may be fatal). The ackee fruit is eaten in the West Indies where the sickness may reach epidemic proportions at certain times of the year.

245
L-Indospicine; L-2-Amino-6-amidinohexanoic acid

$C_7H_{15}N_3O_2$ Mol. wt 173.22

Found in the leaves and seeds of *Indigofera spicata* and other *Indigofera* spp. (Leguminosae).

Teratogenic, abortifacient, and hepatotoxic activities in animals. It causes cleft palate and dwarfism in rat foetuses, and liver damage in sheep, rabbits and cows.

246
L-Isoleucine

$C_6H_{13}NO_2$ Mol. wt 131.18

As a protein amino acid it is found in all plants.
Essential dietary amino acid.

247
Isowillardiine

$C_7H_9N_3O_4$ Mol. wt 199.17

Occurs in the seeds and seedlings of *Pisum sativum* (Leguminosae).

248
L-Lathyrine; Tingitanine

$C_7H_{10}N_4O_2$ Mol. wt 182.19

Occurs in the seeds of *Lathyrus tingitanus* and other *Lathyrus* spp. (Leguminosae).

249
L-**Leucine**; (S)-2-Amino-4-methylvaleric acid

$C_6H_{13}NO_2$ Mol. wt 131.18

As a protein amino acid it is found in all plants.

Essential dietary amino acid. Excess leucine is involved in the pathogenesis of pellagra.

250
L-**Lysine**; (S)-2,6-Diaminohexanoic acid

$C_6H_{14}N_2O_2$ Mol. wt 146.19

As a protein amino acid it is found in all plants.

Essential dietary amino acid. It has been used to treat herpes simplex lesions.

251
L-**Methionine**

$C_5H_{11}NO_2S$ Mol. wt 149.22

As a protein amino acid it is found in all plants.

Essential dietary amino acid. It is lipotropic, and has been used as an adjuvant in the treatment of liver disease. It enhances the synthesis of glutathione necessary for the detoxification of paracetamol, and is used in the treatment of paracetamol overdose.

252
3-Methylamino-L-alanine

$C_4H_{10}N_2O_2$ Mol. wt 118.14

Found in the seeds and leaves of the primitive order Cycadales, including *Cycas circinalis*.

Growth retardant in rats when given orally; very toxic, causing convulsions and considerable mortality when injected. It contributes to the toxicity of these plants in humans, but is not thought to be responsible for the main toxic effects, which are probably due to cycasin (q.v.).

253
α-**(Methylenecyclopropyl)glycine**

$C_6H_9NO_2$ Mol. wt 127.15

Found in *Litchi sinensis* (Sapindaceae), *Billia hippocastanum* (Hippocastanaceae), and *Acer pseudoplatanus*, the sycamore (Aceraceae).

Hypoglycaemic; similar toxicity to its higher homologue hypoglycin (q.v.).

254
Se-Methyl-L-selenocysteine;
(R)-2-Amino-3-selenomethylpropanoic acid

$C_4H_9NO_2Se$ Mol. wt 182.08

Occurs in the seeds of accumulating plants grown on seleniferous soils, including *Astragalus bisulcatus* (Leguminosae) and *Oonopsis condensata* (Compositae).

Produces selenosis, in a similar manner to selenocystathionine (q.v.), as well as the syndrome known as "blind staggers" in animals.

255
L-Mimosine; Leucenol; Leucenine; Leucaenol; Leucaenine

$C_8H_{10}N_2O_4$ Mol. wt 198.18

Occurs in the seeds and foliage of *Mimosa pudica* and *Leucaena leucocephala*, the jumbie bean, previously *L. glauca* (Leguminosae).

Depilatory: in animals such as horses, sheep, swine (but not ruminants which can detoxify mimosine) it affects growing hair, and animals can become completely bald. Resting hairs are not affected. It has potential as a defleecing agent, but toxic effects include weight loss, general malaise and eye inflammation. It is goitrogenic in calves produced by heifers fed mimosine, and teratogenic in rats. It inhibits DNA synthesis.

256
L-Ornithine; (S)-2,5-Diaminopentanoic acid; (S)-α,δ-Diaminovaleric acid

$C_5H_{12}N_2O_2$ Mol. wt 132.17

Ubiquitous, as a metabolic intermediate.

A nonessential amino acid. It has been used as an anticholesterolaemic, and to treat ammonia intoxication.

257
L-Phenylalanine; β-Phenyl-L-alanine

$C_9H_{11}NO_2$ Mol. wt 165.19

As a protein amino acid, it is found in all plants.

Essential dietary amino acid. It has been used to treat depression. Normally, it is metabolized to tyrosine, except in children with phenylketonuria, in which an enzyme is lacking and dietary levels must be kept low to avoid mental retardation.

258
L-Pipecolic acid; Dihydro-L-baikiaine; L-Homoproline; (S)-Pipecolinic acid

$C_6H_{11}NO_2$ Mol. wt 129.16

Found in the seeds of *Phaseolus vulgaris* (Leguminosae) and subsequently shown to be widespread in both plants and fungi.

Germination inhibitor.

259
L-**Proline**; (S)-2-Pyrrolidine carboxylic acid

C$_5$H$_9$NO$_2$ Mol. wt 115.13

As a protein amino acid, it is found in all plants. It accumulates in some plants as a response to drought or salt stress. Thus the free proline content of the halophyte *Triglochin maritima* (Juncaginaceae) reaches 10–20% of the dry weight.

Nonessential dietary amino acid.

260
S-[(E)-Prop-1-enyl]-L-**cysteine** S-oxide; *trans*-S-(1-Propenyl)cysteine S-oxide; S-(Prop-1-enyl)cysteine sulfoxide

C$_6$H$_{11}$NO$_3$S   Mol. wt 177.22

Found in onions, *Allium cepa* (Liliaceae), and other *Allium* spp.

Flavour ingredient; it is the precursor of propanethial S-oxide, the main lachrymatory principle of onion, which is produced by the action of alliinase when the onion is cut or bruised.

261
L-**Selenocystathionine**

C$_7$H$_{14}$N$_2$O$_4$Se   Mol. wt 269.16

Occurs in the seeds of accumulating plants grown on soils rich in selenium, which substitutes for sulfur. These include *Stanleya pinnata* (Cruciferae); *Lecythis ollaria* (Lecythidaceae), known as coco de mono or monkey nut, *Astragalus* spp., and *Neptunia amplexicaulis* (Leguminosae).

Toxic: causes abdominal pain, nausea, vomiting and diarrhoea and, a week or two later, reversible loss of scalp and body hair. It causes the acute selenium poisoning known as "blind staggers" in animals.

262
L-**Serine**; (S)-2-Amino-3-hydroxypropionic acid; β-Hydroxy-L-alanine

C$_3$H$_7$NO$_3$ Mol. wt 105.10

As a protein amino acid, it is found in all plants.

Dietary amino acid.

263
L-**Threonine**; (2S,3R)-2-Amino-3-hydroxybutyric acid

C$_4$H$_9$NO$_3$ Mol. wt 119.12

As a protein amino acid, it is found in all plants.

Essential dietary amino acid.

264
Tricholomic acid; L-*erythro*-Dihydroibotenic acid

$C_5H_8N_2O_4$ Mol. wt 160.13

Found in *Tricholoma muscarium* and other *Tricholoma* spp. (Russulaceae).

Toxic: excites central neurones. It is lethal to flies. It is used as a flavour enhancer.

265
L-Tryptophan; α-Aminoindole-3-propionic acid

$C_{11}H_{12}N_2O_2$ Mol. wt 204.23

As a protein amino acid, it is found in all plants.

Essential dietary amino acid. It is used clinically to treat depression.

266
L-Tyrosine; β-(*p*-Hydroxyphenyl)-L-alanine

$C_9H_{11}NO_3$ Mol. wt 181.19

As a protein amino acid, it is found in all plants.

Dietary amino acid.

267
L-Valine; (*S*)-2-Aminoisovaleric acid

$C_5H_{11}NO_2$ Mol. wt 117.15

As a protein amino acid, it is found in all plants.

Essential dietary amino acid.

Chapter 11
Amines

Plant amines can be considered simply as the products of decarboxylation of amino acids, although they are mainly synthesized *in vivo* by transamination of the corresponding aldehydes. The most widespread plant amines are conveniently divided into three groups: aliphatic monoamines, aliphatic polyamines (including diamines) and aromatic amines.

Aliphatic monoamines are volatile compounds, ranging from simple compounds like methylamine, CH_3NH_2, to *n*-hexylamine, $CH_3(CH_2)_5NH_2$. They are widely distributed in higher plants and fungi and, when present in any concentration, have an unpleasant fish-like smell. They function in flowers (e.g., in the cow parsley, *Heracleum sphondylium*) as insect attractants by simulating the smell of carrion.

By contrast to the monoamines, diamines and other polyamines are less volatile, although they still possess offensive odours. Widespread polyamines include putrescine, (**297**), agmatine (**268**), spermidine (**300**) and spermine (**301**). There are several others of more limited occurrence, such as cadaverine (**271**) (Smith, 1981). Polyamines are of research interest because of their growth-stimulating activity in relation to their effect on ribosomal RNA.

Polyamines can occur in plants conjugated with hydroxycinnamic acids (q.v.). There are simple ester conjugates (e.g., caffeoylputrescine) and also more complex derivatives. The antifungal hordatines A and B (**282, 283**) of barley are derived from the condensation of agmatine (**268**) with *p*-coumaric and ferulic acids. Polyamine fatty acid conjugates have also been found (Smith, 1977).

The best known aromatic amine from plants is probably mescaline (**287**), the active principle of the flowering heads (peyote) of the cactus *Lophophora williamsii*, and a powerful natural hallucinogenic compound. Indeed, many of the known aromatic amines are physiologically active and, for this reason, they are sometimes classified with the alkaloids. Three substances which are very important in animal physiology (e.g., in brain metabolism) are noradrenaline (**292**), histamine (**281**) and serotonin (**298**); all three occur in plants, noradrenaline, for example, being present in the banana and the potato. Perhaps the most widespread aromatic amine is tyramine (**303**), which has been detected in 15 of 77 plant families surveyed.

All the amines so far mentioned are primary amines, i.e., have the general formula RNH_2. Secondary amines, general formula R_2NH, and tertiary amines, R_3N, are known in plants but they are not very common. A typical secondary amine found in plants is dimethylamine; a typical tertiary amine is hordenine or *N,N*-dimethyltyramine (**284**) which is the principal "alkaloid" of barley.

References

Smith, T.A. (1977). Recent advances in the biochemistry of plant amines. In *Progress in Phytochemistry*, Volume 4 (Reinhold, L., Harborne, J.B. and Swain, T., eds), pp. 27–82. Oxford: Pergamon.
Smith, T.A. (1980). Plant amines. In *Encyclopedia of Plant Physiology*, New Series, Volume 8, *Secondary Plant Products* (Bell, E.A. and Charlwood, B.V., eds), pp. 430–460. Berlin: Springer-Verlag.
Smith, T.A. (1981). Amines. In *The Biochemistry of Plants*, Volume 7 (Conn, E.E., ed.), pp. 249–268. New York: Academic Press.

268
Agmatine; (4-Aminobutyl)guanidine;
1-Amino-4-guanidinobutane

$$\text{H}_2\text{N}\diagdown\text{C}-\text{NH}-[\text{CH}_2]_4-\text{NH}_2$$
$$\text{HN}\diagup$$

$C_5H_{14}N_4$ Mol. wt 130.20

Occurs in soya beans, *Glycine max* (Leguminosae), sesame seeds, *Sesamum indicum* (Pedaliaceae), barley, *Hordeum vulgare* (Graminae), the Indian pea, *Lathyrus sativus* (Leguminosae), and many other plants.

269
***p*-Aminobenzoic acid;** PABA; Bacterial vitamin H[1]; Vitamin B$_x$

$C_7H_7NO_2$ Mol. wt 137.14

Wide occurrence in plants, and particularly abundant in rice polishings.

Component of folic acid and B complex vitamins. It is a sulfonamide antagonist. As a sunscreen, it is used topically as a component of many suntan preparations. It had been used to treat rickettsial infections such as Rocky Mountain spotted fever and typhus, before the advent of broad spectrum antibiotics.

270
Bufotenine; 5-Hydroxy-*N,N*-dimethyltryptamine;
N,N-Dimethylserotonin

$C_{12}H_{16}N_2O$ Mol. wt 204.27

Occurs in the flowers of the reed *Arundo donax* (Gramineae), and in the seeds and leaves of *Piptadenia peregrina* and *P. macrocarpa* (Leguminosae).

Hallucinogenic, and causes other mental effects including anxiety and perceptual disturbances. Autonomic effects are also produced, such as increase in blood pressure and dilatation of the pupil.

271
Cadaverine; Animal coniine; Pentamethylenediamine; Pentane-1,5-diamine

$$\text{NH}_2-[\text{CH}_2]_5-\text{NH}_2$$

$C_5H_{14}N_2$ Mol. wt 102.18

Occurs in the seedlings of a number of leguminous plants, including pea, *Pisum sativum*, subterranean clover, *Trifolium subterraneum*, Indian pea, *Lathyrus sativus*, soybean, *Glycine max* and soybean flour. Also present in stonecrop, *Sedum acre* (Crassulaceae).

Toxic; it is also a skin irritant and sensitizer. In low concentrations it acts as a plant growth stimulant.

272
***N*-Carbamoylputrescine;**
N-Carbamoylbutane-1,4-diamine; *N*-(4-Aminobutyl)urea; 4-Ureidobutylamine

$$\text{H}_2\text{N}\diagdown\text{C}-\text{NH}-[\text{CH}_2]_4-\text{NH}_2$$
$$\text{O}\diagup$$

$C_5H_{13}N_3O$ Mol. wt 131.18

Found in the roots of sea lavender, *Limonium vulgare* (Plumbaginaceae), and probably other plants which produce putrescine (q.v.).

273
D-Cathine; Katine; Norpseudoephedrine; Pseudonorephedrine; ψ-Norephedrine; Nor-ψ-ephedrine

$C_9H_{13}NO$ Mol. wt 151.21

Occurs in the kat (or khat) plant, *Catha edulis*, and in *Maytenus krukovii* (Celastraceae), and in mother liquors from *Ephedra* spp. (Ephedraceae) after production of ephedrine (q.v.).

Central nervous system stimulant, mild euphoriant, and anorexic.

274
D-Cathinone; (S)-2-Amino-1-oxo-1-phenylpropane

$C_9H_{11}NO$ Mol. wt 149.19

Occurs in the kat (or khat) plant, *Catha edulis* and in *Maytenus krukovii* (Celastraceae).

Similar to cathine (q.v.): it is a central nervous system stimulant, mild euphoriant, and anorexic.

275
***N,N*-Dimethyltryptamine**; 3-(2-Dimethylaminoethyl)indole; DMT

$C_{12}H_{16}N_2$ Mol. wt 188.27

Found in the leaves of *Prestonia amazonica* (Apocynaceae), the flowers of the reed *Arundo donax* (Gramineae); the seeds and leaves of *Piptadenia peregrina* and in cowhage, *Mucuna pruriens* (Leguminosae).

Psychotomimetic activity including hallucinations, anxiety and perceptual distortions; it also causes autonomic effects such as hypertension and pupillary dilatation.

276
Dopamine; 4-(2-Aminoethyl)benzene-1,2-diol; 4-(2-Aminoethyl)pyrocatechol; 3,4-Dihydroxyphenethylamine; α-(3,4-Dihydroxyphenyl)-β-aminoethane; 3-Hydroxytyramine

$C_8H_{11}NO_2$ Mol. wt 153.18

High concentrations are found in banana peel (6–12 μmol per g fresh weight), and in *Musa sapientum* (Musaceae). It is present in mescal buttons, *Lophophora williamsii* (Cactaceae), broom, *Cytisus scoparius* (Leguminosae), and *Hermidium alipes* (Nyctaginaceae).

A neurotransmitter in the brain; dopamine levels are reduced in patients with parkinsonism. It is a direct acting sympathomimetic with effects on both α- and β-adrenergic receptors. Also, it increases cardiac output. The main use is in the treatment of shock, but it is inactive orally and must be given by dilute intravenous infusion.

277
L-Ephedrine; (1R, 2S)-1-Phenyl-1-hydroxy-2-methylaminopropane

$C_{10}H_{15}NO$ Mol. wt 165.24

Occurs in the leaves and stems of Ma-Huang, obtained from *Ephedra* spp.: *E. sinica, E. equisetina, E. gerardiana* and others (Ephedraceae). It co-occurs with the D-isomer, pseudoephedrine.

Sympathomimetic activity with direct and indirect effects on both α- and β-adrenergic receptors; it produces peripheral vasoconstriction and raises blood pressure. Also, it is a central nervous system stimulant. The main uses are as a bronchodilator and respiratory stimulant in asthma and as a vasoconstrictor of mucous membranes in rhinitis and sinusitis. Pseudoephedrine is broadly similar in activity.

278
Galegine; N-(3,3-Dimethylallyl)guanidine; Isoamyleneguanidine; (3-Methylbut-2-enyl)guanidine

$C_6H_{13}N_3$ Mol. wt 127.19

Occurs in the leaves, flowers and seeds of goat's rue, or French lilac, *Galega officinalis* (Leguminosae), and in *Verbesina encelioides* (Compositae).

Toxic: affects mitochondrial function.

279
Gramine; 3-(Dimethylaminomethyl)indole; Donaxine

$C_{11}H_{14}N_2$ Mol. wt 174.25

Occurs in the leaves of some varieties of barley, *Hordeum vulgare*, the reed *Arundo donax*, reed canary grass *Phalaris arundinacea*, and other Gramineae spp., the silver maple, *Acer saccharinum*, and *A. rubrum* (Aceraceae), and some *Lupinus* spp. (Leguminosae).

Insect feeding inhibitor. It is partly responsible for the toxic condition in sheep, known as "Phalaris staggers".

280
Gyromitrin; Acetaldehyde formylmethylhydrazone; Acetaldehyde methylformylhydrazone

$C_4H_8N_2O$ Mol. wt 100.12

Found in the false morel mushroom, *Gyromitra esculenta* (or *Helvella esculenta*.)

Causes gastroenteritis about 2–6 hours after eating, with vomiting, abdominal cramps, lassitude, headache, cyanosis, jaundice, convulsions and coma. The poison causing the effects is thought to be monomethylhydrazine, formed by hydrolysis of gyromitrin.

281
Histamine; β-Aminoethylglyoxaline; 4-(2-Aminoethyl)imidazole; Ergamine; 4-Imidazoleethylamine; 5-Imidazoleethylamine; 1H-Imidazole-4-ethanamine; 2-(Imidazol-4-yl)ethylamine

$C_5H_9N_3$ Mol. wt 111.15

Widespread occurrence, e.g., banana fruit, *Musa sapientum* (Musaceae), several Chenopodiaceae spp. including spinach, *Spinacea oleracea*, insectivorous plants such as sundews, *Drosera* spp. (Droseraceae) and pitcher plants, *Nepenthes* spp. (Nepenthaceae), and *Sarracenia* (Sarraceniaceae); it also occurs in the stinging hairs of the nettle, *Urtica dioica* (Urticaceae) and many other families.

Vasodilator, irritant, and bronchoconstrictor, but also a mediator of inflammatory processes and hypersensitivity; it is present in all mammalian tissues. The main use is as a diagnostic agent for gastric secretion and circulatory disorders, but it is also applied topically to chilblains and similar conditions.

282
Hordatine A

$C_{28}H_{38}N_8O_4$ Mol. wt 550.67

Occurs in the seedlings of barley, *Hordeum vulgare* (Gramineae).

Antifungal activity, e.g., against *Monilinia fructicola* and other pathogenic fungi.

283
Hordatine B

$C_{29}H_{40}N_8O_5$ Mol. wt 580.70

Occurs in the seedlings of barley, *Hordeum vulgare* (Gramineae).

Antifungal activity against *Monilinia fructicola*.

284
Hordenine; 4-(2-(Dimethylamino)ethyl)phenol; N,N-Dimethyltyramine

$C_{10}H_{15}NO$ Mol. wt 165.24

Widespread occurrence, e.g., in germinating barley, *Hordeum vulgare*, and other grasses, such as the reed canary grass, *Phalaris arundinacea*; it is also found in several Cactaceae spp., including *Ariocarpus scapharostrus* and *A. kotschoubeyanus*

Insect feeding inhibitor.

285
Indole; 2,3-Benzopyrrole

C_8H_7N Mol. wt 117.15

Occurs in the flowers of *Sauromatum guttatum, Arum maculatum, Dracunculus vulgaris* and *Amorphophallus* spp. (Araceae), and in the flowers of *Jasminum* (Oleaceae) and of *Citrus* spp. (Rutaceae).

Insect attractant. It has an unpleasant faecal odour.

286
Indoleacetic acid; Auxin; Heteroauxin; IAA; Indol-3-ylacetic acid; 1*H*-Indole-3-acetic acid

$C_{10}H_9NO_2$ Mol. wt 175.19

Occurs in trace amounts in the growing parts of all higher plants: normally present in the free and conjugated forms (e.g., as the glucose ester, etc.).

Plant hormone and growth regulator.

287
Mescaline; Mezcaline

$C_{11}H_{17}NO_3$ Mol. wt 211.26

Found in the flowering heads of peyote, or mescal buttons, *Lophophora williamsii* and other cacti, including *Trichocereus pachanoi* (Cactaceae).

Psychotomimetic. It is a central nervous system depressant and hallucinogenic in high doses.

288
5-Methoxy-*N*,*N*-dimethyltryptamine; *O*-Methylbufotenine

$C_{13}H_{18}N_2O$ Mol. wt 218.30

Found in the pasture grasses *Phalaris arundinacea* and *P. tuberosa* (Gramineae), and in *Desmodium pulchellum* (Leguminosae).

Toxic. It is partly responsible for the condition in sheep known as "Phalaris staggers" when the concentration is high. It is psychotomimetic, and similar in effect to bufotenine (q.v.).

289
***N*-Methylmescaline**

$C_{12}H_{19}NO_3$ Mol. wt 225.29

Occurs in cacti, including *Lophophora williamsii* and, unusually, *Alhagi pseudoalhagi* (Leguminosae).

Psychotomimetic: similar to mescaline (q.v.).

290
Muscarine; L-(+)-Muscarine

$[C_9H_{20}NO_2]^+$ Mol. wt 174.27

Occurs in the fly agaric mushroom, *Amanita muscaria* (Amanitaceae), other fungi including *Inocybe obscuroides, I. napipes, I. imbrina* and other *Inocybe* spp. (Cortinariaceae), and some *Clitocybe* spp. (Tricholomataceae).

Cholinergic, acting at the receptors designated as "muscarinic". Symptoms of poisoning occur within 30–120 minutes of ingestion: blurred vision, contraction of pupil, excessive perspiration, salivation and lachrimation, slowing of the heart, lowering of blood pressure, congestion of pulmonary circulation, and breathing difficulties. (LD_{50} intravenously in mice 0.23 mg kg^{-1}.) The specific antidote is atropine (q.v.).

291
Muscimol; Agarin;
5-Aminomethyl-3-hydroxyisoxazole; Pantherine

$C_4H_6N_2O_2$ Mol. wt 114.11

Occurs in the fly agaric mushroom, *Amanita muscaria*, and the panther cap, *A. pantherina* (Amanitaceae).

Induces hallucinations, delirium, muscular spasm, and sleep. Death from ingesting these mushrooms is uncommon but not unknown; they are sometimes taken for their hallucinogenic effects. Muscimol mimics the effects of the neurotransmitter GABA. It is responsible for the insecticidal activity of the mushrooms.

292
L-Noradrenaline; L-Norepinephrine

$C_8H_{11}NO_3$ Mol. wt 169.18

Found in banana fruit, *Musa sapientum* (Musaceae), *Passiflora quadrangularis* (Passifloraceae), in many legumes, including *Pisum sativum, Mimosa pudica, Albizia julibrissin, Phaseolus multiflorus* and *Samanea saman*, and in *Portulaca oleracea* (Portulacaceae).

A sympathomimetic neurohormone with mainly α-adrenergic activity; inactive by mouth but, when given by injection, produces vasoconstriction, reduced blood flow to the brain, kidney, liver, skin and skeletal muscle and dilatation of the pupil. It is used to treat peripheral vasomotor collapse, some hypotensive states and acute cardiac infarction.

293
Octopamine; *p*-Hydroxyphenylethanolamine;
Norsympathol; Norsynephrine

$C_8H_{11}NO_2$ Mol. wt 153.18

Found in the cactus *Coryphantha macromeris* var. *runyonii*, in *Capsicum frutescens* (Solanaceae), in *Citrus* spp. (Rutaceae), and in *Cyperus rotundus* and *C. papyrus* (Cyperaceae).

294
Phenethylamine; β-Phenylethylamine;
β-Aminoethylbenzene; 1-Amino-2-phenylethane;
Benzeneethanamine; PEA

$C_8H_{11}N$ Mol. wt 121.18

Occurs in the oil of bitter almonds, obtained from *Prunus dulcis* var. *amara* (Rosaceae), banana fruit, *Musa sapientum* (Musaceae), and algae of the Rhodophyceae.

Skin irritant and sensitizer. It has a fishy odour, and is found in normal human urine.

295
Psilocin; Psilocyn

$C_{12}H_{16}N_2O$ Mol. wt 204.27

Found in the sacred mushroom of Mexico, Teonanacatl, *Psilocybe mexicana*, and other blue-staining *Psilocybes* known as "magic mushrooms" (Strophariaceae).

Hallucinogenic, with an unusually pleasant sensation of intellectual and physical relaxation, involving distortions of time and space perception. The effects are due to its structural similarity to serotonin (q.v.). It is readily oxidized to a blue pigment when the mushroom is bruised.

296
Psilocybin; Indocybin

$C_{12}H_{17}N_2O_4P$ Mol. wt 284.26

Occurrence as for psilocin (q.v.).

Psychotomimetic. Its properties are as for psilocin (q.v.), which is converted into psilocybin by metabolic phosphorylation in the body.

297
Putrescine; Butane-1,4-diamine; Tetramethylenediamine

$NH_2-[CH_2]_4-NH_2$

$C_4H_{12}N_2$ Mol. wt 88.15

Present in all plant cells. The content has been shown to increase in higher plants, for example barley, *Hordeum vulgare* (Gramineae) under conditions of potassium deficiency, and is affected by levels of other trace elements.

Thought to be a protein synthesis and growth regulator. The main use is as a tool in biochemical research.

298
Serotonin; 5-HT; 5-Hydroxytryptamine

$C_{10}H_{12}N_2O$ Mol. wt 176.22

Found in banana fruit, *Musa sapientum* (Musaceae), tomatoes, *Lycopersicon esculentum* (Solanaceae), the stinging hairs of the nettle, *Urtica dioica* (Urticaceae), and those of cowhage, *Mucuna pruriens* (Leguminosae).

Neurotransmitter in the central nervous system. It does not pass the blood–brain barrier, so its precursor, 5-hydroxytryptophan, is used clinically to treat depression.

299
Skatole; 3-Methyl-1*H*-indole

C_9H_9N Mol. wt 131.18

Found in beetroot, *Beta vulgaris* (Chenopodiaceae), in *Nectandra* wood (Lauraceae), and in the vile odours of inflorescences of *Arum* spp. (Araceae).

Found in faeces, and is responsible for the characteristic odour.

300
Spermidine

$NH_2-[CH_2]_4-NH-[CH_2]_3-NH_2$

$C_7H_{19}N_3$ Mol. wt 145.25

Occurs universally, either free or conjugated with fatty acids or as cinnamic acid derivatives, e.g., in oats, *Avena sativa* (Gramineae), tobacco, *Nicotiana tabacum*, tomatoes, *Lycopersicon esculentum* (Solanaceae), soybeans, *Glycine max* (Leguminosae), and many others.

Concerned with growth regulation; increased levels found in plant and animal tumours. It inhibits fungal spore germination in *Penicillium digitatum*. The main use is as a tool in biochemical research.

301
Spermine; Gerontine; Musculamine; Neuridine

$$NH_2-[CH_2]_3-NH-[CH_2]_4-NH-[CH_2]_3-NH_2$$

$C_{10}H_{26}N_4$ Mol. wt 202.35

Occurs in all plants: sources include oats, *Avena sativa*, barley, *Hordeum vulgare* (Gramineae), and Jerusalem artichoke, *Helianthus tuberosus* (Compositae).

May be involved in protein synthesis and growth regulation. It inhibits fungal spore germination in *Penicillium digitatum*. The main use is as a tool in biochemical research.

302
Tryptamine; 3-(2-Aminoethyl)indole

$C_{10}H_{12}N_2$ Mol. wt 160.22

Widespread occurrence, e.g., tomatoes, *Lycopersicon esculentum*, potatoes, *Solanum tuberosum*, and tobacco, *Nicotiana tabacum* (Solanaceae), *Piptadenia peregrina* (Leguminosae), barley seedlings, *Hordeum vulgare*, maize, *Zea mays* (Gramineae), cucumber, *Cucumis sativus* (Cucurbitaceae), and others.

Precursor of indoleacetic acid (q.v.).

303
Tyramine; 4-Hydroxyphenethylamine; Tyrosamine

$C_8H_{11}NO$ Mol. wt 137.18

Widespread occurrence, e.g., the cacti *Lophophora williamsii* and *Trichocereus pachanoi*, barley, *Hordeum vulgare*, and ryegrass, *Lolium multiflorum* (Gramineae), and mistletoe, *Viscum album* (Viscaceae).

A sympathomimetic agent with indirect adrenergic activity.

Chapter 12
Cyanogenic Glycosides

The fact that bitter almonds from *Prunus amygdalus* trees produce a poison, hydrogen cyanide (HCN), has been known since antiquity. Cyanide occurs in the intact nut in bound form as a cyanogenic glycoside. The release of the "smell of bitter almonds" occurs only when the tissue is damaged, either through enzymatic (β-glucosidase) or non-enzymatic hydrolysis of the glycoside. The immediate hydrolytic product is a cyanohydrin, which spontaneously decomposes to HCN and an aldehyde or ketone. The ease of detection of HCN either by its smell or by its reaction with moist picrate paper (yellow to brownish-red) has meant that many plants have been surveyed for cyanogenesis or the release of HCN. At least 1000 species representing 100 families and 500 genera have this character (Hegnauer, 1977). The bound toxins have been characterized in only some of these plants, and about 40 cyanogenic glycosides have been identified (Seigler, 1977; Evered and Harnett, 1988). Cyanogens are not only widely (albeit sporadically) distributed within vascular plants, including ferns, but they also occur in defence secretions of centipedes, millipedes and moths.

Most of the known glycosides are formed biosynthetically from protein amino acids, and are best classified according to their various amino acid precursors. Thus, dhurrin (**311**) is formed from tyrosine, prunasin (**323**) from phenylalanine, linamarin (**315**) from valine, lotaustralin (**317**) from isoleucine, and cardiospermin (**309**) from leucine. A distinctive group of glycosides present in plants of the Passifloraceae have a cyclopentene nucleus (volkenin, **327**), and are probably derived from the non-protein amino acid cyclopentenylglycine.

The most common sugar of cyanogenic glycosides is glucose, but the disaccharide gentiobiose (q.v.) is present in amygdalin (**305**), the disaccharide primeverose in lucumin (**318**) and the disaccharide vicianose in vicianin (**326**). Several glycosides are known in two forms as stereoisomers: one such pair is (*R*)-prunasin (**323**), characteristic of *Prunus*, and (*S*)-sambunigrin, from the elderberry, *Sambucus*. The most commonly occurring glycosides are linamarin and lotaustralin which, although of different biosynthetic origin, are nearly always found together.

The main function of cyanogenesis in plants is a protective one, against herbivory. Because of the metabolic cost of synthesis from protein amino acids cyanogenesis is sometimes variable within a plant population, with some members being cyanogenic and others acyanogenic. This is true of clover, *Trifolium repens*, and birdsfoot trefoil, *Lotus corniculatus*. Many herbivores have developed the ability to detoxify cyanide, either by the addition of sulfur to form thiocyanate (CNS^-) or by reaction with cysteine to produce β-cyanoalanine. Nevertheless, cyanogenesis still provides plants with a useful defence against browsing insects, the glycosides themselves being bitter-tasting, and the ketones formed on hydrolysis often being as toxic as the cyanide that is produced simultaneously (Spencer, 1988). Cyanogenic plants are still a significant hazard to unadapted farm animals, and humans are not immune from the cyanide content of cassava and certain other tropical food plants.

References

Chemistry and Distribution

Hegnauer, R. (1977). Cyanogenic glycosides as systematic markers in the Tracheophyta. *Plant Systematics and Evolution*, Supplement 1, 191–210.

Nahrstedt, A. (1987). Recent developments in the chemistry, distribution and biology of the cyanogenic glycosides. *Annual Proceedings of the Phytochemical Society of Europe* **27**, 213–234.

Seigler, D.S. (1977). Naturally occurring cyanogenic glycosides. *Progress in Phytochemistry*, **4**, 83–120.

Biological Activities

Poulton, J.E. (1983). Cyanogenic compounds in plants and their toxic effects. In *Plant and Fungal Toxins* (Keeler, R.F. and Tu, A.T., eds), pp. 117–160. New York: Marcel Dekker.

Spencer, K.V. (ed.) (1988). *Chemical Mediation in Coevolution*. San Diego: Academic Press.

Vennesland, B., Conn, E.E., Knowles, C.J., Westly, J. and Wissing, F. (eds) (1981). *Cyanide in Biology*. London: Academic Press.

Evered, D. and Harnett, S. (1988). *Cyanide Compounds in Biology*, CIBA Foundation Symposium Volume No. 140. Chichester: Wiley.

304
Acalyphin

$C_{14}H_{20}N_2O_9$ Mol. wt 360.32

Found in all parts of *Acalypha indica* (Euphorbiaceae); it co-occurs with *N*-methyl-3-cyanopyridones, which derive from nicotinic acid.

Toxic to all life.

305
Amygdalin; Amygdaloside; Mandelonitrile-β-gentiobioside

$C_{20}H_{27}NO_{11}$ Mol. wt 457.44

First isolated from bitter almonds, seeds of *Prunus amygdalus* (Rosaceae); the sugar moiety is gentiobiose; this is typical for seeds of many members of Rosaceae, and trace amounts may be present in the fleshy parts of fruits, in bark and leaves; also it is found in subterranean parts of *Gerbera jamesonii* cultivars (Compositae) together with vicianin (q.v.) and prunasin (q.v.). An isomer of amygdalin with sophorose (q.v.) instead of gentiobiose (q.v.) as the sugar moiety occurs together with prunasin in the leaves of *Perilla frutescens* (Labiatae).

Toxic; it has been used, interchangeably with mandelonitrile-β-glucuronide, as the drug "laetrile" for treating cancer, but is ineffective. Amygdalin is responsible for the bitterness and toxicity of the seeds of bitter almond (100 µmol g^{-1}) and of apricots (20–80 µmol g^{-1}). Minimum lethal dose to humans of hydrogen cyanide is 0.5–3.5 mg per kg body weight.

306
***Anthemis* glycoside A**

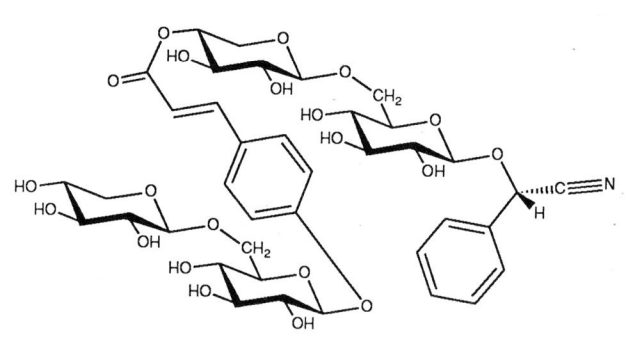

$C_{39}H_{49}NO_{21}$ Mol. wt 867.83

Found in cyanogenic achenes of *Anthemis altissima* but probably also present in other *Anthemis* spp. (Compositae); it occurs together with *Anthemis* glycoside B (q.v.) and epilucumin.

Toxic.

307
***Anthemis* glycoside B**

$C_{34}H_{41}NO_{17}$ Mol. wt 735.71

Found in cyanogenic achenes of *Anthemis altissima* but probably also present in other *Anthemis* spp. (Compositae); it occurs together with *Anthemis* glycoside A (q.v.) and epilucumin.

Toxic.

308
Cardiospermin

$C_{11}H_{17}NO_7$ Mol. wt 275.26

Found in the leaves of *Cardiospermum grandiflorum* and *C. hirsutum* and of *Heterodendron oleaefolium* (both Sapindaceae); it occurs as the *p*-hydroxybenzoate and *p*-hydroxycinnamate (esterified through the aglucone hydroxyl) in *Sorbaria arborea* (Rosaceae); cardiospermin sulfate accompanies cardiospermin in the leaves and stems of *Cardiospermum grandiflorum*.

Toxic.

309
Deidaclin

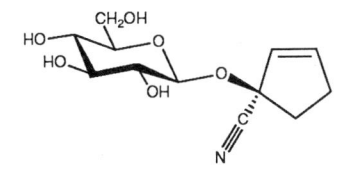

$C_{12}H_{17}NO_6$ Mol. wt 271.28

Found in the leaves of *Deidamia clematoides*, also in *Passiflora coriaceae*, *Tetrapathea tetandra*, *Adenia fruticosa*, *A. spinosa*, and *A. globosa* (all Passifloraceae).

Toxic.

310
Dhurrin; *p*-Hydroxymandelonitrile glucoside

$C_{14}H_{17}NO_7$ Mol. wt 311.30

First isolated from *Sorghum* cultivars and later shown to be present in some other grasses (Gramineae); it occurs also rather erratically in dicotyledonous taxa, e.g., *Trochodendron aralioides* (Trochodendraceae), *Borago officinalis* and *Papaver nudicaulis* (Papaveraceae), *Platanus* spp. (Platanaceae) and the leaves of *Macadamia ternifolia* (Proteaceae). The (*R*) epimer, called taxiphyllin, was first isolated from *Phyllanthus gasstroemeri* (Euphorbiaceae); it occurs in several Gramineae and is relatively widespread in dicotyledons (e.g., in *Liriodendron tulipifera* (Magnoliaceae).

Toxic. The high dhurrin content of young sorghum seedlings has led to accidental cattle poisoning but, fortunately, sorghum seeds, which are widely eaten in India and Africa, are non-cyanogenic. Taxiphyllin, present in young bamboo shoots, has been responsible for several cases of human cyanide poisoning.

311
***p*-Glucosyloxymandelonitrile;**
p-Glucosyloxybenzaldehyde cyanohydrin

$C_{14}H_{17}NO_7$ Mol. wt 311.30

Found in *Goodia latifolia* (Leguminosae), and in *Nandina domestica* (Berberidaceae) as such and as the 4'-caffeic acid ester.

Toxic: it can release hydrogen cyanide without the intervention of a β-glucosidase.

312
Gynocardin

$C_{12}H_{17}NO_8$ Mol. wt 303.28

Found in the seeds of *Gynocardia odorata* and the leaves and seeds of *Pangium edule* (Flacourtiaceae), in the pericarps and immature seeds of *Carpotroche brasiliensis* (Flacourtiaceae), and in several other genera of the family.

Toxic.

313
Heterodendrin; Dihydroacacipetalin

$C_{11}H_{19}NO_6$ Mol. wt 261.28

Occurs in the leaves of South African *Acacia* spp. (Leguminosae) and the leaves of *Heterodendron oleaefolium* (Sapindaceae); it is also a minor glucoside in the leaves of *Sorbaria arborea* (Rosaceae). The epimer, (R)-epiheterodendrin, occurs in *Acacia globulifera*, in the dried leaves of barley, *Hordeum vulgare* (Gramineae), and in some *Passiflora* spp. (Passifloraceae).

Toxic.

314
Linamarin; Manihotoxine; Phaseolunatin

$C_{10}H_{17}NO_6$ Mol. wt 247.25

Found in the seedlings of *Linum usitatissimum* (Linaceae), in several spp. of Compositae and Euphorbiaceae, in the genus *Linum* (predominantly in blue flowering spp.), the genus *Acacia* and the tribes Loteae and Phaseoleae of Leguminosae, and in some *Passiflora* spp. (Passifloraceae). Usually, it co-occurs with lotaustralin (q.v.).

Toxic. It is sequestered, with lotaustralin, from the food plant *Lotus corniculatus*, by the five-spot burnet moth, *Zygaena trifolii*. This is the major toxin of cassava, *Manihot esculentum*, through the release of free cyanide. Although acute poisoning from cassava consumption is rare, more serious long term effects occur in people, due to the thiocyanate formed during detoxification.

315
Linustatin

$C_{16}H_{27}NO_{11}$ Mol. wt 409.40

Found in the seed meal of flax, *Linum usitatissimum* (Linaceae), where it occurs together with neolinustatin (q.v.). It occurs in a few *Passiflora* spp. (Passifloraceae). Gentiobiose (q.v.) is the sugar moiety.

Toxic.

316
Lotaustralin

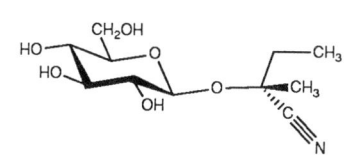

$C_{11}H_{19}NO_6$ Mol. wt 261.28

Found in *Lotus australis* and *Trifolium repens*, in other Leguminosae taxa, e.g., *Dorycnium, Ornithopus* and *Tetragonolobus*, in *Haloragis erecta* and probably other Haloragidaceae. It also occurs in *Linum usitatissimum* (Linaceae), some *Passiflora* spp. (Passifloraceae), some *Triticum* spp. (Gramineae), e.g., *T. monococcum*, and in trace to appreciable amounts in many spp. containing linamarin. The epimer (S)-epilotaustralin co-occurs in *Triticum dicoccum* and in some *Passiflora* spp.

Toxic: activity similar to that of linamarin (q.v.).

317
Lucumin; Lucuminoside; Prunasin xyloside

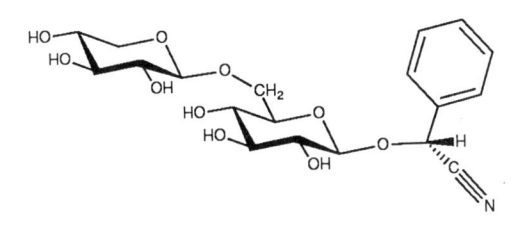

$C_{19}H_{25}NO_{10}$ Mol. wt 427.42

Occurs in the seeds of *Calocarpum sapota* (=*Lucuma mammosa*) (Sapotaceae). The sugar moiety is primverose. It probably occurs in many other cyanogenic Sapotaceae genera. The epimer, (S)-epilucumin, occurs in *Anthemis altissima* (Compositae) together with *Anthemis* glycosides A and B (q.v.).

Toxic.

318
Neolinustatin

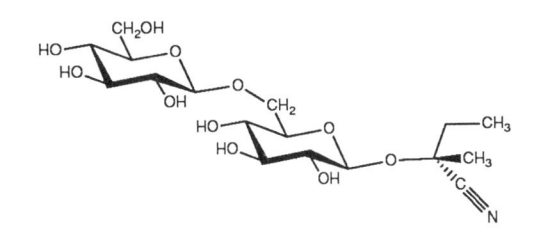

$C_{17}H_{29}NO_{11}$ Mol. wt 423.43

Found in the seed meal of flax, *Linum usitatissimum* (Linaceae), where it occurs together with linustatin (q.v.); it also occurs in a few *Passiflora* spp. (Passifloraceae). Gentiobiose (q.v.) is its sugar moiety.

Toxic.

319
Proacaciberin; Proacacipetalin 6′-arabinoside

$C_{16}H_{25}NO_{10}$ Mol. wt 391.38

Found together with proacacipetalin (q.v.) in the pods of *Acacia sieberiana* (Leguminosae). It is a diglycoside with vicianose (q.v.) as the sugar moiety.

Toxic.

320
Proacacipetalin

$C_{11}H_{17}NO_6$ Mol. wt 259.26

Occurs in many spp. of section Gummiferae of the genus *Acacia*, and has not been reported from other sources. It is converted to the isomer acacipetalin during the isolation process. The epimer, (R)-epiproacacipetalin, occurs in *Acacia globulifera* (Leguminosae), and is easily converted during isolation to 3-hydroxyepiheterodendrin, due to the addition of water across the double bond.

Toxic.

321
Proteacin

$C_{20}H_{27}NO_{12}$ Mol. wt 473.43

Only known from the seed kernels of bitter Macadamia nuts, *Macadamia ternifolia* (Proteaceae).

Toxic.

322
Prunasin; Mandelonitrile glucoside

$C_{14}H_{17}NO_6$ Mol. wt 295.30

Occurs in the leaves and bark of *Prunus* taxa, rather erratically in vascular plants, in some ferns, e.g., *Pteridium aquilinum*, *Cystopteris* spp., and many dicotyledonous taxa, e.g., many Rosaceae-Prunoideae and Maloideae and a few Rosaceae-Spiraeoideae. It is also known from some spp. of Compositae, Leguminosae, Myrtaceae, Myoporaceae and Scrophulariaceae. The epimer, (S)-sambunigrin, occurs in *Sambucus nigra* (Caprifoliaceae), *Acacia glaucescens* (Leguminosae) and *Ximenia americana* (Olacaceae).

Toxic.

323
Sarmentosin epoxide

$C_{11}H_{17}NO_8$ Mol. wt 291.26

Occurs in *Sedum cepaea* (Crassulaceae).

Toxic. This is not a typical cyanogenic glycoside: the glycosylated cyanhydrin structure is lacking; for generating hydrogen cyanide spontaneously the epoxy ring has to be hydrolysed by an epoxyhydrolase.

324
Triglochinin

$C_{14}H_{17}NO_{10}$ Mol. wt 359.30

Found in arrow grass, *Triglochin maritima* (Juncaginaceae), in other Juncaginaceae, in some Gramineae, in many representatives of Magnoliidae *sensu* Cronquist, in several Euphorbiaceae – Phyllanthoideae, in some *Campanula* spp., e.g., *C. cochleariifolia* and *C. rotundifolia* (Campanulaceae), and in many Araceae. It is often accompanied by dhurrin (q.v.) or taxiphyllin (see dhurrin).

Toxic. Cattle and sheep poisoning has been recorded, after the ingestion of arrow grass.

325
Vicianin; Mandelonitrile vicianoside

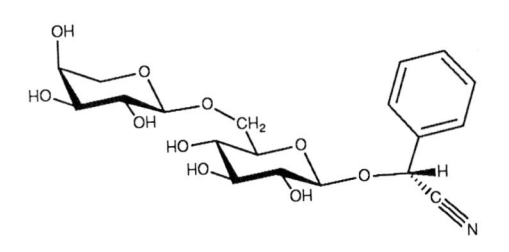

$C_{19}H_{25}NO_{10}$ Mol. wt 427.42

Occurs in the seeds of *Vicia angustifolia* and *V. sativa* (Leguminosae). The sugar moiety is vicianose. It is found together with amygdalin and prunasin in subterranean parts of *Gerbera jamesonii* cultivars (Compositae), and occurs in the fronds of several spp. of the fern genus *Davallia*.

Toxic.

326
Volkenin; Barterin; Epitetraphyllin B

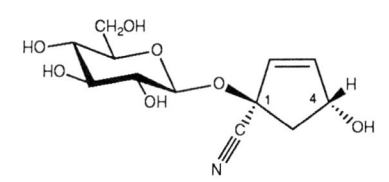

$C_{12}H_{17}NO_7$ Mol. wt 287.27

The first 1-cyano-1,4-dihydroxycyclopentenyl-1-β-glucoside to be isolated from *Barteria fistulosa* (Passifloraceae). Volkenin is the (1R,4R) form, but other epimers, namely tetraphyllin B (1S,4S), taraktophyllin (1R,4S) and epivolkenin (1S,4R), have been isolated from *Adenia, Passiflora, Tetrapathaea* spp. (Passifloraceae) and *Hydnocarpus* spp. (Flacourtiaceae).

Toxic.

327
Zierin; *m*-Hydroxysambunigrin

$C_{14}H_{17}NO_7$ Mol. wt 311.30

Occurs in the leaves and twigs of *Zieria laevigata* (Rutaceae) and is probably also present in other cyanogenic *Zieria* species; it also occurs in some Danish *Sambucus nigra* (Caprifoliaceae) populations, in leaves of *Aster ptarmicoides* (Compositae), and in full grown plants of *Oxytropis campestris* (Leguminosae). It is found as zierin xyloside, a diglycoside with primverose as the sugar moiety, in achenes of *Xeranthemum cylindraceum* (Compositae-Cardueae). The epimer, (*R*)-holocalin, was first isolated from the seeds of *Holocalyx balansae* (Leguminosae) and also occurs in *Sambucus nigra* (as above) and in *Chlorophytum capense* (Liliaceae).

Toxic.

Chapter 13
Glucosinolates

The glucosinolates (or mustard oil glycosides) are a group of bound toxins, like the cyanogenic glycosides (q.v.), which are not toxic *per se* but, on enzymatic hydrolysis, release the volatile, evil-smelling, acrid-tasting mustard oils. Release occurs when plant tissue is crushed, since thioglucosidases, called myrosinases, always accompany glucosinolates in the plants where they occur. The major hydrolysis product is the related isothiocyanate, RCNS, formed by rearrangement, but small amounts of thiocyanate RCNS and nitrile RCN may also be produced.

Chemically, the glucosinolates are nitrogen sulfur compounds with the same basic formula $RC(SGlc)NOSO_3^-$, one sulfur atom being conjugated to glucose as the *S*-glucoside and the other sulfur being present as the oxygen-linked sulfate anion. They may occur naturally as the potassium salts, although this has only been established in a few cases. About 80 glucosinolates are known (Fenwick *et al.*, 1983); the majority are aliphatic, the remainder being benzyl or indole derivatives. The best known aliphatic derivatives are sinigrin (R=allyl, **353**) and glucocapparin (R=methyl, **332**) while a typical benzyl derivative is sinalbin (**352**) and an indole derivative is glucobrassicin. Glucosinolates are formed from protein amino acids in the plant, although a carbon chain elongation may be a necessary intermediary step. Sinigrin, for example, is formed from methionine via its higher homologue, 2-amino-5-(methylthio)hexanoic acid (Kjaer and Larsen, 1973).

Glucosinolates are characteristic constituents of the mustard family, the Cruciferae, and they have been detected in every one of some 400 species that have been tested (Kjaer, 1976). They occur, for example, in such well known crops as cabbage, cauliflower, cress, mustard, rape and turnip. These glycosides are also uniformly present in other families in the same order, Capparales, to which the Cruciferae belongs, namely in the Capparaceae, Moringaceae, Resedaceae, Stegnospermaceae and Tovariaceae. There are several unrelated occurrences, notably in *Carica* (Caricaceae), *Limnanthes* (Limnanthaceae) and *Tropaeolum* (Tropaeolaceae). The seeds are often a rich source, although the compounds are normally distributed throughout the plant.

The major function of glucosinolates in plants is as feeding deterrents, against both insects (Chew, 1988) and mammals. However, specialist insects, such as the cabbage white butterfly, *Pieris brassicae*, become adapted to their toxic effects and use the presence of mustard oils in their host plants as feeding cues and oviposition guides. Toxic effects can be observed in unadapted insects, e.g., a parsley plant infiltrated with sinigrin is lethal to a caterpillar of the black swallowtail, *Papilio polyxenes*. Regular dietary intake of quantities of glucosinolate causes toxic effects in farm animals, especially damage to the thyroid, liver and kidneys. The same effects may occur in human, although it has rarely been observed in practice. The effect on the thyroid gland is due to the thiocyanate formed by rearrangement of the unstable isothiocyanate. Isothiocyanates are themselves irritating to the skin and mucous membrane, and are also active against microorganisms (Van Etten and Tookey, 1979).

Glucosinolates can be analysed, after purification on an anion exchange resin, either by thin layer or high performance liquid chromatography. Alternatively, the related isothiocyanates (or thiocyanates), obtained by steam distillation of crushed plant tissue, can be separated and determined by gas liquid chromatography. For the latest methods of analysis, see Fenwick *et al.* (1983).

References

Biochemistry and Analysis

Fenwick, G.R., Heaney, R.K. and Mullin, W.J. (1983). Glucosinolates and their breakdown products in food and food plants. *Critical Reviews in Food Science and Nutrition* **18**, 123-201.

Kjaer, A. and Larsen, P.O. (1973). Non-protein amino acids, cyanogenic glycosides and glucosinolates. In *Biosynthesis*, Volume 2 (Geissman, T.A., ed.), pp. 71-105. London: The Chemical Society.

Larsen, P.O. (1981). Glucosinolates. In *The Biochemistry of Plants*, Volume 7 (Conn, E.E., ed.), pp. 501-525. New York: Academic Press.

Natural Distribution

Ettlinger, M.G. and Kjaer, A. (1968). Sulphur compounds in plants. *Recent Advances in Phytochemistry* **1**, 59-144.

Kjaer, A. (1976). Glucosinolates in the Cruciferae. In *The Biology and Chemistry of the Cruciferae* (Vaughan, J.G., Macleod, A.J. and Jones, B.M.G., eds), pp. 207-220. London: Academic Press.

Biological Activities

Chew, F.S. (1988). Biological effects of glucosinolates. In *Biologically Active Natural Products for Use in Agriculture* (Cutler, H.G., ed.), pp. 155-181. Washington DC: American Chemical Society.

Van Etten, C.H. and Tookey, H.L. (1979). Chemistry and biological effects of glucosinolates. In *Herbivores: their Interaction with Secondary Metabolites* (Rosenthal, G.A. and Janzen, D.H., eds), pp. 471-501. New York: Academic Press.

328
Glucoalyssin; 5-(Methylsulfinyl)pentylglucosinolate

$[C_{13}H_{24}NO_{10}S_3]^-$ Mol. wt 450.51

Found in Chinese cabbage, pak-choi, *Brassica campestris* var. *chinensis*, turnip, *B. campestris* var. *rapifera*, and *Alyssum* spp. (all Cruciferae).

Flavour component, together with its decomposition products formed during cooking.

329
Glucoberteroin; 5-(Methylthio)pentylglucosinolate

$[C_{13}H_{24}NO_9S_3]^-$ Mol. wt 434.54

Found in horseradish, *Armoracia lapathifolia*, radish, *Raphanus sativus*, Chinese cabbage, pe-tsai, *Brassica pekinensis*, and swede, *B. napus* var. *napobrassica* (all Cruciferae).

Flavour component, together with its decomposition products. The hydrolysis product, 5-(methylthio)pentylisothiocyanate, is pungent and volatile.

330
Glucobrassicanapin; 4-Pentenylglucosinolate

$[C_{12}H_{20}NO_9S_2]^-$ Mol. wt 386.43

Found in horseradish, *Armoracia lapathifolia*, brown mustard, *Brassica juncea*, wasabi, Japanese horseradish, *Wasabi japonica*, rape cress, cole cress, *Brassica napus*, swede, *B. napus* var. *napobrassica*, turnip, *B. campestris* ssp. *rapifera*, and sea rocket, *Cakile maritima* (all Cruciferae).

Flavour component, together with enzymatic hydrolysis products.

331
Glucobrassicin; Indol-3-ylmethylglucosinolate

$[C_{16}H_{19}N_2O_9S_2]^-$ Mol. wt 447.47

Common in the Cruciferae, e.g., in radish, *Raphanus sativus*, rape cress and cole cress, *Brassica napus*, cabbage, *B. oleracea* var. *capitata*, and Brussels sprouts, *B. oleracea* var. *gemmifera*.

Flavour component together with decomposition products produced during cooking. Hydrolysis produces an unstable isothiocyanate which liberates free thiocyanate, SCN^-.

332
Glucocapparin; Methylglucosinolate

$[C_8H_{14}NO_9S_2]^-$ Mol. wt 332.34

Common in the Cruciferae, e.g., in horseradish, *Armoracia lapathifolia*, wasabi, Japanese horseradish, *Wasabi japonica*, and cauliflower, *Brassica oleracea* var. *botrytis* subvar. *cauliflora*. It is also present in plants of the Capparaceae.

Flavour component together with its decomposition products formed during cooking. The hydrolysis product, methyl isothiocyanate, is lachrimatory, and is partly responsible for the characteristic flavour of horseradish.

333
Glucocheirolin; 3-(Methylsulfonyl)propylglucosinolate

$[C_{11}H_{20}NO_{11}S_3]^-$ Mol. wt 438.49

Found in horseradish, *Armoracia lapathifolia*, cauliflower, *Brassica oleracea* var. *botrytis* subvar. *cauliflora*, swede, *B. napus* var. *napobrassica*, and turnip, *B. campestris* var. *rapifera* (all Cruciferae).

Flavour component together with breakdown products formed during cooking. One of these, 3-(methylsulfonyl)propylisothiocyanate, is goitrogenic and cytotoxic in animals.

334
Glucocleomin; 2-Methyl-2-hydroxybutylglucosinolate

$[C_{12}H_{22}NO_{10}S_2]^-$ Mol. wt 404.45

Occurs in *Cleome* spp. (Capparaceae).

335
Glucocochlearin; Glucojiabutin; (1*S*)-1-Methylpropylglucosinolate

$[C_{11}H_{20}NO_9S_2]^-$ Mol. wt 374.42

Found in horseradish, *Armoracia lapathifolia* (Cruciferae).

Flavour component, together with its breakdown products.

336
Glucoconringiin; 2-Methyl-2-hydroxypropylglucosinolate

$[C_{11}H_{20}NO_{10}S_2]^-$ Mol. wt 390.42

Found in *Conringia* spp. (Cruciferae).

337
Glucoerucin; 4-(Methylthio)butylglucosinolate

$[C_{12}H_{22}NO_9S_3]^-$ Mol. wt 420.51

Common in the Cruciferae, e.g., in cabbage, *Brassica oleracea* var. *capitata*, Brussels sprouts, *B. oleracea* var. *gemmifera*, kohlrabi, *B. napus* var. *napobrassica*, and turnip, *B. campestris* ssp. *rapifera*.

Contributes significantly to flavour and odour, together with breakdown products formed during cooking. The aglucone, erucin, attracts and stimulates egg-laying in the cabbage root fly, *Delia brassicae*.

338
Glucoerysolin; 4-(Methylsulfonyl)butylglucosinolate

$[C_{12}H_{22}NO_{11}S_3]^-$ Mol. wt 452.51

Found in swede, *Brassica napus* var. *napobrassica*, and turnip, *B. campestris* ssp. *rapifera* (Cruciferae).

Flavour component. The aglucone, erysoline, has antifungal and antibacterial activity and is cytotoxic in animals.

339
Glucoiberin; 3-(Methylsulfinyl)propylglucosinolate

$[C_{11}H_{20}NO_{10}S_3]^-$ Mol. wt 422.49

Occurs in candytuft, *Iberis amara*, horseradish, *Armoracia lapathifolia*, kohlrabi, *Brassica oleracea* var. *gongylodes*, cabbage *B. oleracea* var. *capitata*, Brussels sprouts, *B. oleracea* var. *gemmifera*, cauliflower, *B. oleracea* var. *botrytis* subvar. *cauliflora*, and kale, *B. oleracea* var. *sabauda* (all Cruciferae).

Flavour component, together with hydrolysis products formed during cooking. The hydrolysis product 3-(methylsulfinyl)propylisothiocyanate is cytotoxic in animal tests.

340
Glucoiberverin; 3-(Methylthio)propylglucosinolate

$[C_{11}H_{20}NO_9S_3]^-$ Mol. wt 406.49

Found in the seeds of candytuft, *Iberis amara*, horseradish, *Armoracia lapathifolia*, radish, *Raphanus sativus*, cabbage, *Brassica oleracea* var. *capitata*, Chinese cabbage, pak-choi, *B. campestris* var. *chinensis*, cauliflower, *B. oleracea* var. *botrytis* subvar. *cauliflora*, mustard spinach, *B. juncea*, and mustard greens, *B. hirta* (= *Sinapis alba*) (all Cruciferae).

Flavour component, together with the breakdown products formed during cooking. 3-(Methylthio)propyl isothiocyanate, the hydrolysate, is steam-volatile and pungent.

341
Glucolepidiin; Ethylglucosinolate

$[C_9H_{16}NO_9S_2]^-$ Mol. wt 346.37

Occurs in garden cress, *Lepidium sativum*, and horseradish, *Armoracia lapathifolia* (Cruciferae).

Flavour component. The hydrolysis product, ethyl isothiocyanate, is extremely pungent, and garlic-like.

342
Glucolimnanthin; *m*-Methoxybenzylglucosinolate

$[C_{15}H_{20}NO_{10}S_2]^-$ Mol. wt 438.46

Found in the seeds of *Limnanthes douglasii*, and other *Limnanthes* spp. (Limnanthaceae).

Odour principle. Hydrolysis leads to the formation of a volatile, pungent isothiocyanate.

343
Gluconapin; But-3-enylglucosinolate

$[C_{11}H_{18}NO_9S_2]^-$ Mol. wt 372.40

Widespread occurrence in the Cruciferae: e.g., in brown and black mustard, *Brassica juncea* and *B. nigra*, horseradish, *Armoracia lapathifolia*, and wasabi, Japanese horseradish, *Wasabi japonica*.

Flavour component; the hydrolysis product, but-3-enylisothiocyanate, is aromatic and pungent.

344
Gluconapoleiferin; 2-Hydroxypent-4-enylglucosinolate

$[C_{12}H_{20}NO_{10}S_2]^-$ Mol. wt 402.43

Found in Chinese cabbage, pak-choi, *Brassica campestris* ssp. *chinensis*, and pe-tsai, *B. pekinensis*, turnip, *B. campestris* ssp. *rapifera*, swede, *B. napus* var. *napobrassica*, and brown mustard, *B. juncea* (all Cruciferae).

Flavour component, together with the enzymatic breakdown products.

345
Gluconasturtiin; 2-Phenylethylglucosinolate

$[C_{15}H_{20}NO_9S_2]^-$ Mol. wt 422.46

Very widespread occurrence in the Cruciferae: e.g., in water cress, *Nasturtium officinale*, garden cress, *Lepidium sativum*, and black, brown and white mustard, *Brassica nigra*, *B. juncea* and *B. hirta* (=*Sinapis alba*).

Flavour component. The hydrolysis product, 2-phenylethyl isothiocyanate, has the aroma of water cress, and produces a tingling sensation on the tongue; it is cytotoxic in animals.

346
Glucoputranjivin; Isopropylglucosinolate

$[C_{10}H_{18}NO_9S_2]^-$ Mol. wt 360.39

Occurs in white, brown and black mustard, *Brassica hirta* (=*Sinapis alba*), *B. juncea* and *B. nigra*, horseradish, *Armoracia lapathifolia*, wasabi, Japanese horseradish, *Wasabi japonica*, Chinese cabbage, pak-choi, *B. campestris* ssp. *chinensis*, and Brussels sprouts, *B. oleracea* var. *gemmifera* (all Cruciferae).

Flavour component. The breakdown product, isopropyl isothiocyanate, is pungent.

347
Glucoraphanin; 4-(Methylsulfinyl)butylglucosinolate

$[C_{12}H_{22}NO_{10}S_3]^-$ Mol. wt 436.51

Found in radish, *Raphanus sativus*, Brussels sprouts, *Brassica oleracea* var. *gemmifera*, cabbage, *B. oleracea* var. *capitata*, and broccoli, *B. oleracea* var. *botrytis* subvar. *cymosa* (all Cruciferae).

Flavour component. The aglycone, 4-(methylsulfinyl)butyl isothiocyanate (sulforaphane), has considerable antifungal and antimicrobial activities.

348
Glucoraphenin;
4-(Methylsulfinyl)but-3-enylglucosinolate

$[C_{12}H_{20}NO_{10}S_3]^-$ Mol. wt 434.50

Found in radish, *Raphanus sativus* (Cruciferae).

Flavour component.

349
Glucotropaeolin; Benzylglucosinolate

$[C_{14}H_{18}NO_9S_2]^-$ Mol. wt 408.44

Occurs in nasturtium, Indian cress, *Tropaeolum majus* (Tropaeolaceae), horseradish, *Armoracia lapathifolia*, and other Cruciferae; it occurs also in the dry latex of papaya, *Carica papaya* (Caricaceae).

Flavour component. It is a feeding stimulant for the cabbage white butterfly, *Pieris brassicae*. Benzyl isothiocyanate and benzyl thiocyanate are hydrolysis products and inhibit the neoplastic effects of some carcinogens in animals. Benzyl isothiocyanate has strong antibacterial and antifungal activities. Benzyl thiocyanate is thought to be goitrogenic in large amounts.

350
Neoglucobrassicin;
1-Methoxy-3-indolylmethylglucosinolate;
N-Methoxy-3-indolylmethylglucosinolate

$[C_{17}H_{21}N_2O_{10}S_2]^-$ Mol. wt 477.50

Occurs in rape and cole cress, *Brassica napus*, Brussels sprouts, *B. oleracea* var. *gemmifera*, cauliflower, *B. oleracea* var. *botrytis* subvar. *cauliflora*, broccoli, *B. oleracea* var. *botrytis* subvar. *cymosa*, kohlrabi, *B. oleracea* var. *gongylodes*, and hedge mustard, *Sisymbrium officinale* (all Cruciferae).

Flavour component, together with the breakdown products. Hydrolysis yields the thiocyanate ion, SCN^-.

351
Progoitrin; (2R)-2-Hydroxybut-3-enylglucosinolate

$[C_{11}H_{18}NO_{10}S_2]^-$ Mol. wt 388.40

Found in most brassicas, including rapeseed, *B. napus* var. *napus*, brown mustard, *B. juncea*, and cabbage, *B. oleracea* var. *capitata* (Cruciferae). Epiprogoitrin ((2S)-2-hydroxybut-3-enylglucosinolate) is found in *Crambe abyssinica* (Cruciferae).

Flavour component, together with the breakdown products. It is the precursor of (R)-5-vinyl-2-oxazolidinethione, "goitrin", which is responsible for the bitter taste found in some frozen *Brassica* vegetables, and is goitrogenic, inhibiting the incorporation of iodine into thyroxine precursors and interfering with thyroxine secretion. It stimulates feeding in the diamond back moth, *Plutella maculipennis*, and egg laying in the cabbage root fly, *Delia brassicae*. Both goitrin and epigoitrin are toxic in animals but are nonteratogenic in rats.

352
Sinalbin; Glucosinalbin; *p*-Hydroxybenzylglucosinolate

$[C_{14}H_{18}NO_{10}S_2]^-$ Mol. wt 424.44

Found in white mustard, *Brassica hirta* (=*Sinapis alba*); radish, *Raphanus sativus*, and charlock, *Sinapis arvensis* (Cruciferae).

Flavour component, together with the breakdown products. It decomposes to form the thiocyanate ion, SCN^-. Also, it stimulates egg laying in the cabbage maggot, *Hymlemya brassicae*.

353
Sinigrin; Allylglucosinolate; Prop-2-enylglucosinolate

$[C_{10}H_{16}NO_9S_2]^-$ Mol. wt 358.38

Common in the Cruciferae, e.g., in horseradish, *Armoracia lapathifolia*, black, white and brown mustards, *Brassica nigra*, *B. hirta* and *B. juncea*, wasabi, Japanese horseradish, *Wasabi japonica*, and cabbage, *B. oleracea* var. *capitata*.

Flavour component, together with the breakdown products. Allyl isothiocyanate, formed during hydrolysis, is volatile and intensely pungent, and responsible for the biting taste of mustard. Both sinigrin and allyl isothiocyanate attract and stimulate egg laying in the cabbage maggot, *Hymlemya brassicae*, and in the cabbage white butterfly, *Pieris brassicae*, and have been used as attractants in field traps for cabbage root fly, *Delia brassicae*. However, they are feeding deterrents for many other insects.

354
(*N*-Sulfoindol-3-yl)methylglucosinolate

$[C_{16}H_{18}N_2O_{12}S_3]^{2-}$ Mol. wt 526.53

Occurs in woad, *Isatis tinctoria* (Cruciferae).

Hydrolysis liberates the thiocyanate ion, SCN^-.

Chapter 14
Purines and Pyrimidines

The purines and pyrimidines present in plants can be considered under four headings. First of all, there are the well known bases of nucleic acids: the purines, adenine (**355**) and guanine (**362**); and the pyrimidines, cytosine (**359**), uracil (**372**) (only in RNA) and thymine (**371**) (only in DNA). These are common to all living tissues. These five bases, besides occurring in the structures of the nucleic acids, also occur at least in trace amounts, bound in low molecular weight form in plants. Each base can occur with the appropriate sugar ribose or deoxyribose attached (as nucleosides) or with sugar and phosphate (as nucleotides). Certain nucleotides, such as adenosine triphosphate (**356**), have very important functions in primary metabolism. Others, such as uridine diphosphate glucose (**373**), are involved in carbohydrate metabolism (Miflin, 1981).

Second, there are a number of unusual bases found in plants which are closely related in structure to the nucleic acid bases. These are present either bound in nucleic acids, as is 5-methylcytosine (**367**), in the DNA of wheat germ, or in low-molecular-weight form. Examples of the latter are the two pyrimidine glycosides, vicine (**374**) and convicine (**358**), which are found in certain legume seeds of the genera *Vicia* and *Pisum*. Another example is the nonprotein amino acid lathyrine, which has a pyrimidyl ring in its structure and which also occurs in legumes, in *Lathyrus* seed (see Amino acids) (Wang, 1973).

A third group of bases of importance in plants is the methylated purines such as theobromine (**369**) and caffeine (**357**), which are highly valued as stimulants. They occur in relatively high concentrations in tea, coffee and cocoa, but are also found as trace constituents in a number of other plants. Caffeine and related purines are sometimes classified with the Alkaloids (q.v.). There is good evidence in the case of caffeine that its distribution in the coffee plant is correlated with a protective function for it against herbivory (Frischknecht *et al.*, 1986).

The fourth and final group is also purines, this time substituted in the 6-position, and are known collectively as cytokinins. They comprise an important group of plant growth regulators, and are primarily responsible for initiation of cell division during growth. The first cytokinin to be described was kinetin (**365**), actually a breakdown product of an animal nucleic acid preparation discovered during plant tissue culture studies. Subsequently, naturally occurring cytokinins were detected in a number of plants and presumably they are of universal occurrence in the plant kingdom. The first natural cytokinin to be discovered was zeatin (**375**) from *Zea mays*, but a number of other closely related substances, including dihydrozeatin (**361**), N^6-methylaminopurine (**366**) and N^6-(Δ^2-isopentenyl)adenine (**364**), have since been implicated as cytokinins in other plants (for review, see MacMillan, 1980).

Purines and pyrimidines are colourless crystalline relatively stable compounds. They are generally only slightly soluble in water, being more soluble in either weak acid or alkaline solution. Distribution studies have been handicapped by the fact that there is no specific colour reagent (like ninhydrin for the amino acids) which allows for their easy detection in micro amounts in plant extracts. Once isolated, purines and pyrimidines, however, can be satisfactorily characterized by their UV spectra, particularly since these undergo specific shifts according to the pH of the solution. This is because purines and pyrimidines can exist in more than one tautomeric form; in basic solution, the hydroxyl groups will ionize and this form will be stabilized whereas, in acid solution, the amino groups will ionize and the keto tautomer will be stabilized.

References

Frischknecht, P.M., Ulmer-Dufek, J. and Baumann, T.W. (1986). Purine alkaloid formation in buds and developing leaflets of *Coffea arabica*: expression of an optimal defence strategy. *Phytochemistry* **25,** 613–616.

MacMillan, J. (ed.) (1980). *Hormonal Regulation of Development. Aspects of Plant Hormones.* Berlin: Springer-Verlag.

Miflin, B.J. (ed.) (1981). *Biochemistry of Plants (Amino Acids and Derivatives*, Volume 5). New York: Academic Press.

Wang, D. (1973). Purines and pyrimidines. In *Phytochemistry*, Volume 2 (Miller, L.P., ed.), pp. 61–117. New York: Van Nostrand-Reinhold.

355
Adenine; 1*H*-Purin-6-amine; 6-Aminopurine; 6-Amino-1*H*-purine; 6-Amino-3*H*-purine; 6-Amino-9*H*-purine; 1,6-Dihydro-6-iminopurine; 3,6-Dihydro-6-iminopurine

$C_5H_5N_5$ Mol. wt 135.13

Universal in living tissues as a constituent of nucleic acids, especially as the D-riboside, adenosine; it occurs in coenzymes such as coenzyme A, nicotinamide adenine dinucleotide (NAD), flavin adenine dinucleotide (FAD), and adenosine mono-, di- and triphosphates (q.v.) (AMP, ADP and ATP).

Essential in metabolism as a constituent of the above compounds. The main uses are in biochemical research on heredity, viral diseases and cancer.

356
Adenosine triphosphate; ATP; Adenosine 5′-(tetrahydrogen phosphate); Adenosine 5′-triphosphoric acid

$C_{10}H_{16}N_5O_{13}P_3$ Mol. wt 507.20

Universal in living tissues.

Coenzyme involved in the transfer of phosphate bond energy, for example in the contraction of skeletal muscle, where ATP is hydrolysed to the diphosphate, ADP. It is used in biochemical research and in industry to inhibit enzymatic browning of raw potatoes, apples and other fruit and vegetables.

357
Caffeine; Coffeine; Guaranine; 3-Methyltheobromine; Thein; Theine; 1,3,7-Trimethylxanthine

$C_8H_{10}N_4O_2$ Mol. wt 194.20

Occurs in the coffee bean, *Coffea* spp. (Rubiaceae), tea leaves, *Camellia sinensis* (Theaceae), maté, *Ilex paraguayensis* (Aquifoliaceae), guarana, *Paullinia cupana* (Sapindaceae), and cola, *Cola acuminata* (Sterculiaceae).

Central nervous system stimulant, causing wakefulness and increased mental activity, diuretic, cardiac and respiratory stimulant. It is used in many proprietary preparations containing aspirin and paracetamol to enhance analgesic activity.

358
Convicine

$C_{10}H_{15}N_3O_8$ Mol. wt 305.25

Found in the seeds of vetch, *Vicia sativa* (Leguminosae).

A nucleoside.

359
Cytosine; 4-Amino-2-hydroxypyrimidine; 4-Aminopyrimidin-2-ol

$C_4H_5N_3O$ Mol. wt 111.11

Universally distributed in nature, especially as a constituent of nucleic acids, as the riboside, cytidine, and as cytidine monophosphate CMP (2′- and 3′-cytidylic acid), ribonuclease inhibitors.

Essential in metabolism as a constituent of the above compounds.

360
DNA; Deoxyribonucleic acid; Desoxyribonucleic acid

R = purine or pyrimidine base

Occurs in all chromosomes.

Essential carrier of genetic information for inheritance and the direction of cell development.

361
Dihydrozeatin

$C_{10}H_{15}N_5O$ Mol. wt 221.27

Widespread occurrence in plants.

Plant growth regulator.

362
Guanine; 2-Aminohypoxanthine

$C_5H_5N_5O$ Mol. wt 151.13

Universal occurrence in living tissues as a constituent of nucleic acids, especially as the riboside, guanosine, and as guanosine mono-, di-, and triphosphates.

Essential in metabolism as a constituent of the above compounds.

363
Hypoxanthine; 1,7-Dihydro-6*H*-purin-6-one; Oxypurine; Purin-6(1*H*)-one

$C_5H_4N_4O$ Mol. wt 136.12

Widespread occurrence in the plant kingdom; formed in the body from the breakdown of nucleic acids. The 9-β-D-riboside is inosine (hypoxanthosine), and is found in sugar beet, *Beta vulgaris* (Chenopodiaceae).

Nucleic acid base of limited distribution. Hypoxanthine breaks down to form uric acid in the body; inhibition of this process forms the basis of the treatment of gout. Inosine activates cellular functions. The complex inosine pranobex is immunostimulatory, and has been used to treat viral diseases.

364
N^6-(Δ^2-Isopentenyl)adenine

$C_{10}H_{13}N_5$ Mol. wt 203.25

Widespread occurrence in plants.

Plant growth regulator. The nucleoside has also been found in the transfer RNA (q.v.) of some plants and microorganisms.

365
Kinetin; 6-Furfurylaminopurine; N^6-Furfuryladenine

$C_{10}H_9N_5O$ Mol. wt 215.22

Derived artefactually from the breakdown of DNA during autoclaving.

Plant growth regulator, active at very low concentrations but only in the presence of indoleacetic acid (q.v.). It is used in microbial culture. The name kinetin is also used as a synonym for the enzyme hyaluronidase.

366
6-Methylaminopurine; N^6-Methyladenine

$C_6H_7N_5$ Mol. wt 149.16

A minor constituent of nucleic acids. Also, it is produced as an artefact during alkaline extraction of 1-methyladenine compounds from transfer RNA (see RNA).

Forms a nucleoside, replacing adenine, in some types of nucleic acids.

367
5-Methylcytosine

$C_5H_7N_3O$ Mol. wt 125.13

A constituent of DNA from some sources, e.g., wheatgerm, where approximately 25% of the cytosine (q.v.) is substituted by 5-methylcytosine.

A nucleoside in some plants.

368
RNA; Ribonucleic acid

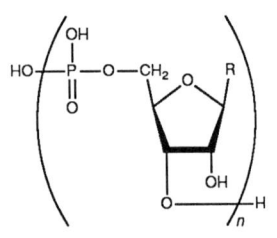

R = purine or pyrimidine base

Universal occurrence in cell nuclei and cytoplasm.

Directly involved in protein synthesis. Ribosomal RNA is involved in ribosomal protein synthesis. Messenger RNA acts as a template, one strand of DNA complementing one strand of mRNA to translate the genetic code to protein synthesis. Each transfer RNA molecule is specific for an amino acid; recognizing, binding and placing the amino acid in the correct position in the polypeptide chain being synthesized, and attaching it to the ribosome.

369
Theobromine; 3,7-Dimethylxanthine

$C_7H_8N_4O_2$ Mol. wt 180.17

Found in cocoa, *Theobroma cacao* (Sterculiaceae), tea, *Camellia sinensis* (Theaceae), guarana, *Paullinia cupana* (Sapindaceae), and cola, *Cola acuminata* (Sterculiaceae).

Diuretic, cardiac stimulant, vasodilator, and smooth muscle relaxant. It also has some central nervous system activity, but less so than caffeine (q.v.).

370
Theophylline; 1,3-Dimethylxanthine; Theocin

$C_7H_8N_4O_2$ Mol. wt 180.17

Occurs in small amounts in tea, *Camellia sinensis* (Theaceae), and guarana, *Paullinia cupana* (Sapindaceae).

Diuretic of short duration, and a cardiac stimulant, with only slight central nervous system stimulation. It relaxes involuntary muscle and, therefore, is used widely for the prophylaxis and treatment of bronchospasm associated with asthma, emphysema and chronic bronchitis.

371
Thymine; 2,4-Dihydroxy-5-methylpyrimidine; 5-Methyluracil

$C_5H_6N_2O_2$ Mol. wt 126.12

Universal occurrence in nature, especially as the deoxyriboside thymidine, a constituent of nucleic acids.

Essential in metabolism as a nucleotide, and is used in biochemical research.

372
Uracil; 2,4-Dioxopyrimidine;
2-Hydroxypyrimidine-4(1*H*)-one;
2-Hydroxypyrimidine-4(3*H*)-one;
4-Hydroxypyrimidine-4(3*H*)-one; 2,4-Pyrimidinediol;
Pyrimidine-2,4(1*H*,3*H*)-dione

$C_4H_4N_2O_2$ Mol. wt 112.09

Widespread occurrence in living tissues as the riboside, uridine, a component of nucleic acids; and as uridine diphosphate glucose (q.v.).

Essential in metabolism as a constituent of the above compounds. It is used in biochemical research.

373
Uridine diphosphate glucose; UDPG;
Uridine-5′-diphosphoglucose; Co-galactoisomerase;
Co-waldenase

$C_{15}H_{24}N_2O_{17}P_2$ Mol. wt 566.32

Widespread occurrence in nature.

Involved in carbohydrate metabolism: the coenzyme of the galactowaldenase system which catalyses the conversion of galactose 1-phosphate into glucose 1-phosphate.

374
Vicine; Divicine-β-glucoside

$C_{10}H_{16}N_4O_7$ Mol. wt 304.27

Found in the seeds of vetch, *Vicia sativa* (Leguminosae).

A nucleoside.

375
Zeatin; *trans*-Zeatin

$C_{10}H_{13}N_5O$ Mol. wt 219.25

Widespread occurrence in plants, together with *cis*-ribosylzeatin.

Plant growth regulator.

Chapter 15
Proteins and Peptides

The proteins in plants, as in other organisms, are high-molecular-weight polymers of amino acids. The amino acids are arranged in a given linear order, as determined by the triplet base code of the DNA in the nucleus, and each protein has a specific amino acid sequence. In the simplest cases, a protein may consist of a single chain of polypeptide; two or more identical chains may, however, form a higher-molecular-weight hydrogen-bonded aggregate *in vivo*. Because of the stereochemistry of the peptide bond, a polypeptide does not usually have a random form; the chain is often coiled in the form of a helix and this, in turn, can fold on itself and adopt a particular three-dimensional shape. These are described as the secondary and tertiary structures of the protein. Many plant proteins are, in fact, roughly rounded in shape and, hence, are called globular proteins.

Proteins which contain other structural elements besides amino acids are classified as conjugated proteins. The nature of the linkage between the polypeptide chain and the other structural moiety is not always precisely known. A simple conjugated protein is one which contains a metal such as calcium, iron, copper or molybdenum, complexed in its structure (e.g., calmodulin, **381**, ferredoxin, **387**). Other conjugated proteins contain lipid, phosphate, carbohydrate or nucleic acid. Chromoproteins have a chromophoric group attached and are coloured. The respiratory enzyme, cytochrome *c* (**385**), for example, is yellow and consists of a polypeptide chain of 112 amino acid residues linked to a haem (iron-porphyrin) chromophore. Proteins with sugars attached, linked covalently via, e.g., serine residues, are known as glycoproteins. Within this group are the phytohaemagglutinins or lectins (e.g., favin, **386**), which are distinguished by their ability to cause the clumping of red blood cells.

Proteins vary in their molecular weight from 12 000 in cytochrome *c*, and 60 000 in alcohol dehydrogenase to 500 000 in urease. Molecular weights of over 1 000 000 are also known. Proteins separate according to molecular weight when they are subjected to gel filtration on a column of Sephadex, a cross-linked dextran. Separation of proteins by gel electrophoresis is also partly determined by their molecular size, although the charge properties of the protein are equally important in determining their mobility on the gel. The net charge on a protein depends on the number of basic and acidic amino acids present in the polypeptide chain. Most plant proteins have more acidic than basic amino acids and thus have a net negative charge and move towards the anode.

Many proteins are also enzymes, catalysing particular steps in either primary or secondary metabolism, and can clearly be distinguished from other proteins by their enzymatic properties (e.g., rubisco, **407**). Some plant proteins such as papain (**397**) are well known and valued because of their enzymatic proteolytic activity, i.e., their ability to hydrolyse other proteins. Certain proteins, found particularly in legume seeds, are likewise notorious because of their ability to inhibit the activity of mammalian proteases, such as trypsin. The kunitz inhibitor of soybean (**392**) is but one example.

Plant proteins have a few special features which distinguish them from proteins in animals. They are, in general, relatively low by comparison in their content of sulfur amino acids (methionine and cysteine). They also occasionally contain amino acids other than the so-called 20 protein amino acids. Plant (but not animal) cytochrome *c* may have the compound trimethyllysine present as one of its amino acid constituents. Again, hydroxyproline is found as an amino acid constituent of the cell wall protein in many plant tissues. From the nutritional viewpoint, some cereal proteins and notably the prolamins are undesirable because of their relatively low content of the basic amino acid lysine. This problem has been rectified by breeding cereal varieties, the so-called 'high lysine' strains, in which there is a better balance of protein types and in which prolamin synthesis is partly suppressed. Another feature of certain plant proteins is their deleterious effect on mammalian systems; two examples are the toxic abrin (**376**) from *Abrus precatorius*, and ricin (**406**) from castor bean, one of the most poisonous plant substances known. Uniquely, plant proteins can occasionally taste very sweet, see, e.g., monellin (**396**), and thaumatin (**409**).

Although the amino acid composition of the total leaf protein is practically identical whatever the plant source, plants do vary in their protein make-up. Variation in primary sequence has been extensively explored in the case of three electron-carrier proteins in plants, namely cytochrome *c*, ferredoxin and plastocyanin. Partial or complete sequence data are also available for some enzymes (e.g. rubisco), for protease inhibitors and for storage proteins (see Boulter and Parthier, 1982).

By contrast with the ubiquity of proteins in plant tissues, peptides are rarely found. The only universally distributed compound is the tripeptide glutathione (**389**). Some nonprotein Amino acids (q.v.) occur in dipeptide form, e.g., as the γ-glutamyl derivatives. Otherwise, there are larger peptides, such as viscotoxin (**412**) from mistletoe, with 46 amino acid residues, which is really a small protein.

There are many general reference books on proteins, which may deal *inter alia* with those specifically present in plants. Books on plant proteins include those of Boulter and Parthier (1982), Harborne and van Sumere (1975), and Norton (1978). There is also a book specifically on seed proteins (Daussant *et al.*, 1983).

References

Boulter, D. and Parthier, B. (eds) (1982). *Structure, Biochemistry and Physiology of Proteins. Encylopedia of Plant Physiology*, New Series, Volume 14A. Berlin: Springer-Verlag.
Daussant, J., Mosse, J. and Vaughan, J. (eds) (1983). *Seed Proteins*. London: Academic Press.
Harborne, J.B. and van Sumere, C.F. (eds) (1975). *Chemistry and Biochemistry of Plant Proteins*. London: Academic Press.
Liener, I.E. (ed.) (1969). *Toxic Constituents of Plant Foodstuffs*. London: Academic Press.
Norton, G. (ed.) (1978). *Plant Proteins*. London: Butterworths.

376
Abrin

Mol. wt Abrins: 63 000–67 000
Abrus agglutinin: 134 900

All are glycoproteins (but not metalloproteins) and are lectins. Abrins are monovalent, composed of two polypeptide chains joined by a disulfide bridge, the smaller chain being the A chain, the other, containing more sugar residues, the B chain. *Abrus* agglutinin is a bivalent tetramer.

Abrin occurs in the seeds of Jequirity, *Abrus precatorius* (Leguminosae).

All are potent haemagglutinins; abrins are extremely toxic: as little as one seed can cause fatal poisoning. Abrins are more toxic to tumour cells than normal cells. The A chain is an enzyme that inhibits protein synthesis. *Abrus* agglutinin is nontoxic in animal cells. They are used experimentally in cancer research.

377
Actinidin

An acidic protein (pI 3.1) consisting of 220 amino acid residues with three disulfide bridges and one free (active site) sulfhydryl group.

Actinidin occurs in the Chinese gooseberry, the fruit of *Actinidia chinensis* (Actinidiaceae).

It is a sulfhydryl protease.

378
α-Amanitin

$C_{39}H_{54}N_{10}O_{13}S$ Mol. wt 903.00

Found in the death cap mushroom, *Amanita phalloides*, and also *A. verna, A. bisporigera, A. ocreata* and, to some extent, the destroying angel mushroom, *A. virosa* (Amanitaceae). α-Amanitin is one of a number of structurally similar amatoxins present in these fungi.

Very poisonous, 10–15 times more toxic than phalloidin (q.v.). It affects the kidney and liver and has a delayed onset of action: at least 12 h after ingestion. Symptoms include nausea, vomiting, convulsions, colic, severe diarrhoea; then apparent recovery for up to 5 days, followed by hepatitis, renal failure, coma and vascular collapse. It is used as a tool in biochemical research.

379
Bowman-Birk inhibitor, soybean

Mol. wt ~ 8000

Composed of a single polypeptide chain of 71 residues, cross-linked by 7 disulfide bonds.

Occurs in soybeans, *Glycine max* (Leguminosae).

Protease inhibitor. It is active against trypsin and chymotrypsin.

380
Bromelain; Bromelin

Mol. wt ~ 33 000

Stem bromelain is a basic glycoprotein; fruit bromelain is acidic.

Found in pineapple, *Ananas comosus* (Bromeliaceae).

Proteolytic enzymes. They are used medicinally as an anti-inflammatory agent in soft tissue oedema and injury. Further uses are to tenderize meat and to chill-proof beer.

381
Calmodulin

Mol. wt ~ 15 000–19 000

An acidic monomer consisting of mainly acidic amino acids, almost always including a trimethylated lysyl residue but lacking cysteine, hydroxyproline and tryptophan, and with a low tyrosine content. There are four homologous internal amino acid sequences, called domains, each one being a calcium ion binding site.

Found in most plant and animal cells. Has been characterized from several plant species, e.g., spinach, *Spinacia oleracea* (Chenopodiaceae).

A calcium-dependent protein responsible for activating and regulating a number of enzymes involved in basic cell functions, such as contractile processes, protein phosphorylation and many metabolic reactions.

382
Chymopapain

Mol. wt Chymopapain A, B: ~ 35 000

A basic protein with four components, A, B, C, D.

Found in the latex of papaya, *Carica papaya* (Caricaceae).

Sulfhydryl proteases similar to papain (q.v.). Chymopapain has been used medicinally to treat spinal disc ruptures, and for wound healing. In the food industry it is used for tenderizing meat.

383
Concanavalin A

A nonglycoprotein lectin existing in two pH-dependent forms, each composed of identical subunits of molecular weight 25 500. Below pH 5.5 it is predominantly dimeric and above pH 7 it is tetrameric. Each subunit is a single polypeptide chain of 237 amino acids containing binding sites for Mn^{2+} and Ca^{2+} ions as well as for sugars.

Found in the jack bean, *Canavalia ensiformis* (Leguminosae).

Agglutinates a variety of cells. It inhibits the growth of tumour cells in experimental animals and restores a normal growth pattern to virus-transformed fibroblasts in tissue culture. A further use is as a probe in biochemical research.

384
Cyclosporin A

$C_{62}H_{111}N_{11}O_{12}$ Mol. wt 1202.65

Occurs in *Tolypocladium inflatum* (= *Trichoderma polysporum*) and other fungi imperfecti.

Immunosuppressant. A major use is to prevent rejection of implanted organs such as heart and kidney.

385
Cytochrome *c*

A single polypeptide chain of 103–113 residues with no disulfide bridges and a covalently bound haem group linked by thioether bridges to two cysteine residues at positions 14 and 17. Sequences vary slightly with source.

Occurs in the cells of all aerobic organisms; first sequenced in higher plants from maize, *Zea mays* (Gramineae).

The catalyst of respiration: part of the electron-transport chain linking substrates with oxygen.

386
Favin

Mol. wt ~ 51 000

A glycoprotein consisting of two pairs of chains, the α and β chains, of molecular weight approximately 2×5600 and 2×20 000, respectively.

Occurs in broad beans, *Vicia faba* (Leguminosae).

Agglutinates cells. It is a lectin with sugar specificity to α-D-mannose and α-D-glucose but with no metal binding sites.

387
Ferredoxin

Mol. wt ~ 11 000

Ferredoxins from higher plants are small non-haem iron and sulfur proteins. They have a 2Fe+2S active centre. The amino acid sequences vary according to source, but have between 93 and 98 residues, of which 26 are invariant. Four are cysteine residues at positions 39, 44, 47 and 77, which are required for the binding of the 2 Fe atoms involved in the catalytic site of the protein.

Ubiquitous in bacteria, algae and higher plants. Some plants, e.g., the horsetail, *Equisetum telmateia* (Equisetaceae) can contain two different ferredoxins in their tissues.

An electron carrier in a variety of biological processes, including photosynthetic electron transport, respiration and nitrogen fixation.

388
Ficin; Ficus protease

The complete sequence has not yet been established: a protein of estimated molecular weight 23 800–25 500 with a free sulfhydryl group.

Occurs in the latex of the fig, *Ficus glabrata*, and other *Ficus* spp. (Moraceae).

Protease. It is used in the brewing industry as a chill-proofing agent, in cheesemaking as a substitute for rennet, to produce peptones, as a meat tenderizer, and in the textile industry for shrinkproofing wool, removing gelatine from sized thread and as a spot remover.

389
L-Glutathione; Glutathione-SH; *N*-(*N*-L-γ-Glutamyl-L-cysteinyl)glycine

$C_{10}H_{17}N_3O_6S$ Mol. wt 307.33

Occurs in all living cells; it may be isolated from yeast.

Component of normal metabolism. It is used medically to improve liver function in alcoholism and hepatitis, and to aid detoxification in cases of poisoning by substances which block the thiol groups of some enzymes.

390
Gluten

A mixture of proteins consisting mainly of gliadins (prolamins, composed of up to 43% glutamine) and glutenins in variable amounts, together with small amounts of albumins and globulins. The gliadins can be separated in fractions α, β, γ, and ω, each containing many components.

Occurs in wheat flour and, to a lesser extent, in barley, maize, oats, rye and other cereals (Gramineae).

Gluten causes retention of carbon dioxide in bread dough during yeast fermentation, causing the bread to rise and giving a spongy texture. It has been used as an adhesive, and as a flour substitute. Patients with coeliac disease are sensitive to gluten, mainly due to the gliadin content. Fractions α, β and, to a lesser extent, γ are toxic in coeliac disease; only the ω fraction is harmless. Treatment is with gluten-free diets, avoiding, particularly, wheat products.

391
Histones

Small basic proteins of 5 main types, designated histone 1, 2, 3, 4, and 5, with various subtypes. Pea histone H4 contains 102 residues as a single polypeptide chain.

Associated with nuclear DNA in the chromosomes of plants.

Chromosomal proteins thought to act as repressors of template activity of DNA.

392
Kunitz inhibitor, soybean

Mol. wt ~ 21 000

Consists of a single polypeptide chain of 181 residues, with 2 disulfide bonds. The active site is arginine–isoleucine–arginine–phenylalanine.

Found in soybeans, *Glycine max* (Leguminosae).

Protease inhibitor: inhibits trypsin.

393
Leghaemoglobin; Leghemoglobin; Legoglobin

Similar to animal globins in many ways but with a molecular weight about a quarter that of haemoglobin: sequence determined by the host plant.

Found in the nitrogen fixing root nodules of leguminous plants formed after infection with *Rhizobium* bacteria. Different types of leghaemoglobin may be present in the same nodule.

Ensure an adequate supply of oxygen to the symbiotic bacterium at low concentrations of free oxygen. They have a much greater affinity for oxygen than haemoglobin and are also able to bind long chain carboxylic acids.

394
Lentil lectin; Lentil agglutinin

Mol. wt ~ 46 000

A glycoprotein consisting of two pairs of chains, the α and β chains, of approximate molecular weights 2×5700 and $2 \times 17\,500$, respectively.

Occurs in lentils, *Lens culinaris* (Leguminosae).

Agglutinates cells, including erythrocytes.

395
Mistletoe lectins

Glycoproteins: lectin I, composed of 4 chains, molecular weight 115 000; lectin II, composed of 2 chains, molecular weight 60 000; and lectin III, composed of 2 chains, molecular weight 50 000. Lectins II and III are not monomers of lectin I.

Found in the vegetative parts of mistletoe, *Viscum album* (Viscaceae).

Highly toxic. They react with human erythrocytes without specificity to blood groups A, B and O. Sugar specificity: lectin I, D-galactose, lectin II, D-galactose and *N*-acetyl-D-galactosamine; lectin III, *N*-acetyl-D-galactosamine.

396
Monellin

A protein composed of two nonidentical chains of 50 and 42 residues respectively, with no disulfide bonds and no bound carbohydrate.

Occurs in serendipity berries, *Dioscoreophyllum cumminsii* (Menispermaceae).

Sweetener: over 10^5 times sweeter than sucrose on a molecular basis.

397
Papain; Vegetable pepsin

Mol. wt ~ 23 400

A folded polypeptide chain of 212 residues.

Found in the latex of papaya, *Carica papaya* (Caricaceae).

Proteolytic enzyme. It is used in the food industry to tenderize meat and clarify drinks, and it has been used medicinally both to aid digestion and, applied topically to infected wounds, to aid sloughing and prevent adhesions.

398
Papaya peptidases A and B

Peptides which, so far, have been only partially sequenced.

Found in the latex of papaya, *Carica papaya* (Caricaceae).

Sulfhydryl proteases.

399
Peanut lectin; Peanut agglutinin; PNA

Glycoprotein which has been partially sequenced and which shows some homology with other legume lectins.

Found in peanuts, *Arachis hypogaea* (Leguminosae).

Agglutinates cells, but not blood group specific. It does not react with normal human peripheral lymphocytes, but does agglutinate immature thymocytes. In acute, but not chronic, leukaemia, PNA reacts with peripheral lymphocytes. It has been concluded that a peanut lectin receptor is exposed in immature and acute leukaemia cells but masked in mature and chronic leukaemia cells. This may be of diagnostic and even therapeutic value. Sugar specificity α- and β-galactose.

400
Phalloidin

$C_{35}H_{48}N_8O_{11}S$ Mol. wt 788.89

Occurs in the death cap mushroom, *Amanita phalloides, A. virosa,* and possibly other *Amanita* spp. (Amanitaceae). Phalloidin is one of a number of similar toxic peptides present in *Amanita* spp.

Toxic: less poisonous but much quicker acting than amanitin (q.v.), 1–2 h in mice; some uncertainty exists as to whether or not it is orally toxic. It reacts with the structural protein actin, and is severely hepatotoxic.

401
Phoratoxin

A small basic protein, with 46 amino acid residues.

Occurs in Californian mistletoe, *Phoradendron tomentosum* (Viscaceae).

Toxic to heart muscle.

402
Phytohaemagglutinin; Kidney bean lectin(s); Phasin(s); PHA

Glycoproteins consisting of two pairs of polypeptide chains composed of glutamic acid, aspartic acid, serine, alanine, tyrosine, lysine, and arginine residues, and 4–10% carbohydrate. Total molecular weight about 120 000; each chain about 30 000.

Occurs in kidney beans, *Phaseolus vulgaris* (Leguminosae).

Agglutinates cells, including erythrocytes, but not blood group specific. Sugar specificity is a complex oligosaccharide.

403
Plastocyanin

Mol. wt ~ 11 000

A monomeric protein with no bound carbohydrate, and one copper atom per molecule. Sequence of 99 amino acids varies according to source.

Ubiquitous in plants and algae. Has been completely sequenced from, e.g., the potato, *Solanum tuberosum* (Solanaceae) and the lettuce *Lactuca sativa* (Compositae).

A photosynthetic electron transfer protein.

404
Purothionins

Low molecular weight proteins of very similar composition, consisting of a single chain of 45 residues with four disulfide bonds.

Extracted from wheat flour as lipoprotein complexes; in hexaploid, tetraploid and diploid spp. of *Triticum* (Gramineae).

Toxic. They bind to cell membranes and inhibit sugar incorporation.

405
Ragweed pollen allergen Ra5

A single polypeptide chain of 45 residues, including eight cysteine residues constituting four disulfide bonds, and no detectable carbohydrate.

A minor allergen from ragweed, *Ambrosia eliator* (Compositae).

Allergen. A major use is in genetic studies of allergy, because of its simple molecular structure.

406
Ricin

The term ricin was originally used for the agglutinating extract. It consists of at least four lectins; two agglutinins (RCA), RCL I and RCL II, and two toxins, ricin D (RCL III) and RCL IV.

All are glycoproteins consisting of two different polypeptide chains held together by disulfide bonds. The agglutinins occur as a tetramer of two 30 000 and two 35 000 molecular weight subunits, and the toxins as dimers of 30 000 and 33 000 molecular weight chains. The sequence for ricin D has been determined.

Found in castor beans, *Ricinus communis* (Euphorbiaceae).

Ricin is one of the most toxic substances known. The agglutinins are nontoxic but ricin D inactivates ribosomes and binds to the cell surface. It has antitumour activity, and is used experimentally in cancer research and to investigate cell surface properties.

407
Rubisco; Ribulose 1,6-bisphosphate carboxylase-oxygenase; Fraction I protein

Mol. wt ~ 560 000

A large protein consisting of eight large (molecular weight ~ 56 000) and eight small (molecular weight ~ 12 000) subunits. The large subunit (475 amino acid residues) of *Zea mays* (Gramineae) and the small subunit (120 residues) of *Spinacia oleracea* (Chenopodiaceae) have been sequenced.

Universally present in the chloroplasts of all green plants. Makes up about 50% of the soluble leaf protein and can be obtained crystalline from *Nicotiana tabacum* (Solanaceae) leaf. Described as the world's most abundant protein.

The key enzyme of photosynthesis, catalysing the carboxylation of ribulose 1,6-bisphosphate to yield 2 molecules of phosphoglyceric acid, which is then converted to sugar. Nutritionally, it provides a good balance of essential amino acids.

408
Soybean lectin; Soybean agglutinin; SBA

Mol. wt ~ 120 000

A glycoprotein consisting of two pairs of chains, the α and β chains, each of molecular weight approximately 30 000.

Found in soyabean, *Glycine max* (Leguminosae).

Agglutinates cells. Sugar specificity is N-acetyl-D-galactosamine.

409
Thaumatin

Thaumatin I is a single polypeptide chain of 207 residues with 8 disulfide bonds. Thaumatin II is also entirely proteinaceous but the sequence is not yet known.

Occurs in the fruit of *Thaumatococcus daniellii* (Marantaceae).

Sweetener: over 10^5 times sweeter than sucrose on a molecular basis, but is unstable on heating.

410
Trichosanthin

Mol. wt ~ 24 000

A single polypeptide chain of 220 amino acid residues, pI 9.4, containing 8 segments of α-helix, 13 strands of β-helix and some extended chains.

Occurs in tubers of *Trichosanthes kirilowii* (Cucurbitaceae).

Abortifacient used clinically in China.

411
***Vicia cracca* lectins**

The one-chain lectin, *Vicia cracca* Gal/NAc lectin, consists of four subunits each of molecular weight about 33 000, total 125 000; the two-chain lectin, *Vicia cracca* Man/Glc lectin, consists of two pairs of chains of molecular weight 2×5700 and $2 \times 17\,500$, totalling 46 400.

Found in the seeds of tufted vetch, *Vicia cracca* (Leguminosae).

The one-chain lectin is blood group A specific, binding to α-N-acetyl-D-galactosamine; the two-chain lectin is unspecific in the ABO system and binds to mannose/glucose.

412
Viscotoxin

A mixture of at least three components, A2, A3 and B (sometimes classified as I, II (=B), III (=A2) and IVb (=A3)), each small basic protein composed of 46 amino acid residues and of molecular weight ~ 5000.

Occurs in mistletoe, *Viscum album* (Viscaceae).

Toxic to heart muscle and cytotoxic. It inhibits DNA synthesis.

413
Viscumin

Mol. wt ~ 60 000

A protein composed of two chains with some disulfide links: the A chain, of molecular weight about 29 000, and the B chain, of about 34 000. It is a monomer at low concentrations and a dimer at high concentrations.

Occurs in mistletoe, *Viscum album* (Loranthaceae).

Cytotoxic. It inhibits protein synthesis in cell-free systems. Toxicity may be prevented by pretreatment with galactose, lactose and calcium ionophore A23187.

414
Wheat germ agglutinin; WGA

Structurally, it is a dimer except at low pH, where it dissociates into two identical subunits or protomers of molecular weight 17 000. It consists of a single polypeptide chain of 164 residues, particularly rich in half-cystine residues, which are present in each protomer as 16 disulfide bonds. There are four distinct domains, A, B, C and D, each extensively linked by four disulfide bridges.

Found in wheat germ, *Triticum aestivum* (Gramineae).

Agglutinates a variety of animal cells, including both malignant and normally dividing cells. Sugar specificity is N-acetyl-D-glucosamine and its $(1\rightarrow 4)$ linked oligomers.

Chapter 16
Miscellaneous Nitrogen Compounds

A number of the vitamins contain nitrogen in their structures, particularly those of the B group, and are included in this section. They are usually present in plants in low amounts, and play a role in plant metabolism as well as providing a dietary source of these essential components of human health. The vitamins listed are biotin, cyanocobalamin, folic acid, nicotinamide, nicotinic acid, pantothenic acid, pyridoxine, riboflavin and thiamine. For non-nitrogenous vitamins, see under appropriate sections: e.g., ascorbic acid (vitamin C) can be found under organic acids.

Two of the most abundant nitrogenous compounds of plants are chlorophylls a and b (**419, 420**), the two essential catalysts of photosynthesis which are universally present in green plants. The nitrogen of chlorophyll is in the form of a pyrrole ring, four of which are united to give the porphyrin nucleus to which a magnesium atom is chelated. While chlorophylls a and b are the major pigments of higher plants, ferns and mosses, they are accompanied or replaced by chlorophylls c, d and e in algae (Vernon and Seeley, 1966). A related tetrapyrrole-based pigment is phytochrome (**441**), which differs in having a protein component.

Many of the remaining compounds included in this section are plant toxins. There are aliphatic nitro compounds such as miserotoxin (**437**) and aromatic nitro derivatives, represented here by aristolochic acid, **416** (Mix *et al.*, 1982). The azo derivatives cycasin (**424**) and macrozamin (**435**) are potentially dangerous because of their carcinogenic properties (Hooper, 1978). The anthranilic-based avenalumin (**417**) and dianthalexin (**425**) are notable for being phytoalexins and having antifungal activity. The latter property is also associated with hydroxamic acids, of which DIMBOA glucoside (**426**) is one example (Niemeyer, 1988).

References

Mix, D.B., Guinaudeau, H. and Shamma, M. (1982). The aristolochic acids and aristolactams. *Journal of Natural Products* **45**, 657–666.
Hooper, P.J. (1978). Cycad poisoning in Australia. In *Effects of Poisonous Plants on Livestock* (Keeler, R.F., van Kempen, K.R. and James, L.F.) (eds), pp. 337–348. New York: Academic Press.
Niemeyer, H.M. (1988). Hydroxamic acids, defence chemicals in the Gramineae. *Phytochemistry* **27**, 3349–3358.
Robinson, F.A. (1973). Vitamins. In *Phytochemistry*, Volume 3. (Miller, L.P., ed.), pp. 195–220. New York: Van Nostrand–Reinhold.
Vernon, L.P. and Seeley, G.R. (1966). *The Chlorophylls*. New York: Academic Press.

415
Agaritine

$C_{12}H_{17}N_3O_4$ Mol. wt 267.29

Occurs in the commercial edible mushroom *Agaricus bisporus* and in ten other *Agaricus* spp.: *argentatus, campestris, comptulis, crocodilinus, edulis, hortensis, micromegathus, patersonii, perrarus* and *xanthodermus*.

A weak mutagenic effect has been reported in a spot test in *Salmonella typhimurium*.

416
Aristolochic acid; Aristolochic acid-I; Aristolochic acid A

$C_{17}H_{11}NO_7$ Mol. wt 341.28

Found in Birthwort, *Aristolochia clematis*, Indian birthwort, *A. indica*, long birthwort, *A. longa*, and wild ginger, *Asarum canadense* (Aristolochiaceae). Co-occurs with several closely related structures, which vary in the number and position of hydroxy or methoxy substituents.

Anti-inflammatory, antifertility and antibiotic activity in animals, and it is an *in vitro* tumour growth inhibitor; also it is carcinogenic in animals. It is sequestered and stored from the food plants *Aristolochia* spp. by the butterfly *Battus archidamus*.

417
Avenalumin I

$C_{16}H_{11}NO_4$ Mol. wt 281.27

Formed as a phytoalexin in *Avena sativa* (Gramineae) in response to microbial infection.

Antifungal activity.

418
Biotin; Vitamin H; Coenzyme R

$C_{10}H_{16}N_2O_3S$ Mol. wt 244.32

Present in minute amounts in every living cell.

Coenzyme for carboxylation during metabolism of proteins and carbohydrates.

419
Chlorophyll a

$C_{55}H_{72}MgN_4O_5$ Mol. wt 893.51

Universal green pigment, with chlorophyll b (q.v.), in all higher plants. Occurs in the chloroplast, loosely associated with protein.

Essential catalyst for photosynthesis. It has been employed as a deodorant (in a wick type freshener), and is used both as a colourant in food and perfumery, and also for dyeing leather.

420
Chlorophyll b

$C_{55}H_{70}MgN_4O_6$ Mol. wt 907.50

Universal green pigment, with chlorophyll a (q.v.), in all higher plants. The ratio of chlorophyll a to b is usually 3:1.

Essential catalyst for photosynthesis. Its uses are as for chlorophyll a (q.v.).

421
Cibarian; 1,6-Di(3-nitropropanoyl)-β-D-glucopyranoside

$C_{12}H_{18}N_2O_{12}$ Mol. wt 382.29

Found in *Astragalus cibarius, A. canadensis* var. *brevidens* and var. *mortonii, A. falcatus* and *A. flexuosus* (Leguminosae).

Toxic.

422
Coronarian; 2,6-Di(3-nitropropanoyl)-α-D-glucopyranoside

$C_{12}H_{18}N_2O_{12}$ Mol. wt 382.29

Found in *Astragalus cibarius, A. falcatus* and *A. flexuosus* (Leguminosae).

Toxic.

423
Cyanocobalamin; Vitamin B_{12};
5,6-Dimethylbenzimidazolyl cyanocobamide

$C_{63}H_{88}CoN_{14}O_{14}P$　　　　　Mol. wt 1355.40

Found in minute amounts in higher plants (compared with animal sources, where it is produced by bacteria in the colon).

Haematopoietic vitamin, used to treat pernicious anaemia, where absorption through intestinal wall is prevented by lack of Castle's Intrinsic Factor.

424
Cycasin; β-D-Glucosyloxyazoxymethane;
Methylazoxymethanol-β-D-glucoside

$C_8H_{16}N_2O_7$　　　　　Mol. wt 252.23

Occurs in the seeds of the sago-palm, *Cycas circinalis* and *C. revoluta* (Cycadaceae).

Responsible for the toxicity of cycad palms to livestock. Cycasin is hepatotoxic, carcinogenic and teratogenic in cattle and sheep. It is nontoxic to insects and is sequestered from cycad food plants by the caterpillar of the hairstreak butterfly, *Eumaeus atala*, and stored in the adult for defensive purposes.

425
Dianthalexin

$C_{14}H_9NO_3$　　　　　Mol. wt 239.23

Formed as a phytoalexin in carnation tissues, *Dianthus caryophyllus* (Caryophyllaceae) in response to microbial infection.

Antifungal activity.

426
Dianthramine

$C_{14}H_{11}NO_6$　　　　　Mol. wt 289.25

Formed as a phytoalexin in carnation tissues, *Dianthus caryophyllus* (Caryophyllaceae) in response to microbial infection.

Antifungal activity.

427
DIMBOA glucoside

$C_{15}H_{17}NO_{10}$ Mol. wt 371.31

Found in the leaves of wheat, *Triticum aestivum*, and maize, *Zea mays*, and other related cereals (Gramineae). Several closely similar hydroxamic acid glucosides co-occur in these plants. It is very readily hydrolysed during extraction to free DIMBOA, which in turn decomposes with loss of formic acid to the corresponding benzoxazolinone.

In part responsible for the resistance in cereal leaves to insect and aphid attack as well as to microbial infection. Also, it shows allelopathic effects in noncereals.

428
Folic acid; Pteroylglutamic acid

$C_{19}H_{19}N_7O_6$ Mol. wt 441.41

Found in green leaves, especially spinach, *Spinacea oleracea* (Chenopodiaceae), and grasses (Gramineae).

Haematopoietic vitamin, reduced in the body to tetrahydrofolate, a coenzyme for various processes including synthesis of purine and pyrimidine nucleotides and therefore DNA (q.v.). Deficiency causes megaloblastic anaemia. Folic acid is usually given in conjunction with cyanocobalamin (q.v.) or hydroxocobalamin to treat deficiency states, and sometimes prophylactically, with iron, in pregnancy.

429
6-Hydroxykynurenic acid

$C_{10}H_7NO_4$ Mol. wt 205.17

Occurs in tobacco, *Nicotiana tabacum* (Solanaceae), *Thapsia villosa* (Umbelliferae) and *Ginkgo biloba* (Ginkgoaceae). It appears to be widespread in plants as a trace constituent, especially in Labiatae, Scrophulariaceae and Solanaceae.

430
Indican; 3-(β-Glucosido)indole; Indoxyl-β-D-glucoside; Plant indican

$C_{14}H_{17}NO_6$ Mol. wt 295.30

Found in *Indigofera* spp. (Leguminosae), and *Polygonum tinctorum* (Polygonaceae).

Colourless precursor of the dark blue indigo, converted to the dye after hydrolysis of the glucoside and subsequent oxidation to isatin, and dimerization. Indigo is probably the oldest known naturally occurring colouring matter, and is still used for dyeing cloth.

431
Indole-3-acetonitrile; 3-Cyanomethyl-1*H*-indole; 3-Indolylacetonitrile

$C_{10}H_8N_2$ Mol. wt 156.19

Found in cabbage, *Brassica oleracea*, and other Cruciferae.

Naturally occurring plant growth hormone, which functions following enzymatic hydrolysis to the acid.

432
Indole-3-carboxaldehyde; 3-Formylindole; Indole-3-aldehyde; 3-Aldehydoindole

C_9H_7NO Mol. wt 145.16

Occurs in the stembark of *Murraya exotica* (Rutaceae), barley seedlings, *Hordeum vulgare* (Gramineae), and in the cotton plant, *Gossypium barbadense* (Malvaceae).

433
Isotan B

$C_{14}H_{15}NO_7$ Mol. wt 309.28

Occurs in the first year leaves of the biennial *Isatis tinctoria*, woad (Cruciferae).

Fermentation yields indoxyl, which is rapidly oxidized by air to the dye indigo. Its use as a pigment by the ancient Britons was recorded by Julius Caesar.

434
Karakin;
1,2,6-Tri(3-nitropropanoyl)-β-D-glucopyranoside

$C_{15}H_{21}N_3O_{15}$ Mol. wt 483.35

Occurs in *Astragalus canadensis* var. *brevidens* and var. *mortonii*, *A. cibarius*, *A. falcatus* and *A. flexuosus* (Leguminosae).

Toxic.

435
Macrozamin;
Methylazoxymethanol-β-D-primeveroside

$C_{13}H_{24}N_2O_{11}$ Mol. wt 384.34

Found in *Macrozamia spiralis* (Zamiaceae).

Carcinogenic; its toxic properties are similar to those of cycasin (q.v.).

436
Methoxybrassinin

$C_{12}H_{14}N_2OS_2$ Mol. wt 266.39

Formed as a phytoalexin in Chinese cabbage, *Brassica campestris*, and in rape *Brassica napus* (Cruciferae) in response to microbial infection. Several related structures are also similarly formed in these crucifers as phytoalexins.

Antifungal activity.

437
Miserotoxin; 3-Nitro-1-propyl-β-D-glucopyranoside

$C_9H_{17}NO_8$ Mol. wt 267.24

Occurs in *Astragalus miser* var. *oblongifolia*, and some other, but not all, varieties, *A. atropubescens, A. pterocarpus, A. tetrapleurus,* and *A. toanus* (Leguminosae).

Toxin causing methaemoglobinaemia, particularly in livestock.

438
Nicotinamide; Niacinamide; Nicotinic acid amide; Nicotyl amide; 3-Pyridine carboxylic acid amide

$C_6H_6N_2O$ Mol. wt 122.13

Widespread occurrence in plants.

Vitamin or enzyme co-factor. It is used in the same way as nicotinic acid (q.v.) to prevent pellagra, but has no vasodilator action or effect on serum lipids.

439
Nicotinic acid; Niacin, Pyridine carboxylic acid; Pyridine 3-carboxylic acid

$C_6H_5NO_2$ Mol. wt 123.11

Found in rice polishings, potatoes, legumes, and cereals, and in minute amounts in all living cells.

Vitamin or enzyme co-factor. Deficiency causes pellagra, characterized by loss of appetite, weakness, diarrhoea, dermatitis and mental changes. It has a vasodilator effect and is used to treat vascular disorders, e.g., chilblains, frostbite, and Menières disease. It also lowers serum cholesterol in large doses.

440
Pantothenic acid

$C_9H_{17}NO_5$ Mol. wt 219.24

Rice bran, molasses and fresh vegetables are rich sources, but it is present in all living tissues.

Component of co-enzyme A, and a nutritional factor, often described as a vitamin of the B group. A deficiency syndrome has been described with symptoms of fatigue, headache and sleep disturbance.

441
Phytochrome

Universally present in all green plants in two interchangeable forms: P_r with a UV maximum at 660 nm and P_{fr} with a maximum at 730 nm. Closely related pigments, the phycobilins, are present in several algal groups.

The major light receptor pigment of plants.

442
Pyridoxine; Adermine; Vitamin B_6

$C_8H_{11}NO_3$ Mol. wt 169.18

Rice, bran cereals and legumes are good sources, but it is universally present in trace amounts.

Co-enzyme, vitamin, converted in the body to pyridoxal phosphate, which is the co-enzyme for amino acid decarboxylase and transaminase. It is used to treat deficiency states, such as pyridoxine-dependent convulsions of infancy and some types of depression and pre-menstrual syndrome, and those due to drug therapy, e.g., during isoniazid treatment of tuberculosis.

443
Riboflavin; Lactoflavin; Vitamin B_2

$C_{17}H_{20}N_4O_6$ Mol. wt 376.38

Occurs in leafy vegetables and malted barley, and present in small amounts in all cells, mainly in combined forms, e.g., flavin adenine dinucleotide, FAD (q.v.) and other prosthetic groups of flavoprotein enzymes, and riboflavin 5-phosphate.

Vitamin, co-enzyme. The flavoprotein enzymes are necessary for the oxidation of carbohydrates, amino acids, aldehydes, and other substances.

444
Ruscopine

$C_{16}H_{26}N_4O_2$ Mol. wt 306.41

Found in the thistle, *Carduus acanthoides* (Compositae).

445
Thiamine; Aneurine; Vitamin B_1

$[C_{12}H_{17}N_4OS]^+$ Mol. wt 265.36

Good sources include peas, beans and other legumes, oatmeal, whole wheat, rice husks, and most green leafy vegetables.

Vitamin or enzyme co-factor. It is associated with carbohydrate metabolisms by combining with pyrophosphoric acid to produce co-carboxylase. Deficiency causes fatigue, anorexia, gastrointestinal disturbances and, in severe cases, beri-beri.

446
Trigonelline; Caffearine; Caffearine; Gynesine; Nicotinic acid *N*-methylbetaine; Trigenolline

$C_7H_7NO_2$ Mol. wt 137.14

Occurs in the seeds of fenugreek, *Trigonella foenum-graecum*, and alfalfa, *Medicago sativa* (Leguminosae), coffee beans, *Coffea* spp. (Rubiaceae), the seeds of *Strophanthus* spp. (Apocynaceae), and cannabis, *Cannabis sativa* (Cannabinaceae), and yarrow herb, *Achillea millefolium* (Compositae).

Hypoglycaemic in animals. It is an important flavour precursor in coffee.

PART III
Alkaloids

Alkaloids

General Introduction

The alkaloids are basic plant substances. They normally contain a nitrogen atom in their chemical structures as part of a heterocyclic ring. They possess an array of structural diversity and physiological activity unrivalled by any other group of natural products. They number about ten thousand structures; on average, one new alkaloid is described in the scientific literature every day.

Originally, alkaloids were thought to be unique to the plant kingdom; in recent times, however, they have been detected in some animals, e.g., in the toxic secretions of fire ants, ladybirds and toads. Their major occurrence is still in the flowering plants, and 40% of all plant families have at least one alkaloid-bearing member. Their distribution is decidedly uneven: they may be universal in some families (e.g., Papaveraceae), common in others (e.g., Amaryllidaceae, Rutaceae) and rare in yet others (e.g., Umbelliferae). They can be present throughout the plant or, alternatively, restricted to certain tissues such as the root or bark. The concentration can vary from a small fraction of 0.1% to as much as 12% dry weight.

Alkaloids are principally of interest to humans because of their medicinal properties, and many are widely used as drugs, e.g., atropine, codeine, morphine and vincristine. Some are very poisonous, e.g., coniine and strychnine, and others have hallucinogenic effects, e.g., cocaine and muscimol. The importance of alkaloids in plant metabolism has been much debated. There is, however, increasing evidence that they can both protect the plant against herbivory and, at the same time, serve as a storage form of nitrogen.

Classification is usually based on the type of ring system present (e.g., pyrrolidine, piperidine, etc.) and on the biosynthetic origin from one or other of the protein amino acids. Some structures fit into more than one category, and a rigid hierarchal classification is difficult to devise. In this chapter, fourteen groups of alkaloids are separately enumerated. There is also a miscellaneous section, covering those structures which do not fit in easily anywhere else. Some groups of alkaloid can be defined by their botanical occurrences. The betalain alkaloids are found only in the plant order Centrospermae, while the Lycopodium alkaloids are confined to the club mosses and the Amaryllidaceae alkaloids to plants of that one family. Some alkaloid groups have relatively simple structures, such as the pyrrolidines and piperidines (and the related pyridines), while others are quite complex (e.g., the indole, isoquinoline and quinoline classes).

Certain alkaloid groups are clearly defined by their chemistry, e.g., the peptide, pyrrolizidine, quinolizidine and tropane classes. Other groups are partly terpenoid in origin. These are conveniently classified, according to the number of isoprene units present, as monoterpene (2), sesquiterpene (3), diterpenoid (4) or steroidal (6). Representatives of other alkaloid groups (e.g., pyrazines, oxazoles) can be found in the miscellaneous section. Other bases, sometimes included with the alkaloids, are listed in this Dictionary in chapter 2 with other nitrogen-containing compounds. These include the protoalkaloids or amines and such pseudoalkaloids as the purines and pyrimidines.

Alkaloids are detected in plant extracts by a number of well known colour tests. Some are general (Dragendorff, Mayer) while others are more specific (Ehrlich for ergot alkaloids). Methods of chromatographic purification and structure determination vary according to the class of alkaloid being studied (Waterman, 1992), and are referred to briefly under each alkaloid type. Alkaloids are known to occur *in vivo* both as the free base and as the *N*-oxide. The *N*-oxide may be the form in which the alkaloid is transported around the plant. In practice, alkaloids are normally isolated as the free bases, but the fact that they can occur as *N*-oxides should not be forgotten.

A common characteristic of alkaloid names is that they usually terminate in "ine". The name may be derived from the plant genus where it occurs, e.g., atropine from *Atropa*, or from the plant species, e.g., cocaine from *Erythroxylon coca*. Other names may be derived from that of a plant drug, e.g., ergotamine from ergot, or from a physiological activity, e.g., emetine from its emetic effects.

The alkaloid literature is extensive. The most comprehensive and up-to-date single reference work is the dictionary of Southon and Buckingham (1989). This replaces earlier dictionaries of Glasby (1978) and Raffauf (1970). The classic textbook on plant alkaloids is that of Henry, which appeared in its fourth edition in 1949. The best modern account of the chemistry, biosynthesis and pharmacology is the monograph by Cordell (1981). Two books which concentrate on alkaloid biochemistry and biology are the works of Robinson (1981) and of Waller and Nowacki (1978).

There are three major continuing series of review publications on the alkaloids. Pride of place must go to the Manske/Rodrigo/Brossi series, simply entitled *The Alkaloids*, which began in 1950 and is still being produced on a regular basis about once a year. A second series was instituted by the Royal Society of Chemistry in 1971 as a Specialist Periodical Report on alkaloids, edited first by Saxton (1971–5) and then by Grundon (1976–83). This is now continued in journal form, published bimonthly, as *Natural Product Reports*. A third series on chemical and biological perspectives and with a wider appeal is edited by Pelletier. This began in 1983, and so far seven volumes have appeared.

References

Cordell, G. (1981). *Introduction to Alkaloids: A Biogenetic Approach*. New York: Wiley-Interscience.

Glasby, J.S. (1978). *Encyclopedia of the Alkaloids*, in 3 Volumes. New York: Plenum Press.

Grundon, M.F. (1976–1983). *The Alkaloids*, Specialist Periodical Reports, Volumes 6–13. London: Royal Society of Chemistry.

Henry, T.A. (1949). *The Plant Alkaloids*, 4th Edition. London: Churchill.

Manske, R.H.F., Rodrigo, R., Brossi, A. and Cordell, G.A. (eds) (1950–). *The Alkaloids*, Volumes 1–41. New York: Academic Press.

Pelletier, S.W. (ed.) (1983–). *Alkaloids: Chemical and Biological Perspectives*, Volumes 1–6, New York: Wiley; Volume 7, Berlin: Springer-Verlag.

Raffauf, R.F. (1970). *Handbook of Alkaloids and Alkaloid-containing Plants*. New York: Wiley-Interscience.

Robinson, T. (1981). *The Biochemistry of Alkaloids*, 2nd Edition. Berlin: Springer-Verlag.

Saxton, J.E. (1971–1975). *The Alkaloids*, Specialist Periodical Reports, Volumes 1–5. London: Royal Society of Chemistry.

Southon, I.W. and Buckingham, J. (1989). *Dictionary of Alkaloids*, in 2 Volumes. London: Chapman & Hall.

Waller, G.R. and Nowacki, E.K. (1978). *Alkaloid Biology and Metabolism in Plants*. New York: Plenum Press.

Waterman, P.G. (ed.) (1992). *Methods in Plant Biochemistry*, Volume 8, *The Alkaloids*. London: Academic Press.

Chapter 17
Amaryllidaceae alkaloids

These alkaloids, as their name implies, are restricted in their occurrence to this one family of plants, and no other class of alkaloid has been reported from these sources. The daffodil family, with 75 genera and 1100 species, includes narcissi, snowdrops and amaryllis. The characteristic alkaloids are mainly reported from the bulbs, but they do occur in the aerial parts. About 275 alkaloids have been described so far and 15 are listed here. They share a common 15-carbon nucleus, based on an aromatic C_6-C_1 unit that is linked to a reduced aromatic C_6-C_2 unit. Their biosynthesis involves the phenolic coupling of two aromatic amino acids, with one of the two nitrogen atoms being retained in the process. Belladine (**452**) in demethylated form can be regarded as a precursor of most of the other alkaloids of this type.

There are several different ring systems represented among the Amaryllidaceae alkaloids. Three common ones are apparent in the structures of galanthamine (**460**), lycorine (**466**) and narciclasine (**467**). Several of the alkaloids are known to occur in glycosidic form, e.g., kalbreclasine (**463**), and the glucoside of narciclasine (**467**).

Accidental human poisoning has occurred as a result of mistaking daffodil bulbs for onions, and many of the alkaloids of this family are relatively toxic. Galanthamine has been used medically in Russia to treat various nervous diseases, and the same alkaloid is reported to exhibit analgesic activity comparable to that of morphine. Several other alkaloids, including narciclasine, show some promise as anti-cancer agents.

References

In *The Alkaloids* (Manske, R.H.F. and Brossi, A., eds) (1952) Volume 2, pp. 331–352; (1960) Volume 6, pp. 289–413; (1968) Volume 11, pp. 302–406; (1975) Volume 15, pp. 83–164; (1987) Volume 30, pp. 252–376.

447
Acetylcaranine; Bellamarine

Skeletal type: galanthan

$C_{18}H_{19}NO_4$ Mol. wt 313.36

Occurs in the bulbs of *Amaryllis belladonna* hybrid, *Crinum macrantherum* and *Ammocharis coranica* (Amaryllidaceae).

In vitro antineoplastic activity to P388 leukaemia; it is a uterine stimulant.

448
3-Acetylnerbowdine

Skeletal type: crinane

$C_{19}H_{23}NO_6$ Mol. wt 361.40

Occurs in the bulbs of *Boophane disticha* and *Nerine crispa* (Amaryllidaceae).

Highly toxic alkaloid, and a constituent of arrow poisons which are used in South Africa for hunting; the bulb extract shows muscle relaxant, atropine-like properties.

449
Albomaculine

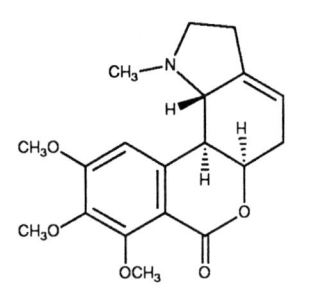

Skeletal type: lycoranan

$C_{19}H_{23}NO_5$ Mol. wt 345.40

Isolated from the bulbs of *Haemanthus albomaculatus* (Amaryllidaceae).

Natives of South Africa have employed extracts of *Haemanthus* as topical treatment of such diverse afflictions as leprosy, ulcers, febrile colds, asthma, coughs and wounds.

450
Amaryllisine

Skeletal type: crinane

$C_{18}H_{23}NO_4$ Mol. wt 317.39

Found in *Brunsvigia rosea* (Amaryllidaceae).

Alkaloids of this family show a diversity of pharmacodynamic activity on the cardiovascular and central nervous systems (e.g., antitumour effect; analgesic activity) in experiments, but they are much too toxic to warrant further studies.

451
Ambelline

Skeletal type: crinane

$C_{18}H_{21}NO_5$ Mol. wt 331.37

Occurs in *Amaryllis belladonna*, *Boophane fischeri*, *Crinum laurentii* and *Hippeastrum* spp. (Amaryllidaceae).

Analgesic action and duration of effect approach those of morphine and codeine, but it is too toxic to be used in medicine.

452
Belladine

Skeletal type: *N*-benzylphenethylamine

$C_{19}H_{25}NO_3$ Mol. wt 315.42

Found in the bulbs of *Amaryllis belladonna* hybrid and *Crinum powellii* (Amaryllidaceae).

453
Brunsvigine

Skeletal type: pancracine

$C_{16}H_{17}NO_4$ Mol. wt 287.32

Isolated, together with brunsvinine, crinamine and lycorine (q.v.), from *Brunsvigia cooperi* (Amaryllidaceae).

454
Candimine

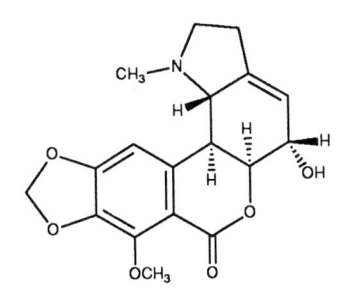

Skeletal type: lycoranan

$C_{18}H_{19}NO_6$ Mol. wt 345.36

Found in the bulbs of *Hippeastrum candidum* (Amaryllidaceae).

Toxic.

455
Caranine

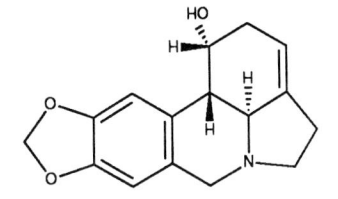

Skeletal type: galanthan

$C_{16}H_{17}NO_3$ Mol. wt 271.32

Occurs in the bulbs of *Amaryllis belladonna*, *Ammocharis coranica* and *Clivia defixum* (Amaryllidaceae).

Toxic to animals, killing them by respiratory paralysis. The analgesic action and duration of effect approach those of morphine and codeine.

456
Caribine

Skeletal type: galanthan

$C_{19}H_{22}N_2O_3$ Mol. wt 326.40

Found in *Hymenocallis arencola* (Amaryllidaceae).

Possible antiviral and antineoplastic activities.

457
Carinatine; 10-Demethylgalanthine

Skeletal type: galanthan

$C_{17}H_{21}NO_4$ Mol. wt 303.36

Found in the fresh bulbs of *Zephyranthes carinata* (Amaryllidaceae).

Similar biological activities to caribine (q.v.).

458
Cavinine; 1,2-β-Epoxyambelline

Skeletal type: crinane

$C_{18}H_{21}NO_6$ Mol. wt 347.37

Found in the bulbs of *Hippeastrum punecium* (Amaryllidaceae).

Immunostimulant. It produces moderate activation of mouse spleen lymphocytes.

459
Crinamine

Skeletal type: crinane

$C_{17}H_{19}NO_4$ Mol. wt 301.35

Occurs in *Crinum asiaticum* and other *Crinum* spp. (Amaryllidaceae).

Toxic (LD_{50} orally in dogs 10 mg kg^{-1}); also, it is a powerful transient hypotensive agent in dogs. It shows respiratory depressant activity.

460
Galanthamine; Galantamine; Lycorimine

Skeletal type: galanthaman

$C_{17}H_{21}NO_3$ Mol. wt 287.36

Occurs in the following Amaryllidaceae genera: *Crinum, Galanthus, Hippeastrum, Hymenocallis, Leucojum, Lycoris, Narcissus, Pancratium,* and *Ungernia*.

Inhibits cholinesterase activity (reversible); it exhibits strong analgesic activity comparable to that of morphine, and has been used in Russia in the treatment of nervous diseases. It shows potent insecticidal activity against the yellow butterfly, *Eurema hecabe mandarina*.

461
Hippeastrine; Trisphaerine; Trispherine

Skeletal type: lycoranan

$C_{17}H_{17}NO_5$ Mol. wt 315.33

Occurs in the bulbs of *Clivia miniata, Crinum amabile, C. latifolium, Hippeastrum* spp., *Lycoris radiata, Sternbergia lutea* and *Ungernia vvedenskyi* (Amaryllidaceae).

Feeding inhibitor for the fifth instar larva of the yellow butterfly, *Eurema hecabe mandarina*.

462
Haemanthamine; 3-Epicrinamine; Hemanthidine; Natalensine

Skeletal type: crinane

$C_{17}H_{19}NO_4$ Mol. wt 301.35

First isolated from the *Haemanthus* hybrid "King Albert"; also occurs in twenty other Amaryllidaceae genera.

Slight hypotensive activity. Extracts of the bulbs are used by Africans as topical treatment for diverse afflictions.

463
Kalbreclasine; 2-*O*-β-D-Glucopyranosylnarciclasine

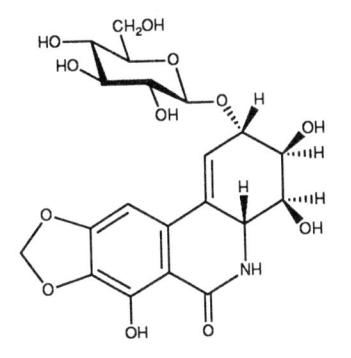

Skeletal type: phenanthridine

$C_{20}H_{23}NO_{12}$ Mol. wt 469.41

Occurs in the fresh roots of the flowering bulbs of *Haemanthus kalbreyeri* (Amaryllidaceae).

Causes significant mitogenic activation of splenic lymphocytes, characteristic of immunostimulants. It is reported that extracts of its roots, bulbs and flowers are used in India as a popular medicine in the treatment of the common cold, coughs, and asthma, and in the healing of wounds.

464
Lycorenine

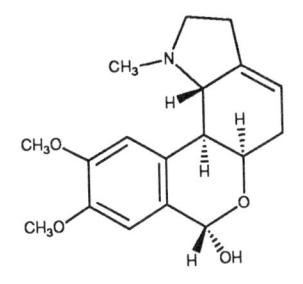

Skeletal type: lycoranan

$C_{18}H_{23}NO_4$ Mol. wt 317.39

Occurs in bulbs of *Lycoris radiata* and in many other species in the Amaryllidaceae.

Antifeedant against the larva of the yellow butterfly, *Eurema hecabe mandarina*.

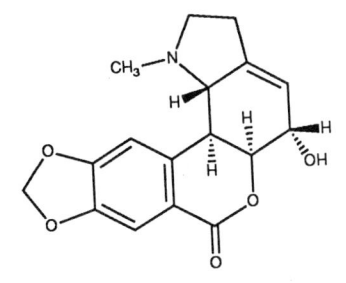

465
Lycoricidine; Margetine

Skeletal type: phenanthridine

$C_{14}H_{13}NO_6$ Mol. wt 291.27

Occurs in *Lycoris radiata* and *L. sanguinea*, and many other Amaryllidaceae spp.

Antifeedant against the larva of the yellow butterfly *Eurema hecabe mandarina* (see also lycorenine). It is a cytotoxic agent, and shows plant growth inhibitory activity.

466
Lycorine; Narcissine; Galanthidine

Skeletal type: galanthan

$C_{16}H_{17}NO_4$ Mol. wt 287.32

Occurs in *Lycoris radiata*. It is the most widespread of all the Amaryllidaceae alkaloids, and occurs free in the (−) form (illustrated) and in a conjugated form (linked to glucose, acetic acid or fatty acids).

Shows antiviral activity against poliomyelitis, coxsackie and herpes type I viruses. It possesses plant growth inhibitory properties by suppressing cell division and cell elongation, and inhibits protein synthesis in eukaryotic cells by inhibiting peptide bond formation. It is highly toxic, and frequently responsible for accidental human poisoning by *Narcissus* spp. (LD_{50} in dogs 41 mg kg^{-1}).

467
Narciclasine; Lycoricidinol

Skeletal type: phenanthridine

$C_{14}H_{13}NO_7$ Mol. wt 307.27

Found in *Narcissus* spp., *Haemanthus kalbreyeri* and *Lycoris longituba* (Amaryllidaceae).

Antitumour agent, exerting an antimitotic effect by immediately terminating protein synthesis in eukaryotic cells. It inhibits HeLa cell growth and stabilizes HeLa cell polysomes *in vivo*. Also, it is a potent antifeedant against the larva of the yellow butterfly, *Eurema hecabe mandarina* (see also lycorenine).

468
Narwedine; Galanthaminone

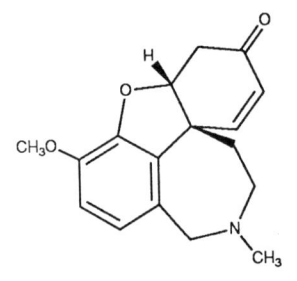

Skeletal type: galanthaman

$C_{17}H_{19}NO_3$ Mol. wt 285.35

Occurs in the bulbs of the snowdrop *Galanthus nivalis*, and of *Lycoris guangxiensis*, *Ungernia severtzovii*, *U. victoris* and *U. vvedenskyi* (Amaryllidaceae).

Increases the amplitude and frequency of respiratory movements, and increases the amplitude and decreases the frequency of cardiac contractions. It potentiates the analgesic effects of morphine. Also, it shows hypotensive activity.

469
Pancratistatin

Skeletal type: phenanthridine

$C_{14}H_{15}NO_8$ Mol. wt 325.28

Found in the bulbs of *Pancratium littorale* and *Zephyranthes grandiflora* (Amaryllidaceae).

Shows anticancer activity when tested against lymphocytic P388 leukaemia, PS system and M5076 ovary sarcoma.

470
Pretazettine; Isotazettine

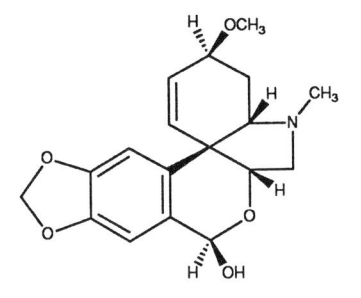

Skeletal type: tazettine

$C_{18}H_{21}NO_5$ Mol. wt 331.37

Occurs in *Leucojum aestivum*, *Narcissus tazetta*, *Pancratium biflorum*, *Lycoris radiata* and *Zephyranthes carinatus* (Amaryllidaceae).

Significant antitumour activity; it inhibits HeLa cell growth as well as protein synthesis in eukaryotic cells. Therapeutic activity has been shown against Rauscher leukaemia virus.

Chapter 18
Betalain alkaloids

The betalains are a class of nitrogen-containing water-soluble plant pigments comprising two groups: the purple betacyanins (e.g., amaranthin, **471**) and the yellow betaxanthins (e.g., indicaxanthin, **479**). They are formed biosynthetically from the amino acid 3,4-dihydroxyphenylalanine (dopa), which is converted to betalamic acid (**472**), a key precursor of both groups of pigment. Condensation of betalamic acid with cyclodopa gives rise to the betacyanin skeleton, while condensation of betalamic acid with aliphatic protein amino acids yields the betaxanthins. For example, betalamic acid and proline yield indicaxanthin, a yellow betaxanthin (**479**) from *Opuntia* flowers. The betacyanins may exist and can co-occur in two epimeric forms. In beetroot, for example, there is betanin (**474**) and the C_{15} epimer isobetanin (**480**). These are both glucosides, and the majority of betacyanin pigments occur with glycosidic or acylated glycosidic attachment.

Betalains are mutually exclusive in their natural occurrence with the purple anthocyanin pigments (q.v.), and are found in families of only one higher plant order, the Centrospermae. Even here, they are not universally present, since they are replaced by anthocyanins in the families Caryophyllaceae and Molluginaceae. However, betalains are widely present as flower, fruit or leaf pigments in all members of the Cactaceae and other families of this order. The purple vegetable beetroot, *Beta vulgaris*, is often misquoted in textbooks as having anthocyanin pigmentation; in fact, a mixture of purple betacyanins (e.g., betanin, **474**; prebetanin, **489**) and yellow betaxanthins (e.g., vulgaxanthins-I and -II, **490, 491**) are present. Pigments related to the betalains are also present in entirely unrelated sources, in the fruiting bodies of the fly agaric, *Amanita muscaria*, and in *Hygrocybe* spp. Muscapurpurin (**487**) and musca-aurins-I and -II (**485, 486**) are typical pigments from these fungi.

Betalain pigments are readily distinguished from other classes of water soluble plant pigment by their instability in hot acid solution, by their visible spectral properties and by their electrophoretic mobilities. Although they are alkaloidal in nature, they are nontoxic to humans and are readily metabolized on ingestion. A proportion (14% of the UK population) of human subjects, however, are unable to completely metabolize dietary betacyanin (e.g., from beetroot) and suffer from "beeturia" as a result.

References

Piattelli, M. (1976). Betalains. In *Chemistry and Biochemistry of Plant Pigments* (Goodwin, T.W., ed.), 2nd Edition, Volume 1, pp. 560–596. London: Academic Press.

Reznik, H. (1980). Betalains. In *Pigments in Plants* (Czygan, F.C., ed.), 2nd Edition, pp. 370–392. Stuttgart: Gustav Fischer.

471
Amaranthin

$C_{30}H_{34}N_2O_{19}$ Mol. wt 726.60

Occurs in the leaves of *Amaranthus caudatus* and *A. tricolor*, in the inflorescences of *Celosia cristata* (all Amaranthaceae) and in the leaves of *Atriplex hortense* and of *Chenopodium amaranticolor* (both Chenopodiaceae).

Purple pigment.

472
Betalamic acid

$C_9H_9NO_5$ Mol. wt 211.18

Occurs in numerous spp. of the betalain-producing families of the Order Centrospermae: e.g., *Beta vulgaris* (Chenopodiaceae), *Celosia cristata* (Amaranthaceae), and *Portulaca grandiflora* (Portulacaceae).

Pigment. It is a key intermediate in the biogenesis of all betalains.

473
Betanidin

$C_{18}H_{16}N_2O_8$ Mol. wt 388.34

Occurs together with isobetanidin (q.v.) in the flowers of *Mesembryanthemum edule* (Aizoaceae) and of *Portulaca grandiflora* (Portulacaceae).

Purple pigment.

474
Betanin; 5-*O*-β-D-Glucopyranosylbetanidin; Phytolaccanin

$C_{24}H_{26}N_2O_{13}$ Mol. wt 550.49

Found in the flowers of *Carpobrotus acinaciformis*, *Drosanthemum floribundum* and *Mesembryanthemum* spp. (all Aizoaceae), *Opuntia bergeriana* and other *Opuntia* spp. (Cactaceae), and the fruits of *Phytolacca americana* (Phytolaccaceae) and *Portulaca grandiflora* (Portulacaceae).

Purple pigment. It is used as a colouring agent in the food industry.

475
Bougainvillein-r-I; 5-*O*-β-Sophorosylbetanidin

$C_{30}H_{36}N_2O_{18}$ Mol. wt 712.63

Occurs in the red-violet bracts of the cultivar *Bougainvillea* "Mrs Butt" (Nyctaginaceae). The 6-*O*-β-sophoroside of betanidin and of its C-15 epimer isobetanidin and also the 6-*O*-rhamnosylsophoroside of betanidin and of isobetanidin have been isolated from the bracts of *Bougainvillea glabra* var. *sanderiana*.

Purple pigment.

476
Dopaxanthin

$C_{18}H_{17}N_2O_8$ Mol. wt 389.35

Found in the flowers of *Glottiphyllum longum* (Aizoaceae).

Yellow pigment.

477
Gomphrenin-I; 6-*O*-β-D-Glucopyranosylbetanidin

$C_{24}H_{26}N_2O_{13}$ Mol. wt 550.49

Found in the inflorescences of *Gomphrena globosa* (Amaranthaceae), together with gomphrenin-II and small amounts of amaranthin (q.v.), isoamaranthin and celosianin. Gomphrenin-II is the C-15 diastereoisomer of gomphrenin-I (compare betanin and isobetanin, q.v.).

Purple pigment.

478
Gomphrenin-V; *trans*-Feruloylgomphrenin-I

$C_{34}H_{34}N_2O_{16}$ Mol. wt 726.66

In addition to gomphrenins-I and -II (q.v.), five other gomphrenins occur in the inflorescences of *Gomphrena globosa* (Amaranthaceae); they are *trans*-feruloyl, *trans*-*p*-coumaroyl or *cis*-*p*-coumaroyl derivatives of gomphrenin-I or -II.

Purple pigment.

479
Indicaxanthin

$C_{14}H_{16}N_2O_6$ Mol. wt 308.30

Found in the fruit of *Opuntia ficus-indica* (Cactaceae) and the flowers of *Mirabilis jalapa* (Nyctaginaceae).

Yellow pigment.

480
Isobetanin; 5-*O*-β-D-Glucopyranosylisobetanidin

$C_{24}H_{26}N_2O_{13}$ Mol. wt 550.49

Occurs in the fruits of *Opuntia vulgaris* and *O. ficus-indica* (Cactaceae), in the leaves of *Phyllocactus hybridus* (Cactaceae) and in the flowers of *Mesembryanthemum edule* and *M. conspicuum* (Aizoaceae).

Purple pigment.

481
Miraxanthin-I

$C_{14}H_{18}N_2O_7S$ Mol. wt 358.38

Occurs in the flowers of *Mirabilis jalapa* (Nyctaginaceae).

Yellow pigment.

482
Miraxanthin-II

$C_{13}H_{14}N_2O_8$ Mol. wt 326.27.

Occurs in the flowers of *Mirabilis jalapa* (Nyctaginaceae).

Yellow pigment.

483
Miraxanthin-III

$C_{17}H_{17}N_2O_5$ Mol. wt 329.34

Found in the flowers of *Mirabilis jalapa* (Nyctaginaceae).

Yellow pigment.

484
Miraxanthin-V

$C_{17}H_{18}N_2O_6$ Mol. wt 346.35

Found in the flowers of *Mirabilis jalapa* (Nyctaginaceae).

Yellow pigment.

485
Musca-aurin-I

$C_{14}H_{13}N_3O_8$ Mol. wt 351.27

Occurs in the red cap of fly agaric *Amanita muscaria* (Agaricaceae).

Yellow pigment.

486
Musca-aurin-II

$C_{18}H_{16}N_2O_{10}$ Mol. wt 420.34

Found in the red cap of fly agaric *Amanita muscaria* (Agaricaceae).

Yellow pigment.

487
Muscapurpurin

$C_{18}H_{14}N_2O_{10}$ Mol. wt 418.32

Found in the red cap of fly agaric *Amanita muscaria* (Agaricaceae).

Purple pigment.

488
Portulaxanthin

$C_{14}H_{16}N_2O_7$ Mol. wt 324.30

Occurs in the flowers of *Portulaca grandiflora* (Portulacaceae).

Yellow pigment.

489
Prebetanin; Betanin sulfate

$C_{24}H_{26}N_2O_{16}S$ Mol. wt 630.55

Found in the beetroot *Beta vulgaris* (Chenopodiaceae) and in the fruits of *Phytolacca decandra* (Phytolaccaceae).

Purple pigment.

490
Vulgaxanthin-I

$C_{14}H_{17}N_3O_7$ Mol. wt 339.30

Found in the roots of *Beta vulgaris* (Chenopodiaceae) and in the flowers of *Mirabilis jalapa* (Nyctaginaceae).

Yellow pigment.

491
Vulgaxanthin-II

$C_{14}H_{16}N_2O_8$ Mol. wt 340.30

Occurs in the beetroot *Beta vulgaris* (Chenopodiaceae).

Yellow pigment.

Chapter 19
Diterpenoid Alkaloids

Although not at present of therapeutic significance, these alkaloids have been extensively studied because of their interesting chemical structures, and some 450 compounds have now been described. They are conveniently considered under five headings: C_{19} alkaloids, C_{20} alkaloids, *Erythrophleum* alkaloids, bisditerpenoids, and miscellaneous. The largest group is the C_{19} alkaloids, represented by aconitine (**494**) and many related structures, where the trivial name (e.g., falaconitine, **518**) often indicates the relationship with the parent compound. These compounds can be very poisonous. The next group is the C_{20} alkaloids, represented by two related bases, atisine (**498**) and veatchine (**550**). These alkaloids often occur as esters, e.g., lindheimerine (**530**), and are not as toxic as aconitine. The *Erythrophleum* alkaloids are represented by cassaidine (**506**), and characteristically contain their nitrogen atom in an aliphatic side chain. Plants containing these alkaloids are particularly poisonous to farm animals. A typical bisditerpenoid is staphidine (**546**), and an example of a miscellaneous diterpenoid alkaloid is ryanodine (**539**).

Diterpene alkaloids occur mainly in the family Ranunculaceae, notably in species of *Aconitum* and *Delphinium*. They have also been obtained from *Garrya* (Garryaceae), *Icacina* (Icacinaceae), *Inula* (Compositae), *Spiraea* (Rosaceae), *Anopterus* (Saxifragaceae) and *Erythrophleum* (Leguminosae). Interest in these compounds has been based on the use of *Aconitum* roots and leaves as poisons, and in native remedies, for treating hypertension, neuralgia and rheumatism. *Aconitum* species are among the most poisonous plants known, and 2–5 mg of a pure alkaloid can be a fatal dose in humans. They act by slowing the heart rate and lowering the blood pressure. Some *Delphinium* species, such as *D. nelsonii*, growing in pastures in the Western United States are eaten by cattle, and the resulting toxicosis, which can lead to death, is due to the alkaloid content (Benn and Jacyno, 1983).

The detection and analysis of diterpenoid alkaloids in plants has been made easier recently by the publication of a comprehensive listing of their spectral data (Atta-ur-Rahman, 1990).

References

Atta-ur-Rahman (1990). *Handbook of Natural Products Data* Volume 1. *Diterpenoid and Steroidal Alkaloids*. Amsterdam: Elsevier.

Benn, M.H. and Jacyno, J.M. (1983). The toxicology and pharmacology of diterpenoid alkaloids. In *Alkaloids: Chemical and Biological Perspectives* (Pelletier, S.W., ed.), pp. 153–210, New York: Wiley.

In *The Alkaloids* (Manske, R.H.F. and Brossi, A., eds) (1954) Volume 4, pp. 265–274; (1960) Volume 7, pp. 473–504; (1968) Volume 10, pp. 287–305; (1970) Volume 12, pp. 2–206; (1979) Volume 17, pp. 1–103; (1981) Volume 18, pp. 99–216; (1988) Volume 34, pp. 95–179.

492
Acetylbrowniine; Browniine 14-acetate

$C_{27}H_{43}NO_8$ Mol. wt 509.65

Occurs in the aerial parts of *Delphinium oreophilum* and the seeds of *Consolida ambigua* (Ranunculaceae).

Some activity in the guinea-pig ileum assay with a response at a concentration of 2×10^{-4}M.

493
Aconifine; 10-Hydroxyaconitine; Nagarine

$C_{34}H_{47}NO_{12}$ Mol. wt 661.76

Occurs in the roots of *Aconitum karakolicum* and of *A. nagarum* (Ranunculaceae).

The pharmacology most closely resembles that of aconitine in its spectrum of effects, but it has only about half that alkaloid's toxicity in mice.

494
Aconitine; Acetylbenzoylaconine

$C_{34}H_{47}NO_{11}$ Mol. wt 645.76

Occurs mainly in the tuberous roots of *Aconitum napellus* (Ranunculaceae), but all parts of the plant are poisonous. It also occurs in a great many other *Aconitum* spp.

Very potent and quick-acting poison, causing slowing of the heart rate and lowering of blood pressure; formerly used as an antineuralgicum but now rarely used internally. Several *Aconitum* spp. are used in China, after processing, for different purposes, e.g., as a local anaesthetic. Absorption through the skin can be fatal.

495
Ajaconine

$C_{22}H_{33}NO_3$ Mol. wt 359.51

Occurs in the seeds of *Delphinium ajacis*, in the roots of *D. tatsienense* and in the whole plant of *D. viriscens* (Ranunculaceae).

Poisonous and insecticidal.

496
Anopterine

$C_{31}H_{43}NO_7$ Mol. wt 541.69

Occurs in the leaf and bark of *Anopterus macleayanus* and the bark of *A. glandulosus* (Saxifragaceae), along with the related anopterimine and anopterimine *N*-oxide.

Antitumour activity in mouse leukaemia assays with P-388 and KB-systems.

497
Anthranoyllycoctonine; Inuline

$C_{32}H_{46}N_2O_8$ Mol. wt 586.73

Occurs in the roots of *Delphinium barbeyi* (Ranunculaceae) and other *Delphinium* spp., and in *Inula royleana* (Compositae).

Toxic in animals: partial loss of motor control and respiratory paralysis, culminating in death preceded by violent convulsions. In mice it has some potency as a neuromuscular blocking agent.

498
Atisine; Anthorine

$C_{22}H_{33}NO_2$ Mol. wt 343.51

Major constituent of the roots of *Aconitum heterophyllum* (Ranunculaceae).

Relatively nontoxic, compared with aconitine. It is used in India as a febrifuge under the name "utees", "atees" or "atis". Cardiovascular activity has been reported.

499
Avadharidine; Avadkharidine; Awadcharidine

$C_{36}H_{51}N_3O_{10}$ Mol. wt 685.83

Occurs in the roots of *Aconitum orientale, A. finetianum* and *Delphinium cashmirianum* (Ranunculaceae).

Possesses curare-like properties; it has a negligible effect on the heart rate of anaesthetized rabbits in doses up to 5 mg kg^{-1} intravenously.

500
Beiwutine

$C_{32}H_{46}NO_{12}$ Mol. wt 646.73

Occurs with four known alkaloids – aconitine (q.v.), deoxyaconitine, hypaconitine (q.v.) and mesaconitine (q.v.) – in *Aconitum kusnezoffi* (Ranunculaceae).

Analgesic activity.

501
Bikhaconitine

$C_{36}H_{51}NO_{11}$ Mol. wt 673.81

Occurs in the roots of *Aconitum spicatum, A. ferox* and *A. violaceum* (Ranunculaceae).

Highly toxic, producing respiratory paralysis and a direct action upon the heart, terminating in ventricular fibrillation. The plant is used extensively as a poison in India ("bikh" meaning "poison").

502
Browniine

$C_{25}H_{41}NO_7$ Mol. wt 467.61

Occurs in *Aconitum* and *Delphinium* spp., e.g., *D. brownii* (Ranunculaceae).

Relatively nontoxic. It is moderately active in depressing the response of the phrenic nerve-diaphragm preparation of the rat, and also the response of the guinea-pig ileum preparation.

503
Cammaconine

$C_{23}H_{37}NO_5$ Mol. wt 407.56

Occurs in the tuberous roots of *Aconitum variegatum* (Ranunculaceae) together with talatizamine (q.v.).

504
Cardiopetalidine

$C_{21}H_{33}NO_4$ Mol. wt 363.50

Occurs in *Delphinium cardiopetalum* (Ranunculaceae), along with cardiopetaline, the 7-deoxy derivative.

505
Cashmiradelphine; Septentriodine

$C_{37}H_{52}N_2O_{11}$ Mol. wt 700.84

Occurs in the roots of *Delphinium cashmirianum* and *Aconitum septentrionale* (Ranunculaceae).

No arrhythmogenic activity, in contrast to, e.g., aconitine and lappaconitine (q.v.).

506
Cassaidine

$C_{24}H_{41}NO_4$ Mol. wt 407.60

Occurs in the bark of *Erythrophleum guineense* (Leguminosae), together with cassaine (q.v.).

Toxic, with a digitalis-like effect on the heart, and a very strong local anaesthetic action.

507
Cassaine

$C_{24}H_{39}NO_4$ Mol. wt 405.58

Occurs in the bark of *Erythrophleum guineense* (Leguminosae).

Convulsant action. It is digitalis-like, with cardiotonic and cardiotoxic activities. It has only moderate anaesthetic activity.

508
Condelphine; 14-Acetylisotalatizidine

$C_{25}H_{39}NO_6$ Mol. wt 449.59

Occurs in a number of *Delphinium* spp., e.g., *D. confusum* and *D. denudatum* (Ranunculaceae), and *Aconitum delphinifolium* (Ranunculaceae).

Said to have similar pharmacological properties to those of methyllycaconitine (q.v.); it is a curare-like neuromuscular blocker and a hypotensive agent. It has been used clinically in the USSR in the treatment of various neurological disorders.

509
Cuauchichicine

$C_{22}H_{33}NO_2$ Mol. wt 343.51

Occurs in the bark of *Garrya laurifolia* (Garryaceae) and *G. ovata* var. *lindheimeri*, along with garryfoline (q.v.), lindheimerine (q.v.) and ovatine (q.v.).

In mice the most toxic of the veatchine (q.v.)-like alkaloids; it causes tremors, lowers the blood pressure and increases the heart rate.

510
Delcorine

$C_{26}H_{41}NO_7$ Mol. wt 479.62

Occurs in *Delphinium corumbosum* (Ranunculaceae).

Intravenous doses of 5–15 mg kg^{-1} in cats and dogs cause a fall in blood pressure and a temporary respiratory depression. It inhibits the contractions of isolated rat and rabbit intestine, and induces contractions of the guinea-pig uterus; it is a ganglionic blocking agent.

511
Delcosine; Iliensine; Ilienzine

$C_{24}H_{39}NO_7$ Mol. wt 453.58

Occurs in *Aconitum ibukiense*, *Delphinium ajacis* and *D. consolida* (Ranunculaceae).

Poisonous to cold-blooded animals, but less toxic to warm-blooded creatures; the acute toxicity is about three times that of lycoctonine (q.v.). In anaesthetized cats, doses of 10 mg kg^{-1} produce a substantial fall in blood pressure. It reduces the response of an electrically stimulated sciatic nerve–gastrocnemius muscle preparation. It is also reported to be insecticidal.

512
Delphinine

$C_{33}H_{45}NO_9$ Mol. wt 599.73

Occurs in the seeds of *Delphinium staphisagria* and in the roots of *Atragene sibirica* (Ranunculaceae).

The physiological effects of delphinine are closely similar to those of aconitine (q.v.). The most characteristic symptom of poisoning in animals seems to be respiratory depression, which is the primary cause of death; in general, the toxicity seems to be slightly lower than that of aconitine.

513
Delsoline; Acomonine; 14-*O*-Methyldelcosine

C$_{25}$H$_{41}$NO$_7$ Mol. wt 467.61

Occurs in *Delphinium consolida* and *Aconitum monticola* (Ranunculaceae).

Parenteral administration to mice causes weakness in the extremities, clonic convulsions and respiratory depression. The blood pressure of anaesthetized cats and dogs is lowered by doses of 5–15 mg kg^{-1}. It is reported to show insecticidal activity.

514
Deltaline; Eldeline

C$_{27}$H$_{41}$NO$_8$ Mol. wt 507.63

Occurs in several *Delphinium* spp., e.g., *D. barbeyi*, *D. occidentale* and *D. elatum* (Ranunculaceae).

Intravenous administration of 20 mg kg^{-1} to cats lowered the blood pressure, but respiration was not affected.

515
Denudatine

C$_{22}$H$_{33}$NO$_2$ Mol. wt 343.51

Occurs in certain *Delphinium* spp., e.g., the roots of *D. denudatum* (Ranunculaceae).

No effect on blood pressure or respiration in anaesthetized dogs; it markedly inhibits the contraction of isolated rabbit duodenal strip.

516
Elatine

C$_{38}$H$_{50}$N$_2$O$_{10}$ Mol. wt 694.83

Occurs in *Delphinium elatum* (Ranunculaceae).

Pharmacology similar to that of methyllycaconitine (q.v.). It has been used in the USSR for the treatment of neurological disorders. When compared with tubocurarine (q.v.), elatine proved to be 7–8 times less toxic but has a 5–6 times wider therapeutic spectrum.

517
Erythrophleguine; 6α-Hydroxycassamine

$C_{25}H_{39}NO_6$ Mol. wt 449.59

Occurs in the bark of *Erythrophleum guineense* (Leguminosae).

Digitalis-like activity: reduction of the heart frequency, intensification of heart contraction and diureses. It has cardiotonic activity.

518
Falaconitine; Pyropseudoaconitine

$C_{34}H_{47}NO_{10}$ Mol. wt 629.76

Occurs in the roots of *Aconitum falconeri* (Ranunculaceae).

In small animals, the symptoms of poisoning generally resemble those of aconitine: guinea-pigs showed paresis and clonic movements; death was preceded by convulsions.

519
Finaconitine

$C_{32}H_{44}N_2O_{10}$ Mol. wt 616.72

Occurs in the roots of *Aconitum finetianum* (Ranunculaceae).

Analgesic activity.

520
Garryine

$C_{22}H_{33}NO_2$ Mol. wt 343.51

Occurs in the bark of *Garrya veatchii* (Garryaceae).

Poisonous effects on mice are virtually identical with those obtained with veatchine (q.v.). In the anaesthetized cat it decreases the blood pressure and the heart rate.

521
Heteratisine

C$_{22}$H$_{33}$NO$_5$ Mol. wt 391.51

Occurs in *Aconitum heterophyllum* and *A. zeravschanicum* (Ranunculaceae).

Pharmacology shows features in common with that of napelline (q.v.), i.e., brief hypertension and disturbed respiration.

522
Hetisine; Delatine

C$_{20}$H$_{27}$NO$_3$ Mol. wt 329.44

Occurs in the roots of *Aconitum heterophyllum* and *Delphinium cardinale* (Ranunculaceae).

Nontoxic.

523
Hypaconitine; 3-Deoxymesaconitine

C$_{33}$H$_{45}$NO$_{10}$ Mol. wt 615.73

Occurs in numerous *Aconitum* spp., e.g., *Aconitum callianthum*, *A. carmichaeli*, and *A. napellus* (Ranunculaceae).

Pharmacological effects are similar to those of aconitine (q.v.) and mesaconitine (q.v.) but it is 5–8 times less effective, among others, in analgesic activity; its anti-inflammatory action is similar to that of aconitine.

524
Icaceine

C$_{22}$H$_{33}$NO$_4$ Mol. wt 375.51

Occurs in the leaves and roots of *Icacina güssfeldtii* (Icacinaceae) along with icacine (q.v.) and de-*N*-methylicaceine.

Root decoction of *Icacina güssfeldtii* used in different regions of tropical Africa in popular medicine as an anticonvulsant, probably due to the alkaloid content.

525
Icacine

C$_{22}$H$_{31}$NO$_6$ Mol. wt 405.50

Occurs in the leaves and roots of *Icacina güssfeldtii* (Icacinaceae), along with icaceine (q.v.) and de-*N*-methylicaceine.

Use described under icaceine.

526
Indaconitine; 15-Deoxyaconitine

$C_{34}H_{47}NO_{10}$ Mol. wt 629.76

First isolated from *Aconitum chasmanthum* and subsequently found in other *Aconitum* spp., e.g., *A. ferox,* and *A. falconeri* (Ranunculaceae).

Acute toxicity towards small animals very close to that of aconitine (q.v.).

527
Jesaconitine

$C_{35}H_{49}NO_{12}$ Mol. wt 675.78

Occurs in *Aconitum fischeri, A. subcuneatum* and *A. sachalinense* (Ranunculaceae).

Possesses pharmacological properties similar to those of aconitine (q.v.); it is highly toxic to mammals and affects blood pressure, heart rate and respiration. Small doses increase the contractions of a mammalian intestine preparation.

528
Karakoline; Karacoline; Carmichaeline

$C_{22}H_{35}NO_4$ Mol. wt 377.53

Occurs in the tubers of *Aconitum carmichaeli, A. karakolicum* and in *Delphinium pentagynum* (Ranunculaceae).

Pharmacological activity closely resembles that of talatizamine; the acute toxicity is about twice that of talatizamine (q.v.) in mice. Intraperitoneal injection into mice produces muscular weakness, and death from respiratory depression. Intravenous administration to cats or dogs lowers blood pressure for 15–20 min.

529
Lappaconitine

$C_{32}H_{44}N_2O_8$ Mol. wt 584.72

Occurs in the tubers and aerial parts of *Aconitum excelsum, A. orientale, A. ranunculaefolium, A. septentrionale*, and other *A.* spp. (Ranunculaceae).

Highly toxic though 40 times less toxic than aconitine intravenously in mice; toxic doses produce respiratory paralysis and have a direct action upon the heart, often terminating in ventricular fibrillation.

530
Lindheimerine

$C_{22}H_{31}NO_2$ Mol. wt 341.50

Occurs in *Garrya ovata* var. *lindheimeri* (Garryaceae) along with ovatine (q.v.).

531
Lycaconitine; *N*-Succinylanthranoyllycoctonine

$C_{36}H_{48}N_2O_{10}$ Mol. wt 668.79

Occurs in *Aconitum lycoctonum* and *Delphinium cashmirianum* (Ranunculaceae).

Little effect on the heart rate of anaesthetized rabbits in doses up to 5 mg kg^{-1} intravenously; doses above 0.25 mg kg^{-1} produce a small rise in blood pressure.

532
Lycoctonine; Delsine; Royline

$C_{25}H_{41}NO_7$ Mol. wt 467.61

Occurs in *Aconitum lycoctonum*, and in various *Delphinium* spp., e.g., *D. consolida* (Ranunculaceae), and in *Inula royleana* (Compositae).

Causes a sudden fall in blood pressure when given intravenously to cats in doses of 2–5 mg kg^{-1}; there is a hypotensive response in anaesthetized cats with intravenous doses of 5–15 mg kg^{-1}.

533
Mesaconitine; 3α-Hydroxyhypaconitine

$C_{33}H_{45}NO_{11}$ Mol. wt 631.73

Predominant alkaloid in certain subspp. of *Aconitum napellus*, the traditional source of European aconitine (q.v.) (Ranunculaceae); it is widely present in other *Aconitum* spp.

Pharmacological properties approach those of aconitine (q.v.), but with minor differences in potency.

534
Methyllycaconitine; Delartine; Delsemidine

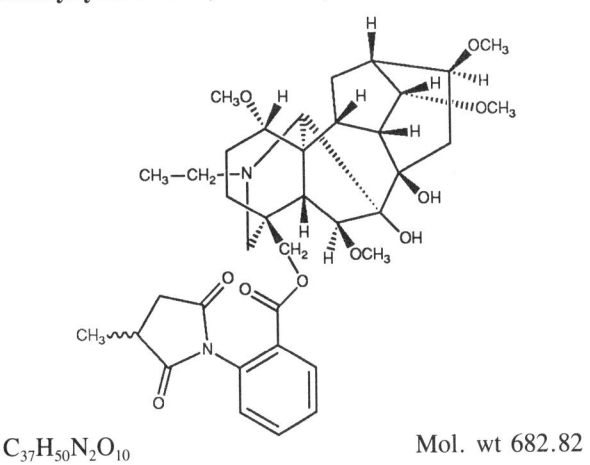

$C_{37}H_{50}N_2O_{10}$ Mol. wt 682.82

Found in numerous *Delphinium* spp., e.g., the roots of *Delphinium elatum* (Ranunculaceae).

Potent neuromuscular poison in mammals with classical curariform activity.

535
Napelline; Luciculine

$C_{22}H_{33}NO_3$ Mol. wt 359.51

Occurs in several *Aconitum* spp., including *A. napellus* (Ranunculaceae).

Causes cardiovascular effects in cats, including brief lowering of blood pressure and disturbed respiration.

536
Norerythrostachaldine; 19-Oxonorcassaidine

$C_{23}H_{37}NO_5$ Mol. wt 407.56

Found in the bark of *Erythrophleum chlorostachys* (Leguminosae) along with its naturally occurring 3β-acetate (at C-3).

High cytotoxicity against human nasopharyngeal cancer cells.

537
Ovatine; (15β)-Veatchine acetate

$C_{24}H_{35}NO_3$ Mol. wt 385.55

Found in *Garrya ovata* var. *lindheimeri* (Garryaceae) along with lindheimerine (q.v.) and garryfoline (q.v.).

Crude extracts from the bark and leaves of the above plant show antitumour activity *in vivo*, probably due to the two diterpenoid alkaloids, ovatine and lindheimerine (q.v.).

538
Pseudaconitine

$C_{36}H_{51}NO_{12}$ Mol. wt 689.81

Occurs in the tubers of *Aconitum ferox* and other *Aconitum* spp. (Ranunculaceae), e.g., *A. falconeri*, and *A. spictatum*.

About twice as toxic in small animals and birds as aconitine and slightly more so than bikhaconitine (q.v.). It is more effective as a respiratory depressant than aconitine, but the cardiovascular potencies of the two alkaloids are very similar.

539
Ryanodine

$C_{25}H_{35}NO_9$ Mol. wt 493.56

Found in *Ryania speciosa* (Flacourtiaceae).

Insecticide against the European cornborer. It inhibits the binding of calcium to muscle protein, and retards circulation by vascular constriction.

540
Septentrionine

$C_{38}H_{54}N_2O_{11}$ Mol. wt 714.86

Found in the roots of *Aconitum septentrionale* and *A. barbatum* (Ranunculaceae).

541
Songorine; Napellonine; Bullatine G

$C_{22}H_{31}NO_3$ Mol. wt 357.50

Found in *Aconitum karakolicum, A. soongaricum* and *A. monticola* (Ranunculaceae).

Acute toxicity at a relatively low dose in mice: doses of 20 mg kg^{-1} cause a fall in blood pressure attributed to its ganglion-blocking properties. Larger doses in mice and rats lead to a decrease in motor activity, to respiratory difficulties, tremor, increased tone in the skeletal musculature and clonicotonic convulsions.

542
Spiradine A

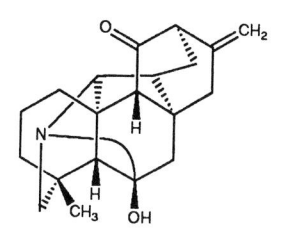

$C_{20}H_{25}NO_2$ Mol. wt 311.43

Spiraea japonica var. *fortunei* (Rosaceae) yields a series of closely related alkaloids, spiradines A–G, which also occur in *Thalictrum sessile* (Ranunculaceae) along with the alkaloids spiredine (q.v.), spirasine I–III (q.v.), thalifarazine, magnoflorine (q.v.) and berberine (q.v.).

543
Spiramine A

$C_{24}H_{33}NO_4$ Mol. wt 399.54

Found in *Spiraea japonica* var. *acuminata* (Rosaceae) together with related alkaloids spiramine B, C and D.

544
Spirasine I

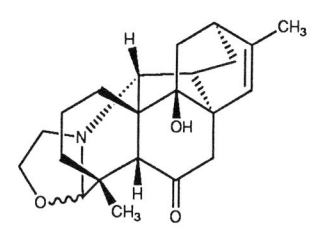

$C_{22}H_{29}NO_3$ Mol. wt 355.48

Occurs in *Spiraea japonica* var. *fortunei* (Rosaceae) along with spirasine II–VIII, spiradine A–G (q.v.) and spiredine (q.v.), and in *Thalictrum sessile* (Ranunculaceae) along with spiradine A, spirasine III and spiredine.

545
Spiredine

$C_{22}H_{27}NO_3$ Mol. wt 353.47

Occurs in *Spiraea japonica* var. *fortunei* (Rosaceae) along with spiradine A and other related alkaloids, and in the fresh roots of *Thalictrum sesseli* (Ranunculaceae) along with thalicsessine (q.v.) and thalisiline.

546
Staphidine

$C_{42}H_{58}N_2O$ Mol. wt 606.93

Occurs in the seeds of *Delphinium staphisagria* (Ranunculaceae), along with eight closely related bisterpenoid alkaloids.

Extracts of the above seeds are used as an external parasiticide.

547
Talatizamine; Talatisamine

$C_{24}H_{39}NO_5$ Mol. wt 421.58

Occurs in *Aconitum* spp. such as *A. talassicum, A. nemorum* and *A. carmichaeli* (Ranunculaceae).

Produces brief hypotension in small animals at doses of 5–15 mg kg^{-1} but no effect observed on heart rhythm and no central effects on conditioned reflexes.

548
Thalicsessine

$C_{22}H_{27}NO_4$ Mol. wt 369.47

Occurs in the fresh roots of *Thalictrum sessile* (Ranunculaceae).

Anti-inflammatory and analgesic activities.

549
Tricornine; 18-*O*-Acetyllycoctonine; Lycoctonine 18-acetate

$C_{27}H_{43}NO_8$ Mol. wt 509.65

Occurs in *Delphinium tricorne* (Ranunculaceae).

Low doses produce only a slight enhancement of the response in the rat to the phrenic nerve–diaphragm preparation; higher doses cause an initial enhancement followed by depression.

550
Veatchine

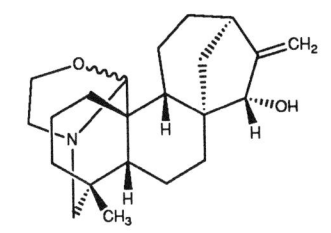

$C_{22}H_{33}NO_2$ Mol. wt 343.51

Occurs in *Garrya* spp., e.g., *G. veatchii* (Garryaceae). Garryfoline, the 15-epimer, occurs in *G. laurifolia* and *G. ovata* var. *lindheimeri*.

The toxicity is relatively high in comparison with other alkanolamines of either the C_{19} or C_{20} type; toxic symptoms include gasping, convulsions and respiratory failure. Garryfoline is slightly less toxic than veatchine.

Chapter 20
Indole Alkaloids

The indole alkaloids comprise the second largest single group of plant bases known today. Many are of clinical value in medicine, e.g., the tranquillizer reserpine (**646**) from *Rauwolfia serpentina* and used for treating hypertension. These alkaloids are derived biosynthetically from the protein amino acid tryptophan, the indole nucleus of which can be seen in every structure. There are a small number of simple indoles, such as harman (**622**) and hypaphorine (**625**). However, the majority of these alkaloids are derived from tryptophan and a C_{10} monoterpenoid precursor secologanin (q.v.). Condensation of these two different classes of precursor produces a range of complex skeletal types, with anything from four to seven ring systems. There are tetracyclic structures such as agroclavine (**555**), pentacyclic structures such as ajmalicine (**556**), hexacycles such as apodine (**565**) and heptacyclic compounds such as brucine (**582**).

Indole alkaloids can undergo dimerization during biosynthesis, so that dimeric structures are formed as in amataine (**562**) or vincristine (**664**). Functional groups present in these alkaloids include hydroxy, methoxy, aldehyde and carbomethoxy substitution, but glycosides are almost unknown. The biosynthetic pathway to the indole alkaloids has been actively pursued, largely by its study in plant cell culture. Most of the enzymes of biosynthesis leading to yohimbine-type bases (e.g., ajmalicine, **556**) have been characterized.

Some of the simple indoles have a wide distribution in the angiosperms. Thus, harman alkaloids have been detected in more than ten plant families, as distantly related as the Chenopodiaceae and Zygophyllaceae. By contrast, the secologanin-based indoles are restricted in their occurrence to three related families: the Apocynaceae, Loganiaceae and Rubiaceae. At least 1200 structures occur throughout many species and genera of these families. A notable source is the Madagascar periwinkle, *Vinca rosea* (or *Catharanthus roseus*), the roots and leaves of which have yielded over 60 alkaloids. Two of these (vinblastine and vincristine) at least are important in medicine for treating leukaemia.

The ergot alkaloids are exceptional in occurring both in higher plants and in fungi. The major source is the fungal pathogen of rye, *Claviceps purpurea*, but they also occur in the seeds of the morning glory plant, *Ipomoea tricolor* from the Convolvulaceae. The hallucinogenic properties of the synthetic derivative, lysergic acid diethylamide (LSD), are notorious, because of the harmful disorientating effects. There are, however, many valuable drugs among the natural indoles of ergot. Ergometrine (**608**) is indispensable for relaxing the uterine muscle during childbirth, while ergotamine (**610**) relieves the symptoms of migraine.

As with other plant alkaloids, some indole alkaloids can be very toxic. Strychnine (**653**) is well known as a poison weapon and as a means of killing rats but, in small doses, it can be used as a general stimulant. Physostigmine (**642**) from the calabar bean is also toxic, but this does not prevent its use in ophthalmology to return the pupil to normal after administration of atropine.

An excellent general reference to the indole alkaloids is the review volume of Phillipson and Zenk (1980); methods of chromatography and analysis are reviewed by Verpoorte in that volume. Extensive coverage of these alkaloids is also available in Cordell (1981). There are also monographs on individual groups of alkaloid, e.g., on the ergot group by Bove (1970).

References

Chemistry and Distribution

Bove, F.J. (1970). *The Story of Ergot*. Basel: Karger Verlag.
Cordell, G. (1981). *Introduction to the Alkaloids: A Biogenetic Approach*. New York: Wiley.
Phillipson, J.D. and Zenk, M.H. (eds) (1980). *Indole and Biogenetically Related Alkaloids*. London: Academic Press.
In *The Alkaloids* (Manske, R.H.F. and Brossi, A., eds) (1965) Volume 8, whole volume; (1967) Volume 9, pp. 467–482; (1970) Volume 12, pp. 207–244; (1973) Volume 14, pp. 157–179; (1980) Volume 20, pp. 1–295; (1985) Volume 26, pp. 1–51; (1986) Volume 27, pp. 1–268; (1987) Volume 30, pp. 223–249; (1988) Volume 34, pp. 211–239; (1989) Volume 36, pp. 1–68, 135–170; (1990) Volume 37, whole volume; Volume 38, pp. 1–156.

551
1-Acetylaspidoalbidine

$C_{21}H_{26}N_2O_2$ Mol. wt 338.45

Found in leaves and branches of *Vallesia dichotoma* and *V. glabra* (Apocynaceae).

552
Adifoline; Adinine

$C_{22}H_{20}N_2O_7$ Mol. wt 424.42

Found in the heartwood of *Adina cordifolia* (Rubiaceae).

553
Affinine

$C_{20}H_{24}N_2O_2$ Mol. wt 324.43

Occurs in *Peschiera affinis* (Apocynaceae) together with affinisine (q.v.).

Produces gross behavioural changes in the mouse: slight central nervous system depression, delayed intention tremors, ataxia, hypothermia and bradypnoea.

554
Affinisine

$C_{20}H_{24}N_2O$ Mol. wt 308.42

Found in *Alstonia macrophylla*, *Peschiera affinis* and *Tabernaemontana fuchsiaefolia* (Apocynaceae).

Produces central nervous system depression, lachrymation, and tremors in the mouse; it has moderate analgesic properties in the rat.

555
Agroclavine

$C_{16}H_{18}N_2$ Mol. wt 238.33

Occurs in the mycelium of the fungus *Claviceps purpurea*, and also in *Cuscuta*, *Ipomoea* and *Rivea* spp. (Convolvulaceae).

Highly active *in vivo* on the uteri of the rabbit and guinea-pig; it can undergo stepwise oxidation to lysergic acid, which forms a peptide linkage with a variety of amino acids, to yield the therapeutically useful ergot alkaloids.

556
Ajmalicine; Raubasine; Tetrahydroserpentine; δ-Yohimbine; Vinceine; Vincaine

$C_{21}H_{24}N_2O_3$ Mol. wt 352.44

Found in the bark of *Corynanthe johimbe* (Rubiaceae), and the roots of many *Rauwolfia* spp., e.g., *R. serpentina*, and in *Vinca rosea* (Apocynaceae).

Tranquillizer, antihypertensive, and used to improve cerebral blood circulation.

557
Ajmaline; Rauwolfine; Raugalline

$C_{20}H_{26}N_2O_2$ Mol. wt 326.44

Occurs in *Melodinus balansae*, in *Tonduzia longifolia*, and in the ground roots of *Rauwolfia serpentina* (all Apocynaceae). The 17-epimer is called epiajmaline or sandwicine, and occurs in *R. sandwicensis*.

Coronary dilating and antiarrhythmic activities.

558
Akuammicine

$C_{20}H_{22}N_2O_2$ Mol. wt 322.41

Occurs in the (−)-form, together with the racemate (pseudoakuammicine), in the seeds and aerial parts of *Picralima klaineana* and in the roots of *Catharanthus roseus* (Apocynaceae).

Gonadotrophic (follicular stimulation) activity.

559
Akuammidine; Rhazine

$C_{21}H_{24}N_2O_3$ Mol. wt 352.44

Major alkaloid of both *Picralima klaineana* and *Rhazya stricta* (Apocynaceae); it is also found in many other Apocynaceae genera such as *Vinca*, *Voacango*, *Melodinus*, *Vallesia* and *Aspidosperma*.

Hypotensive; it is also a skeletal muscle relaxant, and has local anaesthetic action. Its local anaesthetic activity is said to be about three times as potent as cocaine.

560
Akuammine; Vincamajoridine

$C_{22}H_{26}N_2O_4$ Mol. wt 382.46

Found in the seeds of *Picralima klaineana* (Apocynaceae).

Augments the hypertensive effects of adrenaline, and has local anaesthetic action, almost equal to that of cocaine; the seeds are used by the inhabitants of Ghana in place of quinine.

561
Alstonine

$C_{21}H_{20}N_2O_3$ Mol. wt 348.41

Occurs in *Alstonia constricta*, *Rauwolfia hirsuta*, *R. obscura*, *R. vomitoria* and in *Vinca rosea* (all Apocynaceae).

Neoplasm-inhibiting activity.

562
Amataine; Grandifoline; Subsessiline

$C_{43}H_{48}N_4O_6$ Mol. wt 716.89

Found in the root bark of *Hedranthera barteri* (Apocynaceae).

This plant is used in Nigeria as a children's laxative; its roots, rhizomes, stems and leaves are mildly cardiotonic.

563
Angustine

$C_{20}H_{15}N_3O$ Mol. wt 313.36

Found in the leaves of *Strychnos angustiflora* (Loganiaceae).

Toxic.

564
Antirhine; Rhazinine

$C_{19}H_{24}N_2O$ Mol. wt 296.41

Occurs in *Antirhea putaminosa* (Rubiaceae), and in *Catharanthus longifolius* and *Rhazya stricta* (Apocynaceae).

565
Apodine

$C_{21}H_{22}N_2O_4$ Mol. wt 366.42

Found in *Tabernaemontana armenica* and *T. apoda* (Apocynaceae).

566
Apovincamine

$C_{21}H_{24}N_2O_2$ Mol. wt 336.44

Found in *Tabernaemontana riedelii* and *T. rigida* (Apocynaceae).

Most common medicinal use of many *Tabernaemontana* spp. (roots and latex) is based on their antimicrobial action against infections or wounds; apovincamine shows purgative and febrifuge activity, as well as central nervous system action.

567
(−)-Apparicine; Gomezine; Pericalline; Tabernoschizine

$C_{18}H_{20}N_2$ Mol. wt 264.37

Both the (+) and (−) forms occur in *Aspidosperma dasycarpon*; the (−) form is present in *Tabernaemontana pachysiphon* and *T. divaricata* (all Apocynaceae).

The (−) form shows antimicrobial activity against *Shigella*, *Salmonella*, *Pseudomonas*, *Escherichia*, *Proteus*, *Staphylococcus* and *Corynebacterium*. It shows pronounced analeptic properties, and is active against polio virus.

568
Aricine; Quinovatine; Cinchovatine; Heterophylline

$C_{22}H_{26}N_2O_4$ Mol. wt 382.46

Found in the bark of *Cinchona pelletierana* (Rubiaceae), in *Rauwolfia canescens* (Apocynaceae) and in *Aspidosperma marcgravianum* (Apocynaceae).

It does not seem to share the hypotensive and sedative activities of reserpine (q.v.), although it is similar in structure.

569
Aspidoalbine

C₂₄H₃₂N₂O₅ — Mol. wt 428.53

Occurs in the stem bark of both *Aspidosperma album* and *A. spruceanum* (Apocynaceae).

570
Aspidodasycarpine

C₂₁H₂₆N₂O₄ — Mol. wt 370.45

Occurs in the aerial bark of *Aspidosperma dasycarpon* and *A. cuspa* (Apocynaceae).

The leaves and bark are used in some parts of Venezuela as a febrifuge.

571
Aspidofractine

C₂₂H₂₆N₂O₃ — Mol. wt 366.46

Found in *Aspidosperma refractum* and *A. populifolium* (Apocynaceae).

572
Aspidospermatine

C₂₁H₂₆N₂O₂ — Mol. wt 338.45

Found in *Aspidosperma quebrachoblanco* (Apocynaceae).

Used in Argentina, South and Central America under the name "quebracho blanco" and "quebracho colorada" against dyspnoea.

573
Aspidospermine

C₂₂H₃₀N₂O₂ — Mol. wt 354.50

Common in *Aspidosperma* and *Vallesia* spp. (Apocynaceae), e.g., *A. quebrachoblanco*, *A. quirandy*, *V. dichotoma* and *V. glabra*.

Pharmacological properties include diuretic and respiratory stimulant activities.

574
Auricularine

$C_{33}H_{42}N_4$ Mol. wt 494.73

Occurs in *Hedyotis auricularia* (Rubiaceae).

Popular remedy in tropical and subtropical regions in early cholera, also dysentery, colitis and other abdominal complaints. The herb is employed as a prophylactic against enteric fever.

575
Baloxine

$C_{21}H_{24}N_2O_4$ Mol. wt 368.44

Found in the leaves of *Melodinus balansae* (Apocynaceae).

576
Bleekerine; 10,11-Dimethoxyalstonine

$C_{23}H_{24}N_2O_5$ Mol. wt 408.46

Found in the stem-bark and wood of *Bleekeria vitiensis* (Apocynaceae).

577
Bonafousine

$C_{35}H_{40}N_4O_3$ Mol. wt 564.73

Found in the leaves of *Bonafousia tetrastachya* and *Tabernaemontana siphilitica* (Apocynaceae).

578
Borrecapine

$C_{20}H_{24}N_2O$ Mol. wt 308.43

Occurs in *Borreria capitata* (Rubiaceae).

Substitute for ipecacuanha, being emetic, expectorant and astringent.

579
Borreline

$C_{17}H_{18}N_2O$ Mol. wt 266.35

Occurs in *Borreria capitata* (Rubiaceae) and other *Borreria* spp.

Properties similar to those of borrecapine (q.v.).

580
Borreverine

$C_{32}H_{40}N_4$ Mol. wt 480.70

Occurs in the leaves and twigs of *Flindersia fournieri* (Rutaceae) and in *Borreria verticillata* (Rubiaceae).

581
Brevicolline

$C_{17}H_{19}N_3$ Mol. wt 265.36

Occurs, together with the related brevicarine, in *Carex brevicollis* (Cyperaceae).

Significant effect on uterine contractability in animals.

582
Brucine; 10,11-Dimethoxystrychnine

$C_{23}H_{26}N_2O_4$ Mol. wt 394.48

Found in the bark, wood and seeds of several *Strychnos* spp., e.g., *S. ignatii*, *S. nux vomica*, and *S. aculeata* (Loganiaceae).

Central nervous system stimulant, resembling strychnine (q.v.), but less toxic. It is used in analytical chemistry for separating racemic mixtures. The lethal dose in humans is about 200 mg.

583
Cabucine; 10-Methoxyajmalicine

$C_{22}H_{26}N_2O_4$ Mol. wt 382.46

Occurs in the root and bark of *Cabucala madagascariensis* (Apocynaceae).

584
Calebassine; *C*-Calebassine; *C*-Curarine II; *C*-Strychnotoxine; *C*-Toxiferine II

$[C_{40}H_{48}N_4O_2]^{2+}$ Mol. wt 616.85

Occurs in various *Strychnos* spp., including *S. trinervis, S. mittschelichii, S. solimoensana* and *S. divaricans* (Loganiaceae).

Highly toxic. It is a neuromuscular blocking agent. A component of Calabash curare, it is an arrow poison.

585
Callichiline

$C_{42}H_{48}N_4O_5$ Mol. wt 688.87

Occurs, together with vobtusine (q.v.), in the root bark of *Callichilia subsessilis* and *C. barteria*, and also in *Conopharyngia* and *Voacanga* spp. (all Apocynaceae).

586
Calligonine; Elaeagnine

$C_{12}H_{14}N_2$ Mol. wt 186.26

Major alkaloid from the root of *Calligonum minimum* (Polygonaceae), from *Elaeagnus angustifolia* (Elaeagnaceae), *Petalostylis labicheoides* (Leguminosae) and *Banisteriopsis argentea* (Malphigiaceae).

Causes substantial and lasting depression of blood pressure, comparable with reserpine (q.v.).

587
Calycanthidine

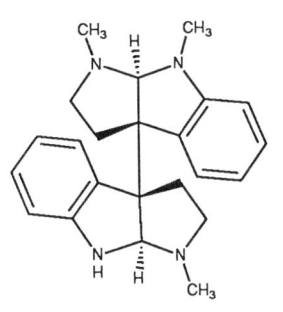

$C_{23}H_{28}N_4$ Mol. wt 360.51

Found in the seeds of *Calycanthus floridus* (Calycanthaceae).

Fruits of the above cause sheep and cattle intoxication in Tennessee, USA, probably the result of another alkaloid, calycanthine (q.v.).

588
Canthin-6-one; Canthinone

$C_{14}H_8N_2O$ Mol. wt 220.23

Occurs in the wood and the leaves of *Pentaceras australis* and *Zanthoxylum suberosum* (Rutaceae), and in *Picrasma crenata* (Simaroubaceae).

Strong antimicrobial activity against *Staphylococcus aureus* and *Mycobacterium smegmatis*. It is cytotoxic to guinea-pig keratinocytes.

589
Caracurine V

$C_{38}H_{40}N_4O_2$ Mol. wt 584.77

Found in the stem bark of *Strychnos dolichothyrsa* (Loganiaceae).

Toxic. It is a neuromuscular blocking agent.

590
Carapanaubine; Vinine

$C_{23}H_{28}N_2O_6$ Mol. wt 428.49

Occurs in the dried bark of *Aspidosperma carapanauba* and *Vinca pubescens* (both Apocynaceae).

591
Catharanthine

$C_{21}H_{24}N_2O_2$ Mol. wt 336.44

Found in *Vinca rosea* (Apocynaceae).

Hypoglycaemic activity.

592
Catharine

$C_{46}H_{54}N_4O_{10}$ Mol. wt 822.97

Found in *Vinca rosea* (Apocynaceae).

593
Chanoclavine-I; Chanoclavine; Secaclavine

$C_{16}H_{20}N_2O$ Mol. wt 256.35

Occurs in *Rivea corymbosa*, *Ipomoea argyrophylla*, *I. violacea* and *I. tricolor* (Convolvulaceae). These form the drug "ololiuqui" used by Central American Indians. Also, it is present in ergot, the fungus *Claviceps purpurea* (Hypocreaceae), which grows on rye and other Gramineae, and in *Aspergillus fumigatus* (Hyphomycetes). The C-10 epimer (chanoclavine-II) is also present in the same sources.

Hallucinogen. It is the major constituent of ololiuqui.

594
Coronaridine; Carbomethoxyibogamine

$C_{21}H_{26}N_2O_2$ Mol. wt 338.45

Found in *Tabernaemontana coronaria* and *Tabernanthe iboga* (Apocynaceae).

Diuretic and oestrogenic activities in animals; it is cytotoxic *in vitro*.

595
Cryptolepine

$C_{16}H_{12}N_2$ Mol. wt 232.29

Found in *Cryptolepis triangularis* and *C. sanguinolenta* (Asclepiadaceae).

Hypotensive activity.

596
***C*-Curarine**

$[C_{40}H_{44}N_4O]^{2+}$ Mol. wt 596.82

Occurs in *Strychnos* spp., including *S. froesii*, *S. mittschelichii*, *S. solimoensana* and *S. divaricans* (Loganiaceae).

Extremely toxic. It is a neuromuscular blocking agent. Curarine is responsible for the paralytic effects of Amazonian arrow poisons, and is a component of Calabash curare.

597
Deserpidine; Canescine; 11-Desmethoxyreserpine; Harmonyl; Raunormine; Recanescine; Reserpidine

$C_{32}H_{38}N_2O_8$ Mol. wt 578.67

Occurs in the roots of *Rauwolfia canescens*, *R. cubana*, and *R. littoralis* (Apocynaceae).

Antihypertensive; tranquillizer.

598
Eburnamine; Hunteriline; Pleiocarpinidine

$C_{19}H_{24}N_2O$ Mol. wt 296.42

Found in *Hunteria eburnea* and *Amsonia tabernaemontana* (Apocynaceae).

Occurs in *Alstonia* spp. (formerly *Echites*), including *A. scholaris*, *A. spectabilis* and *A. spatulata* (Apocynaceae).

Highly toxic when taken orally. It is hypotensive. *Alstonia scholaris* bark is used as a febrifuge.

599
Eburnamonine; Huntericine

$C_{19}H_{22}N_2O$ (−) form   Mol. wt 294.40

The (+) form occurs in *Hunteria eburnea*, while the (−) form (illustrated, known as vincamone or viburnine) occurs in the lesser periwinkle, *Vinca minor* (Apocynaceae).

Vasodilator. The (−) form is used as a drug for stimulating muscle activity.

600
Echitamine; Ditaine

$[C_{22}H_{29}N_2O_4]^+$ Mol. wt 385.49

601
Ellipticine

$C_{17}H_{14}N_2$ Mol. wt 246.31

Occurs in *Aspidosperma subincanum*, *Bleekeria vitiensis* and *Ochrosia elliptica* (Apocynaceae).

Antitrypanosomal *in vitro* against *Trypanosoma cruzi*. It shows broad antitumour activity.

602
Elymoclavine

$C_{16}H_{18}N_2O$ Mol. wt 254.33

Found in ergot, the fungus *Claviceps purpurea* (Hypocreaceae), which grows on rye and other Gramineae, including *Elymus mollis* and *Pennisetum typhoideum*; it is also found in *Aspergillus fumigatus* of the Hyphomycetes.

Prolactin release inhibitor. It is a central nervous system excitatory agent.

603
Epivoacorine

$C_{43}H_{52}N_4O_6$ Mol. wt 720.92

Found in the trunk bark of *Tabernaemontana brachyantha* (Apocynaceae). The 20-epimer, voacorine, occurs in the same source and in *Voacanga africana* (Apocynaceae).

Cytotoxic *in vitro*.

604
Ergine; Lysergic acid amide; Lysergamide

$C_{16}H_{17}N_3O$ Mol. wt 267.33

Found in *Rivea corymbosa, Ipomoea argyrophylla, I. violacea* and *I. tricolor* (Convolvulaceae), and related plants. The drug "ololiuqui" is obtained from these, and used by Central American Indians. Ergine is produced by hydrolysis of ergot alkaloids, from the fungus ergot, *Claviceps purpurea* (Hypocreaceae), c.f., ergotamine. The C-8 isomer is known as erginine or isoergine.

Hallucinogen. It is the principal active constituent of ololiuqui.

605
Ergocornine

$C_{31}H_{39}N_5O_5$ Mol. wt 561.69

Found in ergot, *Claviceps purpurea*, and other *Claviceps* spp. (Hypocreaceae), a fungus growing on rye and other cereals. The C-8 isomer also occurs in ergot and is known as ergocorninine.

Inhibits prolactin release, preventing implantation and lactation. For the effects of ergot poisoning or ergotism, see ergotamine.

606
Ergocristine

$C_{35}H_{39}N_5O_5$ Mol. wt 609.73

Found in ergot, *Claviceps purpurea*, and other *Claviceps* spp. (Hypocreaceae), a fungus growing on rye and other cereals. Ergocristinine is the C-8 epimer.

Inhibits prolactin release, preventing implantation and lactation. For the effects of ergot poisoning or ergotism, see ergotamine.

607
Ergocryptine; Ergokryptine

$C_{32}H_{41}N_5O_5$ Mol. wt 575.72

Found in ergot, *Claviceps purpurea*, and other *Claviceps* spp. (Hypocreaceae), a fungus growing on rye and other cereals. α-Ergocryptine (illustrated) yields L-leucine on hydrolysis, while the closely similar β-ergocryptine yields L-isoleucine.

Inhibits prolactin release, preventing implantation and lactation. The 2-bromo derivative of α-ergocryptine, bromocryptine, is used clinically as a prolactin inhibitor and to treat Parkinson's disease. For the effects of ergot poisoning or ergotism, see ergotamine.

608
Ergometrine; Ergobasine; Ergonovine; Ergostetrine; Ergotocin; D-Lysergic acid L-propanolamide

$C_{19}H_{23}N_3O_2$ Mol. wt 325.41

Found in ergot, *Claviceps purpurea*, and other *Claviceps* spp. (Hypocreaceae), a fungus growing on rye and other cereals. It is also found in *Ipomoea argyrophylla*, and related Convolvulaceae. The C-8 epimer is known as ergometrinine or ergonovinine.

Potent oxytocic. It is used clinically to treat post-partum haemorrhage. Also, it inhibits prolactin release, preventing implantation and lactation. For the effects of ergot poisoning or ergotism, see ergotamine.

609
Ergosine

C₃₀H₃₇N₅O₅   Mol. wt 547.66

Found in ergot, *Claviceps purpurea* (Hypocreaceae), a fungus growing on rye and other cereals, and in *Ipomoea argyrophylla* (Convolvulaceae), and related spp. The C-8 epimer is known as ergosinine. An equimolar racemic mixture is sometimes called ergoclavine.

Prolactin release inhibitor, preventing implantation and lactation. For the effects of ergot poisoning or ergotism, see ergotamine.

610
Ergotamine

C₃₃H₃₅N₅O₅   Mol. wt 581.68

Found in ergot, *Claviceps purpurea*, and *C. paspali* (Hypocreaceae), a fungus growing on rye and other cereals. The C-8 epimer is known as ergotaminine.

Highly toxic. It is a potent vasoconstrictor, and a weak oxytocic. Clinically, it is used to treat migraine and, in obstetrics, used as a haemostatic. It is one of the main alkaloids responsible for the gangrenous infections of the extremities, due to the loss of blood supply, known in the Middle Ages as St Anthony's Fire. These plagues, or ergotism, were caused by flour made from infected rye and the symptoms included hallucinations, delirium and convulsions.

611
Eseramine

C₁₆H₂₂N₄O₃   Mol. wt 318.38

Found in the calabar bean, the seeds of *Physostigma venenosum* (Leguminosae).

Anticholinesterase, with activity similar to but much weaker than that of physostigmine (q.v.).

612
Eseridine; Eserine aminoxide; Eserine oxide; Geneserine; Physostigmine aminoxide; Physostigmine oxide

C₁₅H₂₁N₃O₃   Mol. wt 291.35

Found in the calabar bean, the seeds of *Physostigma venenosum* (Leguminosae).

Anticholinesterase, with activity similar to that of physostigmine (q.v.). It is used for treating gastro-intestinal disorders and chronic dermatoses.

613
Evodiamine

$C_{19}H_{17}N_3O$ Mol. wt 303.37

Occurs in *Evodia rutaecarpa* (Rutaceae) and *Araliopsis tabouensis* (Araliaceae).

Diuretic and diaphoretic agent.

614
Gabunamine; *N*-Dimethylconoduramine

$C_{42}H_{50}N_4O_5$ Mol. wt 690.89

Found in the stem bark of *Tabernaemontana johnstonii* (Apocynaceae).

Antitumour activity. It is cytotoxic to P-388 lymphocytic leukaemia cells *in vitro*.

615
Gabunine

$C_{42}H_{50}N_4O_5$ Mol. wt 690.89

Occurs in *Tabernaemontana johnstonii*, *T. holstii* and *Gabunia odoratissima* (=*T. odoratissima*) (Apocynaceae).

Antitumour activity. It is cytotoxic to P-388 lymphocytic leukaemia cells *in vitro*.

616
Geissoschizoline; 16α-Curan-17-ol; Pereirine

$C_{19}H_{26}N_2O$ Mol. wt 298.43

Found in *Geissospermum vellosii* (Apocynaceae).

617
Geissospermine

$C_{40}H_{48}N_4O_3$ Mol. wt 632.85

Found in *Geissospermum vellosii* and *G. sericeum* (Apocynaceae).

618
Gelsemicine

$C_{20}H_{26}N_2O_4$ Mol. wt 358.44

Occurs in the roots and rhizomes of the Carolina or yellow jessamine, *Gelsemium sempervirens*, and in *Mostuea brunonis* (Loganiaceae).

Highly toxic. It is a central nervous system stimulant. Small doses stimulate respiration; large doses cause respiratory paralysis.

619
Gelsemine

$C_{20}H_{22}N_2O_2$ Mol. wt 322.41

Found in the roots and rhizomes of the Carolina or yellow jessamine, *Gelsemium sempervirens* (Loganiaceae).

Highly toxic: frequent cause of human poisoning. It is a central nervous system stimulant.

620
Haplophytine

$C_{37}H_{40}N_4O_7$ Mol. wt 652.76

Occurs in the Mexican cockroach plant, *Haplophyton cimicidum* (Apocynaceae).

Insecticidal to a wide range of insects.

621
Harmaline; 3,4-Dihydroharmine; Harmidine

$C_{13}H_{14}N_2O$ Mol. wt 214.27

Occurs in the seeds of *Peganum harmala* (Zygophyllaceae), *Banisteria caapi* (Malpighaceae), and *Passiflora incarnata* (Passifloraceae).

Hallucinogenic. It causes ataxia and excitation tremors in animals. At one time it was used to treat Parkinson's disease.

622
Harman; Aribine; Loturine; 1-Methyl-β-carboline; Passiflorin

$C_{12}H_{10}N_2$ Mol. wt 182.23

Found in the passion flower, *Passiflora incarnata* (Passifloraceae), *Singickia* (=*Arariba*) *rubra* (Rubiaceae), *Zygophyllum fabago* (Zygophyllaceae), and *Symplocos racemosa* (Symplocaceae). Present in tobacco smoke.

Motor depressant at low doses in animals; causes convulsions at high doses. It is a plant growth inhibitor, and a cytotoxic intercalating agent.

623
Harmine; Banisterine; Leucoharmine; Telepathine; Yageine

$C_{13}H_{12}N_2O$ Mol. wt 212.25

Occurs in *Peganum harmala* (Zygophyllaceae), *Banisteria caapi* (Malpighaceae), and the passion flower, *Passiflora incarnata* (Passifloraceae).

Central nervous system stimulant. It is hallucinogenic at high doses.

624
11-Hydroxycanthin-6-one; Amarorine

$C_{14}H_8N_2O_2$ Mol. wt 236.23

Found in the stem bark of *Quassia kerstingii* and *Amaroria soulameoides* (Simaroubaceae).

Active *in vitro* against the P388 lymphocytic leukaemia cell line.

625
Hypaphorine

$C_{14}H_{18}N_2O_2$ Mol. wt 246.31

Occurs in the seeds of *Erythrina hypaphorus* and *Pterocarpus officinalis* (Leguminosae).

Convulsive poison. It is a feeding deterrent for the seed eating rodent *Liomys salvini*.

626
Ibogaine; 12-Methoxyibogamine

$C_{20}H_{26}N_2O$ Mol. wt 310.44

Occurs in *Tabernanthe iboga* and *Voacanga thouarsii* (Apocynaceae).

Central nervous system activity; it is a hallucinogen and an anticonvulsant.

627
Ibogamine

$C_{19}H_{24}N_2$ Mol. wt 280.42

The parent compound of a series of alkaloids found in iboga, *Tabernanthe iboga*, and related spp. (Apocynaceae).

Moderate cytotoxic activity. It also shows brachycardiac and hypotensive activities.

628
Isoajmaline

$C_{18}H_{26}N_2O_2$ Mol. wt 302.42

Occurs in the roots of *Rauwolfia serpentina* (Apocynaceae).

Causes initial central nervous stimulation followed by depression; it decreases blood pressure and also stimulates uterine contractions.

629
Leurosidine; Vinrosidine; VRD; 4′α-Vincaleukoblastine

$C_{46}H_{58}N_4O_9$ Mol. wt 811.00

Found in the Madagascar periwinkle, *Vinca rosea* (Apocynaceae).

Antineoplastic agent.

630
Leurosine; Vinleurosine; VLR

$C_{46}H_{56}N_4O_9$ Mol. wt 808.98

Found in the Madagascar periwinkle, *Vinca rosea* and in *Catharanthus* spp. (Apocynaceae).

Antineoplastic agent.

631
Mahanimbine

$C_{23}H_{25}NO$ Mol. wt 331.46

Occurs in the stem bark of the Indian curry leaf plant, *Murraya koenigii*, and of *M. exotica* (Rutaceae).

632
Melosatin A

$C_{21}H_{23}NO_4$ Mol. wt 353.42

Found in *Melochia tomentosa* (Sterculiaceae).

Yellow pigment. The plant is tumourigenic.

633
Melosatin B

$C_{19}H_{19}NO_2$ Mol. wt 293.37

Found in *Melochia tomentosa* (Sterculiaceae).

The plant is tumourigenic.

634
Mesembrenone

$C_{17}H_{21}NO_3$ Mol. wt 287.36

Occurs in *Sceletium* spp. (formerly *Mesembryanthemum*), including *S. expansum*, *S. tortuosum* and *S. anatomicum* (Aizoaceae).

Sceletium leaves are chewed by Namaqualand bushmen and fermented to form a narcotic, stimulant drug preparation known as "channa".

635
Mesembrine; Mesembranone

$C_{17}H_{23}NO_3$ Mol. wt 289.36

Found in *Sceletium* spp. (formerly *Mesembryanthemum*), including *S. expansum*, *S. tortuosum*) and *S. anatomicum* (Aizoaceae).

Narcotic, cocaine-like stimulant. *Sceletium* leaves are chewed by Namaqualand bushmen and fermented to form a narcotic, stimulant drug preparation known as "channa".

636
Mesembrinol; Mesembranol

$C_{17}H_{25}NO_3$ Mol. wt 291.39

Found in *Sceletium* spp. (formerly *Mesembryanthemum*), including *S. expansum*, *S. tortuosum* and *S. anatomicum* (Aizoaceae).

Sceletium leaves are chewed by Namaqualand bushmen and fermented to form a narcotic, stimulant drug preparation known as "channa".

637
Mitragynine

$C_{23}H_{30}N_2O_4$ Mol. wt 398.51

Found in *Mitragyna speciosa* (Rubiaceae).

Central nervous system depressant; antitussive; analgesic. Leaves of *M. speciosa* are sometimes smoked for their narcotic action.

638
Mitraphylline

$C_{21}H_{24}N_2O_4$ Mol. wt 368.44

Occurs in *Mitragyna rubrostipula*, *M. macrophylla* and *Uncaria kawakamii* (Rubiaceae). Several isomers, the uncarines A–F, occur in related species of these two genera.

Hypotensive; it is a weak central nervous system depressant.

639
Murrayanine

$C_{14}H_{11}NO_2$ Mol. wt 225.25

Occurs in the stem bark of the Indian curry leaf plant, *Murraya koenigii*, and the root bark of *Clausena heptaphylla* (Rutaceae).

640
Ochrolifuanine A

C$_{29}$H$_{34}$N$_4$ Mol. wt 438.62

Occurs in *Ochrosia lifuana, O. miana, O. confusa* and *Dyera costulata* (Apocynaceae). One of four stereoisomeric forms present in these plants.

The root barks have been used as an ingredient of arrow poisons.

641
Olivacine; Guatambuinine

C$_{17}$H$_{14}$N$_2$ Mol. wt 246.31

Occurs in *Aspidosperma nigricans* (Apocynaceae).

Antitrypanosomal *in vitro* against *Trypanosoma cruzi* epimastigotes. Strong cytotoxic activity against human carcinomas.

642
Physostigmine; Eserine; Physosterine; Physostol

C$_{15}$H$_{21}$N$_3$O$_2$ Mol. wt 275.35

Main alkaloid of the calabar bean, the seeds of *Physostigma venenosum* (Leguminosae).

Anticholinesterase. It has wide ranging parasympathetic activity when taken internally, and may be used to counteract the effects of anticholinergics such as atropine (q.v.). Large doses can be fatal. The main use is in the form of eye drops as a miotic.

643
Physovenine

C$_{14}$H$_{18}$N$_2$O$_3$ Mol. wt 262.31

Found in the calabar bean, the seeds of *Physostigma venenosum* (Leguminosae).

Anticholinesterase. It is similar in activity to physostigmine (q.v.), but not in clinical use.

644
Psychotridine

C$_{55}$H$_{62}$N$_{10}$ Mol. wt 863.17

Found in *Psychotria beccaroides* (Rubiaceae).

Trypanocidal *in vitro*.

645
Rescinnamine; Reserpinine

$C_{35}H_{42}N_2O_9$ Mol. wt 634.74

Occurs in *Rauwolfia* spp., including *R. serpentina*, *R. vomitoria* and *R. nitida* (Apocynaceae).

Antihypertensive; it is a tranquillizer.

646
Reserpine

$C_{33}H_{40}N_2O_9$ Mol. wt 608.70

Occurs in *Rauwolfia serpentina* and *R. vomitoria* (Apocynaceae); it is present in many other species and genera of this family.

Antihypertensive and tranquillizer, in clinical usage. It is possibly carcinogenic.

647
Rhynchophylline; Rhyncophylline; Mitrinermine

$C_{22}H_{28}N_2O_4$ Mol. wt 384.48

Found in *Ourouparia* (formerly *Uncaria*) *rhynchophylla*, *Mitragyna rotundifolia* and *M. africana* (Rubiaceae).

Antipyretic; hypotensive.

648
Roxburghine B

$C_{31}H_{32}N_4O_2$ Mol. wt 492.63

Occurs in the leaves of *Uncaria gambir* (Rubiaceae). One of six stereoisomeric forms present in *Uncaria* spp.

649
Rutaecarpine; Rutecarpine; Rhetine

$C_{18}H_{13}N_3O$ Mol. wt 287.32

Found in *Evodia rutaecarpa* and *Hortia arborea* (Rutaceae).

Hypotensive.

650
Sarpagine; Raupine

$C_{19}H_{22}N_2O_2$ Mol. wt 310.40

Found in snakeroot, *Rauwolfia serpentina* (Apocynaceae).

Nicotine antagonist, but with no human medicinal usage.

651
Sempervirine; Sempervirene; Sempervine

$C_{19}H_{16}N_2$ Mol. wt 272.35

Occurs in the roots and rhizomes of the Carolina or yellow jessamine, *Gelsemium sempervirens*, and in *Mostuea buchholzii* (Loganiaceae).

Antitumour activity. *G. sempervirens* is highly poisonous; however, this is due to the gelsemine (q.v.) content.

652
Serpentine

$C_{21}H_{20}N_2O_3$ Mol. wt 348.41

Occurs in *Rauwolfia serpentina, R. beddomei, R. fruticosa, Vinca major* and *V. rosea* (Apocynaceae).

Antihypertensive. It also shows antitumour activity.

653
Strychnine

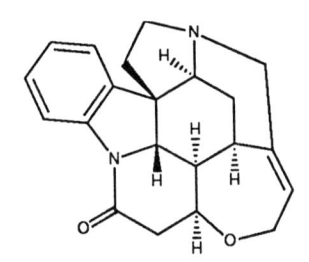

$C_{21}H_{22}N_2O_2$ Mol. wt 334.42

Occurs in nux-vomica, the seeds of *Strychnos nux-vomica*, ignatius beans, from *S. ignatii*, and others such as *S. tieute* and *S. triplinervia* (Loganiaceae). The richest known source (6.6% dry weight) is the bark of *Strychnos icaja*.

Central nervous system stimulant; respiratory stimulant. It is highly toxic and used as a rodenticide. Nux-vomica extracts were formerly used as a nervous tonic and appetite stimulant.

654
Tabernamine

$C_{40}H_{48}N_4O_2$ Mol. wt 616.85

Occurs in the stem bark of *Tabernaemontana johnstonii* (Apocynaceae).

Antitumour activity. It is cytotoxic to P-388 lymphocytic leukaemia cells *in vitro*.

655
Tabernanthine; 13-Methoxyibogamine

$C_{20}H_{26}N_2O$ Mol. wt 310.44

Occurs in *Tabernanthe iboga*, and some spp. of *Conopharingia, Tabernaemontana* and *Stemmadenia* (all Apocynaceae).

Central nervous system activity; it interacts with benzodiazepine receptors.

656
Tabersonine

$C_{21}H_{24}N_2O_2$ Mol. wt 336.44

Widely distributed in the Apocynaceae; e.g., in *Amsonia tabernaemontana* and *Voacanga africana*.

Used in the synthesis and biosynthesis of other alkaloids, such as vindoline, vincamine and vinblastine (q.v). It shows about a quarter of the hypotensive activity of reserpine (q.v.).

657
Toxiferine I; *C*-Toxiferine I; Toxiferine V; Toxiferine XI

$[C_{40}H_{46}N_4O_2]^{2+}$ Mol. wt 614.84

A component of calabash curare, found in *Strychnos* spp., including *S. toxifera* and *S. froesii* (Loganiaceae).

Neuromuscular blocking agent, six to eight times more potent than the isoquinoline alkaloid tubocurarine (q.v.). Calabash curare is an arrow poison.

658
Trichotomine

$C_{30}H_{20}N_4O_6$ Mol. wt 532.52

Occurs in the fruits of *Clerodendron trichotomum* (Verbenaceae).

Bronchodilator; hypotensive; sedative. It is a blue pigment.

659
Tubulosine

$C_{29}H_{37}N_3O_3$ Mol. wt 475.64

Occurs in *Pogonopus tubulosus*, *Psychotria granadensis* and *Cephaelis ipecacuanha* (Rubiaceae).

Amoebicidal. It is highly toxic and shows antitumour activity.

660
Usambarensine

$C_{29}H_{28}N_4$ Mol. wt 432.57

Found in the root bark of *Strychnos usambarensis* (Loganiaceae).

Antimuscarinic effects on isolated rat intestinal muscle. *S. usambarensis* root bark is used as an arrow poison.

661
Usambarine

$C_{30}H_{34}N_4$ Mol. wt 450.63

Occurs in the root bark of *Strychnos usambarensis* (Loganiaceae).

S. usambarensis root bark is used as an arrow poison.

662
Vinblastine; Vincaleukoblastine; VLB

$C_{46}H_{58}N_4O_9$ Mol. wt 811.00

Occurs in the leaves of the periwinkle. *Vinca rosea* (Apocynaceae).

Antitumour activity: in widespread clinical usage, especially to treat certain types of leukaemia and Hodgkin's disease.

663
Vincamine; Vincamarine; Minorine

$C_{21}H_{26}N_2O_3$ Mol. wt 354.45

Occurs in the greater periwinkle, *Vinca major* (Apocynaceae).

Vasodilator. Muscle relaxant derivatives have been prepared. It is used in a variety of disorders to increase cerebral blood circulation.

664
Vincristine; Leurocristine; 22-Oxovincaleukoblastine; VCR; LCR

$C_{46}H_{56}N_4O_{10}$ Mol. wt 824.98

Found in the Madagascar periwinkle, *Vinca rosea* (Apocynaceae).

Antitumour agent used particularly for acute lymphocytic leukaemia in childhood.

665
Vindoline

$C_{25}H_{32}N_2O_6$ Mol. wt 456.55

Found in the madagascar periwinkle, *Vinca rosea* and *V. rusilla* (Apocynaceae).

Used in the synthesis of vinblastine (q.v.), of which it constitutes half the molecule.

666
Voacamine; Voacanginine

$C_{43}H_{52}N_4O_5$ Mol. wt 704.92

Found in many spp. of *Voacanga*, including *V. africana*, *V. globosa*, *V. grandiflora* and *V. thouarsii*, and of *Tabernaemontana*, including *T. arborea*, *T. australis* and *T. oppositifolia*; it is also found in *Hedranthera barteri* (all Apocynaceae).

Cytotoxic *in vitro*. It shows antibacterial activity, especially against Gram-positive species.

667
Vobasine

$C_{21}H_{24}N_2O_3$ Mol. wt 352.44

Occurs in *Voacanga africana* and *Tabernaemontana odoratissima* (Apocynaceae).

Weakly sedative, antipyretic and analgesic in mice.

668
Vobtusine; Papuanine

$C_{43}H_{50}N_4O_6$ Mol. wt 718.90

Widespread in the genus *Voacanga* (Apocynaceae).

669
Vomicine; Struxine; Strychnicine

$C_{22}H_{24}N_2O_4$ Mol. wt 380.45

Occurs in the seeds and leaves of *Strychnos nux-vomica* (Loganiaceae).

The toxicity of *S. nux-vomica* is due to strychnine (q.v.) and not to this alkaloid.

670
Yohimbine; Aphrodine; Corynine; Hydroergotocin; Quebrachine

$C_{21}H_{26}N_2O_3$ Mol. wt 354.45

Found in yohimbe bark, *Pausinystalia yohimbe* (Rubiaceae) and in *Rauwolfia serpentina* (Apocynaceae). Two stereoisomers, α- and β-yohimbine, occur in the same and related plants.

Toxic. It is an α-adrenergic blocking agent, a serotonin antagonist, and a mydriatic. It was once considered an aphrodisiac, due to its use in veterinary medicine. The main clinical use is as an antidepressant.

Chapter 21
Isoquinoline Alkaloids

The isoquinolines are the largest single group of plant alkaloids. Their structures are based on the tetrahydroisoquinoline nucleus, which can be seen in simple compounds such as the cactus alkaloid carnegine (**710**). There is a considerable array of more complex structural types, which can all be derived biosynthetically from phenylalanine and/or tyrosine. The major skeletal types dealt with in this section are the aporphines (actinodaphnine, **671**), the benzophenanthridines (angoline, **685**), the bisbenzylisoquinolines (adiantifoline, **672**), the cularines (cularine, **720**), the emetines (emetine, **727**), the morphinans (codeine, **716**), the pavines (amurensine, **682**) and the rhoeadines (alpinine, **681**).

Chemically, these alkaloids are aromatic bases, with varying numbers of hydroxyl, methoxyl and methylenedioxy substituents. Many are highly *O*-methylated, see, e.g., alpinine (**681**) with five methoxyl groups. The ring nitrogen atom may carry a hydrogen, as in annonaine (**688**) or be *N*-methylated, as in apoglaziovine (**689**). Alkaloids with the nitrogen in the quaternary form are also fairly frequent (e.g., bocconine, **701**). Compounds with sugar attachment, as glycosides, are however, rare (ipecoside, **745**). The majority of isoquinoline alkaloids are tetracyclic, as in the aporphine series, but a few are pentacyclic, as in the emetines. The bisbenzylisoquinolines are distinguished from all other isoquinolines by their dimeric nature and their formation from two benzylisoquinoline nuclei joined by a diphenyl or a diphenyl ether linkage. Some have one bridging linkage (adiantifoline, **672**), others have two (aromoline, **692**) and a few have three (tiliacorine, **796**).

The most widely distributed isoquinolines are the aporphines and protoberberines, which are present in many species of the Magnoliales (e.g., Magnoliaceae), the Ranunculales (e.g., Ranunculaceae), the Papaverales (e.g., Papaveraceae) and the Rutales (e.g., Rutaceae). Other classes tend to have a more discrete distribution pattern. Simple isoquinolines are mainly found in Cactaceae and Leguminosae, while the morphinans are characteristic of the Papaveraceae. The cularines are confined to the Fumariaceae, while the Erythrina alkaloids are restricted to *Erythrina* in the Leguminosae and *Cocculus* in the Menispermaceae. In plant families rich in alkaloids, several classes of isoquinoline may occur together. This is particularly true of the poppy family, where morphinans, rhoeadines, aporphines and protoberberines may all be found (Hegnauer, 1966).

The most notable of these alkaloids, and the most notorious, is morphine, the major base in the opium poppy latex. It is still the most effective of painkillers available in medical practice. At the same time, it is addictive, and the mis-use of the diacetate, heroin, as a narcotic and euphoric drug can lead to the early death of the addict. Other medicinally useful drugs are papaverine (**774**), employed as a vasodilator, boldine (**702**) and codeine (**716**). Many others have a variety of valuable pharmacological properties, but have yet to be developed as commercial drugs.

The isoquinolines are detected in plant extracts by the standard procedures used with other alkaloid types. They are further identified on the basis of their spectral properties. The aporphines for example, are readily characterized by means of their ultraviolet spectral properties and their carbon-13 NMR data. The quantification and detection of morphine in plant material have been much studied from the forensic viewpoint (Hashimoto *et al.*, 1988). Three major compilations on the isoquinoline alkaloids are the works of Shamma (1972), Shamma and Moniot (1978) and Phillipson *et al.* (1985).

References

Chemistry and Biology
Shamma, M. (1972). *The Isoquinoline Alkaloids*. New York: Academic Press.
Shamma, M. and Moniot, J.L. (1978). *Isoquinoline Alkaloid Research 1972–1977*. New York: Plenum Press.
Phillipson, J.D., Roberts, M.F. and Zenk, M. H. (eds) (1985). *The Chemistry and Biology of the Isoquinoline Alkaloids*. Berlin: Springer-Verlag.

Analysis
Hashimoto, Y., Kawanishi, K. and Moriyasu, M. (1988). Forensic chemistry of alkaloids by chromatographic analysis. In *The Alkaloids* (Brossi, A., ed.), Volume 32, pp. 1–78. New York: Academic Press.

Distribution
Hegnauer, R. (1966). Comparative phytochemistry of alkaloids. In *Comparative Phytochemistry* (Swain, T., ed.), pp. 211–230. London: Academic Press.

In *The Alkaloids*. (Manske, R.H.F. and Brossi, A., eds) (1967) Volume 9, pp. 2–40; pp. 41–116; pp. 117–133; pp. 483–514; (1971) Volume 13, pp. 165–188; pp. 189–212; (1975) Volume 15, pp. 207–261; (1977) Volume 16, pp. 250–392; (1983) Volume 21, pp. 255–328; (1985) Volume 24, pp. 153–252; (1985) Volume 24, pp. 253–286.

671
Actinodaphnine

Skeletal type: aporphine

$C_{18}H_{17}NO_4$ Mol. wt 311.34

Found in the bark of *Actinodaphne hookeri*, and the bark and wood of *Laurus nobilis* and *Litsea sebifera* (all Lauraceae), in the leaves of *Hernandia cordigera* and in *Illigera luzonensis* (both Hernandiaceae).

Antimicrobial activity.

672
Adiantifoline

Skeletal type: bisbenzylisoquinoline

$C_{42}H_{50}N_2O_9$ Mol. wt 726.88

Occurs in the roots and aerial parts of *Thalictrum minus* var. *adiantifolium* (Ranunculaceae).

Too toxic for clinical trials. Intravenous administration to rabbits produces a brief hypotensive response.

673
(+)-Adlumine

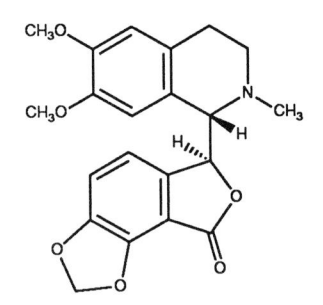

Skeletal type: phthalideisoquinoline

$C_{21}H_{21}NO_6$ Mol. wt 383.41

Occurs in *Adlumia fungosa* (Papaveraceae). The (−) form occurs in *Corydalis scouleri*, *C. sempervirens* and *C. ophiocarpa*, and the (±) form occurs in *C. rosea* (all Papaveraceae).

Activity as a convulsant, cardiac depressor and uterine and intestinal stimulant similar to but weaker than that of hydrastine.

674
Aknadicine

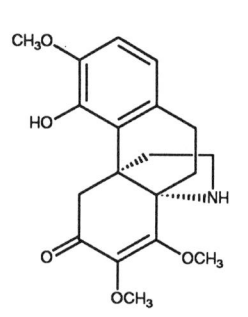

Skeletal type: hasubanan

$C_{19}H_{23}NO_5$ Mol. wt 345.40

Two closely related alkaloids, aknadicine and aknadinine, occur in the roots and rhizomes of *Stephania hernandifolia* (Menispermaceae).

Antimicrobial activity, and used for various types of fever, diarrhoea and urinary diseases. It has some reputation in folk medicine.

675
Alamarine

Skeletal type: isoquinonaphthyridine

$C_{19}H_{18}N_2O_4$ Mol. wt 338.36

Found in the seeds of *Alangium lamarckii* (Alangiaceae).

Not recognized in Western medicine. In India, the seeds are used as a remedy against leprosy and various skin diseases, as an anthelmintic, as a laxative, and as a tonic.

676
Alangicine

Skeletal type: emetine

$C_{28}H_{36}N_2O_5$ Mol. wt 480.61

Found in the root bark of *Alangium lamarckii* (Alangiaceae).

The plant is used in native medicine as a tonic and laxative, anthelmintic and antiprotic.

677
Alangimarine

Skeletal type: isoquinonaphthyridine

$C_{19}H_{16}N_2O_3$ Mol. wt 320.35

Occurs in the seeds of *Alangium lamarckii* (Alangiaceae).

The plant is used in native medicine as a tonic and laxative, vermifuge and antiprotic.

678
Alangimarckine

Skeletal type: tubulosine

$C_{29}H_{37}N_3O_3$ Mol. wt 475.64

Occurs in the leaves of *Alangium lamarckii* (Alangiaceae).

The plant is used in India for skin diseases, and as an anthelmintic, tonic and laxative.

679
Alangiside

Skeletal type: ipecoside

$C_{25}H_{31}NO_{10}$ Mol. wt 505.53

Found in the unripe fruit, leaves and roots of *Alangium lamarckii* (Alangiaceae).

A. lamarckii is a medicinal plant indigenous to India and a rich source of alkaloids, structurally related to the ipecac bases (e.g., ipecoside); the juice of roots is said to be anthelmintic, and a purgative, and is used for dropsical cases.

680
Allocryptopine; β-Homochelidonine; α-Fagarine

Skeletal type: protopine

$C_{21}H_{23}NO_{5}$ Mol. wt 369.42

Occurs occasionally in combination with the β stereoisomer, in a great many Papaveraceae genera (*Bocconia, Chelidonium, Corydalis, Dicentra, Eschscholtzia, Glaucium*, and *Sanguinaria*) and, exceptionally, in *Zanthoxylum* (Rutaceae).

Oxytocic agent.

681
Alpinine; *O*-Methylalpinigenine

Skeletal type: rhoeadine

$C_{23}H_{29}NO_{6}$ Mol. wt 415.49

Occurs together with alpinigenine in *Papaver alpinum* (Papaveraceae), and in other *Papaver* spp.

682
Amurensine; Xanthopetaline

Skeletal type: pavine

$C_{19}H_{19}NO_{4}$ Mol. wt 325.37

Occurs, together with amurensinine (=*O*-methyl ether of amurensine), in several *Papaver* spp., e.g., *P. alpinum* and *P. nudicaule* var. *amurense* (Papaveraceae).

Ameliorates pain; it is also used as an expectorant and as a tranquillizer.

683
Amurine

Skeletal type: morphinan

$C_{19}H_{19}NO_4$ Mol. wt 325.37

Found in *Papaver auranticum* and *P. nudicaule* var. *amurense* (Papaveraceae).

Uses as for amurensine (q.v.).

684
Ancistrocladine

Skeletal type: acetogenic isoquinoline

$C_{25}H_{29}NO_4$ Mol. wt 407.51

Occurs in the leaves, bark and roots of *Ancistrocladus hamatus, A. heyneanus* and *A. congolensis* (Ancistrocladaceae), together with stereoisomers hamatine, ancistrocladinine and ancistrocladisine.

The roots are boiled and used for dysentery in Burma, Siam, and Banka; in Siam, very young leaves are used for flavouring.

685
Angoline; 9-Methoxychelerythrine

Skeletal type: benzophenanthridine

$C_{22}H_{21}NO_5$ Mol. wt 379.42

Found in *Fagara angolensis* (Rutaceae) and other *Fagara* spp., together with nitidine (q.v.), and also in *Bocconia arborea* (Papaveraceae).

686
Ankorine

Skeletal type: ipecoside

$C_{19}H_{29}NO_4$ Mol. wt 335.45

Found in the leaves of *Alangium lamarckii* and *A. kurzii* (Alangiaceae).

The plant has been used in India in the treatment of leprosy and skin diseases; the leaves are used as a poultice to relieve rheumatic pains; the alkaloids have been found to be hypotensive and have interesting cardiovascular properties.

687
Annolobine; Anolobine; Analobine

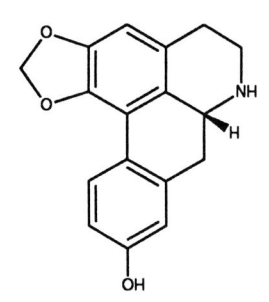

Skeletal type: aporphine

$C_{17}H_{15}NO_3$ Mol. wt 281.31

Occurs in *Asimina triloba* (Annonaceae).

The aporphines display a wide range of pharmacodynamic activities although only a few have commercial use.

688
(−)-**Annonaine**: Anonaine

Skeletal type: aporphine

$C_{17}H_{15}NO_2$ Mol. wt 265.31

Found in *Annona reticulata* (Annonaceae) and *Nelumbo nucifera* (Nymphaeaceae).

Insecticidal and antimicrobial activities.

689
(−)-**Apoglaziovine**

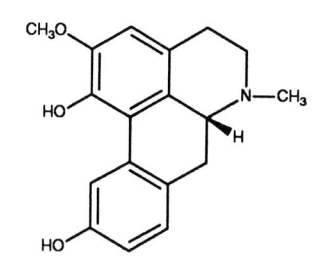

Skeletal type: aporphine

$C_{18}H_{19}NO_3$ Mol. wt 297.36

Found in the leaves of *Ocotea glaziovii* (Lauraceae) together with glaziovine (q.v.). The (+) form is from *O. variabilis* and the (±) form from *O. glaziovii*.

Hypotensive agent and tranquillizer.

690
(−)-**Argemonine**

Skeletal type: pavine

$C_{21}H_{25}NO_4$ Mol. wt 355.44

Found in *Argemone* spp., e.g., *A. mexicana* (Papaveraceae), and in *Thalictrum* spp., e.g., *T. thunbergii*, *T. minus*, *T. rugosum* (Ranunculaceae). The (+) form occurs in *Leontice smirnovii* (Berberidaceae).

Weak pain relieving and antiarrhythmic activities.

691
Armepavine

Skeletal type: benzylisoquinoline

$C_{19}H_{23}NO_3$ Mol. wt 313.40

Occurs in *Papaver caucasicum* and *P. persicum* (Papaveraceae), *Euonymus europaea* (Celastraceae), *Rhamnus frangula* (Rhamnaceae) and *Nelumbo nucifera* (Nymphaeaceae).

Convulsive and irritant; it produces cardiac irregularity.

692
(+)-Aromoline; Thalicrine

Skeletal type: bisbenzylisoquinoline

$C_{36}H_{38}N_2O_6$ Mol. wt 594.71

Occurs in *Daphnandra aromatica* (Monimiaceae) and *Thalictrum thunbergii* and other *Thalictrum* spp. (Ranunculaceae). The (−) form, also called macolidine, occurs in *Abuta grisebachii* (Menispermaceae).

Antimicrobial action; it is a membrane stabilizer. It has a hypotensive action in dogs.

693
Atheroline

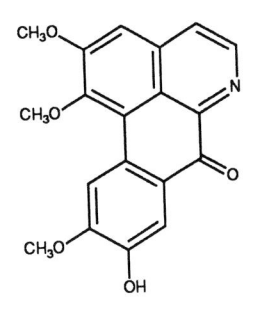

Skeletal type: aporphine

$C_{19}H_{15}NO_5$ Mol. wt 337.34

Found in *Atherosperma moschatum* and in the bark of *Nemuaron vieillardii* (both Monimiaceae).

694
Atherospermidine; Psilopine

Skeletal type: aporphine

$C_{18}H_{11}NO_4$ Mol. wt 305.29

Found in *Atherosperma moschatum* (Monimiaceae) and in *Guatteria psilopus* (Annonaceae).

695
(+)-Atherospermoline

Skeletal type: bisbenzylisoquinoline

$C_{36}H_{38}N_2O_6$ Mol. wt 594.71

Occurs in *Atherosperma moschatum* (Monimiaceae). The (−) form is called krukovine and is from both *Abuta splendida* and *Pycnarrhena longifolia* (Monispermaceae).

696
(+)-Bebeerine; Curine; Chondodendrine

Skeletal type: bisbenzylisoquinoline

$C_{36}H_{38}N_2O_6$ Mol. wt 594.71

Found in the stems and roots of *Chondodendron platyphyllum*, *C. candicans*, and *C. tomentosum* (Menispermaceae), the stems of *Cissampelos pareira* (Menispermaceae), the bark of *Nectandra rodioei* (Lauraceae), and the leaves of *Buxus sempervirens* (Buxaceae). Occurs in the (+) and (−) forms; the racemic form is called hayatine.

Cardiac action; it is an antimalarial.

697
Berbamine; Berbenine

Skeletal type: bisbenzylisoquinoline

$C_{37}H_{40}N_2O_6$ Mol. wt 608.74

Found in *Atherosperma moschatum* (Monimiaceae), *Berberis thunbergii*, *B. vulgaris*, and *Mahonia aquifolium* (Berberidaceae), and *Pycnarrhena manillensis* and *Stephania sasakii* (Menispermaceae).

Tumour inhibiting and antibiotic activity; it is a strong curarizing agent and is highly toxic. It has spasmolytic and vasodilatating action in animal experiments.

698
Berberastine

Skeletal type: protoberberine

$[C_{20}H_{18}NO_5]^+$ Mol. wt 352.37

Occurs in the rhizomes of *Coptis japonica*, the roots of *Hydrastis canadensis*, the stems of *Zanthorrhiza simplicissima* (all Ranunculaceae), and in the roots of *Berberis laurina* (Berberidaceae).

Activity and uses are similar to those of berberine (q.v.).

699
Berberine; Umbellatine

Skeletal type: protoberberine

$[C_{20}H_{18}NO_4]^+$ Mol. wt 336.37

Found in many plant families, including Annonaceae (*Coelocline*), Berberidaceae (*Berberis, Mahonia, Nandina*), Menispermaceae (*Archangelisia*), Papaveraceae (*Argemone, Chelidonium, Corydalis*), Rutaceae (*Evodia, Toddalia, Zanthoxylum*), and Ranunculaceae (*Coptis, Thalictrum*). It is the yellow pigment in the stem bark and leaves of the common barberry, *Berberis vulgaris*.

Moderately toxic (LD_{50} in humans 27.5 mg kg^{-1}), causing cardiac damage, dyspnoea, and lowered blood pressure; it is used as a bitter stomachic, antimalarial, antimicrobial, antipyretic, anthelmintic and cytotoxic agent.

700
(+)-Bicuculline

Skeletal type: phthalideisoquinoline

$C_{20}H_{17}NO_6$ Mol. wt 367.36

Occurs in the entire plant of *Adlumia fungosa, Corydalis thalictrifolia* and *C. incisa* (Papaveraceae). The diastereoisomer adlumidine also occurs in the same plants.

Effective γ-aminobutyric acid antagonist and is widely used in neurological research.

701
Bocconine; Chelirubine

Skeletal type: benzophenanthridine

$[C_{21}H_{16}NO_5]^+$ Mol. wt 362.37

Occurs in *Bocconia cordata* (Papaveraceae), and is widely present in Papaveraceae and Fumariaceae.

Nematocidal activity; it is also used as a local anaesthetic.

702
Boldine

Skeletal type: aporphine

$C_{19}H_{21}NO_4$ Mol. wt 327.38

Found in the leaves of *Boldea fragrans, Peumus boldus*, and *Monimia rotundifolia* (Monimiaceae), and in *Litsea laurifolia, L. turfosa* and *Sassafras albidem* (Lauraceae). Also recorded in the Annonaceae (*Desmos*), Magnoliaceae (*Liriodendron*), Rhamnaceae (*Retanilla*) and Atherospermataceae (*Laurelia*).

Ingredient in choleretics and laxatives, and used in cases of hepatic dysfunction and cholelithiasis.

703
Bracteoline

Skeletal type: aporphine

$C_{19}H_{21}NO_4$ Mol. wt 327.38

Found in *Papaver bracteatum* and *P. orientale* (Papaveraceae).

704
Bulbocapnine; *N*-Methyl-launobine

Skeletal type: aporphine

$C_{19}H_{19}NO_4$ Mol. wt 325.37

Found in most *Corydalis* spp., e.g., *C. bulbosa, C. cava, C. decumbens*, and *C. solida*, and in *Fumaria officinalis, Glaucium flavum*, and *G. pulchrum* (all Papaveraceae).

Causes catalepsy and sedative effects, and also potentiation of some hypnotics such as pentobarbital.

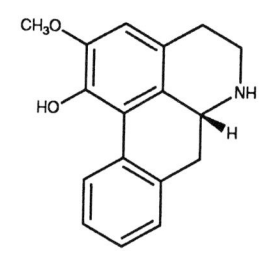

705
(−)-Caaverine

Skeletal type: aporphine

$C_{17}H_{17}NO_2$ Mol. wt 267.33

Occurs in *Symplocos celastrinea* (Symplocaceae), *Liriodendron tulipifera* (Magnoliaceae), *Isolona pilosa* (Annonaceae), and *Ocotea glaziovii* (Lauraceae).

Extracts of the bark of the trees growing in Brazil and North Paraguay are toxic to animals.

706
Calafatimine

Skeletal type: bisbenzylisoquinoline

$C_{38}H_{40}N_2O_7$ Mol. wt 636.75

Found in the roots of *Berberis buxifolia* (Berberidaceae), together with the structurally related alkaloid calafatine.

Weak tumour-inhibiting activity compared with other bisbenzylisoquinolines.

707
(−)-Canadine

Skeletal type: protoberberine

$C_{20}H_{21}NO_4$ Mol. wt 339.40

(+)-Canadine occurs in several *Corydalis* spp., e.g., *C. cava*; (−)-canadine occurs in *C. cheilantheifolia* (Papaveraceae), *Fagara rhoifolia* (Rutaceae), *Hydrastis canadensis* (Ranunculaceae), and *Zanthoxylum brachyacanthum* and *Z. veneficium* (Rutaceae).

Hypotensive activity.

708
Cancentrine

Skeletal type: morphinan

$C_{36}H_{33}N_2O_7$ Mol. wt 605.67

Yellow base found in *Dicentra canadensis* (Fumariaceae).

709
Capaurine; Capaurimine 10-methyl ether

Skeletal type: protoberberine

$C_{21}H_{25}NO_5$ Mol. wt 371.44

Found in *Corydalis aurea, C. micrantha, C. montana*, and *C. pallida* (Fumariaceae).

Uterine stimulant activity.

710
(−)-Carnegine; Pectenine

Skeletal type: simple isoquinoline

$C_{13}H_{19}NO_2$ Mol. wt 221.30

Found in *Carnegiea gigantea* and *Cereus pectenaboriginum* (Cactaceae).

Provokes convulsions in warm-blooded animals.

711
Caseadine

Skeletal type: protoberberine

C$_{20}$H$_{23}$NO$_4$ Mol. wt 341.41

Found, together with caseamine and caseanadine, in *Corydalis caseana* (Fumariaceae).

712
Cassyfiline; Cassythine

Skeletal type: aporphine

C$_{19}$H$_{19}$NO$_5$ Mol. wt 341.37

Occurs in the stems of *Cassytha filiformis* (Lauraceae).

Tetanic action on animals.

713
(+)-Cassythicine; *N*-Methylactinodaphnine

Skeletal type: aporphine

C$_{19}$H$_{19}$NO$_4$ Mol. wt 325.37

Found in *Cassytha melantha* and *C. glabella* (Lauraceae); the (−) form is found in *Annona glabra* (Annonaceae).

Antimicrobial and cytotoxic activities.

714
Cephaeline

Skeletal type: emetine

C$_{28}$H$_{38}$N$_2$O$_4$ Mol. wt 466.63

After emetine (q.v.) this is the most important alkaloid in the roots of *Cephaelis ipecacuanha* (Rubiaceae). It is also found in *Alangium lamarekii* (Alangiaceae).

Emetic, expectorant and anti-amoebic activities; the roots are used medicinally in South Brazil.

715
Cepharanthine

Skeletal type: bisbenzylisoquinoline

$C_{37}H_{38}N_2O_6$ Mol. wt 606.73

Occurs in the roots of *Stephania cepharantha* and *S. sasakii* (Menispermaceae).

Effective against human tuberculosis and leprosy, although it has now been superseded by synthetic drugs.

716
Codeine; 3-*O*-Methylmorphine

Skeletal type: morphinan

$C_{18}H_{21}NO_3$ Mol. wt 299.37

Occurs in opium, the dried latex of *Papaver somniferum* (Papaveraceae).

Spasmolytic, narcotic analgesic, and antitussive. It is used extensively for pain, diarrhoea, and as a cough suppressant.

717
Cularicine

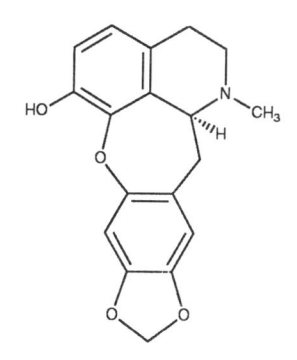

Skeletal type: cularine

$C_{18}H_{17}NO_4$ Mol. wt 311.34

Occurs in *Corydalis clariculata* (Fumariaceae).

Cytotoxic.

718
Cularidine; O^7-Demethylcularine

Skeletal type: cularine

$C_{19}H_{21}NO_4$ Mol. wt 327.38

Found in *Corydalis claviculata* and *Dicentra cucullaria* (Fumariaceae).

Cytotoxic.

719
Cularimine; Alkaloid F30

Skeletal type: cularine

$C_{19}H_{21}NO_4$ Mol. wt 327.38

Found in *Dicentra eximia* (Fumariaceae).

Antineoplastic.

720
Cularine

Skeletal type: cularine

$C_{20}H_{23}NO_4$ Mol. wt 341.41

Occurs in *Berberis valdiviana* (Berberidaceae), and in *Corydalis claviculata* and *Dicentra eximia* (Fumariaceae).

Stimulates the uterus, and increases heart tone and contractility. It is hypotensive in rabbits, and anaesthetic to rabbit cornea. Also, it shows cytotoxic activity.

721
Daphnandrine; *O*-Methyldaphnoline

Skeletal type: bisbenzylisoquinoline

$C_{36}H_{38}N_2O_6$ Mol. wt 594.71

Found in the bark of *Daphnandra micrantha*, and *Doryphora aromatica* (Monimiaceae).

Antiprotozoal activity *in vitro*.

722
Daphnoline; Trilobamine

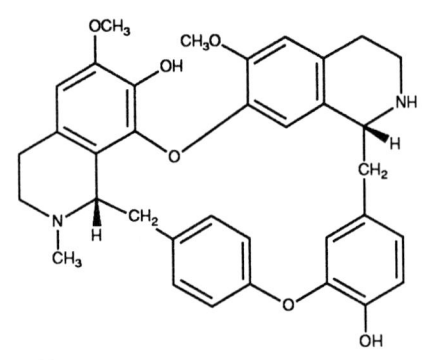

Skeletal type: bisbenzylisoquinoline

$C_{35}H_{36}N_2O_6$ Mol. wt 580.69

Found in *Daphnandra micrantha* (Monimiaceae), and in *Cocculus trilobus* (Menispermaceae).

Vasodilator, oedematous agent, central nervous system depressant and respiratory paralytic.

723
Dauricine

Skeletal type: bisbenzylisoquinoline

$C_{38}H_{44}N_2O_6$ Mol. wt 624.78

Occurs in *Menispermum dauricum, M. canadense*, and probably other Menispermaceous plants.

Anaesthetic and anti-inflammatory activities in animals. It is a weak curarizing agent, and is toxic (LD_{50} intraperitoneally in mice 6 mg kg^{-1}).

724
(+)-Demethylcoclaurine

Skeletal type: bisbenzylisoquinoline

$C_{16}H_{17}NO_3$ Mol. wt 271.32

Found in *Aconitum japonicum* (Ranunculaceae). The racemic form is known as higenamine.

Cardiac stimulant.

725
Dihydrosanguinarine

Skeletal type: benzophenanthridine

$C_{20}H_{15}NO_4$ Mol. wt 333.35

Found in *Fumaria parviflora, F. vaillantii, Corydalis gigantea, C. ledebouriana*, and other Fumariaceae, in *Eschscholzia californica* (Papaveraceae) and in *Pteridophyllum* spp. (Sapindaceae).

Activity and uses similar to those of sanguinarine (q.v.). Many *Fumaria* and *Corydalis* spp. are used as antiseptics and anti-inflammatory agents.

726
Emetamine

Skeletal type: emetine

$C_{29}H_{36}N_2O_4$ Mol. wt 476.62

Found in the ipecacuanha, *Cephaelis* (=*Uragoga*) *ipecacuanha* and *C. acuminata* (Rubiaceae).

Ipecacuanha is used as an expectorant, emetic and amoebicide; however, these effects are all attributable to emetine (q.v.).

727
Emetine; Cephaeline methyl ether

Skeletal type: emetine

$C_{29}H_{40}N_2O_4$ Mol. wt 480.65

Occurs in all varieties of ipecacuanha, from *Cephaelis* (=*Uragoga*) *ipecacuanha*, and *C. acuminata* (Rubiaceae).

Expectorant, emetic, antiamoebic, cytotoxic. Ipecacuanha extracts are used widely in cough mixtures, and in higher doses, in cases of poisoning and overdosage, to produce vomiting. Emetine is used to treat amoebic dysentery, despite the gastrointestinal effects. It has antiviral and anticancer activity, probably due to its inhibition of DNA and other protein synthesis, but has not yet proved useful clinically for treating advanced carcinomas. The lethal dose is about one gram in humans; it is a cumulative poison over short periods.

728
Erysonine

Skeletal type: erythrina

$C_{17}H_{19}NO_3$ Mol. wt 285.34

Occurs in many *Erythrina* spp., including *E. caribea* and *E. melanacantha* (Leguminosae).

Neuromuscular blocking agent.

729
Erysotrine

Skeletal type: erythrina

$C_{19}H_{23}NO_3$ Mol. wt 313.40

Found in *Erythrina suberosa* and many other *Erythrina* spp. (Leguminosae).

Neuromuscular blocking agent. The leaves and bark of *Erythrina suberosa* have antitumour activity.

730
Erythratidine

Skeletal type: erythrina

$C_{19}H_{25}NO_4$ Mol. wt 331.41

Occurs in many *Erythrina* spp., including *E. caribea* and *E. melanacantha* (Leguminosae).

Curare-like neuromuscular blocking agent.

731
α-Erythroidine

Skeletal type: erythrina

$C_{16}H_{19}NO_3$ Mol. wt 273.34

Found in many *Erythrina* spp. (Leguminosae).

Curare-like neuromuscular blocking agent.

732
β-Erythroidine

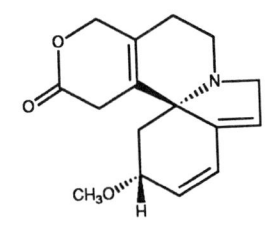

Skeletal type: erythrina

$C_{16}H_{19}NO_3$ Mol. wt 273.34

Occurs in many *Erythrina* spp. (Leguminosae).

Orally active neuromuscular blocking agent, hypnotic, respiratory depressant and hypotensive. Dehydro-β-erythroidine is approximately five times more potent. Both have been used as curare substitutes. It is toxic (LD_{50} intraperitoneally in mice 29.5 mg kg^{-1}).

733
Fagaridine

Skeletal type: benzophenanthridine

$[C_{20}H_{16}NO_4]^+$ Mol. wt 334.36

Found in *Zanthoxylum* (=*Fagara*) spp., including *Z. tessmanii* (Rutaceae).

734
Fagaronine

Skeletal type: benzophenanthridine

$[C_{21}H_{20}NO_4]^+$ Mol. wt 350.4

Occurs in the roots of *Zanthoxylum zanthoxyloides* (=*Fagara xanthoxylum*) and other Rutaceae spp.

Antitumour activity. It inhibits reverse transcriptase activity of various RNA oncogenic viruses. Also, it is bactericidal.

735
Fetidine; Foetidine

Skeletal type: bisbenzylisoquinoline

$C_{40}H_{46}N_2O_8$ Mol. wt 682.82

Found in the aerial parts of *Thalictrum foetidum* (Ranunculaceae).

Hypotensive, and anti-inflammatory. It depresses nervous activity in mice.

736
Fumaricine

Skeletal type: benzylisoquinoline

$C_{21}H_{23}NO_5$ Mol. wt 369.42

Found in fumitory, *Fumaria officinalis* (Fumariaceae).

Fumaria officinalis is used as a herbal medicine to treat skin infections, liver complaints, and inflammation, and as a tonic.

737
Gigantine

Skeletal type: simple isoquinoline

$C_{13}H_{19}NO_3$ Mol. wt 237.30

Occurs in the giant or saguaro cactus, *Carnegeia gigantea* (Cactaceae).

Thought to be hallucinogenic in animals, but unconfirmed in humans.

738
Glaucine; Boldine dimethyl ether

Skeletal type: aporphine

$C_{21}H_{25}NO_4$ Mol. wt 355.44

Found in 11 plant families, including *Glaucium flavum* (Papaveraceae), *Dicentra eximia* and *Corydalis ambigua* (Fumariaceae), and *Beilschmiedia podagrica* (Lauraceae).

Antitussive. It inhibits respiration, lowers blood pressure and blood glucose levels in animals. It is toxic (LD_{50} intravenously 4.8 mg kg^{-1}).

739
Glaziovine

Skeletal type: aporphine

$C_{18}H_{19}NO_3$ Mol. wt 297.36

Found in *Ocotoea glaziovii* (Lauraceae), and in *Meconopsis cambrica* (Papaveraceae).

Antidepressant. It shows weak anticancer activity.

740
Gyrocarpine

Skeletal type: bisbenzylisoquinoline

$C_{37}H_{40}N_2O_6$ Mol. wt 608.74

Found in the stem bark of *Gyrocarpus americanus* (Hernandiaceae).

Antitrypanosomal against *Leishmania* spp.

741
Hasubanonine; *O*-Methylaknadinine

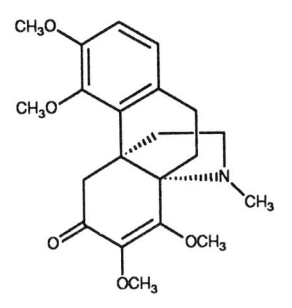

Skeletal type: hasubanan

$C_{21}H_{27}NO_5$ Mol. wt 373.45

Found in *Stephania japonica* (Menispermaceae).

742
Heliamine

Skeletal type: simple isoquinoline

$C_{11}H_{15}NO_2$ Mol. wt 193.25

Occurs in cacti *Pachycereus weberi*, *P. pringlei*, *P. pecten-aboriginum*, and *Carnegeia gigantea* (Cactaceae).

Inhibits the growth of sarcoma 45 in rats.

743
Hernandezine; Thalicsimine; Thaliximine

Skeletal type: bisbenzylisoquinoline

$C_{39}H_{44}N_2O_7$ Mol. wt 652.80

Found in *Stephania hernandiifolia* (Menispermaceae), and *Thalictrum simplex* (Ranunculaceae).

Strong anti-inflammatory activity. In rats, it inhibits conditioned avoidance reactions and reflexes associated with movement and eating. *Stephania hernandiifolia* has been used as a fish poison.

744
(−)-α-Hydrastine

Skeletal type: phthalideisoquinoline

$C_{21}H_{21}NO_6$ Mol. wt 383.41

Occurs in golden seal, *Hydrastis canadensis* (Ranunculaceae).

Haemostatic, and antiseptic. Formerly, it has been used to treat uterine haemorrhage.

745
Ipecoside

Skeletal type: terpenoisoquinoline

$C_{27}H_{35}NO_{12}$ Mol. wt 565.58

Found in ipecacuanha, from *Cephaelis* (=*Uragoga*) *ipecacuanha*, and *C. acuminata* (Rubiaceae).

746
Isoboldine; Aurotensine

Skeletal type: aporphine

$C_{19}H_{21}NO_4$ Mol. wt 327.38

Found in 13 plant families, including *Corydalis cava* and other *Corydalis* spp., *Glaucium flavum* and other *Glaucium* spp. (Papaveraceae), *Peumus boldo* (Monimiaceae), and *Litsea glutinosa* var. *glabraria* (Lauraceae).

Insect feeding inhibitor. *Peumus boldo* is used as a choleretic and slimming aid.

747
Isochondrodendrine; Isobeberine

Skeletal type: bisbenzylisoquinoline

$C_{36}H_{38}N_2O_6$ Mol. wt 594.71

Found in *Chondrodendron tomentosum*, pareira, and *Sciadotenia toxifera* (Menispermaceae), in *Guatteria megalophylla* (Annonaceae), and in *Heracleum wallichi* (Umbelliferae).

Muscle relaxant. The juice of plants containing it are constituents of "curare", the arrow poisons used in South America.

748
Isococculidine; *O*-Methylisococculine

Skeletal type: erythrina

$C_{18}H_{23}NO_2$ Mol. wt 285.39

Found in the leaves of *Cocculus laurifolius* (Menispermaceae).

Neuromuscular blocking agent.

749
Isocorydine; Artabotrine; Luteanine

Skeletal type: aporphine

$C_{20}H_{23}NO_4$ Mol. wt 341.41

Found in 12 plant families, including *Corydalis cava* (Fumariaceae), *Arbabotrys suaveolens* (Annonaceae), several *Papaver* and *Glaucium* spp. (Papaveraceae), and *Phoebe clemensii* (Lauraceae).

Anti-adrenergic. It is a sedative in animals, and cataleptic at high doses. It is toxic (LD_{50} intraperitoneally in rats 10.9 mg kg^{-1}).

750
Isothebaine; 1-Hydroxy-2,11-dimethoxyaporphine

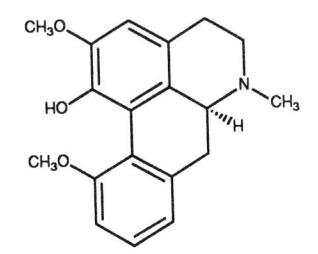

Skeletal type: aporphine

$C_{19}H_{21}NO_3$ Mol. wt 311.38

Found in *Papaver orientale* and *P. pseudo-orientale* (Papaveraceae).

Depresses respiration and heart rate, and decreases motor activity. It is anti-inflammatory and analgesic in animals. It is toxic (LD_{50} in humans 26 mg kg^{-1}).

751
Jatrorrhizine; Jateorrhizine; Neprotine

Skeletal type: berberine

$[C_{20}H_{20}NO_4]^+$ Mol. wt 338.39

Occurs in the calumba root, *Jateorrhiza palmata* (Menispermaceae), in many *Berberis* and *Mahonia* spp. (Berberidaceae), in *Thalictrum* spp. (Ranunculaceae), and in *Michelia* spp. (Magnoliaceae).

Antifungal and antiprotozoal. It is a sedative and hypotensive in animals.

752
(−)-Laudanidine; Laudanine

Skeletal type: benzylisoquinoline

$C_{20}H_{25}NO_4$ Mol. wt 343.43

The (−) form occurs in opium, from *Papaver somniferum* (Papaveraceae). The (+) form occurs in *Machilus obovatifolia* (Lauraceae) and in *Thalictrum dasycarpum* (Ranunculaceae). The (±) form, called laudanine, is in *P. somniferum* and also in *Xylopia pancheri* (Annonaceae).

Toxic. In animals, large doses have a strychnine-like effect, causing convulsions and paralysis.

753
Laudanosine; Laudanine methyl ether

Skeletal type: benzylisoquinoline

$C_{21}H_{27}NO_4$ Mol. wt 357.45

Found in opium, from *Papaver somniferum* (Papaveraceae).

Toxic; strong tetanic poison. It depresses blood pressure and causes convulsions in animals.

754
Liriodenine; Spermatheridine

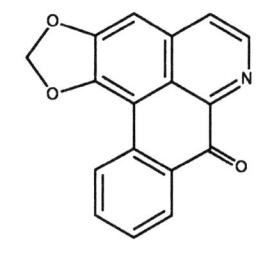

Skeletal type: benzylisoquinoline

$C_{17}H_9NO_3$ Mol. wt 275.27

Occurs widely in 13 plant families, including *Liriodendron tulipifera*, and *Magnolia obovata* (Magnoliaceae).

Cytotoxic to human nasopharyngeal carcinoma cells *in vitro*. It is also antifungal.

755
Lophocerine

Skeletal type: simple isoquinoline

$C_{15}H_{23}NO_2$ Mol. wt 249.36

Found in *Lophocereus schottii* (Cactaceae).

756
Lophophorine; *N*-Methylanhalonine

Skeletal type: simple isoquinoline

$C_{13}H_{17}NO_3$ Mol. wt 235.29

Occurs in peyote (mescal buttons), *Lophophora williamsii* (Cactaceae).

Respiratory stimulant and convulsive agent. It is toxic (LD_{50} intravenously in rabbits 15–20 mg kg^{-1}).

757
(+)-Luguine

Skeletal type: benzophenanthridine

$C_{19}H_{13}NO_5$ Mol. wt 335.32

Found in the roots of *Glaucium flavum* var. *vestitum* (Papaveraceae).

758
Macarpine

Skeletal type: benzophenanthridine

$[C_{22}H_{18}NO_6]^+$ Mol. wt 392.39

Found in *Macleaya cordata*, *Eschscholtzia douglasii* and *Stylophorum diphyllum* (Papaveraceae).

759
Macoline

Skeletal type: bisbenzylisoquinoline

$[C_{37}H_{41}N_2O_6]^+$ Mol. wt 609.75

Occurs in the stem wood of *Abuta grisebachii* (Menispermaceae).

Muscle relaxant. The juice of the plant was mixed into a type of curare by South American Indians, and used for hunting.

760
Magnoflorine; Thalictrine; Escholine; Corytuberine

Skeletal type: aporphine

$[C_{20}H_{24}NO_4]^+$ Mol. wt 342.42

The most widespread aporphine alkaloid, found in many Papaveraceae and Fumariaceae spp., e.g. *Argemone, Papaver, Eschscholzia, Glaucium, Meconopsis, Dicranostigma, Chelidonium* and *Corydalis*; it is also found in *Magnolia grandiflora* (Magnoliaceae), most *Zanthoxylum* spp. (Rutaceae), *Thalictrum thunbergii* (Ranunculaceae), *Aristolochia thunbergii* (Aristolochiaceae), and *Croton cumingii* (Euphorbiaceae).

Weak neuromuscular blocking agent. It is a hypotensive in rodents.

761
Mecambrine; Fugapavine

Skeletal type: proaporphine

$C_{18}H_{17}NO_3$ Mol. wt 295.34

Occurs in *Papaver fugax, P. dubium* and *Meconopsis cambrica* (Papaveraceae).

Increases blood pressure, stimulates respiration slightly and produces bradycardia in animals. It is weakly anti-inflammatory. Large doses cause convulsions, and it is toxic. (LD_{50} in mice 4.1 mg kg^{-1}).

762
(+)-Mecambroline; Isofugapavine

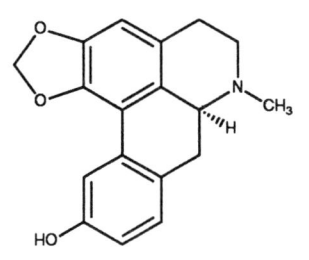

Skeletal type: aporphine

$C_{18}H_{17}NO_3$ Mol. wt 295.34

Occurs in the Welsh poppy, *Meconopsis cambrica*, and in *Papaver fugax* (Papaveraceae). The (−) form is found in *Phoebe chemensii* (Lauraceae).

Hypotensive in rodents.

763
(+)-O-Methylthalicberine; Thalmidine

Skeletal type: bisbenzylisoquinoline

$C_{38}H_{42}N_2O_6$ Mol. wt 622.77

Found in the stems, leaves and roots of *Thalictrum* spp., including *T. minus* (Ranunculaceae), and in *Berberis laurina* (Berberidaceae).

Weak antitumour agent. It is an anti-inflammatory in animals.

764
Morphine; Morphia

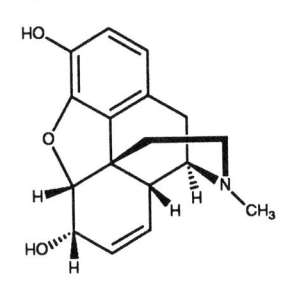

Skeletal type: morphinan

$C_{17}H_{19}NO_3$ Mol. wt 285.35

Found in opium, from *Papaver somniferum* (Papaveraceae).

Powerful analgesic, spasmolytic, narcotic, sedative, gastric sedative, and antitussive. Prolonged use leads to habituation. It is used extensively for pain relief, especially in terminal care. The lethal dose in humans lies between 1 and 10 mg. The drug of addiction is the diacetate, heroin.

765
Narceine

Skeletal type: ring-opened isoquinoline

$C_{23}H_{27}NO_8$ Mol. wt 445.48

Found in opium, from *Papaver somniferum* (Papaveraceae).

Antitussive like codeine (q.v.), but with no analgesic activity. It stimulates respiration, lowers blood pressure and stimulates intestinal peristalsis in animals.

766
α-Narcotine; (−)-α-Narcotine; Noscapine; Narcosine; Methoxyhydrastine; Opianine

Skeletal type: phthalideisoquinoline

$C_{22}H_{23}NO_7$ Mol. wt 413.43

Found in opium, from *Papaver somniferum* (Papaveraceae). It readily epimerizes to the artefact (−)-β-narcotine during its isolation.

Antitussive, and spasmolytic.

767
Narcotoline; Desmethylnarcotine

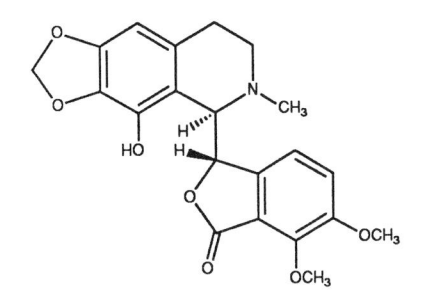

Skeletal type: phthalideisoquinoline

$C_{21}H_{21}NO_7$ Mol. wt 399.41

Found in opium, from *Papaver somniferum* (Papaveraceae).

Respiratory stimulant, and spasmolytic.

768
Neopine; β-Codeine

Skeletal type: morphinan

$C_{18}H_{21}NO_3$ Mol. wt 299.37

Found in opium, from *Papaver somniferum* (Papaveraceae); it is also found in *P. bracteatum*.

Activity and uses similar to those of codeine (q.v.), analgesic, and spasmolytic.

769
Nitidine; Angolinine

Skeletal type: benzophenanthridine

$[C_{21}H_{18}NO_4]^+$ Mol. wt 348.38

Found in prickly ash bark, *Zanthoxylum americanum*, and in *Z. clava-herculis* (Rutaceae), and occurs widely in *Zanthoxylum* and *Fagara* spp.

Antitumour properties, but too toxic for clinical use. It inhibits reverse transcriptase activity of various RNA oncogenic viruses.

770
Obaberine; O-Methyloxyacanthine

Skeletal type: bisbenzylisoquinoline

$C_{38}H_{42}N_2O_6$ Mol. wt 622.77

Occurs in *Berberis* spp. (Berberidaceae), *Thalictrum lucidum* (Ranunculaceae), and *Albertisia papuana* (Lauraceae).

Antitrypanosomal against *Leishmania* spp.

771
Ochotensine

Skeletal type: spirobenzylisoquinoline

$C_{21}H_{21}NO_4$ Mol. wt 351.41

Found in *Corydalis ochotensis* and *C. sibirica* (Fumariaceae).

772
Oxyacanthine; Vinetine

Skeletal type: bisbenzylisoquinoline

$C_{37}H_{40}N_2O_6$ Mol. wt 608.74

Found in the roots of *Berberis vulgaris*, and *Mahonia acanthifolia* (Berberidaceae), and also reported in the Menispermaceae, Magnoliaceae and Ranunculaceae.

Antimicrobial against *Bacillus subtilis* and *Colpidium colpoda*. It is an adrenaline antagonist and a vasodilator.

773
Palmatine; Calystigine

Skeletal type: benzophenanthridine

$[C_{21}H_{22}NO_4]^+$ Mol. wt 352.41

Occurs in calumba root, *Jateorrhiza palmata* (Menispermaceae), and is widely distributed in the Berberidaceae (in most *Berberis* and *Mahonia* spp.) and the Papaveraceae.

Antiarrhythmic, positively inotropic, adrenocorticotrophic, anticholinesterase, analgesic and bactericidal.

774
Papaverine

Skeletal type: benzylisoquinoline

$C_{20}H_{21}NO_4$ Mol. wt 339.40

Found in opium, *Papaver somniferum* (Papaveraceae), and in *Rauwolfia serpentina* (Apocynaceae).

Smooth muscle relaxant, cerebral vasodilator, and antitussive. It is used extensively as a spasmolytic and in cough medicines.

775
(−)-Pellotine; *N*-Methylanhalonidine; Peyotline

Skeletal type: simple isoquinoline

$C_{13}H_{19}NO_3$ Mol. wt 237.30

Found in peyote, pellote or mescal buttons, *Lophophora williamsii* (Cactaceae).

Peyote is psychotomimetic and hallucinogenic, due mainly to the presence of mescaline (q.v.) and *N*-methylmescaline.

776
Phaeantharine

Skeletal type: bisbenzylisoquinoline

$[C_{39}H_{40}N_2O_6]^{2+}$ Mol. wt 632.76

Found in the bark of *Phaeanthus ebracteolatus* (Annonaceae).

Anticancer activity in animals.

777
Pilocereine

Skeletal type: isoquinoline trimer

$C_{45}H_{65}N_3O_6$ Mol. wt 744.04

Occurs in *Cephalocereus* (=*Pilocereus*) *sargentianus* and *Lophocereus schotti* (Cactaceae).

778
(+)-Pronuciferine; Miltanthin

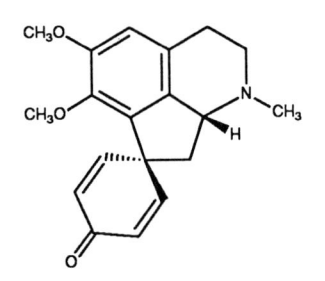

Skeletal type: proaporphine

$C_{19}H_{21}NO_3$ Mol. wt 311.38

Wide occurrence, e.g., several *Papaver* spp. (Papaveraceae), *Croton linearis* (Euphorbiaceae), *Stephania glabra* (Menispermaceae), and the "sacred lotus", *Nelumbo nucifera* (Nelumbonaceae).

Mild local anaesthetic activity; it has a synergistic effect with acetylcholine on guinea-pig ileum.

779
Protopine; Fumarine; Macleyine; Corydalis C; Corydinine; Biflorine

Skeletal type: benzylisoquinoline

$C_{20}H_{19}NO_5$ Mol. wt 353.38

Widespread occurrence in the Papaveraceae and Fumariaceae, including *Papaver somniferum*, the opium poppy, *Fumaria officinalis*, fumitory, and *Argemone mexicana*, the prickly poppy.

Smooth muscle relaxant and sedative in animals. It shows bactericidal activity against Gram-positive bacteria.

780
Psychotrine

Skeletal type: emetine

$C_{28}H_{36}N_2O_4$ Mol. wt 464.61

Occurs in ipecacuanha root, *Cephaelis* (=*Uragoga*) *ipecacuanha*, and *C. acuminata* (Rubiaceae); and in the seed, root bark and stem bark of *Alangium lamarckii* (Alangiaceae).

Ipecacuanha is used medicinally for its expectorant, emetic and anti-amoebic activities, attributed mainly to emetine (q.v.).

781
Pukateine; *N*-Methyloboranine

Skeletal type: aporphine

$C_{18}H_{17}NO_3$ Mol. wt 295.34

Found in the bark of *Laurelia novaezelandiae* (Lauraceae).

782
(+)-Reticuline; Coclanoline

Skeletal type: benzylisoquinoline

$C_{19}H_{23}NO_4$ Mol. wt 329.40

Found in opium, the dried latex of *Papaver somniferum*, and other *Papaver* spp., in *Annona glabra* (Annonaceae), and in *Cryptocarpa odorata* (Lauraceae).

The (−) form is found in *Romneya coulteri* (Papaveraceae).

Precursor of many aporphine and morphinane alkaloids.

783
Rhoeadine; Rheadine

Skeletal type: rhoeadan

$C_{21}H_{21}NO_6$ Mol. wt 383.40

Occurs in the seed capsules of the corn poppy, *Papaver rhoeas* (Papaveraceae), and is present in other *Papaver* spp.

Slightly cytotoxic to ascites tumour cells *in vitro*. Large doses cause spasms in animals. *Papaver rhoeas* is used as a sedative and mild expectorant in folk medicine. It is toxic (LD_{50} intraperitoneally in rats 530 mg kg^{-1}).

784
Rodiasine; 6'-O-Methylphlebicine

Skeletal type: bisbenzylisoquinoline

$C_{38}H_{42}N_2O_6$ Mol. wt 622.77

Found in the bark and seeds of *Ocotea venenosa* (Lauraceae).

Neuromuscular blocking agent; it is a muscle-relaxant similar to (+)-tubocurarine (q.v.). The plant has been used as a curare ingredient in South American arrow poisons.

785
(−)-**Salsoline**

Skeletal type: simple isoquinoline

$C_{11}H_{15}NO_2$ Mol. wt 193.25

Occurs in *Salsola richteri* (Chenopodiaceae).

Lethal to mice when given intravenously, but does not appear to be toxic orally.

786
(−)-**Salsolinol**

Skeletal type: simple isoquinoline

$C_{10}H_{13}NO_2$ Mol. wt 179.22

Occurs in banana fruit, *Musa paradisiaca* (Musaceae), and in cocoa powder, *Theobroma cacao* (Sterculiaceae).

Dopamine antagonist *in vivo* (in cell culture and in radioreceptor assays).

787
(+)-**Salutaridine**; Floripavine

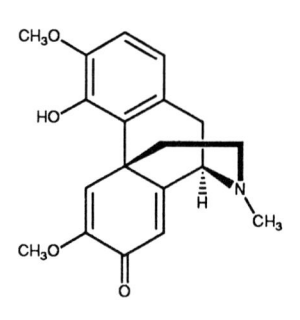

Skeletal type: morphinan

$C_{19}H_{21}NO_4$ Mol. wt 327.38

Occurs in many *Papaver* spp., including *P. somniferum*, *P. orientale*, and *P. bracteatum* (Papaveraceae), and in *Croton* spp., including *C. salutaris* and *C. balsamifera* (Euphorbiaceae). The (−) isomer is called sinoacutine.

An important intermediate in the biosynthesis of morphine (q.v.). It shows antitumour activity against Walker 256 carcinosarcoma.

788
Sanguinarine; Pseudochelerythrine

Skeletal type: benzophenanthridine

$[C_{20}H_{14}NO_4]^+$ Mol. wt 332.34

Widespread occurrence in the Papaveraceae, including opium from *Papaver somniferum*, *Dicentra spectabilis*, *D. peregrina* and *Chelidonium majus*, and in the Fumariaceae, including bloodroot, *Sanguinaria canadensis*, and fumitory, *Fumaria officinalis*. It is also found in *Zanthoxylum* spp. (Rutaceae), and *Pteridophyllum* spp. (Sapindaceae).

Antibacterial, cytotoxic, and anti-inflammatory. A major use is in dentrifices and mouthwashes because of its antiplaque activity. It has a positive inotropic effect on the heart. It inhibits various enzymes including ATP-ase, diamine oxidase and some aminotransferases. It causes glaucoma at high doses over a prolonged period. Also, it is toxic (LD_{50} in mice 19.4 mg kg^{-1}).

789
Sinomenine; Cucoline

Skeletal type: morphinan

$C_{19}H_{23}NO_4$ Mol. wt 329.40

Found in *Sinomenium acutum* (Menispermaceae).

Antitussive; it is also weakly analgesic. It is an abortifacient in large doses.

790
(+)-Tetrandrine

Skeletal type: bisbenzylisoquinoline

$C_{38}H_{42}N_2O_6$ Mol. wt 622.77

Found in *Stephania tetrandra*, *S. discolor*, *Cyclea peltata*, and *Cissampelos pareira* (Menispermaceae). The (−) isomer is called phaeanthine.

Analgesic, antipyretic, and anti-inflammatory. *Cissampelos pareira* is used as a native remedy in Uruguay for fertility control and snakebite.

791
Thalicarpine; Thaliblastine

Skeletal type: bisbenzylisoquinoline

$C_{41}H_{48}N_2O_8$ Mol. wt 696.85

Found in *Thalictrum dasycarpum*, *T. flavum*, and *T. polygamum* (Ranunculaceae).

Hypotensive, vasodilatory, antimicrobial, and antitumour. It is toxic (LD_{50} intravenously in mice 58.6 mg kg^{-1}).

792
Thalidasine

Skeletal type: bisbenzylisoquinoline

$C_{39}H_{44}N_2O_7$ Mol. wt 652.80

Found in *Thalictrum dasycarpum* (Ranunculaceae).

Antimicrobial: active *in vivo* against *Mycobacterium smegmatis*.

793
Thalmine

Skeletal type: bisbenzylisoquinoline

$C_{37}H_{40}N_2O_6$ Mol. wt 608.74

Occurs in *Thalictrum* spp. (Ranunculaceae).

Anti-inflammatory, in several experimental models of inflammation, in animals. Antitumour activity is shown against ascites lymphoma in rats and mice.

794
Thalsimine

Skeletal type: bisbenzylisoquinoline

$C_{38}H_{40}N_2O_7$ Mol. wt 636.75

Found in *Thalictrum simplex* and *T. rugosum* (Ranunculaceae).

Affects the nervous system: inhibits conditioned avoidance reactions and reflexes associated with eating and movement in rats, and reduces the time taken for dogs to run through a labyrinth. It shows weak antitumour activity.

795
Thebaine; Paramorphine

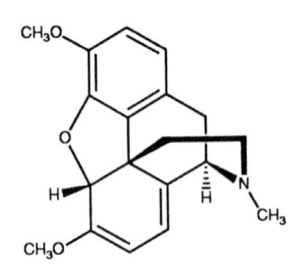

Skeletal type: morphinan

$C_{19}H_{21}NO_3$ Mol. wt 311.38

Minor constituent of opium from *Papaver somniferum*; it occurs in some strains of *P. bracteatum* (Papaveraceae).

Weak narcotic and analgesic, more toxic than morphine (q.v.). Large doses causes convulsions.

796
Tiliacorine

Skeletal type: bisbenzylisoquinoline

$C_{36}H_{36}N_2O_5$ Mol. wt 576.70

Found in the bark and root of *Tiliacora acuminata* and *T. racemosa* (Menispermaceae).

Antimalarial activity against *Plasmodium falciparum* in vivo.

797
Trilobine

Skeletal type: bisbenzylisoquinoline

$C_{35}H_{34}N_2O_5$ Mol. wt 562.67

Found in the roots of *Cocculus trilobus* and *C. sarmentosus*, and in the stems of *Anisocycla grandidieri* (Menispermaceae).

Toxic (LD_{50} in man 500 mg kg^{-1}).

798
(+)-Tubocurarine

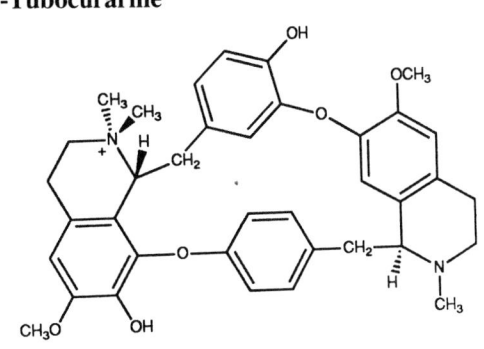

Skeletal type: bisbenzylisoquinoline

$[C_{37}H_{42}N_2O_6]^+$ Mol. wt 610.76

Occurs in pareira bark, *Chondrodendron tomentosum*, and other *Chondrodendron* spp. (Menispermaceae). The (−) form is also found naturally in *C. tomentosum*, but is much less active than the (+) form.

Skeletal muscle relaxant used to paralyse muscles during surgical operations. An important ingredient of the South American arrow poisons known as "curare".

799
Xylopine

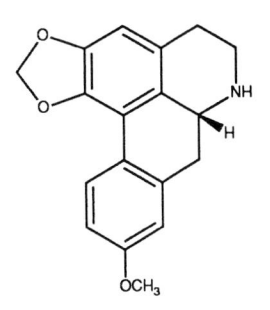

Skeletal type: aporphine

$C_{18}H_{17}NO_3$ Mol. wt 295.34

Found in *Xylopia discreta* and *X. papuana* (Annonaceae).

Sedative, and analgesic.

800
Xylopinine

Skeletal type: aporphine

$C_{21}H_{25}NO_4$ Mol. wt 355.44

Found in *Xylopia discreta* and *X. buxifolia* (Annonaceae), and in *Duguetia* spp. (Menispermaceae).

Adrenergic α-blocker.

Chapter 22
Lycopodium alkaloids

The club mosses, and especially those classified in the genus *Lycopodium*, are prolific producers of a series of alkaloids that are essentially unique to these lowly plants. The Lycopodium alkaloids are related to the quinolizidines (q.v.) but have unusual tetracyclic ring systems which can be seen in the structures of annotinine (**804**) and lycopodine (**818**) on the one hand and in the structures of cernuine (**806**) and lycodine (**814**) on the other. There are about 150 compounds, which are derived from one or other of these four representative substances. Structural variation is mainly provided by skeletal rearrangements (see, e.g., fawcettimine, **808**) and by the positioning of methyl, hydroxy and other substituents.

As a class, these alkaloids are moderately toxic but, owing to the inedible nature of the club mosses, no fatalities appear to have been recorded in either humans or farm animals. They are pharmacologically active and have the ability to stimulate respiratory activities and to paralyse the central nervous system. Although none of these alkaloids has so far been used in modern medicine, several species of club moss have been employed in herbal practice for the treatment of skin disorders.

References

Blumenkopf, T.A. and Heathcock, C.H. (1985). Synthesis of *Lycopodium* alkaloids. In *Alkaloids, Chemical and Biological Perspectives*, Volume 3 (Pelletier, S.W., ed.), pp. 185–240, Chichester: Wiley.

In *The Alkaloids* (Manske, R.H.F. and Brossi, A., eds) (1955) Volume 5, pp. 259–300; (1968) Volume 10, pp. 306–382; (1973) Volume 14, pp. 347–406; (1985) Volume 26, pp. 241–298.

801
Acrifoline

$C_{16}H_{23}NO_2$ Mol. wt 261.37

Found in *Lycopodium selago* and *L. annotinum* var. *acrifolium* (Lycopodiaceae).

Moderately toxic.

802
Annofoline

$C_{16}H_{25}NO_2$ Mol. wt 263.38

Found in *Lycopodium annotinum* and other *Lycopodium* spp. (Lycopodiaceae).

803
Annotine

$C_{16}H_{21}NO_3$ Mol. wt 275.35

Found as the main alkaloid in *Lycopodium annotinum* (Lycopodiaceae) together with many related alkaloids.

This club moss has been used in Chinese herbal medicine since the seventeenth century for the treatment of skin disorders and as a tonic.

804
Annotinine

$C_{16}H_{21}NO_3$ Mol. wt 275.35

Found in *Lycopodium annotinum* (Lycopodiaceae).

For the uses of this club moss see annotine.

805
Carolinianine

$C_{16}H_{24}N_2O_2$ Mol. wt 276.38

Found in *Lycopodium carolinianum* var. *affine* (Lycopodiaceae).

Moderately toxic.

806
Cernuine

$C_{16}H_{26}N_2O$ Mol. wt 262.40

Found in *Lycopodium cernuum* and *L. carolinianum* (Lycopodiaceae).

Slightly toxic.

807
Fawcettidine

$C_{16}H_{23}NO$ Mol. wt 245.37

Found in *Lycopodium fawcettii* and *L. phlegmaria* (Lycopodiaceae).

808
Fawcettimine

$C_{16}H_{25}NO_2$ Mol. wt 263.38

Found in *Lycopodium fawcettii* and *L. clavatum* var. *inflexum* (Lycopodiaceae).

809
Flabellidine

$C_{18}H_{28}N_2O$ Mol. wt 288.44

Found in *Lycopodium paniculatum* (Lycopodiaceae).

810
Inundatine

$C_{16}H_{23}NO_2$ Mol. wt 261.37

Found in *Lycopodium inundatum* (Lycopodiaceae).

811
Lucidine B; Serratanine A

$C_{30}H_{49}N_3O$ Mol. wt 467.74

Found in *Lycopodium serratum* and *L. lucidum* (Lycopodiaceae).

812
Luciduline

$C_{13}H_{21}NO$ Mol. wt 207.32

Found in *Lycopodium lucidulum* (Lycopodiaceae).

813
Lycocernuine

$C_{16}H_{26}N_2O_2$ Mol. wt 278.40

Found in *Lycopodium cernuum* and *L. carolinianum* (Lycopodiaceae).

814
Lycodine

$C_{16}H_{22}N_2$ Mol. wt 242.37

Found in *Lycopodium annotinum* (Lycopodiaceae).

Some *Lycopodium* spp. are used as a remedy for skin disorders in herbal medicine.

815
Lycofawcine

$C_{18}H_{29}NO_4$ Mol. wt 323.44

Found in *Lycopodium fawcettii* (Lycopodiaceae).

Some *Lycopodium* spp. are used as a remedy for skin disorders in herbal medicine.

816
Lycoflexine; Lycobergine

$C_{17}H_{25}NO_2$ Mol. wt 275.39

Found in *Lycopodium clavatum* var. *inflexum*, *L. carolinum* and *L. inundatum* (Lycopodiaceae).

817
Lyconnotine

$C_{17}H_{25}NO_3$ Mol. wt 291.39

Found in *Lycopodium annotinum* (Lycopodiaceae).

818
Lycopodine

C$_{16}$H$_{25}$NO   Mol. wt 247.38

Found in *Lycopodium complanatum* and *L. clavatum* (Lycopodiaceae).

Toxic. It is used as a remedy for skin disorders in herbal medicine. It causes paralysis in frogs, and causes uterine contractions and an increase in the peristalsis of the small intestine in rabbits, rats and guinea-pigs.

819
Magellanine

C$_{17}$H$_{25}$NO$_2$ Mol. wt 275.39

Found in *Lycopodium magellanicum* (Lycopodiaceae).

820
Megastachine

C$_{20}$H$_{29}$NO$_3$ Mol. wt 331.46

Found in *Lycopodium megastachyum* (Lycopodiaceae).

821
α-Obscurine

C$_{17}$H$_{26}$N$_2$O   Mol. wt 274.41

Found in *Lycopodium clavatum, L. magellanicum, L. sitchense* and *L. thyoides* (Lycopodiaceae).

822
β-Obscurine

C$_{17}$H$_{24}$N$_2$O   Mol. wt 272.39

Found in *Lycopodium clavatum* (Lycopodiaceae).

823
Oxolucidine B; Serratanine B

C$_{30}$H$_{49}$N$_3$O$_2$ Mol. wt 483.74

Found in *Lycopodium serratum* and *L. lucidum* (Lycopodiaceae).

824
Phlegmarine

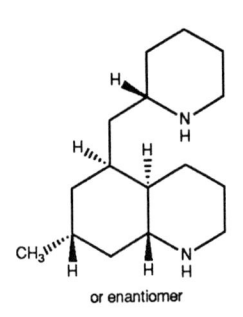

$C_{16}H_{30}N_2$ Mol. wt 250.43

Found in *Lycopodium phlegmaria* (Lycopodiaceae).

825
Sauroxine

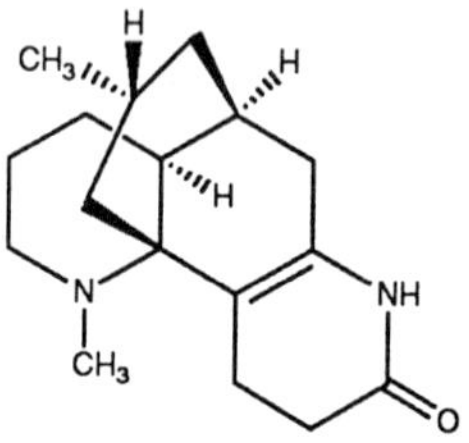

$C_{17}H_{26}N_2O$ Mol. wt 274.41

Found in *Locopodium saururus* (Lycopodiaceae).

826
Selagine

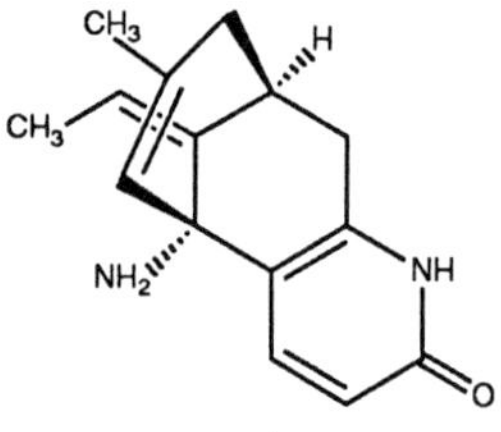

$C_{15}H_{18}N_2O$ Mol. wt 242.32

Found in *Lycopodium selago*, *L. erythraeum*, *L. gnidioides*, and *L. saururus* (Lycopodiaceae).

827
Serratanidine

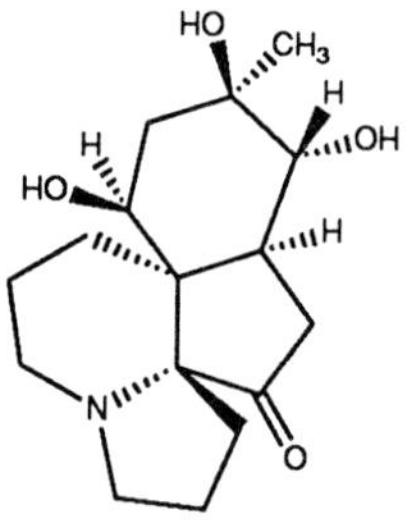

$C_{16}H_{25}NO_4$ Mol. wt 295.38

Found in *Lycopodium serratum* (Lycopodiaceae).

828
Serratine

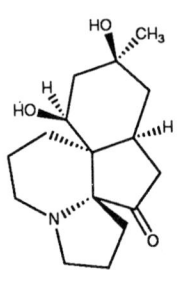

$C_{16}H_{25}NO_3$ Mol. wt 279.38

Found in *Lycopodium serratum* (Lycopodiaceae).

829
Spirolucidine

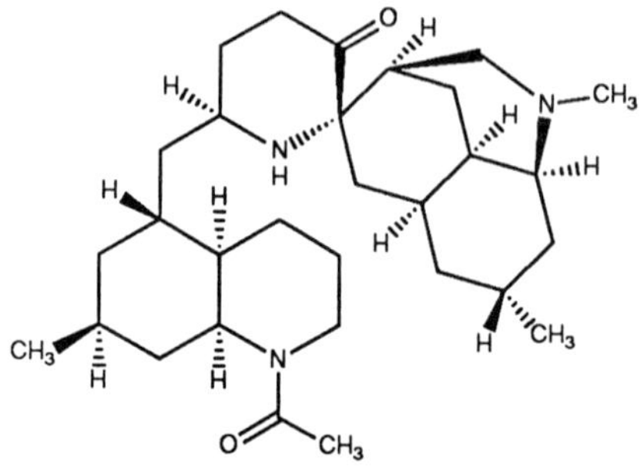

$C_{30}H_{49}N_3O_2$ Mol. wt 483.74

Found in *Lycopodium lucidum* (Lycopodiaceae).

Chapter 23
Monoterpene and Sesquiterpene alkaloids

The monoterpene alkaloids are a small group of plant alkaloids, formed biosynthetically from the iridoids loganin and secologanin (q.v.) by condensation with ammonia. This reaction can sometimes occur during the isolation of iridoids from plants, so that gentianine (**846**) can be formed from swertiamarin or gentiopicroside (q.v.) as an artefact of using ammonium hydroxide in the extraction medium. There is no doubt that these alkaloids do genuinely occur in plants, and especially in members of the genus *Gentiana* (Gentianaceae). Gentian roots have been widely used in Europe as a source of tonics, and the activity of such extracts is due to the presence of gentianaine (**844**), gentianine (**846**) or gentioflavine (**847**).

Of the other monoterpene alkaloids, arenaine (**832**) is notable for its guanidino substitution (cf., arginine). Chaksine (**839**) is also guanidino substituted, and is further distinguished by its macrocyclic dimeric structure. Actinidine (**831**) is interesting because of its biological properties. Like nepetalactone (q.v.), it has a remarkable attractant effect on most members of the cat family.

Sesquiterpene alkaloids, which are derived biosynthetically from a C_{15} farnesol precursor, can be divided into three groups according to plant source. There are tetracyclic alkaloids such as dendrobine (**840**), which characterize orchid plants of the large genus *Dendrobium*. Then there are alkaloids like deoxynupharidine (**841**), which are found in the water lily family, in *Nuphar japonicum* and *N. luteum*. Some sulfur-containing alkaloids, e.g., thiobinupharidine (**854**) are also in this group. Finally, there are the Celastraceae alkaloids, represented here by cassinine (**836**), catheduline E2 (**837**), celapanine (**838**) and wilfordine (**856**). They are all distinctive in their highly oxygenated structures. Although formally derived from a sesquiterpene precursor, their biosynthesis has yet to be examined. They are found in the spindle tree, *Euonymus europaeus*, and in the leaves of khat, *Catha edulis*, a common drug plant of the Arabian world.

References

In *The Alkaloids* (Manske, R.H.F. and Brossi, A., eds) (1967) Volume 9, pp. 441–465; (1977) Volume 16, pp. 181–214; 215–248; 432–510; (1989) Volume 35, pp. 215–257.

830
Acanthicifoline

$C_{10}H_{12}N_2O_2$ Mol. wt 192.22

Found in *Acanthus ilicifolius* (Acanthaceae).

831
Actinidine

$C_{10}H_{13}N$ Mol. wt 147.22

The (+) form is found only in the leaves of *Tecoma stans* (Bignoniaceae), in which it co-occurs with boschniakine (q.v.). The (−) form occurs in the leaves and galls of *Actinidia polygama* and *A. arguta* (Actinidiaceae), the leaves of *Tecoma radicans* (Bignoniaceae), and in the roots of *Valeriana officinalis* (Valerianaceae).

Causes excitation in cats and other Felidae. It also occurs in the defensive secretions of certain ants and of rove beetles, so is presumably toxic to their animal predators.

832
Arenaine

$C_{11}H_{17}N_3O$ or enantiomer   Mol. wt 207.28

Found in the seeds of *Plantago arenaria* (Plantaginaceae), in traces, along with narcotine (q.v.).

833
Bakankoside; Bakankosine; Bacancosin

$C_{16}H_{23}NO_8$ Mol. wt 357.37

Found in the seeds of *Strychnos vacacoua* (Loganiaceae), called bakanko in Madagascar.

Although the plant belongs to the genus *Strychnos*, the seeds do not contain strychnine.

834
Boschniakine; Indicaine

$C_{10}H_{11}NO$ Mol. wt 161.21

Occurs in *Boschniaka rossica* (Orobanchaceae), *Tecoma stans* and *T. radicans* (Bignoniaceae), *Plantago sempervirens* (Plantaginaceae) and *Pedicularis* spp. (Scrophulariaceae).

835
Cantleyine

$C_{11}H_{13}NO_3$ Mol. wt 207.23

Found in the trunk bark of *Cantleya corniculata* (Icacinaceae), in *Dipsacus azureus* (Dipsacaceae), in *Strychnos nux-vomica* (Loganiaceae) and in several *Jasminum* spp. (Oleaceae). Cantleyine is easily formed in certain other species as an artefact from appropriate heterosides on treatment with ammonia.

836
Cassinine

$C_{44}H_{51}NO_{17}$ Mol. wt 865.90

Found in the root bark of *Cassine matabelica* (Celastraceae).

Weak tranquillizing effect in animals.

837
Catheduline E2; Catheduline 2

$C_{38}H_{40}N_2O_{11}$ Mol. wt 700.75

One of many structurally related alkaloids isolated from *Catha edulis* (Celastraceae).

The leaves are used as a tea (Khat) or for chewing, in Arabia and adjacent territories, owing to the presence of ephedrine-like bases.

838
Celapanine

$C_{30}H_{35}NO_{10}$ Mol. wt 569.62

One of five structurally closely related alkaloids isolated from *Celastrus paniculata* (Celastraceae).

The oil from the seeds is a powerful stimulant, also employed for relieving rheumatic pains and for paralysis. The juice from the leaves is used in the sub-Himalayan region as an antidote for opium poisoning.

839
Chaksine

$C_{22}H_{38}N_6O_4$ Mol. wt 450.59

Found in the seeds, leaves and roots of *Cassia absus* (Leguminosae), together with isochaksine.

Causes respiratory paralysis in mice, vasoconstriction in rats and contraction of the ileum in guinea-pigs.

840
Dendrobine

$C_{16}H_{25}NO_2$ Mol. wt 263.38

Found in *Dendrobium nobile* and other *Dendrobium* spp. (Orchidaceae).

Lowers blood pressure and has weak antipyretic and analgesic actions; in large doses, cardiac activity is reduced. Intravenous injection produces convulsions.

841
Deoxynupharidine; α-Nupharidine

$C_{15}H_{23}NO$ Mol. wt 233.36

Found in the rhizomes of *Nuphar japonicum* and *N. luteum* and in various *Nymphaea* spp. (Nymphaeaceae).

Rhizomes used in folk medicine in eastern Europe and Asia for sedative and narcotic purposes.

842
Enicoflavine

$C_{10}H_{13}NO_4$ Mol. wt 211.22

Found in *Enicostemma hyssopifolium* (Gentianaceae). On standing at room temperature for 3 weeks, it converts by ring closure to gentianine (q.v.).

843
Gentianadine

$C_8H_7NO_2$ Mol. wt 149.15

Found in the aerial parts of *Gentiana turkestanorum, G. olgae* and *G. olivieri* (Gentianaceae).

Very mildly toxic. It shows hypothermic, hypotensive, anti-inflammatory and muscular relaxant actions, like gentianine (q.v.).

844
Gentianaine; Gentiocrucine

$C_6H_7NO_3$ Mol. wt 141.13

Occurs in *Enicostemma hyssopifolium, Gentiana caucasa, G. olgae, G. kaufmanniana, G. olivieri* and *G. turkestanorum* (Gentianaceae).

Low anti-inflammatory activity.

845
Gentianamine

$C_{11}H_{11}NO_3$ Mol. wt 205.22

Found in *Gentiana olivieri* and *G. turkestanorum* (Gentianaceae).

Anti-inflammatory activity.

846
Gentianine

$C_{10}H_9NO_2$ Mol. wt 175.19

Found in *Gentiana kirilowi*, other *Gentiana* spp., and in *Swertia* spp. (Gentianaceae). Occasionally, it is isolated as an artefact when ammonia is used to obtain a crude alkaloid fraction from a plant extract.

Low toxicity. It stimulates central nervous system activity at low doses, but in higher doses it has a paralysing effect. It exerts hypotensive, anti-inflammatory and muscular relaxant action. The tonic effects of the gentian liquor are probably due to the hypotensive effects of this alkaloid.

847
Gentioflavine

$C_{10}H_{11}NO_3$ Mol. wt 193.21

Found in *Gentiana asclepiadea* and other *Gentiana* spp., and also from *Erythrea centaurium*, *Swertia connata* and other *Swertia* spp. (Gentianaceae).

848
N-(p-Hydroxyphenethyl)actinidine

$[C_{18}H_{22}NO]^+$ Mol. wt 268.38

Occurs in the roots of *Valeriana officinalis* (Valerianaceae).

Highly active inhibitor of cholinesterase activity.

849
Jasminine

$C_{11}H_{12}N_2O_3$ Mol. wt 220.23

Found in *Jasminum gracile* and *J. lineare* together with jasminidine, also in *Ligustrum novoguineense*, *Olea paniculata* and *Syringa vulgaris* (all Oleaceae).

850
Rhexifoline

$C_{11}H_{13}NO_3$ Mol. wt 207.23

Occurs in the seeds and the blossoms of *Castilleja rhexifolia* (Scrophulariaceae).

Larval feeding stimulant for the plume moth, *Platyptilia pica* (Lepidoptera).

851
β-Skytanthine

$C_{11}H_{21}N$ Mol. wt 167.30

Found in *Skytanthus acutus* (Apocynaceae).

Tremorigenic agent of low toxicity.

852
Tecomine; Tecomanine

$C_{11}H_{17}NO$ Mol. wt 179.26

Found in *Tecoma stans* (Bignoniaceae), and also in *Calopogonium stans* and *C. fulva* (Leguminosae).

Hypoglycaemic activity in fasting rabbits at 20 mg kg^{-1} intravenously and 50 mg kg^{-1} orally. In Mexico it is used for treating diabetes mellitus.

853
Tecostanine

$C_{11}H_{21}NO$ Mol. wt 183.30

Occurs in *Tecoma stans* (Bignoniaceae) together with tecostidine, the related pyridine derivative.

Similarly to tecomine (q.v.), lowers blood sugar levels in animals, e.g., rabbits.

854
Thiobinupharidine; Thionuphlutine A

$C_{30}H_{42}N_2O_2S$ Mol. wt 494.75

Found in the rhizomes of *Nuphar luteum* (Nymphaeaceae).

Weak antibacterial activity.

855
Valerianine

$C_{11}H_{15}NO$ Mol. wt 177.25

Found in the roots of *Valeriana officinalis* (Valerianaceae).

856
Wilfordine

$C_{43}H_{49}NO_{19}$ Mol. wt 883.87

Occurs in the seeds of *Tripterygium wilfordii* and of *Euonymus alatus* (Celastraceae).

Insecticidal activity.

Chapter 24
Peptide alkaloids

About 130 plant alkaloids are known which are formed by the linking of protein amino acids through peptide bonding, followed by cyclization. They are macrocyclic, usually with 13-, 14- or 15-membered rings and are known more correctly as cyclopeptides. Typical examples are zizyphine A (**879**) with a 13-membered ring, frangulanine (**867**) with a 14-membered ring and mucronine A (**873**) with a 15-membered ring. Almost all have an aromatic styrylamide group within the structure, formed by the decarboxylation of a phenylalanine unit.

These alkaloids occur regularly in the Rhamnaceae and have also been found less frequently in members of the Celastraceae, Pandaceae, Rubiaceae, Sterculiaceae and Urticaceae. Some related peptides are known in the fungi, e.g., in *Amanita phalloides*, but these are listed in this Dictionary in Chapter 2, Peptides and proteins.

Peptide alkaloids are not readily detected in plant surveys, because they are only very weak bases and lack characteristic ultraviolet spectral properties. They are, however, readily analysed by mass spectrometry, since they undergo characteristic fragmentation with sequential release of the different amino acid residues present. Little is known of their biological activities, apart from the fact that plants containing them have been used as herbal medicines. Some of them do possess antibiotic properties.

References

In *The Alkaloids* (Manske, R.H.F. and Brossi, A., eds) (1975) Volume 15, pp. 165–205; (1985) Volume 26, pp. 299–326.

857
Adouétine X; Ceanothamine B

$C_{28}H_{44}N_4O_4$ Mol. wt 500.69

Occurs in *Waltheria americana* (Sterculiaceae), and *Ceanothus americanus* (Rhamnaceae).

858
Adouétine Y

$C_{34}H_{40}N_4O_4$ Mol. wt 568.72

Occurs in *Waltheria americana* (Sterculiaceae), and *Ceanothus americanus* (Rhamnaceae).

859
Adouétine Z

$C_{42}H_{45}N_5O_5$ Mol. wt 699.86

Most complex of the four peptide alkaloids isolated from *Waltheria americana* (Sterculiaceae).

Sedative and hypotensive activity. In the Ivory Coast it is used as a febrifuge.

860
Americine

$C_{31}H_{39}N_5O_4$ Mol. wt 545.69

Found in the dried ground root of *Ceanothus americanus* (Rhamnaceae).

861
Amphibine A; Discarine A

$C_{35}H_{43}N_5O_4$ Mol. wt 573.74

Occurs in the bark of *Ziziphus amphibia* (Rhamnaceae) together with eight closely related peptide alkaloids.

Antibiotic properties. It is weakly active against lower fungi and Gram positive bacteria.

862
Amphibine B

$C_{39}H_{47}N_5O_5$ Mol. wt 665.84

Occurs in *Ziziphus amphibia* and *Z. mauritania* (Rhamnaceae), accompanied by the closely related amphibines C and D.

863
Aralionine A

$C_{34}H_{38}N_4O_5$ Mol. wt 582.71

Found in the leaves of *Araliorhamnus vaginata* (Rhamnaceae), which contains minor alkaloids aralionine B and C in the bark.

Antibiotic properties. It is weakly active against lower fungi and Gram positive bacteria.

864
Canthiumine

$C_{33}H_{36}N_4O_4$ Mol. wt 552.68

Found in *Canthium euryoides* (Rubiaceae).

Antibiotic properties against Gram positive bacteria.

865
Ceanothine B

$C_{29}H_{36}N_4O_4$ Mol. wt 504.64

Major constituent of the root bark of *Ceanothus americanus* (Rhamnaceae) together with four minor peptide alkaloids.

866
Crenatine A

$C_{34}H_{40}N_4O_4$ Mol. wt 568.72

Occurs in *Discaria crenata* (Rhamnaceae).

867
Frangulanine; Ceanothamine A

$C_{28}H_{44}N_4O_4$ Mol. wt 500.69

Occurs in frangula, or alder buckthorn, *Rhamnus frangula*, and *Ceanothus americanus* (Rhamnaceae).

Weak antibiotic activity against lower fungi and Gram positive bacteria. Frangula bark is used more often for its content of anthraquinone glycosides, which are cathartic.

868
Hymenocardine

$C_{37}H_{50}N_6O_6$ Mol. wt 674.85

Occurs in *Hymenocardia acida* (Euphorbiaceae).

869
Integerrenine

$C_{31}H_{42}N_4O_4$ Mol. wt 534.71

Occurs in *Ceanothus integerrimus* (Rhamnaceae).

870
Integerressine

$C_{33}H_{38}N_4O_4$ Mol. wt 554.70

Occurs in *Ceanothus integerrimus* (Rhamnaceae).

871
Integerrine

$C_{35}H_{39}N_5O_4$ Mol. wt 593.73

Occurs in *Ceanothus integerimus* (Rhamnaceae).

Weak antibiotic activity against Gram positive bacteria and lower fungi. It inhibits photophosphorylation in plants.

872
Lasiodine A

$C_{39}H_{49}N_5O_7$ Mol. wt 699.86

Occurs in *Lasiodiscus marmoratus* (Rhamnaceae).

Weak antibiotic activity against Gram positive bacteria and lower fungi. It inhibits photophosphorylation in plants.

873
Mucronine A

$C_{29}H_{38}N_4O_4$ Mol. wt 506.65

Occurs in *Ziziphus mucronata* and *Z. abyssinica* (Rhamnaceae).

Weak antibiotic activity against Gram positive bacteria and lower fungi. *Ziziphus mucronata* is used in central and southern Africa as a remedy for diarrhoea and dysentery.

874
Mucronine B

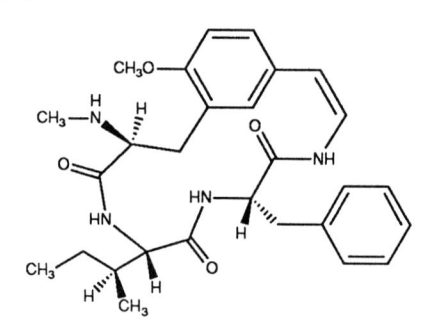

$C_{28}H_{36}N_4O_4$ Mol. wt 492.62

Occurs in *Ziziphus mucronata* (Rhamnaceae).

875
Nummularine F

$C_{23}H_{32}N_4O_4$ Mol. wt 428.54

Occurs in *Ziziphus nummularia* and *Z. oenoplia* (Rhamnaceae).

876
Pandamine

$C_{31}H_{44}N_4O_5$ Mol. wt 552.72

Occurs in *Panda oleosa* (Pandaceae).

Weak antibiotic activity against Gram positive bacteria and lower fungi. It inhibits photophosphorylation in plants.

877
Sativanine B

$C_{30}H_{38}N_4O_4$ Mol. wt 518.66

Occurs in *Ziziphus sativa* (Rhamnaceae).

878
Scutianine F

$C_{38}H_{45}N_5O_5$ Mol. wt 651.81

Occurs in *Scutia buxifolia* (Rhamnaceae).

879
Zizyphine A; Ziziphine A; Zizyphine

$C_{33}H_{49}N_5O_6$ Mol. wt 611.79

Occurs in *Ziziphus (=Zizyphus) oenoplia* (Rhamnaceae).

Weak antibiotic activity against lower fungi and Gram positive bacteria. It inhibits photophosphorylation in plants.

880
Zizyphine F

$C_{32}H_{47}N_5O_6$ Mol. wt 597.76

Occurs in *Ziziphus nummularia* and *Z. oenoplia* (Rhamnaceae).

Chapter 25
Pyrrolidine and Piperidine alkaloids

Plant alkaloids which have a nitrogen atom either in a 5-membered ring (pyrrolidines) or in a 6-membered ring (piperidines) are included in this section. Those with a nitrogen atom shared between a 5- and a 6-membered ring (indolizidines) are also included, but quinolizidines (a nitrogen shared between two 6-membered rings) and pyrrolizidines (a nitrogen shared between two 5-membered rings) are listed elsewhere. One further group included here are the pyridine alkaloids, many of which have additional pyrrolidine or piperidine rings in their structures.

Hygrine (**910**) is an example of a simple pyrrolidine, while cuscohygrine (**902**) is a base with two pyrrolidine rings in its structure. Codonopsine (**897**) is a pyrrolidine with an aromatic substituent, while betonicine (**893**) is one which occurs as a quaternary ammonium salt. Stachydrine (**934**), another quaternary salt, is notable for its relatively widespread occurrence, especially in plants of the Capparidaceae and Labiatae.

The best known piperidine alkaloid is coniine (**899**), the poisonous principle of hemlock, which occurs in equilibrium with its dehydro derivative, γ-coniceine (**898**). The *Lobelia* alkaloids are piperidines with two aromatic substituents attached adjacent to the nitrogen atom (see lobeline, **916**). The piperidine alkaloid pinidine (**924**) is of interest because of its occurrence in pine trees (*Pinus*); otherwise, alkaloids are very rare in conifers (but see also taxine, in *Taxus*). Piperidine dimers are not uncommon, as in anaferine (**885**), in which the two 6-membered rings are joined by a 3-carbon moiety, and in carpaine (**894**), where the two rings are linked together in a large macrocyclic structure.

Indolizidines are represented here by castanospermine (**896**) and swainsonine (**935**), of current medical interest because of their possible use in the treatment of the human disease AIDS. Two other alkaloids with similar biological properties are the polyhydroxypiperidines deoxymannojirimycin (**903**) and DMDP (**905**). These four and other similar polyhydroxy compounds have been referred to as "sugar-shaped" alkaloids (Fellows, 1986), because of their structural resemblances to monosaccharides (q.v.), and the fact that they are potent inhibitors of sugar-hydrolysing enzymes.

The most familiar pyridine alkaloid is nicotine (**919**), the structure of which is based on the linking of pyridine and pyrrolidine rings. Although nicotine occurs in a number of unrelated plants, the major source is *Nicotiana*, so that it and related structures are usually referred to as tobacco alkaloids. Anabasine (**884**) is a tobacco alkaloid where a pyridine ring is linked to a piperidine rather than a pyrrolidine nucleus. It may be noted that addiction to tobacco smoking (and to nicotine, the active principle) can be treated to some extent by taking the piperidine derivative lobeline (**916**) instead. There are simple pyridine alkaloids as well, such as arecaidine (**888**) and arecoline (**889**), which occur in the nuts of *Areca catechu*, or betel. This is the chewing "product" of many Asian people, but its well known carcinogenic effects are probably due to the tannins of *Areca* rather than to its alkaloids.

References

Indolizidine Alkaloids

Elbein, A.D. and Molyneux, R.J. (1987). The chemistry and biochemistry of simple indolizidine and related polyhydroxyalkaloids. In *Alkaloids, Chemical and Biological Perspectives*, Volume 5 (Pelletier, S.W., ed.), pp. 1-54, New York: Wiley.

Fellows, L.E. (1986). The biological activity of polyhydroxyalkaloids from plants. *Pesticide Science* 17, 602-606.

In *The Alkaloids* (Manske, R.H.F. and Brossi, A., eds) (1986) Volume 28, pp. 183-308.

Piperidine Alkaloids

In *The Alkaloids* (Manske, R.H.F. and Brossi, A., eds) (1985) Volume 26, pp. 89-184.

Pyridine Alkaloids

von Euler, V.S. (ed.) (1965). *Tobacco Alkaloids and Related Compounds*. Oxford: Pergamon Press.

Foder, G.B. and Colasanti, B. (1985). The pyridine and piperidine alkaloids: chemistry and pharmacology. In *Alkaloids, Chemical and Biological Perspectives*, Volume 3 (Pelletier, S.W., ed.), pp. 1-90, New York: Wiley.

In *The Alkaloids* (Manske, R.H.F. and Brossi, A., eds) (1950) Volume 1, pp. 165-270; (1960) Volume 6, pp. 123-144; (1968) Volume 11, pp. 459-501; (1985) Volume 26, pp. 89-184.

Pyrrolidine Alkaloids

In *The Alkaloids* (Manske, R.H.F. and Brossi, A., eds) (1950) Volume 1, pp. 91-106; (1960) Volume 6, pp. 31-34; (1986) Volume 27, pp. 270-322.

881
Adenocarpine; Teidine; Orensine

C₁₉H₂₄N₂O   (+) form   Mol. wt 296.41

Major component in several *Adenocarpus* spp., e.g., *A. hispanicus* and *A. intermedius* (Leguminosae).

882
Alexine

C₈H₁₄NO₄   Mol. wt 188.21

Found in *Alexa leiopetala* (Leguminosae).

α-Glucosidase inhibitor.

883
(+)-Ammodendrine; *N*-Acetyltetrahydroanabasine

C₁₂H₂₀N₂O   Mol. wt 208.30

(±)-Ammodendrine obtained with the (+) form (=sphaerocarpine) from *Ammodendron conollyii* and from several *Sophora* spp., e.g., *S. franchetiana* (Leguminosae).

Toxic to many insects.

884
(−)-Anabasine; Nicotinimine

C₁₀H₁₄N₂   Mol. wt 162.24

Occurs either in the (−) form or as the racemate (neonicotine) in *Anabasis aphylla* (Chenopodiaceae), *Nicotiana acuminata* and other *Nicotiana* spp., *Duboisia myoporoides* (Solanaceae), *Zinnia elegans*, *Zollikoferia eliquiensis* (Compositae), and *Alangium* spp., e.g., *A. kurzii* and *A. platanifolium* (Alangiaceae).

Acute and subacute toxicity. It has respiratory muscle-stimulating action. A major use is to discourage smoking of tobacco. It has insecticidal properties.

885
(−)-Anaferine

C₁₃H₂₄N₂O   Mol. wt 224.34

Occurs in the roots of *Withania somnifera* (Solanaceae), together with anahygrine, the *N*-methyl derivative.

The root drug is described in early and modern Indian literature as a curative for many ills, and sedative properties are stressed.

886
Anatabine

$C_{10}H_{12}N_2$ Mol. wt 160.22

Most abundant of the minor alkaloids present in the roots of *Nicotiana tabacum* (Solanaceae).

887
Anibine

$C_{11}H_9NO_3$ Mol. wt 203.20

Found in the sawdust of several *Aniba* spp. (Lauraceae), the South American rosewood tree.

888
Arecaidine; Arecaine

$C_7H_{11}NO_2$ Mol. wt 141.17

One of several alkaloids obtained from the seeds of *Areca catechu* (Palmae), betel nut.

Not markedly toxic. The seeds are used medicinally, being diaphoretic, miotic, teniacide, astringent, and they are used in veterinary medicine as a laxative and teniafuge.

889
Arecoline

$C_8H_{13}NO_2$ Mol. wt 155.20

Most important alkaloid of *Areca catechu* (Palmae), betel nut.

Exhibits markedly toxic properties. It shows a parasympathetic stimulant action, and some anthelmintic activity. It is a teniacide for animals, a cathartic in horses, and a ruminatoric in cattle.

890
Astrocasine

$C_{20}H_{26}N_2O$ Mol. wt 310.44

Found in *Astrocasia phyllantoides* (Euphorbiaceae) together with astrophylline (q.v.), a related alkaloid.

891
Astrophylline

$C_{19}H_{26}N_2O$ Mol. wt 298.43

Found in *Astrocasia phyllantoides* (Euphorbiaceae), together with astrocasine (q.v.).

892
Australine

$C_8H_{15}NO_4$ Mol. wt 189.22

Occurs in the seeds of *Castanospermum australe* (Leguminosae), together with castanospermine (q.v.).

Specific inhibitor of exo-1,4α-glucosidase.

893
(−)-Betonicine; 4-Hydroxyproline betaine; Achillein

$C_7H_{13}NO_3$ Mol. wt 159.19

Found in *Betonica officinalis*, *Marrubium vulgare* and *Stachys sylvatica* (Labiatae), and in *Achillea moschata* and *A. millefolium* (Compositae). The (+) enantiomer is called combretin A and the racemate combretin B.

Anti-inflammatory activity.

894
Carpaine

$C_{27}H_{50}N_2O_4$ Mol. wt 466.71

Occurs in the seeds and leaves of the pawpaw, *Carica papaya*, and of *Vasconcellea hastata* (Caricaceae).

Cardiotonic agent and a potent amoebicide.

895
Cassine

$C_{18}H_{35}NO_2$ Mol. wt 297.49

Occurs in the leaves of *Cassia excelsa*, *C. carnaval*, and *Prosopis ruscifolia* (Leguminosae).

896
Castanospermine

$C_8H_{15}NO_4$ Mol. wt 189.22

Found in the seeds of *Castanospermum australe* (Leguminosae).

Unripe seeds cause severe gastrointestinal irritation and sometimes death when eaten by horses and cattle; Australian Aborigines use the seeds as food after soaking in water and roasting. Castanospermine is toxic to insects and is a potent inhibitor of α- and β-glucosidases. It also reduces the ability of the human immunodeficiency virus (HIV) to infect cultured cells, and has potential for treating AIDS.

897
Codonopsine

$C_{14}H_{21}NO_4$ Mol. wt 267.33

Found in *Codonopsis clematidea* (Campanulaceae).

Decreases blood pressure in cats at doses of 20 mg kg^{-1}.

898
γ-Coniceine; 2,3,4,5-Tetrahydro-6-propylpyridine

$C_8H_{15}N$ Mol. wt 125.22

Co-occurs with coniine (q.v.) in hemlock, *Conium maculatum* (Umbelliferae). It is also found in *Aloe gililandii*, *A. ballyi* and *A. ruspoliana* (Liliaceae).

Like coniine (q.v.), it is highly toxic and teratogenic.

899
(+)-Coniine; (S)-2-Propylpiperidine

$C_8H_{17}N$ Mol. wt 127.23

Found in the leaves and seeds of the hemlock, *Conium maculatum* (Umbelliferae). Also, it is produced in the pitcher plant, *Sarracenia flava* (Sarraceniaceae), to paralyse insect prey.

Extremely toxic, causing paralysis of motor nerve endings. Teratogenic effects have been recorded after cows and swine have eaten hemlock during pregnancy. Hemlock was used by the Greeks to execute criminals. It has a mousy odour.

900
Cryptophorine

$C_{17}H_{27}NO$ Mol. wt 261.41

Occurs in *Bathiorhamnus cryptophorus* (Rhamnaceae).

901
Cucurbitine; 3-Amino-β-proline

$C_5H_{10}N_2O_2$ Mol. wt 130.15

Occurs in the seeds of *Cucurbita moschata* (Cucurbitaceae).

Inhibits the growth of *Schistosoma japonica* and has promise in the treatment of schistosomiasis.

902
Cuscohygrine; Cuskhygrine; Hellaridine

$C_{13}H_{24}N_2O$ Mol. wt 224.35

Common in the Solanaceae, e.g., in *Hyoscyamus niger*, *Atropa belladonna* (roots) and in *Datura* and *Scopolia* spp. It co-occurs with cocaine (q.v.) in *Erythroxylum coca* (Erythroxylaceae), and is also found in *Convolvulus erinaceus* (Convolvulaceae).

Inhibitor of 2,4-dinitrofluorobenzene-induced hypersensitivity in mice. Several of the above plants are used in folk medicine as sedatives or narcotics.

903
Deoxymannojirimycin; DMJ

$C_6H_{13}NO_4$ Mol. wt 163.18

Found in the seeds of *Lonchocarpus sericeus* and *L. costaricensis* (Leguminosae). The isomer at position 4 is known as deoxynojirimycin (DNJ) and occurs in *Morus* spp. (Moraceae) and in various *Bacillus* spp.

While DMJ is a mannosidase inhibitor, DNJ inhibits α-glucosidase. DNJ is used as an additive in animal feedstuff.

904
Dioscorine

$C_{13}H_{19}NO_2$ Mol. wt 221.30

Occurs in the tubers of *Dioscorea hirsuta* and *D. hispida* (Dioscoreaceae).

905
DMDP; 2,5-Dihydroxymethyl-3,4-dihydroxypyrrolidine

$C_6H_{13}NO_4$ Mol. wt 163.18

Occurs in the leaves of *Derris elliptica* and the seeds of *Lonchocarpus sericeus* (Leguminosae).

Potent inhibitor of viral glycoprotein processing glucosidase I and of insect α- and β-glucosidases. It is a locust antifeedant and toxic to armyworms.

906
Gerrardine

$C_{11}H_{19}NO_2S_4$ Mol. wt 325.54

Found in *Cassipourea gerrardii* and *C. guianensis* (Rhizophoraceae).

Antibacterial activity, e.g., against *Salmonella*, and antifungal activity against *Candida albicans*.

907
Girgensonine

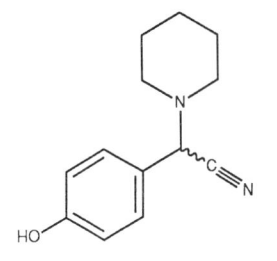

$C_{13}H_{16}N_2O$ Mol. wt 216.29

Found in the green tissues of *Girgensohnia oppositifolia* (Chenopodiaceae).

908
Guvacine; Δ^3-Tetrahydronicotinic acid

$C_6H_9NO_2$ Mol. wt 127.15

Occurs in *Areca catechu* (Palmae), betel nut, where it co-occurs with the methyl ester, guvacoline.

Potent inhibitor of 4-aminobutanoic acid.

909
3-Hydroxystachydrine

$C_7H_{13}NO_3$ Mol. wt 159.19

Characteristic alkaloid of the Capparidaceae, e.g., in *Boscia senegalensis*, *Capparis biloba* and *Steriphoma paradoxum*.

910
(−)-Hygrine

$C_8H_{15}NO$ Mol. wt 141.22

An intermediate in tropane alkaloid (q.v.) synthesis, and commonly found in plants which contain these alkaloids, e.g., in *Atropa* (Solanaceae), *Erythroxylum* (Erythroxylaceae), and *Convolvulus* (Convolvulaceae). Also, it is found in *Dendrobium* (Orchidaceae) and *Cochlearia* (Cruciferae).

911
(−)-Hygroline

$C_8H_{17}NO$ Mol. wt 143.23

Occurs in *Erythroxylum coca* (Erythroxylaceae) and *Cochlearia arctica* (Cruciferae).

912
Isolobinine

or enantiomer

$C_{18}H_{25}NO_2$ Mol. wt 287.41

Found in *Lobelia inflata* (Campanulaceae), where it occurs with its stereoisomer, lobinine.

Anti-asthmatic, antitussive and hypertensive activities.

913
Juliflorine; Juliprosopine

$C_{40}H_{75}N_3O_2$ Mol. wt 630.06

Found in *Prosopis juliflora* (Leguminosae).

Antibacterial and antifungal activities.

914
Lobelanidine

$C_{22}H_{24}NO_2$ Mol. wt 334.44

Occurs in *Lobelia inflata*, *L. hassleri*, *Isotoma longiflora* and *Laurentia longiflora* (Campanulaceae), and in *Sedum acre* (Crassulaceae).

Poisonous.

915
Lobelanine; 8,10-Diphenyllobelidione

$C_{22}H_{25}NO_2$ Mol. wt 335.45

Occurs in *Lobelia inflata* (Campanulaceae).

916
(−)-Lobeline

$C_{22}H_{27}NO_2$ Mol. wt 337.47

Found in *Lobelia inflata*, *L. nicotianaefolia* and *L. hassleri*, and from the seeds of *Campanula medium* (all Campanulaceae).

Used in anti-smoking preparations. The racemate, known as lobelidine, is an analeptic.

917
(+)-N-Methylconiine

$C_9H_{19}N$ Mol. wt 141.26

Found in hemlock, *Conium maculatum* (Umbelliferae).

Toxic.

918
Myosmine

$C_9H_{10}N_2$ Mol. wt 146.19

Occurs in *Nicotiana tabacum* and other *Nicotiana* spp. (Solanaceae). Found in tobacco smoke.

919
Nicotine

$C_{10}H_{14}N_2$ Mol. wt 162.24

Occurs in quantity in *Nicotiana tabacum* and other *Nicotiana* spp. (Solanaceae). It is a trace component in *Asclepias syriaca* (Asclepiadaceae), *Lycopodium* spp., *Equisetum arvense* (Equisetaceae) and *Sedum acre* (Crassulaceae).

Highly toxic, causing respiratory paralysis; the fatal human dose is about 50 mg. This is the addictive component of tobacco, with tranquillizing properties. It is used as a horticultural insecticide.

920
Nicotyrine; β-Nicotyrine

$C_{10}H_{10}N_2$ Mol. wt 158.20

Obtained from cured tobacco, *Nicotiana tabacum* (Solanaceae).

921
Nigrifactin

$C_{12}H_{17}N$ Mol. wt 175.28

Mould metabolite from *Streptomyces*. It is very unstable.

Shows antihistamine activity and affects blood pressure.

922
(+)-Nornicotine; 3-(2-Pyrrolidinyl)pyridine

$C_9H_{12}N_2$ Mol. wt 148.21

Occurs as the (+) form (illustrated) in *Duboisia hopwoodii* and as the (−) form in *Nicotiana tabacum* (Solanaceae).

923
(−)-Pelletierine; Isopelletierine; 8-Methylnorlobelone

$C_8H_{15}NO$ Mol. wt 141.22

Found in the pomegranate plant, *Punica granatum* (Punicaceae), *Duboisia myoporoides* (Solanaceae) and *Sedum acre* (Crassulaceae). It has been reported as the (−) form from pomegranate but racemizes rapidly during isolation. The *N*-methyl derivative, methyl isopelletierine, occurs in pomegranate, in *Sedum* spp. and in *Lupinus formosus* (Leguminosae).

Highly toxic to tapeworms and has been used as an anthelmintic.

924
Pinidine;
(2*R*,6*R*)-2-Methyl-6-(*E*-prop-1-enyl)piperidine

$C_9H_{17}N$ Mol. wt 139.24

Found in *Pinus sabiniana*, *P. jeffreyii* and *P. torreyana* (Pinaceae).

925
Piperine; 1-Piperinoylpiperidine

$C_{17}H_{19}NO_3$ Mol. wt 285.35

Occurs in pepper, *Piper nigrum* and other *Piper* spp. (Piperaceae). Piperine is the *E,E* form (shown) and its isomers chavicine (*Z,Z* form), isochavicine (*E,Z* form) and isopiperine (*Z,E* form) are also present in pepper.

Responsible for the hot taste of pepper. It is used as an insecticide, and also to impart a pungent taste to brandy.

926
Piplartine; Piperlongumine

$C_{17}H_{19}NO_5$ Mol. wt 317.35

Found in the roots of *Piper longum* and *P. sylvaticum* (Piperaceae).

Effective for the treatment of asthma and chronic bronchitis.

927
(+)-Prosopinine

$C_{18}H_{35}NO_3$ Mol. wt 313.49

Found in the leaves of *Prosopis africana* (Leguminosae). It co-occurs with the 6-epimer, prosophylline, which is present as the racemate.

928
Pseudoconhydrine; ψ-Conhydrine; 5-Hydroxy-2-propylpiperidine

$C_8H_{17}NO$ Mol. wt 143.23

Minor component of hemlock, *Conium maculatum* (Umbelliferae).

Toxic.

929
Ricinine

$C_8H_8N_2O_2$ Mol. wt 164.17

Occurs in the seeds of castor oil, *Ricinus communis* (Euphorbiaceae).

Mildly toxic. A highly toxic protein, ricin (q.v.), is also present in castor oil seeds.

930
Ruspolinone

$C_{14}H_{19}NO_3$ Mol. wt 249.31

Occurs in the seeds of *Ruspolia hypercrateriformis* (Acanthaceae).

931
(−)-Santiaguine

$C_{38}H_{48}N_4O_2$ Mol. wt 592.83

Found in *Adenocarpus commutatus* and *A. grandiflorus* (Leguminosae). It can exist in four isomeric forms, which have all been found in *Adenocarpus* spp.

932
(−)-Sedamine; 8-Phenyllobelol

$C_{14}H_{21}NO$ Mol. wt 219.33

Occurs in *Sedum acre*, *S. lydium* and other *Sedum* spp. (Crassulaceae).

933
Slaframine

$C_{10}H_{18}N_2O_2$ Mol. wt 198.27

Toxin of the fungus *Rhizoctonia leguminicola*, which infects red clover, *Trifolium repens* (Leguminosae).

Responsible for "black patch" disease in dairy cattle, horses and sheep eating infected clover. The disease is characterized by excessive salivation, diarrhoea and anorexia. Slaframine is a strong parasympathomimetic agent, and has been suggested as a therapeutic treatment for cystic fibrosis.

935
Swainsonine

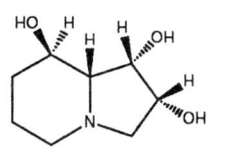

$C_8H_{15}NO_3$ Mol. wt 173.22

Occurs in *Swainsona canescens*, *S. luteola* and *S. galagifolia* (Leguminosae). It co-occurs with slaframine (q.v.) in various fungi, e.g. *Rhizoctonia leguminicola* and *Metarrhizium* sp.

Toxic to livestock feeding on *Swainsona* plants, causing symptoms similar to the genetic disorder mannosidosis, and eventually death. Is an α-mannosidase inhibitor.

934
Stachydrine; Cadabine

$C_7H_{13}NO_2$ Mol. wt 143.19

Common in the Capparidaceae, e.g., *Capparis tomentosa*, and the Labiatae, e.g., *Stachys betoniciflora*. Also, it is found in *Achillea millefolium* (Compositae), *Desmodium triflorum* (Leguminosae), and *Asphodelus microcarpus* (Liliaceae).

Systolic depressant. *Capparis* plants are widely used in the treatment of rheumatism.

Chapter 26
Pyrrolizidine alkaloids

The pyrrolizidine or *Senecio* alkaloids are probably the most poisonous single group of alkaloids. They are hepatotoxins and cause death and liver damage not only in humans but also in many farm animals. Some 50% of all livestock deaths arising from the consumption of harmful plants can be attributed to these bases. Chemically, the pyrrolizidines are readily distinguished by the presence of a fused two 5-membered ring system, which has a bridge nitrogen shared between the two rings. Most are esters of amino alcohols (or necines) such as supinidine (**973**), retronecine (**961**) or crotonecine (**942**). The esters can be classified as monoesters (amabiline, **936**), diesters (echimidine, **944**) or cyclic esters (anacrotine, **937**). The necine bases are biosynthesized from the amino acid ornithine (q.v.) via the amines putrescine and homospermidine.

The major occurrence of these alkaloids is in plants of the cosmopolitan genus *Senecio* (Compositae); probably all 1500 species contain them in varying amounts. The protection from animal grazing enjoyed by the common ragwort, *Senecio jacobeae*, is due to its poisonous alkaloid content. Other genera in the Compositae in the same tribe, the Senecioneae, usually have these alkaloids. Certain genera of two other plant families, the Leguminosae (e.g., *Crotalaria*) and the Boraginaceae (e.g., *Borago, Symphytum*) are also rich in pyrrolizidines. They are synthesized in the roots and are then transported and distributed throughout the aerial parts. These alkaloids are also found in a number of Lepidoptera as pheromones or defence agents, but they are not synthesized by the animal: they are obtained from plant sources by pharmacophagous means (Boppre, 1986).

Pyrrolizidines often occur naturally as the *N*-oxides and are usually converted to the free bases during the isolation procedure. Some alkaloids (e.g., platyphylline, **960**) have no double bonds in the necine ring system, and these are not toxic. All the others show some toxicity; this increases with ester substitution and the cyclic esters (e.g., senecionine, **969**) are the most poisonous. Human beings are at risk as a result of taking herbal medicines or bush teas, based, e.g., on *Crotalaria* plants, or from consuming comfrey (*Symphytum* spp.) as a salad. While toxic effects are mainly centred on the liver, the damage can spread to the lungs.

Analytical tests are available for the specific detection of these alkaloids in plant materials. They are based on the oxidation to pyrrole derivatives and the production of a magenta colour with a modified Ehrlich's reagent (Mattocks and Jukes, 1987). Because of variations in the solubility properties of these alkaloids, they are best isolated by ion-exchange chromatography. They are characterized by spectral measurements, especially by carbon-13 nuclear magnetic resonance spectroscopy (Roder, 1990). The major source of information on pyrrolizidines is the book of Mattocks (1986); other recent reviews are those of Wrobel (1985) and Robins (1989).

References

Chemistry and Toxicology

Mattocks, A.R. (1988). *Chemistry and Toxicology of Pyrrolizidine Alkaloids*. London: Academic Press.

Mattocks, A.R. and Jukes, R. (1987). Improved field tests for toxic pyrrolizidine alkaloids. *Journal of Natural Products*, **50**, 161–166.

Robins, D.J. (1989). Pyrrolizidine alkaloids. *Natural Product Reports*, **6**, 221–239.
Roder, E. (1990). Carbon-13 NMR spectroscopy of pyrrolizidine alkaloids. *Phytochemistry*, **29**, 11–30.
Wrobel, J. (1985). Pyrrolizidine alkaloids. In *The Alkaloids*, Volume 26 (Brossi, A., ed.); New York: Academic Press.

Biology

Boppre, M. (1986). Insects pharmacophagously utilizing defensive plant chemicals (pyrrolizidine alkaloids). *Naturwissenschaften* **73**, 17–26.

936
Amabiline

$C_{15}H_{25}NO_4$ Mol. wt 283.37

Found in *Cynoglossum australe* (Boraginaceae).

Hepatotoxic activity.

937
Anacrotine; Crotalaburnine

$C_{18}H_{25}NO_6$ Mol. wt 351.41

Found in several *Crotalaria* spp., mainly in *C. laburnifolia*, *C. anagyroides* and *C. incana* (Leguminosae).

Hepatotoxin and pneumotoxin.

938
O^7-**Angelylheliotridine**; Rivularine

$C_{13}H_{19}NO_3$ Mol. wt 237.30

Ester alkaloid found in *Senecio rivularis* (Compositae) but also in *Cynoglossum* spp. (Boraginaceae).

Produces chronic liver disease in sheep.

939
Angularine; 13,19-Didehydrorosmarinine

$C_{18}H_{25}NO_6$ Mol. wt 351.40

Occurs in *Senecio angulatus* (Compositae).

Hepatotoxic alkaloid causing necrosis of the liver in cattle.

940
Auriculine

$C_{31}H_{45}NO_8$ Mol. wt 559.71

Occurs in *Liparis auriculata* and *L. loeselii* (Orchidaceae).

Slightly hepatotoxic.

941
Clivorine

$C_{21}H_{27}NO_7$ Mol. wt 405.45

Occurs in *Anchusa officinalis* (Boraginaceae), and *Ligularia clivorum*, *L. dentata* and *L. elegans* (Compositae).

Hepatocarcinogenic. It is mutagenic in the Ames and DNA repair tests.

942
Crotanecine

$C_8H_{13}NO_3$ Mol. wt 171.20

Thought to be present as the free necine base in plants containing its esters, such as anacrotine (q.v.), in various *Crotalaria* spp. (Leguminosae).

Nonhepatotoxic unless esterified.

943
Doronine

$C_{21}H_{30}NO_8Cl$ Mol. wt 459.93

Occurs in *Doronicum macrophyllum* and *Senecio othonnae* (Compositae).

Toxicity not yet established, but hepatotoxicity suspected.

944
Echimidine; 7-Angelyl-9-echimidinylretronecine

$C_{20}H_{31}NO_7$ Mol. wt 397.48

Found in "Paterson's curse", *Echium plantagineum*, and honey derived from the nectar of this plant, in *E. italicum*, and *E. lycopsis*, in comfrey, *Symphytum officinale*, and tuberous comfrey, *S. tuberosum*, *S. caucasium*, and *S. orientale* (all Boraginaceae).

Hepatotoxic in animals; it is mutagenic to *Drosophila* chromosomes.

945
Europine; 9-Lasiocarpylheliotridine

$C_{16}H_{27}NO_6$ Mol. wt 329.40

Occurs in *Heliotropium arbainense, H. europeum, H. maris-mortui, H. rotundifolium* and *Trichodesma africana* (Boraginaceae).

Acutely toxic. It is hepatotoxic and anticholinergic in rats.

946
Fulvine

$C_{16}H_{23}NO_5$ Mol. wt 309.37

Occurs in *Crotalaria crispa, C. fulva, C. madurensis* and *C. paniculata* (Leguminosae).

Hepatotoxic, pneumotoxic, and mutagenic to *Drosophila* chromosomes.

947
Heliosupine; Cynoglossophine; 7-Angelyl-9-echimidinylheliotridine

$C_{20}H_{31}NO_7$ Mol. wt 397.48

Occurs in hound's tongue, *Cynoglossum officinale, C. australe,* and *C. pictum,* in viper's bugloss, *Echium vulgare,* in *Heliotropium supinum, Symphytum asperum,* and comfrey, *S. officinale* (all Boraginaceae).

Hepatotoxic.

948
Heliotridine

$C_8H_{13}NO_2$ Mol. wt 155.20

The most common necine base in American and Indian *Heliotropium* spp., probably found free together with the esters, e.g., lasiocarpine and heliosupine (q.v.).

Hepatotoxic.

949
Heliotrine; 9-Heliotrylheliotridine

$C_{16}H_{27}NO_5$ Mol. wt 313.40

Occurs in many spp. of heliotrope, including *Heliotropium europaeum, H. arbainense, H. curassavicum* and *H. indicum* (Boraginaceae).

Highly toxic. It produces neuromuscular block and respiratory failure in animals. It is hepatotoxic, and impairs mitochondrial respiratory activity in rat liver; also it is teratogenic in rats. It is mutagenic in *Drosophila* and other test systems, including the Ames test.

950
Indicine

$C_{15}H_{25}NO_5$ Mol. wt 299.37

Found in the heliotropes *Heliotropium indicum* and *H. amplexicaule* (Boraginaceae).

Low toxicity in animals compared with most pyrrolizidine alkaloids. The *N*-oxide has been used in clinical trials as an anticancer agent; the mode of action is unknown but is not dependent on reduction to indicine.

951
Integerrimine; Squalidine

$C_{18}H_{25}NO_5$ Mol. wt 335.41

Occurs in *Senecio integerrimus, S. squalidus, S. alpinus, S. brasiliensis, Cacalia hastata*, and *Petasites hybridus* (Compositae), and in *Crotalaria brevifolia*, and *C. incarnata* (Leguminosae). It is an isomer of senecionine (q.v.).

Toxic. It is mutagenic to *Drosophila*. This alkaloid is collected by adult Monarch butterflies, together with other *Senecio* alkaloids such as senecionine (q.v.), from *Senecio* plants and utilized for pheromone production and defence against bird predation.

952
Intermedine; 9-(+)-Trachelanthylretronecine

$C_{15}H_{25}NO_5$ Mol. wt 299.37

Found in Russian comfrey, *Symphytum × uplandicum*, in *Amsinckia intermedia* and *A. lycopsoides*, and in *Trichodesma africana* (Boraginaceae), and in *Conoclinum coelastinum* (Compositae). It is diastereoisomeric with indicine (q.v.).

Hepatotoxic.

953
Isatidine; Retrorsine *N*-oxide

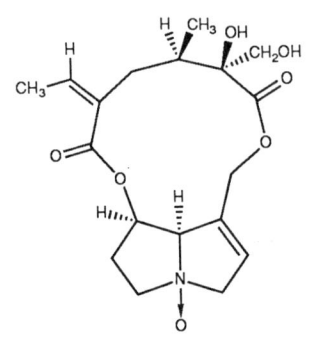

$C_{18}H_{25}NO_7$ Mol. wt 367.41

Found in *Senecio longilobus* and other *Senecio* spp. (Compositae), as for retrorsine (q.v.), to which it is converted during extraction.

Hepatotoxic, causing veno-occlusive disease in humans. It is carcinogenic in animals. It is far less toxic than retrorsine (q.v.), except when given orally, where it is converted by gut enzymes to retrorsine base.

954
Jacobine

$C_{18}H_{25}NO_6$ Mol. wt 351.41

Found in the tansy ragwort, *Senecio jacobaea*, and other *Senecio* spp., including *S. alpinus, S. cineraria* and *S. incarnus* (Compositae).

Toxic; it is potent inducer of hepatic epoxide hydrolase, and anticholinergic in rats. Also, it is mutagenic to plant cell chromosomes.

955
Lasiocarpine; Lassiocarpine;
7-Angelyl-9-lasiocarpylheliotridine

$C_{21}H_{33}NO_7$ Mol. wt 411.50

Found in *Heliotropium lasiocarpum, H. hirsutum, H. arbainense*, and *H. europaeum*, in *Lappula intermedia*, in comfrey, *Symphytum officinale*, and *S. asperum* (all Boraginaceae).

Hepatotoxic. It is carcinogenic to rat liver, skin, and intestine. Also, it is mutagenic in all systems tested, including the Ames test, *Drosophila*, DNA repair test and cultured cell tests.

956
Lycopsamine; 9-Viridiflorylretronecine

$C_{15}H_{25}NO_5$ Mol. wt 299.37

Occurs in *Parsonsia eucalyptophylla* and *P. straminea* (Apocynaceae), in *Amsinckia hispida* and *A. intermedia*, in *Anchusa arvensis (Lycopsis arvensis), Heliotropium steudneri, Messerschmidia sibirica*, Russian comfrey, *Symphytum × uplandicum*, and borage, *Borago officinale* (all Boraginaceae), and in *Eupatorium compositifolium* (Compositae). It is a stereoisomer of intermedine (q.v.).

Hepatotoxic. It is nonmutagenic in the Ames test.

957
Monocrotaline; Crotaline

$C_{16}H_{23}NO_6$ Mol. wt 325.37

Occurs in *Crotalaria crispata, C. paulina, C. quinquefolia, C. stipularia*, and many other *Crotalaria* spp. (Leguminosae), and in *Lindelofia spectabilis* (Boraginaceae).

Hepatocarcinogenic and pneumotoxic. It is mutagenic to cultured mammalian cells, in the DNA repair test and to *Drosophila* chromosomes. This is the causative agent of accidental poisoning due to contamination in bread by alkaloid-containing material.

958
Otonecine

C$_9$H$_{15}$NO$_3$ Mol. wt 185.23

Thought to occur as the free necine base in plants containing the esters, such as senkirkine (q.v.), in various *Senecio* spp. (Compositae).

Hepatotoxic.

959
Petasitenine; Fukinotoxin

C$_{19}$H$_{27}$NO$_7$ Mol. wt 381.43

Found in *Petasites japonicus* and *P. hybridus* (Compositae).

Hepatocarcinogenic. It is mutagenic to cultured mammalian cells, and in the Ames and DNA repair tests.

960
Platyphylline

C$_{18}$H$_{27}$NO$_5$ Mol. wt 337.42

Occurs in *Senecio platyphyllus* and other *Senecio* spp., in *Petasites laevigatus*, and in *Adenostyles* spp. (all Compositae).

Nontoxic in all systems tested: nonhepatotoxic; nonalkylating; nonmutagenic; no effect on mitosis. It is anticholinergic in animals. In the USSR it has been used for treating peptic ulceration.

961
Retronecine

C$_8$H$_{13}$NO$_2$ Mol. wt 155.20

Occurs in *Senecio pseudo-orientalis*, and other *Senecio* spp. (Compositae), some *Crotalaria* spp. (Leguminosae), and some spp. of Boraginaceae. The 7-*O*-angelyl ester occurs in *Senecio sylvaticus*. It is the most common base form of pyrrolizidine alkaloids.

Hepatotoxic. It is dehydrogenated after ingestion to the related pyrrole, which is more toxic because it binds to the DNA in the liver.

962
Retrorsine; β-Longilobine

C$_{18}$H$_{25}$NO$_6$ Mol. wt 351.41

Occurs in *Senecio retrorsus*, *S. vulgaris*, *S. filaginoides*, *S. phillipicus*, *S. grisebachii*, and other *Senecio* spp. (Compositae), and in *Crotalaria usaramoenensis* and *C. spartioides* (Leguminosae).

Hepatotoxic and pneumotoxic. It is mutagenic to *Drosophila* chromosomes, and in the Ames test.

963
Riddelline; Riddelliine; 18-Hydroxyseneciphylline

$C_{18}H_{23}NO_6$ Mol. wt 349.39

Found in the tansy ragwort, *Senecio vulgaris*, *S. riddellii*, *S. aegypticus*, *S. ambrosioides* and *S. eremophilus* (Compositae), and in *Crotalaria juncea* (Leguminosae). The alkaloid content of *S. riddellii* is exceptionally high, some plants containing 10–18% of the leaf dry weight.

Hepatotoxic. *S. longilobus*, which contains riddelline, has caused fatal veno-occlusive disease. Carcinogenicity tests have been inconclusive. It is anticholinergic in animals.

964
Rinderine; 9-(+)-Trachelanthylheliotridine

$C_{15}H_{25}NO_5$ Mol. wt 299.37

Found in *Rindera baldschuanica* and *Solenanthus circinatus* (Boraginaceae), and in *Eupatorium altissimum* and *E. cannabinum* (Compositae).

Hepatotoxic, and pneumotoxic.

965
Rosmarinine

$C_{18}H_{27}NO_6$ Mol. wt 353.42

Occurs in *Senecio rosmarinifolius* (Compositae).

Nonhepatotoxic and nonmutagenic in cell transformation tests. Dehydration gives senecionine (q.v.).

966
Sarracine; Mikanoidine; 7-Angelyl-9-sarracinylplatynecine

$C_{18}H_{27}NO_5$ Mol. wt 337.42

Occurs in *Senecio sarracinus*, and *S. sylvaticus* (Compositae).

Anticholinergic in animals. It is nonhepatotoxic, in common with other saturated necines, and has been used in the USSR for treating peptic ulceration.

967
Senaetnine

$C_{20}H_{23}NO_7$ Mol. wt 389.41

Occurs in *Senecio aetnensis* (Compositae).

Causes damage to pulmonary vascular tissue in rats but without hepatotoxicity.

968
Senampeline A

$C_{25}H_{31}NO_8$ Mol. wt 473.52

Occurs in *Senecio aetnensis*, *S. aucheri* and *S. pterophorus* (Compositae).

Toxicity not established, but suspected to be hepatotoxic.

969
Senecionine; 12-Hydroxysenecionane-11,16-dione

$C_{18}H_{25}NO_5$ Mol. wt 335.41

Found in the common groundsel, *Senecio vulgaris*, the tansy ragwort, *S. jacobaea*, and many other *Senecio* spp. (Compositae). It is also present in *Castilleja rhexifolia* (Scrophulariaceae), a hemiparasite, which obtains it by absorption from its host plant *S. triangularis*.

Hepatotoxic, pneumotoxic, and genotoxic. It is mutagenic to *Drosophila* chromosomes, but not in the Ames test. Also, it is anticholinergic in rats.

970
Seneciphylline; Jacodine; α-Longilobine

$C_{18}H_{23}NO_5$ Mol. wt 333.39

Found in *Senecio platyphyllus*, *S. phillipicus* and many other *Senecio* spp. (Compositae), and in *Crotalaria juncea* (Leguminosae).

Hepatotoxic, cardiotoxic, and pneumotoxic. It is anticholinergic in rats, and a cause of grazing toxicity in farm animals.

971
Senecivernine

C₁₈H₂₅NO₅   Mol. wt 335.41

Occurs in *Senecio vernalis* and *S. seratophiloides* (Compositae).

Hepatotoxic.

972
Senkirkine; Renardine

C₁₉H₂₇NO₆   Mol. wt 365.43

Found in *Emilia sonchifolia, Brachyglottis repanda, Farfugium japonicum, Petasites albus, P. hybridus, Senecio jacobaea, S. kirkii, S. illinitus, S. renardii, S. vernalis,* and *Tussilago farfara* (Compositae), and in *Crotalaria laburnifolia* (Leguminosae).

Hepatocarcinogenic in animals. It is mutagenic in the Ames test, and to cultured cells and *Drosophila* chromosomes.

973
Supinidine

C₈H₁₃NO   Mol. wt 139.20

Necine base probably found free in plants containing the esters, e.g., supinine (q.v.), in some *Heliotropium* spp. (Boraginaceae).

974
Supinine

C₁₅H₂₅NO₄   Mol. wt 283.37

Occurs in *Heliotropium indicum, H. supinum, Tournefortia sarmentosa,* and *T. zeylanicum* (Boraginaceae), and in *Eupatorium cannabinum, E. serotinum* and *E. stoechadosum* (Compositae).

Hepatotoxic. It is anticholinergic in animals, and produces chromosomal aberrations in plant cells.

975
Symlandine; 7-Angelyl-9-(−)-viridiflorylretronecine

C₂₀H₃₁NO₆   Mol. wt 381.48

Found in Russian comfrey, *Symphytum × uplandicum* (Boraginaceae).

Toxicity not established, but suspected to be hepatotoxic.

976
Symphytine; 7-Tiglyl-9-(−)-viridiflorylretronecine

$C_{20}H_{31}NO_6$ Mol. wt 381.48

Found in comfrey, *Symphytum officinale*, Russian comfrey, *S. × uplandicum*, *S. orientale* and the water forget-me-not, *Myosotis scorpioides* (Boraginaceae).

Hepatocarcinogenic.

977
Triangularine; 6-Angelyl-9-sarracinylretronecine

$C_{18}H_{25}NO_5$ Mol. wt 355.41

Occurs in *Senecio triangularis* (Compositae) and *Alkanna tinctoria* (Boraginaceae).

Toxicity not established, but suspected to be hepatotoxic.

978
Tussilagine

$C_{10}H_{17}NO_3$ Mol. wt 199.25

Occurs in coltsfoot, *Tussilago farfara* (Compositae).

Toxicity not established. It did not cause chromosome damage in cultured human lymphocytes.

979
Uplandicine; 7-Acetyl-9-echimidinylretronecine

$C_{17}H_{27}NO_7$ Mol. wt 357.40

Occurs in Russian comfrey, *Symphytum × uplandicum* (Boraginaceae).

Toxicity not established, but suspected to be hepatotoxic.

980
Usaramine; Mucronatine; Usuramine

$C_{18}H_{25}NO_6$ Mol. wt 351.41

Found in *Crotalaria brevifolia*, *C. incana*, *C. intermedia*, *C. mucronata*, and *C. usaramoensis* (Leguminosae).

Hepatotoxic, pneumotoxic, and hypotensive.

Chapter 27
Quinoline alkaloids

Quinoline alkaloids are derived biosynthetically from anthranilic (2-aminobenzoic) acid and are based on a bicyclic system in which a benzene and a pyridine ring are fused together. Echinopsine (**1004**) is an example of a simple quinoline, while japonine (**1023**) has a similar structure but is substituted by methoxyl and phenyl groups. Attachment of a furan ring to the pyridine nucleus, as in dictamnine (**1001**) gives rise to the furoquinolines, while similar attachment of a pyran ring produces an alkaloid such as flindersine (**1012**).

Attachment of a benzene ring to a quinoline gives rise to the acridine skeleton and a series of alkaloids exemplified by acronycine (**983**) and evoxanthidine (**1007**). Acridine dimers are known, as in the case of atalanine (**990**) from *Atalantia ceylonica*. In yet another group, the quinoline alkaloids contain two nitrogen atoms in the heterocyclic ring and are known as quinazolines. Two examples are aniflorine (**985**) and febrifugine (**1010**).

Structural variation is due to the attachment of varying numbers of hydroxyl, methoxyl or methylenedioxy groups to the quinoline nucleus. Isoprenyl substitution can also occur, as in atalaphylline (**991**), a diprenylated acridine. The nitrogen atom of the quinoline nucleus is sometimes methylated, and this can lead to the existence of quaternary forms. Several alkaloids, e.g., balfourodinium (**993**), are known naturally as the quaternary salts.

The major occurrence in nature of the quinoline alkaloids is in the plant family Rutaceae (Price, 1963; Waterman and Grundon, 1985). Compounds such as dictamnine (**1001**), γ-fagarine (**1009**) and skimmianine (**1043**) are widespread among these plants. A typical rutaceous source is the rue, *Ruta graveolens*, from which thirty quinoline alkaloids have been identified. While the acridines and furoquinolines are largely confined to the Rutaceae, the quinazolines have a wider distribution. They have been reported from members of Acanthaceae, Malvaceae, Saxifragaceae and Zygophyllaceae as well as in the Rutaceae. It may also be noted that the simple isoquinoline echinopsine (**1004**) was found in the Compositae, so this group of alkaloids may occasionally be found in yet other plant families.

Although quinoline alkaloids exhibit a range of pharmacological activities, few are used in medicine at the present time. Cinchonidine (**997**) and quinine (**1038**) from the bark of the cinchona tree are, however, well known for their antimalarial properties. Quinoline alkaloids are relatively easily detected in plant extracts either by standard alkaloid tests or by means of their aromatic absorption. The furoquinolines, in particular, have characteristic ultraviolet spectra, and can then be identified further by proton nuclear magnetic resonance spectral measurements.

References

Acheson, R.M. (1956). *Acridines*. New York: Interscience.
Price, J.R. (1963). Distribution of alkaloids in Rutaceae. In *Chemical Plant Taxonomy* (T. Swain, ed.), pp. 429–452. London: Academic Press.
Waterman, P.G. and Grundon, M.F. (eds) (1985). *Chemistry and Chemical Taxonomy of the Rutales*. London: Academic Press.

In *The Alkaloids* (Manske, R.H.F. and Brossi, A., eds) (1952) Volume 2, pp. 353–368; (1955) Volume 3, pp. 1–118; (1960) Volume 7, pp. 229–252; (1967) Volume 9, pp. 223–267; (1973) Volume 14, pp. 181–223; (1979) Volume 17, pp. 105–198; (1983) Volume 21, pp. 1–28; (1986) Volume 29, pp. 99–140; (1988) Volume 32, pp. 341–439; (1988) Volume 34, pp. 331–398.

981
Acronidine

$C_{18}H_{17}NO_4$ Mol. wt 311.34

Found in the leaves of *Acronychia baueri*, *Melicope leptococca* and *Sarcomelicope leiocarpa* (Rutaceae).

982
Acronycidine

$C_{15}H_{15}NO_5$ Mol. wt 289.29

Occurs in the leaves of *Acronychia baueri*, and the bark of *Melicope fareana* (Rutaceae).

Central nervous system depressant.

983
Acronycine

$C_{20}H_{19}NO_3$ Mol. wt 321.38

Occurs in *Acronychia baueri*, *A. haplophylla*, and *Melicope leptococca* (Rutaceae).

Potent antitumour agent; possesses a broad-spectrum activity against experimental neoplasts. It is carcinogenic.

984
Acrophylline

$C_{17}H_{17}NO_3$ Mol. wt 283.33

Occurs in the leaves and bark of *Acronychia haplophylla* (Rutaceae), together with acronycine (q.v.).

985
Aniflorine

$C_{20}H_{21}N_3O_3$ Mol. wt 351.41

Occurs in the leaves and branches of *Anisotes sessiliflorus* (Acanthaceae).

986
Anisessine

$C_{20}H_{19}N_3O_3$ Mol. wt 349.39

Found in *Anisotes sessiliflorus* (Acanthaceae).

987
Anisotine

C₂₀H₁₉N₃O₃   Mol. wt 349.39

Found in *Anisotes sessiliflora* and *Adhatoda vasica* (Acanthaceae).

988
Arborine; Glycosine

C₁₆H₁₄N₂O   Mol. wt 250.30

Major alkaloid in the leaves of *Glycosmis arborea*, in *Ruta graveolens* and in the fruit of *Zanthoxylum budrunga* (all Rutaceae).

Inhibits peripheral action of acetylcholine and acts as a central hypotensive agent. It shows an inhibitory action against pituitrin on rat uterus.

989
Arborinine

C₁₆H₁₅NO₄   Mol. wt 285.30

Found in *Glycosmis arborea, Fagara leprieurii, Euodia xanthoxyloides, Ruta graveolens, Teclea natalensis* and other Rutaceae, mostly together with arborine (q.v.).

Spasmolytic activity.

990
Atalanine; *trans*-Atalantine

C₃₄H₃₀N₂O₉   Mol. wt 610.63

Occurs in *Atalantia ceylanica* (syn. *A. wightii*) (Rutaceae), together with ataline, an alkaloid similar in structure.

Decoction of the leaves of the above plant is applied for itching and other skin complaints.

991
Atalaphylline

C₂₃H₂₅NO₄   Mol. wt 379.46

Found in the root bark of *Atalantia monophylla* (Rutaceae).

992
Balfourodine

C₁₆H₁₉NO₄   (+) form   Mol. wt 289.34

Occurs in *Balfourodendron riedelianum* and *Ptelea trifoliata* (Rutaceae).

993
Balfourodinium

$[C_{17}H_{22}NO_4]^+$ Mol. wt 304.37

Occurs in the (+) form (shown) in the bark of *Balfourodendron riedelianum*, and in the (−) form in the leaves, roots and stems of *Choisya ternata* (both Rutaceae).

Plant growth inhibitor.

994
Bucharaine

$C_{19}H_{25}NO_4$ Mol. wt 331.42

Found in *Haplophyllum bucharicum* (Rutaceae).

Hypothermic activity. It suppresses aggressive responses to electrical stimulation in rats.

995
Camptothecin; Camptothecine

$C_{20}H_{16}N_2O_4$ Mol. wt 348.36

Occurs in the fruits, stemwood and bark of *Camptotheca acuminata* (Nyssaceae) and *Mappia foetida* (Icacinaceae).

High toxicity in both animals and humans. It has pronounced antitumour and antileukaemia activities, and is widely used in China in the treatment of various forms of cancer.

996
Casimiroin

$C_{12}H_{11}NO_4$ Mol. wt 233.23

Occurs in the seeds, trunk and root bark of *Casimiroa edulis* (Rutaceae).

997
Cinchonidine; Cinchocatine; Cinchonan-9-ol; α-Quinidine

$C_{19}H_{22}N_2O$ Mol. wt 294.40

Occurs in *Cinchona succirubra* and *C. tucujensis*, and *Remijia* spp. (Rubiaceae), and in the leaves of *Olea europaea* and *Ligustrum vulgare* (Oleaceae). Its stereoisomer, cinchonine, occurs in most of these spp., including *Cinchona officinalis*.

Antimalarial activity; it is a resolving agent.

998
Cusparine

$C_{19}H_{17}NO_3$ Mol. wt 307.35

Found in the bark of *Galipea officinalis*, angostura bark (Rutaceae).

Antispasmodic properties. Angostura bark has antipyretic, antiperiodic, antidysenteric and bitter tonic properties. In pharmacy it is therefore used in bitter tonics, but not in the drinks called "angostura bitters". In folk medicine, angostura serves as a febrifuge, an antidiarrhoeic and as a bitter tonic.

999
Deoxypeganine; Deoxyvasicine

$C_{11}H_{12}N_2$ Mol. wt 172.23

Occurs in the aerial parts of *Peganum harmala* and in the leaves of *P. nigellastrum* (Zygophyllaceae).

Cholinergic activity.

1000
Deoxyvasicinone

$C_{11}H_{10}N_2O$ Mol. wt 186.22

Found in the leaves of *Peganum nigellastrum* and in the aerial parts and seeds of *P. harmala* (Zygophyllaceae), in the roots of *Adhatoda vasica* (Acanthaceae), in *Linaria* spp. (Scrophulariaceae), and in *Mackinlaya macrosciadia* (Araliaceae).

Cholinergic activity.

1001
Dictamnine; Dictamine

$C_{12}H_9NO_2$ Mol. wt 199.21

Occurs widely within the Rutaceae: in the roots of *Dictamnus albus*, in *Adiscanthus fusciflorus*, *Aegle marmelos*, *Afraegle paniculata*, and *Esenbeckia* spp., and in the genera *Flindersia*, *Geijera*, *Glycosma*, *Haplophyllum*, *Ruta* and *Zanthoxylum*.

Antibacterial and antifungal activities, and DNA-binding effects. Also, it is a strong muscle contractant.

1002
Dubamine

$C_{16}H_{11}NO_2$ Mol. wt 249.27

Occurs in *Haplophyllum dubium* and *H. latifolium* (Rutaceae).

Hydrochloride is a strong antimicrobial agent.

1003
Dubinidine

$C_{15}H_{17}NO_4$ Mol. wt 275.31

Found in the leaves of *Haplophyllum dubium* (Rutaceae).

Central nervous system depressant (sedative). It has hypothermic activity.

1004
Echinopsine

C₁₀H₉NO   Mol. wt 159.19

Found in the seeds of *Echinops ritro* and other *Echinops* spp. (Compositae).

1005
Echinorine

[C₁₁H₁₂NO]⁺   Mol. wt 174.22

Found in the fruits of *Echinops ritro*, in the leaves of *E. commutatus* and in *E. sphaerocephalus* (Compositae).

1006
Evoprenine

C₂₀H₂₁NO₄   Mol. wt 339.40

Found in *Euodia alata* (Rutaceae).

1007
Evoxanthidine

C₁₅H₁₁NO₄   Mol. wt 269.26

Occurs in *Euodia xanthoxyloides* (Rutaceae).

1008
Evoxine; Haploperine

C₁₈H₂₁NO₆   Mol. wt 347.37

Occurs in the leaves of *Orixa japonica*, in *Euodia xanthoxyloides*, in *Haplophyllum* spp., in *Teclea boiviniana* and in *Monnieria trifolia* (all Rutaceae).

Sedative and spasmolytic effects. Also moderate insect antifeedant activity.

1009
γ-Fagarine; Aegelenine; Haplophine; 8-Methoxydictamnine

C₁₃H₁₁NO₃   Mol. wt 229.24

Widespread occurrence within the Rutaceae: in *Haplophyllum* spp., *Fagara coco*, *Aegle marmelos*, *Dictamnus angustifolius*, and *Flindersia* and *Zanthoxylum* spp.

Anti-arrhythmic principle of *F. coco*.

1010
Febrifugine; β-Dichroine; γ-Dichroine; Dichroine B

$C_{16}H_{19}N_3O_3$ Mol. wt 301.35

Occurs in the roots of *Dichroa febrifuga* and in *Hydrangea* spp. (Saxifragaceae).

Antipyretic, emetic and antimalarial activities. Also, it is used as a coccidostat in veterinary practice. It is a hundred times more active than quinine (q.v.) as an antimalarial drug, but its use as such is limited by its toxicity.

1011
Flindersiamine

$C_{14}H_{11}NO_5$ Mol. wt 273.25

Occurs in *Flindersia collina* and *F. maculosa*, *Araliopsis soyauxii* and *Teclea verdoorniana* (Rutaceae).

Phototoxic against *Saccharomyces cerevisiae* and *Candida albicans*.

1012
Flindersine

$C_{14}H_{13}NO_2$ Mol. wt 227.27

Found in *Haplophyllum perforatum* and *Flindersia australis* (Rutaceae).

1013
Furofoline I; Furofoline; Furacridone

$C_{16}H_{11}NO_3$ Mol. wt 265.27

Found in the root and stem bark of *Glycosmis citrifolia* and in the stem bark of *Ruta graveolens* (both Rutaceae).

1014
Galipine

$C_{20}H_{21}NO_3$ Mol. wt 323.40

Occurs in the bark of *Cusparia febrifuga*, angostura (Rutaceae).

Antispasmodic activity.

1015
Glycophymoline

$C_{16}H_{14}N_2O$ Mol. wt 250.30

Found in the flower heads of *Glycosmis pentaphylla* (Rutaceae).

1016
Glycosminine; Glycophymine

C₁₅H₁₂N₂O Mol. wt 236.28

Found in the leaves and flowers of *Glycosmis arborea* (=*pentaphylla*) (Rutaceae).

Inhibits specifically serine protease and human leukocyte elastase.

1017
Graveoline; Rutamine

C₁₇H₁₃NO₃ Mol. wt 279.30

Occurs in *Ruta graveolens* and *R. bracteosa* (Rutaceae).

1018
Haplophyllidine

C₁₈H₂₃NO₄ Mol. wt 317.39

Found in the seeds of *Haplophyllum perforatum* (Rutaceae).

Hypnotic synergist and strong central nervous system depressant.

1019
Haplopine

C₁₃H₁₁NO₄ Mol. wt 245.24

Found in *Monnieria trifolia, Zanthoxylum cuspidatum, Melicope lasioneura* and *Haplophyllum robustum* (all Rutaceae). It is present as the α-mannoside, glycoperine, in *H. perforatum*.

Phototoxic activity against *Saccharomyces cerevisiae* in long wavelength UV.

1020
Hydroquinidine; Hydrochinene; Hydroconchinene; Hydroconchinine; Quinotidine

C₂₀H₂₆N₂O₂ Mol. wt 326.44

Found in the bark of *Cinchona officinalis* as minor constituent, and also in *Remijia pedunculata* (both Rubiaceae); it is also found in the bark and leaves of *Aspidiosperma marcgravianum* (Apocynaceae).

Antimalarial activity.

1021
Isodictamnine

C₁₂H₉NO₂   Mol. wt 199.21

Occurs in the root bark of *Dictamnus albus* and in the trunk bark of *Helietta longifoliata* (both Rutaceae).

Phototoxic to bacteria and yeasts.

1022
Isofebrifugine

C₁₆H₁₉N₃O₃   Mol. wt 301.35

Occurs in *Dichroa febrifuga* and *Hydrangea umbellata* (Saxifragaceae).

Anti-moth activity.

1023
Japonine

C₁₈H₁₇NO₃   Mol. wt 295.34

Found in the leaves of *Orixa japonica* (Rutaceae).

1024
Kokusaginine; 6,7-Dimethoxydictamnine

C₁₄H₁₃NO₄   Mol. wt 259.27

Common within Rutaceae, e.g., in spp. of *Euodia, Orixa, Haplophyllum, Melicope* and *Acronychia*.

Moderate insect antifeedant activity, and phototoxic to yeasts (*Candida albicans, Saccharomyces cerevisiae*). It enhances noradrenaline and dopamine levels in the mouse brain.

1025
Lunacridine

C₁₇H₂₃NO₄   Mol. wt 305.37

Occurs in the bark of *Lunasia costulata, L. amara* and *L. quercifolia* (Rutaceae).

Hypotensive agent.

1026
Lunacrine

$C_{16}H_{19}NO_3$ Mol. wt 273.34

Occurs in the bark of *Lunasia costulata* and *L. amara* (Rutaceae).

Hypotensive agent, causing transient fall in blood pressure in cats. It is toxic (LD_{50} intravenously in mice 80 mg kg^{-1}).

1027
Lunamarine

$C_{18}H_{15}NO_4$ Mol. wt 309.33

Found in *Lunasia amara* (Rutaceae).

Weak smooth muscle stimulant, stimulating isolated rabbit intestine and uterus. Also, it shows hypotensive activity, causing a transient fall in blood pressure in cats.

1028
Maculine; 6,7-Methylenedioxydictamnine

$C_{13}H_{19}NO_4$ Mol. wt 253.30

Occurs in *Esenbeckia litoralis*, *Helietta parvifolia*, *Sargentea greggii*, *Teclea simplicifolia*, *Flindersia dissosperma* and *F. maculosa* (Rutaceae).

1029
Maculosidine; 6,8-Dimethoxydictamnine

$C_{14}H_{13}NO_4$ Mol. wt 259.27

Found in *Eriostemon* spp. and in *Flindersia maculosa* (Rutaceae).

Phototoxic to the yeasts *Candida albicans* and *Saccharomyces cerevisiae*.

1030
Maculosine

$C_{17}H_{17}NO_6$ Mol. wt 331.33

Occurs in the bark of *Flindersia maculosa* (Rutaceae).

1031
Melicopicine

$C_{18}H_{19}NO_5$ Mol. wt 329.36

Occurs in *Acronychia baueri*, *Teclea boiviniana*, *Melicope fareana* and *M. leptococca* (Rutaceae).

1032
Melicopine

$C_{17}H_{15}NO_5$ Mol. wt 313.31

Found in *Melicope fareana*, *Euodia alata*, *Acronychia baueri*, *Teclea natalensis* and *Bauerella simplicifolia* (Rutaceae).

Found in *Ptelea trifoliata*, *Xylocarpus granatum*, *Atalantia roxburghiana* and *Fagara chalybaea* (Rutaceae).

Antifeedant activity against beetles. Also, it shows antimicrobial action against the yeast *Candida albicans*.

1033
Melochinone

$C_{22}H_{21}NO_2$ Mol. wt 331.42

Occurs in *Melochia tomentosa* (Sterculiaceae).

1034
N-Methylflindersine

$C_{15}H_{15}NO_2$ Mol. wt 241.29

1035
O-Methylptelefolonium; Ptelefolonium

$[C_{18}H_{22}NO_4]^+$ Mol. wt 316.38

Occurs in the leaves of *Ptelea trifoliata*, and also in callus cultures of the same plant (Rutaceae).

The chloride inhibits growth of plant tissues and seedlings. It is also cytotoxic against plant and animal tumours, and shows antimicrobial and antifungal activities.

1036
Peganine; Linarine; Vasicine

$C_{11}H_{12}N_2O$ Mol. wt 188.23

The (−) form (illustrated) is found in the seeds of *Peganum harmala* (Zygophyllaceae), in the leaves of *Adhatoda vasica* (Acanthaceae), in the roots of *Sida cordifolia* (Malvaceae) and in *Lunaria* spp. (Cruciferae). The (±) form is found as a major alkaloid from *Anisotes sessiliflorus* (Acanthaceae).

Hypotensive effects, bronchodilatory activity, and respiratory stimulant. It also stimulates smooth muscle, is an abortifacient, an anthelmintic, and shows uterotonic effects.

1037
Pteleatine

$[C_{16}H_{20}NO_4]^+$ Mol. wt 290.34

Found in leaves, stems and branches of *Ptelea trifoliata* (Rutaceae).

The chloride shows antimicrobial activity against *Mycobacterium smegmatis*, *Staphylococcus aureus*, and *Candida albicans*.

1038
Quinine

$C_{20}H_{24}N_2O_2$ Mol. wt 324.43

Found in the bark of *Cinchona officinalis* and in many other *Cinchona* spp., and also in *Remijia pedunculata* (Rubiaceae). It occurs together with its stereoisomer quinidine (syn: β-quinine; pitayine; conquinine; conchinine).

Both isomers are important antimalarial drugs; used also against *Plasmodium falcicarpum*, which is usually drug-resistant. Quinine is weakly active as a cardiac depressant and partly used as a clinical antifibrillatory agent. Quinine acts as a stimulant for horses and is used in horse doping. It has abortifacient and spermicidal properties and is used as a bitter stomachic, analgesic and antipyretic in veterinary practice. Quinidine has antitumour and immunosuppressive activities. Quinine is one of the bitterest substances known and is significantly bitter to humans at a molar concentration of 1×10^{-5}. It is widely used in the food industry as a bittering agent (e.g., in bitter lemon drinks).

1039
Ribalinium

$[C_{16}H_{20}NO_4]^+$ Mol. wt 290.34

Occurs in leaves of *Ruta graveolens* and in the trunk bark of *Balfourodendron riedelianum* (Rutaceae).

The ribalinium salt is moderately active against *Mycobacterium smegmatis*.

1040
Robustine

$C_{12}H_9NO_3$ Mol. wt 215.21

Found in *Dictamnus caucasicus*, *Haplophyllum robustum*, *Thamnosma montana*, *Toddalia aculeata*, and *Zanthoxylum* spp. (Rutaceae).

Potentiates effects of barbiturate.

1041
Rutacridone

$C_{18}H_{18}NO_3$ Mol. wt 296.35

Found in *Ruta graveolens* and *R. chalepensis* (Rutaceae).

1042
Rutacridone epoxide

$C_{19}H_{17}NO_4$ Mol. wt 323.35

Found in the roots and callus tissue of *Ruta graveolens* (Rutaceae).

Antimicrobial and antibacterial activities.

1043
Skimmianine; Chloroxylonine; 7,8-Dimethoxydictamnine; β-Fagarine; Pentaphylline

$C_{14}H_{13}NO_4$ Mol. wt 259.27

Fairly common occurrence within Rutaceae: in *Skimmia japonica, Dictamnus albus, Essenbeckia flava, Fagara manschurica, Ruta graveolens*, and in the genera *Flindersia, Glycosmis, Haplophyllum, Zanthoxylum* and *Murraya*.

Analgesic, anticonvulsant and antipyretic activities. It potentiates the effects of barbiturate and acts as a central nervous system depressant. Also, it is phototoxic to the yeasts *Candida albicans* and *Saccharomyces cerevisiae*.

1044
Tryptanthrine; Couroupitine A

$C_{15}H_8N_2O_2$ Mol. wt 248.24

Occurs in the leaves of *Strobilanthes cusia* (Acanthaceae), in *Polygonum tinctorum* (Polygonaceae), and in *Couroupita guaianensis* (Lecythidaceae). Also, it has been isolated from the yeast *Candida albicans*, grown in the presence of tryptophan.

Antimicrobial and antifungal activities against a range of dermatophytes. *Strobilanthes cusia* is used as a folk remedy for athlete's foot in Japan.

1045
Vasicinol; 7-Hydroxypeganine

$C_{11}H_{12}N_2O_2$ Mol. wt 204.23

Found in the roots, leaves and seeds of *Adhatoda vasica* (Acanthaceae), and in the roots of *Sida cordifolia* (Malvaceae).

Antifertility effects, and active as an insect antifeedant. It shows moderate anticholinesterase activity, and is active as a histamine antagonist, a cardiac depressant and as a transient hypotensive agent.

1046
Vasicinone

$C_{11}H_{10}N_2O_2$ Mol. wt 202.22

The (−) form shown is found in the seeds and aerial parts of *Peganum harmala* and in the leaves of *P. nigellastrum* (Zygophyllaceae); it is also found in the roots of *Sida cordifolia* (Malvaceae), and in the leaves of *Adhatoda vasica* (Acanthaceae). Further sources are *Biebersteinia multifida* (Biebersteiniaceae) and *Nitraria sibirica* (Zygophyllaceae). The (±) form occurs in the inflorescence of *Adhatoda vasica*.

Anthelminthic and antifertility activities. It shows bronchodilatory and weakly hypotensive ionotropic action. It is also an antifeedant.

1047
Veprisinium

$[C_{18}H_{24}NO_5]^+$ Mol. wt 334.40

Occurs in the stem bark of *Vepris louisii* (Rutaceae).

Antibacterial activity against *Staphylococcus aureus*. In Cameroon, the plant extract is used for the treatment of skin diseases of bacterial origin.

Chapter 28
Quinolizidine alkaloids

Quinolizidine alkaloids are chemically defined by the presence of a structural unit in which a nitrogen atom occupies a central position in two fused cyclohexane rings, as in lupinine (**1068**). They can also be defined botanically by their occurrence in plants of the Leguminosae and, particularly, in the genus *Lupinus*, and they are often referred to as the lupin alkaloids. Structurally related but distinctively different quinolizidine alkaloids occur in *Lycopodium* and are known as Lycopodium alkaloids (q.v.), while a few more complex compounds of this type may be found under isoquinoline and indole alkaloids (q.v.).

There are about 170 quinolizidine alkaloids in legumes (Kinghorn and Balandrin, 1984) and these are readily classified according to the number of rings present. A typical bicyclic structure is lupinine (**1098**), while there are tricyclic (e.g., angustifoline, **1052**) and tetracyclic compounds (e.g., anagyrine, **1051**). Some of these alkaloids occur with aromatic ester attachment; cinegalline (**1060**) is such an ester of 13-hydroxylupanine (**1063**). A few hexacyclic alkaloids are also known, especially in the genus *Ormosia*, substances such as jamine (**1064**) and panamine (**1073**). Finally, several dimeric quinolizidines have been found, such as argentine (**1054**) and dimethamine (**1062**). Quinolizidine alkaloids are synthesized in the leaf from lysine and cadaverine as precursors; an important intermediate is 17-oxosparteine, which is the immediate precursor of lupanine (**1067**) and sparteine (**1078**).

Although universally present in lupins, a genus of some 200 species, these alkaloids are regularly found in over 40 related genera of legumes. They occur, for example, in a number of well known trees and shrubs, such as broom, laburnum and gorse. They have been responsible for much accidental poisoning in sheep, cattle and humans from the consumption of such plants, and particularly their seeds. Cytisine and anagyrine are more acutely toxic than lupanine and sparteine, and they are also damaging in livestock due to their teratogenic effects. A few of these alkaloids are potentially useful in medicine; retamine, for example, has strong hypotensive activity. The chemical ecology of lupin alkaloids has been extensively studied (Wink, 1987), and there is good evidence that they are of protective value against herbivory in the plants in which they occur.

Quinolizidine alkaloids can be detected after thin layer chromatography by the colours produced by Munier's modification of the Dragendorff reagent. They are commonly identified and quantified by combined gas chromatography/mass spectrometry, but a range of other techniques is also available for their analysis (Kinghorn and Balandrin, 1984).

References

Kinghorn, A.D. and Balandrin, M.F. (1984). Quinolizidine alkaloids of the Leguminosae. In *Alkaloids, Chemical and Biological Perspectives*, Volume 2 (Pelletier, W.S., ed.), pp. 105–147. Chichester: Wiley.

Smolenski, S.J., Kinghorn, A.D. and Balandrin, M.F. (1981). Toxic constituents of legume forage plants, *Economic Botany* **35**, 321–355.

Wink, M. (1987). *Defensive Role of Quinolizidine Alkaloids*. American Chemical Society Symposium Series 330.
In *The Alkaloids* (Manske, R.H.F. and Brossi, A., eds) (1953) Volume 3, pp. 119–199; (1960) Volume 7, pp. 253–317; (1967) Volume 9, pp. 175–221; (1987) Volume 31, pp. 116–192.

1048
Albine; Dehydroalbine

$C_{14}H_{20}N_2O$ Mol. wt 232.33

Found in the seeds of *Lupinus albus* (Leguminosae).

1049
Aloperine

$C_{15}H_{24}N_2$ Mol. wt 232.37

Found in the seeds and aerial parts of *Sophora alopecuroides* (Leguminosae).

Causes paralysis of the central nervous system and of respiratory organs.

1050
Ammothamnine; Oxymatrine; Matrine *N*-oxide

$C_{15}H_{24}N_2O_2$ Mol. wt 264.37

Found in *Ammothamnus lehmanni, A. songorica* and *Sophora flavescens* (Leguminosae).

Toxic.

1051
Anagyrine; Rhombinine; Monolupine

$C_{15}H_{20}N_2O$ Mol. wt 244.34

Found in the seeds of *Anagyris foetida, Ulex europaeus* and *Thermopsis chinensis*; it has been reported also in some spp. of *Cytisus, Genista, Lupinus, Sophora* and *Ammodendron* (all Leguminosae).

Highly toxic and teratogenic. It is responsible for "crooked calf" disease caused by ingestion of leguminous plants. It is a cardiotonic agent, inducing tachycardia.

1052
Angustifoline; Jamaicensine

$C_{14}H_{22}N_2O$ Mol. wt 234.34

Found in *Lupinus angustifolius, L. polyphyllus, L. albus, Ormosia jamaicensis* and *O. panamensis* (Leguminosae).

1053
Aphylline; 10-Oxosparteine

$C_{15}H_{24}N_2O$ (+) form   Mol. wt 248.36

Found in the leaves of *Anabasis aphylla* (Chenopodiaceae); the (−) form is found in *Argyrolobium megorhizum* (Leguminosae).

Insecticidal properties.

1054
Argentine; 3,3′-Carbonylbiscytisine

$C_{23}H_{26}N_4O_3$ Mol. wt 406.49

Found in *Ammodendron longiracemosum*, *Sophora griffithii* and *Thermopsis alterniflora* (Leguminosae).

1055
Argyrolobine

$C_{15}H_{22}N_2O_2$ Mol. wt 262.36

Found in the whole plant of *Argyrolobium megarhizum* (Leguminosae).

1056
Baptifoline; 13-Hydroxyanagyrine

$C_{15}H_{20}N_2O_2$ Mol. wt 260.34

Found in the seeds of *Baptisia perfoliata*, *B. minor*, *Sophora alopecuroides* and *S. flavescens* (Leguminosae), and in *Caulophyllum thalictroides* (Berberidaceae).

1057
Cadiamine

$C_{15}H_{26}N_2O_3$ Mol. wt 282.39

Found in *Cadia purpurea* (Leguminosae).

1058
Calpurnine; Oroboidine; 13-Hydroxylupanine 2-pyrrolecarboxylate

$C_{20}H_{27}N_3O_3$ Mol. wt 357.46

Found in *Calpurnia lasiogyne*, *C. aurea*, *Virgilia capensis* and *Cadia ellisiana* (Leguminosae).

Highly toxic to fish and mice. It shows hypotensive properties, and is anti-arrhythmic.

1059
Caulophylline; Methylcytisine

$C_{12}H_{16}N_2O$ Mol. wt 204.27

Found in *Cytisus laburnum* and *Spartium junceum*, and regularly present in many related plants in the Leguminosae.

Toxic. It is repellent to snails.

1060
Cinegalline

$C_{23}H_{30}N_2O_6$ Mol. wt 430.51

Found in *Genista cinerea* (Leguminosae).

1061
Cytisine; Baptitoxine; Citisine; Sophorine; Ulexine

$C_{11}H_{14}N_2O$ Mol. wt 190.25

Found in the seeds of the laburnum tree, *Laburnum anagyroides*; it is also found widely in *Anagyris*, *Baptisia*, *Cytisus*, *Genista*, *Sophora* and *Thermopsis* spp. (all Leguminosae).

Highly toxic (LD_{50} 18 mg kg^{-1} intraperitoneally in mice). It is the cause of poisoning in humans and animals following ingestion of laburnum seeds. Also, it is teratogenic in rabbits and poultry. It is a respiratory stimulant, with a nicotine-like activity, and it is hallucinogenic.

1062
Dimethamine

$C_{24}H_{32}N_4O_2$ Mol. wt 408.55

Found in the aerial parts of *Thermopsis alterniflora* (Leguminosae).

1063
13-Hydroxylupanine; 13-Hydroxy-2-oxosparteine; Octalupine

$C_{15}H_{24}N_2O_2$ Mol. wt 264.37

Found in *Cytisus scoparius*, *Genista cinerea*, *Thermopsis cinerea* and in several *Cadia* and *Lupinus* spp. (Leguminosae). It is often present with ester substitution (e.g., of benzoic, tiglic or cinnamic acid) at the 13-hydroxyl group.

Anti-arrhythmic and hypotensive.

1064
Jamine; Homo-ormosanine

$C_{21}H_{35}N_3$ (−) form   Mol. wt 329.53

The (−) form occurs in *Ormosia costulata*, while the (±) form is found in seeds of *O. panamensis* and *O. jamaicensis* (Leguminosae).

1065
Lamprolobine

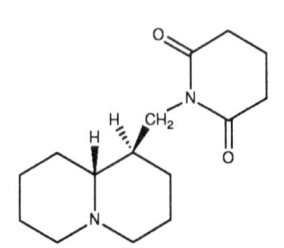

$C_{15}H_{24}N_2O_2$ Mol. wt 264.37

Found in *Lamprolobium fruticosum* and *Lupinus holosericeus* (Leguminosae).

1066
Leontiformine

$C_{15}H_{26}N_2O$ Mol. wt 250.39

Found in *Leontice leonopetalum* (Berberidaceae).

1067
Lupanine; 2-Oxo-11α-sparteine

$C_{15}H_{24}N_2O$ Mol. wt 248.37

Regularly present in over 12 genera of legumes: e.g., in broom, *Cytisus scoparius*, and in several *Lupinus* spp. It is found in the (+) form (illustrated) and in the (−) form, and as the racemate.

Toxic, being responsible for cattle poisoning. It is anti-arrhythmic, hypotensive and hypoglycaemic.

1068
Lupinine

$C_{10}H_{19}NO$ Mol. wt 169.27

Found in *Anabasis aphylla* (Chenopodiaceae), and in *Lupinus luteus* and *L. palmeri* (Leguminosae).
Orally toxic. It is an insect antifeedant, and a growth inhibitor for the grasshopper *Melanoplus bivittatus*.

1069
Matrine; Lupanidine

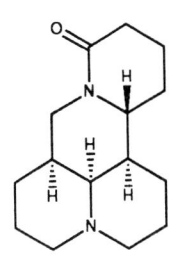

$C_{15}H_{24}N_2O$ Mol. wt 248.37

Found in *Euchresta horsfeldii, Goebelia pachycarpa, Vexibia pachycarpa* and numerous *Sophora* spp. (Leguminosae). Five stereoisomers of this structure have been found in these plants.

Anti-ulcer, antitumour and antibacterial activities.

1070
Multiflorine

$C_{15}H_{22}N_2O$ Mol. wt 246.36

Found in *Cadia ellisiana, Lupinus multiflorus* and *L. albus* (Leguminosae).

Central nervous system depressant.

1071
Nuttalline; 4β-Hydroxylupanine

$C_{15}H_{24}N_2O_2$ Mol. wt 264.37

Occurs in *Lupinus nuttallii, L. hartwegi* and *L. sericeus* (Leguminosae).

1072
(−)-Ormosanine; Piptamine

$C_{20}H_{35}N_3$ Mol. wt 317.52

Found in *Podopetalum ormondii*, *Ormosia semicastrata* and *O. jamaicensis* (Leguminosae).

1073
Panamine

$C_{20}H_{33}N_3$ Mol. wt 315.51

Occurs in *Ormosia* spp., e.g., *O. coccinea* and *O. jamaicensis* (Leguminosae).

1074
Piptanthine

$C_{20}H_{35}N_3$ (−) form   Mol. wt 317.52

The (−) form occurs in *Piptanthus nanus*, and the (±) form occurs in the stems and leaves of *Hovea linearis*, *Ormosia semicastrata* and *Templetonia retusa* (Leguminosae).

Similar pharmacological activity to that of sparteine (q.v.) but of lower potency.

1075
Retamine; 12α-Hydroxysparteine

$C_{15}H_{26}N_2O$ Mol. wt 250.39

Occurs in *Genista sphaerocarpa*, *G. junceum* and other *Genista* spp. (Leguminosae).

Powerful hypotensive agent; also it is a uterotonic and a diuretic.

1076
Rhombifoline; *N*-(3-Butenyl)cytisine

$C_{15}H_{20}N_2O$ Mol. wt 244.34

Found in *Genista tinctoria*, *Ammodendron longiracemosum* and *Thermopsis rhombifolia* (Leguminosae).

1077
Sophoramine

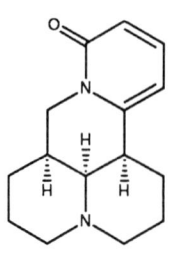

$C_{15}H_{20}N_2O$ Mol. wt 244.34

Found in *Sophora alopecuroides*, *S. flavescens* and *pachycarpa* (Leguminosae).

1078
(−)-Sparteine; Lupinidine

$C_{15}H_{26}N_2$ Mol. wt 234.39

Probably the most commonly occurring alkaloid of this class, present in many *Baptisia, Cytisus, Lupinus* and *Sarothamnus* spp. (Leguminosae). It occurs in both the (+) form (pachycarpine), e.g., in *Lupinus pusillus*, and the (−) form (lupinidine), e.g., in *Piptanthus nanus*.

The (−) form is an oxytocic agent, which has been used in treatment of cardiac insufficiency. Also, it is a diuretic and a hypoglycaemic. It is toxic to livestock and to many insects, but is a feeding stimulant to the broom aphid, *Acyrthosiphon spartii*.

The (−) form (illustrated) occurs in seeds of *Sophora secundiflora* and *Thermopsis lanceolata*. The (+) form, also known as hexalupine or isoanagyrine, occurs in *Lupinus caudatus* and *L. corymbosus* (all Leguminosae).

1081
Tinctorine; Alteramine

$C_{15}H_{20}N_2O$ Mol. wt 244.34

Occurs in *Baptisia australis, Genista tinctoria* and *Thermopsis alterniflora* (Leguminosae).

1079
Templetine

$C_{20}H_{35}N_3$ Mol. wt 317.52

Found in the leaves of *Templetonia retusa* (Leguminosae).

1082
Tsukushinamine A

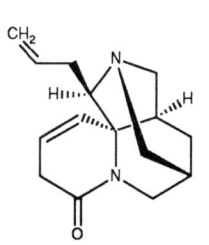

$C_{15}H_{20}N_2O$ Mol. wt 244.34

Occurs, together with two stereoisomers, in *Sophora franchetiana* (Leguminosae).

1080
Thermopsine

$C_{15}H_{20}N_2O$ Mol. wt 244.34

Chapter 29
Steroidal alkaloids

These alkaloids are triterpenoid in origin, being derived biosynthetically from six isoprene units. They might equally well be placed close to the phytosterols (q.v.), except that they do contain nitrogen and hence are bases, like the other alkaloids. The nitrogen may be part of a ring system, usually incorporated at a late stage in biosynthesis, as in the structure of solanidine (**1101**) or it may be as an *N*-methyl substituted amino group, as in the structure of cyclobuxine D (**1086**).

Steroidal alkaloids are found in four unrelated plant families: the Apocynaceae, Buxaceae, Liliaceae and Solanaceae. Each family has a different set of alkaloids in its tissues. Many of the names given to these bases indicate the particular plant sources. thus, the alkaloids of *Buxus* (Buxaceae) that are listed here are buxamine (**1083**), cyclobuxine D (**1086**), cycloprotobuxine C (**1088**) and cyclovirobuxine C (**1089**). They are characterized chemically by their substitution with a methyl, methylene, hydroxymethyl or alkoxymethyl group at C-4 and a methyl group at C-14. They are known to be antimalarial and antitubercular in their pharmacological activities, and they are also purgatives.

The steroidal alkaloids of the Solanaceae are glycosidic, derived from such well known bases as solanidine (**1101**) in the potato and tomatidine (**1106**) in the tomato. For example, α-solanine (**1102**) is a trisaccharide derivative of solanidine, while tomatine (**1107**) is a tetrasaccharide derivative of tomatidine. They are both widely present in *Solanum* and *Lycopersicon* spp., respectively. They are poisonous and both human and cattle deaths have occurred following ingestion of such alkaloid-containing plant materials. While the potato tuber normally contains only small amounts (less than 20 mg per 100 g fresh weight), its sprouts, leaves and flowers can contain lethal concentrations. Certain of these alkaloids have antifeedant properties (e.g., against the Colorado potato beetle) and others are antifungal, so that they probably have a protective role in nature against insect and microbial attack (Roddick, 1986).

From the viewpoint of human medicine, the most important steroidal alkaloids are those from the genus *Veratrum* (Liliaceae), since several (e.g., protoverine, **1099**) can be used for treating hypertension. These alkaloids fall into two groups according to the degree of oxygenation. The jervatrum alkaloids contain one to four oxygen atoms, and occur in the free state or as monoglycosides (e.g., jervine, **1094** and veratramine, **1109**). The ceveratrum bases, by contrast, contain seven to nine oxygen atoms and occur esterified, as in the case of protoveratrines A and B (**1097, 1098**). These alkaloids are noted for the teratogenic effects they produce in livestock feeding on pastures containing *Veratrum* species. Other Liliaceous genera containing this type of alkaloid are *Amianthum*, *Fritillaria*, *Schoenocoulon* and *Zigadenus*.

The last group of steroidal bases considered here is those from the Apocynaceae, from species of *Funtumia*, *Holarrhena* and *Malonetia*. Representative examples are conessine (**1085**), funtumine (**1091**), irehine (**1093**) and kurchessine (**1095**). These alkaloids are of medicinal interest, largely because they provide suitable starting materials for the partial synthesis of useful steroids, such as the adrenocortical hormone, aldosterone.

References

Atta-ur-Rahman (1990). *Handbook of Natural Products Data*, Volume 1, *Diterpenoid and Steroidal Alkaloids*. Amsterdam: Elsevier.

Roddick, J.G. (1986). Steroidal alkaloids of the Solanaceae. In *Solanaceae, Biology and Systematics* (D'Arcy, W.G., ed.), pp. 201–222. New York: Columbia University Press.

In *The Alkaloids* (Manske, R.H.F. and Brossi, A., eds) (1953) Volume 3, pp. 247–312; (1960) Volume 7, pp. 319–342, 343–361, 363–417; (1967) Volume 9, pp. 305–426; (1968) Volume 10, pp. 1–192, 193–285; (1973) Volume 14, pp. 1–82; (1981) Volume 19, pp. 81–192; (1988) Volume 32, pp. 79–240.

1083
Buxamine E; Buxamine

$C_{26}H_{44}N_2$ Mol. wt 384.65

Occurs in the leaves and seeds of *Buxus* spp., notably, *B. sempervirens*. Various *N*-methyl derivatives (buxamines A, B, C and G) co-occur in these plants.

Buxus alkaloids are strongly purgative.

1084
α-Chaconine

$C_{45}H_{73}NO_{14}$ Mol. wt 852.09

Found in *Solanum chacoense, S. nigrum* and *S. tuberosum* (Solanaceae), and also in *Notholiron hyacinthinum* and *Veratrum stenophyllum* (Liliaceae).

Toxic constituent of potato tubers. *Solanum chacoense* extracts impart resistance to Colorado beetle feeding.

1085
Conessine; Neriine; Rochessine; Roquessine; Wrightine

$C_{24}H_{40}N_2$ Mol. wt 356.60

Found in *Holarrhena* spp., notably, in *H. pubescens* (Kurchibark), *H. antidysenterica, H. africana, H. congolensis, H. febrifuga, H. floribunda* and *H. waltsbergii* (Apocynaceae).

Anti-amoebic drug for the treatment of dysentery.

1086
Cyclobuxine D; Cyclobuxine

$C_{25}H_{42}N_2O$ Mol. wt 386.63

Occurs in the leaves and seeds of *Buxus sempervirens, B. harlandi, B. hyrcana, B. microphylla* and *B. wallichiana* (Buxaceae). Cyclobuxine B and pseudocyclobuxine D are also present, and are *N*-methyl derivatives.

Strongly purgative; it is also anti-inflammatory and hypertensive.

1087
Cyclopamine; 11-Deoxyjervine

$C_{27}H_{41}NO_2$ Mol. wt 411.63

Found in *Veratrum californicum* and *V. album* (Liliaceae).

Severe teratogen, responsible for cyclopic malformation in sheep grazing on *V. californicum*.

1088
Cycloprotobuxine C

$C_{27}H_{48}N_2$ Mol. wt 400.70

Occurs in *Buxus sempervirens, B. balearica* and *B. malayana* (Buxaceae). It co-occurs with the closely related cycloprotobuxines A, D and F.

Purgative.

1089
Cyclovirobuxine C

$C_{27}H_{48}N_2O$ Mol. wt 416.69

Occurs in *Buxus malayana, B. sempervirens, B. argentea, B. microphylla* and *B. wallichiana* (Buxaceae). It co-occurs with the related cyclovirobuxines D and F.

Anti-arrhythmic agent. *Buxus* spp. are notably purgative.

1090
Demissine; β-Demissidine glycoside

$C_{50}H_{83}NO_{20}$ Mol. wt 1018.22

Found widely in *Solanum* spp., notably, *Solanum demissum, S. cartilobum, S. chacoense, S. commersonii, S. decemlineata, S. deporum, S. horoviteii, S. jamesii* and *S. juzepczukii*. It is present also in *Lycopersicon pimpinellifolium*.

Haemolytic activity. On hydrolysis, it gives demissidine, for use as a cholinesterase inhibitor or as a repellent to the cataract beetle.

1091
Funtumine

$C_{21}H_{35}NO$ Mol. wt 317.52

Occurs in trees of the genus *Funtumia*, notably, in *F. elastica* and *F. latifolia* (Apocynaceae). It occurs also in *Holarrhena febrifuga* and *H. congolensis*.

Hypocholesterolaemic agent and respiratory stimulant.

1092
Germine

$C_{27}H_{43}NO_8$ Mol. wt 509.65

Found in *Zigadenus venenosus*, *Veratrum viride* and other *Veratrum* spp. (Liliaceae).

Potent hypotensive agent with characteristic action on the heart, causing irregularity and prolongation of the beat. Ester derivatives have been used medicinally for treating myasthenia gravis.

1093
Irehine; *N,N*-Dimethylholafebrine; Buxomegine

$C_{23}H_{38}NO$ Mol. wt 344.56

Occurs in the bulbs of *Narcissus* spp. (Amaryllidaceae). Other sources are *Funtumia elastica* (Apocynaceae) and *Buxus sempervirens* (Buxaceae).

Starting base for synthesizing medicinally useful steroids.

1094
Jervine; Jervanin-11-one

$C_{27}H_{39}NO_3$ Mol. wt 425.62

Found in the rhizomes of *Veratrum* spp., notably *V. album*, *V. grandiflorum* and *V. viride* (Liliaceae).

Antibacterial activity. It is used as a starting material for C-nor-D-homosteroids. *Veratrum album* has shown teratogenic activity to ewes.

1095
Kurchessine; Irehdiamine I; Saraconinine

$C_{25}H_{44}N_2$ Mol. wt 372.64

Occurs in *Holarrhena antidysenterica* and *H. pubescens* (Apocynaceae).

Anti-amoebic compound used to treat dysentery.

1096
Paravallarine

$C_{22}H_{33}NO_2$ Mol. wt 343.51

Occurs in *Paravallaris microphylla*; the 3-epimer occurs in *Kibatalia gitingensis* (Apocynaceae).

Paravallarine is used as starting point for the synthesis for various hydroxy derivatives and of the 3- or 20-epimers. Alkaloids from *Kibatalia* are poisonous to fish.

1097
Protoveratrine A; Protoalba

$C_{41}H_{63}NO_{14}$ Mol. wt 793.96

Found in *Veratrum alba*, white hellibore (Liliaceae), together with protoveratrine B (q.v.).

Antihypertensive drug, profound in high dosages. *Veratrum* alkaloids may cause nausea. Protoverine is obtained by hydrolysis of protoveratrine A.

1098
Protoveratrine B; Neoprotoveratrine; Veratetrine

$C_{41}H_{63}NO_{15}$ Mol. wt 809.96

Occurs in white hellibore, *Veratrum alba* (Liliaceae), together with protoveratrine A (q.v.).

Hypertensive drug, profound in high dosages. *Veratrum* alkaloids may cause nausea.

1099
Protoverine; 6α-Hydroxygermine

$C_{27}H_{43}NO_9$ Mol. wt 525.65

Occurs in *Veratrum alba* (Liliaceae) in esterified form.

Veratrum alkaloids are hypertensive and show evidence of teratogenicity.

1100
Solamargine; Solasodine glycoside

$C_{45}H_{73}NO_{15}$ Mol. wt 868.09

Occurs in the leaves and fruits of *Solanum marginatum*, *S. melongena* and *S. scabrum* (Solanaceae).

1101
Solanidine; Purapuridine; Solancarpidine; Solamidine T

$C_{27}H_{43}NO$ Mol. wt 397.65

Found in *Solanum* spp., notably, in *S. nigrum*, black nightshade (Solanaceae). It is a hydrolysis product of solanine (q.v.). Potato sprouts can contain 0.008% of solanidine.

Toxic constituent of potato tubers. It is a starting point for synthesis of steroidal drugs.

1102
α-Solanine; Solatunine

$C_{45}H_{73}NO_{15}$ Mol. wt 868.07

Occurs together with closely related glycosides β- and γ-solanine in *Solanum tuberosum* (potato), about 0.04% in fresh sprouts; it also occurs in *Solanum nigrum*, woody nightshade, and in *Lycopersicon esculentum*, tomato (all Solanaceae).

The hydrochloride has been used as an agricultural pesticide (LD_{50} intraperitoneally in mice 42 mg kg^{-1}). Inhibition of oxygen intake of mouse ascites tumour reported. Human toxicity symptoms include vomiting, diarrhoea, hallucination and coma; oral ingestion of 2.8 mg kg^{-1} is toxic to humans.

1103
Solasodine; Purpapuridine; Solancarpidine; Solasodine-S

$C_{27}H_{43}NO_2$ Mol. wt 413.65

Widespread in glycosidic combination in *Solanum* spp., notably, in *S. laciniatum* (1 to 2% of solasodine) and *S. nigrum* (Solanaceae).

Solasodine (like diosgenin) is a viable starting material for producing adrenocortical and glucocortical steroids, which are used for contraception or as anti-inflammatory agents. It is teratogenic in rats and guinea-pigs.

1104
Solasonine; Purapurine; Solanine-S; Solasodamine

$C_{45}H_{73}NO_{16}$ Mol. wt 884.09

Widely distributed, especially in the fruits, of the *Solanum* genus, notably, *S. aviculare*, kangaroo apple, *S. sodomeum*, *S. torvum*, *S. viarum*, *S. xanthocarpum*, *S. incanum* and *S. melongena* (Solanaceae).

Solanum sodomeum is lethal to cockroaches.

1105
Terminaline

$C_{23}H_{41}NO_2$ Mol. wt 363.59

Found in *Pachysandra terminalis* (Buxaceae).

Base for synthesizing other steroids.

1106
Tomatidine

$C_{27}H_{45}NO_2$ Mol. wt 415.66

Occurs in the roots of *Lycopersicon esculentum*, Rutgers tomato plant, and *Solanum demissum*. It is also widely present in glycosidic form in other *Lycopersicon* and *Solanum* spp. (Solanaceae).

Antifungal and cytostatic activities. It is a cholinesterase inhibitor and clinically active against forms of dermatitis. Also, it is repellent to the Colorado potato beetle.

1107
Tomatine; α-Tomatine; Lycopericin

$C_{50}H_{83}NO_{21}$ Mol. wt 1034.22

Found in the leaves and fruits of tomato, *Lycopersicon esculentum* (Solanaceae). Also, it is present in wild *Lycopersicon* spp. and in several *Solanum* spp.

Inhibits growth of bacteria and fungi. It is used as an insecticide, notably as a repellent for the Colorado beetle. Tomatine precipitates steroids and is proposed as an alternative to digitonin. Some antihistamine activity has been reported.

1108
Veracevine; Protocevine

$C_{27}H_{43}NO_8$ Mol. wt 509.65

Occurs in the seeds of *Schoenocoulon officinale* (Liliaceae).

Used as an insecticide in veterinary medicine. It is a source of cevine by epimerization or of veratridine (weakly hypertensive) by methoxybenzoylation.

1109
Veratramine

$C_{27}H_{39}NO_2$ Mol. wt 409.62

Component of the coarse rhizomes of *Veratrum grandiflorum* and *Veratrum viride* (Liliaceae).

Antihypertensive drug. *Veratrum viride* extracts have veterinary use as circulatory depressants, emetics and parasiticides.

1110
Verticine; Apoverticine; Peimine

$C_{27}H_{45}NO_3$ Mol. wt 431.66

Occurs in the bulbous portions of *Fritillaria* spp., notably, in *F. verticillata*, *F. thunbergii* and *F. roylei* (Liliaceae), admixed with other alkaloids.

Fritillaria alkaloids are used for treatment of chest ailments.

1111
Zygadenine; 7-Deoxygermine

$C_{27}H_{43}NO_7$ Mol. wt 493.65

Found in *Zigadenus gramineus*, *Z. venenosus* and *Veratrum album* (Liliaceae).

Poisonous. Antitumour and antihypertensive activities have been shown.

Chapter 30
Tropane alkaloids

These alkaloids in their simplest form are aliphatic nitrogen compounds, based on the hydroxytropane skeleton, as can be seen in tropine (**1142**) or its stereoisomer ψ-tropine (**1137**). This skeleton may be substituted by further hydroxyl groups (e.g., as in meteloidine, **1133**, a trihydroxytropane) or by epoxide groups (as in apohyoscine, **1115**). Most are further substituted at a hydroxyl group by an aliphatic or aromatic acid. Typical esters are atropine (**1116**) where the esterifying acid is tropic, cocaine (**1123**) where the acid is benzoic and meteloidine (**1133**) where the acid is tiglic. The tropane skeleton in these alkaloids is derived biosynthetically from the amino acid ornithine, while the tropic acid moiety in, for example, atropine is formed from the aromatic amino acid, phenylalanine.

Tropane alkaloids occur widely in the plant family Solanaceae, particularly in those species which have long been utilized for their medicinal, hallucinogenic or poisonous properties (Evans, 1979). A typical source is the deadly nightshade, *Atropa belladonna*, which contains atropine (**1116**), hyoscyamine (**1130**) and apoatropine (**1114**) as major constituents. The alkaloids are formed in the roots and transported as the *N*-oxides to the aerial parts, where they accumulate in varying amounts in leaves, flowers and fruits. Other solanaceous sources include species of *Datura, Duboisia, Hyoscyamus, Mandragora* and *Scopolia*. A second family rich in tropanes is the Erythroxylaceae, which includes the coca plant, *Erythroxylum coca*. The leaves of coca are a major source of the alkaloid cocaine (**1123**), currently one of the most widely used drugs of abuse (or addiction) for its hallucinogenic effects. Tropane alkaloids have a limited occurrence in at least eight other, unrelated, plant families: Convolvulaceae, Cruciferae, Dioscoreaceae, Elaeocarpaceae, Euphorbiaceae, Orchidaceae, Proteaceae and Rhizophoraceae (Romeike, 1978). They also occur rarely in algae (anatoxin *a* **1112**).

Cocaine can be detected in plant extracts by a colour reaction with cobalt thiocyanate but, for the others, a modified Dragendorff's reagent or iodoplatinate is used. A variety of TLC, GLC and HPLC techniques has been developed for quantifying atropine and cocaine and their respective derivatives in plant extracts (Hashimoto *et al.*, 1988). In spite of their illicit use as stimulants, these alkaloids are valuable medicinally, especially as local anaesthetics in ophthalmology and for controlling travel sickness. The seeds of thornapple and of belladonna are poisonous because of their high alkaloid content, and eating the fruits can lead to human fatalities.

References

General

Clarke, R.L. (1977). Chemistry of the tropane alkaloids. In *The Alkaloids*, Volume 16 (Manske, R.H.F., ed.), pp. 83–180. London: Academic Press.
Fodor, G. and Dharanipragada, R. (1988). Tropane alkaloids. *Natural Product Reports* **5**, 67–72.

Distribution

Evans, W.C. (1979). Tropane alkaloids of the Solanaceae. In *The Biology and Taxonomy of the Solanaceae* (Hawkes, J.G., Lister, R.N. and Skelding, A.D., eds), pp. 241–254. London: Academic Press.

Romeike, A. (1978). Tropane alkaloids: occurrence and systematic importance in angiosperms. *Botaniska Notiser*, **131**, 85–96.

Detection

Hashimoto, Y., Kawanishi, K. and Mariyasu, M. (1988). Forensic chemistry of alkaloids by chromatographic analysis. In *The Alkaloids*, Volume 32 (Brossi, A., ed.), pp. 1–78. San Diego: Academic Press.

1112
Anatoxin a; "Very fast death" factor

$C_{10}H_{15}NO$ Mol. wt 165.24

Found in the alga *Anabaena flos-aquae* (Cyanophyceae).

Highly toxic; this "very fast death" factor is released from the blue-green alga into surrounding water (e.g., reservoir) during growth.

1113
Anisodamine; 6-Hydroxyhyoscyamine

$C_{17}H_{23}NO_4$ Mol. wt 305.38

Occurs in *Datura* spp., as a possible precursor of hyoscyamine and hyoscine (q.v.) (Solanaceae).

Used as a biochemical tool in studying the mode of action of anaesthetics, because of its effects on liposome formation. It reduces acute myocardial infarction in rabbits, and prevents haemorrhagic shock in cats and endotoxin-induced shock in dogs.

1114
Apoatropine; Atropamine; Atropyltropeine

$C_{17}H_{21}NO_2$ Mol. wt 271.36

Occurs in the roots of *Atropa belladonna* (Solanaceae).

Antispasmodic agent. In relatively small doses, it causes respiratory arrest in animals.

1115
Apohyoscine; 6,7β-Epoxy-3α-atropyloxytropane

$C_{17}H_{19}NO_3$ Mol. wt 285.35

Found in the aerial parts of *Datura meteloides*, *Duboisia myoporoides* and *Hyoscyamus niger* (Solanaceae).

1116
Atropine; Tropine tropate

$C_{17}H_{23}NO_3$ Mol. wt 289.38

Found in *Atropa belladonna, Datura stramonium*, and other Solanaceae, especially in the roots.

Highly toxic. It has anticholinergic activity, and causes blurred vision, suppressed salivation, vasodilation, excitement and delirium. It is used in anaesthesia. The lethal dose is 100 mg in humans. See also hyoscyamine.

1117
Belladonnine; Tropylisatropate

$C_{34}H_{42}N_2O_4$ Mol. wt 542.72

Found in the berries of *Hyoscyamus niger* (α form), *Physochliana alaica* (β form) and *Atropa belladonna* (Solanaceae).

This is also a thermal dimerization product of apoatropine (q.v.), so that there is some question as to whether its occurrence may be artefactual. It is found in both the *trans* (1*RS*,4*SR*) form (α-belladonnine) and *cis*(1*RS*,4*RS*) form (β-belladonnine).

Local anaesthetic potentiator.

1118
Bellendine; 10-Methylpyranotropane

$C_{12}H_{15}NO_2$ Mol. wt 205.26

Occurs in the upper leaves and flowers of *Darlingia darlingiana* and *Bellendena montana* (Proteaceae).

1119
Benzoylecgonine

$C_{16}H_{19}NO_4$ Mol. wt 289.34

Found in *Erythroxylum coca* and other *Erythroxylum* spp. (Erythroxylaceae).

Metabolite of cocaine (q.v.) in humans.

1120
Benzoyltropein; Tropine benzoate

$C_{15}H_{19}NO_2$ Mol. wt 245.32

Occurs in *Erythroxylum coca* (Erythroxylaceae), *Crossostylis ebertii* and *Bruguiera sexangula* (Rhizophoraceae).

The parent compound, tropine (q.v.), is poisonous.

1121
Brugine

$C_{12}H_{19}NO_2S_2$ Mol. wt 273.42

Found in the bark of *Bruguiera sexangula* and *B. exaristata* (Rhizophoraceae).

1122
Cinnamoylcocaine; Cinnamoylcocaine; Cinnamoylmethylecgonine; Ecgonine cinnamate methyl ester

$C_{19}H_{23}NO_4$ Mol. wt 329.40

Found in coca leaves, *Erythroxylum coca* and, particularly, Javanese coca, *E. truxillense* (Erythroxylaceae).

Usually hydrolysed as a crude extract and converted to cocaine (q.v.).

1123
Cocaine; Benzoylmethylecgonine; Methyl benzoylecgonine

$C_{17}H_{21}NO_4$ Mol. wt 303.36

Found in coca leaves, *Erythroxylum coca*, and other *Erythroxylum* spp. (Erythroxylaceae).

Central nervous system stimulant and narcotic, subject to widespread abuse. It is a local anaesthetic used mainly in ophthalmology, and a mydriatic.

1124
Convolamine; *O*-Veratroyltropine

$C_{17}H_{23}NO_4$ Mol. wt 305.38

Found in *Convolvulus pseudocantabricus*, *C. lineatus* and *C. subhirsutus* (Convolvulaceae).

1125
Convoline; 3α-Veratroyl-*N*-hydroxynortropane

$C_{16}H_{21}NO_5$ Mol. wt 307.35

Occurs in *Convolvulus krauseanus* (Convolvulaceae).

1126
Convolvine; O-Veratroylnortropine

$C_{16}H_{21}NO_4$ Mol. wt 291.35

Widespread in the Solanaceae, including *Datura, Physalis* and *Salpichroa*, in *Convolvulus* spp. (Convolvulaceae), and also in the mangrove, *Bruguiera* (Rhizophoraceae).

1127
Darlingine

$C_{13}H_{17}NO_2$ Mol. wt 219.29

Found in the upper stems and leaves of *Darlingia ferruginea* and *D. darlingiana* (Proteaceae).

1128
Ecgonine

$C_9H_{15}NO_3$ Mol. wt 185.23

Occurs in coca leaves, *Erythroxylum coca*, and other *Erythroxylum* spp. (Erythroxylaceae).

Highly toxic by inhalation. It is a starting material for the manufacture of cocaine. It is also a topical anaesthetic.

1129
Hyoscine; 6,7-Epoxytropine tropate; Scopine tropate; Scopolamine

$C_{17}H_{21}NO_4$ Mol. wt 303.36

Found in henbane, *Hyoscyamus niger, Datura metel, D. innoxia, Scopolia carniolica, Anthocercis viscosa, A. fasciculata* (Solanaceae).

Anticholinergic, with both central and peripheral actions. A particular use is as an antispasmodic, for motion sickness. It is also used as a pre-operative medication to sedate, reduce secretions, produce amnesia, assist the induction of anaesthesia and reduce some of its side-effects.

1130
Hyoscyamine; Daturine; Duboisine

$C_{17}H_{23}NO_3$ Mol. wt 289.38

Occurs in henbane and Egyptian henbane, *Hyoscyamus niger* and *H. muticus*, deadly nightshade, *Atropa belladonna*, thornapple, *Datura stramonium, Duboisia myoporoides*, and other Solanaceae.

Anticholinergic with actions similar to but more potent than those of atropine (q.v.), which is the racemate; it causes central nervous system depression followed by stimulation. It is a mydriatic, anti-emetic, antispasmodic, and antisecretory for saliva, perspiration and gastric secretions, and therefore useful as a premedication before anaesthesia; also, it has been used to treat symptoms of Parkinson's disease. Toxic to humans.

1131
Knightinol

$C_{17}H_{23}NO_3$ Mol. wt 289.38

Occurs in *Knightia strobilina* (Proteaceae).

1132
Littorine

$C_{17}H_{23}NO_3$ Mol. wt 289.38

Found in *Anthocercis littorea*, and other *Anthocercis* spp., and *Datura sanguinea* (Solanaceae).

The racemate is a mydriatic.

1133
Meteloidine; 6,7-Dihydroxy-3-tiglyloxytropane; 3-(3,6,7-Tropanetriol) tiglate

$C_{13}H_{21}NO_4$ Mol. wt 255.32

Occurs in *Datura meteloides, D. ferox* and some *Anthocercis* spp. (Solanaceae), and *Erythroxylum* spp. (Erythroxylaceae).

1134
Norhyoscyamine; Pseudohyoscyamine; Solandrine; 1-Tropic acid 3α-nortropanyl ester

$C_{16}H_{21}NO_3$ Mol. wt 275.35

Found in Egyptian henbane, *Hyoscyamus muticus*, and other Solanaceae.

1135
Physoperuvine; 1-Hydroxytropane

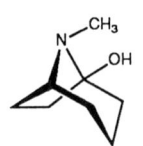

$C_8H_{15}NO$ Mol. wt 141.22

Occurs in the Cape gooseberry, *Physalis peruviana* (Solanaceae). The free base can exist in two tautomeric forms.

Moderately toxic.

1136
Pseudopelletierine; 9-Methyl-3-granatanone; *N*-Methylgranatonine; ψ-Pelletierine; Pseudopunicine Pelletierine; Pseudopunicine

$C_9H_{15}NO$ Mol. wt 153.23

Found in pomegranate root bark, *Punica granatum* (Punicaceae).

1137
Pseudotropine; ψ-Tropine; 3-Pseudotropanol; 3β-Tropanol

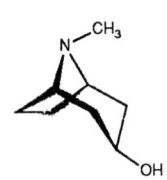

$C_8H_{15}NO$ Mol. wt 141.22

Occurs in *Scopolia carniolica* and probably other Solanaceae.

1138
Scopoline; Oscine; 3α,6α-Epoxy-7β-hydroxytropane

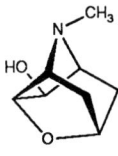

$C_8H_{13}NO_2$ Mol. wt 155.20

Found in *Solandra grandiflora* (Solanaceae). It is a decomposition product of hyoscine (scopolamine) (q.v.).

The name scopoline or scopolin (q.v.) is also the name of a coumarin glycoside.

1139
Strobamine

$C_{17}H_{19}NO_2$ Mol. wt 269.35

Found in the leaves of *Knightia strobilina* (Proteaceae).

1140
Tigloidine; 3β-Tigloyloxytropane; Tiglylpseudotropeine

$C_{13}H_{21}NO_2$ Mol. wt 223.32

Occurs in *Duboisia myoporoides* and *Datura innoxia* (Solanaceae).

Central nervous system depressant, and mild anticholinergic. It has been used to treat muscular rigidity and Parkinson's disease.

1141
Tropacocaine; Tropacaine; Benzoylpseudotropeine; Pseudotropine benzoate

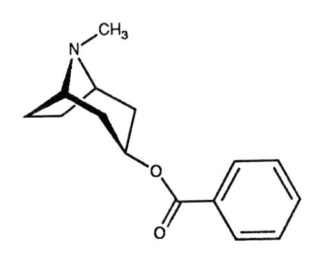

$C_{15}H_{19}NO_2$ Mol. wt 245.32

Found in Javanese and Peruvian coca leaves, *Erythroxylum truxillense* (Erythroxylaceae) and *Peripentadenia mearsii* (Elaeocarpaceae).

Poisonous.

1142
Tropine; 2,3-Dihydro-3α-hydroxytropidine;
1α*H*,5α*H*-Tropan-3α-ol

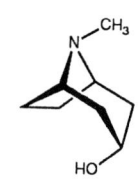

$C_8H_{15}NO$ Mol. wt 141.22

Found in *Scopolia carniolica* and *Atropa belladonna* (Solanaceae).

Highly toxic.

1143
Valeroidine;
(3*S*,6*S*)-3α-Isovaleroxy-6β-hydroxytropane

$C_{13}H_{23}NO_3$ Mol. wt 241.33

Found in *Duboisia myoporoides* and other *Duboisia* and *Datura* spp. (Solanaceae).

Chapter 31
Miscellaneous alkaloids

This section includes alkaloids which are difficult to classify elsewhere and those which have skeletal types of which there are only a few examples. Among minor classes of alkaloid are the oxazoles, e.g., annuloline (**1151**), and the pyrazines, e.g., aspergillic acid (**1152**). Some 14 alkaloids with flavonoid (q.v.) substituents are known, ficine (**1173**) from the fig being one example.

Several alkaloids listed here have complex structures which are unique to particular plant families. There is decaline (**1169**) from the Lythraceae, elaeocarpine (**1171**) from Elaeocarpaceae and himbacine (**1177**) from Himantandraceae. There is also maytansine (**1184**), which occurs characteristically in Celastraceae; other Celastraceae bases with different structures are listed with the monoterpene and sesquiterpene alkaloids.

Some alkaloids in this section are very toxic, such as taxine A (**1195**) from the yew. This is also true of mycotoxins such as verruculotoxin (**1199**), which can get into the human food chain via the peanut. Several alkaloids have very promising anticancer activity, notably cephalotoxine (**1160**), cryptopleurine (**1166**) and harringtonine (**1176**). In addition, the diterpenoid alkaloid taxol (**1196**) from the yew is currently under clinical trial for the treatment of breast cancer.

References

Cordell, G.A. (1981). *Introduction to Alkaloids: A Biogenetic Approach*. New York: Wiley.

1144
Acanthoidine

$C_{16}H_{26}N_4O_2$ Mol. wt 306.41

Occurs, together with the closely related acanthoine, in the dried stalks of *Carduus acanthoides* (Compositae) and probably also in other *Carduus* spp.

Hypotensive activity in humans.

1145
***O*-Acetylcypholophine**

$C_{20}H_{20}N_2O_4$ Mol. wt 360.46

Found in the leaves and stems of *Cypholophus friesianus* (Urticaceae), together with cypholophine.

Cytotoxic and antimicrobial activities.

1146
Acutumidine; *N*-Noracutumine

$C_{18}H_{22}NO_6Cl$ Mol. wt 383.83

Found in the leaves of *Menispermum dauricum* and *Sinomenium acutum* (Menispermaceae).

Similar to quinine, small amounts stimulate the uterus in animals.

1147
Alchorneine

$C_{12}H_{19}N_3O$ Mol. wt 221.31

Occurs in *Alchornea floribunda* and *A. hirtella* (Euphorbiaceae).

Spasmolytic agent in dogs. It is a ganglioplegic parasympathomimetic agent, and a strong vagolytic agent and inhibitor of intestinal peristalsis.

1148
Alchornine

$C_{11}H_{17}N_3O$ Mol. wt 207.28

Found in the bark and leaves of the rain-forest tree *Alchornea javanensis* (Euphorbiaceae).

Similar activity to that of alchorneine (q.v.).

1149
Anantine

$C_{15}H_{15}N_3O$ Mol. wt 253.31

Found in the leaves of *Cynometra ananta*, and the root bark, trunk bark and leaves of *Cynometra lujae* (Leguminosae).

1150
Androcymbine

$C_{21}H_{25}NO_5$ Mol. wt 371.44

Together with colchicine, found in the leaves of *Androcymbium melanthoides* var. *stricta* (Liliaceae).

Activity very similar to that of colchicine (q.v.).

1151
Annuloline

$C_{20}H_{19}NO_4$ Mol. wt 337.38

Found in the leaves of annual rye grass, *Lolium multiflorum* (Gramineae).

Fungistatic activity.

1152
Aspergillic acid

$C_{12}H_{20}N_2O_2$ Mol. wt 224.31

Occurs in the fungus *Aspergillus flavus* (Hyphomycetes).

Antibiotic activity. Its odour is similar to that of black walnuts.

1153
Balsoxine

$C_{17}H_{15}NO_3$ Mol. wt 281.31

Found in *Amyris balsamifera* (Rutaceae).

1154
Calycanthine

$C_{22}H_{26}N_4$ Mol. wt 346.48

Occurs in the seeds of *Calycanthus glaucus, C. floridus, C. occidentalis* and *C. praecox* (Calycanthaceae), and also in the dried stems and leaves of *Palicourea alpina* (Rubiaceae).

Highly toxic: can cause violent convulsions, paralysis, and cardiac depression. It shows uterine stimulant activity.

1155
Candicine; Maltotoxin

$[C_{11}H_{18}NO]^+$ Mol. wt 180.27

Found in the young roots of *Trichocereus candidans* and *T. lampochlorus* (Cactaceae), *Fagara* spp. (Rutaceae), *Hordeum vulgare* (Gramineae), *Magnolia grandiflora* (Magnoliaceae), *Phellodendron amurense* (Rutaceae), and *Lysichitum camtschatcense* (Araceae).

Nicotine-like action on the nervous system. In animals it provokes hypertension, while large doses have a curare-like action.

1156
Cannabisativine

$C_{21}H_{39}N_3O_3$ Mol. wt 381.56

Occurs in the roots and leaves of *Cannabis sativa* (Cannabinaceae).

Shows no cannabimetic activity. Also, there are no gross signs of toxicity when injected in mice.

1157
Casimiroedine

$C_{21}H_{27}N_3O_6$ Mol. wt 417.47

Principal alkaloid from the seeds of *Casimiroa edulis* (Rutaceae).

Cytostatic activity.

1158
Celabenzine; Celabenzene

$C_{23}H_{29}N_3O_2$ Mol. wt 379.51

Occurs in *Tripterygium wilfordii* (Celastraceae), together with the related celafurine, and from *Maytenus mossambicensis* (Celastraceae).

Insecticidal activity.

1159
Cephalomannine; Taxol B

$C_{45}H_{53}NO_{14}$ Mol. wt 831.93

Found in all parts of the yew tree, *Taxus baccata* (Taxaceae).

Antileukaemic and antitumour activities. It is cytotoxic in cell culture experiments.

1160
Cephalotaxine

$C_{18}H_{21}NO_4$ Mol. wt 315.37

Occurs in *Cephalotaxus harringtonia*, *C. fortunei* and *C. wilsoniana* (Cephalotaxaceae).

Antileukaemic activity.

1161
Codonocarpine

$C_{26}H_{31}N_3O_5$ Mol. wt 465.56

Found in the bark of *Codonocarpus australis* (Gyrostemonaceae).

1162
Colchicine

$C_{22}H_{25}NO_6$ Mol. wt 399.45

Occurs in *Colchicum autumnale*, other *Colchicum* spp., and in *Gloriosa superba* (Liliaceae).

Highly toxic, with a lethal dose of 10 mg in humans. It is also an irritant, a carcinogen and a teratogen, but has been used to relieve the pain of acute gout. Plant scientists use it to induce polyploidy in plants.

1163
Crinasiadine

$C_{14}H_9NO_3$ Mol. wt 239.23

Occurs in the flowering bulbs of *Crinum asiaticum* (Amaryllidaceae).

Bacteriostatic and tumour-inhibiting activities.

1164
Crinasiatine

$C_{22}H_{17}NO_4$ Mol. wt 359.39

Occurs in the flowering bulbs of *Crinum asiaticum* (Amaryllidaceae).

Bacteriostatic and tumour-inhibiting activities.

1165
Cryogenine; Vertine

$C_{26}H_{29}NO_5$ Mol. wt 435.53

Found in *Decodon verticillatus*, *Heimia salicifolia*, *H. myrtifolia*, and *Lagerstroemia fauriei* (Lythraceae).

Anti-inflammatory, hyperglycaemic, sedative and hypotensive activities.

1166
Cryptopleurine

$C_{24}H_{27}NO_3$ Mol. wt 377.49

Found in *Cryptocarya pleurosperma* (Lauraceae), *Boehmeria cylindrica* (Urticaceae), and *Cissus rheifolia* (Vitidaceae).

Antiviral activity against herpes virus. Also, it shows potent cytotoxic activity against human nasopharyngeal cells; however, it lacks antitumour properties.

1167
Cynometrine

$C_{16}H_{19}N_3O_2$ Mol. wt 285.35

Occurs in the leaves of *Cynometra ananta* and *C. lujae*, and in the stem bark and seeds of *C. hankei* (Leguminosae).

1168
Damascenine; Nigelline; Methyl damasceninate

$C_{10}H_{13}NO_3$ Mol. wt 195.22

Found in the seeds of *Nigella damascena* and *N. arvensis* (Ranunculaceae).

Antipyretic activity. It inhibits induced oedema formation in rat paw. This is the odoriferous principle of *Nigella* seeds.

1169
Decaline

$C_{26}H_{31}NO_5$ Mol. wt 437.54

Occurs in *Decodon verticillatus* (Lythraceae).

1170
Elaeocarpidine; Eleocarpidine

$C_{17}H_{21}N_3$ Mol. wt 267.38

Occurs in the leaves of *Elaeocarpus polydactylus*, *E. sphaericus* and *E. densiflorus* (Elaeocarpaceae).

1171
(+)-Elaeocarpine; Eleocarpine

$C_{16}H_{19}NO_2$ Mol. wt 257.34

Occurs in the leaves of *Elaeocarpus polydactylus, E. sphaericus* and *E. ganitrus* (Elaeocarpaceae).

1172
Elaeokanine C; Eleokanine C

$C_{12}H_{21}NO_2$ Mol. wt 211.31

Occurs in the leaves of *Elaeocarpus kaniensis* (Elaeocarpaceae).

1173
Ficine

$C_{20}H_{19}NO_4$ Mol. wt 337.38

Occurs in the wild fig, *Ficus pantoniana* (Moraceae).

1174
Gliotoxin; Aspergillin

$C_{13}H_{14}N_2O_4S_2$ Mol. wt 326.40

Antibiotic produced by the fungi *Gliocladium fimbriatum, Aspergillus fumigatus*, and various *Penicillium* spp.

Antiviral and immunomodulating activities. It has been used as an agricultural fungicide.

1175
Halfordinol

$C_{14}H_{10}N_2O_2$ Mol. wt 238.25

Found in the bark of *Halfordia scleroxyla* (Rutaceae).

1176
Harringtonine

$C_{28}H_{37}NO_9$ Mol. wt 531.61

Occurs in *Cephalotaxus harringtonia, C. fortunei* and *C. hainensis* (Cephalotaxaceae).

Shows antitumour activity, and has been used clinically to treat acute myelocytic leukaemia. It acts as a protein and DNA biosynthesis inhibitor.

1177
Himbacine

$C_{22}H_{35}NO_2$ Mol. wt 345.53

Occurs in the bark of *Himantandra* (syn. *Galbulimima*) *baccata* and *H. belgraveana* (Himantandraceae).

Antispasmodic activity.

1178
Homaline

$C_{30}H_{42}N_4O_2$ Mol. wt 490.69

Found in the leaves of *Homalium pronyense* and other *Homalium* spp. (Flacourtiaceae).

1179
Ibotenic acid

$C_5H_6N_2O_4$ Mol. wt 158.12

Found in the fly agaric, *Amanita muscaria*, and the panther cap, *A. pantherina* (Amanitaceae).

Toxic precursor of the hallucinogenic principle, muscimol (q.v.), of this mushroom. It undergoes decarboxylation *in vivo*, when the fly agaric is eaten, and muscimol is excreted in the urine. Such urine is drunk for its hallucinogenic effects. Ibotenic acid is used commercially as a synergist for the flavour additive, monosodium glutamate. Also, it is lethal to flies.

1180
Isoficine

$C_{20}H_{19}NO_4$ Mol. wt 337.38

Occurs as a minor alkaloid in *Ficus pantoniana* (Moraceae).

1181
Lilaline

$C_{20}H_{17}NO_7$ Mol. wt 383.36

Occurs in the aerial parts of *Lilium candidum* (Liliaceae).

1182
Lunarine

$C_{25}H_{31}N_3O_4$ Mol. wt 437.54

Found in the seeds of *Lunaria biennis* and *L. rediviva* (Cruciferae).

1183
Lythramine

$C_{29}H_{37}NO_5$ Mol. wt 479.62

Found in *Lythrum anceps* (Lythraceae).

1184
Maytansine

$C_{34}H_{46}N_3O_{10}Cl$ Mol. wt 692.22

Occurs in the fruits of *Maytenus ovatus* and *M. senata* (Celastraceae). The stem and wood of *Putterlickia verrucosa* (Celastraceae) are the richest known sources (12 mg kg^{-1}).

Antileukaemic, cytotoxic, antitubulin and antimitotic activities. This is one of the most potent plant anticancer agents discovered, but its harmful side-effects have so far prevented its use in clinical practice.

1185
Mycosporine; Mycosporine 1; Mycosporin

$C_{11}H_{19}NO_6$ Mol. wt 261.28

Found in the fungi *Stereum hirsutum* (Stereaceae) and *Nectria galligena* (Hypocreales).

1186
Palustrine

$C_{17}H_{31}N_3O_2$ Mol. wt 309.46

Occurs in *Equisetum arvense*, *E. palustre* and *E. silvaticum* (Equisetaceae).

1187
Phyllospadine

$C_{21}H_{21}NO_6$ Mol. wt 383.41

Found in the seagrass *Phyllospadix iwatensis* (Zosteraceae).

1188
Pilocarpine

$C_{11}H_{16}N_2O_2$ Mol. wt 208.26

Found in *Pilocarpus microphyllus* and other *Pilocarpus* spp. (Rutaceae).

Stimulates the parasympathetic nerve endings, increasing thereby salivatory, gastric and lachrymal secretions. It is used in the treatment of glaucoma.

1189
Pilosine; Carpidine

$C_{16}H_{18}N_2O_3$ Mol. wt 286.33

Occurs in *Pilocarpus microphyllus* (Rutaceae).

Narcotic to mammals. It resembles pilocarpine (q.v.) in its pharmacological properties, but less active.

1190
Pithecolobine

$C_{22}H_{46}N_4O$ Mol. wt 382.64

The major component of a mixture of analogues occurring in *Samanea (Pithecolobium) saman* (Leguminosae).

Toxic.

1191
Pleurostyline

$C_{25}H_{29}N_3O_2$ Mol. wt 403.53

Occurs in *Pleurostylia africana* (Celastraceae).

1192
Securinine

$C_{13}H_{15}NO_2$ Mol. wt 217.27

Occurs in the leaves, roots and stems of *Securinega suffruticosa* and *Phyllanthus discoides* (Euphorbiaceae) and the bark of *Securidaca longepedunculata* (Leguminosae). It co-occurs with the 2-epimer, allosecurinine, in *Securinega suffruticosa*. The optical antipodes, virosecurinine and viroallosecurinine, remarkably, co-occur in a related plant, *Securinega virosa*.

Stereospecific γ-aminobutyric acid receptor antagonist. It is a central nervous system stimulant with strychnine-like activity and increase of blood pressure. A medical use is in the treatment of paralysis following infectious disease.

1193
Shihunine

$C_{12}H_{13}NO_2$ Mol. wt 203.24

Occurs in the orchids *Dendrobium lohohense* and *D. pierardii* (Orchidaceae).

D. lohohense is a component of the Chinese drug, "shi-hu", so that shihunine may be pharmacologically active.

1194
Sporidesmin; Sporidesmin A

$C_{18}H_{20}N_3O_6S_2Cl$ Mol. wt 473.96

Found in the fungus *Pithomyces chartarum*: one of several closely related alkaloids present.

Shows antitumour activity. It causes facial eczema in sheep grazing on infected pastures in New Zealand.

1195
Taxine A

$C_{35}H_{47}NO_{10}$ Mol. wt 641.76

Major constituent of the alkaloid mixture from the yew *Taxus baccata* (Taxaceae).

Main causal agent of cattle poisoning by yew leaves.

1196
Taxol; Taxol A

$C_{47}H_{51}NO_{14}$ Mol. wt 853.93

Found in the stem bark of *Taxus brevifolia* and *T. cuspidata*; it is also present in trace amount in other *Taxus* spp., including *T. baccata* (Taxaceae).

Toxic (LD_{50} orally in the dog 9 mg kg^{-1}). It is an antileukaemic and antitumour agent, and is undergoing clinical trials for treating breast cancer.

1197
(−)-Tylocrebine

$C_{24}H_{27}NO_4$ Mol. wt 393.49

The (+) form occurs in *Ficus septica* (Moraceae), while the (−) form (illustrated) is present in *Tylophora crebriflora* (Asclepiadaceae).

Toxic and powerful vesicants. The (−) form possesses antitumour activity in the mouse, and has central nervous system toxicity in humans.

1198
Tylophorine

$C_{24}H_{27}NO_4$ Mol. wt 393.49

Occurs in *Tylophora asthmatica, Cynanchum vincetoxicum, Pergularia pallida*, and *Vincetoxicum officinale* (Asclepiadaceae), and *Ficus septica* (Moraceae).

Weak antitumour activity. It is a powerful vesicant. It is highly toxic to frogs but shows low toxicity to mice.

1199
Verruculotoxin

$C_{15}H_{20}N_2O$ Mol. wt 244.34

Mycotoxin from the fungus *Penicillium verruculosum* (Hyphomycetes).

The toxic metabolite produced in infected green peanuts, *Arachis hypogaea* (LD_{50} in one-day-old chicks 20 mg kg^{-1}). Toxicity is characterized by ataxia, prostration and loss of muscle control.

1200
Withasomnine

$C_{12}H_{12}N_2$ Mol. wt 184.24

Found in the roots of *Withania somnifera* (Solanaceae).

PART IV
Phenolics

Phenolics

General Introduction

Phenolic compounds or polyphenols are plant substances which possess in common an aromatic ring bearing one or more hydroxyl groups. They range from simple substances, such as arbutin, eugenol, hydroquinone, khellin and myristicin, to the more complex structures of griseofulvin, podophyllotoxin, procyanidin, rotenone, tetrahydrocannabinol and usnic acid. The majority of phenolics are water-soluble, occurring naturally combined with sugar in glycosidic form; they are then located within the plant cell in the central vacuole. Other phenolics are lipophilic, with many of the phenolic groups masked by O-methylation, and they are present in the cell cytoplasm or else at the surface of plants in waxes and bud exudates.

There are about eight thousand naturally occurring plant phenolics and about half this number are flavonoids. These flavonoids have closely related structures, based on the C_{15} heterocyclic nucleus of flavone and varying chiefly in the number of phenolic, methoxyl and other substituents. They are derived biosynthetically from the union of aromatic (hydroxycinnamyl coenzyme A ester) and aliphatic (malonyl coenzyme A) precursors. There are many structurally simpler phenols, such as the phenolic acids, phenylpropanoids and phenolic quinones. Several important groups of plant polymeric materials are phenolic: the lignins, melanins and tannins. Finally, phenolic groups are occasionally found in alkaloids and terpenoids; such compounds will be found chiefly in other chapters in this Dictionary.

Phenolics are classified according to structural complexity and biosynthetic origin, but no ideal system is available. The simpler classes are the phenols themselves, phenolic acids and phenolic ketones. The phenylpropanoids, based on a C_6-C_3 nucleus, are a large group, and there are various derived phenylpropanoids which can be distinguished: the coumarins, chromones and chromenes, the benzofurans and the dimeric lignans. Three further classes of phenolic considered here are the xanthones (with a C_6-C_1-C_6 skeleton), the stilbenoids (with a C_6-C_2-C_6 skeleton) and the quinones. These last-mentioned are pigments, which share a common quinone nucleus, and which are further subdivided into benzoquinones, naphthoquinones and anthraquinones.

The flavonoids may be found under five separate headings: the anthocyanins and anthochlors, which are red-to-blue and yellow flower pigments, respectively; the minor flavonoids, which include flavanones, dihydroflavonols and dihydrochalcones; the flavones and flavonols, the most widely occurring and structurally variable flavonoids; the isoflavonoids, a distinctive class found mainly in one plant family, the Leguminosae; and the tannins, which are characterized by their affinity to bind with protein. Among the tannins are both flavonoids (the proanthocyanidins or flavolans) and simpler phenolics based on gallic acid (the gallo- and ellagi-tannins).

Phenolics contribute significantly to the colour, taste and flavour of many foods and drinks. Some flavonoids are valued pharmacologically for their anti-inflammatory and antihepatotoxic properties. Certain isoflavonoids exert oestrogenic effects in mammals, while others are insecticidal or piscicidal. The anthocyanins have a clearly defined function in flowers and fruits in attracting pollinators and seed dispersal agents to the plant. The more abundant phenolics in leaves and other tissues are often distasteful to animals, and serve a useful role in deterring herbivory.

No comprehensive dictionary of plant phenolics has appeared since Karrer (1958). The first modern monograph on phenolic compounds appeared 25 years ago (Harborne, 1964). Symposium proceedings provide more up-to-date information on these compounds (Swain *et al.*, 1979; van Sumere and Lea, 1985). The proceedings of the Groupe Polyphenols, which are published biennially, also contain new data on plant phenolics (see Anon, 1988). Methodology is covered in Pridham (1964), Ribereau-Gayon (1972) and Harborne (1984, 1989). For a concise modern account of phenolic biochemistry, see Harborne (1980). The recent monograph of Haslam (1989), entitled *Plant Polyphenols*, is not a general review but concentrates on the plant tannins. Other monographs concerned with individual classes of phenolic may be found listed under the appropriate sections.

References

Anon (1988). Proceedings of the Fourteenth International Conference of the Groupe Polyphenols. Ontario, Canada: St. Catherines.

Harborne, J.B. (1964). *The Biochemistry of Phenolic Compounds*. London: Academic Press.

Harborne, J.B. (1980). Plant phenolics. In *Encyclopedia of Plant Physiology*, New Series, Volume 8, *Secondary Plant Products* (Bell, E.A. and Charlwood, B.V.) (eds), pp. 328–402. Berlin: Springer-Verlag.

Harborne, J.B. (1984). *Phytochemical Methods*, 2nd Edition. London: Chapman & Hall.

Harborne, J.B. (1989). *Methods in Plant Biochemistry*, Volume 1, *Plant Phenolics* (Dey, P. and Harborne, J.B.) (series eds). London: Academic Press.

Haslam, E. (1989). *Plant Polyphenols: Vegetable Tannins Revisited*. Cambridge University Press.

Pridham, J.B. (1964). *Methods in Polyphenol Chemistry*. Oxford: Pergamon Press.

Ribereau-Gayon, P. (1972). *Plant Phenolics*. Edinburgh: Oliver and Boyd.

Swain, T., Harborne, J.B. and van Sumere, C.F. (1979). *Recent Advances in Phytochemistry*, Volume 12, *Biochemistry of Plant Phenolics*. New York: Plenum Press.

van Sumere, C.F. and Lea, P.J. (1985). *Phytochemical Society of Europe Annual Proceedings*, Volume 25, *The Biochemistry of Plant Phenolics*. Oxford: Clarendon Press.

Chapter 32
Anthocyanins and anthochlors

There are three main classes of flavonoid pigment providing flower colour in plants: the anthocyanins, which contribute red to blue colours throughout the angiosperms; and the anthochlor chalcones and the aurones, which have a more restricted occurrence in yellow petals. The most common anthocyanidin (or anthocyanin aglycone) is cyanidin (**1208**) and the other 15 known anthocyanidins differ only in the number and position of methoxyl or hydroxyl groups (delphinidin, **1213**, is the 5′-hydroxy derivative, etc.). These 16 anthocyanidins occur in nature with sugar attachment as glycosides or acylated glycosides. Cyanidin alone occurs in 76 different glycosidic combinations (see, e.g., cyanin, **1212**). Both aliphatic and aromatic acyl groups may be attached to the same anthocyanin (see, e.g., malonylawobanin, **1222**).

Anthocyanins provide colour in the flowers, fruits and leaves of all angiosperm families, except those which have betalains (q.v.) and are included in the order Centrospermae. Anthocyanins are also found in gymnosperms, ferns and occasionally in mosses. The distribution of anthocyanins (and anthochlors, see below) in plants is listed in detail in Harborne (1967), Harborne *et al.* (1975), Harborne and Mabry (1982) and Harborne (1988). Anthocyanin structure is related to *in vivo* petal colour, so that red flower colour is usually based on cyanidin and blue on delphinidin (Brouillard, 1988). The use of anthocyanins as natural food colourants has been proposed in recent years to replace the synthetic and possibly carcinogenic dyes in present use (Timberlake and Henry, 1986).

The yellow anthochlor pigments — the chalcones and aurones — are so described because, when flowers containing them are fumed with ammonia, they change to red. Yellow flowers containing only carotenoids (q.v.) are unchanged after this treatment. Anthochlor-containing flowers are found in such families as the Caryophyllaceae (*Dianthus*), Compositae (*Coreopis* and *Dahlia*), Rosaceae (*Potentilla*), and Scrophulariaceae (*Antirrhinum*). The chalcones are readily oxidized both *in vivo* and *in vitro* to aurones and, indeed, the two classes of pigment are often found together in yellow flowers. Two structural series can be discerned: those with trihydroxy substitution in the left-hand ring, e.g., chalconaringenin (**1207**) and aureusidin (**1203**), and those with dihydroxy substitution, e.g., butein (**1205**) and sulphuretin (**1232**). Chalcones, unlike aurones, also occur sporadically in other parts of plants. Among the 200 known chalcones, significant numbers have isopentenyl substitution (e.g., xenognosin A, **1233**).

The anthocyanins and anthochlors can be characterized initially by simple chromatographic and absorption spectroscopic procedures (Harborne, 1967). Some of the more complex structures require more sophisticated methods of analysis (Harborne, 1989).

References

Chemistry and Natural Occurrence

Harborne, J.B. (1967). *Comparative Biochemistry of the Flavonoids*, London, Academic Press.
Harborne, J.B., Mabry, T.J. and Mabry, H. (eds) (1975). *The Flavonoids*, London: Chapman and Hall.

Harborne, J.B. and Mabry, T.J. (1982). *The Flavonoids — Advances in Research*, London: Chapman and Hall.

Harborne, J.B. (1988). *The Flavonoids — Advances in Research since 1980*, London: Chapman and Hall.

Harborne, J.B. (1994). *The Flavonoids — Advances in Research since 1986*, London: Chapman and Hall.

Function

Brouillard, R. and Dangles, O. (1994). Flavonoids and flower colour. In *The Flavonoids — Advances in Research since 1986* (Harborne, J.B., ed.), pp. 565–588, London: Chapman and Hall.

Cody, V., Middleton, E. and Harborne, J.B. (1986). *Plant Flavonoids in Biology and Medicine*, New York: Alan R. Liss.

Timberlake, C.F, and Henry, B.S. (1986). Plant pigments as natural food colours. *Endeavour*, New Series, **10,** 31–36.

Methodology

Harborne, J.B. (ed.) (1989). *Methods in Plant Biochemistry*, Volume 1, *Plant Phenolics*, London: Academic Press.

1201
Abyssinone VI

Skeletal type: chalcone

$C_{25}H_{28}O_4$ Mol. wt 392.50

Found in the roots of the East African medicinal plant *Erythrina abyssinica* (Leguminosae).

Antipeptic activity. Strong inhibitory activity is shown against rabbit platelet aggregation.

1202
Apigeninidin; 5,7,4′-Trihydroxyflavylium

Skeletal type: anthocyanidin

$[C_{15}H_{11}O_4]^+$ Mol. wt 255.25

Occurs in glycosidic form in the fronds of many ferns, e.g., *Adiantum* and *Dryopteris* spp. (Pteridophyta), and as the 5-*O*-glucoside in the petals of *Reichsteineria cardinalis* (=*Sinningia cardinalis*, Gesneriaceae).

Orange-yellow flower pigment.

1203
Aureusidin; 4,6,3′,4′-Tetrahydroxyaurone

Skeletal type: aurone

$C_{15}H_{10}O_6$ Mol. wt 286.25

Occurs as the 6-glucoside, aureusin, in the petals of the snapdragon, *Antirrhinum majus* (Scrophulariaceae), as the 6-glucuronide in liverworts, e.g., *Marchantia polymorpha* (Hepaticae), and as the 4-glucoside (cernuoside) in *Chirita micromusa* and *Petrocosmea kerrii* (Gesneriaceae).

Yellow flower pigment. Aurones are potent inhibitors of iodothyronine-deiodinase in rat liver microsomal membranes.

1204
Bracteatin; 4,6,3′,4′,5′-Pentahydroxyaurone

Skeletal type: aurone

$C_{15}H_{10}O_7$ Mol. wt 302.25

Occurs as the 6-glucoside in the petals of the snapdragon, *Antirrhinum majus* (Scrophulariaceae), and as the 4-glucoside in the flowers of *Helichrysum bracteatum* (Compositae).

Activity similar to that of aureusidin (q.v.).

1205
Butein; 2′,4′,3,4-Tetrahydroxychalcone

Skeletal type: chalcone

$C_{15}H_{12}O_5$ Mol. wt 272.26

Occurs widely as the 4′-glucoside, coreopsin, in yellow flowered Compositae, e.g., in *Coreopsis douglasii* and *Bidens* spp. See also isobutrin.

Yellow flower pigment. It completely uncouples the oxidative phosphorylation in mitochondria of mungbean hypocotyls and potato tubers. Also, it is an inhibitor of NADH oxidase and succinoxidase activities. See also chalconaringenin.

1206
Carthamone

Skeletal type: chalcone

$C_{21}H_{20}O_{11}$ Mol. wt 448.39

Found in safflower, *Carthamus tinctorius* (Compositae).

The flowers of *C. tinctorius* were formerly used in rouge and for dyeing food.

1207
Chalconaringenin; 2′,4′,6′,4-Tetrahydroxychalcone

Skeletal type: chalcone

$C_{15}H_{12}O_5$ Mol. wt 272.26

Exists only in conjugated form, e.g., as the 2′-glucoside, isosalipurposide, which occurs in the bark of the willow *Salix purpurea* (Salicaceae), and in the petals of yellow carnations (*Dianthus*, Caryophyllaceae), in *Helichrysum* spp. (Compositae) and in *Paeonia* spp. (Paeoniaceae).

Yellow flower pigment. Isosalipurposide in flowers also provides a nectar guide to pollinating insects, since it absorbs UV light and appears to the pollinators as "bee-purple" stripes or patterns in the flowers. Chalcones are also potent iodothyronine-deiodinase inhibitors in rat hepatocytes.

1208
Cyanidin; 3,5,7,3′,4′-Pentahydroxyflavylium

Skeletal type: anthocyanidin

$[C_{15}H_{11}O_6]^+$ Mol. wt 287.25

Found in glycosidic form, frequently acylated, in one or more organs (flowers, fruits, leaves, and tubers) of nearly all green plants. In addition to the cyanidin glycosides described in the following four entries, there are many more examples, e.g., the 3-arabinoside in the leaves of the cocoa plant, *Theobroma cacao* (Sterculiaceae), the 3-sophoroside-5-glucoside in the pods of the pea, *Pisum sativum* (Leguminosae), and the 3-sambubioside in *Sambucus nigra* berries (Caprifoliaceae).

Widespread red plant pigment. Elderberries and red cabbage, which both contain many different cyanidin glycosides, are potential sources of natural food colourings.

1209
Cyanidin 3-*O*-galactoside; Idaein; Idein

Skeletal type: anthocyanin

$[C_{21}H_{21}O_{11}]^+$ Mol. wt 449.40

Occurs in the leaves of the copper beech, *Fagus sylvatica* (Fagaceae), the kernel skin of the pistachio nut, *Pistacia vera* (Anacardiaceae), and the berries of *Vaccinium* spp. (Ericaceae).

Red fruit pigment. The anthocyanin extract of *V. myrtillus* berries, which contains the 3-galactosides of cyanidin, delphinidin, peonidin, petunidin and malvidin and the 3-glucosides and 3-arabinosides of the same anthocyanidins, has anti-inflammatory activity, and is used in the prevention of capillary fragility.

1210
Cyanidin 3-*O*-glucoside; Asterin; Chrysanthemin; Kuromanin

Skeletal type: anthocyanin

$[C_{21}H_{21}O_{11}]^+$ Mol. wt 449.40

This is the commonest of all anthocyanins, occurring in many flowers, e.g., of *Aster* cultivars; in fruits, e.g., raspberries, *Rubus idaeus* (Rosaceae); in tubers, e.g., of the yam, *Dioscorea alata* (Dioscoreaceae); and in leaves, particularly in autumnal ones.

Red plant pigment. It is a larval growth inhibitor for the tobacco budworm, *Heliothis virescens*.

1211
Cyanidin 3-*O*-rutinoside; Antirrhinin; Keracyanin; Sambucin

Skeletal type: anthocyanin

$[C_{27}H_{31}O_{15}]^+$ Mol. wt 595.55

Found in the magenta flowers of the snapdragon, *Antirrhinum majus* (Scrophulariaceae) and the petals of *Potentilla atrosanguinea* (Rosaceae), and in the leaves, stems, fruits and reproductive organs of Araceae, e.g., in the spadix and spathe of the cuckoo pint, *Arum maculatum*.

Red plant pigment.

1212
Cyanin; Cyanidin 3,5-di-*O*-glucoside; Shisonin A

Skeletal type: anthocyanin

$[C_{27}H_{31}O_{16}]^+$ Mol. wt 611.55

Occurs in the fruits of elderberry, *Sambucus* spp. (Caprifoliaceae) and in the leaves of many Bromeliaceae, e.g., the pineapple, *Ananas comosus*, in which it co-occurs with the 3,5,3'-triglucoside. It occurs as the succinic acid ester in the petals of the cornflower, *Centaurea cyanus* (Compositae) and also occurs, in the acylated form, in red cabbage, *Brassica oleracea* (Cruciferae).

Potential source of natural red food colouring (see cyanidin).

1213
Delphinidin; 3,5,7,3′,4′,5′-Hexahydroxyflavylium

Skeletal type: anthocyanidin

$[C_{15}H_{11}O_7]^+$ Mol. wt 303.25

Found in glucosidic form in many blue and purple flowers and fruits, e.g., as the 3,5-diglucoside in flowers of *Delphinium consolida* (Ranunculaceae) and *Verbena hybrida* (Verbenaceae), where it co-occurs with the 3-glucoside, as the 3-rhamnoside in *Plumbago rosea* petals (Plumbaginaceae), and the 3-rutinoside in potato tubers, *Solanum tuberosum* (Solanaceae). See also gentiodelphin, malonylawobanin and delphinidin 3,3′,5′-tri-*O*-glucoside.

Mauve plant pigment.

1214
Delphinidin 3,3′,5′-tri-*O*-glucoside

Skeletal type: anthocyanin

$[C_{33}H_{41}O_{22}]^+$ Mol. wt 789.69

Occurs in various acylated forms (ternatins A–F) in the flowers of *Clitoria ternatea* (Leguminosae).

These pigments are used in Malaysia to colour rice cakes blue.

1215
Gentiodelphin

Skeletal type: anthocyanin

$[C_{51}H_{53}O_{28}]^+$ Mol. wt 1113.99

Found in the petals of *Gentiana makinoi* (Gentianaceae).

Mauve to blue flower pigment.

1216
Heavenly blue anthocyanin; Peonidin 3-sophoroside-5-glucoside tri(caffeoylglucose) ester

Skeletal type: anthocyanin

$[C_{79}H_{91}O_{45}]^+$ Mol. wt 1760.60

Occurs in the petals of the morning glory, *Ipomoea purpurea* "heavenly blue" (Convolvulaceae).

Blue flower pigment.

1217
Hirsutidin; 3,5,4′-Trihydroxy-7,3′,5′-trimethoxyflavylium

Skeletal type: anthocyanidin

$[C_{18}H_{17}O_7]^+$ Mol. wt 345.33

Occurs as the 3,5-diglucoside (hirsutin) in the flowers of *Primula hirsuta* and other *Primula* spp. (Primulaceae).

Mauve flower pigment.

1218
6-Hydroxycyanidin; 3,5,6,7,3′,4′-Hexahydroxyflavylium

Skeletal type: anthocyanidin

$[C_{15}H_{11}O_7]^+$ Mol. wt 303.25

Occurs as the 3-rutinoside and 3-glucoside in the flowers of *Alstroemeria* cultivars (Alstroemeriaceae).

Red flower pigment.

1219
Isobutrin; Butein 3,4′-di-*O*-glucoside

Skeletal type: chalcone *O*-glycoside

$C_{27}H_{32}O_{15}$ Mol. wt 596.55

Found in the petals of *Butea monosperma* and *B. frondosa* (Leguminosae).

Yellow flower pigment. It shows antihepatotoxic activity.

1220
Isoliquiritigenin; 2′,4′,4-Trihydroxychalcone

Skeletal type: chalcone

$C_{15}H_{12}O_4$ Mol. wt 256.26

Found as the 4-glucoside in liquorice root, *Glycyrrhiza glabra* (Leguminosae), and as the 4′-glucoside and 4′-diglucoside in yellow-flowered varieties of *Dahlia variabilis* (Compositae).

Yellow flower pigment. It inhibits rat liver mitochondrial monoamine oxidase, and it completely uncouples the oxidative phosphorylation in mitochondria of mungbean hypocotyls and potato tubers.

1221
Luteolinidin; 5,7,3′,4′-Tetrahydroxyflavylium

Skeletal type: anthocyanidin

$[C_{15}H_{11}O_5]^+$ Mol. wt 271.25

Occurs in glycosidic form in the fronds of many ferns, e.g., in *Adiantum, Azolla, Dryopteris* and *Pteris* spp. (Pteridophyta). The 5-*O*-glucoside is present in petals of *Rechsteineria cardinalis* (Gesneriaceae).

Orange-red flower pigment.

1222
Malonylawobanin; Delphinidin 3-(6″-*p*-coumaroylglucose)-5-(6‴-malonylglucoside)

Skeletal type: anthocyanin

$[C_{39}H_{39}O_{22}]^+$ Mol. wt 859.73

Occurs in the flowers of *Commelina communis* (Commelinaceae) and of bluebells, *Hyacinthoides nonscripta* (Liliaceae).

Blue flower pigment.

1223
Malvidin; 3,5,7,4′-Tetrahydroxy-3′,5′-dimethoxyflavylium

Skeletal type: anthocyanidin

$[C_{17}H_{15}O_7]^+$ Mol. wt 331.31

Found as the 3-glucoside in grapes, *Vitis vinifera* cultivars (Vitaceae). The 3-rhamnoside occurs in the petals of sweet pea, *Lathyrus odoratus* (Leguminosae) and of *Anagallis arvensis* f. *coerulea* (Primulaceae), and the 3-galactoside in the berries of *Vaccinium uliginosum* (Ericaceae).

Mauve plant pigment. The pomace of wine grapes and grape juice lees, which contains malvidin 3-glucoside and sometimes the 3,5-diglucoside, is used as a natural food colouring.

1224
Malvin; Malvidin 3,5-diglucoside

Skeletal type: anthocyanin

$[C_{29}H_{35}O_{17}]^+$ Mol. wt 655.60

Found in the petals of the mallow, *Malva sylvestris* (Malvaceae) as the malonate. Malvin and acetylated derivatives are also found in wine grapes, *Vitis* spp. (Vitaceae).

Activity similar to that of malvidin (q.v.).

1225
Maritimetin; 6,7,3',4'-Tetrahydroxyaurone

Skeletal type: aurone

$C_{15}H_{10}O_6$ Mol. wt 286.25

Occurs as the 6-glucoside, maritimein, in the yellow flowers of several Compositae, e.g., in *Coreopsis maritima* and *C. bigelovii*. Maritimein often co-occurs with the chalcone glycoside marein (see okanin).

Activity similar to that of aureusidin (q.v.).

1226
Monardaein; Monardein; Pelargonidin 3-(6″-*p*-coumaroylglucoside)-5-(4‴,6‴-dimalonylglucoside)

Skeletal type: anthocyanin

$C_{42}H_{41}O_{23}$ Mol. wt 913.77

Found in the flowers of bergamot, *Monarda didyma* and of *Salvia splendens* (Labiatae).

Scarlet flower pigment.

1227
Okanin; 2',3',4',3,4-Pentahydroxychalcone

Skeletal type: chalcone

$C_{15}H_{12}O_6$ Mol. wt 288.26

Occurs as the 4'-O-glucoside, marein, in yellow flowers of several Compositae, e.g., in *Coreopsis* and *Bidens* spp. Marein often co-occurs with the aurone glycoside maritimein (see maritimetin).

Yellow flower pigment. It completely uncouples the oxidative phosphorylation in mitochondria of mungbean hypocotyls and potato tubers. See also chalconaringenin.

1228
Pelargonidin; 3,5,7,4′-Tetrahydroxyflavylium

Skeletal type: anthocyanidin

$[C_{15}H_{11}O_5]^+$ Mol. wt 271.25

Occurs as the 3-glucoside in the fruit of the wild strawberry, *Fragaria vesca* (Rosaceae), the 3-galactoside in beech leaves, *Fagus sylvatica* (Fagaceae), the 3-rutinoside in *Antirrhinum majus* petals, the 3-lathyroside in *Lathyrus odoratus* petals, etc. See also pelargonin and monardaein.

Scarlet plant pigment. It shows antiviral activity.

1229
Pelargonin; Pelargonidin 3,5-di-*O*-glucoside

Skeletal type: anthocyanin

$[C_{27}H_{31}O_{15}]^+$ Mol. wt 595.55

Occurs in the petals of *Pelargonium zonale* (Geraniaceae), of red *Gladiolus* cultivars (Iridaceae), of *Commiphora muhul* (Burseraceae), etc. Acylated derivatives of pelargonin occur in red *Dahlia* cultivars (Compositae) and in *Monarda didyma* (Labiatae). See monardaein.

Scarlet to pink flower pigment.

1230
Peonidin; 3,5,7,4′-Tetrahydroxy-3′-methoxyflavylium

Skeletal type: anthocyanidin

$[C_{16}H_{13}O_6]^+$ Mol. wt 301.28

Occurs in glycosidic form in flowers and fruits: e.g., as the 3-galactoside, 3-arabinoside and 3-glucoside in the berries of *Vaccinium* spp. (Ericaceae), and as the 3,5-diglucoside in flowers of species of *Rosa* (Rosaceae), *Pelargonium* (Geraniaceae) and *Paeonia* (Paeoniaceae). See also heavenly blue anthocyanin.

Activity similar to that of cyanidin 3-galactoside (q.v.).

1231
Petunidin; 3,5,7,4′,5′-Pentahydroxy-3′-methoxyflavylium

Skeletal type: anthocyanidin

$[C_{16}H_{13}O_7]^+$ Mol. wt 317.28

Found in glycosidic form in flowers and fruits: e.g., as the 3-glucoside in the petals of *Primula sinensis*, the 3-galactoside and 3-arabinoside in the berries of *Vaccinium* spp. (Ericaceae), the 3-rhamnoside in the petals of *Lathyrus odoratus* (Leguminosae), and the 3,5-diglucoside (petunin) in the flowers of *Anchusa* spp. (Boraginaceae).

Mauve to blue flower pigment. Activity similar to that of cyanidin 3-galactoside (q.v.).

1232
Sulphuretin; Sulfuretin; 6,3′,4′-Trihydroxyaurone

Skeletal type: aurone

$C_{15}H_{10}O_5$ Mol. wt 270.25

Occurs in the wood of *Cotinus* (Anacardiaceae), and as the 6-glucoside, sulphurein, in the flowers of several genera of Compositae, e.g., *Coreopsis, Bidens, Helianthus*, and in yellow varieties of *Dahlia*.

Activity similar to that of aureusidin (q.v.).

1233
Xenognosin A

Skeletal type: 1,3-diphenylpropene

$C_{16}H_{16}O_3$ Mol. wt 256.30

This chalcone-like substance is produced, together with the closely related obtustyrene (which lacks the B-ring hydroxyl) and 4,4′-dihydroxy-2′-methoxychalcone, by copper(II)-stressed peas, *Pisum sativum* (Leguminosae).

Stress metabolite.

Chapter 33
Benzofurans

The benzofuran nucleus, in which a benzene ring and a furan ring are fused together, is present in many plant products, and some of these are classified elsewhere in this Dictionary as coumarins and flavonoids (q.v.). The compounds dealt with in this section are either simple benzofurans, e.g., dehydrotremetone (**1237**), or dibenzofurans, e.g., cotonefuran (**1236**), where two benzene rings are fused on either side of a central furan ring. They are generally substituted with hydroxyl, methoxyl or ketonic groups. Compounds in which the 2,3-double bond is reduced are also known, e.g., albafuran (**1234**) and tremetone (**1250**).

The biosynthesis of benzofurans may occur by more than one pathway. Simple benzofurans, such as dehydrotremetone, are probably formed in plants by the condensation of a polyketide-derived phenol with a terpenoid precursor, isopentenyl pyrophosphate, and subsequent ring closure on the phenolic group. By contrast, the 2-arylbenzofurans (e.g., moracin A, **1242**) are clearly related biosynthetically to the stilbenoids (q.v.) and may co-occur with them in some plants (e.g., in *Morus* spp.).

The best known source of benzofurans is the family Compositae, in which substances like tremetone are present; other sources include members of the Leguminosae, Moraceae, Myrtaceae and Rosaceae. These aromatic compounds are present occasionally in fungi, but are more characteristically present in lichens. Usnic acid (**1251**) is but one of a number of lichen benzofurans (Culberson, 1969; Culberson and Elix, 1989).

A biological property many benzofurans share is antifungal activity. This includes both constitutive natural products such as sainfuran (**1247**) and substances formed after plants have suffered microbial infection, i.e., phytoalexins. Benzofuran phytoalexins have now been recorded in Leguminosae (vignafuran, **1252**), Moraceae (moracin A, **1242**) and Rosaceae (cotonefuran, **1236**). Others are toxic to insects, fish, cattle and humans. Tremetone (**1250**), for example, ingested by cattle from plants such as white snake root, *Eupatorium rugosum*, causes "trembling" symptoms which may lead to death. Also it passes into the cow's milk, producing a "milk sickness" in humans who drink it.

There is no simple colour reaction available for detecting benzofurans in plant extracts. However, many may be recognized by their fluorescence in ultraviolet light, while a few such as usnic acid are yellow in colour. They also have characteristic spectral absorption properties. Methods for analysing lichen benzofurans are reviewed by Culberson and Elix (1989).

References

Chemistry
Dean, F.M. (1963). *Naturally Occurring Oxygen Ring Compounds*. London: Butterworths.

Distribution and Analysis
Culberson, C.F. (1969). *Chemical and Botanical Guide to Lichen Products*. Chapel Hill: University of North Carolina Press.
Culberson, C.F. and Elix, J.A. (1989). Lichen substances. In *Methods in Plant Biochemistry*, Volume 1, *Plant Phenolics* (Harborne, J.B., ed.), pp. 509–536. London: Academic Press.

1234
Albafuran A

$C_{24}H_{26}O_4$ Mol. wt 378.47

Found in the shoot epidermis of *Morus alba* (Moraceae) together with an isomer, albafuran B.

Antifungal activity.

1235
Albanol A; Mulberrofuran G

$C_{34}H_{26}O_8$ Mol. wt 562.58

Found in the root bark of *Morus lhou* and *M. alba* (Moraceae).

Hypotensive activity.

1236
Cotonefuran

$C_{15}H_{14}O_6$ Mol. wt 290.27

A phytoalexin produced in necrotic wood of *Cotoneaster lactea* (Rosaceae).

Antifungal activity against *Cladosporium cucumerinum*.

1237
Dehydrotremetone

$C_{13}H_{12}O_2$ Mol. wt 200.24

Found in *Eupatorium urticaefolium* and *Haplopappus heterophyllus* (=*Isocoma wrightii*) (both Compositae).

Bacteriostatic activity. It is toxic to goldfish. Also, it is jointly responsible for milk sickness after human consumption of milk from cattle feeding on *Eupatorium urticaefolium*. See also toxol and tremetone.

1238
Griseofulvin; Curling factor; Spirofulvin; Grifulvin

$C_{17}H_{17}O_6Cl$ Mol. wt 352.78

Metabolite of the moulds *Penicillium griseofulvin* and *P. janczewskii*.

Antifungal activity. It is effective against ringworm when taken orally.

1239
6-Hydroxytremetone

$C_{13}H_{14}O_3$ Mol. wt 218.26

Found in many Compositae, e.g., in *Clibadium armanii*, *Encelia ventuorum*, *Fluorensia heterolepis*, *Hemizonia congesta*, and *Ligularia intermedia*, and in *Helianthella*, *Austrobrickellia*, *Senecio*, *Abrotanella*, *Helichrysum* and *Tagetes* spp.

Toxic to goldfish.

1240
Lithospermic acid

$C_{27}H_{22}O_{12}$ Mol. wt 538.47

Found in the roots of various Boraginaceae, e.g., in *Lycopus europaeus*, *L. virginicus*, *Lithospermum ruderale*, *L. officinale*, *Symphytum officinale*, *Anchusa officinale* and *Echium vulgare*.

Reputed to have contraceptive properties in humans, but this remains to be confirmed.

1241
6-*O*-Methyleuparin; Euparin methyl ether

$C_{14}H_{14}O_3$ Mol. wt 230.27

Found in *Encelia californica* (Compositae).

Moderately phototoxic against bacteria and yeasts; it is active against *Pseudomonas fluorescens*, which is usually unaffected by psoralen and other photosensitizers.

1242
Moracin A

$C_{16}H_{14}O_5$ Mol. wt 286.29

Occurs in the cortex and phloem tissues of shoots from the mulberry, *Morus alba* (Moraceae), infected with the fungus *Fusarium solanii* f.sp. *mori*. One of 26 closely related benzofurans (moracins A–Z) formed in *Morus alba* as phytoalexins.

Antifungal activity.

1243
Mulberrofuran A

$C_{25}H_{28}O_4$ Mol. wt 392.50

Found in the root bark of *Morus alba* (Moraceae).

Inhibits formation of cyclooxygenase products of the arachidonate metabolism *in vitro*. Also, it is active against Gram-positive bacteria.

1244
Mulberrofuran C; Moruberofuran C

$C_{41}H_{42}O_9$ Mol. wt 678.79

Found in the root bark of *Morus bombycis* (Moraceae).

Hypotensive action in animals.

1245
α-Pyrufuran

$C_{15}H_{14}O_5$ Mol. wt 274.28

Occurs in the sapwood of *Pyrus communis* (Rosaceae) infected with *Chondrostereum purpureum*.

Antifungal activity against *Cladosporium cucumerinum*.

1246
ψ-Rhodomyrtoxin

$C_{24}H_{28}O_7$ Mol. wt 428.49

Occurs in the fruits of *Rhodomyrtus macrocarpa* (Myrtaceae).

Consumption of the fruits of *Rhodomyrtus* causes blindness and poisoning of livestock, but there is no verification that this compound is responsible. It is toxic to mice.

1247
Sainfuran

$C_{16}H_{14}O_5$ Mol. wt 286.29

Found in the roots of *Onobrychis viciifolia* (Leguminosae), together with its methyl ether, methylsainfuran. Also, it is found in roots of *Hedysarum polybotris* (Leguminosae).

Insect antifeedant. It shows antifungal activity against *Cladosporium cladosporioides*.

1248
Toxol

$C_{13}H_{14}O_3$ Mol. wt 218.26

Occurs in *Haplopappus heterophyllus* (=*Isocoma wrightii*) and *Morithamnus crassua* (both Compositae).

Bacteriostatic and antitumour activities. *Haplopappus heterophyllus* is also claimed to be responsible for causing milk sickness (together with *Eupatorium* spp.) as described for dehydrotremetone (q.v.).

1249
Toxyl angelate

$C_{18}H_{20}O_4$ Mol. wt 300.36

Found in *Haplopappus heterophyllus* (=*Isocoma wrightii*), *H. tenuisectus*, *Bahianthus viscidus* and *Morithamnus crassua*. The stereoisomer occurs in *Liatris* spp. (all Compositae).

Moderately active against P-388 lymphocytic leukaemia tumours.

1250
Tremetone

$C_{13}H_{14}O_2$ Mol. wt 202.26

Occurs in many Compositae, e.g., in *Eupatorium urticaefolium*, *E. rugosum*, *Haplopappus heterophyllus* and in *Ageratina*, *Brickellia*, *Liatris*, *Ligularia*, *Baccharis* and *Grindelia* spp.

Toxic to goldfish. It is jointly responsible for milk sickness in humans after consumption of milk from cattle feeding on *Eupatorium urticaefolium*. See also dehydrotremetone.

1251
Usnic acid; Usninic acid; Usnein; Usniacin

$C_{18}H_{16}O_7$ (−) form   Mol. wt 344.33

Found in a number of lichens, e.g., *Usnea*, *Ramalina*, *Evernia*, *Parmelia*, *Lecanora* and *Cladonia* spp. It occurs as the (+) form in *Usnea*, as the (−) form in *Cladonia* and also, rarely, as the racemate (in *Cetraria islandica*). The lichen *Haematomma coccineum* contains as much as 20% of its dry weight as usnic acid.

Antibacterial activity, especially against the tuberculosis bacterium, *Mycobacterium tuberculosis*. It is ineffective against fungi.

1252
Vignafuran

$C_{16}H_{14}O_4$ Mol. wt 270.29

Formed as a phytoalexin in the hypocotyls of *Lablab niger* and in the leaves of *Vigna unguiculata* (Leguminosae).

Fungitoxic.

Chapter 34
Chromones and chromenes

The chromones comprise a small group of naturally occurring plant phenolics which have an aromatic ring fused to a 6-membered heterocyclic pyrone ring. The biosynthetic origins of these simple chromones are not always clear. Those which have a 2-methyl substituent (e.g., biflorin, **1254**) are probably of acetate origin, formed by the polyketide pathway. Even when a 2-methyl group is absent, e.g., in the case of the 3-ethylchromone lathodoratin (**1269**), the aromatic ring is still of acetate origin while the rest of the carbon skeleton is derived from the protein amino acid isoleucine. The chromones are formally related to the flavonoids (q.v.) (or 2-phenylchromones) and isoflavonoids (3-phenyl-chromones), but these two classes of natural product are dealt with separately in this Dictionary. Some chromones have isopentenyl substitution in addition (cimifugin, **1258**) and isopentenyl groups are clearly involved in the biosynthesis of the naturally occurring chromenes. These mainly have 2,2-dimethyl substitution, e.g., encecalin (**1262**), indicating the origin of the heterocyclic part of the structure from an isoprenyl unit.

The best known chromone is undoubtedly the furochromone khellin (**1267**), which has been used therapeutically as a vasodilator in the treatment of angina pectoris. The plant source of khellin, the seeds of *Ammi visnaga*, also contains a number of related compounds with similar activity. The chromenes are mainly of biological interest because of their insecticidal activity. The precocenes 1 and 2 are of special note because, as anti-juvenile hormone agents, they interfere with larval growth in a number of insect species, causing precocious metamorphosis.

References

Dean, F.M. (1963). *Naturally Occurring Oxygen Ring Compounds*. London: Butterworths.
Staal, G.B. (1986). Anti-juvenile hormone agents. *Annual Reviews of Entomology*, **31**, 391–429.

1253
Aurasperone D

$C_{31}H_{24}O_{10}$ Mol. wt 556.53

Occurs in mango fruit, *Mangifera indica* (Anacardiaceae), infected with *Aspergillus niger*.

Depressant effects on central nervous system in animals, leading to death by respiratory failure. Stored foods can become contaminated with *Aspergillus niger* – a serious health hazard to people and animals.

1254
Biflorin

$C_{16}H_{18}O_9$ Mol. wt 354.32

Root constituent of *Pancratium biflorum* (Amaryllidaceae).

Phosphodiesterase inhibitor. Biflorin is also the name of an *o*-quinone found in *Copraria biflora*.

1255
Butyrylmallotochromene

$C_{26}H_{30}O_8$ Mol. wt 470.53

Found in the pericarp of *Mallotus japonicus* (Euphorbiaceae).

Moderately cytotoxic against KB-cell lines. The bark of *Mallotus* is used as medication for ulcers and cancer in Japan.

1256
Cannabichromene

$C_{21}H_{30}O_2$ Mol. wt 314.47

Occurs in *Cannabis sativa* (Cannabaceae).

Anti-inflammatory activity; it protects erythrocytes from hypotonic lysis.

1257
Capillarisin

$C_{16}H_{12}O_7$ Mol. wt 316.27

Found in the aerial parts of *Artemisia capillaris* (Compositae).

Choleretic activity.

1258
Cimifugin

$C_{16}H_{18}O_6$ Mol. wt 306.32

Found in the roots and rhizomes of *Ledebouriella seseloides* (=*Siler divaricatum*) (Umbelliferae), in *Cimicifuga* spp. and *Eranthis pinnatifida* (both Ranunculaceae), and also as a glucoside in *Angelica japonica* (Umbelliferae).

Hypotensive activity in animals; it shows increasing coronary flow in isolated guinea-pig heart. The roots of *Ledebouriella* are used as a diaphoretic, an analgesic, and an antipyretic in Chinese medicine.

1259
5,7-Dihydroxychromone

$C_9H_6O_4$ Mol. wt 178.15

Found in *Cneorum tricoccum* and *C. pulverulentum* (Cneoraceae), in the root bark of *Morus alba* (Moraceae), in the seeds of *Silybum marianum* (Compositae), and in the shells of *Arachis hypogaea* (Leguminosae). Possibly of artefactual origin, it is formed during extraction by cleavage of the 2-phenyl substituent of 5,7-dihydroxyflavanones (q.v.) present in the plant.

Active against Gram-positive bacteria. *Cneorum* plants are used as a rubifacient and an antifebrifuge.

1260
8-(3,3-Dimethylallyl)spatheliachromene; Cneorum chromone B

$C_{20}H_{22}O_4$ Mol. wt 326.40

Occurs in the leaves of *Cneorum tricoccum* and *C. pulverulentum* (Cneoraceae), and in *Spathelia sorbifolia* (Rutaceae).

Inhibitory action against HeLa cells; also, it shows antibacterial activity.

1261
Drummondin A

$C_{26}H_{30}O_8$ Mol. wt 470.53

Found in the roots of *Hypericum drummondii* (Guttiferae) together with the closely related drummondins B, C and F.

All drummondins have antimicrobial activity and are cytotoxic to P-388, KB and human cancer cell lines.

1262
Encecalin

$C_{14}H_{16}O_3$ Mol. wt 232.28

Occurs in *Lagascea rigida, Encelia californica, Hemizonia congesta, Eupatorium glandulosum* and *Ageratina scorodonoides* (all Compositae).

Insecticidal and phototoxic activity (against bacteria and yeasts).

1263
Eupatoriochromene; 7-Hydroxyencecalin; 7-Demethylencecalin

$C_{13}H_{14}O_3$ Mol. wt 218.26

Occurs in *Lagascea rigida, Verbesina alternifolia, Encelia californica, Eupatorium riparium, Austrobrickellia patens* and *Ageratina aromatica* (all Compositae). Also, it is a flower constituent of *Arnica sachalinensis* (Compositae).

Phototoxic activity against bacteria and yeasts.

1264
Flindersiachromone; 2-(2-Phenylethyl)chromone

$C_{17}H_{14}O_2$ Mol. wt 250.30

Found in the wood of *Aquillaria* spp., agarwood (Thymeleaceae) and in *Flindersia laevicarpa* (Rutaceae).

Contributes to the odour of agarwood when it is burned as incense. Its resinous wood is used as an aphrodisiac, diuretic and stimulant, and wood decoctions are used in Chinese herbal medicine.

1265
Frutinone A

$C_{16}H_8O_4$ Mol. wt 264.24

Found in the leaves of *Polygala fruticosa* (Polygalaceae).

Antifungal activity against *Cladosporium cucumerinum*.

1266
Isobutyrylmallotochromene

$C_{26}H_{30}O_8$ Mol. wt 470.52

Found in the pericarp of *Mallotus japonicus* (Euphorbiaceae).

Cytotoxic activity against KB cell lines. The bark of *Mallotus* is used as medication for ulcers and cancer in Japan.

1267
Khellin; Kellin; Kelamin; Kelicor; Eskel; Gynokhellan; Visammin

$C_{14}H_{12}O_5$ Mol. wt 260.25

Found in the seeds of *Ammi visnaga* (Umbelliferae).

Used medicinally as a vasodilatory agent; it exhibits antiviral phototoxicity and moderate action against bacteriophage T4, M13. Also, it is a phosphodiesterase inhibitor. Further activities are anthelmintic, antispasmodic, anti-arteriosclerotic, anti-asthmatic, bronchodilatory, antidiabetic and anti-ulcerogenic.

1268
Khellol glucoside; Khellinin

$C_{19}H_{20}O_{10}$ Mol. wt 408.37

Found in the seeds of *Ammi visnaga* (Umbelliferae), and in *Eranthis hyemalis* (Ranunculaceae).

1269
Lathodoratin; 5,7-Dihydroxy-3-ethylchromone

$C_{11}H_{10}O_4$ Mol. wt 206.20

Found in the leaflets of *Lathyrus odoratus* (Leguminosae), after fungal infection, as a phytoalexin. Co-occurs with the 7-methyl ether.

Fungicidal activity against *Cladosporium herbarum*.

1270
Mallotochromene

$C_{24}H_{26}O_8$ Mol. wt 442.47

Found in the pericarp of *Mallotus japonicus* (Euphorbiaceae).

Antileukaemic activity against KB cell lines. The bark of *Mallotus* is used as a medication for ulcers and cancer in Japan.

1271
2-(4-Methoxyphenethyl)chromone

$C_{18}H_{16}O_3$ Mol. wt 280.33

Occurs in agarwood, *Aquillaria* spp. (Thymeleaceae).

Contributes to the odour of agarwood when it is burned as an incense. Its resinous wood is used as an aphrodisiac, a diuretic and a stimulant, and wood decoctions are used in Chinese herbal medicine.

1272
5-O-Methylalloptaeroxylin; Methylallopteroxylin; Perforatin A

$C_{16}H_{16}O_4$ Mol. wt 272.30

Found in the leaves of *Ptaeroxylon obliquum* and in *Cedrelopsis grevei* (Ptaeroxylaceae). It also occurs in *Neochamaelea pulverulenta* (=*Cneorum pulverulenta*) (Cneoraceae) and in *Harrisonia perforata* (Simaroubaceae).

Antihypertensive activity in rats. *Ptaeroxylon* is used in Southern Africa as a traditional medicine. The powdered wood is pungent and irritating, causing violent sneezing.

1273
5-O-Methylvisamminol

$C_{16}H_{18}O_5$ Mol. wt 290.32

Occurs in the roots and rhizomes of *Ledebouriella seseloides* (=*Siler divaricatum*) (Umbelliferae).

Hypotensive action in animals. It increases blood flow in isolated guinea-pig heart. *Ledebouriella* is called fang feng in Chinese medicine, and is used as a diaphoretic, an analgesic and an antipyretic.

1274
Precocene 1; 7-Methoxy-2,2-dimethylchromene

$C_{12}H_{14}O_2$ Mol. wt 190.24

Found in *Ageratina aromatica* (Compositae).

Active as an antijuvenile hormone in insects.

1275
Precocene 2; Ageratochromene; 6,7-Dimethoxy-2,2-dimethylchromene

$C_{13}H_{16}O_3$ Mol. wt 220.27

Found in *Ageratum houstonianum*, in *Ageratina aromatica* and in *Senecio longifolius* (all Compositae).

Insecticidal activity, and active as an antijuvenile hormone in insects.

1276
Ptaerochromenol; Pterochromenol

$C_{15}H_{14}O_5$ Mol. wt 274.28

Occurs in *Cneorum tricoccum* and *C. pulverulentum* (Cneoraceae), and also in the heartwood of *Ptaeroxylon obliquum* (Ptaeroxylaceae), where it co-occurs with the 5-methyl ether.

Active against Gram-positive bacteria. *Cneorum* is used as a rubifacient and antifebrifuge, and *Ptaeroxylon* is used in Southern Africa as a traditional medicine.

1277
Ptaeroglycol; Cneorum chromone F; Pteroglycol

$C_{15}H_{14}O_6$ Mol. wt 290.28

Occurs in the heartwood of *Cedrelopsis grevei* and *Ptaeroxylon obliquum* (Ptaeroxylaceae), and in *Cneorum tricoccum* and *C. pulverulentum* (Cneoraceae).

Antibacterial activity; it is also active against HeLa cells.

1278
Pulverochromenol

$C_{20}H_{22}O_4$ Mol. wt 326.40

Found in *Cneorum tricoccum* and *C. pulverulentum* (Cneoraceae).

Inhibitory to HeLa cells; it shows antibacterial activity.

1279
Quinquangulin

$C_{16}H_{14}O_5$ Mol. wt 286.29

Occurs in the roots of *Cassia quinquangulata* (Leguminosae).

Moderately cytotoxic in the P-388 lymphocytic leukaemia cell system.

1280
Rubrofusarin

$C_{15}H_{12}O_5$ Mol. wt 272.26

Occurs in the mycelium of *Fusarium graminearum* and *F. culmorum* as a red pigment. It occurs also in the roots of *Cassia tora* and *C. quinquangulata* (Leguminosae), and in mango fruit, *Mangifera indica* (Anacardiaceae) infected with *Aspergillus niger*.

Moderately cytotoxic in the P-388 lymphocytic leukaemia cell system. Also, it acts as a depressant on the central nervous system in animals. Infected mango fruits may be toxic to man.

1281
Spathelia bischromene; Cneorum chromone A

$C_{20}H_{20}O_4$ Mol. wt 324.38

Occurs in *Cneorum tricoccum* and *C. pulverulentum* (Cneoraceae), and in *Spathelia glabrescens* and *S. sorbifolia* (Rutaceae).

Active against HeLa cells; also it shows antibacterial activity.

1282
Visnagin; Visnagidin; Visnacorin;
5-Methoxy-2-methylfuranochromone

$C_{13}H_{10}O_4$ Mol. wt 230.22

Occurs in the fruits of *Ammi visnaga* (Umbelliferae).

Phototoxic activities against green algae and bacteriophages T4, M13 and it shows antiviral phototoxicity.

Chapter 35
Coumarins

The 700 or more plant coumarins can all be derived from the parent compound coumarin itself (**1295**), which has a characteristic odour of newmown hay and which occurs widely, usually in bound form. There are three major classes: the simple hydroxycoumarins, such as umbelliferone (**1338**) and esculetin (**1305**); the furanocoumarins, exemplified by angelicin (**1286**); and the pyranocoumarins, such as decursinol (**1300**). Furanocoumarins are further subdivided into linear (6,7-disubstituted, e.g., bergapten, **1289**) and angular (7,8-disubstituted, e.g., pimpinellin, **1325**) derivatives. There are furanocoumarins with additional prenyl substitution (e.g., alloimperatorin, **1284**) and a number of more complex coumarins (e.g., the mycotoxin, aflatoxin B1, **1283**) which are generally of microbial origin. Yet another minor class is the 4-phenylcoumarins (e.g., mammeisin, **1316**) which, alternatively, may be classified with the flavonoids (q.v.) as neoflavonoids.

The biosynthesis of coumarins normally proceeds from *p*-hydroxycinnamic acid, which then undergoes *o*-hydroxylation and ring closure to umbelliferone (**1338**). Isoprenylation of this key intermediate in either the 6- or 8-position, followed by ring closure and oxidative or other modification, gives rise to the majority of known derivatives.

Simple hydroxycoumarins and their glucosides are found regularly in higher plants, having been recorded in about 100 families (Murray *et al.*, 1982). The furano- and pyranocoumarins, which usually occur in the free state in fruits and roots, are mainly restricted to two large plant families, the Umbelliferae and Rutaceae, where they occur widely. They have been recorded occasionally in six other families: the Compositae, Leguminosae (*Psoralea*), Moraceae, Pittosporaceae, Solanaceae and Thymeleaceae. Furanocoumarins have been reported additionally as phytoalexins (antifungal agents) in two of these families, the Umbelliferae and Compositae.

Furanocoumarins are more biologically active than simple coumarins. In particular, they exhibit phototoxicity (Towers, 1980) and many are allergenic (Mitchell and Rook, 1979). They exhibit a range of toxic effects on insects, though some are able to detoxify them (Berenbaum and Zangerl, 1988). To mammals and humans, the most dangerous coumarins are the aflatoxins, which are hepatotoxic, and the substance dicoumarol (**1301**), which is a blood anticoagulant.

Coumarins are aromatic lactones, have characteristic ultraviolet spectra, and can be identified on the basis of spectral and chromatographic measurements (Ibrahim, 1989). They normally fluoresce in solution in ultraviolet light, the fluorescence sometimes being enhanced by treatment with alkali. Such fluorescence is important in the detection of the highly poisonous aflatoxins, which can contaminate stored food products such as wheat and groundnuts following infection with *Aspergillus* (Smith and Moss, 1985).

The key reference to the occurrence, chemistry and biochemistry of the coumarins is the book of Murray *et al.* (1982).

References

Murray, R.D.H., Mendez, J. and Brown, S.A. (1982). *The Natural Coumarins*. Chichester: Wiley.

Nielsen, B.E. (1971). Coumarin patterns in the Umbelliferae. In *The Biology and Chemistry of the Umbelliferae* (Heywood, V.H., ed.), pp. 325–336. London: Academic Press.

Biological Activity

Berenbaum, M.R. and Zangerl, A.R. (1988). Stalemates in the coevolutionary Arms Race: Synthesis, synergisms and sundry other sins. In *Chemical Mediation of Coevolution* (Spencer, K.C.), pp. 113–132. San Diego: Academic Press.

Mitchell, J.C. and Rook. A. (1979). *Botanical Dermatology*, Vancouver: Greenglass.

Smith, J.E. and Moss, M.O. (1985). *Mycotoxins, Formation, Analysis and Significance*, Chichester: Wiley.

Towers, G.H.N. (1980). Photosensitisers from plants and their photodynamic action. *Progress in Phytochemistry*, **6,** 183–202.

Identification

Ibrahim, R.K. (1989). Phenylpropanoids. In *Methods in Plant Biochemistry*, Volume I, *Plant Phenolics* (Harborne, J.B., ed.), pp. 75–112. London: Academic Press.

1283
Aflatoxin B₁

$C_{17}H_{12}O_6$ Mol. wt 312.28

Found in *Aspergillus flavus* (Ascomycetes) and *A. parasiticus*. Aflatoxin B₁ is only one metabolite of the aflatoxin series, of which 14 or more are known as naturally occurring compounds.

Extremely toxic and carcinogenic to animals and humans, affecting DNA, RNA and protein synthesis as well as lipid metabolism.

1284
Alloimperatorin; Prangenidine

$C_{16}H_{14}O_4$ Mol. wt 270.29

Found in the roots of *Prangos pabularia*, in the seeds of *Heracleum nepalense*, in the fruits of *Ammi majus* and of *Selinum monnieri*, and in the aerial parts of *Smyrniopsis armena* (all Umbelliferae). It is found also in *Aegle marmelos*, *Poncirus trifoliata*, and *Thamnosma montana* (Rutaceae) and in *Zea mays* (Gramineae).

Exhibits piscicidal activity, and acts as a growth inhibitor of microorganisms.

1285
Ammoresinol

$C_{24}H_{30}O_4$ Mol. wt 382.50

Occurs in the resinous juice of *Dorema ammoniacum* (Umbelliferae).

Bacteriostatic activity, e.g., against *Bacillus megatherium*.

1286
Angelicin; Isopsoralen

$C_{11}H_6O_3$ Mol. wt 186.17

Found in the roots of *Angelica archangelica*, *Heracleum* spp. and *Selinum vaginatum* (Umbelliferae). Further, it accumulates in the seeds of *Psoralea corylifolia* (Leguminosae) and *Castanopsis indica* (Fagaceae), and *Ficus nitida* (Moraceae).

Slight photosensitizing activity; recently, it was also shown to have spasmolytic activity. The name angelicin has also been used as a synonym of β-sitosterol, a plant steroid.

1287
Archangelicin

$C_{24}H_{26}O_7$ Mol. wt 426.47

Occurs in the roots of *Angelica archangelica*, *A. keiskei*, *A. longeradiata* and *Cnidium japonicum* (Umbelliferae).

Spasmolytic activity.

1288
Athamantin

$C_{24}H_{30}O_7$ Mol. wt 430.50

Found in the roots and seeds of *Athamanta oreoselinum* and in *Ammi visnaga*; it also occurs in *Angelica sylvestris*, *Athamanta cretensis*, *Libanotis transcaucasica*, *Peucedanum* spp. and *Seseli libanotis* (all Umbelliferae).

Spasmolytic activity.

1289
Bergapten; Bergaptene; Heraclin; Majudin; 5-Methoxypsoralen

$C_{12}H_8O_4$ Mol. wt 216.20

Widespread occurrence among Rutaceae, in the oil of bergamot, *Citrus bergamia*, and in essential oils of fruits of *Fagara* spp. and *Ruta graveolens*. Occurrence is also common in the Umbelliferae: in the genera *Heracleum*, *Ligusticum*, *Angelica*, *Ammi*, *Seseli*, *Levisticum*, *Pimpinella* and *Petroselinum*. The leaves of *Ficus carica* (Moraceae) and members of the Pittosporaceae are other sources. It has been found in the aerial parts of the tomato *Lycopersicon esculentum* (Solanaceae).

Used in the treatment of leukoderma (vitiligo) and psoriasis, but it is less effective than psoralen (q.v.). Bergapten in combination with UV causes impairment of DNA synthesis. It has been used as an ingredient in suntan preparations, but was recently shown to have lethal and clastogenic effects on mammalian cells in tissue culture, and lethal and mutagenic photosensitization effects on bacteria. Is also toxic to fish, to toads, and to the snails which carry the *Schistosoma* parasite.

1290
Byakangelicin

$C_{17}H_{18}O_7$ Mol. wt 334.33

Occurs in the roots and seeds of *Angelica glabra* and the seeds of *Heracleum* spp. (Umbelliferae). It is also present in the essential oil of *Citrus* spp. (Rutaceae).

Inhibits human chorionic gonadotrophin in laboratory animals.

1291
Calophyllolide

$C_{26}H_{24}O_5$ Mol. wt 416.48

Occurs in the nuts of *Calophyllum inophyllum* (Guttiferae).

Anti-inflammatory activity. Its reported anticoagulant activity is still disputed.

1292
Chalepensin; Xylotenin

$C_{16}H_{14}O_3$ Mol. wt 254.29

Present in all parts (root, stem, leaf) of *Ruta graveolens* (Rutaceae). *Psilopeganum sinense* and *Boenninghausenia* spp. are other taxa of the Rutaceae accumulating this compound.

Causes antifertility activity in rats at nontoxic dosages.

1293
Chartreusin; Antibioticum X-465 A; Lambdamycin

$C_{32}H_{32}O_{14}$ Mol. wt 640.61

Produced by *Streptomyces chartreusis* (from African soil), and by *Streptomyces* spp. (Actinomycetes) from American soil.

Antibiotic activity; it acts as an inhibitor of DNA and RNA polymerase of leukaemia cells.

1294
Cichoriin; 6,7-Dihydroxycoumarin 7-glucoside

$C_{15}H_{16}O_9$ Mol. wt 340.29

Found in the flowers of *Cichorium intybus* and further Compositae such as *Artemisia, Centaurea, Koelpinia, Launea* and *Sonchus* spp. It is also accumulated in *Fraxinus* spp. (Oleaceae).

Antifeedant activity against locusts.

1295
Coumarin; 2*H*-1-Benzopyran-2-one; 1,2-Benzopyrone; *cis-o*-Coumarinic acid lactone; Tonka bean camphor; Coumarone

$C_9H_6O_2$ Mol. wt 146.15

One of the most widespread coumarins in the plant kingdom, being present in most families of angiosperms. Among the gymnosperms it is known from the Pinaceae and, among the ferns, from the Polypodiaceae. The occurrence is not limited to any one part of the plant. Often it occurs *in vivo* in bound glucosidic form. The free coumarin is released during tissue damage. Coumarin is responsible, for example, for the smell of newmown hay.

Coumarin has a haemorrhagic effect, and it causes liver damage in rats and dogs. Thus, it is used as a rat poison. It has piscicidal, antifungal and antitumour activities. Coumarin inhibits the growth of *Escherichia coli* and larval development of houseflies.

1296
Coumermycin A₁; Notomycin A₁; Sugordomycin D$_{1\alpha}$

$C_{55}H_{59}N_5O_{20}$ Mol. wt 1110.11

Produced by cultures of *Streptomyces rishiriensis* (Actinomycetales).

Antibiotic activity, inhibiting DNA synthesis of *Staphylococcus aureus*. In general, its effects are similar to those of novobiocin (q.v.).

1297
Daphnetin

$C_9H_6O_4$ Mol. wt 178.15

Occurs in the fungus *Puccinia graminis* (Basidiomycetes). Among angiosperms, it occurs in seeds of *Euphorbia lathyris* (Euphorbiaceae), in aerial parts and roots of *Cicer arietinum* and in *Tetragonolobus foenum-graecum* (Leguminosae). It is widely present in the Thymeleaceae, in genera such as *Arthrosolen* and *Daphne*, where it occurs as the 7-glucoside, daphnin.

Antibacterial activity against most Gram-positive and some Gram-negative bacteria.

1298
Decuroside III; Nodakenetin cellobioside

$C_{26}H_{34}O_{14}$ Mol. wt 570.56

Occurs in the roots of *Peucedanum decursivum* (Umbelliferae), as a cellobioside.

Inhibits human platelet aggregation *in vitro*. Another nodakenetin glycoside of *P. decursivum*, called decuroside IV, exhibits the same activity.

1299
Decursin

$C_{19}H_{20}O_5$ Mol. wt 328.37

Occurs in the roots of *Angelica decursiva* and of *A. gigas* (Umbelliferae).

Decreases the rate of heart beat of cultured myocardial cells.

1300
Decursinol

$C_{14}H_{14}O_4$ Mol. wt 246.27

Found mainly in the roots of *Angelica decursiva, A. gigas*, and *Seseli grandivittatum* (all Umbelliferae), and in *Aegle marmelos* (Rutaceae).

Decreases the rate of heart beat of cultured myocardial cells.

1301
Dicoumarol; Dicumarol; Dicumol; Dicoumarin; Dufalone; Melitoxin

$C_{19}H_{12}O_6$ Mol. wt 336.31

Originates from *o*-coumaric acid via 4-hydroxycoumarin in decomposing hay of *Anthoxanthum* (Gramineae) and *Melilotus* (Leguminosae) spp.

Haemorrhagic and anticoagulant activities. It is an inhibitor of various enzymes. Haemorrhagic disorders and even death may be caused by cattle eating "spoiled sweet clover" containing dicumarol. Related compounds (e.g., warfarin sodium) are used as anticoagulants in medicine, or to kill rodents.

1302
Dihydrosamidin

$C_{21}H_{12}O_7$ Mol. wt 376.33

Found in the fruit and flowers of *Ammi visnaga* (Umbelliferae).

Vasodilatory agent; it is used in human medicine.

1303
Disenecionyl *cis*-khellactone; *cis*-Khellactone disenecioate

$C_{24}H_{26}O_7$ Mol. wt 426.47

Occurs in the roots of *Seseli* spp. such as *S. incanum* and *S. libanotis* (Umbelliferae).

Spasmolytic and coronary vasodilatory activities.

1304
(3′*R*,4′*R*)-3′-Epoxyangeloyloxy-4′-acetoxy-3′,4′-dihydroseselin

$C_{21}H_{22}O_8$ Mol. wt 402.41

Occurs in the roots of *Angelica shikokiana* (Umbelliferae).

Anti-allergic activity. The roots of *Angelica shikokiana* are used in folk medicine against hyperlipidaemia, diabetes, inflammatory and allergic diseases.

1305
Esculetin; Aesculetin; Cichorigenin; Esculetol; 6,7-Dihydroxycoumarin

$C_9H_6O_4$ Mol. wt 178.15

Relatively widespread among angiosperm families: e.g., in the seeds of *Euphorbia lathyris* (Euphorbiaceae), in the bark of *Fraxinus* spp. (Oleaceae), and in the wood of *Aesculus turbinata* (Hippocastanaceae).

Bacteriostatic and antifungal activity.

1306
Esculin; Aesculin; Crataegin; Polychrom; Esculoside; Bicolorin

$C_{15}H_{16}O_9$ Mol. wt 340.29

This 6-glucoside of esculetin (q.v.) is widely occurring; e.g., in the bark of *Aesculus hippocastanum* (Hippocastanaceae), in the bark of *Crataegus oxyacantha* (Rosaceae), in the bark of *Fraxinus* spp. (Oleaceae) and in the leaves of *Bursaria spinosa* (Pittosporaceae).

Inhibits chemically induced carcinogenic action. It is a growth inhibitor of *Bacillus subtilis*.

1307
Fraxetin; 7,8-Dihydroxy-6-methoxycoumarin

$C_{10}H_8O_5$ Mol. wt 208.17

Found in *Fraxinus* spp. (Oleaceae), *Echites hirsuta* (Apocynaceae), *Aesculus turbinata* and *A. hippocastanum* (Hippocastanaceae) and also in *Vestia lycioides* (Solanaceae). It is further obtained as the aglycone from fraxin (q.v.) after hydrolysis.

1308
Fraxin; Fraxetin-8-glucoside; Fraxoside; Paviin

$C_{16}H_{18}O_{10}$ Mol. wt 370.32

Occurs in most *Fraxinus* spp. (Oleaceae), in *Aesculus hippocastanum* (Hippocastanaceae), in Caprifoliaceae (*Diervilla* spp., *Symphoricarpus* spp.), as well as in Tiliaceae (*Tilia* spp.), in Calycanthaceae (*Chimonanthus fragrans*) and in Campanulaceae (*Campanula* spp.).

1309
Heliettin; Chalepin

$C_{19}H_{22}O_4$ Mol. wt 314.39

Occurs in the bark of *Helietta longifolia* and in the aerial parts of *Ruta chalepensis* (both Rutaceae).

Tumour-inhibitor *in vitro*.

1310
Herniarin; Ayapanin

$C_{10}H_8O_3$ Mol. wt 176.17

One of the more common coumarins, it occurs in the fungus *Puccinia graminis* (Basidiomycetes), in angiosperm families such as Gramineae, Caryophyllaceae (e.g., *Herniaria glabra*, *H. hirsuta*), Compositae (*Eupatorium ayapana*), Labiatae, Leguminosae, Moraceae, Rosaceae, Rutaceae (e.g., essential oil of *Ruta montana*) and Solanaceae. It is reported as a phytoalexin in the leaves of *Apium graveolens* (Umbelliferae).

Antifungal and antibacterial activities.

1311
Imperatorin; Marmelosin; Ammidin; 8-Isoamylenoxypsoralen

$C_{16}H_{14}O_4$ Mol. wt 270.29

Common occurrence in the roots and seeds of the Umbelliferae, e.g., in *Angelica*, *Heracleum* and *Pastinaca* spp. It further occurs in the seeds of *Citrus meyeri* and *Aegle marmelos* (Rutaceae) and in the leaves of *Fragaria* spp. (Rosaceae).

Weak activity against HeLa-cell proliferation, and antimutagenic properties. The compound has some piscicide activity and is toxic to toads.

1312
Isopimpinellin

$C_{13}H_{10}O_5$ Mol. wt 246.22

Found in Umbelliferae genera such as *Pimpinella* (roots), *Angelica* (roots), *Heracleum* (roots), *Pastinaca* (seeds), *Ferula* and *Seseli* (seeds), and in Rutaceae, particularly in lime oil (*Citrus aurantifolia*), the leaves of *Kimmia laureola*, the fruit of *Luvunga* sp. and in *Flindersia bennettiana* (wood, bark, leaves). It has also been isolated from the aerial parts of *Trichocline incana* (Compositae).

Tuberculostatic activity (against *Myobacterium tuberculosis*). Also, it shows toxicity against the host intermediates (snails) of *Schistosoma*, and has antifungal and piscicidal effects.

1313
Isosamidin

$C_{21}H_{22}O_7$ Mol. wt 386.41

Found mainly in Umbelliferae genera, accumulating in the stems of *Seseli gummiferum*, in the aerial parts of *S. tortuosum* and in the aerial parts and roots of *Zizia aptera*.

Spasmolytic and coronary vasodilatory activities.

1314
Libanorin; Senecionyl dihydrooreoselol

C₁₉H₂₀O₅   Mol. wt 328.37

Found in the roots of *Libanotis schrenkiana* and in the fruit of *Peucedanum oreoselinum* (both Umbelliferae).

Spasmolytic and coronary vasodilatory activities.

1315
Luvangetin

C₁₅H₁₄O₄   Mol. wt 258.28

Occurs in the heartwood of *Chloroxylon swietenia*, the leaves of *Hesperethusa crenulata* and the fruit of *Luvunga scandens* (all Rutaceae). Also isolated from the heartwood of *Brosimum rubescens* (Moraceae).

Moderate antifungal activity.

1316
Mammeisin; Mammea-compound A/AA

C₂₅H₂₆O₅   Mol. wt 406.48

Found in *Mammea africana* and *M. thwailesii* (Guttiferae). One of a number of related 4-phenylcoumarins present together in *Mammea* spp.

In vitro antitumour activity.

1317
Micromelin; Micromelumin

C₁₅H₁₂O₆   Mol. wt 288.26

Occurs in the bark of *Caesaria graveolens* (*Samydra glabra*) (Samydraceae). It also occurs in stems and leaves of *Micromelum integerrimum* (Rutaceae).

Antitumour activity reported.

1318
Nodakenetin

$C_{14}H_{14}O_4$ Mol. wt 246.27

Occurs in the roots of *Peucedanum decursivum*, the fruit and roots of *Heracleum sprengleianum* (Umbelliferae), in the leaves of *Scaevola frutescens* (Goodeniaceae) and in the Rutaceous genera *Chloroxylon* and *Ptelea*.

Inhibits human platelet aggregation *in vitro*.

1319
Nodakenin; Nodakenetin glucoside

$C_{20}H_{24}O_9$ Mol. wt 408.41

Occurs in the roots of *Peucedanum decursivum* and of *Sphenosciadium capitellatum* (Umbelliferae).

Inhibits human platelet aggregation *in vitro*.

1320
Novobiocin; Streptonivicin; Albamycin; Biotexin; Cardelmycin; Cathocin; Cathomycin; Inamycin; Spheromycin; Vulcamicina; Vulkamycin

$C_{31}H_{36}N_2O_{11}$ Mol. wt 612.64

Found in *Streptomyces spheroides* and *S. niveus* (Actinomycetales).

Plasma protein binder (e.g., albumin). It has antibacterial and antiviral activities. In addition, it exhibits efficacy in canine respiratory infections, and is used as an antimicrobial agent in veterinary medicine.

1321
Osthol; 8-(3-Methyl-2-butenyl)herniarin

$C_{15}H_{16}O_3$ Mol. wt 244.29

Occurs in the roots and rhizomes of *Peucedanum ostruthium, Angelica archangelica* and *Prangos pabularia* (Umbelliferae). It also accumulates in the wood, leaves and bark of *Flindersia bennettiana* and other spp. of this genus as well as in *Citrus, Clausenia, Creoridium* and *Haplophyllum* spp. (all Rutaceae).

Antimutagenic activity. It also has antimalarial properties.

1322
Ostruthin

$C_{19}H_{22}O_3$ Mol. wt 298.39

Found in the rhizomes of *Peucedanum ostruthium* (Umbelliferae), and it is also known from genera of the Rutaceae, particularly occurring in the leaves and twigs of *Eriostemon tomentellus* and in *Luvunga eleutherandra*.

Antibacterial (*Staphylococcus aureus*), antifungal (*Saccharomyces cerevisiae*) and piscicidal activities. Also, it has antimalarial properties.

1323
Peucedanin; Oreoselone methyl ether

C₁₅H₁₄O₄   Mol. wt 258.28

Occurs in the rhizomes of *Peucedanum officinale* and the fruit of *Prangos pabularia* (Umbelliferae). Other sources within the Umbelliferae include *Anthriscus cerefolium*, *Meum athamanticum*, *Myrrhis odorata*, *Pastinaca silvestris* and *Daucus carota*.

Antineoplastic activity.

1324
Peucenidin

C₂₁H₂₂O₇   Mol. wt 386.41

Found in the roots of *Libanotis pyrenaicum*, the leaves of *Peucedanum bourgaei* and the fruit of *P. oreoselinum* (all Umbelliferae).

Spasmolytic and coronary vasodilatory activities.

1325
Pimpinellin

C₁₃H₁₀O₅   Mol. wt 246.22

Occurs in the roots of Umbelliferae, e.g., *Pimpinella saxifraga* and *Heracleum sphondylium*, in the fruits of *Heracleum* spp., and *Angelica genuflexa* (Umbelliferae), in *Cyperus papyrus* (Cyperaceae), and in *Artemisia canariensis* (Compositae).

Tuberculostatic activity (against *Mycobacterium tuberculosis*).

1326
Prangolarine; (+)-Oxypeucedanin

C₁₆H₁₄O₅   Mol. wt 286.29

Occurs in the Umbelliferae, in the roots of *Peucedanum palustre*, *Prangos pabularia* and *Angelica archangelica*, and in the fruits of *Agasyllis latifolia*, *Coelopleurum lucidum* var. *gmelinii* and *Cymopterus longipes*.

Piscicidal and antimicrobial activities.

1327
Psoralen; Ficusin

C₁₁H₆O₃   Mol. wt 186.17

Occurs in the seeds of *Psoralea* spp. and *Coronilla glauca* (Leguminosae), and in the essential oil of *Phebalium argenteum* (Rutaceae), and in the wood of *Xanthoxylum flavum* (Rutaceae), and in *Ficus carica* (Moraceae).

Photosensitizing activity. Also, it is active against *Mycobacterium tuberculosis*. In the presence of ultraviolet light, it causes impairment of DNA synthesis.

1328
Pteryxin

$C_{21}H_{22}O_7$ Mol. wt 386.41

Found in the roots of *Angelica ursina*, *Libanotis lehmannia*, *Pteryxia terebinthina*, *Seseli eriocephalum* and *Zizia aptera* (all Umbelliferae).

Spasmolytic and coronary vasodilatory activities.

1329
Rutamarin; Chalepin acetate

$C_{21}H_{24}O_5$ Mol. wt 356.42

Occurs in the leaves of *Boenninghausenia japonica*, the heartwood of *Chloroxylon swietenia* and in *Ruta graveolens* (all Rutaceae).

Antitumour activity against HeLa-cells, blocking DNA synthesis.

1330
Rutarin; Campesenin

$C_{20}H_{24}O_{10}$ Mol. wt 424.41

Found in the aerial parts and roots of *Ruta graveolens* and *R. chalepensis* (Rutaceae).

Antifungal and moderate antibacterial activities.

1331
Samidin

$C_{21}H_{22}O_7$ Mol. wt 386.41

Occurs in the fruit of *Ammi visnaga* and the aerial parts of *Seseli tortuosum* (Umbelliferae).

Vasodilatory activity.

1332
Scoparone; Dimethylesculetin; Esculetin dimethyl ether; Di-*O*-methylesculetin; Dimethoxycoumarin

$C_{11}H_{10}O_4$ Mol. wt 206.20

Found in the fruit and flowerheads of *Artemisia scoparia* (Compositae). Further sources are the wood of *Fagaria macrophylla*, the bark of *Zanthoxylum schinifolium* (both Rutaceae), the leaves of *Dendrobium thyrsiflorum* (Orchidaceae) and the resin of *Ferula oopoda* (Umbelliferae). Also, it is found in the bark of *Liriodendron tulipifera* (Magnoliaceae) and in *Gentiana kuroo* (Gentianaceae).

Antihepatotoxic activity. *In vitro* tests indicate effects on heart beat activity, with stimulation at lower concentrations and inhibition at higher concentrations. It also produces a significant fall of blood pressure. Scoparone affects calcium mobilization in vascular smooth muscle, inhibiting calcium release and extracellular influx.

1333
Scopoletin; 6-Methoxyumbelliferone;
β-Methylesculetin; Chrysatropic acid; Gelseminic acid

$C_{10}H_8O_4$ Mol. wt 192.17

Widespread occurrence, including *Atropa belladonna* (Solanaceae), and *Gelsemium sempervirens* (Loganaceae); the bark of *Prunus serotina* (Rosaceae), the roots of *Ipomoea orizabensis* and *Convolvulus scammonia* (both Convolvulaceae), the bark of *Diospyros maritima* (Ebenaceae), the flower heads of *Artemisia afra* (Compositae), and *Nerium odorum* (Apocynaceae).

Hypotensive activity in animals. It also exhibits spasmolytic, antibacterial and antifungal activities. In plants, it acts as a bud growth inhibitor of *Pisum sativum* and a stimulator of germination in *Striga asiatica*.

1334
Scopolin; Murrayin

$C_{16}H_{18}O_9$ Mol. wt 354.32

Occurs in the roots of *Scopolia japonica* and in *Nicotiana tabacum* (Solanaceae). Further sources are *Nerium odorum*, the heartwood of *Rhus lanceolata* (Anacardiaceae), the genera *Hedera* (Araliaceae), *Swertia* (Gentianaceae), *Anthemis* and *Artemisia* (Compositae), *Baptisia* (Leguminosae) and *Celtis* (Ulmaceae).

1335
Seselin; Amyrolin

$C_{14}H_{12}O_3$ Mol. wt 228.25

Found in the Rutaceae, in the roots of *Citrus aurantium*, in the bark, leaves and wood of *Flindersia bennettiana*, in *Haplophyllum dubium* and in *Myrtopsis macrocarpa*. In the Umbelliferae, it occurs in *Apium leptophyllum*, *A. graveolens*, *Foeniculum vulgare* and *Seseli indicum*.

Antifungal activity against *Curvularia lunata* and *Aspergillus niger*.

1336
5,6,7-Trimethoxycoumarin

$C_{12}H_{12}O_5$ Mol. wt 236.23

Occurs in *Diosma pilosa* (Rutaceae), in the roots of *Pelargonium reniforme* (Geraniaceae) and in the stems of *Aleurites fordii* (Euphorbiaceae).

Antimalarial activity.

1337
4,5′,8-Trimethylpsoralen; Trioxsalen; Trioxalen; TMP

$C_{14}H_{12}O_3$ Mol. wt 228.25

Produced as a phytoalexin by celery plants (*Apium graveolens*, Umbelliferae) infected by the fungus *Sclerotina sclerotiorum*.

Causes dermatitis in humans.

1338
Umbelliferone; Skimmetin; Hydrangin; Dichrin A; 7-Hydroxycoumarin

$C_9H_6O_3$ Mol. wt 162.15

Widespread occurrence, e.g., in the resin of various Umbelliferae: *Ferula, Apium, Pimpinella* and *Heracleum* spp. It is present in the flowers of *Hydrangea paniculata* (Saxifragaceae), the bark of *Aegle marmelos*, the peel of *Citrus grandis*, grapefruit (Rutaceae) and the roots of *Atropa belladonna* (Solanaceae).

Antifungal and antibacterial activities. It is used in sunscreen lotions and creams.

1339
Visnadin; Cardine; Carduben; Provismine; Vibeline; Visnamine

$C_{21}H_{24}O_7$ Mol. wt 388.42

Occurs in the seeds of *Ammi visnaga* (Umbelliferae). Further Umbelliferae genera containing this compound are *Anethum, Ferula* and *Phloiodicarpus*.

Spasmolytic and coronary vasodilator activities.

1340
Xanthotoxin; Ammoidin; 8-Methoxypsoralen; Methoxsalen

$C_{12}H_8O_4$ Mol. wt 216.20

Produced as a phytoalexin by celery plants *Apium graveolens* (Umbelliferae) after infection with the fungus *Sclerotina sclerotiorum*. Also, it occurs naturally in *Pastinaca sativa* and *Ammi majus* (seeds), *Angelica archangelica* (seeds), *A. officinalis* (roots) and *Heracleum sphondylium* (roots and herb) (Umbelliferae). It is common in the Rutaceae in genera such as *Ruta* and *Fagara*.

Causes dermatitis in man. It is tuberculostatic, and is used as a pigmentation agent in the treatment of leukodermia (vitiligo) and psoriasis. Furthermore, it exhibits toxicity towards intermediate hosts (snails) of *Schistosoma*, and has piscicidal and antibacterial effects. Also, it is toxic to toads. In combination with UV, it causes impairment of DNA synthesis.

1341
Xanthotoxol

$C_{11}H_6O_4$ Mol. wt 202.17

Occurs in the seeds of *Angelica archangelica, Pastinaca sativa* and *Heracleum lanatum* var. *asiaticum* (all Umbelliferae), in the fruit of *Cnidium monnieri* (Umbelliferae), and in the seeds of *Poncirus trifoliata* (Rutaceae).

Inhibits HeLa cell proliferation.

1342
Xanthyletin

$C_{14}H_{12}O_3$ Mol. wt 228.25

Found in the bark of *Xanthoxylum americanum*, the roots of *X. ailanthoides*, the fruit of *Luvunga scandens*, the wood of *Chloroxylon swietenia*, and in *Citrus aurantifolia* (all Rutaceae). Also, it is found in the wood of *Brosimum* sp. (Moraceae).

Antitumour and antibacterial activities.

Chapter 36
Minor flavonoids

The term "minor flavonoid" is a convenient one for describing those flavonoids which have a more limited natural distribution than the more widespread anthocyanins, flavones and flavonols. They are mainly derived from the reduced flavan skeleton that can be seen in 7,4′-dihydroxy-8-methylflavan (**1362**). They can be further categorized as flavanones, dihydrochalcones, dihydroflavonols, flavan-3-ols, flavan-3,4-diols and flavans. They are clearly distinct from the anthocyanins, flavones and flavonols (q.v.) which have a central pyran ring at a higher level of oxidation. Indeed, flavanones such as naringenin (**1391**) and dihydroflavonols such as aromadendrin (**1347**) are the immediate biosynthetic precursors of flavones and flavonols, respectively. Dihydrochalcones (e.g., asebogenin, **1348**) are ring-opened flavanones, with two aromatic rings linked by a central three carbon unit. They are structurally related to both the flavanones and the chalcones (q.v.) but are more restricted in their natural occurrence than either of these two flavonoid classes.

The best known minor flavonoids are the flavanones and the flavanon-3-ols or dihydroflavonols. Two flavanones, naringenin (**1391**) and eriodictyol (**1368**) are relatively common in nature, while isoprenylated derivatives, e.g., abyssinone V (**1344**), are correspondingly rare. Flavanone glycosides containing a neohesperidose sugar unit are notable for their taste properties. Naringin (**1392**) and eriocitrin (**1367**) are major contributors to the bitterness of the citrus fruits where they occur. Not all flavanones are bitter, since some can be bitter-sweet or sweet. For example, a related structure to neohesperidin (**1396**), namely the corresponding dihydrochalone, is used in the food industry as a nonsugar sweetener. Sweetness is also a property of some dihydroflavonols and their 3-acetates, see, e.g., 6-methoxytaxifolin (**1389**).

The three dihydroflavonols corresponding to kaempferol, quercetin and myricetin, namely aromadendrin (**1347**), taxifolin (**1412**) and ampelopsin (**1346**) are relatively widespread in nature, especially in woody plants. They may be accompanied in wood and bark by flavan-3,4-diols, e.g., fisetinidol-4β-ol (**1371**), and flavan-3-ols, e.g., (+)-catechin (**1355**). These 3,4-diols and 3-ols also act as precursors of anthocyanins and condensed tannins (q.v.).

Two substances, (+)-catechin (**1355**) and silybin (**1408**) from *Silybum marianum* are used as drugs because of their anti-inflammatory and anti-hepatotoxic properties. Several other compounds have medicinal promise; e.g., sanggenon C (**1405**) is hypotensive. Many structures have antimicrobial activity, and several have been characterized as phytoalexins, e.g., betagarin (**1350**) from infected sugar beet and broussin (**1351**) from the paper mulberry tree.

Flavanones and dihydroflavonols can be recognized in plant extracts by simple colour tests based on reduction with magnesium or zinc in acid solution. Other classes are less easily detected in direct extracts, but most have distinctive ultraviolet spectra. Methods of isolation and characterization are reviewed by Grayer (1988).

References

Chemistry

Harborne, J.B. (ed.) (1988). *The Flavonoids: Advances in Research since 1980*. London: Chapman and Hall.

Harborne, J.B. (1994). *The Flavonoids: Advances in Research since 1986*. London: Chapman and Hall.

Methodology

Grayer, R.J. (1989). Flavanoids In *Methods in Plant Biochemistry*, Volume 1, *Plant Phenolics* (Harborne, J.B., ed.), pp. 283–323, London: Academic Press.

Biological Activities

Gao, F., Wang, H., Mabry, T.J. and Kinghorn, A.D. (1990). Dihydroflavonol sweeteners and other constituents from *Hymenoxys turneri*. *Phytochemistry*, **29,** 2865–2869.

Nomura, T. (1988). Phenolic compounds of the mulberry tree and related plants. *Progress in the Chemistry of Organic Natural Products*, **53,** 87–201.

Saini, K.S. and Ghosal, S. (1984). Naturally occurring flavans unsubstituted in the heterocyclic ring. *Phytochemistry*, **23,** 2415–2421.

1343
Abyssinone I

Skeletal type: flavanone

$C_{20}H_{18}O_4$ Mol. wt 322.36

Occurs in the roots of the East African medicinal plant *Erythrina abyssinica* (Leguminosae).

Antimicrobial activity against the bacteria *Staphylococcus aureus* and *Bacillus subtilis* and the fungi *Sclerotinia libertiana* and *Mucor mucedo*.

1344
Abyssinone V

Skeletal type: flavanone

$C_{25}H_{28}O_5$ Mol. wt 408.50

Occurs in the roots of the East African medicinal plant *Erythrina abyssinica* (Leguminosae).

Antibacterial activity against *Staphylococcus aureus*, *Bacillus subtilis* and *Micrococcus lysodeikticus*.

1345
Afzelechin; 3,5,7,4'-Tetrahydroxyflavan

Skeletal type: flavan-3-ol

$C_{15}H_{14}O_6$ Mol. wt 290.27

Found in the wood of *Eucalyptus calophylla* (Myrtaceae), the heartwood of *Nothofagus fusca* (Fagaceae), the root of *Saxifraga ligulata* (Saxifragaceae) and the fruit of *Juniperus communis* (Cupressaceae).

1346
Ampelopsin; Dihydromyricetin

Skeletal type: dihydroflavanol

$C_{15}H_{12}O_8$ Mol. wt 320.26

Widespread occurrence in the wood and bark of conifer trees, e.g., in pine and cedar bark. Also, it is found in woody angiosperms, e.g., in the leaves of *Rhododendron cinnabarinum* (Ericaceae) and the wood of *Cercidiphyllum japonicum* (Cercidiphyllaceae).

Anti-oxidant

1347
Aromadendrin; Dihydrokaempferol;
3,5,7,4′-Tetrahydroxyflavanone

Skeletal type: dihydroflavonol

$C_{15}H_{12}O_6$ Mol. wt 288.26

Widespread occurrence in the free state, especially in the wood of trees. Many glycosides occur: e.g., the 3-rhamnoside, engeletin, which is found in *Eucalyptus* spp. (Myrtaceae), *Nothofagus* spp. (Fagaceae) and *Exocarpos* spp. (Santalaceae), the 3-glucoside, which is found in *Dalbergia sericea* (Leguminosae) and in *Podocarpus* spp. (Podocarpaceae, Coniferales), and the 7-galactoside, which is found in the bark of *Sapium sebiferum* (Euphorbiaceae).

Some antifungal activity. The 7-galactoside has antimicrobial activity against Gram-negative bacteria, e.g., *Klebsiella pneumoniae*, and against the fungus *Microsporum gypseum*.

1348
Asebogenin; Phloretin 4′-methyl ether

Skeletal type: dihydrochalcone

$C_{16}H_{16}O_5$ Mol. wt 288.30

Found in *Iryanthera laevis* (Myristicaceae) and in the ferns *Notholaena sulphurea* (Polypodiaceae) and *Pityrogramma calomelanos* (Adiantaceae). The 2′-glucoside, asebotin, occurs in spp. of *Kalmia*, *Pieris*, and *Rhododendron* (Ericaceae).

1349
Auriculoside

Skeletal type: flavan *O*-glycoside

$C_{22}H_{26}O_{10}$ Mol. wt 450.45

Found in *Acacia auriculiformis* (Leguminosae).

Minor central nervous system depressant with some antistress/anti-anxiety activity.

1350
Betagarin;
5,2′-Dimethoxy-6,7-methylenedioxyflavanone

Skeletal type: flavanone

$C_{18}H_{16}O_6$ Mol. wt 328.33

Induced as a phytoalexin in sugar beet, *Beta vulgaris* (Chenopodiaceae) infected with the leaf-spot fungus *Cercospora beticola*.

Antifungal activity.

1351
Broussin; 7-Hydroxy-4′-methoxyflavan

Skeletal type: flavan

$C_{16}H_{16}O_3$ Mol. wt 256.30

Produced as a phytoalexin from wounded xylem tissue of the paper mulberry tree, *Broussonetia papyrifera* (Moraceae).

Significant antimicrobial activity against *Bipolaris leersiae*.

1352
Broussonin C

Skeletal type: 1,3-diphenylpropane

$C_{20}H_{24}O_3$ Mol. wt 312.41

Phytoalexin from diseased shoot cortical tissues and from wounded xylem tissue of the paper mulberry tree, *Broussonetia papyrifera* (Moraceae).

Antifungal activity.

1353
Butin; 7,3′,4′-Trihydroxyflavanone

Skeletal type: flavanone

$C_{15}H_{12}O_5$ Mol. wt 272.26

Occurs in many spp. of the Leguminosae and in *Mangifera indica* (Anacardiaceae). The 7-glucoside, isocoreopsin, occurs in *Coreopsis, Cosmos, Dahlia* and *Bidens* spp. (Compositae), and the 3′-glucoside occurs in *Butea monosperma* (Leguminosae). See also butrin.

1354
Butrin; Butin 7,3′-di-*O*-glucoside

Skeletal type: flavanone *O*-glycoside

$C_{27}H_{32}O_{15}$ Mol. wt 596.55

Found in the flowers of *Butea monosperma* (Leguminosae).

Antihepatotoxic activity, but less strong than that of its chalcone isomer isobutrin (q.v.). In India, flowers of *B. monosperma* are used in the treatment of hepatic disorders and viral hepatitis.

1355
(+)-Catechin; Catechinic acid; Catechol; Catechuic acid; (+)-Cyanidanol; (+)-Cyanidan-3-ol

Skeletal type: flavan-3-ol

$C_{15}H_{14}O_6$ Mol. wt 290.28

Widespread occurrence in nature, especially in woody plants, e.g., in willow catkin, *Salix caprea* (Salicaceae).

Biologically highly active. It is used as a haemostatic drug, and in the treatment of various liver diseases, especially acute hepatitis. It shows strong liver-protective and potent antiperoxidative activities, so that it may act as a "radical scavenger" by neutralizing free radicals produced by hepatotoxic substances. However, prolonged treatment with (+)-catechin can induce several adverse reactions, most of them immunomediated, such as haemolysis, acute renal failure and skin rashes.

1356
Catechin 7-*O*-β-D-xyloside

Skeletal type: flavan-3-ol *O*-glycoside

$C_{20}H_{22}O_{10}$ Mol. wt 422.40

Occurs in the elm *Ulmus americana* (Ulmaceae).

Feeding stimulant for the smaller European elm bark beetle, *Scolytus multistriatus*.

1357
Davidigenin; 2′,4′,4-Trihydroxydihydrochalcone

Skeletal type: dihydrochalcone

$C_{15}H_{14}O_4$ Mol. wt 258.28

The 2′-glucoside, davidioside, occurs in several spp. of *Viburnum*, e.g., *V. davidii* (Caprifoliaceae), and the 4′-glucoside, confusoside, occurs in *Symplocos confusa* (Symplocaceae).

1358
Diffutin

Skeletal type: flavan *O*-glycoside

$C_{23}H_{28}O_{10}$ Mol. wt 464.48

Found in *Canscora diffusa* (Gentianaceae).

C. diffusa is used in India in some mental disorders such as melancholia. Diffutin is probably its active principle, since it has a mild central nervous system depressant activity which is associated with antistress and anti-anxiety activities. Nontoxic in dogs up to 500 mg kg^{-1}.

1359
7,4′-Dihydroxyflavan

Skeletal type: flavan

$C_{15}H_{14}O_3$ Mol. wt 242.28

Induced as a phytoalexin in the bulb of the daffodil, *Narcissus pseudonarcissus* (Amaryllidaceae), and in diseased shoot cortical tissue of the paper mulberry tree, *Broussonetia papyrifera* (Moraceae).

Fungitoxic activity against *Botrytis cinerea*, and also against germinated spores of *B. cinerea* in liquid culture (ED_{50} 65 µg ml^{-1}). It shows antibacterial activity against plant pathogens such as *Corynebacterium betae* and *C. fascians*.

1360
2′,6′-Dihydroxy-4′-methoxydihydrochalcone

Skeletal type: dihydrochalcone

$C_{16}H_{16}O_4$ Mol. wt 272.30

Occurs in *Lindera umbellata* (Lauraceae) and in the leaf exudates of ferns, e.g., *Adiantum sulphureum* and certain *Pityrogramma* and *Notholaena* spp.

Allelopathic activity: inhibits the germination of spores of the fern *Pityrogramma calomelanos* at concentrations of 5×10^{-5} M and higher.

1361
7,3′-Dihydroxy-4′-methoxy-8-methylflavan

Skeletal type: flavan

$C_{17}H_{18}O_4$ Mol. wt 286.33

Found in the bulbs of *Lycoris radiata* (Liliaceae).

Antifeedant activity against larvae of the butterfly *Eurema hecabe mandarina*.

1362
7,4′-Dihydroxy-8-methylflavan

Skeletal type: flavan

$C_{16}H_{16}O_3$ Mol. wt 256.30

Found in the resin of the dragon tree, *Dracaena draco* (Agavaceae), and induced as a phytoalexin in bulbs of the daffodil *Narcissus pseudonarcissus* (Amaryllidaceae).

Fungitoxic activity against *Botrytis cinerea*, and also against germinated spores of *B. cinerea* in liquid culture (ED_{50} 32 µg ml^{-1}). It shows antibacterial activity against the plant pathogens *Corynebacterium betae* and *C. fascians*.

1363
Dracorubin

Skeletal type: biflavanoid

$C_{32}H_{24}O_5$ Mol. wt 488.54

Found in dragon's blood resin, from the dragon tree, *Dracaena draco* (Agavaceae). This resin also yields many other monomeric and dimeric flavans which are easily oxidized to form stable, intensely red-coloured quinone methides.

Dragon's blood resin is used in red varnishes.

1364
(−)-Epicatechin

Skeletal type: flavan-3-ol

$C_{15}H_{14}O_6$ Mol. wt 290.28

Very widespread occurrence, e.g., in hawthorn, *Crataegus monogyna* (Rosaceae), in the bark of *Pterocarpus* spp. (Leguminosae), in *Aesculus californica* (Hippocastanaceae), and in *Podocarpus nagi* (Podocarpaceae, Gymnospermae).

Antibacterial, anti-anaphylactic, antihyperglycaemic, anti-inflammatory, antimutagenic and antiperoxidative activities.

1365
ent-Epicatechin

Skeletal type: flavan-3-ol

$C_{15}H_{14}O_6$ Mol. wt 290.28

Occurs in the fruits and leaves of the Palmae (e.g., *Chamaerops humilis*), in the leaves of *Uncaria gambir* (Rubiaceae) and in the roots of *Polygonum multiflorum* (Polygonaceae).

1366
Epigallocatechin 3-gallate

Skeletal type: flavan-3-ol

$C_{22}H_{18}O_{11}$ Mol. wt 458.39

Occurs as the major phenolic (10% dry wt) of the tea leaf *Camellia sinensis* (Theaceae), together with several closely related catechins. Also present in the bark of *Hamamelis virginiana* (Hamamelidaceae) and the leaf of *Davidsonia pruriens* (Davidsoniaceae).

Undergoes oxidation during the processing of fresh tea leaves yielding products (theaflavins and thearubigins) which provide the colour, taste and flavour of the tea beverage.

1367
Eriocitrin; Eriodictyol 7-*O*-rutinoside

Skeletal type: flavanone *O*-glycoside

$C_{27}H_{32}O_{15}$ Mol. wt 596.55

Occurs in *Citrus* spp. and hybrids (Rutaceae), *Mentha piperita* (Labiatae) and *Myoporum tenuifolium* (Myoporaceae).

Activity similar to that of neoeriocitrin (q.v.).

1368
Eriodictyol; 5,7,3′,4′-Tetrahydroxyflavanone

Skeletal type: flavanone

$C_{15}H_{12}O_6$ Mol. wt 288.26

Widespread occurrence, e.g., in *Eriodictyon californicum* (Hydrophyllaceae), and many Compositae, Leguminosae and Labiatae. The 7-glucoside is found in *Lophophytum leandri* (Balanophoraceae), and the 7-rhamnoside (eriodictin) in *Citrus* spp. (Rutaceae). See also eriocitrin and neoeriocitrin.

Antibacterial activity against *Pseudomonas maltophilia* and *Enterobacter cloacae*. It shows growth inhibitory activity against larvae of *Heliothis zea*, and induces nodulation gene expression in *Rhizobium leguminosarum* in the symbiosis between this bacterium and its legume host, *Pisum sativum*.

1369
Farrerol

Skeletal type: flavanone

$C_{17}H_{16}O_5$ Mol. wt 300.32

The 5,7-diglucoside occurs in several *Rhododendron* spp., e.g., *R. farreri* (Ericaceae) and the 7-glucoside occurs in *Cyrtomium* spp. (Pteridophyta).

Expectorant activity; farrerol seems to act directly on the mucous membrane of the respiratory tract and increases the secretion of respiratory tract fluid. It is used to treat chronic bronchitis.

1370
Fisetinidol; 3,7,3′,4′-Tetrahydroxyflavan

Skeletal type: flavan-3-ol

$C_{15}H_{14}O_5$ Mol. wt 274.28

Found in the heartwood of *Acacia mearnsii* and *Colophospermum mopane* (Leguminosae), and in the wood of *Virola minutifolia* and *V. elongata* (Myristicaceae).

1371
Fisetinidol-4β-ol; Fisetini-3,4-diol

Skeletal type: flavan-3,4-diol

$C_{15}H_{14}O_6$ Mol. wt 290.28

Found in the heartwood of *Guibourtia coleosperma*, *Colophospermum mopane* and many *Acacia* spp., and in the root of *Neorautanenia amboensis* (Leguminosae). Fisetinidol-4β-ol is one of six epimers which occur naturally, and are termed leucofisetinidins.

1372
Fustin; Dihydrofisetin

Skeletal type: dihydroflavonol

$C_{15}H_{12}O_6$ Mol. wt 288.26

Common occurrence in the Leguminosae; also it is found in *Rhus* and *Schinopsis* (Anacardiaceae), *Platanus* (Platanaceae) and *Tilia* spp. (Tiliaceae). Fustin 3-glucoside occurs in *Baptisia* spp. (Leguminosae).

Antibacterial activity against *Pseudomonas maltophilia* and *Enterobacter cloacae*. It shows antiviral activity against Herpes simplex type 1 virus. It is an inhibitor of NADH-oxidase and succinoxidase enzyme systems.

1373
Garbanzol; 3,7,4′-Trihydroxyflavanone

Skeletal type: dihydroflavonol

$C_{15}H_{12}O_5$ Mol. wt 272.26

Found in chick pea or garbanzo bean, *Cicer arietinum* (Leguminosae), shepherd's purse, *Capsella bursa-pastoris* (Cruciferae), and *Rhus* spp. (Anacardiaceae). The 3-glucoside, lecontin, occurs in *Baptisia* spp. (Leguminosae).

1374
Glabranin

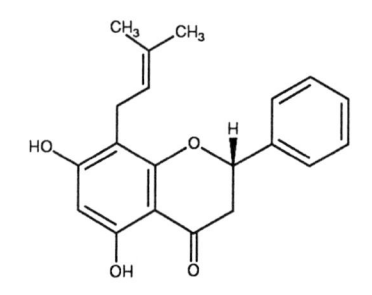

Skeletal type: flavanone

$C_{20}H_{20}O_4$ Mol. wt 324.38

Found in American liquorice, *Glycyrrhiza lepidota*, and in *Piscidia erythrina* (Leguminosae).

Broad-spectrum antimicrobial activity.

1375
Glepidotin B

Skeletal type: dihydroflavonol

$C_{20}H_{20}O_5$　　　　　　　　　　Mol. wt 340.38

Found in American liquorice, *Glycyrrhiza lepidota* (Leguminosae).

Broad-spectrum antimicrobial activity.

1376
Glycyphyllin; Phloretin 2′-*O*-rhamnoside

Skeletal type: dihydrochalcone *O*-glycoside

$C_{21}H_{24}O_9$　　　　　　　　　　Mol. wt 420.42

Occurs in some *Smilax* spp. (Smilacaceae).

Bitter-sweet taste.

1377
Hesperetin; Eriodictyol 4′-methyl ether

Skeletal type: flavanone

$C_{16}H_{14}O_6$　　　　　　　　　　Mol. wt 302.29

Occurs in *Citrus* spp. and hybrids (Rutaceae). The 5-glucoside occurs in the peach, *Prunus persica* (Rosaceae), and the 7-rhamnoside in *Cordia obliqua* (Boraginaceae). See also hesperidin and neohesperidin.

Feeding deterrent activity against the aphids *Schizaphis graminum* and *Myzus persicae*. It induces nodulation gene expression in *Rhizobium leguminosarum* in the symbiosis between this bacterium and its legume host, *Pisum sativum*. Also, it stimulates RNA synthesis in isolated rat liver nuclei *in vitro*. Some antibacterial and antiviral activities are shown.

1378
Hesperidin; Hesperetin 7-*O*-rutinoside; Ciratin

Skeletal type: flavanone *O*-glycoside

$C_{28}H_{34}O_{15}$　　　　　　　　　　Mol. wt 610.58

Occurs in many Rutaceae, e.g., *Citrus* spp. and *Poncirus trifoliata*, and in Labiatae, e.g., *Mentha* and *Hyssopus* spp.

One of the flavonoids in *Citrus* leaf which acts as an oviposition stimulant for the butterflies *Papilio xuthus* and *P. protenor*.

1379
Homoeriodictyol: Eriodictyol 3′-methyl ether

Skeletal type: flavanone

$C_{16}H_{14}O_6$ Mol. wt 302.29

Occurs in *Eriodictyon* spp. (Hydrophyllaceae), and in *Artemisia campestris* and *Tanacetum sibiricum* (Compositae).

Feeding deterrent activity against the aphids *Schizaphis graminum* and *Myzus persicae*.

1380
7-Hydroxyflavan

Skeletal type: flavan

$C_{15}H_{14}O_2$ Mol. wt 226.28

Induced as a phytoalexin in the bulbs of the daffodil *Narcissus pseudonarcissus* (Amaryllidaceae).

Fungitoxic activity against *Botrytis cinerea*, and also against germinated spores of *B. cinerea* in liquid culture (ED_{50} 22 µg ml^{-1}). It shows antibacterial activity against the plant pathogens *Corynebacterium betae* and *C. fascians*.

1381
Isochamaejasmin; 3,3″-Binaringenin

Skeletal type: biflavanone

$C_{30}H_{22}O_{10}$ Mol. wt 542.51

Found in the roots of *Stellera chamaejasme* (Thymelaeaceae).

Prevents both initiation and promotion in the process of chemical carcinogenesis.

1382
Isosakuranetin; Citrifoliol; Naringenin 4′-methyl ether

Skeletal type: flavanone

$C_{16}H_{14}O_5$ Mol. wt 286.29

Found in, e.g., *Artemisia campestris*, *Wyethia* spp. and other Compositae, from Betulaceae, and some ferns. The 7-glucoside, isosakuranin, occurs in *Prunus* spp. (Rosaceae); the 7-rutinoside, didymin, in satsuma orange and other *Citrus* hybrids (Rutaceae) and in *Monarda didyma* (Labiatae). See also poncirin.

1383
Isouvaretin

Skeletal type: dihydrochalcone

$C_{23}H_{22}O_5$ Mol. wt 378.43

Occurs in the roots of *Uvaria angolensis* (Annonaceae).

Antimicrobial and cytotoxic activities.

1384
Kazinol A

Skeletal type: flavan

$C_{25}H_{30}O_4$ Mol. wt 394.52

Found in the cortex of the paper mulberry tree, *Broussonetia papyrifera* (Moraceae).

1385
Kolaflavanone

Skeletal type: biflavanone

$C_{31}H_{24}O_{12}$ Mol. wt 588.53

Occurs in *Garcinia kola* (Guttiferae).

Anti-inflammatory, anticholesteraemic and anticapillary fragility activities. It shows antihyperglycaemic activity on alloxanated rats. Also, it shows antihepatotoxic activity against phalloidin and *Amanita phalloides* poisoning.

1386
Liquiritigenin; 7,4'-Dihydroxyflavanone

Skeletal type: flavanone

$C_{15}H_{12}O_4$ Mol. wt 256.26

Occurs in many spp. of Leguminosae, e.g., *Cicer arietinum*. It is induced as a phytoalexin in *Medicago sativa* and *M. lupulina* (Leguminosae) following inoculation with *Helminthosporium carbonum*. The 4'-O-glucoside (liquiritin) and 4'-rhamnosylglucoside (rhamnoliquiritin) occur in liquorice, *Glycyrrhiza* spp. (Leguminosae), whereas the 7-glucoside and 7-diglucoside occur in *Dahlia variabilis* (Compositae).

Weak antifungal activity. It shows haemoglobin induction activity. It inhibits rat liver mitochondrial monoamine oxidase *in vitro* and, hence, acts on the central nervous system.

1387
Manniflavanone

Skeletal type: biflavanone

$C_{30}H_{22}O_{13}$ Mol. wt 590.51

Occurs in the bark of *Garcinia mannii* (Guttiferae).

Inhibitor of aldose reductase. It is used for the treatment of diseases resulting from disorders of vascular permeability and fragility, and in the prevention of the complications of diabetes mellitus.

1388
6-Methoxyaromadendrin 3-*O*-acetate

Skeletal type: dihydroflavonol

$C_{18}H_{16}O_8$ Mol. wt 360.33

Occurs in the aerial parts of *Hymenoxys turneri* (Compositae).

Sweet tasting substance: about 20 times sweeter than sucrose.

1389
6-Methoxytaxifolin

Skeletal type: dihydroflavonol

$C_{16}H_{14}O_8$ Mol. wt 334.29

Found in the aerial parts of *Hymenoxys turneri* (Compositae), where it occurs with the 3-*O*-acetate.

Sweet tasting substance: about 12 times sweeter than sucrose, whereas the 3-acetate is 25 times sweeter.

1390
2′-*O*-Methylodoratol

Skeletal type: α-hydroxydihydrochalcone

$C_{18}H_{20}O_5$ Mol. wt 316.36

Induced as a phytoalexin in the cotyledons of the sweet pea, *Lathyrus odoratus* (Leguminosae), by means of the fungus *Phytophthora megasperma* var. *sojae* or mercuric acetate as elicitors.

Fungitoxic activity.

1399
Pinobanksin; 3,5,7-Trihydroxyflavanone

Skeletal type: dihydroflavonol

$C_{15}H_{12}O_5$ Mol. wt 272.26

Found in the heartwood of *Pinus banksiana* (Pinaceae), and e.g., in *Baccharis oxydonta* (Compositae), *Eremophila alternifolia* (Myoporaceae) and *Polygonum nodosum* (Polygonaceae).

1400
Pinocembrin; 5,7-Dihydroxyflavanone

Skeletal type: flavanone

$C_{15}H_{12}O_4$ Mol. wt 256.26

Very widespread occurrence, e.g., in *Pinus cembra* (Pinaceae), *Glycyrrhiza* spp. (Leguminosae), *Prunus* spp. (Rosaceae) and *Helichrysum* spp. (Compositae).

Antimicrobial activity; it inhibits the development of *Bacillus subtilis* in a concentration of 3 μg ml^{-1}. Also, it is effective against *Candida albicans*, *Saccharomyces cerevisiae* and *Cryptococcus neoformans* in concentrations of 0.1–3 mg ml^{-1}.

1401
Poncirin; Citrifolioside; Isosakuranetin 7-*O*-neohesperidoside

Skeletal type: flavanone *O*-glycoside

$C_{28}H_{34}O_{14}$ Mol. wt 594.58

Found in *Citrus*, *Eremocitrus* and *Microcitrus* spp. (Rutaceae), and in *Calamintha nepeta* and *Acinos* spp. (Labiatae).

Bitter taste (1/5 as bitter as quinine). The corresponding rutinoside, didymin, is tasteless.

1402
5′-Prenylhomoeriodictyol; Sigmoidin B 4′-methyl ether

Skeletal type: flavanone

$C_{21}H_{22}O_6$ Mol. wt 370.41

Occurs in the stem bark of *Erythrina berteroana* (Leguminosae).

Antifungal activity against *Cladosporium cucumerinum*.

1403
Prunin; Naringenin 7-*O*-glucoside

Skeletal type: flavanone

$C_{21}H_{22}O_{10}$ Mol. wt 434.41

Found in many gymnosperms, e.g., some *Abies*, *Pinus* and *Podocarpus* spp., in the tomato, *Lycopersicon esculentum* (Solanaceae), and in the bark of *Prunus persica* (Rosaceae).

Antifungal activity against the woodrot fungus *Sporotrichum pulverulentum*.

1404
Sakuranetin; Naringenin 7-methyl ether

Skeletal type: flavanone

$C_{16}H_{14}O_5$ Mol. wt 286.29

Occurs in *Artemisia campestris*, *Baccharis* spp. and other Compositae, in *Juglans* spp. (Juglandaceae), *Betula* spp. (Betulaceae), etc. The 5-glucoside, sakuranin, occurs in *Prunus puddum* (Rosaceae).

Surface fungitoxin on blackcurrant leaves, *Ribes nigrum* (Grossulariaceae).

1405
Sanggenon C

Skeletal type: flavanone

$C_{40}H_{36}O_{12}$ Mol. wt 708.73

Found in the root bark of *Morus mongolica* (Moraceae).

Marked hypotensive activity. It shows antimicrobial activity against *Staphylococcus aureus*, *Bacillus subtilis*, *Trichophyton mentagrophytes* and *Pyricularia oryzae*. Also, it inhibits the specific binding of the tumour-promoting agent 12-*O*-tetradecanoylphorbol 13-acetate to mouse skin, and inhibits the activation of protein kinase C in a dose-dependent manner.

1406
Sanggenon D

Skeletal type: flavanone

$C_{40}H_{36}O_{12}$ Mol. wt 708.73

Found in the root bark of *Morus mongolica* (Moraceae).

Hypotensive activity. It shows antimicrobial activity against *Staphylococcus aureus*, *Bacillus subtilis*, *Trichophyton mentagrophytes*, and *Pyricularia oryzae*, but weaker than that of sanggenon C (q.v.). It inhibits the tumour-promoting activity of teleocidin on mouse-skin, and the activation of protein kinase C.

1407
Silandrin

Skeletal type: flavanolignan

$C_{25}H_{22}O_9$ Mol. wt 466.45

Occurs in the fruit of white-flowered plants of *Silybum marianum* (Compositae).

Strong antihepatotoxic activity.

1408
Silybin

Skeletal type: flavanolignan

$C_{25}H_{22}O_{10}$ Mol. wt 482.45

Occurs in the fruit of violet-flowered plants of *Silybum marianum* (Compositae).

Silybin (formerly called silymarin) is one of the components of "silymarin group", a mixture of *S. marianum* flavanolignans having strong antihepatotoxic activity, especially against phalloidin. They are used medicinally to treat liver disease and cases of *Amanita* poisoning. They increase the RNA-synthesis in isolated rat liver nuclei *in vitro*. As a consequence, the formation of ribosomes is accelerated and protein synthesis is increased.

1409
Silychristin; Silymarin II

Skeletal type: flavanolignan

$C_{25}H_{22}O_{10}$ Mol. wt 482.45

Found in the fruit of violet-flowered plants of *Silybum marianum* (Compositae).

One of the components of the "silymarin group", which is a mixture of *S. marianum* flavanolignans which have liver-protective activity. See also silybin.

1410
Sophoranone

Skeletal type: flavanone

$C_{30}H_{36}O_4$ Mol. wt 460.62

Occurs in the roots of *Sophora subprostrata* (Leguminosae), together with the corresponding chalcone isomer, sophoradin.

Sophoranone and sophoradin have antigastric ulcer activity.

1411
Strobopinin; 6-Methylpinocembrin

Skeletal type: flavanone

$C_{16}H_{14}O_4$ Mol. wt 270.29

Found together with the 8-*C*-methyl isomer, cryptostrobin, in the wood of many *Pinus* spp. (Pinaceae). Both occur in the heartwood of *P. krempfii*, which also contains the 6,8-di-*C*-methyl ether, demethoxymatteucinol. Both isomers are present in *Comptonia peregrina* (Myrtaceae), while strobopinin occurs on its own in the fern *Pityrogramma pallida* (Adiantaceae).

1412
Taxifolin; Dihydroquercetin; Distylin; 3,5,7,3′,4′-Pentahydroxyflavanone

Skeletal type: dihydroflavonol

$C_{15}H_{12}O_7$ Mol. wt 304.26

Widespread occurrence, e.g., in many Coniferae, in *Acacia catechu* (Leguminosae), *Polygonum nodosum* (Polygonaceae), and *Salix capraea* (Salicaceae). The 3-rhamnoside, astilbin, occurs in *Astilbe* spp. (Saxifragaceae), and the 3- and 7-galactosides in *Rhododendron* spp. (Ericaceae).

Antibacterial, antifungal, antiviral, anti-inflammatory, antihepatotoxic and anti-oxidant activities. It stimulates RNA synthesis in isolated rat liver nuclei *in vitro*. It inhibits NADH-oxidase, succinoxidase and Δ^5-lipoxygenase. Also, it inhibits the growth of *Heliothis zea* larvae, as does astilbin.

1413
Taxifolin 3-*O*-acetate

Skeletal type: dihydroflavonol

$C_{17}H_{14}O_8$ Mol. wt 346.30

Found in *Hymenoxys turneri, Tessaria dodoneifolia, Baccharis varicans* and *Inula viscosa* (Compositae).

Intensely sweet (80 times sweeter than sucrose).

1414
Tephrowatsin A

Skeletal type: flavan-4-ol

$C_{22}H_{26}O_4$ Mol. wt 354.45

Occurs in the leaves of *Tephrosia watsoniana* (Leguminosae).

1415
Theasinensin A

Skeletal type: biflavan-3-ol galloyl ester

$C_{44}H_{34}O_{22}$ Mol. wt 914.76

Found in green tea, *Camellia (Thea) sinensis* (Theaceae).

One of the natural precursors of the theaflavins, the pigments of black tea.

1416
Uvaretin

Skeletal type: dihydrochalcone

$C_{23}H_{22}O_5$ Mol. wt 378.43

Found in the roots of *Uvaria angolensis* (Annonaceae).

Antimicrobial and cytotoxic activities.

Chapter 37

Flavones and Flavonols

Flavones and flavonols are very widely distributed in plants, both as co-pigments to anthocyanins in petals, and also in the leaves of higher plants. Like the anthocyanins (q.v.), they occur most frequently in glycosidic combination. They are also present in the free state (or in *O*-methylated form) on leaf surfaces, in fruits and in bud exudates. Although several hundred flavonol aglycones are known, only three are at all common: kaempferol (**1469**), quercetin (**1492**) and myricetin (**1478**). The other known flavonols are mostly simple structural variants on the common flavonols and are of limited natural occurrence. In the case of quercetin, a number of *O*-methylated derivatives are known, the 3′-methyl ether (isorhamnetin, **1464**) and the 5-methyl ether (azaleatin, **1426**) being but two examples. Addition of a hydroxyl in the 8-position to the structure of quercetin gives gossypetin (**1453**), one of the few flavonols which are pigments in their own right, providing yellow flower colour in the primrose, and in the cotton plant. Isoprenyl derivatives are known, e.g., glepidotin A (**1452**), but are not common.

There is a considerable range of flavonol glycosides present in plants. One hundred and thirty-five different glycosides of quercetin alone have been described. By far the commonest is quercetin 3-rutinoside, known as rutin (**1500**), which is of pharmaceutical interest in relation to the treatment of capillary fragility in man.

Flavones differ from flavonols only in lacking a 3-hydroxyl substitution; this affects their UV absorption and chromatographic mobility and colour reactions, so that simple flavones can be distinguished from flavonols on these bases. There are only two common flavones, apigenin (**1421**) and luteolin (**1474**), corresponding in hydroxylation pattern to kaempferol and quercetin. The flavone tricetin (**1510**), corresponding to myricetin, is known, but it is of very rare occurrence. More common are two methyl ethers: chrysoeriol (**1433**), the 3′-methyl ether of luteolin, and tricin (**1511**), the 3′,5′-dimethyl ether of tricetin.

Flavones occur as glycosides, but the range of different glycosides is less than in the case of the flavonols. A common type is the 7-glucoside, exemplified by luteolin 7-glucoside (**1475**). Flavones, unlike flavonols, also occur, remarkably, with sugar bound by a carbon–carbon bond. A series of such glycosylflavones has been described, one example being orientin (**1484**), the 8-*C*-glucoside of luteolin. The carbon–carbon bond is very resistant to acid hydrolysis, so that it is relatively easy to distinguish these *C*-glycosides from *O*-glycosides, which are more readily hydrolysed.

One other structural variant in the flavone series must be mentioned, the biflavonyl. These dimeric compounds are formed by carbon–carbon or carbon–oxygen coupling between two flavone (usually apigenin) units, e.g., amentoflavone (**1420**) and hinokiflavone (**1456**), respectively. Methyl ethers are common, e.g., ginkgetin (**1451**), a dimethyl ether of amentoflavone, occurs in autumnal leaves of *Ginkgo biloba*. Biflavonyls occur almost exclusively in the gymnosperms, but they have been found occasionally in angiosperms, in families such as the Anacardiaceae, Casuarinaceae and Guttiferae.

References

Chemistry and Distribution
Geissman, T.A. (ed.) (1962). *The Chemistry of Flavonoid Compounds*. Oxford: Pergamon Press.
Harborne, J.B., Mabry, T.J. and Mabry, H. (1975). *The Flavonoids*. London: Chapman and Hall.
Harborne, J.B. and Mabry, T.J. (1982). *The Flavonoids: Advances in Research*. London: Chapman and Hall.
Harborne, J.B. (1988). *The Flavonoids: Advances in Research since 1980*. London: Chapman and Hall.
Harborne, J.B. (1994). *The Flavonoids: Advances in Research since 1986*. London: Chapman and Hall.

Identification
Harborne, J.B. (ed.) (1989). *Methods in Plant Biochemistry*, Volume 1, *Plant Phenolics*. London: Academic Press.

Biological Properties
Cody, V., Middleton, E. and Harborne, J.B. (1986). *Plant Flavonoids in Biology and Medicine*. New York: Alan R. Liss.
Gabor, M. (1986). *The Pharmacology of Benzopyrone Derivatives and Related Compounds*. Budapest: Akademiai Kiado.

1417
Acacetin; Apigenin 4′-methyl ether

Skeletal type: flavone

$C_{16}H_{12}O_5$ Mol. wt 284.27

The aglycone occurs in leaf bud excretions of Betulaceae, on the leaf surface of certain Compositae, and in the farinose exudate on fronds of some ferns. The 7-glucoside occurs in the leaves of *Tilia japonica* (Tiliaceae); the 7-glucuronide in the thistle *Cirsium arvense* (Compositae), and the 7-rutinoside (linarin or acaciin) in the flowers of *Linaria vulgaris* (Scrophulariaceae).

Inhibitor of lens aldose reductase, iodothyronine deiodinase, and histamine release from rat peritoneal mast cells. It shows contact sensitizing (allergenic) activity.

1418
Acerosin

Skeletal type: flavone

$C_{18}H_{16}O_8$ Mol. wt 360.33

Occurs in *Iva acerosa* (Compositae), in the aerial parts of *Helianthus strumosus* (Compositae), in *Gardenia* spp. (Rubiaceae), and in the seeds of *Vitex negundo* (Verbenaceae).

Causes disruption of later stages of spermatogenesis in dogs.

1419
Agathisflavone; 6,8″-Biapigenin

Skeletal type: biflavone

$C_{30}H_{18}O_{10}$ Mol. wt 538.47

Found in *Agathis dammara* and *Araucaria bidwillii* (Araucariaceae, Coniferales).

Inhibits cyclic nucleotide phosphodiesterase.

1420
Amentoflavone; 3′,8″-Biapigenin

Skeletal type: biflavone

$C_{30}H_{18}O_{10}$ Mol. wt 538.47

Very common occurrence in the Coniferales, e.g., in *Podocarpus montanus* (Podocarpaceae), and also in other Gymnospermae, e.g., in *Cycas revoluta* (Cycadaceae). It occurs sporadically in the Angiospermae, e.g., in *Rhus succedanea* (Anacardiaceae).

Potent inhibitor of nucleotide phosphodiesterase. It shows antifungal activity: it inhibits the growth of *Aspergillus fumigatus, Botrytis cinerea* and *Trichoderma glaucum*.

1421
Apigenin; 5,7,4'-Trihydroxyflavone

Skeletal type: flavone

$C_{15}H_{10}O_5$ Mol. wt 270.25

The aglycone can be found on the leaf surface of certain plants, e.g., members of the Labiatae, and also in the farinose exudate on fern fronds. Many glycosides are known, e.g., the 5-glucoside in leaves of *Amorpha fruticosa* (Leguminosae), the 7-glucuronide in the flowers of *Erigeron annuus* (Compositae), and the 4'-glucoside in the petals of *Dahlia variabilis* (Compositae). See also apigenin 7-glucoside and apiin.

Antibacterial, anti-inflammatory, diuretic and hypotensive activities. It inhibits many enzymes. Also, it promotes smooth muscle relaxation. It acts as a nodulation signal to the bacterium *Rhizobium leguminosarum* in the pea.

1422
Apigenin 7,4'-dimethyl ether

Skeletal type: flavone

$C_{17}H_{14}O_5$ Mol. wt 298.30

Widespread occurrence as an aglycone in many plants families, e.g., in the leaf resin of *Cistus* spp. (Cistaceae), in the roots of *Rhus undulata* (Anacardiaceae), and as an external flavonoid on *Thymus piperella* (Labiatae), *Striga asiatica* (Scrophulariaceae) and *Andrographis paniculata* (Acanthaceae).

Anti-inflammatory and antigastric ulcer activities.

1423
Apigenin 7-O-glucoside; Cosmosiin

Skeletal type: flavone *O*-glycoside

$C_{21}H_{20}O_{10}$ Mol. wt 432.39

Widespread occurrence, e.g., in the flowers of *Cosmos bipinnatus* (Compositae).

Acts as a nodulation signal to *Rhizobium leguminosarum* in the symbiosis between this bacterium and its legume host, the pea.

1424
Apiin; Apioside; Apigenin 7-apiosylglucoside

Skeletal type: flavone *O*-glycoside

$C_{26}H_{28}O_{14}$ Mol. wt 564.51

Occurs in the seeds and leaves of parsley, *Petroselinum crispum*, and celery, *Apium graveolens* (Umbelliferae).

Inhibitor of lens aldose reductase.

1425
Axillarin; Quercetagetin 3,6-dimethyl ether

Skeletal type: flavonol

$C_{17}H_{14}O_8$ Mol. wt 346.30

Found in the aerial parts of chamomile, *Matricaria chamomilla*, in *Achillea* and *Artemisia* spp. and in many other Compositae, and in *Didierea* spp. (Didiereaceae).

Antiviral activity. It inhibits lens aldose reductase.

1426
Azaleatin; Quercetin 5-methyl ether

Skeletal type: flavonol

$C_{16}H_{12}O_7$ Mol. wt 316.27

Found in many *Rhododendron* spp., often as the 3-rhamnoside (azalein), e.g., in the flowers of *R. mucronatum* (Ericaceae). The 3-galactoside occurs in the leaves of *Eucryphia glutinosa* (Eucryphiaceae).

1427
Baicalein; 5,6,7-Trihydroxyflavone

Skeletal type: flavone

$C_{15}H_{10}O_5$ Mol. wt 270.25

Occurs in the roots and leaves of many *Scutellaria* spp. (Labiatae), often as the 7-glucuronide, baicalin (q.v.). The 7-rhamnoside occurs in *S. galericulata* and the 6-glucoside and 6-glucuronide in the leaves of *Oroxylum indicum* (Bignoniaceae).

Inhibits glyoxylase-I and platelet lipoxygenase. It shows anti-promotion activity in carcinogenesis, and anti-inflammatory, anti-allergic and diuretic activities. Baicalein 6-phosphate inhibits immediate hypersensitivity reactions. High concentrations of baicalein prolong the clotting time of fibrinogen by thrombin.

1428
Baicalein 5,6,7-trimethyl ether

Skeletal type: flavone

$C_{18}H_{16}O_5$ Mol. wt 312.33

Found on the surface of the leaves of *Helichrysum nitens* (Compositae).

Antifungal activity.

1429
Baicalin; Baicalein 7-*O*-glucuronide

Skeletal type: flavone *O*-glycoside

$C_{21}H_{18}O_{11}$ Mol. wt 446.38

Occurs in the roots of *Scutellaria baicalensis* (Labiatae).

Diuretic and anti-allergic activities but, in high concentrations (0.1–1 mM), it prolongs the clotting time of fibrinogen by thrombin.

1430
Carlinoside

Skeletal type: flavone C-glycoside

$C_{26}H_{28}O_{15}$ Mol. wt 580.51

Found in *Carlina vulgaris* (Compositae), in the rice plant, *Oryza sativa* (Gramineae), in *Gibasis schiedeana* (Commelinaceae) and in the liverwort, *Takakia lepidozioides* (Hepaticae).

Carlinoside is one of eight flavone C-glycosides present in the rice plant as a mixture which acts as a feeding attractant to plant hoppers feeding on rice phloem.

1431
Chrysin; 5,7-Dihydroxyflavone

Skeletal type: flavone

$C_{15}H_{10}O_4$ Mol. wt 254.25

Occurs free in many plants, e.g., the leaf buds of *Populus* spp. (Salicaceae), the heartwood of *Pinus* spp. (Pinaceae), and the leaves of *Escallonia* spp. (Saxifragaceae). The 5-glucoside (toringin) occurs in the bark of *Docyniopsis tschonoski* (Rosaceae), the 7-glucoside in the wood of *Prunus aequinoctialis* (Rosaceae) and the 7-glucuronide in the leaves *Scutellaria galericulata* (Labiatae).

Anti-inflammatory and antibacterial activities. It inhibits iodothyronine deiodinase, lens aldose reductase, and histamine release from rat peritoneal mast cells. It also induces oestrogen synthetase and haemoglobin.

1432
Chrysin 5,7-dimethyl ether

Skeletal type: flavone

$C_{17}H_{14}O_4$ Mol. wt 282.30

Occurs in the leaves of *Helichrysum nitens* (Compositae) and the rhizomes of *Boesenbergia pandurata* (Zingiberaceae).

Strong antifungal activity.

1433
Chrysoeriol; Luteolin 3'-methyl ether

Skeletal type: flavone

$C_{16}H_{12}O_6$ Mol. wt 300.27

Occurs as an external flavonoid in many plant families, e.g., on the leaves of *Coleus amboinicus* (Labiatae), and in the farinose exudate of some fern fronds, e.g., *Notholaena californica* (Pteridophyta). Glycosides are also present in many plants, e.g., the 7-glucoside in pear leaves, *Pyrus* (Rosaceae), the 7-glucuronide in the petals of *Antirrhinum majus* (Scrophulariaceae), the 7-rutinoside in the flowers of chamomile, *Matricaria chamomilla* (Compositae) and the 7-apiosylglucoside in celery seed, *Apium graveolens* (Umbelliferae).

1434
Chrysosplenetin; Polycladin; Quercetagetin 3,6,7,3'-tetramethyl ether

Skeletal type: flavonol

$C_{19}H_{18}O_8$ Mol. wt 374.35

Occurs in the leaves of *Chrysosplenium tosaense* and *C. japonicum* (Saxifragaceae), in *Digitalis thapsii* (Scrophulariaceae), and in many Compositae, e.g., chamomile, *Matricaria chamomilla*.

Antiviral activity.

1435
Chrysosplenol C; Quercetagetin 3,7,3'-trimethyl ether

Skeletal type: flavonol

$C_{18}H_{16}O_8$ Mol. wt 360.33

Found in the aerial parts of *Chrysosplenium maximowiczii, C. oppositifolium*, and *C. alternifolium* (Saxifragaceae), and in *Parthenium* spp. (Compositae).

Antiviral activity against rhinovirus type 2.

1436
Cirsilineol

Skeletal type: flavone

$C_{18}H_{16}O_7$ Mol. wt 344.33

Found on the leaf surface of many Labiatae, e.g., thyme, *Thymus vulgaris, Salvia tomentosa* and some *Sideritis* spp. It is also found in the Compositae.

Cirsilineol and its 8-methoxy derivative (also present in thyme) show spasmolytic activity.

1437
Cirsiliol

Skeletal type: flavone

$C_{17}H_{14}O_7$ Mol. wt 330.30

Occurs in the aerial parts of many Compositae, e.g., *Cirsium lineare*, and Labiatae, e.g., *Salvia officinalis*.

Potent and relatively selective inhibitor of arachidonate 5-lipoxygenase.

1438
Cupressuflavone; 8,8″-Biapigenin

Skeletal type: biflavone

$C_{30}H_{18}O_{10}$ Mol. wt 538.47

Found in several plants of the Coniferales, e.g., *Agathis palmerstoni* and *Araucaria bidwillii* (Araucariaceae), *Cupressus arizonica* and *Juniperus horizontalis* (Cupressaceae).

Inhibits cyclic nucleotide phosphodiesterases.

1439
Datiscetin; 3,5,7,2′-Tetrahydroxyflavone

Skeletal type: flavonol

$C_{15}H_{10}O_6$ Mol. wt 286.25

The 3-glucoside occurs in *Datisca cannabina* (Datiscaceae).

Strong antibacterial activity against *Pseudomonas maltophilia* and *Enterobacter cloacae*.

1440
5-Deoxykaempferol; 3,7,4′-Trihydroxyflavone

Skeletal type: flavonol

$C_{15}H_{10}O_5$ Mol. wt 270.25

Occurs in glycosidic form in many Leguminosae, e.g., as the 3-glucoside in *Baptisia leucantha*, the 7-glucoside and 7-rutinoside in *B. lecontei*, and the 4′-glucoside in *Trifolium subterraneum*.

1441
Diosmetin; Luteolin 4′-methyl ether

Skeletal type: flavone

$C_{16}H_{12}O_6$ Mol. wt 300.27

Found free on the aerial parts of many Compositae, e.g., the flower heads of *Arnica* spp., many Labiatae, e.g., *Salvia tomentosa*, and some Scrophulariaceae, e.g., *Stemodia viscosa*. Occurs as the 7-glucoside in *Mentha spicata* (Labiatae) and as the 7-glucuronide in the leaves of *Pedalium murex* (Pedaliaceae). See also diosmin.

1442
Diosmin; Diosmetin 7-O-rutinoside

Skeletal type: flavone O-glycoside

$C_{28}H_{32}O_{15}$ Mol. wt 608.56

Occurs in the leaves of *Diosma crenulata* (Rutaceae).

Anti-inflammatory activity. It increases skin capillary resistance in both intensity and duration.

1443
Eupatilin

Skeletal type: flavone

$C_{18}H_{16}O_7$ Mol. wt 344.33

Found on the aerial parts of several Compositae, e.g., *Eupatorium semiserratum* and *Artemisia rubripes*, and in the Labiatae, e.g., *Sideritis tomentosa*. It is also found in the fruit peel of *Citrus reticulata* (Rutaceae).

Selectively inhibits 5-lipoxygenase of cultured mastocytoma cells.

1444
Fisetin; 5-Deoxyquercetin

Skeletal type: flavonol

$C_{15}H_{10}O_6$ Mol. wt 286.25

Occurs as a glycoside in many Leguminosae, e.g., as the 3-glucoside and the 3-rutinoside in *Trifolium subterraneum*. The 7-glucuronide occurs in the heartwood of *Rhus succedanea* (Anacardiaceae).

Inhibits iodothyronine deiodinase, succinoxidase, protein kinase C, Δ^5-lipoxygenase, and NADH-oxidase. It suppresses basophil histamine release, ionophore-induced arachidonic acid release and metabolism, and release of oxidants by human neutrophils. Also, it inhibits smooth muscle contraction. It shows both antibacterial activity, and a moderate sensitizing (allergenic) activity.

1445
Fisetin 8-C-glucoside

Skeletal type: flavonol C-glycoside

$C_{21}H_{20}O_{11}$ Mol. wt 448.39

Found in the heartwood of *Pterocarpus marsupium* (Leguminosae), together with 5-deoxykaempferol 8-C-glucoside.

1446
Flavone

Skeletal type: flavone

$C_{15}H_{10}O_2$ Mol. wt 222.25

Occurs in farina on the leaves of *Primula pulverulenta* and *Dionysia* spp. (Primulaceae), and in whole plants of *Pimelea decora* and *P. simplex* (Thymelaeaceae).

Inhibits Δ^5-lipoxygenase, and aggregation of human platelets, due to inhibition of cyclo-oxygenase. It also shows inhibitory activity against basophil histamine release, weak antibacterial activity against *Pseudomonas maltophilia*, and moderate contact sensitizing (allergenic) activity.

1447
Galangin; 3,5,7-Trihydroxyflavone

Skeletal type: flavonol

$C_{15}H_{10}O_5$ Mol. wt 270.25

Occurs in bud excretions of Salicaceae and Betulaceae, on the leaves of *Escallonia* spp. (Saxifragaceae), in the farinose exudate on fronds of various ferns, and on the leaves of Labiatae. The 3-rutinoside is present in the leaves of *Datisca cannabina* (Datiscaceae).

Antibacterial activity against the skin bacterium *Staphylococcus epidermis* and against *Pseudomonas maltophilia* and *Enterobacter cloacae*. It is a potent inhibitor of bull seminal cyclo-oxygenase activity.

1448
Galangin 3,5,7-trimethyl ether

Skeletal type: flavonol

$C_{18}H_{16}O_5$ Mol. wt 312.33

Occurs in the rhizomes of *Boesenbergia pandurata* (Zingiberaceae) and on the leaves of *Helichrysum nitens* (Compositae).

Antifungal activity.

1449
Genkwanin; Apigenin 7-methyl ether

Skeletal type: flavone

$C_{16}H_{12}O_5$ Mol. wt 284.27

Found in the flowers of *Daphne genkwa* (Thymelaeaceae). The aglycone occurs on the leaves of Labiatae, in bud excretions of Betulaceae and in the farinose exudate of some ferns. The 5- and the 4'-glucosides are found in the bark of *Prunus serrata* and *P. puddum*, respectively (Rosaceae).

1450
Geraldone; 5-Deoxychrysoeriol

Skeletal type: flavone

$C_{16}H_{12}O_5$ Mol. wt 284.27

Occurs in the roots of *Trifolium subterraneum* (Leguminosae), and in *Salvertia convallariodora* (Vochysiaceae), in which the 7-glucoside is also present.

Acts as a nodulation signal to *Rhizobium trifolium* in the symbiosis between this bacterium and its legume host, clover (sp. of *Trifolium*).

1451
Ginkgetin; Amentoflavone 7,4′-dimethyl ether

Skeletal type: biflavone

$C_{32}H_{22}O_{10}$ Mol. wt 566.53

Found in gymnosperms, e.g., in the Cycadales (e.g., in *Zamia angustifolia*), Taxales (e.g., in *Taxus* spp.), Coniferales (e.g., in *Dacrydium* spp.) and Ginkgoales (e.g., in *Ginkgo biloba*).

1452
Glepidotin A; 8-Prenylgalangin

Skeletal type: flavonol

$C_{20}H_{18}O_5$ Mol. wt 338.36

Occurs in the whole plant of *Glycyrrhiza lepidota* (Leguminosae).

Active against the microorganisms *Staphylococcus aureus*, *Mycobacterium smegmatitis*, *Candida albicans* and *Klebsiella pneumoniae*, but not against *Escherichia coli* or *Salmonella gallinarum*.

1453
Gossypetin; 8-Hydroxyquercetin

Skeletal type: flavonol

$C_{15}H_{10}O_8$ Mol. wt 318.25

The 3-glucoside occurs in the flowers of *Hibiscus tiliaceus* (Malvaceae), the 3-galactoside in the petals of *Lotus corniculatus* (Leguminosae) and the 7-glucoside (gossypitrin) in the petals of *Chrysanthemum segetum* (Compositae). See also gossypin.

Yellow flower pigment. Gossypetin and its 8-rhamnoside show antibacterial activity against *Pseudomonas maltophilia* and *Enterobacter cloacae*.

1454
Gossypin; Gossypetin 8-O-glucoside

Skeletal type: flavonol O-glycoside

$C_{21}H_{20}O_{13}$　　　　　　　Mol. wt 480.39

Occurs in *Gossypium indicum* and the flowers of *Hibiscus vitifolius* (Malvaceae).

Yellow flower pigment. Analgesic activity, anti-inflammatory activity, and antigastric ulcer activity (reduces the degree of ulceration, the free and total acidity, and the volume of gastric content).

1455
Herbacetin; 8-Hydroxykaempferol

Skeletal type: flavonol

$C_{15}H_{10}O_7$　　　　　　　Mol. wt 302.27

The aglycone is found in the flowerheads of *Eupatorium gracile* (Compositae), the 3-glucuronide in the petals of *Meconopsis paniculata* (Papaveraceae), the 7-glucoside in the petals of *Gossypium herbaceum* and *G. indicum* (Malvaceae), and the 8-glucoside in the flowers of *Althaea rosea* (Malvaceae).

Yellow flower pigment.

1456
Hinokiflavone

Skeletal type: biflavone

$C_{30}H_{18}O_{10}$　　　　　　　Mol. wt 538.47

Occurs in many Gymnospermae, e.g., in *Cycas revoluta* (Cycadales), *Cupressus funebris* and *Podocarpus macrophyllus* (Coniferales). It also occurs in the water fern *Selaginella tamariscina* (Selaginellales).

Inhibits cyclic nucleotide phosphodiesterases.

1457
Hispidulin; Dinatin; Scutellarein 6-methyl ether

Skeletal type: flavone

$C_{16}H_{12}O_6$　　　　　　　Mol. wt 300.27

Found in the leaves of many Compositae, Labiatae and Hydrophyllaceae, and also in the leaves of *Digitalis orientalis* (Scrophulariaceae) and the green peel of *Citrus sudachii* (Rutaceae).

Antihepatotoxic activity. It inhibits aggregation of human platelets to various agonists (considerably more potent than theophylline), and stimulates platelet cAMP levels.

1458
6-Hydroxykaempferol; Galetin

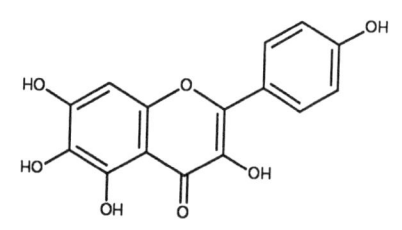

Skeletal type: flavonol

$C_{15}H_{10}O_7$ Mol. wt 302.25

Occurs as the 7-glucoside in the leaves of *Tetragonotheca texana* and *Neurolaena oaxacana* (Compositae). The compound was formerly called galetin, since it was thought to be present in the flowers of *Galega officinalis* (Leguminosae), but this is not the case.

The 3,6-dimethyl ether (from *Acanthospermum australe* and buds of sweet cherry) is a lens aldose reductase inhibitor.

1459
6-Hydroxyluteolin; 5,6,7,3',4'-Pentahydroxyflavone

Skeletal type: flavone

$C_{15}H_{10}O_7$ Mol. wt 302.25

6-Hydroxyluteolin glycosides occur regularly in several related families including the Bignoniaceae, Scrophulariaceae, Labiatae, and Plantaginaceae. For instance, the 7-glucoside is found in *Catalpa bignonioides* leaves (Bignoniaceae) and the 7-glucuronide in *Plantago maritima* leaves (Plantaginaceae).

1460
Hyperin; Hyperoside; Quercetin 3-galactoside

Skeletal type: flavonol *O*-glycoside

$C_{21}H_{20}O_{12}$ Mol. wt 464.39

Commonly occurring: e.g., in leaves of St John's wort, *Hypericum perforatum* (Guttiferae).

Potent inhibitor of lens aldose reductase. It shows antibacterial activity against *Pseudomonas maltophilia*.

1461
Hypolaetin; 8-Hydroxyluteolin

Skeletal type: flavone

$C_{15}H_{10}O_7$ Mol. wt 302.25

Found in *Hypolaena fastigiata* (Restionaceae). The 7-glucoside is in the berries of *Juniperus macropoda* (Cupressaceae); the 8-glucoside in *Sideritis* spp. (Labiatae); the 8-glucuronide and 8,8'-diglucuronide in the liverwort, *Marchantia berteroana* (Hepaticae), and the 8-sulfate in the leaves of *Bixa orellana* (Bixaceae).

Hypolaetin 8-glucoside possesses acute anti-inflammatory activity in rats, and has cytoprotective effects on the rat gastric mucosa. However, its antigastric ulcer activity is less than that of cimetidine.

1462
Isoorientin; Homoorientin; Luteolin 6-C-glucoside

Skeletal type: flavone C-glycoside

$C_{21}H_{20}O_{11}$ Mol. wt 448.39

Widespread occurrence: e.g., in *Polygonum orientale* (Polygonaceae). The 2″-O-glucoside of isoorientin is present in rice, *Oryza sativa* (Gramineae).

Isoorientin 2″-O-glucoside is one of eight flavone C-glycosides present in rice as a mixture which acts as a feeding attractant to plant hoppers feeding on rice phloem.

1463
Isoquercitrin; Quercetin 3-O-glucoside

Skeletal type: flavonol O-glycoside

$C_{21}H_{20}O_{12}$ Mol. wt 464.39

Widespread occurrence: e.g., in the flowers of *Gossypium herbaceum* (Malvaceae), and in the leaves of mulberry, *Morus alba* (Moraceae).

Feeding attractant to the silkworm, *Bombyx mori*, which feeds on the mulberry. It inhibits lens aldose reductase. Also, it shows antibacterial activity against *Pseudomonas maltophilia*.

1464
Isorhamnetin; Quercetin 3′-methyl ether

Skeletal type: flavonol

$C_{16}H_{12}O_7$ Mol. wt 316.27

Very widespread occurrence, both free, e.g., in the flowers of *Arnica* and in the leaves of *Haplopappus* (Compositae), and as a glycoside. The 3-arabinoside (distichin) is present in *Taxodium distichum* leaves (Taxodiaceae), the 3-glucoside in *Argemone mexicana* flowers (Papaveraceae), the 3-galactoside in *Cactus grandiflora* (Cactaceae), and the 5-glucoside in the leaves of *Cotula* spp. (Compositae).

1465
Isoscoparin; Chrysoeriol 6-C-glucoside

Skeletal type: flavone C-glycoside

$C_{22}H_{22}O_{11}$ Mol. wt 462.42

Found in duckweed, *Lemna minor* (Lemnaceae), and in *Potamogeton natans* (Potamogetonaceae). The 2″-O-glucoside of isoscoparin occurs in the rice plant, *Oryza sativa* (Gramineae), together with its 6‴-p-coumaric and 6‴-ferulic acid esters.

Isoscoparin 2″-O-glucoside and its acylated derivatives are among the flavone C-glycosides present in *Oryza* as a mixture which acts as a feeding attractant to plant hoppers which feed on rice phloem.

1466
Isoscutellarein; 8-Hydroxyapigenin

Skeletal type: flavone

$C_{15}H_{10}O_6$ Mol. wt 286.25

Occurs as glycosides in the leaves of *Pinguicula vulgaris* (Lentibulariaceae), and as the 7-(6′′′-acetylallosyl)(1→2)glucoside in the leaves of *Stachys recta* (Labiatae) and of *Veronica filiformis* (Scrophulariaceae).

1467
Isovitexin; Saponaretin; Homovitexin; Apigenin 6-*C*-glucoside

Skeletal type: flavone *C*-glycoside

$C_{21}H_{20}O_{10}$ Mol. wt 432.39

Very widespread occurrence: e.g., in *Vitex lucens* (Verbenaceae).

1468
Kaempferide; Kaempferol 4′-methyl ether

Skeletal type: flavonol

$C_{16}H_{12}O_6$ Mol. wt 300.27

Occurs free in the farinose exudate of fronds of ferns, e.g., *Pityrogramma triangularis* (Adiantaceae), in the leaf buds of Betulaceae and Salicaceae, in the aerial parts of *Baccharis* spp. (Compositae) and in *Linaria dalmatia* (Scrophulariaceae). Glycosides are also common, e.g., the 3-diglucoside in the leaves of *Dillenia indica* (Dilleniaceae).

Inhibits inflammation induced by the tumour promoter 12-*O*-tetradecanoylphorbol-13-acetate.

1469
Kaempferol; 3,5,7,4′-Tetrahydroxyflavone

Skeletal type: flavonol

$C_{15}H_{10}O_6$ Mol. wt 286.25

Very widespread occurrence, both free and bound as glycosides. The 3-arabinofuranoside, juglanin, and 3-rhamnofuranoside occur in the leaves and flowers of *Aesculus hippocastanum* (Hippocastanaceae). The 3-rhamnopyranoside, afzelin, occurs in the heartwood of *Afzelia* spp. (Leguminosae), and the 3-galactoside, trifolin, occurs in the leaves of *Trifolium pratense* (Leguminosae). See also robinin.

Radical scavenger. It shows anti-inflammatory, antibacterial and mutagenic activities. It inhibits the proliferation of rat lymphocytes at a concentration of 10^{-4} M. Also, it inhibits iodothyronine deiodinase, Δ^5-lipoxygenase, and ionophore-induced arachidonic acid release and metabolism.

1470
Kaempferol 3-methyl ether; Isokaempferide

Skeletal type: flavonol

$C_{16}H_{12}O_6$ Mol. wt 300.27

Occurs free in the leaf exudates of many Compositae, in the frond exudates of ferns, in the spines of *Opuntia* spp. (Cactaceae) and in the leaf trichomes of *Solanum sarrachoides* (Solanaceae). The 7-rhamnoside occurs in *Carduus getulus*, and the 7-glucoside in *Haplopappus foliosus* (Compositae).

Antiviral activity.

1471
Kuwanone G

Skeletal type: flavone

$C_{40}H_{36}O_{11}$ Mol. wt 692.73

Found in the root bark of mulberry, *Morus alba* (Moraceae).

Hypotensive activity: lowers blood pressure in rabbits when administered intravenously at a dose of 1.0 mg per kg body weight.

1472
Kuwanone H

Skeletal type: flavone

$C_{45}H_{44}O_{11}$ Mol. wt 760.85

Found in the root bark of mulberry, *Morus alba* (Moraceae).

Hypotensive activity.

1473
Lucenin-2; Luteolin 6,8-di-C-glucoside

Skeletal type: flavone C-glycoside

$C_{27}H_{30}O_{16}$ Mol. wt 610.54

Widespread occurrence: e.g., *Vitex lucens* (Verbenaceae), *Spergularia rubra* (Caryophyllaceae), *Tragopogon* spp. (Compositae) and *Sophora* spp. (Leguminosae).

1474
Luteolin; 5,7,3',4'-Tetrahydroxyflavone

Skeletal type: flavone

$C_{15}H_{10}O_6$ Mol. wt 286.25

Very widespread occurrence, especially as the 7-glucoside (q..v) and 7-glucuronide, e.g., in the petals of *Antirrhinum majus* (Scrophulariaceae). The 7-galactoside and 7-rutinoside occur in *Capsella bursa-pastoris* (Cruciferae), the 3'-glucoside in *Dracocephalum thymiflorum* (Labiatae), and the 4'-glucoside in the flowers of *Spartium junceum* (Leguminosae). The aglycone is also very common, especially in leaf exudates.

Anti-inflammatory and antibacterial activities. It inhibits iodothyronine deiodinase, protein kinase C, NADH-oxidase, succinoxidase, lens aldose reductase, etc. It acts as a nodulation signal to the bacterium *Rhizobium leguminosarum* in pea roots and the bacterium *R. meliloti* in lucerne.

1475
Luteolin 7-O-glucoside

Skeletal type: flavone *O*-glycoside

$C_{21}H_{20}O_{11}$ Mol. wt 448.39

Very widespread occurrence, e.g., in *Humulus japonicus* (Cannabidaceae), and *Salix* spp. (Salicaceae). Luteolin 7-(6″-malonylglucoside) occurs in the leaves of the carrot, *Daucus carota* (Umbelliferae).

Feeding attractant to the beetle *Chrysomela vigintipunctata* feeding on *Salix*. The malonate ester acts as a contact oviposition stimulant for females of the butterfly *Papilio polyxenes*.

1476
Morin; 3,5,7,2',4'-Pentahydroxyflavone

Skeletal type: flavonol

$C_{15}H_{10}O_7$ Mol. wt 302.25

Found in many *Morus* spp., e.g., *M. alba*, and other Moraceae, e.g., *Chlorophora tinctoria* and *Artocarpus integrifolia*.

Antiviral and antibacterial activities, and contact sensitizing (allergenic) potency. It inhibits Δ^5-lipoxygenase, iodothyronine deiodinase, lens aldose reductase, and ionophore-induced arachidonic acid release and metabolism. Feeding attractant to the silkworm, *Bombus mori*, which feeds on *Morus alba*.

1477
Morusin

Skeletal type: flavone

$C_{25}H_{24}O_6$ Mol. wt 420.47

Main phenolic constituent from the root bark of the mulberry tree, *Morus alba* (Moraceae).

Inhibits the tumour-promoting activity of teleocidin.

1478
Myricetin; 3,5,7,3′,4′,5′-Hexahydroxyflavone

Skeletal type: flavonol

$C_{15}H_{10}O_8$ Mol. wt 318.25

Occurs free in the heartwood of *Soymidia febrifuga* (Meliaceae) and the aerial parts of *Haplopappus canescens* (Compositae). Glycosides are widespread, e.g., the 3-glucoside occurs in the petals of *Primula sinensis* (Primulaceae), the 3-galactoside in the leaves of *Camellia sinensis* (Theaceae), and the 3-arabinoside in the berries of *Vaccinium macrocarpon* (Ericaceae). See also myricitrin.

Strong antigonadotropic activity. It shows antibacterial activity against *Pseudomonas maltophilia* and *Enteromorpha cloacae*. It inhibits Δ^5-lipoxygenase, NADH-oxidase, succinoxydase, and ionophore-induced arachidonic acid release and metabolism in rat peritoneal macrophages.

1479
Myricitrin; Myricetin 3-*O*-rhamnoside

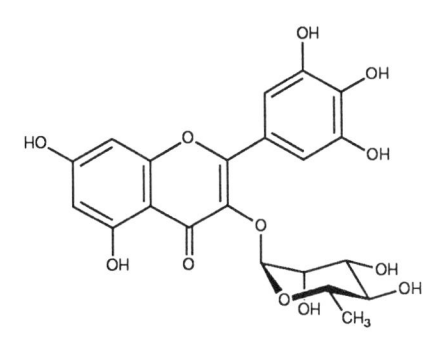

Skeletal type: flavonol *O*-glycoside

$C_{21}H_{20}O_{12}$ Mol. wt 464.39

Found in the bark of *Myrica rubra* (Myricaceae).

Inhibits inflammation induced by the tumour promoter, 12-*O*-tetradecanoylphorbol-13-acetate. It shows antibacterial activity against *Pseudomonas maltophilia* and *Enteromorpha cloacae*.

1480
Neocarlinoside

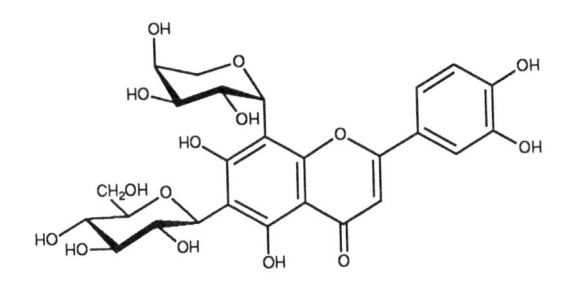

Skeletal type: flavone *C*-glycoside

$C_{26}H_{28}O_{15}$ Mol. wt 580.51

Occurs in the leaves of rice, *Oryza sativa* (Gramineae).

Neocarlinoside is one of the flavone *C*-glycosides present in *O. sativa* as a mixture which acts as a feeding attractant to plant hoppers feeding on rice phloem.

1481
Neoschaftoside

Skeletal type: flavone *C*-glycoside

$C_{26}H_{28}O_{14}$ Mol. wt 564.51

Occurs in *Catananche cerulea* (Compositae), in the leaves of rice, *Oryza sativa* (Gramineae), and many other sources.

Neoschaftoside is one of the flavone-*C*-glycosides present in *O. sativa* the mixture of which acts as a feeding attractant to plant hoppers feeding on rice phloem.

1482
Nevadensin

Skeletal type: flavone

$C_{18}H_{16}O_7$ Mol. wt 344.33

Found in *Iva nevadensis* (Compositae), *Lysionotus pauciflora* (Gesneriaceae), the leaves and glandular trichomes of *Helianthus* spp. (Compositae), and the leaves and flowers of *Ocimum canum* (Labiatae).

Antituberculostatic activity, inhibiting *Bacillus tuberculosis* at a concentration of 0.2 mg ml^{-1} *in vitro*.

1483
Norwogonin; 5,7,8-Trihydroxyflavone

Skeletal type: flavone

$C_{15}H_{10}O_5$ Mol. wt 270.25

Occurs in the roots and other organs of *Scutellaria* spp., e.g., in the leaves of *S. epilobifolia* (Labiatae).

Mutagenic activity in *Salmonella typhimurium* strain TA 100, but hardly any in strain TA 98.

1484
Orientin; Luteolin 8-C-glucoside

Skeletal type: flavone C-glycoside

$C_{21}H_{20}O_{11}$ Mol. wt 448.39

Very widespread occurrence, e.g., in *Polygonum orientale* (Polygonaceae).

Glucosylorientin, which is one of the flavone C-glycosides found in millet grain, *Pennisetum americanum* (Gramineae), is an inhibitor of thyroid peroxidase, and is responsible, together with other phenolics in the grain, for the goitrogenic and antithyroid activity of millet.

1485
Oxyayanin A

Skeletal type: flavonol

$C_{18}H_{16}O_8$ Mol. wt 360.33

Found in ayan or Nigerian satinwood, from the tree *Distemonanthus benthamianus* (Leguminosae).

Allergenic activity: causes contact allergic skin reactions in coffin workers handling ayan wood.

1486
Oxyayanin B

Skeletal type: flavonol

$C_{18}H_{16}O_8$ Mol. wt 360.33

Found in ayan wood from the tree *Distemonanthus benthamianus* (Leguminosae).

Allergenic activity: causes contact allergic skin reactions in coffin workers handling ayan wood.

1487
Pachypodol; Quercetin 3,7,3′-trimethyl ether

Skeletal type: flavonol

$C_{18}H_{16}O_7$ Mol. wt 343.33

Occurs on the leaf surface or in the leaf trichomes of many plants, e.g., of *Solanum pubescens* (Solanaceae), *Begonia glabra* (Begoniaceae), *Jasonia montana* (Compositae) and *Agastache rugosa* (Labiatae).

Antiviral activity.

1488
Patuletin; Quercetagetin 6-methyl ether

Skeletal type: flavonol

$C_{16}H_{12}O_8$ Mol. wt 332.27

Occurs in various organs and/or leaf resins of many Compositae, e.g., in the flowerhead of *Eupatorium gracile* and *Arnica* spp. Patuletin glycosides are also common in the Compositae, e.g., the 3-glucoside and 3-rutinoside in *Hymenoxis scaposa* and 3-glucuronide in *Lasthenia* spp. The 3-galactoside and 3-diglucoside occur in *Ipomopsis aggregata* (Polemoniaceae), and the 7-glucoside in *Prosopis spicigera* (Leguminosae).

1489
Pedalitin

Skeletal type: flavone

$C_{16}H_{12}O_7$ Mol. wt 316.27

Occurs in *Sullivantia* spp. (Saxifragaceae), the liverwort, *Frullania* spp. (Hepaticae), and also the leaves of sesame, *Sesamum indicum* (Pedaliaceae).

Inhibits Δ^5-lipoxygenase.

1490
Primetin; 5,8-Dihydroxyflavone

Skeletal type: flavone

$C_{15}H_{10}O_4$ Mol. wt 254.25

External flavonoid on the leaves and stalks of many *Primula* spp., e.g., *P. modesta* and *P. mistassinica* (Primulaceae).

Contact sensitizing (allergenic) properties.

1491
Quercetagetin; 6-Hydroxyquercetin

Skeletal type: flavonol

$C_{15}H_{10}O_8$ Mol. wt 318.25

Found in the flowers of many spp. of Compositae, both as an aglycone (e.g., from *Eupatorium gracile*) and as glycosides. For instance, the 3-glucoside, tagetiin, and the 7-glucoside occur in the flowers of marigold, *Tagetes erecta*.

Yellow flower pigment. It shows antibacterial activity against *Pseudomonas maltophilia* and *Enterobacter cloaca*.

1492
Quercetin; 3,5,7,3',4'-Pentahydroxyflavone

Skeletal type: flavonol

$C_{15}H_{10}O_7$ Mol. wt 302.25

The commonest flavonoid in higher plants, usually present in glycosidic form (see rutin, isoquercitrin, quercitrin, hyperin and quercimeritrin), but also isolated free from the families Compositae, Passiflorae, Rhamnaceae and Solanaceae.

Inhibits many enzymes, e.g., protein kinase C, lipogenases, lens aldose reductase, 3',5'-cyclic adenosine monophosphate phosphodiesterases. It is a radical scavenger. Quercetin also inhibits smooth muscle contraction, and proliferation of rat lymphocytes. It is antigonadotropic, anti-inflammatory, antibacterial, antiviral and antihepatotoxic, and shows some mutagenic activity and allergenic properties.

1493
Quercetin 3-methyl ether

Skeletal type: flavonol

$C_{16}H_{12}O_7$ Mol. wt 316.27

Found as the aglycone in the leaves, leaf resin, aerial parts and flowerheads of many Compositae and other families, e.g., from *Cistus* spp. (Cistaceae), *Cyperus* spp. (Cyperaceae) and *Opuntia* spp. (Cactaceae). Also common in glycosidic form, e.g., the 7-glucoside in *Artemisia transiliensis* (Compositae).

Antiviral activity.

1494
Quercimeritrin; Quercetin 7-O-glucoside

Skeletal type: flavonol O-glycoside

$C_{21}H_{20}O_{12}$ Mol. wt 464.39

Occurs in the flowers of cotton, *Gossypium hirsutum*, and other *Gossypium* spp. (Malvaceae).

Quercimeritrin, together with quercetin and its 3'-O-glucoside, are feeding stimulants from cotton for the boll weevil, *Anthonomus grandis*.

1495
Quercitrin; Quercetin 3-O-rhamnoside

Skeletal type: flavonol O-glycoside

$C_{21}H_{20}O_{11}$ Mol. wt 448.39

Widespread occurrence, e.g., in the bark of *Quercus tinctoria* (Fagaceae) and in *Polygonum* spp. (Polygonaceae).

Feeding attractant to the beetle *Gastrophysa atrocynea* which feeds on *Polygonum* spp., but a feeding deterrent to the silkworm, *Bombyx mori*. It shows antimutagenic, anti-ulcer, antiviral, antihaemorrhagic and antihepatotoxic activities, as well as antibacterial activity against *Pseudomonas maltophilia* and *Enterobacter cloacae*. It strongly inhibits lens aldose reductase.

1496
Rhamnetin; Quercetin 7-methyl ether

Skeletal type: flavonol

$C_{16}H_{12}O_7$ Mol. wt 316.27

Many glycosides are known: e.g., the 3-glucoside in *Thalictrum foetidum* (Ranunculaceae), the 3-rhamnoside (xanthorhamnin) in the fruit of *Rhamnus cathartica* (Rhamnaceae) and the 3'-glucuronide in *Tamarix aphylla* (Tamaricaceae). The aglycone has been found in the aerial parts of many Compositae and Labiatae, and in the leaf resin of *Cistus* spp. (Cistaceae).

Rhamnetin and its 3-glucoside show antibacterial activity against *Pseudomonas maltophilia* and *Enterobacter cloacae*. It also shows a moderate contact sensitizing (allergenic) capacity.

1497
Robinetin; 5-Deoxymyricetin

Skeletal type: flavonol

$C_{15}H_{10}O_7$ Mol. wt 302.25

Occurs in many Leguminosae, e.g., *Gleditsia monosperma*, *Millettia stuhlmannii*, *Robinia pseudacacia* and *Acacia mearnsii*.

Antibacterial activity against *Pseudomonas maltophilia* and *Enterobacter cloacae*.

1498
Robinin

Skeletal type: flavonol *O*-glycoside

$C_{33}H_{40}O_{19}$ Mol. wt 740.68

Found in the flowers of *Robinia pseudacacia*, and in *Pueraria* spp. and *Vigna* spp. (Leguminosae).

Antibacterial activity against *Pseudomonas maltophilia*.

1499
Robustaflavone; 3′,6″-Biapigenin

Skeletal type: biflavone

$C_{30}H_{18}O_{10}$ Mol. wt 538.47

Found in *Araucaria* spp. (Araucariaceae/Coniferales), *Juniperus* spp. (Cupressaceae/Coniferales) and *Rhus* spp. (Anacardiaceae).

Inhibitor of cyclic nucleotide phosphodiesterases.

1500
Rutin; Quercetin 3-rutinoside; Rutoside

Skeletal type: flavonol *O*-glycoside

$C_{27}H_{30}O_{16}$ Mol. wt 610.54

Very widespread occurrence in higher plants, e.g., in *Polygonum* spp. (Polygonaceae). First isolated from rue, *Ruta graveolens* (Rutaceae).

Radical scavenger. Medicinally, it is used against capillary fragility and varicosis. A more soluble derivative, hydroxyethylrutoside, is also used clinically. It shows antiviral and antibacterial activities, and it inhibits lens aldose reductase and Δ^5-lipoxygenase. It is a feeding attractant to the beetle *Gastrophysa atrocynea*, which feeds on *Polygonum*, but a feeding deterrent to larvae of *Heliothis zea*. Also, it is a contact oviposition stimulant to the butterfly *Papilio xuthus* for laying eggs on citrus leaves.

1501
Schaftoside

Skeletal type: flavone-*C*-glycoside

$C_{26}H_{28}O_{14}$ Mol. wt 564.51

Widespread occurrence, e.g., *Silene schafta* (Caryophyllaceae), *Catananche cerulea* (Compositae) and rice, *Oryza sativa* (Gramineae).

One of the flavone *C*-glycosides present in *O. sativa* as a mixture which acts as a feeding attractant to plant hoppers feeding on rice phloem.

1502
Sciadopitysin

Skeletal type: biflavone

$C_{33}H_{24}O_{10}$ Mol. wt 580.56

Occurs in many Gymnospermae, e.g., in *Encephalartos* spp. and other Zamiaceae (Cycadales), in *Taxus* spp. (Taxales) and in *Ginkgo biloba* (Ginkgoales).

1503
Scullcapflavone II

Skeletal type: flavone

$C_{19}H_{18}O_8$ Mol. wt 374.35

Found in the roots of *Scutellaria baicalensis* (Labiatae).

Cytotoxic activity.

1504
Scutellarein; 6-Hydroxyapigenin

Skeletal type: flavone

$C_{15}H_{10}O_6$ Mol. wt 286.25

The 5-glucuronide occurs in the flowers of *Centaurea depressa* (Compositae), the 7-glucuronide in the flowers and leaves of *Scutellaria polydon* (Labiatae), and the 7-rhamnoside, sorbifolin, in the leaves and flowers of *Sorbaria sorbifolia* (Rosaceae). The aglycone has been isolated from *Asphodeline* spp. (Liliaceae), the leaves of *Digitalis orientalis* (Scrophulariaceae), the leaves of *Pulicaria rivularis* (Compositae), and the roots of *Scutellaria* spp. (Labiatae).

1505
Sexangularetin; Herbacetin 8-methyl ether

Skeletal type: flavonol

$C_{16}H_{12}O_7$ Mol. wt 316.27

Occurs in the flowerheads of *Eupatorium gracile* (Compositae). The 3-rhamnosylglucoside-7-rhamnoside is found in the leafy stems of *Sedum sexangulare* (Crassulaceae).

Yellow flower pigment. It shows mutagenic activity in *Salmonella typhimurium* strain TA 100, but little activity in strain TA 98.

1506
Sinensetin; 5,6,7,3′,4′-Pentamethoxyflavone

Skeletal type: flavone

$C_{20}H_{20}O_7$ — Mol. wt 372.38

Found in the peel of *Citrus* fruits, e.g., *C. sinensis* and *C. aurantium* (Rutaceae), the roots of *Bauhinia championii* (Leguminosae), and the aerial parts of *Chenopodium botrys* (Chenopodiaceae) and of *Kickxia spuria* (Scrophulariaceae).

Antifungal activity.

1507
Swertiajaponin; Isoorientin 7-methyl ether

Skeletal type: flavone C-glycoside

$C_{22}H_{22}O_{11}$ — Mol. wt 462.42

Widespread occurrence, e.g., *Swertia japonica* (Gentianaceae), *Iris nertshinskia* (Iridaceae), *Cephalaria leucantha* (Dipsacaceae) and *Tragopogon* spp. (Compositae).

1508
Tangeretin; Ponkanetin; 5,6,7,8,4′-Pentamethoxyflavone

Skeletal type: flavone

$C_{20}H_{20}O_7$ — Mol. wt 372.38

Occurs in the rind of *Citrus* spp. (Rutaceae).

Antifungal activity. It inhibits the proliferation and invasion of malignant tumour cells *in vitro*, the release of oxidants by human neutrophils, and basophil histamine release. Also, it inhibits smooth muscle contraction. When fed at the rate of 10 mg kg^{-1} each day to female rats during gestation, it caused the death of 83% of the offspring.

1509
Thymonin

Skeletal type: flavone

$C_{18}H_{16}O_8$ — Mol. wt 360.33

Found as an external flavonoid on the leaves on thyme, *Thymus vulgaris*, and other Labiatae.

Inhibits smooth muscle contraction; its spasmolytic activity is higher than that of thymol or carvacrol.

1510
Tricetin; 5,7,3′,4′,5′-Pentahydroxyflavone

Skeletal type: flavone

$C_{15}H_{10}O_7$ Mol. wt 302.25

The 7-glucoside occurs in *Metasequoia glyptostroboides* (Taxodiaceae), and the 3′-glucoside in the leaves of *Thuja occidentalis* (Cupressaceae).

1511
Tricin; Tricetin 3′,5′-dimethyl ether

Skeletal type: flavone

$C_{17}H_{14}O_7$ Mol. wt 330.30

The 5-glucoside and 7-glucuronide occur in *Triticum dicoccum* (Gramineae); the latter glycoside also occurs in alfalfa, *Medicago sativa* (Leguminosae), and the 7-glucoside in rice, *Oryza sativa* (Gramineae). The aglycone is found in the leaves of *Phoenix canariensis* (Palmae) and occurs widespread in the Cyperaceae, e.g., in the stems of *Trichophorum cespitosum*.

Allelopathic agent of the grass *Agropyron repens*, causing inhibition of root growth in competing plant species.

1512
Vicenin-2

Skeletal type: flavone *C*-glycoside

$C_{27}H_{30}O_{15}$ Mol. wt 594.54

Widespread occurrence, e.g., *Vitex lucens* (Verbenaceae), wheat, *Triticum aestivum* (Gramineae), and *Tragopogon* spp. (Compositae) and *Sophora* spp. (Leguminosae).

Contact oviposition stimulant to the butterfly *Papilio xuthus* for laying eggs on *Citrus* leaves.

1513
Violanthin

Skeletal type: flavone *C*-glycoside

$C_{27}H_{30}O_{14}$ Mol. wt 578.54

Found in *Viola tricolor* (Violaceae) and in *Eleusine* spp. and other Gramineae.

1514
Vitexin; Apigenin 8-*C*-glucoside

Skeletal type: flavone *C*-glycoside

$C_{21}H_{20}O_{10}$ Mol. wt 432.39

Widespread occurrence, e.g., *Vitex lucens* (Verbenaceae) and millet, *Pennisetum americanum* (Gramineae).

Potent inhibitor of thyroid peroxidase; vitexin is among the phenolics in millet grain responsible for the goitrogenic and antithyroid activity of *Pennisetum americanum*.

1515
Wogonin; Norwogonin 8-methyl ether

Skeletal type: flavone

$C_{16}H_{12}O_5$ Mol. wt 284.27

Found in the stems of *Anodendron affine* (Apocynaceae) and the roots of *Scutellaria baicalensis* (Labiatae) and other *Scutellaria* spp.

Oestrogenic and anti-implantational activity in the rat.

Chapter 38
Isoflavonoids and neoflavonoids

Isoflavonoids differ from other classes of flavonoid by their greater structural variation, the greater frequency of isoprenoid substitution and the fact that they are usually present in plants in the free state rather than in glycosidic combination. Isoflavones are isomeric with the more widely occurring flavones (q.v.), and genistein (**1534**) is derived biosynthetically by an aryl migration from the same chalcone precursor as that which gives rise to the flavone apigenin. Isoflavonoids are more restricted in nature than flavones and flavonols (q.v.), since they are found regularly in only one subfamily of the Leguminosae, the Papilionoideae. They have been recorded occasionally in a few other families, such as the Compositae, Iridaceae, Myristicaceae and Rosaceae. It is possible that they are more widespread than this, but the lack of a satisfactory method for screening plant tissues for their presence means that our present knowledge of their natural occurrence is imperfect.

Six hundred isoflavonoids have been described (Dewick, 1994), and these are divided into subclasses according to the oxidation level of the central pyran ring. Isoflavones are the most abundant subclass; other subclasses include isoflavanones, pterocarpans, isoflavans and rotenoids. The neoflavonoids (e.g., dalbergin, **1527**) comprise a further small subclass of isomeric structures, many of which can also be described as 4-phenylcoumarins. Neoflavonoids are found in the Leguminosae but, unlike the isoflavonoids, are also reported from the Guttiferae, Rubiaceae and Passifloraceae (Donnelly and Boland, 1994).

Widely occurring isoflavones include daidzein (**1525**), genistein (**1534**), formononetin (**1538**) and biochanin A (**1519**), which are noted for their oestrogenic properties. Other isoflavones differ in the number and position of hydroxyl, methoxyl and methylenedioxy substitution (cf., afrormosin, **1517**). 2′-Hydroxylation is a regular and distinguishing feature in many isoflavones (see e.g., luteone, **1553**, which occurs in lupins). Glycosides are occasionally reported (e.g., genistin, **1536**), as are isopentenyl substituted compounds. Pomiferin (**1569**) from the osage orange has two isopentenyl substituents.

Pterocarpans such as medicarpin (**1555**) and isoflavans such as vestitol (**1588**) are significantly antifungal. Besides occurring constitutively in the heartwood of many legume trees, such compounds are also induced in herbaceous species (e.g., clover) in response to fungal infection, and are then known as phytoalexins (Ingham, 1982, 1983).

The rotenoids have a common tetracyclic ring system, as in rotenone (**1575**), and are distinguished by their insecticidal and piscicidal properties. Powdered derris root, which contains rotenone, is a well known insecticidal dust. Rotenone inhibits the electron transport pathway in mitochondria and is therefore toxic to all forms of life. It can be used safely as a fish poison because the oral toxicity in humans is very low.

Isoflavonoids are detected in plant tissues by their phenolic reactions, and the different subclasses can be distinguished by their ultraviolet spectra. Further details of their identification and characterization can be found in Harborne (1989).

References

Dewick, P.M. (1994). Isoflavonoids. In *The Flavonoids: Advances in Research since 1986* (Harborne, J.B., ed.), pp. 117–238. London: Chapman and Hall.

Donnelly, D.M.X. and Boland, G. (1994). Neoflavonoids. In *The Flavonoids – Advances in Research since 1986* (Harborne, J.B., ed.), pp. 239–258, London: Chapman and Hall.

Harborne, J.B. (1989). *Methods in Plant Biochemistry*, Volume 1, *Plant Phenolics*. London: Academic Press.

Ingham, J.L. (1982). Phytoalexins from the Leguminosae. In *Phytoalexins* (Bailey, J.A. and Mansfield, J.W., eds), pp. 21–80. Glasgow: Blackie.

Ingham, J.L. (1983). Natural occurring isoflavonoids. *Progress in the Chemistry of Organic Natural Products*, **43,** 1–266.

1516
(−)-Acanthocarpan

Skeletal type: pterocarpan

$C_{17}H_{12}O_7$ Mol. wt 328.28

Induced as a phytoalexin in the leaves of *Caragana acanthophylla* and *Tephrosia bidwillii* (Leguminosae).

Antifungal activity.

1517
Afrormosin; Afromosin; Castanin

Skeletal type: isoflavone

$C_{17}H_{14}O_5$ Mol. wt 298.30

Occurs as a constitutive compound in many Leguminosae, e.g., in the heartwood of *Afrormosa elata*. It is induced as a phytoalexin in the leaves of some *Centrosema* spp. (Leguminosae).

Antifungal activity.

1518
Anhydroglycinol

Skeletal type: pterocarpene

$C_{15}H_{10}O_4$ Mol. wt 254.25

Induced as a phytoalexin in *Tetragonolobus maritimus* (Leguminosae).

Antifungal activity.

1519
Biochanin A; Pratensol

Skeletal type: isoflavone

$C_{16}H_{12}O_5$ Mol. wt 284.27

Found in *Cicer arietinum* and *Trifolium*, *Baptisia* and *Dalbergia* spp. (Leguminosae). Also, it is found in the fruit of *Cotoneaster pannosa* (Rosaceae), and the trunkwood of *Virola cadudifolia* (Myristicaceae).

Oestrogenic activity; hypolipidaemic activity.

1520
Cajanin

Skeletal type: isoflavone

$C_{16}H_{12}O_6$ Mol. wt 300.27

Induced as a phytoalexin in pigeon pea, *Cajanus cajan*, in *Canavalia ensiformis*, and in *Centrosema* spp. (Leguminosae).

Antifungal activity.

1521
Cajanol

Skeletal type: isoflavanone

$C_{17}H_{16}O_6$ Mol. wt 316.32

Induced as a phytoalexin in *Cajanus cajan* and *Stizolobium deeringianum* (Leguminosae).

Antifungal activity.

1522
Coumestrol

Skeletal type: coumestan

$C_{15}H_8O_5$ Mol. wt 268.23

Found in *Trifolium* and *Medicago* spp., and induced as a phytoalexin in *Glycine max*, *Phaseolus lunatus*, *P. vulgaris* and *Vigna unguiculata* (Leguminosae). Also, it is found in spinach leaf, *Spinacea oleracea* (Chenopodiaceae).

Oestrogenic activity, and antifungal activity. It is a noncompetitive inhibitor of peroxidase activity.

1523
Cristacarpin; Erythrabyssin I

Skeletal type: pterocarpan

$C_{21}H_{22}O_5$ Mol. wt 354.41

Found in the roots of the East African medicinal plant *Erythrina abyssinica* (Leguminosae).

Antibacterial activity *in vitro* against *Staphylococcus aureus*, *Bacillus subtilis* and *Micrococcus lysodeikticus*.

1524
Cyclokievitone; 1″,2″-Dehydrocyclokievitone

Skeletal type: isoflavanone

$C_{20}H_{18}O_6$ Mol. wt 354.36

Induced as a phytoalexin in *Phaseolus vulgaris* (Leguminosae).

Antifungal activity.

1525
Daidzein

Skeletal type: isoflavone

$C_{15}H_{10}O_4$ Mol. wt 254.25

Occurs widely in the Leguminosae, e.g., in gorse, *Ulex europaeus*, and clover, *Trifolium repens*.

Antifungal activity, and antilipase activity.

1526
Daidzin; Daidzein 7-*O*-glucoside

Skeletal type: isoflavone *O*-glycoside

$C_{21}H_{20}O_9$ Mol. wt 416.39

Occurs in *Glycine max*, *Trifolium pratense* and *Pueraria* spp. and *Baptisia* spp. (Leguminosae).

1527
Dalbergin

Skeletal type: neoflavonoid

$C_{16}H_{12}O_4$ Mol. wt 268.27

Found in the bark, sapwood and heartwood of many *Dalbergia* spp. and *Macherium* spp. (Leguminosae).

1528
(±)-Dalbergioidin

Skeletal type: isoflavanone

$C_{15}H_{12}O_6$ Mol. wt 288.26

Induced as a phytoalexin in *Dolichos biflorus*, *Lablab niger*, *Macrotyloma axillare*, *Phaseolus vulgaris*, and several other Leguminosae.

Antifungal activity.

1529
Dalpanin

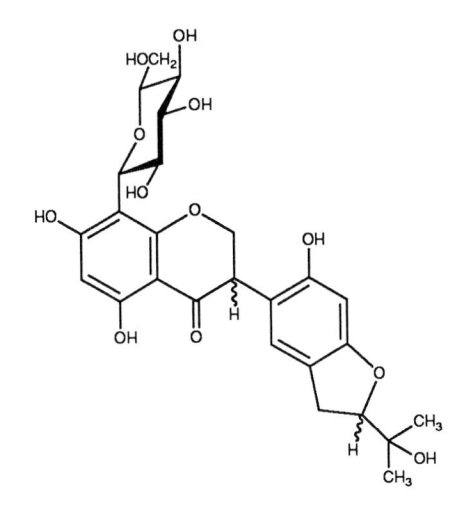

Skeletal type: isoflavanone *C*-glycoside

$C_{26}H_{30}O_{12}$ Mol. wt 534.53

Occurs in the flowers and seeds of *Dalbergia paniculata* (Leguminosae).

1530
Deguelin

Skeletal type: rotenoid

$C_{23}H_{22}O_6$ Mol. wt 394.43

Found in the callus of *Derris elliptica*, and in the roots and seeds of *Lonchocarpus*, *Piscidia* and *Tephrosia* spp. (Leguminosae).

Insecticidal against the larvae of bruchid beetles. It inhibits the activity of plant mitochondria.

1531
(±)-5-Deoxykievitone

Skeletal type: isoflavanone

$C_{20}H_{20}O_5$ Mol. wt 340.38

Induced as a phytoalexin in the pods of *Phaseolus vulgaris* (Leguminosae).

Antifungal activity.

1532
Ferreirin

Skeletal type: isoflavanone

$C_{16}H_{14}O_6$ Mol. wt 302.29

Occurs in the heartwood of *Ferreirea spectabilis* (Leguminosae).

1533
Formononetin; Daidzein 4′-methyl ether; Biochanin B; Neochanin; Pratol

Skeletal type: isoflavone

$C_{16}H_{12}O_4$ Mol. wt 268.27

Occurs widely in the Leguminosae, e.g., in *Cicer arietinum* and in *Trifolium* spp. and *Baptisia* spp.

Antifungal and hypolipidaemic activity. It is responsible for "clover disease", an infertility problem in Australian ewes which have fed on *Trifolium subterraneum*. It is a pro-oestrogen, being converted *in vivo* to the more active isoflavan, equol.

1534
Genistein; Prunetol; Sophoricol; Genisteol

Skeletal type: isoflavone

$C_{15}H_{10}O_5$ Mol. wt 270.25

Occurs in the wood of *Prunus* spp. (Rosaceae), and in broom (*Genista* spp.), clover (*Trifolium* spp.), and many other Leguminosae.

Oestrogenic and antifungal activity. It inhibits soybean lipase activity, and competitively inhibits peroxidase activity.

1535
Genistein 8-*C*-glucoside

Skeletal type: isoflavone *C*-glycoside

$C_{21}H_{20}O_{10}$ Mol. wt 432.39

Occurs in the bark of *Dalbergia nitidula* and the flowers of *Lupinus luteus* (Leguminosae).

Anti-atherosclerotic activity.

1536
Genistin; Genistein 7-*O*-glucoside; Genistoside

Skeletal type: isoflavone *O*-glycoside

$C_{21}H_{20}O_{10}$ Mol. wt 432.39

Occurs in many spp. of Leguminosae, e.g., *Genista tinctoria*, *Ulex nanus*, *Lupinus luteus* and *Glycine max*.

Suppresses the growth of wheat coleoptiles *in vitro*.

1537
Glabridin

Skeletal type: isoflavan

$C_{20}H_{20}O_4$ Mol. wt 324.38

Found in the roots of *Glycyrrhiza glabra* (Leguminosae).

Antimicrobial activity *in vitro* against *Staphylococcus aureus* and *Mycobacterium smegmatis*.

1538
(−)-Glyceollin I

Skeletal type: pterocarpan

$C_{20}H_{18}O_5$ Mol. wt 338.36

Induced as a phytoalexin in some *Glycine* spp. and *Psoralea* spp. (Leguminosae).

Antifungal and antibacterial activity. It is toxic to the root-knot nematode *Meloidogyne incognita*. It inhibits electron transport in isolated mitochondria of soybean, *Glycine max*.

1539
(−)-Glyceollin II

Skeletal type: pterocarpan

$C_{20}H_{18}O_5$ Mol. wt 338.36

Induced as a phytoalexin in many *Glycine* spp., e.g., *G. max* (Leguminosae).

Antifungal and antibacterial activity. It inhibits electron transport in isolated mitochondria of soybean, *Glycine max*.

1540
(−)-Glycinol

Skeletal type: pterocarpan

$C_{15}H_{12}O_5$ Mol. wt 272.26

Induced as a phytoalexin in *Erythrina sandwicensis*, *Pueraria lobata* and *Glycine max* (Leguminosae).

Antifungal activity.

1541
Hildecarpin

Skeletal type: pterocarpan

$C_{17}H_{14}O_7$ Mol. wt 330.30

Found in the roots of *Tephrosia hildebrandtii* (Leguminosae).

Insect antifeedant.

1542
Hispaglabridin A

Skeletal type: isoflavan

C$_{25}$H$_{28}$O$_4$ Mol. wt 392.50

Occurs in the roots of liquorice, *Glycyrrhiza glabra* (Leguminosae).

Antimicrobial activity *in vitro* against *Staphylococcus aureus* and *Mycobacterium smegmatis*.

1543
Homoferreirin

Skeletal type: isoflavanone

C$_{17}$H$_{16}$O$_6$ Mol. wt 316.32

Occurs in *Ferreirea spectabilis*, *Ougeinia dalbergioides* and *Cicer arietinum*, and is induced as a phytoalexin in the leaves of *Argyrocytisus battandieri* (all Leguminosae).

Antifungal activity.

1544
4-Hydroxyhomopterocarpin

Skeletal type: pterocarpan

C$_{17}$H$_{16}$O$_5$ Mol. wt 300.32

Induced as a phytoalexin in the leaves of *Trifolium hybridum* and *T. pallescens* (Leguminosae).

Antifungal activity.

1545
12a-Hydroxyrotenone

Skeletal type: rotenoid

C$_{23}$H$_{22}$O$_7$ Mol. wt 410.43

Found in the roots of *Derris urucu*, *Neorautanenia amboensis*, *Pachyrrhizus erosus* and *Tephrosia* species (Leguminosae).

Insecticidal activity.

1546
Iridin; Irigenin 7-*O*-glucoside

Skeletal type: isoflavone *O*-glycoside

$C_{24}H_{26}O_{13}$ Mol. wt 522.47

Occurs in the rhizomes of *Iris* spp., e.g., the garden iris, *I. germanica*, and orris root, *I. florentina* (Iridaceae).

The aglycone, irigenin, has a stimulatory effect on RNA synthesis. Iridin is a major component of orris root, which has been used in the cosmetics industry.

1548
Irilone

Skeletal type: isoflavone

$C_{17}H_{14}O_6$ Mol. wt 314.30

Found in the rhizomes of *Iris nepalensis* and other *Iris* spp. (Iridaceae).

1548
Irilone

Skeletal type: isoflavone

$C_{16}H_{10}O_6$ Mol. wt 298.26

Occurs in *Iris germanica* (Iridaceae) and red clover, *Trifolium pratense* (Leguminosae).

Suppresses the growth of wheat coleoptiles *in vitro*.

1549
Kievitone; Phaseolus substance II; Vignatin

Skeletal type: isoflavanone

$C_{20}H_{20}O_6$ Mol. wt 356.38

Induced as a phytoalexin in many Leguminosae, e.g., *Dolichos biflorus*, *Lablab niger* and *Phaseolus* spp.

Antibacterial activity against *Pseudomonas*, *Xanthomonas* and *Achromobacter* spp. It shows antifungal activity.

1550
Licoisoflavone A; Phaseoluteone

Skeletal type: isoflavone

$C_{20}H_{18}O_6$ Mol. wt 354.36

Induced as a phytoalexin in *Phaseolus vulgaris* and *Hardenbergia violaceae* (Leguminosae).

Antifungal activity.

1551
Lonchocarpenin

Skeletal type: 3-aryl-4-hydroxycoumarin

$C_{27}H_{28}O_6$ Mol. wt 448.52

Occurs in *Derris scandens* (Leguminosae).

1552
Lotisoflavan

Skeletal type: isoflavan

$C_{17}H_{18}O_5$ Mol. wt 302.33

Induced as a phytoalexin in the leaves of *Lotus angustissimus* and *L. edulis* (Leguminosae).

Antifungal activity.

1553
Luteone

Skeletal type: isoflavone

$C_{20}H_{18}O_6$ Mol. wt 354.36

Occurs in the leaves and fruit of *Lupinus albus* and other *Lupinus* spp.; it is induced as a phytoalexin in *Argyrocytisus battandieri*, *Hardenbergia violacea* and *Laburnum anagyroides* (Leguminosae).

Antifungal activity.

1554
(−)-Maackiain; Inermin; Demethylpterocarpin

Skeletal type: pterocarpan

$C_{16}H_{12}O_5$ Mol. wt 284.27

Occurs in the heartwood of *Maackia amurensis* and many other woody spp. of Leguminosae. Also, it is induced in herbaceous Leguminosae as a phytoalexin, e.g., in *Cicer arietinum*, and in some *Pisum*, *Trifolium* and *Trigonella* spp.

Antifungal activity.

1555
(−)-**Medicarpin**; Demethylhomopterocarpin

Skeletal type: pterocarpan

$C_{16}H_{14}O_4$ Mol. wt 270.29

Occurs regularly in the heartwood of woody legumes, e.g., in *Andira inermis* and *Dalbergia variabilis*, and in leaves of *Osteophleum platyspermum* (Myricaceae). Also, it is induced as a phytoalexin in various *Lathyrus, Medicago, Trifolium* and *Trigonella* spp. (Leguminosae).

Antifungal activity.

1556
Melannin

Skeletal type: neoflavonoid

$C_{16}H_{12}O_5$ Mol. wt 284.27

Found in the heartwood of *Dalbergia melanoxylon* (Leguminosae).

1557
(**R**)-**4-Methoxydalbergione**

Skeletal type: neoflavonoid

$C_{16}H_{14}O_3$ Mol. wt 254.29

Found in the heartwood of *Dalbergia* spp., e.g., *D. retusa* and *D. nigra* (Leguminosae).

1558
Millettone

Skeletal type: rotenoid

$C_{22}H_{18}O_6$ Mol. wt 378.39

Occurs in the seeds of *Milletia dura* and the roots of *Piscidia erythrina* (Leguminosae).

1559
Mucronulatol

Skeletal type: isoflavan

$C_{17}H_{18}O_5$ (−) form   Mol. wt 302.33

Occurs in the wood of *Machaerium mucronulatum* and *Dalbergia variabilis*, and is induced as a phytoalexin in the leaves of *Astragalus* spp., e.g., *A. cicer* (all Leguminosae).

Antifungal activity.

1560
(−)-Nissolin

Skeletal type: pterocarpan

$C_{16}H_{14}O_5$ Mol. wt 286.29

Induced as a phytoalexin in the phyllode of *Lathyrus nissolia* (Leguminosae).

Antifungal activity.

1561
Ononin; Formononetin 7-*O*-glucoside

Skeletal type: isoflavone *O*-glycoside

$C_{22}H_{22}O_9$ Mol. wt 430.42

Found in *Ononis spinosa* and in *Baptisia*, *Thermopsis* and *Trifolium* spp. (Leguminosae).

1562
Orobol; Norsantal; Santol

Skeletal type: isoflavone

$C_{15}H_{10}O_6$ Mol. wt 286.25

Occurs in the roots of *Lathyrus montanus* (*Orobus tuberosus*) and in the leaves and flowers of *Baptisia* spp. (Leguminosae).

Antigonadotrophic activity.

1563
Pachyrrhizone

Skeletal type: rotenoid

$C_{20}H_{14}O_7$ Mol. wt 366.33

Occurs in the seeds of *Pachyrrhizus erosus* (Leguminosae).

Insecticidal activity.

1564
Paniculatin; Genistein 6,8-di-*C*-glucoside

Skeletal type: isoflavone *C*-glycoside

$C_{27}H_{30}O_{15}$ Mol. wt 594.54

Found in the bark of *Dalbergia paniculata* and *D. nitidula* (Leguminosae).

1565
(−)-Phaseollidin

Skeletal type: pterocarpan

$C_{20}H_{20}O_4$ Mol. wt 324.38

Induced as a phytoalexin in *Lablab niger*, *Vigna unguiculata* and *Erythrina* spp. and *Phaseolus* spp. (Leguminosae).

Antibacterial and antifungal activity. It is an insect antifeedant.

1566
(−)-Phaseollin; Phaseolin

Skeletal type: pterocarpan

$C_{20}H_{18}O_4$ Mol. wt 322.36

Induced as a phytoalexin in some *Phaseolus* spp., e.g., *P. vulgaris*, and in *Vigna unguiculata* (Leguminosae).

Antibacterial and antifungal activities. It is an insect antifeedant.

1567
(−)-Phaseollinisoflavan

Skeletal type: isoflavan

$C_{20}H_{20}O_4$ Mol. wt 324.38

Induced as a phytoalexin in *Phaseolus vulgaris* (Leguminosae).

Antifungal activity. It shows antifeedant activity against root-feeding larvae of scarabs *Costelytra zealandica* and *Heteronychus arator*.

1568
Pisatin

Skeletal type: pterocarpan

C$_{17}$H$_{14}$O$_6$ (+) form   Mol. wt 314.30

Induced as a phytoalexin in Leguminosae, for instance the (+) isomer in *Pisum sativum* (pea), and the (−) isomer in *Tephrosia bidwillii*.

Antifungal activity.

1569
Pomiferin

Skeletal type: isoflavone

C$_{25}$H$_{24}$O$_6$ Mol. wt 420.47

Found in the fruit of the osage orange, *Maclura pomifera* (Moraceae), together with the closely related osajin.

Yellow pigment. It shows antimicrobial activity.

1570
Pratensein

Skeletal type: isoflavone

C$_{16}$H$_{12}$O$_6$ Mol. wt 300.27

Found in *Cicer arietinum* and some *Trifolium* spp., e.g., *T. pratense* (Leguminosae).

Hypolipidaemic activity.

1571
Prunetin

Skeletal type: isoflavone

C$_{16}$H$_{12}$O$_5$ Mol. wt 284.27

Occurs in the wood of *Prunus* spp. (Rosaceae), and in *Pterocarpus angolensis* and *Dalbergia miscolobium* (Leguminosae).

1572
Pseudobaptigenin

Skeletal type: isoflavone

C$_{16}$H$_{10}$O$_5$ Mol. wt 282.26

Occurs in *Baptisia*, *Pterocarpus*, *Maackia* and *Dalbergia* spp. (Leguminosae), and was first isolated from the leaves of wild indigo, *Baptisia tinctoria*.

Suppresses the growth of wheat coleoptiles *in vitro*.

1573
Psoralidin

Skeletal type: coumestan

$C_{20}H_{16}O_5$ Mol. wt 336.35

Occurs in the seeds of *Psoralea corylifolia*, and is induced as a phytoalexin in the roots of *Phaseolus lunatus* (Leguminosae).

Antifungal activity.

1574
Puerarin; Daidzein 8-*C*-glucoside

Skeletal type: isoflavone *C*-glycoside

$C_{21}H_{20}O_9$ Mol. wt 416.39

Occurs in *Pueraria* spp., e.g. *P. lobata* (Leguminosae).

Hypotensive, and increases coronary resistance. The roots of *P. lobata*, Radix pueraria, have long been used in China to relieve the symptoms of hypertension and angina pectoris.

1575
Rotenone; Tubotoxin; Nicouline

Skeletal type: rotenoid

$C_{23}H_{22}O_6$ Mol. wt 394.43

Occurs in *Derris*, *Tephrosia*, and *Lonchocarpus* spp., and numerous other Leguminosae. Major sources include the roots of *Derris elliptica* and *Piscidia erythrina*. Also, it is found in the leaves of *Verbascum thapsus* (Scrophulariaceae).

Insecticidal activity, and antiprotozoal. It is toxic (LD_{50} intraperitoneally in mice 2.8 mg kg^{-1}), and more toxic to humans when inhaled than when ingested. Rotenone has been used as a fish poison, as well as an insecticide.

1576
Sainfuran

Skeletal type: 2-arylbenzofuran

$C_{16}H_{14}O_5$ Mol. wt 286.29

Found in the roots of *Onobrychis viciifolia* and *Hedysarum polybotrys* (Leguminosae).

Insect antifeedant.

1577
(−)-Sativan; Sativin

Skeletal type: isoflavan

$C_{17}H_{18}O_4$ Mol. wt 286.33

Occurs in the leaves of *Derris amazonica* and the roots of *Lotus pedunculatus*. Also, it is induced as a phytoalexin in the leaves of some *Lotus, Medicago, Trifolium* and *Trigonella* spp. (Leguminosae).

Antifungal activity.

1578
Sojagol

Skeletal type: coumestan

$C_{20}H_{16}O_5$ Mol. wt 336.35

Induced as a phytoalexin in the leaves of soybean, *Glycine max* (Leguminosae).

Oestrogenic activity, and antifungal activity.

1579
Sophoraisoflavanone A

Skeletal type: isoflavanone

$C_{21}H_{22}O_6$ Mol. wt 370.41

Occurs in the aerial parts of *Sophora tomentosa* (Leguminosae).

Antimicrobial activity *in vitro* against *Staphylococcus aureus, Escherichia coli, Bacillus subtilis, Penicillium citrinum* and *Candida albicans*.

1580
(−)-Sparticarpin

Skeletal type: pterocarpan

$C_{17}H_{16}O_5$ Mol. wt 300.32

Induced as a phytoalexin in the leaves of *Spartium junceum* (Leguminosae).

Antifungal activity.

1581
Sumatrol

Skeletal type: rotenoid

$C_{23}H_{22}O_7$ Mol. wt 410.43

Found in the roots of *Derris malaccensis* and *Piscidia erythrina* (Leguminosae).

Insecticidal activity.

1582
Tectoridin; Tectorigenin 7-*O*-glucoside; Shekanin

Skeletal type: isoflavone *O*-glycoside

$C_{22}H_{22}O_{11}$ Mol. wt 462.42

Occurs in the rhizomes of *Iris tectorum* (Iridaceae), the leaves and stems of *Baptisia* spp., and the wood of *Dalbergia riparia* (Leguminosae).

1583
Tectorigenin

Skeletal type: isoflavone

$C_{16}H_{12}O_6$ Mol. wt 300.27

Occurs in the rhizomes of *Iris germanica* (Iridaceae), in *Baptisia* spp., *Dalbergia* spp. and *Ononis spinosa* (Leguminosae). Also, it is induced as a phytoalexin in the leaves of *Centrosema* spp. (Leguminosae).

Antifungal activity.

1584
Tephrosin

Skeletal type: rotenoid

$C_{23}H_{22}O_7$ Mol. wt 410.43

Found in the roots of *Tephrosia elata*, *Amorpha fruticosa*, *Lonchocarpus longifolius* and *L. spruceanus*, and the seeds of *Piscidia mollis* (Leguminosae).

Insect antifeedant.

1585
Texasin

Skeletal type: isoflavone

$C_{16}H_{12}O_5$ Mol. wt 284.27

Found in the leaves of *Baptisia australis*, and in the heartwood of *Platymiscium praecox* (Leguminosae).

1586
Toxicarol; α-Toxicarol

Skeletal type: rotenoid

$C_{23}H_{22}O_7$ Mol. wt 410.43

Found in the roots of *Tephrosia toxicaria* and *Derris elliptica* (Leguminosae).

Insecticidal activity. Also, rotenoids have been used as fish poisons.

1587
(−)-Variabilin; Homopisatin

Skeletal type: pterocarpan

$C_{17}H_{16}O_5$ Mol. wt 300.32

Occurs in the wood of *Dalbergia variabilis*. Also, it is induced as a phytoalexin in *Lens culinaris, L. nigricans, Caragana* spp. and *Lathyrus* spp. (Leguminosae).

Antifungal activity.

1588
Vestitol

Skeletal type: isoflavan

$C_{16}H_{16}O_4$ (+) form Mol. wt 272.30

(+)-Vestitol occurs in the trunk wood of *Machaerium vestitum* and *Dalbergia variabilis* (Leguminosae), whereas (−)-vestitol is induced as a phytoalexin in herbaceous Leguminosae, e.g., some *Lotus, Medicago, Onobrychis* and *Trifolium* spp.

Antifungal activity.

1589
Vestitone

Skeletal type: isoflavanone

$C_{16}H_{14}O_5$ Mol. wt 286.29

Induced as a phytoalexin in the leaves of *Onobrychis viciifolia* and *Tipuana tipu* (Leguminosae).

Antifungal activity.

1590
Wedelolactone

Skeletal type: coumestan

$C_{16}H_{10}O_7$ Mol. wt 314.26

Found in the leaves of *Wedelia calendulacea* (Compositae), and the heartwood of *Ougeinia dalbergioides* (Leguminosae).

Liver protective activity.

1591
Wighteone; Erythrinin B; Lupinus compound LA-1

Skeletal type: isoflavone

$C_{20}H_{18}O_5$ Mol. wt 338.36

Occurs in the leaves and fruit of *Lupinus albus* and other *Lupinus* spp. Also, it is induced as a phytoalexin in *Argyrocytisus battandieri*, *Laburnum anagyroides*, and *Neonotonia wightii* (all Leguminosae).

Antifungal activity.

# Chapter 39

Lignans

Lignans are chemically related to the polymeric lignins of the plant cell wall and are found mainly in woody tissues. They are dimers of the same phenylpropanoid units that are involved in lignin biosynthesis; two such units are linked through the central carbon atoms of the C_3 side chains to give, e.g., nordihydroguaiaretic acid (**1629**). They are assumed to be formed biosynthetically by stereospecific reductive coupling of phenylpropanoids such as coniferyl and sinapyl alcohols, but experimental evidence for such a pathway is still lacking. Three structural classes of oxygenated lignans can be recognized, in addition to simple lignans: the lignanolides with lactone substitution (e.g., matairesinol, **1626**), monoepoxy lignans (e.g., burseran, **1596**), and bisepoxy-lignans (e.g., pinoresinol, **1638**). Further cyclization within the side chain carbons produces a further large class of cyclolignans (e.g., podophyllotoxin, **1640**), which have a third six-membered ring in their structures. Finally, there are a small number of related neolignans (Gottlieb, 1978), which are formed by unsymmetrical carbon–carbon links in the side chains (e.g., austrobailignan, **1595**).

The majority of the 200 or more naturally occurring lignans (Rao, 1978) occur in the free state in heartwood tissue. Some have been isolated from other plant parts, such as root, leaf and flower, and in such cases they may be found in glycosidic combination. Lignans occur widely in the wood of gymnosperm (e.g., pine) trees, and they have also been recorded in some 50 angiosperm families, where they can be found in both the wood and bark of the trees.

Lignans are of considerable pharmacological interest (Ayres and Loike, 1990), chiefly because they have antitumour and antiviral activity. Podophyllotoxin, which is used externally for destroying warts, has some potential as an anticancer drug (Cassady and Douros, 1980), although the mammalian toxicity may limit its application. Certain other lignans have allergenic, cathartic or cardiovascular effects (MacRae and Towers, 1984).

Lignans usually have one or more free phenolic group, and can be analysed by the same methods as are used for other polyphenols (q.v.). Thus, they respond to phenolic colour tests (e.g., the Maüle reaction for syringyl groups), and their ultraviolet spectra exhibit bathochromic shifts with alkali. The great majority contain methoxy groups in their structures, are relatively nonpolar and may be extracted from plant tissues with other lipophilic constituents. They are then separated and characterized by the standard procedures of natural product chemistry (Rao, 1978).

References

Chemistry and Occurrence

Gottlieb, O.R. (1978). Neolignans. *Fortschritte der Chemische Organische Naturstoffe*, **35**, 1–72.
Rao, C.B.S. (ed.) (1978). *Chemistry of Lignans*. Waltair, India: Andhra University Press.

Biological Properties

Ayres, D.C. and Loike, J.D. (1990). *Lignans: Chemical, biological and clinical properties*. Cambridge University Press.
Cassady, J.M. and Douros, J.D. (eds) (1980). *Anticancer Agents Based on Natural Product Models*. New York: Academic Press.
MacRae, W.D. and Towers, G.H.N. (1984). Biological activities of lignans. *Phytochemistry*, **23**, 1207–1220.

1592
Acanthoside D; (−)-Syringaresinol di-β-D-glucoside

$C_{34}H_{46}O_{18}$ (−) form   Mol. wt 742.74

Occurs in the roots of *Acanthopanax sessiliflorus* and in *Eleutherococcus senticosus* (Araliaceae). The (+) isomer, liriodendrin occurs in the bark of *Liriodendrum tulipifera* (Magnoliaceae) and in the whole plant of *Penstemon deustus* (Scrophulariaceae).

Stress-reducing activity.

1593
1-Acetoxypinoresinol

$C_{22}H_{24}O_8$ Mol. wt 416.43

Occurs free and as the 4′-glucoside in the bark of *Olea europaea* (Oleaceae).

Inhibits cyclic adenosine monophosphate phosphodiesterase.

1594
(−)-Arctigenin

$C_{21}H_{24}O_6$ Mol. wt 372.42

Found in *Trachelospermum asiaticum* (Apocynaceae) and in *Ipomoea cairica* (Convolvulaceae). (+)-Arctigenin occurs in the roots of *Wikstroemia indica* (Thymeleaceae) and, as a glycoside, in the fruits of *Arctium lappa, Lappa minor* and *L. tomentosa* (Compositae).

Inhibitor of cyclic adenosine monophosphate phosphodiesterase; it has cytostatic activity in lymphoma cell systems.

1595
Austrobailignan 1

$C_{21}H_{18}O_7$ Mol. wt 382.37

Found in *Austrobaileya scandens* (Austrobaileyaceae), and also in the twigs of *Amyris pinnata* (Rutaceae).

Antitumour activity.

1596
Burseran

$C_{22}H_{26}O_6$ Mol. wt 386.45

Occurs in the stems and leaves of *Bursera microphylla* (Burseraceae).

Antitumour activity.

1597
Cleistanthin A

$C_{28}H_{28}O_{11}$ Mol. wt 540.53

Found in the heartwood of *Cleistanthus collinus* and *C. patulus* (Euphorbiaceae).

Increases neutrophilic granulocyte count and prevents experimentally induced granulocytopenia.

1598
Cubebin

$C_{20}H_{20}O_6$ Mol. wt 356.38

Occurs in the fruit of *Piper cubeba* (Piperaceae) and in the roots and shoots of *Aristolochia triangularis* (Aristolochiaceae).

Indicated for use as a urinary antiseptic, and recently proved to be effective against gut microsomal monooxygenase of the corn borer (*Ostrinia nubilalis*).

1599
Dehydrodieugenol

$C_{20}H_{22}O_4$ Mol. wt 326.40

Found in the bark of *Litsea turfosa* (Lauraceae).

The extracts of the bark of *L. turfosa* show antifungal activity; its metabolites are believed to possess antitumour activity.

1600
5'-Demethoxydeoxypodophyllotoxin;
5'-Desmethoxydeoxypodophyllotoxin; Morelensin

$C_{21}H_{20}O_6$ Mol. wt 368.39

Occurs in the exudate of *Bursera morelensis* (Burseraceae).

Cytotoxic agent, and shows antitumour activity.

1601
4′-Demethyldeoxypodophyllotoxin

$C_{21}H_{20}O_7$ Mol. wt 384.39

Found in the roots of *Polygala paena*, in the stems, leaves and flowers of *P. macradenia* (Polygalaceae), and in *Hyptis verticillata* (Labiatae). Its 4′-β-D-glucoside is found in the roots of *Podophyllum hexandrum* and *P. peltatum* (Podophyllaceae).

Antimitotic, antileukaemic and antitumour activities.

1602
4′-Demethylpodophyllotoxin

$C_{21}H_{20}O_8$ Mol. wt 400.39

Occurs in Indian podophyllin, i.e., the resin of *Podophyllum hexandrum* (Podophyllaceae), and from *Polygala polygaena* (Polygalaceae).

Antitumour, antimitotic, and cathartic activities.

1603
Denudatin B

$C_{21}H_{23}O_5$ Mol. wt 355.42

Found in the flower buds of *Magnolia fargesii* (Magnoliaceae).

Calcium antagonistic activity on taenia coli of guinea-pig.

1604
Deoxygomisin A; Schisandrin B; Wuweizisu B

$C_{23}H_{28}O_6$ Mol. wt 400.48

Found in the fruit of *Schisandra chinensis* (Schisandraceae).

Antihepatotoxic activity.

1605
Deoxypodophyllotoxin; Anthricin; Hernandion; Silicolin

$C_{22}H_{22}O_7$ Mol. wt 398.42

Found in the roots of *Anthriscus sylvestris* (Umbelliferae), in the seed oil of *Hernandia ovigera* (Hernandiaceae), in the needles of *Juniperus sabina* and *J. silicicola*, in the berries of *Juniperus sabina* var. *tamariscifolia*, and in *Libocedrus* spp. (all Cupressaceae). Also, it is found in *Bursera microphylla*, in the exudate of *B. morelensis* (Burseraceae), and in *Callitris collumellaris* (Cupressaceae).

Antitumour, antimitotic and antiviral (against measles and herpes simplex I) activities.

1606
Dihydroanhydropodorhizol

$C_{22}H_{24}O_7$ Mol. wt 400.43

Occurs in the leaves and stems of *Bursera schlechtendalii* (Burseraceae), and in *Juniperus* spp. (Cupressaceae).

Weak antiviral activity against herpes simplex. Also, it shows antitumour activity.

1607
Dihydrocubebin

$C_{20}H_{22}O_6$ Mol. wt 358.40

Found in the bark, leaves and wood of *Horsfieldia iryaghedhi* (Myristicaceae), in the heartwood of *Cleistanthus collinus* (Euphorbiaceae), and in *Piper guineense* (Piperaceae).

Antimicrobial activity against *Mycobacterium smegmatis*.

1608
Diphyllin

$C_{21}H_{16}O_7$ Mol. wt 380.36

Occurs in the aerial parts of *Haplophyllum hispanicum* (Rutaceae), in *Diphylleia grayi* (Podophyllaceae) and in *Justicia hayatai* (Acanthaceae). It also occurs in the heartwood of *Cleistanthus collinus* (Euphorbiaceae).

Antitumour and piscicidal activities.

1609
Eucommin A;
Medioresinol 4'-*O*-β-D-glucopyranoside

$C_{27}H_{34}O_{12}$ Mol. wt 550.57

Occurs in the bark of *Eucommia ulmoides* (Eucommiaceae), and in *Allamanda neriifolia* (Apocynaceae).

Immunomodulating and anticomplementary activities.

1610
(−)-Eudesmin; L-Eudesmin; Pinoresinol dimethyl ether

$C_{22}H_{26}O_6$ Mol. wt 386.45

Found in kino gum, the exudate of *Eucalyptus hemiphloia* (Myrtaceae). (+)-Eudesmin is present in the wood of *Araucaria angustifolia* (Araucariaceae), in the wood of *Humbertia madagascariensis* (Humbertiaceae), and in the leaves of *Evodia micrococca* var. *micrococca* (Rutaceae). The (+) enantiomer also occurs in the bark of *Litsea gracilipes* (Lauraceae), the flower buds of *Magnolia fargesii* (Magnoliaceae) and in *Haplophyllum* spp. (Rutaceae).

Antitubercular activity *in vitro*, and also calcium antagonistic activity on taenia coli of guinea-pig. Kino gum is used as an astringent and antidiarrhoeal agent.

1611
Eudesobovatol A

$C_{33}H_{44}O_4$ Mol. wt 504.72

Found in the bark of *Magnolia obovata* (Magnoliaceae).

Neurotrophic activity on neuronal cell cultures of foetal rat cerebral hemisphere. The bark of *Magnolia* is used for neurosis and gastrointestinal complaints, and it has depressant effects on the central nervous system.

1612
Fargesone A

$C_{20}H_{24}O_6$ Mol. wt 360.41

Occurs in the flower buds of *Magnolia fargesii* (Magnoliaceae), along with its stereoisomer fargesone B.

Both compounds show calcium antagonistic activity on taenia of guinea-pig.

1613
(+)-Galbacin

$C_{20}H_{20}O_5$ Mol. wt 340.38

Found in the leaves of *Machilus japonica* (Lauraceae), in the roots and shoots of *Aristolochia triangularis* (Aristolochiaceae), and in the leaves of *Virola surinamensis* (Myristicaceae).

Active against the H37Rv strain of *Mycobacterium tuberculosis*. *Aristolochia* is part of a drug mixture used by Brazilian Indians against rheumatic diseases.

1614
Gomisin L₁ methyl ether

$C_{23}H_{28}O_6$ Mol. wt 400.48

Occurs in the fruit of *Schisandra chinensis* (Schisandraceae).

Antihepatotoxic activity. *Schisandra chinensis* fruit is used as an antitussive drug.

1615
Grandisin

$C_{24}H_{32}O_7$ Mol. wt 432.52

Found in the bark of *Litsea grandis* (Lauraceae).

Inhibits platelet activating factor, a lipid mediator of hypersensitivity and inflammation, from binding to its receptor site.

1616
Hinokinin; Cubebinolide

$C_{20}H_{18}O_6$ Mol. wt 354.36

Found in hinoki oil, the extract from red, medium parts of hinoki wood, *Chamaecyparis obtusa* (Cupressaceae).

Insecticide synergist.

1617
cis-Hinokiresinol; Nyasol

$C_{17}H_{16}O_2$ Mol. wt 252.32

Occurs in *Chamaecyparis obtusa* (Cupressaceae), *Araucaria angustifolia* (Araucariaceae), and *Anemarrhena asphodeloides* (Liliaceae).

Potent cyclic adenosine monophosphate phosphodiesterase inhibitor.

1618
Honokiol

$C_{18}H_{18}O_2$ Mol. wt 266.34

Found in the bark of *Magnolia officinalis* and of *M. obovata* (Magnoliaceae).

Sedative and muscle relaxant activity. Also, it shows bactericidal activity (against dental caries).

1619
Justicidin B; Dehydrocollinusin

$C_{21}H_{16}O_6$ Mol. wt 364.36

Found in *Phyllanthus acuminatus* (Euphorbiaceae), *Justicia hayatai* var. *decumbens* (Acanthaceae), *Haplophyllum tuberculatum* and *Boenninghausenia albiflora* (Rutaceae), and *Sesbania drummondii* (Leguminosae).

Piscicidal activity. Also, it shows antiviral activity against murine cytomegalovirus and Sindbis virus.

1620
Kadsurenone

$C_{21}H_{24}O_5$ Mol. wt 356.42

Occurs in the stems of *Piper futukadsura* (Piperaceae).

Inhibits the binding of platelet activating factor to its reception site.

1621
Kadsurin A

$C_{21}H_{24}O_6$ Mol. wt 372.42

Occurs in the stems of *Piper futukadsura* (Piperaceae), and in *Kadsura longipedunculata* (Schizandraceae).

Inhibits the binding of platelet activating factor to its reception site.

1622
Licarin A; (+)-*trans*-Dehydrodiisoeugenol

$C_{20}H_{22}O_4$ Mol. wt 326.40

Found in *Licaria aritu* (Lauraceae), and in *Aristolochia maxima* and *A. taliscana* (Aristolochiaceae). (±)-Dehydrodiisoeugenol occurs in the aril of *Myristica fragrans* (Myristicaceae).

Antimicrobial activity.

1623
Magnolol

$C_{18}H_{18}O_2$ Mol. wt 266.34

Found in the crude drug derived from the bark of *Magnolia officinalis* and/or *M. obovata* (Magnoliaceae). It is also found in the roots of *Sassafras randaiense* (Lauraceae), along with isomagnolol.

Active against cariogenic bacteria and has antidepressant effects on the central nervous system; the ether extract of the crude drug derived from *Magnolia* bark has sedative and muscle relaxant activities.

1624
Magnosalicin

and enantiomer

$C_{24}H_{32}O_7$ Mol. wt 432.52

Occurs in the buds of *Magnolia salicifolia* (Magnoliaceae).

Antagonistic to platelet activating factor. Buds of *Magnolia salicifolia* are used as a medicine for nasal allergy and nasal empyema in traditional medicine. Its chloroform extract has an inhibitory effect on histamine release.

1625
Magnoshinin

and enantiomer

$C_{24}H_{30}O_6$ Mol. wt 414.50

Found in the flower buds of *Magnolia salicifolia* (Magnoliaceae).

Anti-inflammatory activity. The flower buds are used for treating headaches and nasal diseases, and as a tranquillizer.

1626
Matairesinol

$C_{20}H_{22}O_6$ Mol. wt 358.40

Found in the heartwood of *Podocarpus spicata* (Podocarpaceae), in some *Abies* spp. and *Picea* spp. (Pinaceae), in *Trachelospermum asiaticum* var. *intermedium* (Apocynaceae), and in some *Heliopsis* spp. (Compositae).

Active as an insecticide synergist; it shows inhibitory activity against cyclic adenosine monophosphate phosphodiesterase.

1627
Megaphone

$C_{22}H_{30}O_6$ Mol. wt 390.48

Occurs in the roots of *Aniba megaphylla* (Lauraceae), together with its acetate.

Cytotoxic activity.

1628
Neoisostegane

$C_{23}H_{26}O_7$ Mol. wt 414.46

Occurs in *Steganotaenia araliacea* (Umbelliferae).

Cytotoxic and antitumour activities.

1629
Nordihydroguaiaretic acid; NDGA

$C_{18}H_{22}O_4$ Mol. wt 302.37

Found in the resinous exudates of *Larrea* spp., and in the resin of the wood from *Guaiacum sanctum* and *G. officinale* (Zygophyllaceae).

Antitumour, antimicrobial, and antifungal activities. The primary use is as an anti-oxidant in fats and oils (now banned in the USA because suspected of inducing kidney cysts). A 2% solution of Guaiac resin was formerly used for chronic rheumatic diseases, and served also as a reagent for detecting blood in faeces.

1630
Nortrachelogenin; Pinopalustrin

$C_{20}H_{22}O_7$ (+) form   Mol. wt 374.40

The (+) form, wikstromol, occurs in the stem of *Passerina vulgaris* and in *Wikstroemia indica* (Thymeleaceae). The (−) form is found in *Trachelospermum asiaticum* var. *intermedium* (Apocynaceae), and in *Pinus palustris* (Pinaceae). The latter form also occurs in *Daphne odora* (Thymeleaceae) and in *Chaerophyllum maculatum* (Umbelliferae).

Cytotoxic and antileukaemic activities.

1631
Otobain; Otobite

$C_{20}H_{20}O_4$ Mol. wt 324.38

Found in otoba fat, from *Dialeyanthera otoba* (=*Myristica otoba*) and *Myristica simarum* (Myristicaceae).

Serves as a fungicide in otoba butter, and it is used in Colombia in veterinary practice.

1632
α-Peltatin

$C_{21}H_{20}O_8$ Mol. wt 400.39

Occurs in the resin of *Podophyllum peltatum* (Podophyllaceae), and as a glucoside in rhizomes of the same species. β-Peltatin, the 4′-methyl ether, accompanies α-peltatin in the rhizomes.

Antitumour, cathartic and antiviral activities (particularly against measles and herpes simplex I).

1633
β-Peltatin A methyl ether

$C_{23}H_{24}O_8$ Mol. wt 428.45

Occurs in podophyllum resin, from *Podophyllum peltatum* (Podophyllaceae). It also occurs in the needles of *Juniperus sabina* (Cupressaceae), and in *Bursera fagaroides* (Burseraceae).

Antitumour activity.

1634
Phrymarolin I

$C_{24}H_{24}O_{11}$ Mol. wt 488.46

Found in the roots of *Phryma leptostachya* (Phrymaceae).

Synergistic activity with pyrethrin and the pesticide sevin. The roots are used in rural regions of Japan (Kyushu) to exterminate houseflies.

1635
Phyllanthin

$C_{24}H_{34}O_6$ Mol. wt 418.54

Found in the leaves of *Phyllanthus niruri* (Euphorbiaceae).

Bitter taste.

1636
Phyllanthostatin A

$C_{29}H_{30}O_{13}$ Mol. wt 586.56

Occurs in the roots of *Phyllanthus acuminatus* (Euphorbiaceae).

Cytostatic activity.

1637
Picropodophyllin; Picropodophyllic acid lactone

$C_{22}H_{22}O_8$ Mol. wt 414.42

Occurs in the resin of *Podophyllum peltatum* (Podophyllaceae), in the roots of *Diphylleia grayi* (Podophyllaceae), and in *Juniperus sabina* (Cupressaceae). It is isomeric with podophyllotoxin (q.v.) and may be formed during the extraction process from podophyllotoxin β-D-glucoside.

Antiviral activity (against measles and herpes simplex type I).

1638
Pinoresinol

$C_{20}H_{22}O_6$ (+) form Mol. wt 358.40

Found in the resin of *Pinus* spp. and of *Picea jezoensis* (Pinaceae), the wood of *Araucaria angustifolia* (Araucariaceae), the splint wood of *Tsuga heterophylla* (Pinaceae), the bark of *Fraxinus mandshurica* var. *japonica* (Oleaceae), and *Wikstroemia* spp. (Thymeleaceae). (+)-Epipinoresinol occurs in the bark of *Eucommia ulmoides* (Eucommiaceae); (−)-epipinoresinol occurs as the glycoside symplocosin in the bark of *Symplocos lucida* (Symplocaceae).

Inhibits cyclic adenosine monophosphate phosphodiesterase. (+)-Epipinoresinol is an immunomodulating agent (with anticomplementary activity).

1639
Plicatic acid

$C_{20}H_{22}O_{10}$ Mol. wt 422.40

Found in the sawdust of *Thuja plicata* (Cupressaceae).

Allergen in sawdust of *Thuja plicata*, causing asthma and rhinitis.

1640
Podophyllotoxin; Podophyllinic acid lactone

$C_{22}H_{22}O_8$ Mol. wt 414.42

Occurs in the rhizomes (together with the 1-glucoside) of *Podophyllum peltatum*, *P. hexandrum*, and *P. pleianthum* (Podophyllaceae). It also occurs in the roots of *Diphylleia grayi* (Podophyllaceae), in the needles of *Juniperus sabina*, in the shoots of *J. virginiana* and in the needles of *Callitris drummondii* (all Cupressaceae).

Antitumour, antimitotic, cathartic, and antiviral (e.g., against measles and herpes simplex type I) activities.

1641
Podophyllotoxone

$C_{22}H_{22}O_8$ Mol. wt 414.42

Found in the roots of *Podophyllum hexandrum* and *P. peltatum*. Its isomer, isopicropodophyllone, described as found in *P. pleianthum* (Podophyllaceae) is probably an artefact.

Cytotoxic activity.

1642
Podorhizol β-D-glucoside

$C_{28}H_{34}O_{13}$ Mol. wt 578.58

Occurs in *Podophyllum peltatum* and *P. hexandrum* (=*P. emodii*) (Podophyllaceae).

Antimitotic activity.

1643
Prostalidin A

$C_{21}H_{14}O_8$ Mol. wt 394.34

Occurs in *Justicia prostrata* (Acanthaceae).

Mild antidepressant activity, and probably responsible for the use of *Justicia* as an antidepressant drug in popular medicine.

1644
Randainol

$C_{18}H_{18}O_3$ Mol. wt 282.34

Found in the roots of *Sassafras randainense* (Lauraceae).

Slight antimicrobial activity against *Bacillus subtilis* and *Staphylococcus aureus*.

1645
Saucernetin

$C_{22}H_{28}O_5$ Mol. wt 372.47

Found in the leaves and stems of *Saururus cernuus* (Saururaceae).

Antagonistic to platelet activating factor.

1646
Savinin; Hibalactone; Taiwanin B

$C_{20}H_{16}O_6$ (−) form   Mol. wt 352.35

Occurs in the leaves of *Juniperus sabina* (Cupressaceae), and in the wood. Also, it is found in the young leaves of *Chamaecyparis obtusa* var. *breviramea*, in the wood of *Libocedrus formosana* (all Cupressaceae), in the roots of *Ruta graveolens*, in the leaves of *R. microcarpa* and in the twigs of *Amyris pinnata* (all Rutaceae). The (−) form accumulates in the roots of *Acanthopanax sessiliflorus* (Araliaceae). As taiwanin B it was isolated from *Taiwania cryptomerioides* (Taxodiaceae).

Insecticide synergist; it is effective as an emmenagogue, anthelmintic, and antirheumatic.

1647
Schisantherin A; Gomisin-C; Wuweizisu ester

$C_{30}H_{32}O_9$ Mol. wt 536.59

Occurs in the fruits of *Schisandra sphenanthera* and *S. chinensis* (Schisandraceae).

Antihepatotoxic activity.

1648
Sesamin

$C_{20}H_{15}O_6$ (+) form   Mol. wt 351.34

(+)-Sesamin occurs in sesame oil, *Sesamum indicum* and spp. (Pedaliaceae), the bark of *Zanthoxylum acanthapodium* (Rutaceae), the seeds of *Piper longum* (Piperaceae) and the wood of *Paulownia tomentosa* (Scrophulariaceae). (−)-Sesamin occurs in the bark of *Zanthoxylum piperitum*, the leaves of *Ruta montana* (both Rutaceae), *Magnolia mutabilis* (Magnoliaceae), the roots of *Asarum sieboldii* (Aristolochiaceae), and *Eleutherococcus senticosus* (Araliaceae). The racemic mixture, called fagarol, has been isolated from *Fagara* spp. (Rutaceae).

Well known as an insecticide synergist.

1649
Sesamolinol

$C_{20}H_{20}O_7$ Mol. wt 372.38

Occurs in the seeds of *Sesamum indicum* (Pedaliaceae).

Antioxidant properties.

1650
Sesartemin

$C_{23}H_{26}O_8$ Mol. wt 430.46

Occurs in the bark of *Virola elongata* (Myristicaceae), along with its isomer, episesartemin. It is also present in the roots of *Artemisia absinthium* (Compositae).

Effective against the gut microsomal monooxygenase of *Ostrinia nubilalis* (corn borer). The bark resin of *Virola elongata* is used as hallucinogenic snuff and also as an arrow poison by various Indian tribes. Isolation-induced aggression in mice is lowered by sesartemin.

1651
Simplexoside; Piperitol glucoside

$C_{26}H_{30}O_{11}$ Mol. wt 518.53

Found in whole plants of *Justicia simplex* (Acanthaceae).

Mild depressant effects on the central nervous system. Its aglycone acts as an antioxidant in sesame seed, and *Justicia* plants are used as an antistress and antifatigue drug.

1652
Steganacin

$C_{24}H_{24}O_9$ Mol. wt 456.46

Occurs in the wood, stems and bark of *Steganotaenia araliacea* (Umbelliferae).

Inhibits HeLa cell growth, and shows antileukaemic and anticancer activities.

1653
Styraxin

$C_{20}H_{18}O_7$ Mol. wt 370.36

Occurs in the aerial parts of *Styrax officinalis* (Styracaceae).

Antitumour activity. *Styrax* is used for manufacturing fumigating pastilles and powders, in perfumery, as a topical protectant, and as a parasiticide in veterinary medicine.

1654
Surinamensin

$C_{22}H_{28}O_6$ Mol. wt 388.47

Occurs in the leaves of *Virola surinamensis* (Myristacaceae) along with its demethoxy derivative, virolin.

Antischistosomal activity.

1655
(+)-Syringaresinol; Lirioresinol B

$C_{22}H_{26}O_8$ Mol. wt 418.45

Occurs in the wood of *Populus* spp. (Salicaceae) and in *Wikstroemia* spp. (Thymeleaceae). The racemate occurs in *Fagus sylvatica* (Fagaceae).

Cytotoxic effects.

1656
(+)-Syringaresinol *O*-β-D-glucoside; Acanthoside B

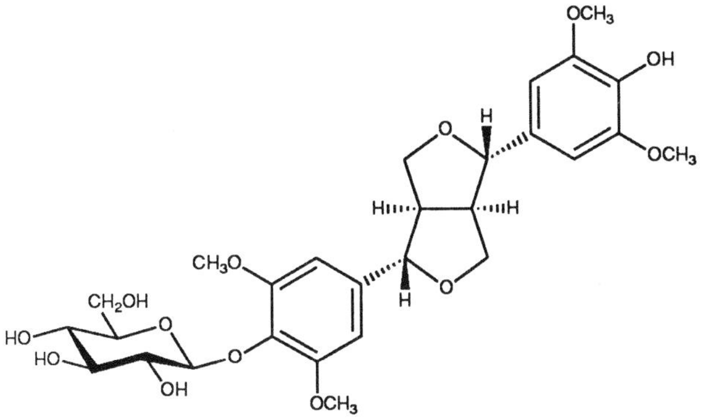

$C_{28}H_{36}O_{13}$ Mol. wt 580.60

Found in the bark of *Eucommia ulmoides* (Eucommiaceae), and in the roots of *Acanthopanax sessiliflorus* (Araliaceae).

Immunomodulating activity.

1657
Trachelogenin

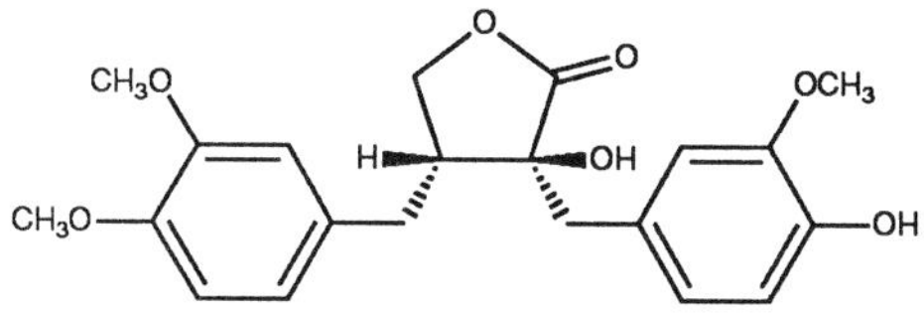

$C_{21}H_{24}O_7$ Mol. wt 388.42

Found in *Trachelospermum asiaticum* var. *intermedium* (Apocynaceae), and in *Ipomoea cairica* (Convolvulaceae).

Calcium antagonistic activity affecting cardiac disorders and hypertension, and cytostatic activity in lymphoma cell systems.

1658
(+)-Veraguensin

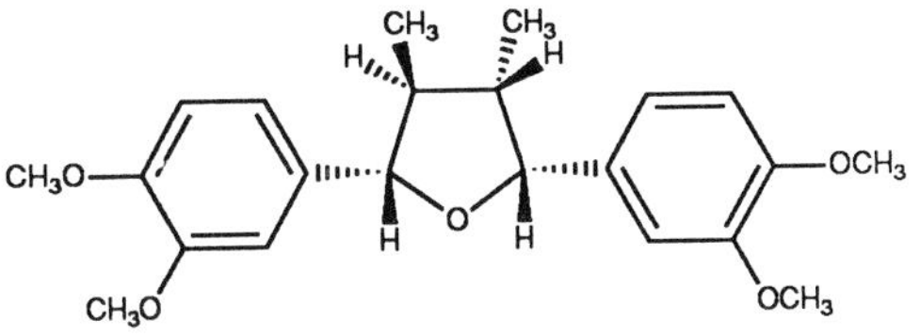

$C_{22}H_{28}O_5$ Mol. wt 372.47

Occurs in the wood of *Ocotea veraguensis* (Lauraceae), the leaves of *Trimenia papuana* (Trimeniaceae), the root bark of *Magnolia acuminata* and the leaves and twigs of *M. liliflora*, the leaves and stems of *Saururus cernuus* (Saururaceae), and the leaves of *Virola surinamensis* (Myristicaceae).

Antagonistic to platelet activating factor.

1659
Wuweizisu C

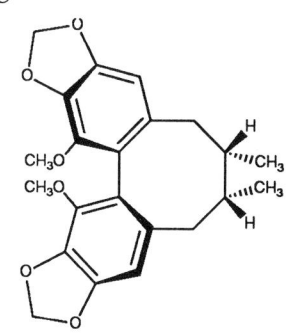

$C_{22}H_{24}O_6$ Mol. wt 384.43

Occurs in the fruit of *Schisandra chinensis* and of *Kadsura longipedunculata* (Schisandraceae).

Antihepatotoxic activity. The plants are used as a treatment for ulcers.

1660
Yangambin

$C_{24}H_{30}O_8$ Mol. wt 446.50

Occurs in the bark of *Virola elongata* (Myristicaceae), along with its isomer, epiyangambin. It also the roots of *Artemisia absinthium* (Compositae).

Lowers isolation-induced aggression in mice. The bark resin of *Virola elongata* is used as hallucinogenic snuff and as an arrow poison by various Indian tribes.

Chapter 40
Phenols and Phenolic Acids

The free phenols (e.g., phenol itself, **1690**), and the phenolic acids (e.g., gentisic acid, **1675**) are best considered together, since they are usually identified together during plant analysis. The acid hydrolysis of plant tissues releases a number of ether soluble phenolic acids, some of which are universal in their distribution. These acids are either associated with lignin at the cell wall or are present elsewhere in the alcohol-insoluble fraction of the leaf; alternatively, they may be present in the alcohol-soluble fraction in glycosidic or ester combination. Universal in angiosperm plants are *p*-hydroxybenzoic acid (**1684**), protocatechuic acid (**1694**), vanillic acid (**1708**) and syringic acid (**1700**). Gentisic acid (2,5-dihydroxybenzoic, **1675**) is relatively widespread, whereas salicylic (2-hydroxybenzoic, **1698**) is uncommon. Gallic acid (**1674**) should also be mentioned, since it is often present in woody plants, in bound form as gallotannin. Also present in acid hydrolysed plant extracts of woody plants is a dimeric condensation product of gallic acid, namely ellagic acid (**1673**), formed from ellagitannins present in direct extracts (see tannins).

By contrast with the phenolic acids, free phenols are relatively rare in plants. Hydroquinone is probably the most widely distributed, occurring for example in a number of plants as the glucoside arbutin (**1664**), named after one of its typical sources *Arbutus unedo*. Other simple phenols such as catechol (**1670**), phloroglucinol (**1691**) and pyrogallol (**1695**) are reported from only a relatively few sources. These simple phenols can occur substituted with aliphatic side chains, and such compounds are almost invariably irritants, responsible for the allergic reactions caused in humans handling such plants. Urushiol III (**1707**), an alkyl catechol, is one of the active principles of the notorious poison ivy plant, which causes much irritation to hikers and other country lovers in North America. Other alkyl phenols cause contact dermatitis, e.g., anacardic acid (**1661**).

A resorcinol derivative well known for its biological activity is Δ^1-tetrahydrocannabinol (**1701**), the hallucinogenic principle of marihuana or cannabis. What is interesting here is that closely related structures, e.g., cannabidiol (**1667**), also present in the cannabis plant, are inactive (Joyce and Curry, 1970). One general property of all phenols is their antimicrobial activity, and phenol itself was the earliest known antiseptic to be used in surgery.

Phenols and phenolic acids are readily detected in plant extracts by thin layer chromatography. They appear as dark absorbing spots in short ultraviolet light, and can be detected specifically by the colour produced after spraying with diazotized *p*-nitroaniline. They also undergo characteristic bathochromic shifts in their UV spectra in the presence of alkali. Their characterization in plants has recently been reviewed by van Sumere (1989).

References

Natural Distribution

Harborne, J.B. and Simmonds, N.W. (1964). The natural distribution of the phenolic aglycones. In *Biochemistry of Phenolic Compounds* (Harborne, J.B.) (ed.), pp. 77–128, London: Academic Press.

Joyce, C.R.B. and Curry, S.H. (eds) (1970). *The Botany and Chemistry of Cannabis*. London: J. & A. Churchill.

Analysis

van Sumere, C.F. (1989). Phenols and phenolic acids. In *Methods in Plant Biochemistry*, Volume 1, *Plant Phenolics* (Harborne, J.B.) (ed.), pp. 29–74. London: Academic Press.

1661
Anacardic acid

$C_{22}H_{36}O_3$ Mol. wt 348.53

Found in the shell liquid of *Anacardium occidentale*, cashew nut (Anacardiaceae). One of four closely related 6-alkylated 2-hydroxybenzoic acids in the shell of the cashew nut (see also ginkgoic acid).

Prostaglandin synthase (enzyme of arachidonic acid metabolism) inhibitor and with antitumour activity. It is responsible for the acute dermatitis caused in humans handling these nuts.

1662
***p*-Anisaldehyde;** Anisic aldehyde; 4-Methoxybenzaldehyde

$C_8H_8O_2$ Mol. wt 136.15

Occurs in the essential oil of the fruit of *Pelea madagascariensis* (Rutaceae), *Agastache rugosa* (Labiatae) and of the leaves of *Magnolia salicifolia* (Magnoliaceae). It also occurs in the essential oils of *Vanilla* spp. (Orchidaceae), *Acacia* spp. (Leguminosae), *Pinus* spp. (Pinaceae), *Cassia* spp. (Leguminosae), *Pimpinella anisum* (Umbelliferae) and *Illicium verum* (Illiciaceae).

Fungistatic activity. Its odour resembles that of coumarin (q.v.); it is used in perfumery and toilet soaps, and in organic syntheses.

1663
Antiarol

$C_9H_{12}O_4$ Mol. wt 184.20

Occurs in the latex of *Antiaris toxicaria* (Moraceae).

1664
Arbutin; Hydroquinone-β-D-glucopyranoside

$C_{12}H_{16}O_7$ Mol. wt 272.26

Found in the leaves of *Arctostaphylos uva-ursi*, *Vaccinium vitis-idaea* (both Ericaceae), *Bergenia crassifolia* (Saxifragaceae), *Pyrus communis* (Rosaceae) and *Origanum majorana* (Labiatae).

Diuretic, antitussive and urinary anti-infective activities. It is easily hydrolysed; the gallotannins in crude extracts prevent hydrolysis, and thus crude extracts are more effective. It inhibits insulin degradation.

1665
Benzoic acid; Dracyclic acid; Phenylformic acid

$C_7H_6O_2$ Mol. wt 122.13

Found in *Dalbergia cochinchinensis* and *D. spruceana* (Leguminosae), *Paeonia albiflora* (Paeoniaceae), and *Uvaria angolensis* (Annonaceae).

Antifungal and choleretic activities. It is used for preserving foods, in the manufacture of dyes, and for curing tobacco. Also, it accumulates as a phytoalexin in apple fruit.

1666
Bilobol; (15:1)-Cardol

$C_{21}H_{34}O_2$ Mol. wt 318.50

Found in the sarcotesta of *Ginkgo biloba* fruit (Ginkgoaceae), and in *Anacardium occidentale* and *Schinus terebinthifolius* (Anacardiaceae).

Antitumour activity against Sarcoma 180 ascites in mice, and a weak antimicrobial agent.

1667
Cannabidiol

$C_{21}H_{30}O_2$ Mol. wt 314.47

Occurs in *Cannabis sativa* var. *indica* (Cannabaceae), in hashish (extract of female flowers) and in marihuana (dried tips of shoots).

Antimicrobial and antibacterial activities. It shows no hallucinogenic effects.

1668
Cannabidiolic acid

$C_{22}H_{30}O_4$ Mol. wt 358.48

Occurs in *Cannabis sativa* var. *indica* (Cannabaceae), in hashish (extract of female flowers) and marihuana (dried tips of shoots).

Sedative properties, but not a hallucinogen.

1669
(15:1)-Cardanol; Ginkgol

$C_{21}H_{34}O$ Mol. wt 302.50

Found in the fruit of *Schinus terebinthifolius*, pink pepper (Anacardiaceae).

Antitumour activity against Sarcoma 180 ascites, and weak antimicrobial activity. It is a skin irritant, and inhibits cyclo-oxygenase and 5-lipoxygenase. It is responsible for the reported toxic effects of pink pepper.

1670
Catechol; Pyrocatechol; Pyrocatechin; 1,2-Dihydroxybenzene

$C_6H_6O_2$ Mol. wt 110.11

Rarely present in plants, but found in the leaves of *Populus* spp. (Salicaceae), in grapefruit, *Citrus paradisi* (Rutaceae) and avocado, *Persea americana* (Lauraceae), and in the leaves of *Gaultheria* spp. (Ericaceae), as the monoglucoside.

Convulsive agent, but may cause eczematous dermatitis; it has been used as a topical antiseptic. Also, it is used in photography, and for dyeing.

1671
p-Cresol; 4-Methylphenol

C_7H_8O Mol. wt 108.14

Occurs in the leaves of *Morus* spp. (Moraceae), in the oil of wood of *Chamaecyparis formosensis* (Cupressaceae), and in the essential oil of *Pimpinella anisum* (Umbelliferae).

Disinfectant agent. Cresols are used in veterinary practice as local antiseptics, parasiticides and disinfectants.

1672
2,6-Dimethoxyphenol

$C_8H_{10}O_3$ Mol. wt 154.17

Found in the stems of *Mucuna birdwoodiana* (Leguminosae).

Inhibits prostaglandin synthase and has an inhibitory effect on rabbit platelet aggregation. The plant source is used in Chinese medicine, promoting blood circulation or relieving stasis.

1673
Ellagic acid; Benzoaric acid; Lagistase

$C_{14}H_6O_8$ Mol. wt 302.20

Occurs free or combined in the plant galls or leaves of many plant families, mostly within the orders Rosidae, Dilleniidae and Hamamelididae. In many cases, it is obtained as a hydrolysis product of ellagitannins (see tannins).

Haemostatic activity. Also, it is a potent antagonist of the mutagenicity of various aromatic hydrocarbons, and a possible prototype of a new class of cancer-preventing drugs.

1674
Gallic acid; 3,4,5-Trihydroxybenzoic acid

$C_7H_6O_5$ Mol. wt 170.13

Relatively widespread occurrence: e.g., *Allanblackia floribunda* and *Garcinia densivenia* (Clusiaceae), *Bridelia micrantha* (Euphorbiaceae), *Caesalpinia sappan* (Caesalpiniaceae), *Dillenia indica* (Dilleniaceae), *Diospyros cinnabarina* (Ebenaceae), *Paratecoma peroba* (Bignoniaceae), *Psidium guajava* and *Syzygium cordatum* (Myrtaceae), *Rhus typhina* (Anacardiaceae), *Tamarix nilotica* (Tamaricaceae) and *Vitis vinifera* (Vitaceae). This is the parent compound of the gallotannins (see tannins).

Antibacterial, antiviral, and antifungal activities; also it shows anti-inflammatory, antitumour, anti-anaphylactic, antimutagenic, choleretic and bronchodilatory activities. It inhibits insulin degradation, and promotes smooth muscle relaxation. Former uses were as an astringent and styptic and, in veterinary practice, as an intestinal astringent. It is a natural inhibitor of flowering present in *Kalanchoe* leaves.

1675
Gentisic acid; 2,5-Dihydroxybenzoic acid; 5-Hydroxysalicylic acid

$C_7H_6O_4$ Mol. wt 154.13

Widespread occurrence: e.g., in the leaves and roots of *Citrus* cultivars (Rutaceae), in the fruit peels of *Vitis vinifera* (Vitaceae), in the tubers of *Helianthus tuberosus* (Compositae), and in the leaves of *Sesamum indicum* (Pedaliaceae). Further sources are *Gentiana* spp. (Gentianaceae), *Pterocarpus santalinus* (Leguminosae), *Eucalyptus grandis* (Myrtaceae), and the Saxifragaceae.

Antibacterial and antiviral activities. It inhibits conidial germination and mycelial growth of *Cryptomeria cubensis* (the cause of a canker disease in eucalypt). The sodium salt shows analgesic and antirheumatic activities.

1676
Geranylhydroquinone

$C_{16}H_{22}O_2$ Mol. wt 246.34

Occurs in the trichomes of *Phacelia ixodes* and *P. crenulata* (Hydrophyllaceae), and in *Cordia eleagnoides* (Boraginaceae).

Antibacterial activity. It shows cancer prevention activities in animals, and causes contact allergies in humans.

1677
Ginkgoic acid; Ginkgolic acid; Romanicardic acid

$C_{22}H_{34}O_3$ Mol. wt 346.51

Occurs in the fruit of *Ginkgo biloba* (Ginkgoaceae) and in the shell liquid of *Anacardium occidentale* (Anacardiaceae).

Antitumour, antimicrobial and molluscicidal activities. It inhibits prostaglandin synthase.

1678
β-Glucogallin; Glucogallic acid; 1-Galloyl-β-D-glucose

$C_{13}H_{16}O_{10}$ Mol. wt 332.27

Found in the flowers of *Rumex tianschanicus*, in *Rheum officinale* (Polygonaceae), and in the flowers of *Paeonia lactiflora* (Paeoniaceae).

1679
Grevillol

$C_{19}H_{32}O_2$ Mol. wt 292.47

Occurs in *Grevillea robusta* and other Proteaceae.

Inhibits 5-lipoxygenase (an enzyme of arachidonic acid metabolism). It is a strong skin irritant, causing contact allergy.

1680
Guaiacol; 2-Methoxyphenol; Methylcatechol

$C_7H_8O_2$ Mol. wt 124.14

Occurs in beechwood tar, *Betula* sp. (Betulaceae), in guaiac resin, *Guaiacum* (Zygophyllaceae), and in various plant oils and saps, e.g., of celery seed, *Apium graveolens* (Umbelliferae).

Expectorant in veterinary practice, nowadays mostly in the form of esters. Earlier it was applied externally to treat eczema and similar skin diseases.

1681
5-(Heptadec-12-enyl)resorcinol

$C_{23}H_{38}O_2$ Mol. wt 346.56

Occurs in the unripe fruit peel of mango, *Mangifera indica* (Anacardiaceae).

Antifungal activity against *Alternaria alternata*, a fungus causing black spot disease of mango.

1682
Hydroquinone; Hydroquinol; Arctuvin; 1,4-Benzenediol; Pyrogentisic acid

$C_6H_6O_2$ Mol. wt 110.11

Found in the leaves of *Protea mellifera* (Proteaceae), the flowers and leaves of *Vaccinium vitis-idaea*, the leaves of *Arbutus unedo* (both Ericaceae), and in the leaf buds of *Pyrus communis* (Rosaceae). Also, it is found in the kernels of *Xanthium canadense* (Compositae), in *Decussocarpus wallichianus* (Podocarpaceae), in the heartwood of *Pinus resinosa* (Pinaceae), and in the essential oils of *Pimpinella anisum*, and *Petroselinum* spp. (Umbelliferae).

Antibacterial, antitumour, antimitotic and hypertensive activities. It is cytotoxic to rat hepatoma cells. Uses include a depigmentor, an antioxidant, and a photographic reducer and developer.

1683
***p*-Hydroxybenzaldehyde;** 4-Formylphenol

$C_7H_6O_2$ Mol. wt 122.13

Widespread occurrence in small amounts: e.g., in *Plocama pendula* (Rubiaceae), and *Pterocarpus marsupium* (Leguminosae). As a hydrolytic product it occurs also in *Lycopodium* spp. (Lycopodiaceae).

1684
***p*-Hydroxybenzoic acid**

$C_7H_6O_3$ Mol. wt 138.13

Very widespread occurrence: e.g., in *Fagara macrophylla* and *Zanthoxylum rubescens* (Rutaceae), in *Paratecoma peroba* and *Tabebuia impetiginosa* (Bignoniaceae), in *Pterocarpus santalinus* (Leguminosae) and in *Vitis vinifera* (Vitaceae).

Antimutagenic and antisickling activities. It stimulates prostaglandin synthase. It is used in organic syntheses, and as an intermediate for dyes and fungicides.

1685
2-Hydroxymethylbenzoic acid

$C_8H_8O_3$ Mol. wt 152.15

Found in the roots of *Fagara zanthoxyloides* (Rutaceae).

Antisickling activity.

1686
Leiocarposide

$C_{27}H_{34}O_{16}$ Mol. wt 614.57

Occurs in *Solidago virgaurea* (Compositae).

Analgesic and antiphlogistic activities.

1687
4-Methylcatechol

$C_7H_8O_2$ Mol. wt 124.14

Occurs in the sapwood of *Picea abies* (Pinaceae) as a phytoalexin.

Inhibits fungal growth.

1688
Orcinol; Orcin; 5-Methylresorcinol

$C_7H_8O_2$ Mol. wt 124.14

Occurs in the flowering tops of *Dittrichia viscosa* (Compositae) and in the leaves of *Erica arborea* (Ericaceae). It also occurs in many species of lichens, but probably as an artefact, formed during extraction from the depsides present.

A component of the antimicrobially active fraction of *Dittrichia viscosa*. It is used as a reagent for pentoses, lignin, beet sugar, etc.

1689
Phenethyl alcohol; Benzyl carbinol; β-Hydroxyethylbenzene

$C_8H_{10}O$ Mol. wt 122.17

Found in the essential oils of *Citrus* spp., *Zanthoxylum hamiltonium* (Rutaceae), in oils of *Heracleum canescens*, *Ligusticum elatum* and *Petroselinum crispum* (all Umbelliferae). Also, it is found in *Pinus* spp. (Pinaceae), in the wood of *Populus tremuloides* (Salicaceae), in *Piper longum* (Piperaceae), *Rosa rugosa* (Rosaceae) and *Tagetes minuta* (Compositae).

Antimicrobial activity. It is used in flavours and perfumery, particularly in rose perfumes.

1690
Phenol; Carbolic acid; Phenylic acid; Hydroxybenzene

C_6H_6O Mol. wt 94.11

Occurs in the essential oil of *Perovskia angustifolia* (Labiatae), in the fruit of *Paedera chinensis* (Rubiaceae), in *Elscholtzia nipponica* (Labiatae), in *Gossypium mexicanum* (Malvaceae) and in *Paeonia albiflora* (Paeoniaceae). Also, it occurs in the wood of *Populus tremuloides* (Salicaceae), and in the leaves of *Morus* spp. (Moraceae).

Toxic to humans. Aqueous solutions are used as a topical anaesthetic; it is antiseptic and antipyruvetic. It is used also in veterinary practice.

1691
Phloroglucinol; Phloroglucin; 1,3,5-Benzenetriol; 1,3,5-Trihydroxybenzene

$C_6H_6O_3$ Mol. wt 126.11

Rare occurrence: reported from the cones of *Sequoia sempervirens* and *S. gigantea* (Taxodiaceae), *Allium cepa* (Liliaceae) and *Lychnis dioica* (Caryophyllaceae). The glucoside, phlorin, occurs in *Citrus paradisi*, *C. sinensis* and *C. limon* (Rutaceae).

Antispasmodic activity. It is used as a reagent for pentoses, lignin, turpentine oil, and free HCl in gastric juice.

1692
Piperonal; Heliotropin

$C_8H_6O_3$ Mol. wt 150.14

Found in the essential oils of the flowers of *Robinia pseudacacia* (Leguminosae), of the leaves of *Doryphora sassafras* (Monimiaceae), and of *Eryngium potericum* (Umbelliferae). Also, it is found in *Heliotropium* spp. (Boraginaceae), in *Vanilla* spp. (Orchidaceae), in extracts of *Viola* spp. (Violaceae), and in *Baccharis rosmarinifolia* (Compositae).

Used in perfumery, in cherry and vanilla flavours, in organic synthesis, and as a pediculicide.

1693
Populin; Populoside; Salicin benzoate

$C_{20}H_{22}O_8$ Mol. wt 390.40

Occurs in the bark and leaves of *Populus tremula* and *P. nigra*, and in the leaves of *P. grandidentata* and *P. tremuloides* (Salicaceae).

1694
Protocatechuic acid

$C_7H_6O_4$ Mol. wt 154.13

Found in *Erica australis* (Ericaceae), *Rosa canina* (Rosaceae), *Picea koraiensis* (Pinaceae), and in several gymnosperms. Also, it is found in *Eucalyptus grandis* (Myrtaceae), in *Picrorhiza kurroa* (Scrophulariaceae), and in ferns.

Antifungal, antihepatotoxic and anti-inflammatory activities. It decreases myocardial oxygen consumption, and leads to myocardial ischaemic improvement. It stimulates prostaglandin synthase. Also, it inhibits the growth of a fungus causing a canker disease in eucalypt.

1695
Pyrogallol; Pyrogallic acid; 1,2,3-Trihydroxybenzene; 1,2,3-Benzenetriol

$C_6H_6O_3$ Mol. wt 126.11

Found in the fruit of *Ceratonia siliqua* (Leguminosae), in the roots of *Statice gmelinii* (Plumbaginaceae), in the roots of *Rheum maximoviczii* (Polygonaceae), in the leaves of *Phyllanthus reticulatus* (Euphorbiaceae), and in those of *Rosa* spp. (Rosaceae). Also, it is found in *Geum urbanum* and *Rubus rigidus* (both Rosaceae).

Antifungal, antimutagenic and antiyeast activities, but little excitatory activity on the central nervous system; it inhibits insulin degradation. It is used as a developer in photography, for the manufacture of dyes, and the staining of leather. Formerly, it was used as an antipsoriatic.

1696
Resorcinol; Resorcin

$C_6H_6O_2$ Mol. wt 110.11

Occurrence rare in nature: e.g., in the needles of *Pinus rigida* (Pinaceae) and, as the dimethyl ether, in the seeds of *Eugenia jambolana* (Myrtaceae).

Keratolytic and antiseborrhoeic agent and, in veterinary practice, a topical antipruritic and antiseptic agent. It is used for tanning, for the manufacture of dyes and explosives, and in cosmetics.

1697
Salicin; Salicoside; Saligenin-β-D-glucopyranoside

$C_{13}H_{18}O_7$ Mol. wt 286.29

Found in the bark of *Populus* spp. and *Salix* spp., and in the female flowers of *Salix* sp. (Salicaceae). Also, it is found in the root bark of *Viburnum prunifolium* (Caprifoliaceae).

Used as an analgesic and, formerly, in veterinary practice, as a bitter stomachic, antirheumatic and analgesic.

1698
Salicylic acid; 2-Hydroxybenzoic acid

$C_7H_6O_3$ Mol. wt 138.13

Occurs free in the upper spadix of the voodoo lily, *Sauromatum guttatum* (Araceae), and as the methyl ester in wintergreen leaves, *Gaultheria procumbens* (Ericaceae) and in the bark of birch, *Betula lenta* (Betulaceae).

In the voodoo lily, it triggers off heat production required for successful fly pollination in this plant. The acetyl derivative of salicylic acid is widely used as a mild painkiller. The free acid has been used as a food preservative and, in medicine, as a topical keratolytic. It can cause skin rashes in sensitive people.

1699
Sesamol

C₇H₆O₃ Mol. wt 138.13

Occurs in the oil of *Sesamum indicum* (Pedaliaceae).

Causes allergic skin reactions in humans.

1700
Syringic acid

C₉H₁₀O₅ Mol. wt 198.18

Widespread occurrence: e.g., in the wood of *Catalpa ovata* (Bignoniaceae); in the leaves of *Impatiens balsamina* (Balsaminaceae), in the root bark of *Ceanothus americanus* (Rhamnaceae), in the leaves and roots of *Citrus* cultivars (Rutaceae), and in the flour of soya bean, *Glycine max* (Leguminosae). Also, it occurs in the leaves of various Saxifragaceae.

1701
Δ¹-Tetrahydrocannabinol; Δ⁹-Tetrahydrocannabinol; Dronabinol

C₂₁H₃₀O₂ Mol. wt 314.47

Occurs in the resin of *Cannabis sativa* (Cannabaceae), and in marihuana (dried tips of shoots).

Active principle of marihuana: anti-inflammatory, anti-emetic and hallucinogenic properties. It is used topically in hypertensive glaucoma. A sterically undefined tetrahydrocannabinol has exhibited antiviral activity against herpes simplex 1 and 2.

1702
Theogallin; 3-Galloylquinic acid

C₁₄H₁₆O₁₀ Mol. wt 344.28

Found in the unprocessed leaves of *Camellia sinensis* (Theaceae).

1703
Trichocarpin

C₂₀H₂₂O₉ Mol. wt 406.40

Found in the bark of *Populus trichocarpa* (Salicaceae).

Fungistatic activity against the phytopathogen *Dothichiza populea*, thus acting as a resistance factor in the bark.

1704
3,3′,4-Tri-O-methylellagic acid

$C_{17}H_{12}O_8$ Mol. wt 344.28

Occurs frequently in genera of the Melastomataceae and Combretaceae, as well as in several Myrtaceae, e.g., *Eugenia*, and was recently isolated from the roots of *Sanguisorba officinalis* (Rosaceae).

Antihaemorrhagic principle of *Sanguisorba officinalis*, a plant used as an antihaemorrhagic agent and as an analgesic and an astringent in Chinese folk medicine.

1705
Turgorin; PLMF 6

$[C_{13}H_{15}O_{13}S]^-$ Mol. wt 411.33

Occurs in *Mimosa pudica* as one of the turgorins. Further sources are *Acacia karroo* and spp., *Albizia julibrissin*, *Gleditsia triacanthos*, and *Robinia pseudacacia* (all Leguminosae), *Oxalis acetosella* (Oxalidaceae), and *Abutilon grandiflorum* (Malvaceae).

One of a number of chemically related plant hormones called turgorins. It is responsible for the induction of leaf movements after perception of external stimulus.

1706
Turricolol E

$C_{21}H_{30}O_3$ Mol. wt 330.47

Occurs in the glandular secretions of *Turricula parryi* (Hydrophyllaceae).

Induces allergic skin reactions and dermatitis in humans.

1707
Urushiol III

$C_{21}H_{32}O_2$ Mol. wt 316.49

Occurs in the irritant oil of poison ivy, *Toxicodendron radicans* (Anacardiaceae). It is one of five closely related alkylcatechols present in this plant.

Provokes heavy allergic skin reactions and inhibits arachidonic acid metabolism. The urushiols are used as anti-allergic agents in hyposensitization therapy.

1708
Vanillic acid

$C_8H_8O_4$ Mol. wt 168.15

Occurs in *Fagara* spp. (Rutaceae), *Alnus japonica* (Betulaceae), *Eleagnus pungens* (Eleagnaceae), *Erica australis* (Ericaceae), *Gossypium mexicanum* (Malvaceae), *Melia azedarach* (Meliaceae), *Panax ginseng* (Araliaceae), *Paratecoma koraiensis* (Bignoniaceae), *Pterocarpus santalinus*, *Rosa canina* (Rosaceae), *Picrorhiza kurroa* (Scrophulariaceae) and *Trachelospermum asiaticum* (Umbelliferae).

Antisickling and anthelmintic activities. *In vitro* tests indicate anti-inflammatory activity.

1709
Vanillin; Methylprotocatechuic aldehyde

$C_8H_8O_3$ Mol. wt 152.15

Widespread occurrence in small amounts, often present together with the glucoside, vanilloside. Found in the pods of *Vanilla planifolia* (Orchidaceae), the bulbs of *Dahlia* spp. (Compositae), the sprouts of *Asparagus* spp. (Liliaceae), and in the beets, *Beta* spp. (Chenopodiaceae). Also, it is found in the essential oils of *Syzygium aromaticum* (Myrtaceae), *Ruta* spp. (Rutaceae), *Spiraea* spp. (Rosaceae) and *Gymnadenia* spp. (Orchidaceae).

Antifungal activity. It is used as a flavouring agent in confectionery, beverages, foods and perfumery, and as a pharmaceutical aid.

Chapter 41
Phenolic ketones

This section includes two related groups of plant phenolics, acetophenones and phloroglucinol derivatives. Most of them could be formed by either of two biosynthetic pathways: from the condensation of four acetate units or by loss of carbon dioxide from a phenylpropanoid (C_6-C_3) precursor, giving in both cases a C_6-C_2 structure. The acetophenones are simple molecules, varying mainly in the number of hydroxyl, methoxyl or other substituents. Picein (**1728**), the glucoside of *p*-hydroxyacetophenone, is a typical member and occurs fairly widely in nature.

The phloroglucinol derivatives listed here are also ketonic but, in general, have more complex structures. Some phloroglucinol derivatives with isopentenyl substitution occur characteristically in hops, *Humulus lupulus*, e.g., humulone (**1720**) and lupulone (**1722**), and provide the bitter flavour of beer. Other acylphloroglucinol derivatives where two (aspidin, **1715**) or three (filixic acid BBB, **1719**) aromatic moieties are linked together occur in a number of ferns, especially in *Aspidium* and *Dryopteris* spp. (Berti and Bottari, 1968). The most complex structure of this group is rottlerin (**1729**), which is a phloracetophenone derivative with a chromene (q.v.) moiety attached.

Some of these phenolic ketones have useful medicinal properties, and several are employed as anthelmintics.

References

Chemistry

Harborne, J.B. (ed.) (1964). *Biochemistry of Phenolic Compounds*. London: Academic Press.

Distribution

Berti, G. and Bottari, F. (1968). Constituents of ferns. *Progress in Phytochemistry* **1**, 589.

1710
Acetophenone; Acetylbenzene; Hypnone;
1-Phenylethanone

C_8H_8O Mol. wt 120.15

Occurs in the essential oils of *Cistus ladaniferus* and *C. creticus* (Cistaceae), of *Stirlingia latifolia* (Proteaceae), *Orthodon linalooliferum* (Labiatae), in the buds of *Populus balsamifera* (Salicaceae) and in *Urtica dioica* (Urticaceae).

Hypnotic activity. It is used in perfumery to impart an orange-blossom-like odour. Also, it is used as a catalyst for the polymerization of olefins.

1711
3-Acetyl-6-methoxybenzaldehyde

$C_{10}H_{10}O_3$ Mol. wt 178.19

Occurs in the leaves of *Encelia farinosa* (Compositae).

Toxic to other plants.

1712
Agrimol C

$C_{39}H_{50}O_{12}$ Mol. wt 710.83

Found in the crude drug of *Agrimonia pilosa* (Rosaceae) along with two related derivatives, agrimol G and F.

All three compounds show antibacterial activity.

1713
Agrimophol

$C_{26}H_{34}O_8$ Mol. wt 474.56

Found in the roots of *Agrimonia pilosa* (Rosaceae).

Antibacterial activity.

1714
Apocynin; Acetovanillone;
4-Hydroxy-3-methoxyacetophenone

$C_9H_{10}O_3$ Mol. wt 166.18

Found in the rhizomes of *Apocynum cannabinum* and *A. androsaemifolium* (Apocynaceae), and in the essential oil of rhizomes of *Iris* spp. (Iridaceae), and in the bulbs of *Buphane disticha* (Amaryllidaceae), *Echinocereus engelmannii* and *Mammillaria runyonii* (both Cactaceae).

1715
Aspidin; Polystichin

$C_{25}H_{32}O_8$ Mol. wt 460.53

Occurs in the roots of the fern *Dryopteris austriaca* and in various *Aspidium* spp. (Pteridophyta).

Anthelmintic activity.

1716
Aspidinol

$C_{12}H_{16}O_4$ Mol. wt 224.26

Found in the ferns *Dryopteris austriaca, Aspidium filix-mas* and *Athyrium filix-femina* (Pteridophyta).

Anthelmintic activity.

1717
3′,4′-Dihydroxyacetophenone

$C_8H_6O_3$ Mol. wt 150.14

Occurs in *Ilex pubescens* var. *glaber* (Aquifoliaceae), in the needles of *Picea maximowiczii* and, as the 3-O-glucoside, pungenin (pungenoside) in the needles of *P. pungens* (Pinaceae).

Inhibits platelet aggregation *in vitro*.

1718
2′,6′-Dimethoxy-4′-hydroxyacetophenone; 4′-Hydroxy-2′,6′-dimethoxyacetophenone

$C_{10}H_{12}O_4$ Mol. wt 196.21

Found as the glucoside in the bulbs of *Pancratium biflorum* (Amaryllidaceae).

Inhibits prostaglandin synthetase, 5-lipoxygenase, and the viability of Ehrlich ascites tumour cells.

1719
Filixic acid BBB; Filixic acid; Filicin; Filicic acid

$C_{36}H_{44}O_{12}$ Mol. wt 668.75

Occurs in the rhizomes of the male fern *Dryopteris filix-mas* (Polypodiaceae). Originally, filixic acid was thought to be a single compound, but later work showed that it consists of a mixture of six closely related phloroglucinol derivatives, of which filixic acid BBB is a major component.

Anthelmintic in veterinary practice.

1720
Humulone; Humulon; α-Lupulic acid

$C_{21}H_{30}O_5$ Mol. wt 362.47

Found in hops *Humulus lupulus* (Moraceae).

Contributes to the bitter taste of beer; it has antibiotic activity.

1721
α-Kosin

$C_{25}H_{32}O_8$ Mol. wt 460.53

Found in the female flowers of *Hagenia abyssinica* (=*Brayera anthelmintica*) (Rosaceae), together with the isomer β-kosin. Older literature claims incorrectly that the kosins are artefacts of protokosin.

Anthelmintic activity.

1722
Lupulone; β-Lupulic acid

$C_{26}H_{38}O_4$ Mol. wt 414.59

Found in hops *Humulus lupulus* (Moraceae).

Contributes to the bitter taste of beer; it has antibiotic activity, but some toxicity (LD_{50} in rats 1.8 g kg^{-1}).

1723
Mallotophenone

$C_{21}H_{24}O_8$ Mol. wt 404.42

Found in the pericarp of *Mallotus japonicus* (Euphorbiaceae).

Cytotoxic against mouse leukaemia L-5178Y cells and the KB cell system *in vitro*.

1724
Multifidol; (2-Methylbutyryl)phloroglucinol

$C_{10}H_{15}O_4$ Mol. wt 199.23

Occurs in the latex of *Jatropha multifida* (Euphorbiaceae), together with its β-D-glucoside.

Both compounds exhibit immunomodulatory activity; the latex of *Jatropha* is used in Asia for the treatment of wounds and skin infections.

1725
Paeonol; 2′-Hydroxy-4′-methoxyacetophenone

$C_9H_{10}O_3$　　　　　　　　　Mol. wt 166.18

Occurs in the roots of *Paeonia moutan* (Paeoniaceae) and in the essential oils of *Xanthorrhoea arborea* and *X. reflexa* (Xanthorroeaceae). Also, it occurs in *Bathysa meridionalis* (Rubiaceae), *Morus alba* (Moraceae), *Betula platyphylla* var. *japonica* (Betulaceae), in the roots of *Primula viscosa* and *P. auricula* (Primulaceae), and in *Cynanchum paniculatum* (Asclepiadaceae).

Stress prevention activity *in vivo*. Also, it shows haemostatic and antiaggregation actions on platelets *in vitro*.

1726
Paeonolide

$C_{20}H_{28}O_{12}$　　　　　　　　Mol. wt 460.44

Found in *Paeonia suffruticosa* and *P. arborea* (Paeoniaceae).

Inhibitory effect on blood platelet aggregation; it is responsible for the effects of the extract of *Paeonia suffruticosa*.

1727
Paeonoside

$C_{15}H_{20}O_8$　　　　　　　　Mol. wt 328.33

Occurs in *Paeonia suffruticosa* and *P. arborea* (Paeoniaceae).

Inhibitory effect on blood platelet aggregation. Paeonoside is also the name of a flavonoid found in *Solanum, Sophora* and *Trigonella* spp.

1728
Picein; *p*-Hydroxyacetophenone glucoside

$C_{14}H_{18}O_7$　　　　　　　　Mol. wt 298.30

Found in the twigs, bark and roots of *Salix* spp. (Salicaceae), the needles of *Picea excelsa* (Pinaceae), and the leaves of *Homogyna alpina* (Compositae). The aglycone occurs in the buds of *Populus balsamifera* (Salicineae).

1729
Rottlerin; Mallotoxin

$C_{30}H_{28}O_8$ Mol. wt 516.55

Unstable pinkish-brown pigment of "Kamala", from the fruit glands of *Mallotus philippensis* (syn. *Rottleria tinctoria*) (Euphorbiaceae).

Toxic. The anthelmintic activity is applied in veterinary practice. In India, it is used for the treatment of tumours, and commercially as a dye for silk.

Found as the glucoside in the bulbs of *Pancratium biflorum* (Amaryllidaceae). Also, it is found in *Xanthoxylum piperitum* and *X. alatum* (Rutaceae), in *Artemisia brevifolia* (Compositae), and *Hippomane mancinella, Sebastiania schottiana* and *Sapium sebiferum* (Euphorbiaceae).

Inhibits prostaglandin synthetase, 5-lipoxygenase, and the viability of Ehrlich ascites tumour cells. Xanthoxylin is also the name of a furoisocoumarin from *Geranium* spp.

1730
Tricyclodehydroisohumulone

$C_{21}H_{28}O_5$ Mol. wt 360.46

Constituent of beer and stored hops, *Humulus lupulus* (Cannabidaceae).

Used as a bittering agent.

1731
Xanthoxylin; Phloracetophenone 4,6-dimethyl ether; Brevifolin

$C_{10}H_{12}O_4$ Mol. wt 196.21

Chapter 42
Phenylpropanoids

Phenylpropanoids are naturally occurring phenolic compounds which have an aromatic ring to which a three-carbon side chain is attached. They are derived biosynthetically from the aromatic protein amino acid phenylalanine, and they contain one or more C_6-C_3 residues. The most widespread are the hydroxycinnamic acids, and compounds like *p*-coumaric acid (**1753**) and caffeic acid (**1737**) have a central role in phenolic metabolism. Another group is the phenylpropenes, e.g., anethole (**1734**), which contribute to the volatile flavours and odours of plants. The lignins, random polymers which contribute universally to the structure and rigidity of plant cell walls, are also phenylpropanoid in origin (Lewis and Yamamoto, 1990). Lignins are represented here by their immediate precursors, the three hydroxycinnamyl alcohols: *p*-coumaryl alcohol (**1755**), coniferyl alcohol (**1750**) and sinapyl alcohol (**1804**). Two other groups of phenylpropanoids, the coumarins and the lignans, are listed separately in this Dictionary.

Four hydroxycinnamic acids are common in plants: ferulic (**1768**), sinapic (**1802**), caffeic and *p*-coumaric acids. Several others are known, such as isoferulic (3-hydroxy-4-methoxy) (**1778**), *o*-coumaric and *p*-methoxycinnamic acids. These acids usually occur in plants in combined form as esters. Esters with quinic acid (q.v.) are particularly common, and the quinic ester of caffeic acid, chlorogenic acid (**1745**), is almost universal in its distribution. Derivatives with sugars, organic acids, lipids and amines have all been described (Herrmann, 1978). The different forms of caffeic acid conjugated with sugars are reviewed by Molgaard and Ravn (1988). Hydroxycinnamic acids (and their esters) can exist in both *Z* and *E* forms. They usually occur naturally as the *E* form, but undergo isomerization during isolation, and are often isolated as an equilibrium mixture.

Phenylpropenes are normally isolated from plant tissues in the essential oil fraction, together with terpenes (q.v.), and are lipid soluble. Some structures are widespread, such as eugenol (**1765**), a principle in the oil of cloves. Others are restricted to a few families. Anethole (**1734**) occurs in anise and fennel in the Umbelliferae, while myristicin (**1788**) is characteristic of nutmeg (Myristicaceae). Pairs of allyl and propenyl isomers, e.g., eugenol (**1765**) and isoeugenol (**1777**), are known and may occasionally occur together in the same plant.

References

Methods of Characterization
Harborne, J.B. (1989). *Phytochemical Methods*, 2nd Edition, London: Chapman and Hall.
Ibrahim, R. and Barron, D. (1989). Phenylpropanoids. In *Methods in Plant Biochemistry*, Volume 1, *Plant Phenolics* (Harborne, J.B.) (ed.), pp. 75–112. London: Academic Press.

Chemistry and Natural Distribution
Herrmann, K. (1978). Hydroxyzimtsauren und Hydroxybenzoesauren enthaltende Naturstoffe in Pflanzen. *Fortschritte der Chemie organischer Naturstoffe*, **35**, 73–132.
Molgaard, P. and Ravn, H. (1988). Evolutionary aspects of caffeoyl ester distribution in dicotyledons. *Phytochemistry*, **27**, 2411–2421.
Lewis, N.G. and Yamamoto, E. (1990). Lignin: occurrence, biogenesis and degradation. *Annual Reviews of Plant Physiology and Plant Molecular Biology*, **41**, 455–496.

1732
1′-Acetoxychavicol acetate

C$_{13}$H$_{14}$O$_4$ Mol. wt 234.26

Occurs in the essential oil of the rhizomes and in the leaves of *Alpinia galanga* (Zingiberaceae).

Antitumour activity against Sarcoma 180 ascites in mice, and antifungal activity. The rhizomes of *A. galanga* are used for treating fungal skin infections, dysentery, and problems of indigestion. The rhizomes are also used as a spice.

1733
1′-Acetoxyeugenol acetate

C$_{14}$H$_{16}$O$_5$ Mol. wt 264.28

Occurs in the rhizomes of Siamese ginger, *Alpinia galanga* (Zingiberaceae).

Pungent. It shows antitumour activity against Sarcoma 180 ascites in mice.

1734
Anethole; Anise camphor; *p*-Propenylanisole

C$_{10}$H$_{12}$O   Mol. wt 148.21

Found in the essential oils of, e.g., *Pimpinella anisum* and *Foeniculum vulgare* (both Umbelliferae), *Clausenia anisata* and *Pelea christophersenii* (both Rutaceae), *Backhousia anisata* (Myrtaceae) and *Magnolia salicifolia* (Magnoliaceae). Also, it is found in the roots of *Artemisia porrecta* and *Aster tartaricus* (Compositae), in *Juniperus rigida* (Cupressaceae) and other conifers, and in *Illicium anisatum* (Illiciaceae).

Stimulates hepatic regeneration in rats. It shows spasmolytic activity. Uses include as a flavouring agent, as a pharmaceutic aid and for the manufacture of anisaldehyde. In veterinary practice it is used as a carminative. It has a sweet taste.

1735
Apiole; Apiol; Apioline; Parsley camphor

C$_{12}$H$_{14}$O$_4$ Mol. wt 222.24

Occurs in the essential oils of the seeds of parsley *Petroselinum crispum*, the roots of *Crithmum maritimum* (both Umbelliferae), camphor wood, *Cinnamomum camphora* (Lauraceae) and the leaves of *Piper angustifolium* (Piperaceae). Also, it occurs in *Ocotea* spp. (Lauraceae).

Insecticidal and spasmolytic activities. In high doses it may cause short-lived intoxication.

1736
β-Asarone; Asarone; *cis*-Asarone; Asarin; Asarum camphor; Asarabacca camphor

$C_{12}H_{16}O_3$ Mol. wt 208.26

Occurs in the essential oil of the roots of *Acorus calamus* (Araceae), and of *Asarum europaeum* (Aristolochiaceae), and in *Piper angustifolium* (Piperaceae). It occurs as the *cis* isomer (illustrated), known as β-asarone and as the *trans* isomer (α-asarone).

Spasmolytic activity. Also it shows antialgal activity, is an insect chemosterilant and a strong insect attractant. β-Asarone is carcinogenic in animals and calamus oil has been banned in the USA. Therefore drugs containing only α-asarone are preferably used in pharmacy.

1737
Caffeic acid; 3,4-Dihydroxycinnamic acid

$C_9H_8O_4$ Mol. wt 180.16

Widespread occurrence: e.g., in green and roasted coffee beans (*Coffea arabica*) and in the root bark of *Cinchona cuprea* (both Rubiaceae), in *Conium maculatum* (Umbelliferae), and in the resin of various conifers. Also, it occurs in herbaceous plants such as *Digitalis purpurea* (Scrophulariaceae), the leaves and flowers of *Papaver somniferum* (Papaveraceae), the roots of *Taraxacum officinale*, and the flowers of *Anthemis nobilis* and *Achillea millefolium* (Compositae). It often occurs in bound form as chlorogenic acid (q.v.).

Antibacterial, antifungal, antiviral and antioxidant activities. It is an analgesic and an anti-inflammatory agent, with antihepatotoxic, antiulcerogenic and clastogenic activities also. It inhibits platelet aggregation *in vitro* and gonadotropin release, and affects both DNA binding and prostaglandin induction.

1738
Caffeic acid 3-glucoside

$C_{15}H_{18}O_9$ Mol. wt 342.31

Occurs in the leaves of *Populus nigra* (Salicaceae) and the berries of wild *Solanum* spp. (Solanaceae). The isomeric 4-glucoside occurs in the berries of *Vaccinium macrocarpon* (Ericaceae). Both glucosides are formed when caffeic acid is infiltrated into plant tissues, e.g., tomato petioles.

1739
1-Caffeoyl-β-D-glucose; β-D-Glucosyl caffeate

$C_{15}H_{18}O_9$ Mol. wt 342.30

Found in the leaves of *Solanum* spp., *Datura stramonium*, and *D. knightii*, in the flowers of *Petunia hybrida*, in the leaves of *Cestrum newellii*, in the leaves and flowers of *Brunfelsia calycina* (all Solanaceae), in the leaves and flowers of *Begonia* spp. (Begoniaceae), in the flowers of Ranunculaceae and of *Sambucus nigra* (Caprifoliaceae), in *Raphanus sativum* (Cruciferae), in *Anagallis* sp. (Primulaceae) and in *Antirrhinum majus* (Scrophulariaceae).

1740
N-Caffeoylputrescine

$C_{13}H_{18}N_2O_3$ Mol. wt 250.30

Occurs in the flowers of *Nicotiana tabacum*, in the leaves and stems of *Iochroma cyaneum* and in *Petunia hybrida* (both Solanaceae), in the seeds of *Pentaclethra macrophylla* (Leguminosae), in *Salix* spp. (Salicaceae) and in *Persea gratissima* (Lauraceae).

1741
5-O-Caffeoylshikimic acid; Date acid; Dactyliferic acid

$C_{16}H_{16}O_8$ Mol. wt 336.30

Found in fresh dates, *Phoenix dactylifera*, in the flowers of other palms, e.g., *Butia capitata* and *Phoenix canariensis*, and also in *Tsuga canadensis* (Pinaceae) in the gymnosperms.

1742
Carpacin; Isosafrole methyl ether

$C_{11}H_{12}O_3$ Mol. wt 192.22

Occurs in the bark of the carpano tree, *Cinnamomum* sp. (Lauraceae), and in *Justicia prostrata* (Acanthaceae).

Insecticidal activity. It has a weak sedative action. Carpacin potentiates the activity of prostalidins A–C (lignans). *Justicia* plants are used as an antidepressant in Asiatic folk medicine.

1743
(−)-Centrolobine

$C_{20}H_{24}O_3$ Mol. wt 312.41

Found in the heartwood of *Centrolobium robustum* in the (−) form and in *C. tomentosum* in the (+) form (Leguminosae).

Antibiotic activity.

1744
Chicoric acid; Dicaffeoyltartaric acid

$C_{22}H_{18}O_{12}$ Mol. wt 474.39

Found in the leaves of *Cichorium intybus*, *C. endivia*, *Lactuca sativa*, and also in *Echinacea* spp. (all Compositae).

1745
Chlorogenic acid; 3-Caffeoylquinic acid

$C_{16}H_{18}O_9$ Mol. wt 354.32

First found in green coffee beans, *Coffea arabica* (Rubiaceae) but widespread in higher plants, e.g., in the leaves of tea, *Thea sinensis* (Theaceae), and in cacao beans, *Theobroma cacao* (Sterculiaceae) and in many other food plants.

Antibacterial, antimutagenic, antitumour and antiviral activities, plus antioxidant and clastogenic activities. The *trans* isomer acts as an insect oviposition stimulant, and it may also reduce larval growth.

1746
Cinnamaldehyde; Cinnamic aldehyde; Cinnamal; Phenylacrolein

C_9H_8O Mol. wt 132.16

Occurs in the essential oil of the bark of *Cinnamomum zeylanicum* (Lauraceae), the flowers of *Hyacinthus* spp. and *Narcissus* spp. (both Liliaceae), *Lavandula* spp. (Labiatae), in patchouli oil, *Pogostemon cablin* (Labiatae), and in the oil of *Commiphora* spp. (Burseraceae).

Used in flavourings and perfumes. It inhibits the germination of velvetleaf, *Abutilon avicennae* (Malvaceae).

1747
Cinnamic acid

$C_9H_8O_2$ Mol. wt 148.16

Widespread occurrence, free and esterified: e.g., in the essential oils of *Cistus ladaniferus* (Cistaceae), of *Cinnamomum* spp. (Lauraceae), *Alpinia* spp. (Zingiberaceae), *Lilium candidum* (Liliaceae), in various resins, e.g., *Liquidambar orientalis*, styrax (Hamamelidaceae), in the leaves of *Erythroxylum coca* (Erythroxylaceae), in *Globularia* spp. (Globulariaceae), in the roots of *Scrophularia nodosa* (Scrophulariaceae) and in the buds of *Populus* spp. (Salicaceae).

Used for manufacture of esters in the perfume industry. It shows spasmolytic activity. However, it may cause contact dermatitis. As the coenzyme A ester, it is the biosynthetic precursor of hydroxycinnamic acids and other phenylpropanoids.

1748
4′-Cinnamoylmussatioside

$C_{34}H_{44}O_{16}$ Mol. wt 708.73

Found in the bark of *Mussatia* spp. (Bignoniaceae).

Inhibitory action on adenosine diphosphate-induced rat platelet aggregation.

1749
Coniferin; Abietin; Coniferoside; Laricin

$C_{16}H_{22}O_8$ Mol. wt 342.35

Occurs frequently in conifers such as *Abies* spp. and *Larix* spp. (Pinaceae), and also in *Beta* spp. (Chenopodiaceae), *Asparagus* spp. (Liliaceae), *Scorzonera hispanica* (Compositae), *Lonicera* spp. (Caprifoliaceae), and in the bark of *Fraxinus quadrangulata* (Oleaceae).

Precursor of lignin biosynthesis in plants.

1750
Coniferyl alcohol; 4-Hydroxyisoeugenol

$C_{10}H_{12}O_3$ Mol. wt 180.21

Occurs as the aglucone of coniferin (q.v.), and in the cambial tissue of *Pinus strobus* (Pinaceae), in *Vanilla mexicana* (Orchidaceae), and as a phytoalexin from *Linum usitatissimum* (Linaceae) after inoculation with the fungus *Melampsora linii*.

Antifungal activity against *Phytophthora megasperma* and *Cladosporium cucumerinum*, but induces virulence in *Agrobacterium tumefaciens*. It is a precursor of lignin biosynthesis in plants.

1751
Coniferyl aldehyde; Coniferaldehyde; Ferula aldehyde; Ferulaldehyde

$C_{10}H_{10}O_3$ Mol. wt 178.19

Occurs, e.g., in the wood of *Quercus* (Fagaceae), in *Acer saccharinum* (Aceraceae), in *Juglans cinerea* (Juglandaceae) and in various conifers such as *Sequoia* (Taxodiaceae). Also, it occurs in the whole plants of *Senra incana* (Bombacaceae) and as a phytoalexin in *Linum usitatissimum* (Linaceae).

Inhibits prostaglandin synthase and rat ear oedema. Also, it shows antifungal activity.

1752
***o*-Coumaric acid**

$C_9H_8O_3$ Mol. wt 164.16

The *trans* acid (shown) occurs in the leaves of *Gliricidia sepium* and *Dipteryx odorata* (Leguminosae), *Anthoxanthum puelii* (Gramineae), *Martynia annua* (Pedaliaceae) and *Citrus* cultivars (Rutaceae). The *cis* acid is unstable, and isomerizes to coumarin (q.v.). It occurs also as the glucoside, melilotoside, in, e.g., the leaves of *Melilotus alba* (Leguminosae).

1753
***p*-Coumaric acid;** *p*-Hydroxycinnamic acid

$C_9H_8O_3$ Mol. wt 164.16

Widespread occurrence: e.g., in the leaves of *Cananga latifolia* and *Annona muricata* (Annonaceae), in the berries of *Solanum tuberosum* (Solanaceae), in the peel of *Citrus limonum* (Rutaceae), in the flowers of *Beta saccharifera* (Chenopodiaceae), and in the kino gum of *Eucalyptus maculata* (Myrtaceae).

Antifungal and antihepatotoxic activities. It shows cytotoxic activity against P-815 and P-388 tumour cells *in vitro*, thus inhibiting tumour growth. Also, it shows allelopathic activity.

1754
4-*p*-Coumaroylquinic acid

$C_{16}H_{18}O_8$ Mol. wt 338.32

Occurs in apple fruit, *Malus domestica* (Rosaceae). This ester or other isomers have been recorded in a variety of plants.

1755
p-Coumaryl alcohol

$C_9H_{10}O_2$ Mol. wt 150.18

Occurs widely in tissues where lignification is taking place, e.g., in the sapwood of trees and the stems of herbs, etc.

Precursor of lignin biosynthesis in plants.

1756
Curcumin; Diferuloylmethane; Turmeric yellow; Turmeric colour

$C_{21}H_{20}O_6$ Mol. wt 368.39

Found in the roots of the spice plants *Curcuma longa*, *C. aromatica* and *C. xanthorrhiza* (Zingiberaceae).

Anti-inflammatory, cytotoxic and antioxidant activities. It reduces cholesterol level and helps control blood sugar. It is used for the preparation of curcuma paper and for detection of boron. This is the yellow colouring matter of *Curcuma* roots.

1757
1,3-Dicaffeoylquinic acid

$C_{25}H_{24}O_{12}$ Mol. wt 516.47

Occurs in the leaves and roots of *Cynara scolymus* and *C. cardunculus* (Compositae). In the fresh leaves it is present as the 1,3 form; on drying it rearranges to the 1,5 form (cynarin(e)).

Choleretic and antihepatotoxic properties. It inhibits fatty acid mobilization, and is indicated for treatment of hyperlipaemic syndrome, mainly due to lowering the serum cholesterol level.

1758
Diferulic acid

$C_{20}H_{18}O_8$ Mol. wt 386.36

Found in the cell walls of grasses, e.g., *Lolium multiflorum* (Gramineae), bound to carbohydrate.

1759
Dihydrocaffeic acid

$C_9H_{10}O_4$ Mol. wt 182.18

Occurs in beetroot, *Beta vulgaris* (Chenopodiaceae), and, in combined form, in caffeic acid sugar esters, e.g., in echinacoside (q.v.).

1760
Dihydroconiferyl alcohol

$C_{10}H_{14}O_3$ Mol. wt 182.22

Occurs in *Lactuca sativa* (Compositae).

Known as "lettuce cotyledon factor"; it is a synergist of gibberellic acid in inducing elongation of hypocotyls.

1761
Dillapiole; Dill apiole

$C_{12}H_{14}O_4$ Mol. wt 222.22

Found in the essential oil of dill, *Anethum graveolens*, and in the oil of the fruit of *Ligusticum scotinum, Orthodon formosanus* and *Crithmum maritimum* (Umbelliferae). It is also found in the leaves of *Piper aduncum* and *P. novae-hollandae* (Piperaceae), in *Erigeron* spp. (Compositae), in the oil of the leaves and stems of *Laurelia serrata* (Monimiaceae).

Insecticide (synergistic with other insecticides), and molluscicide.

1762
Echinacoside

$C_{35}H_{46}O_{20}$ Mol. wt 786.75

Found in the roots of *Echinacea angustifolia* (Compositae), and in *Cistanche salsa* (Orobanchaceae).

Antihepatotoxic activity.

1763
Elemicin

$C_{12}H_{16}O_3$ Mol. wt 208.26

Occurs in the resin of *Canarium commune* (Burseraceae), in the essential oils of the wood of *Cinnamomum glanduliferum* (Lauraceae), of *Cymbopogon procerus* (Gramineae), of *Boronia pinnata* and *Zieria smithii* (Rutaceae), of *Melaleuca bracteata* and *Backhousia myrtifolia* (Myrtaceae), and in *Aniba* (Annonaceae), *Croton nepetaefolius* (Euphorbiaceae), *Dalbergia spruceata* and *Monopteryx uaucu* (Leguminosae), and in *Daucus carota* (Umbelliferae).

DNA binding activity, and inhibits rabbit platelet aggregation *in vitro*.

1764
Estragole; Esdragole; *p*-Allylanisole; Chavicol methyl ether; Methylchavicol

$C_{10}H_{12}O$ Mol. wt 148.21

Found in the bark of *Persea gratissima* (Lauraceae), in the leaves of *Artemisia dracunculus*, in *Solidago odora* and *Tagetes florida* (Compositae), in the leaves of *Feronia elephantum, Orthodon methylchavicoliferum, Agastache rugosa* and *A. foenicula* (Labiatae), *Dictamnus albus* (Rutaceae), and in the leaves of *Magnolia kobus* (Magnoliaceae), *Croton zehntneri* (Euphorbiaceae), *Monopteryx* (Leguminosae), *Pinus* (Pinaceae) and *Piper betle* (Piperaceae).

Stimulates liver regeneration. It shows hypothermic and DNA binding activities. It is used in perfumes and as a flavour in foods and liqueurs.

1765
Eugenol; Allylguaiacol; Caryophyllic acid; Eugenic acid

$C_{10}H_{12}O_2$ Mol. wt 164.21

Found in the essential oils of *Eugenia caryophyllata* (= *Syzygium aromaticum*) (Myrtaceae), the leaves of *Cinnamomum* spp. and in *Sassafras randainense* (Lauraceae), *Pimentum dioica* (Myrtaceae), *Ocimum* spp. and *Origanum majorana* (Labiatae), *Achillea fragrantissima* and *Artemisia klotzschiana* (Compositae), *Myristica fragrans* (Myristicaceae), *Piper betle* (Piperaceae), and *Rosa rugosa* (Rosaceae).

Anticonvulsant, antimitotic, antioxidant, hypothermic and spasmolytic activities. Also, it shows antiyeast and central nervous system depressant activities. It inhibits prostaglandin synthesis by human colonic mucosa, the metabolism of arachidonic acid by human polymorphonuclear leukocytes, smooth muscle activity *in vitro* (humans and animals), and carrageenan-induced foot inflammation in rats. Also, it inhibits induced platelet aggregation *in vitro*. It is used as an antiseptic and anaesthetic in dentistry.

1766
Eugenol methyl ether; *O*-Methyleugenol

$C_{11}H_{14}O_2$ Mol. wt 178.23

Occurs in *Croton nepetaefolius* (Euphorbiaceae), in *Dacrydium frankenii*, *Nectandra polita* and *Ocotea pretiosa* (Lauraceae), and, as a phytoalexin, in *Pinus sylvestris* (Pinaceae).

Spasmolytic, hypothermic and skeletal muscle relaxant activities. It shows DNA binding effects, and is a central nervous system depressant.

1767
Fagaramide

$C_{14}H_{17}NO_3$ Mol. wt 247.30

Occurs in the root bark of *Fagara xanthoxyloides* and of *F. macrophylla* (Rutaceae), and in the stem bark of *Anthocleista djalonensis* and *A. vogelli* (Loganiaceae).

Anti-inflammatory and molluscicidal activities. It inhibits insect growth, and prostaglandin synthesis *in vitro*.

1768
Ferulic acid; Caffeic acid 3-methyl ether; 3-*O*-Methylcaffeic acid

$C_{10}H_{10}O_4$ Mol. wt 194.19

Found first in Asa foetida, the milk sap of the roots of *Ferula foetida* (Umbelliferae), then in many other plant sources: e.g., from the resin of *Pinus laricio* (Pinaceae), the resin of *Opopanax chironicum* (Araliaceae), the wood of *Tsuga heterophylla* (Pinaceae), the wood of *Catalpa ovata* (Bignoniaceae), the leaves of various grasses, e.g., *Oryza sativa*, *Phleum pratense* (Gramineae), the aerial parts of *Ajuga iva* (Labiatae), and the leaves of *Beta vulgaris* (Chenopodiaceae) and of *Periploca graeca* (Periplocaceae).

Antibacterial, antifungal, antihepatotoxic, antioestrogenic, antitumour and antimitotic activities. It stimulates phagocytosis and inhibits platelet aggregation *in vitro*. It acts as a serotonin antagonist and has antiyeast activities. Also, it shows a hydrocholeretic effect. It is used as a food preservative.

1769
Forsythiaside; Forsythoside A

$C_{29}H_{36}O_{15}$　　　　Mol. wt 624.61

Found in the fruits of *Forsythia suspensa* and *F. koreana* (Oleaceae), and in the callus tissue of *Rehmannia glutinosa* (Scrophulariaceae).

Inhibits cyclic adenosine monophosphodiesterase *in vitro* and the formation of 5-lipoxygenase products in rat peritoneal cells and in human leukocytes. It is a strong radical scavenger. It shows antibacterial activity, including plant pathogenic bacteria.

1770
Furcatin; *p*-Vinylphenol apiosylglucoside

$C_{20}H_{28}O_{10}$　　　　Mol. wt 428.44

Occurs in the leaves of *Viburnum furcatum* (Caprifoliaceae).

1771
[6]-Gingerdione

$C_{17}H_{22}O_4$　　　　Mol. wt 290.36

Occurs in the roots of ginger, *Zingiber officinale* (Zingiberaceae), which contain also [10]-gingerdione and the respective dehydro derivatives.

Inhibits prostaglandin biosynthesis *in vitro*.

1772
Gingerenone A

$C_{21}H_{24}O_5$　　　　Mol. wt 356.42

Occurs in the rhizomes of ginger, *Zingiber officinale* (Zingiberaceae).

Moderate antifungal activity *in vitro*.

1773
[6]-Gingerol

$C_{17}H_{26}O_4$　　　　Mol. wt 294.40

Occurs in the rhizomes of ginger, *Zingiber officinale* (Zingiberaceae), where it co-occurs with lesser amounts of [8]-gingerol (side chain with $(CH_2)_6$) and [10]-gingerol (side chain with $(CH_2)_8$).

Antiemetic and antiseratogenic activities. This is the most important pungent principle of ginger. It inhibits cyclo-oxygenase.

1774
Grandidentatin

$C_{21}H_{28}O_9$　　　　Mol. wt 424.46

Found in the bark of *Populus grandidentata* and *P. tremuloides*, and of *Salix peteolaris*, *S. fragilis*, *S. purpurea* and *S. triandra* (Salicaceae).

1775
Hellicoside

$C_{29}H_{36}O_{17}$ Mol. wt 656.61

Found in *Plantago asiatica* (Plantaginaceae).

Inhibits cyclic adenosine monophosphate phosphodiesterase and 5-lipoxygenase. The anti-inflammatory and anti-asthmatic actions of *Plantago* plants are attributed to inhibition of these enzyme activities.

1776
Isochlorogenic acid b; Caffee-tannins

$C_{25}H_{24}O_{12}$ Mol. wt 516.47

Widespread occurrence among Compositae: e.g., *Artemisia, Chrysothamnus paniculatus, Pluchea sagittalis*, as a mixture of 3 isomeric acids: 3,4-, 3,5-, 4,5-dicaffeoylquinic acids. Also, it occurs in *Coffea arabica* and the green beans of *C. robusta* (Rubiaceae). The 3,4-isomer occurs alone in Solanaceae, Polemoniaceae and Convolvulaceae. Unspecified isochlorogenic acids have been reported from *Asarum europaeum* (Aristolochiaceae), *Glycine max* (Leguminosae), fruit of *Prunus laurocerasus* (Rosaceae), and the genera *Dracocephalum* and *Lallemantia* (Labiatae).

Inhibits induced lipid peroxidation in mitochondria and microsomes of liver.

1777
Isoeugenol; 4-Propenylguaiacol

$C_{10}H_{12}O_2$ Mol. wt 164.21

Found in the essential oils of ylang-ylang, *Cananga odorata* (Annonaceae), the fruit of *Myristica fragrans* (Myristicaceae) and the needles of *Juniperus scopulorum* (Cupressaceae).

Inhibits rabbit platelet aggregation *in vitro*. It is used for the manufacture of vanillin.

1778
Isoferulic acid; Hesperitin acid

$C_{10}H_{10}O_4$ Mol. wt 194.19

Occurs in the rhizomes of *Cimicifuga racemosa* (Ranunculaceae), the roots of *Catalpa ovata* (Bignoniaceae) and the leaves of *Tamarix aphylla* (Tamaricaceae).

1779
Isosafrole

$C_{10}H_{10}O_2$ Mol. wt 162.19

Found in the essential oils of *Ligusticum acutilobum* (Umbelliferae), the leaves of *Murraya koenigii* (Rutaceae), and ylang-ylang, *Cananga odorata* (Annonaceae).

Stimulates liver regeneration. It is moderately toxic to humans: similar to safrole (q.v.). Inductor of cytochrome P450. It is used in the manufacture of heliotropin, and to modify oriental perfumes.

1780
Jionoside B$_1$

C$_{37}$H$_{50}$O$_{20}$ Mol. wt 814.81

Occurs in the roots of *Rehmannia glutinosa* (Scrophulariaceae).

Immunosuppressive activity. The roots of *Rehmannia* are used as a tonic, an antianaemic and an antipyretic in Japanese herbal medicine.

1781
Lusitanicoside; Chavicol rutinoside

C$_{21}$H$_{30}$O$_{10}$ Mol. wt 442.47

Occurs in *Cerasus lusitanica* (= *Prunus lusitanicus*) (Rosaceae). The aglycone, chavicol, occurs in the essential oils of leaves of *Piper betle* (Piperaceae), *Barosma venustum*, *Pimenta acris*, *P. racemosa* and *Majorana hortensis* (Labiatae).

1782
Melilotic acid; Dihydro-*o*-coumaric acid; *o*-Hydrocoumaric acid

C$_9$H$_{10}$O$_3$ Mol. wt 166.18

Found in the stem bark of *Cinnamomum cassia* (Lauraceae), and in *Melilotus officinalis*, *M. alba* and *M. caspius* (Leguminosae) as an ester.

Anti-ulcerogenic activity.

1783
***p*-Methoxycinnamaldehyde; PMCA**

C$_{10}$H$_{10}$O$_2$ Mol. wt 162.19

Found in *Ocimum basilicum* (Labiatae), in *Limnophila rugosa* (Scrophulariaceae), *Artemisia dracunculus* and *Sphaeranthus indicus* (Compositae), in *Agastache rugosa* (Labiatae), in *Acorus gramineus* (Araceae), and in *Illicium verum* (Illiciaceae).

Herbicidal activity, and active against germination in *Abutilon avicennae* (Malvaceae).

1784
***p*-Methoxycinnamic acid ethyl ester**; Ethyl *p*-methoxycinnamate

C$_{12}$H$_{14}$O$_3$ Mol. wt 206.24

Found in the rhizomes of *Kaempferia galangae* and of *Hedychium spicatum* (Zingiberaceae).

Cytotoxic against HeLa cells.

1785
Methyl caffeate

$C_{10}H_{10}O_4$ Mol. wt 194.19

Occurs in *Gaillardia pulchella, Tanacetum odessanum, Artemisia apiacea, Pseudostiffita kingii, Bedfordia solicina* and *Gochnatra rusbyana* (Compositae), and as the 4-glucoside, linocaffein, in the fruit of *Linum usitatissimum* (Linaceae).

Antitumour activity against Sarcoma 180, and weak antimicrobial activity.

1786
Methylisoeugenol

$C_{11}H_{14}O_2$ Mol. wt 178.23

Occurs in the essential oil of the roots of *Asarum europaeum* (Aristolochiaceae) and *Acorus calamus* (Araceae), and in the leaf wax of *Daucus carota* (Umbelliferae).

Expectorant, spasmolytic, antihistaminic and antibacterial activities. It is used as a local anaesthetic. It is moderately zoo- and phytotoxic.

1787
Myricoside

$C_{33}H_{43}O_{19}$ Mol. wt 743.71

Found in the roots of *Clerodendron myricoides* (Verbenaceae).

Antifeedant activity against African armyworm.

1788
Myristicin

$C_{11}H_{12}O_3$ Mol. wt 192.22

Found in the seed oil of *Myristica fragrans* (Myristicaceae), in the essential oil of the wood of *Cinnamomum glanduliferum* (Lauraceae), of *Apium graveolens*, of the fruit of *Petroselinum crispum*, of *Levisticum scoticum*, of *Pastinaca sativa* and of *Daucus carota* (Umbelliferae) and *Orthodon* spp. (Labiatae).

Inhibits monoaminoxidase and rabbit platelet aggregation *in vitro*; it shows synergistic activity to insecticides (e.g., xanthotoxin) and spasmolytic effects. It possibly has psychotropic properties.

1789
Orobanchoside; Orobanchin

$C_{28}H_{36}O_{16}$ Mol. wt 628.60

Occurs in *Orobanche rapum-genistae* (Orobanchaceae).

Analgesic and antihypertensive activities. It acts as an agonist of antitremor action of DOPA, and inhibits aldose reductase. It is active against plant pathogenic fungi.

1790
[6]-Paradol

$C_{17}H_{26}O_3$ Mol. wt 278.40

Found in the roots of ginger, *Zingiber officinale* (Zingiberaceae).

Flavour principle of ginger: hot and pungent.

1791
Phaseolic acid; Phaselic acid

$C_{13}H_{12}O_8$ Mol. wt 296.24

Occurs in the leaves of *Phaseolus vulgaris* and *Trifolium pratense* (Leguminosae), and in *Raphanus sativus* (Cruciferae).

1792
Phenethyl caffeate

$C_{17}H_{16}O_4$ Mol. wt 284.32

Occurs in the buds of *Populus* spp. (Salicaceae), and in bee propolis samples.

One of the major contact allergens of propolis.

1793
Plantamajoside

$C_{29}H_{36}O_{16}$ Mol. wt 640.61

Occurs in the leaves of *Plantago major* and *P. asiatica* (Plantaginaceae) and in the callus cultures of *Rehmannia glutinosa* (Scrophulariaceae).

Active against plant pathogenic bacteria. It inhibits 5-lipoxygenase and cyclic adenosine monophosphate diesterase, thus explaining the anti-inflammatory and antiasthmatic activity of *Plantago* plants.

1794
Prenyl caffeate

$C_{14}H_{16}O_4$ Mol. wt 248.28

Found in the buds of *Populus* spp. (Salicaceae), and in bee propolis samples.

One of the major contact allergens in bee propolis.

1795
Purpureaside C

$C_{35}H_{46}O_{20}$ Mol. wt 786.75

Found in the roots of *Rehmannia glutinosa*, and in the leaves of *Digitalis purpurea* (Scrophulariaceae).

Immunosuppressive activity.

1796
Rosmarinic acid

$C_{18}H_{16}O_8$ Mol. wt 360.33

Widespread occurrence within the Labiatae: e.g., in *Rosmarinus officinalis*, *Salvia officinale*, *Melissa officinalis*, *Mentha piperita* and *Teucrium scorodonia*. Also, it occurs in *Symphytum officinale* (Boraginaceae), in some Hydrophyllaceae and Acanthaceae, and in the genera *Anethum*, *Levisticum*, *Sanicula* and *Astrantia major* (Umbelliferae).

Anti-inflammatory activity. It shows antimicrobial activity against plant pathogens. It inhibits gonadotrophin release and adenylate cyclase activities in rat brain preparations and in rat thyroid cells. Also, it shows antiviral activity against herpes simplex.

1797
Safrole; Allylcatechol methylene ether; Shikimol

$C_{10}H_{10}O_2$ Mol. wt 162.19

Found in *Aniba* spp. (Annonaceae), *Cinnamomum camphora*, *Ocotea pretiosa* and *Sassafras albidum* (Lauraceae), *Magnolia salicifolia* (Magnoliaceae), in the leaves of *Illicium religiosum* (Illiciaceae), in the bark of *Nemuaron humboldtii* (Atherospermataceae), in *Eremophila longifolia* (Myoporaceae), in *Ocimum basilicum* (Labiatae), in *Juniperus scopulorum* (Cupressaceae), and in *Myristica fragrans* (Myristicaceae).

Anticonvulsant, DNA-binding, and hypothermic activities. It stimulates liver regeneration in rats, induces cytochrome P450, and acts as a depressant of the central nervous system. It is moderately toxic to man, and a low grade hepatocarcinogen. It inhibits rabbit platelet aggregation *in vitro*. It is used as a topical antiseptic, a pediculicide, a carminative, and also in the manufacture of heliotropin.

1798
Salvianolic acid A

$C_{26}H_{22}O_{10}$ Mol. wt 494.46

Found in the roots of *Salvia multiorrhiza* (Labiatae).

Inhibits pig gastric ATPases, and gastric secretion in rats. It shows anti-ulcer and anticoagulation activities.

1799
Sarisan; Asaricin

$C_{11}H_{12}O_3$ Mol. wt 192.22

Occurs in the leaf oil of *Beilschmiedia miersii* (Lauraceae), in the leaves of *Heteromorpha trifoliata* (Umbelliferae), and in the underground parts of *Asiasarum heterotropoides* (Aristolochiaceae).

Fungicidal activity.

1800
[6]-Shogaol

$C_{17}H_{24}O_3$ Mol. wt 276.38

Occurs in the rhizomes of *Zingiber officinale* and in the seeds of *Amomum melegueta* (Zingiberaceae).

Molluscicidal activity. It is pungent.

1801
Sinapaldehyde

$C_{11}H_{12}O_4$ Mol. wt 208.22

Occurs in the wood of *Acer saccharinum* (Aceraceae), *Juglans nigra* (Juglandaceae), and *Quercus* spp. (Fagaceae), and in the whole plants of *Senra incana* (Bombacaceae).

Inhibits prostaglandin synthetase and rat ear oedema.

1802
Sinapic acid

$C_{11}H_{12}O_5$ Mol. wt 224.22

Relatively widespread occurrence, mostly in bound form: e.g., in the seeds of *Lepidium sativum*, in *Brassica oleracea*, and in the seeds of *Sisymbrium columnae* (Cruciferae), in the rhizomes and leaves of *Aristolochia clematis* and in *Asarum europaeum* (Aristolochiaceae), in the peel of *Citrus limon* (Rutaceae), in *Vitis vinifera* (Vitaceae), in *Erica australis* (Ericaceae), and in *Pterocarpus santalinus* (Leguminosae).

Antibacterial, antifungal and antihepatotoxic activities.

1803
Sinapine; Sinapic acid choline ester

$[C_{16}H_{24}NO_5]^+$ Mol. wt 310.38

Abundant occurrence in the seeds of Cruciferae, e.g., in *Brassica nigra, Sisymbrium columnae, Crambe asiatica, Lepidium sativum* and *Draba nemorosa*.

1804
Sinapyl alcohol; Syringenin

$C_{11}H_{14}O_4$ Mol. wt 210.23

Occurs as the glucoside syringin, which occurs in the leaves and fruit of *Syringa vulgaris*, in *Forsythia suspensa*, in the bark of *Phillyrea* and in other Oleaceae. Also, it occurs in the bark of Caprifoliaceae, in *Paulownia tomentosa* (Scrophulariaceae), and in *Viscum* spp. (Loranthaceae).

Upon release from the host plant it triggers transfer of t-DNA in *Agrobacterium tumefaciens*, thus inducing virulence. It is a precursor of angiosperm lignin.

1805
Sphagnum acid

$C_{11}H_{10}O_5$ Mol. wt 222.20

Occurs in the moss *Sphagnum magellanicum* and other *Sphagnum* spp.

1806
Subaphyllin; Feruloylputrescine

$C_{14}H_{20}N_2O_3$ Mol. wt 264.33

Occurs in the leaves and juice of grapefruit, *Citrus paradisi*, and orange, *Citrus aurantium* (Rutaceae) and in *Salsola subaphylla* (Chenopodiaceae). Also, it occurs in *Ananas comosus* (Bromeliaceae), in *Zea mays* and other Gramineae; in *Gomphrena globosa* (Amaranthaceae), in *Persea gratissima* (Lauraceae) and in *Salix* spp. (Salicaceae).

Hypotensive activity. Occurrence in virus-infected tobacco indicates antiviral effect.

1807
Suspensaside

$C_{29}H_{36}O_{16}$ Mol. wt 640.61

Found in the fruit of *Forsythia suspensa* (Oleaceae).

Antibacterial activity. It inhibits the formation of 5-lipoxygenase products in human leukocytes and in rat peritoneal cells. Also, it is effective in the arachidonate metabolism of leukocytes. It inhibits cyclic adenosine monophosphate phosphodiesterase. It is a strong radical scavenger, indicating activity against asthma and allergic diseases.

1808
1,3,4,5-Tetracaffeoylquinic acid

$C_{43}H_{35}O_{18}$ Mol. wt 839.75

Occurs in the leaves of *Pluchea symphytifolia* (Compositae).

Antimicrobial activity against *Bacillus subtilis*, *Escherichia coli* and *Sarcina citrea*.

1809
Verbascoside; Acteoside; Kusaginin

$C_{29}H_{36}O_{15}$ Mol. wt 624.61

Found in the leaves of *Buddleja globosa, B. officinalis* and in the fruit of *Forsythia* (Oleaceae), in *Verbascum sinuatum* (Scrophulariaceae), and in members of the Orobanchaceae, Gesneriaceae, Acanthaceae, Bignoniaceae, Verbenaceae and Plantaginaceae.

Hypertensive and antihepatotoxic activity. It inhibits formation of 5-lipoxygenase products in human leukocytes, indicating anti-inflammatory properties. Also, it inhibits lens aldose reductase.

Chapter 43
Quinones

The quinone pigments, of which there are at least 1200 structures, range in colour from yellow to almost black, but most are yellow, orange or red. Although they are widely, albeit sporadically, distributed in higher plants, they contribute little to plant colour. Thus, they are either present hidden in bark, heartwood or root or else they are in tissues (e.g., leaves) where their colours are masked by other pigments. By contrast, in bacteria, fungi and lichens they do occasionally provide colour; for example, cultures of *Penicillium* spp. may be pigmented by anthraquinones. These pigments are also found in some arthropods, such as the bean aphid and the cochineal insect, and in sea urchins.

Quinones all contain the same basic chromophore, that of benzoquinone itself, which consists of two carbonyl groups in conjugation with two carbon–carbon double bonds. Quinones can be conveniently divided into three groups of increasing molecular size: the benzoquinones, the naphthoquinones and the anthraquinones. Many are alkylated or substituted by isoprenyl groups and hence are lipophilic (e.g., geranylbenzoquinone, **1850**). Others are hydroxylated, with phenolic properties, and may occur *in vivo* both free and in combined form with sugars (e.g., chrysophanol, **1830**).

Typical benzoquinones are primin (**1874**), the allergen of *Primula* plants, and embelin (**1845**), both of which have aliphatic side chains. Two groups of isoprenoid benzoquinones, which need separate mention, are the plastoquinones, represented here by plastoquinone-9 (**1871**), and the ubiquinones, of which ubiquinone-10 (**1886**) is a common member. These quinones are involved, respectively, in photosynthesis and cellular respiration and, unlike other benzoquinones, are universally distributed in trace amounts in all green plants.

Plant naphthoquinones range from simple structures such as the 5-hydroxy derivative juglone (**1855**), from walnuts, to pigments with isoprenyl attachment such as alkannin (**1813**), a natural dyestuff. There are also a few naphtho-*ortho*-quinones, such as dunnione (**1842**) from *Streptocarpus* and mansonone C (**1862**) from the elm tree. Dimeric structures have been recognized, such as cercosporin (**1824**) from the fungus *Cercospora* and diospyrin (**1840**) from the ebony tree.

Anthraquinones constitute the largest group of plant quinones, and are often red or purple in colour rather than yellow. They are fairly widely distributed, especially in such plant families as the Leguminosae (*Cassia*), Liliaceae (*Aloe*), Polygonaceae (*Rheum*), Rhamnaceae (*Rhamnus*), Rubiaceae (*Rubia*) and Verbenaceae (*Tectona*). Commonly occurring anthraquinones include aloe-emodin (**1815**), chrysophanol (**1830**), emodin (**1846**) and physcion (cf., **1869**). These anthraquinones are known both as *O*-glycosides (e.g., chrysophanol 8-glucoside, **1831**) and, more rarely, as *C*-glycosides (e.g., barbaloin, **1821**). Dimeric anthraquinones include such compounds as cassiamin C (**1823**), sennoside A (**1883**) and the extended quinone hypericin (**1852**) from St. John's Wort.

Quinones are synthesized in plants by any one of four distinct pathways. Anthraquinones, for example, are formed from acetate–malonate precursors in the Polygonaceae and Rhamnaceae, while they are synthesized via *O*-succinylbenzoic acid in the Bignoniaceae and Verbenaceae. The same two pathways are utilized for naphthoquinone synthesis in the Plumbaginaceae (e.g., plumbagin, **1872**) and the Juglandaceae (e.g., juglone, **1855**), respectively. A further pathway, starting from phenylalanine and passing via homogentisic acid, is used for making chimaphylin (**1825**) in the Pyrolaceae, while *p*-hydroxybenzoic acid is the starting material for the production of the isomeric pigments, alkannin **(1813)** and shikonin, in the Boraginaceae.

Natural dyestuffs such as henna and madder are well known to contain quinones. The irritant effects of plant quinones are also well recognized. Many quinone-containing plants from the *Cassia* (senna) and *Aloe* genera are widely used for their purgative properties. Quinones are responsible for skin rashes caused in humans by handling *Primula obconica* leaves (primin) or teak wood (deoxylapachol). Hypericin is a photodynamic pigment, causing a disease in sheep which feed on St. John's wort. A general toxicity to life forms is apparent in many other quinone substances.

Simple colour reactions have been devised to detect quinone pigments in plant extracts. Reversible reduction to a colourless form and restoration of the colour by aerial oxidation in solution is diagnostic. Quinones also give intense bathochromic shifts in colour in the presence of alkali, e.g., from red to purple. They have characteristic spectral properties, with several intense peaks in the ultraviolet and one or two peaks in the visible region. Methods of quinone analysis have been reviewed by Berg and Labadie (1989). The major monographs of Thomson (1971, 1986) should also be consulted.

References

Chemical Analysis and Presence in Cell Culture
Berg, A.J.J. van den and Labadie, R.P. (1989). *Methods in Plant Biochemistry*, Volume 1, *Plant Phenolics* (Harborne, J.B.) (ed.), pp. 451–492. London: Academic Press.

Chemistry and Distribution
Thomson, R.H. (1971). *Naturally Occurring Quinones*, 2nd Edition. London: Academic Press.
Thomson, R.H. (1986). *Naturally Occurring Quinones. III. Recent Advances*. London: Chapman and Hall.

1810
Acamelin

Skeletal type: benzoquinone

$C_{10}H_8O_4$ Mol. wt 192.17

Found in the heartwood of *Acacia melanoxylon* (Leguminosae).

Allergen, causing the contact dermatitis (together with 2,6-dimethoxybenzoquinone) of *Acacia melanoxylon*.

1811
Alizarin; 1,2-Dihydroxyanthraquinone

Skeletal type: anthraquinone

$C_{14}H_8O_4$ Mol. wt 240.22

Occurs in the roots of *Rubia tinctorum*, madder plant, in *Galium* spp., *Asperula odorata*, in the heartwood of *Morinda citrifolia* (all Rubiaceae), *Rheum palmatum* (Polygonaceae) and as a glycoside in *Libertia coerulescens* (Iridaceae).

Orange-red pigment, the principle of madder, one of the most ancient of natural dyestuffs. It is used in the manufacture of acid and chrome dyes for wool, as an acid–base indicator, and as a reagent for aluminium and zinc. It shows antileukaemic activity.

1812
Alizarin 2-methyl ether

Skeletal type: anthraquinone

$C_{15}H_{10}O_4$ Mol. wt 254.25

Occurs in roots of *Rubia cordifolia* and *R. tinctorum*, in the roots and stems of *Morinda umbellata*, in *Asperula odorata* and in *Galium* spp. (all Rubiaceae). Also, it occurs as a phytoalexin in tissue cultures of *Cinchona* spp. (Rubiaceae).

Antimicrobial activity (e.g., against *Bacillus subtilis* and *Escherichia coli*).

1813
Alkannin; Alkanna red; Anchusa acid; Anchusin

Skeletal type: naphthoquinone

$C_{16}H_{16}O_5$ Mol. wt 288.30

Found in the roots of *Alkanna tinctoria*, *Arnebia nobilis*, and *Macrotomia cephalotes*, and as a leaf surface constituent of *Plagiobothrys arizonicus* (all Boraginaceae). Shikonin, the 1′R-isomer, also occurs in the Boraginaceae, in the roots of *Lithospermum erythrorhizon*, in *Echium lycopsis* and in *Onosma caucasicum*.

Both isomers are used as a red dye for cosmetics and food. They have astringent properties, and show a biphasic activity profile: at low concentrations immunomodulatory effects, in higher doses suppressive action in granulocyte and lymphocyte test systems. Shikonin is produced commercially in Japan from *Lithospermum* cell cultures, and is used medicinally and for colouring lipsticks.

1814
Alkannin β,β-dimethylacrylate

Skeletal type: naphthoquinone

$C_{21}H_{22}O_6$ Mol. wt 370.41

Occurs in the roots of *Arnebia nobilis* (Boraginaceae).

Anticancer activity against Walker carcinosarcoma in rats.

1815
Aloe-emodin; Rhabarberone

Skeletal type: anthraquinone

$C_{15}H_{10}O_5$ Mol. wt 270.25

Found in the leaves of *Cassia senna* (Leguminosae) and in some *Rheum* spp. (Polygonaceae), in the tubers of *Asphodelus microcarpus*, in the inflorescences of *Xanthorrhoea australis* in some *Aloe* spp. (Liliaceae), in the leaves of *Oroxylum indicum* (Bignoniaceae), and in the wood of *Tectona grandis*, teak (Verbenaceae).

Cathartic and antileukaemic activities. It is genotoxic in hamster fibroblast mutagenicity assay, and active against *Salmonella typhimurium* strains. Also it helps to protect teak wood against termites.

1816
Anthragallol

Skeletal type: anthraquinone

$C_{14}H_8O_5$ Mol. wt 256.22

Found in the roots of *Relbunium hypocarpium* (Rubiaceae).

Immunosuppressive activity *in vitro*; at higher dosage it shows inhibitory and cytotoxic activity against macrophages, and T- and B-lymphocytes.

1817
Ardisianone

Skeletal type: benzoquinone

$C_{24}H_{38}O_5$ Mol. wt 406.57

Found in the roots of *Ardisia cornudentata* and *A. quinquegona* (Myrsinaceae).

Inhibits the binding of leukotrienes in various receptor assays.

1818
Aristolindiquinone

Skeletal type: naphthoquinone

$C_{12}H_{10}O_4$ Mol. wt 218.21

Occurs in the roots of *Aristolochia indica* (Aristolochiaceae).

Antifertility activity.

1819
Arnebinone

Skeletal type: benzoquinone

$C_{18}H_{22}O_4$ Mol. wt 302.37

Occurs in the roots of *Arnebia euchroma* (Boraginaceae).

Inhibits prostaglandin biosynthesis.

1820
Aurantio-obtusin β-D-glucoside

Skeletal type: anthraquinone

$C_{23}H_{24}O_{12}$ Mol. wt 492.45

Occurs in the seeds of *Cassia obtusifolia* (Leguminosae).

Strong inhibitor of rat platelet aggregation, but the aglycone is only moderately active.

1821
Barbaloin; Aloin

Skeletal type: anthrone

$C_{21}H_{21}O_9$ Mol. wt 417.40

This *C*-glucoside of aloe-emodin occurs in the leaves of several *Aloe* spp. cultivated as drugs, e.g., *A. vera*, *A. ferox* and *A. perryi* (Liliaceae).

Used commercially as a purgative.

1822
Cascaroside A

Skeletal type: anthrone

$C_{27}H_{32}O_{14}$ Mol. wt 580.54

Found in the dried bark of *Rhamnus purshiana* (=*Frangula purshiana*, *Cascara sagrada*) (Rhamnaceae).

Responsible for the cathartic action of *Rhamnus purshiana* bark.

1823
Cassiamin C

Skeletal type: bianthraquinone

$C_{30}H_{18}O_8$ Mol. wt 506.47

One of several bianthraquinones found in the bark of *Cassia siamea* (Leguminosae); it is a dimer of chrysophanol (q.v.).

1824
Cercosporin

Skeletal type: binaphthoquinone

$C_{29}H_{26}O_{10}$ Mol. wt 534.53

Found in cultures of the fungus *Cercospora kikuchi* and numerous other *Cercospora* spp.

A phytotoxin, also exhibiting photodynamic antibacterial activity, which is produced by many *Cercospora* spp. (plant pathogens); this pigment is the effective toxic agent of the fungus.

1825
Chimaphylin

Skeletal type: naphthoquinone

$C_{12}H_{10}O_2$ Mol. wt 186.21

Occurs in *Chimaphila corymbosa*, *Pyrola incarnata* and other members of the Pyrolaceae.

Moderately active as a phagocytose inhibitor of human granulocytes; at low dosage it stimulates phagocytose.

1826
7-Chloroemodin

Skeletal type: anthraquinone

$C_{15}H_9ClO_5$ Mol. wt 304.69

A typical lichen pigment, found, e.g., in *Nephroma laevigatum*, *Lecidea quernea* and *Caloplaca arenaria*.

1827
Chrysazin; Dantron; 1,8-Dihydroxyanthraquinone

Skeletal type: anthraquinone

$C_{14}H_8O_4$ Mol. wt 240.22

Occurs in the roots of *Rheum palmatum* (Polygonaceae), in the leaves and stems of *Xyris semifuscata* (Xyridaceae), and in the tissue cultures of *Cinchona ledgeriana* (Rubiaceae).

At higher dosage, there is immunosuppressive activity in macrophages and lymphocyte cells test systems. It shows cathartic activity, and is indicated as a purgative in veterinary practice.

1828
Chryso-obtusin glucoside

Skeletal type: anthraquinone

$C_{25}H_{28}O_{12}$ Mol. wt 520.50

Found in the seeds of *Cassia obtusifolia* and *C. tora* (Leguminosae).

Strong inhibitor of rat platelet aggregation; the aglycone shows only moderate action.

1829
Chrysophanic acid 9-anthrone

Skeletal type: anthrone

$C_{15}H_{12}O_3$ Mol. wt 240.26

Occurs in the seeds of *Cassia tora, C. siamea*, and in *Ferreirea spectabilis* and *Vatairea quianensis* (all Leguminosae), in *Rumex crispus* (Polygonaceae) and in *Rhamnus purshiana* (Rhamnaceae).

Fungicidal activity, and highly effective inhibitor of dermatophyte growth.

1830
Chrysophanol; Chrysophanic acid; 3-Methylchrysazin; 1,8-Dihydroxy-3-methylanthraquinone

Skeletal type: anthraquinone

$C_{15}H_{10}O_4$ Mol. wt 254.25

Occurs in *Rumex* and *Rheum* spp. (Polygonaceae), in senna leaves, *Cassia senna*, in the fruits and heartwood of *Cassia siamea* (Leguminosae), in *Rhamnus purshiana* (*Cascara sagrada*) (Rhamnaceae), in teak wood, *Tectona grandis* (Verbenaceae), in families like Dipterocarpaceae, Guttiferae, Liliaceae, Simaroubaceae, and in some lichen genera.

Activity against termites in teak wood. It is used as a natural dye.

1831
Chrysophanol 8-glucoside; Chrysophanol 8-*O*-β-D-glucopyranoside

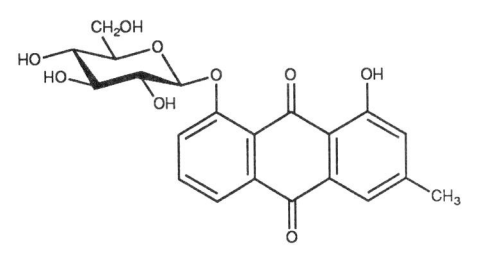

Skeletal type: anthraquinone

$C_{21}H_{20}O_9$ Mol. wt 416.39

Occurs in the rhizomes of *Rheum moorcroftianum* (Polygonaceae) and the flowers of *Woodfordia fruticosa* (Lythraceae).

Inhibits mobilization of human spermatozoa.

1832
Coleone A

Skeletal type: furanonaphthoquinone

$C_{20}H_{22}O_6$ Mol. wt 358.40

Found in the red glandular hairs under the leaves of *Coleus ignarius* (Labiatae).

Antimicrobial activity.

1833
Cornudentanone

Skeletal type: benzoquinone

$C_{22}H_{34}O_5$ Mol. wt 378.51

Occurs in the roots of *Ardisia cornudentata* (Myrsinaceae).

Inhibits binding of leukotrienes in various receptor assays.

1834
Cyperaquinone

Skeletal type: furanobenzoquinone

$C_{14}H_9O_4$ Mol. wt 241.23

Occurs in the roots and rhizomes of *Cyperus haspan*, and other *Cyperus* spp. (Cyperaceae). It co-occurs in these plants with related furanobenzoquinones.

Carmine pigment.

1835
Cypripedin

Skeletal type: benzonaphthoquinone

$C_{16}H_{12}O_5$ Mol. wt 284.27

Found in the leaves of *Cypripedium calceolus* (Orchidaceae).

Responsible for contact dermatitis caused by the leaves of *C. calceolus*.

1836
Deoxylapachol

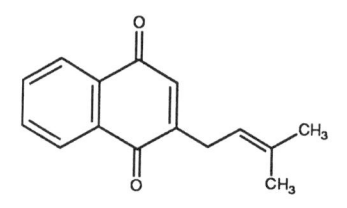

Skeletal type: naphthoquinone

$C_{15}H_{14}O_2$ Mol. wt 226.28

Occurs in the wood of *Catalpa ovata* (Bignoniaceae) and in *Tectona grandis*, teak (Verbenaceae). Major allergen in teak, causing allergic skin reactions. It shows antitermite activity.

1837
Digifferugineol;
1-Hydroxy-2-hydroxymethylanthraquinone

Skeletal type: anthraquinone

$C_{15}H_{10}O_4$ Mol. wt 254.25

Found in the roots of *Streptocarpus dunnii* (Gesneriaceae), in the leaves of *Digitalis ferruginea* (Scrophulariaceae), in the roots of *Morinda parvifolia*, and in tissue cultures of *Cinchona* spp. (both Rubiaceae).

Cytotoxic *in vitro* (KB-human epidermoid carcinoma of nasopharynx). It is a phytoalexin in tissue cultures of *Cinchona* and a main component in a fraction exhibiting antimicrobial activity (e.g., *Bacillus subtilis*, *Escherichia coli*).

1838
1,4-Dihydroxy-2-methylanthraquinone

Skeletal type: anthraquinone

$C_{15}H_{10}O_4$ Mol. wt 254.25

Occurs in teak wood, *Tectona grandis* (Verbenaceae).
Termite repellent.

1839
2,6-Dimethoxybenzoquinone; 2,6-Dimethoxyquinone

Skeletal type: benzoquinone

$C_8H_8O_4$ Mol. wt 168.15

Fairly widespread occurrence in the angiosperms, e.g., in wheat grains, *Triticum vulgare* (Gramineae), the aerial parts of *Adonis vernalis* (Ranunculaceae) and the roots of *Rauwolfia vomitoria* (Apocynaceae).

In vitro cytotoxicity in P-388 lymphocytic leukaemia tests; it is responsible for the contact dermatitis of *Acacia melanoxylon* (together with acamelin, q.v.). It induces haustorial formation in the plant parasite *Striga asiatica*, following its release in trace amounts from the roots of a host plant such as sorghum.

1840
Diospyrin; Euclein

Skeletal type: binaphthoquinone

$C_{22}H_{14}O_6$ Mol. wt 374.35

Occurs in the roots, bark, wood and leaves of *Diospyros* spp. and in *Euclea* spp. (Ebenaceae).

Cytotoxic against Ehrlich ascites carcinoma cells in animals. It is immunostimulating at low doses, but cytotoxic at higher doses. These plants are reported to be useful in the treatment of tumours.

1841
Droserone

Skeletal type: naphthoquinone

$C_{11}H_8O_4$ Mol. wt 204.19

Found in the roots of *Drosera whittakeri* and *D. peltata* (Droseraceae). The 5-methyl ether occurs in the heartwood of *Diospyros melanoxylon* (Ebenaceae).

1842
Dunnione

Skeletal type: α-naphthoquinone

$C_{15}H_{14}O_3$ (+) form Mol. wt 242.28

The (+) form occurs in the orange-red deposit on the undersurface of leaves of *Streptocarpus dunnii* and *S. pole-evansii* (Gesneraceae). The (−) form occurs in *Calceolaria integrifolia* (Scrophulariaceae).

1843
Echinone

Skeletal type: naphthoquinone

$C_{19}H_{20}O_6$ Mol. wt 344.37

Found in the callus cultures of *Echium lycopsis* (Boraginaceae).

Active against Gram-positive bacteria.

1844
Eleutherin

Skeletal type: naphthoquinone

$C_{16}H_{16}O_4$ Mol. wt 272.30

Occurs in the tubers of *Eleutherine bulbosa* (Liliaceae).

Antimicrobial activity (against *Staphylococcus aureus* and *Mycobacterium smegmatis*).

1845
Embelin; Embelic acid

Skeletal type: benzoquinone

$C_{17}H_{26}O_4$ Mol. wt 294.40

Found in the fruit of *Embelia ribes*, in the roots of *Ardisia crenata*, and in the berries of *Myrsine africana* and of *Rapanea* (all Myrsinaceae), and also in the twigs and stems of *Aegiceras corniculatum* (Aegiceraceae).

Potent oral contraceptive, possessing 85% anti-implantation activity in rats. Ammonium embelate is used as an anthelmintic; it may be irritant to mucous membranes, causing violent sneezing.

1846
Emodin; Archin; Frangula emodin; Frangulic acid; Rheum emodin

Skeletal type: anthraquinone

$C_{15}H_{10}O_5$ Mol. wt 270.25

Found in *Rumex* and *Rheum* spp. (Polygonaceae), in the root bark of *Ventilago calyculata*, in *Rhamnus frangula* (both Rhamnaceae), in *Myrsine africana* (Myrsinaceae), in *Psorospermum glaberrimum* (Guttiferae), and in some lichen genera.

Inhibits the growth of crown gall tumours on potato disks; murine antileukaemic (P388) and antitumour (Walker Sarcoma) activities. Also, it is moderately cytotoxic in 3 human tumour cell lines.

1847
Emodin 8-glucoside; Emodin 8-*O*-glucopyranoside

Skeletal type: anthraquinone

$C_{21}H_{20}O_{10}$ Mol. wt 432.38

Occurs in the rhizomes of *Rheum moorcroftianum*, in *Polygonum cuspidatum* (Polygonaceae), and in the stem bark of *Rhamnus frangula* (Rhamnaceae).

Causes immobilization of human spermatozoa.

1848
Frangulin A; Emodin-3-rhamnoside; Rhamnoxanthin

Skeletal type: anthraquinone

$C_{21}H_{20}O_9$ Mol. wt 416.39

Found in the seed, bark and root bark of *Rhamnus cathartica* and *R. frangula* (Rhamnaceae).

Cathartic activity. The synonyms franguloside, avornin and cascarin have been used earlier for both frangulin A and B, since these 2 compounds were believed at one time to be isomeric.

1849
Frangulin B

Skeletal type: anthraquinone

$C_{20}H_{18}O_9$ Mol. wt 402.36

Occurs in the seed, bark and root bark of *Rhamnus cathartica* and *R. frangula* (Rhamnaceae).

Cathartic activity. The synonyms franguloside, avornin and cascarin have been used earlier for both frangulin A and B, since these two compounds were believed at one time to be isomeric.

1850
Geranylbenzoquinone

Skeletal type: benzoquinone

$C_{16}H_{20}O_2$ Mol. wt 244.34

Found in the trichomes of *Phacelia ixodes* (Hydrophyllaceae) and in the wood of *Cordia alliodora* (Boraginaceae).

Elicitor of allergic skin reactions and a potent skin irritant.

1851
2-Hydroxymethylanthraquinone

Skeletal type: anthraquinone

$C_{15}H_{10}O_3$ Mol. wt 238.25

Occurs in the roots of *Morinda parvifolia* (Rubiaceae).

Cytotoxic *in vitro*, and antileukaemic *in vivo* (P-388 lymphocytic leukaemia in mice).

1852
Hypericin; Hypericum red

Skeletal type: bianthraquinone

$C_{30}H_{16}O_8$ Mol. wt 504.46

Found in *Hypericum perforatum* and many other *Hypericum* spp. (Hypericaceae).

Antiretroviral activity *in vitro* and *in vivo*; photosensitizing and antidepressant activities in mammals. It causes facial eczema in sheep grazing on these plants; the photogenic disease produced is called "hypericism".

1853
Isodiospyrin

Skeletal type: binaphthoquinone

$C_{22}H_{14}O_6$ Mol. wt 374.35

Occurs in the roots and stem bark of *Diospyros usamabarensis* (Ebenaceae).

Molluscicidal and antifungal activities; it is cytotoxic.

1854
6-Isohexenyl-α-naphthoquinone

Skeletal type: naphthoquinone

$C_{16}H_{16}O_2$ Mol. wt 240.30

Occurs in the wood of *Radermachera sinica* (Bignoniaceae).

Active against Gram-positive bacteria.

1855
Juglone; Nucin; Regianin; Natural Brown 7

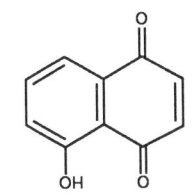

Skeletal type: naphthoquinone

$C_{10}H_6O_3$ Mol. wt 174.16

Found in the stem bark of *Juglans nigra*, in *Juglans regia*, in *Carya ovata*, and in the leaves and nuts of *C. illinoensis* (all Juglandaceae), and in *Lomatia* spp. (Proteaceae).

Antifungal, antiviral (against HSV-1 virus), molluscicidal, and sedative activities in fish and mammals. Also, it is a feeding deterrent to *Scolytus multistriatus*. This is the allelopathic agent of the walnut tree.

1856
Kigelinone

Skeletal type: naphthoquinone

$C_{14}H_{10}O_5$ Mol. wt 258.23

Found in the wood of *Kigelia pinnata* and *Crescentia cujete*, and in the stem bark of *Tabebuia cassinoides* (all Bignoniaceae).

Cytotoxic activity.

1857
Lapachol; Lapachic acid; Greenhartin; Tecomin

Skeletal type: naphthoquinone

$C_{15}H_{14}O_3$ Mol. wt 242.28

Found in the heartwood of the Bignoniaceae, e.g., *Haplophragma adenophyllum* and *Tabebuia rosea*, in the roots of *Kigelia pinnata*, and in *Stereospermum suaveolens* (both Bignoniaceae), in the heartwood of *Hibiscus tiliaceus* (Malvaceae), in the wood of *Diphysa robinoides* (Leguminosae), and in the roots of *Conospermum teretifolium* (Proteaceae).

Antitumour activity in the Walker 256 tumour cell system. It inhibits respiratory processes and is cytotoxic at high doses, but immunostimulating at low doses.

1858
β-Lapachone

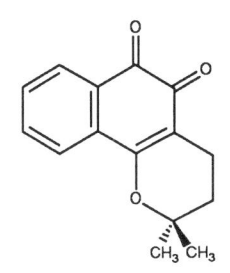

Skeletal type: α-naphthoquinone

$C_{15}H_{14}O_3$ Mol. wt 242.28

Found in *Haplophragma adenophyllum*, in the heartwood of *Tabebuia avellanedae* and *Phyllarthron comorense* (all Bignoniaceae), and in the roots of *Tectona grandis* (Verbenaceae).

Antimicrobial and antitumour activities. Also, it is active against the enzyme reverse transcriptase.

1859
Lawsone; Henna; Isojuglone

Skeletal type: naphthoquinone

$C_{10}H_6O_3$ Mol. wt 174.16

Occurs in the leaves of *Lawsonia inermis* and *L. alba* (Lythraceae), and in *Impatiens balsamina* (Balsaminaceae).

Used as a dye and as a UV screen in therapy, and as cosmetic in African and Eastern countries. Lawsone is the principle of the ancient colouring matter "henna" found on the nails of Egyptian mummies.

1860
Lucidin

Skeletal type: anthraquinone

$C_{15}H_{10}O_5$ Mol. wt 270.25

Occurs in the bark of *Coprosma* spp., in the root bark of *Coelospermum reticulatum*, in the roots of *Asperula odorata, Galium* spp. and *Rubia tinctorum*, and in *Morinda umbellata* (all Rubiaceae).

Active against strains of *Salmonella typhimurium*, and exhibits genotoxicity in the hamster fibroblast-mutagenicity assay.

1861
Lucidin ω-methyl ether

Skeletal type: anthraquinone

$C_{16}H_{12}O_5$ Mol. wt 284.27

Occurs in the roots of *Morinda parvifolia* (Rubiaceae).

Cytotoxic and antileukaemic effects *in vitro*. The roots of *M. parvifolia* are used for treatment of bronchitis and whooping cough in Chinese traditional medicine.

1862
Mansonone C

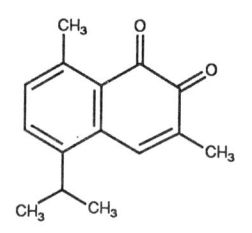

Skeletal type: α-naphthoquinone

$C_{15}H_{16}O_2$ Mol. wt 228.29

One of ten or more mansonones obtained from the heartwood of *Mansonia altissima* (Sterculiaceae), and of various elm spp., e.g., *Ulmus glabra* (Ulmaceae).

The sawdust of *Mansonia* causes sneezing, vertigo and eczema, due to the presence of these quinones, which are also antifungal compounds. Some are produced in elm seedlings as phytoalexins.

1863
9-Methoxy-α-lapachone

Skeletal type: naphthoquinone

C$_{16}$H$_{16}$O$_4$ Mol. wt 272.30

Found in the wood of *Catalpa ovata* (Bignoniaceae).

1864
5-O-Methylembelin

Skeletal type: benzoquinone

C$_{18}$H$_{28}$O$_4$ Mol. wt 308.42

Occurs in the twigs and stems of *Aegiceras corniculatum* (Aegicerataceae).

Piscicidal and antifungal (against *Pythium ultimum*) activities.

1865
Morindone

Skeletal type: anthraquinone

C$_{15}$H$_{10}$O$_5$ Mol. wt 270.25

Found in the roots and heartwood of *Morinda citrifolia* and *M. tinctoria*, in the bark of *Coprosma australis* and in the roots of *Hymenodictyon excelsum* (all Rubiaceae).

Orange-red pigment. Morinda root was an important natural dyestuff in India; *Coprosma* bark was used by the Maoris for dyeing flax.

1866
Naphthazarin

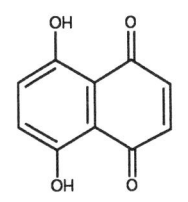

Skeletal type: naphthoquinone

C$_{10}$H$_6$O$_4$ Mol. wt 190.16

Occurs in the wood and bark of *Lomatia obtigua* (Proteaceae), and in the husks of *Juglans mandshurica* var. *sieboldiana* (Juglandaceae).

Moderate molluscicidal activity.

1867
Norobtusifolin; 2-Hydroxychrysophanol

Skeletal type: anthraquinone

C$_{15}$H$_{10}$O$_5$ Mol. wt 270.25

Found in the roots of *Myrsine africana* (Myrsinaceae).

Moderately cytotoxic in 3 human tumour cell lines.

1868
Obtusifolin 2-glucoside

Skeletal type: anthraquinone

$C_{22}H_{22}O_{10}$ Mol. wt 446.42

Occurs in the seeds of *Cassia obtusifolia* (Leguminosae); the aglycone obtusifolin occurs in roots of *Tectona grandis* (Verbenaceae).

Strongly inhibits rat platelet aggregation.

1869
Physcion 8-gentiobioside;
Physcion 8-*O*-β-D-gentiobioside

Skeletal type: anthraquinone

$C_{28}H_{32}O_{15}$ Mol. wt 608.56

Occurs in the roots of *Rheum palmatum* (Polygonaceae), in *Cassia torosa* (Leguminosae) and in *Rhamnus virgata* (Rhamnaceae). The aglycone, physcion, also occurs in many plants, e.g., in heartwood of *Maesopsis eminii* (Rhamnaceae).

Cathartic activity.

1870
Physcion 8-glucoside; Physcionin; Rheochrysin; Physcion 8-*O*-β-D-monoglucoside

Skeletal type: anthraquinone

$C_{22}H_{22}O_{10}$ Mol. wt 446.42

Occurs in the roots of *Rheum palmatum* and in *Polygonum cuspidatum* (Polygonaceae), and in *Rhamnus purshiana* (Rhamnaceae). The isomeric 1-β-D-glucopyranoside occurs in *Cassia occidentalis* (Leguminosae).

Undefined physcion monoglucoside from *Rheum palmatum* (presumably 8-glucoside) exhibits cathartic activity.

1871
Plastoquinone-9

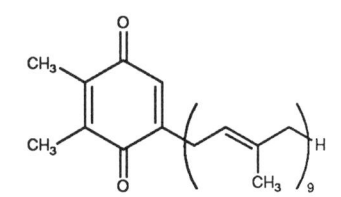

Skeletal type: benzoquinone

$C_{53}H_{80}O_2$ Mol. wt 749.22

Plastoquinones, exemplified here by plastoquinone-9, are constituents of plant chloroplasts.

Associated with photosynthetic and respiratory pathways.

1872
Plumbagin

Skeletal type: naphthoquinone

$C_{11}H_8O_3$ Mol. wt 188.19

Found in the roots of *Plumbago europaea* (Plumbaginaceae), in *Dionaea muscipula* and *Drosera rotundifolia* and other spp. (Droseraceae), in *Aristea, Sisyrynchium* and *Sparaxis* spp. (all Iridaceae), in the bark of *Diospyros* spp. (Ebenaceae), and in the root bark of *Pera ferruginea* (Euphorbiaceae).

Enhances *in vitro* phagocytosis of human granulocytes. It is cytotoxic at high doses and immunostimulating at low doses. Also, it is active as a molluscicidal agent. *Dionaea* plants have been used as an anticancer drug.

1873
Prenylbenzoquinone

Skeletal type: benzoquinone

$C_{11}H_{12}O_2$ Mol. wt 176.22

Occurs in the fruit, leaves and stems of *Phagnalon sordidum* (Compositae).

The contact allergen of *Phagnalon* spp.

1874
Primin

Skeletal type: benzoquinone

$C_{12}H_6O_3$ Mol. wt 198.18

Occurs in the glandular hairs on the leaves of *Primula obconica* and *P. elatior*, in *Anagallis hirtella, Dionysia aretioides* and *Glaux maxima* (all Primulaceae), and in the root bark of *Miconia* spp. (Melastomataceae).

Active as a molluscicide and as a feeding deterrent to insects. It is an allergen of *Primula*, causing skin irritations.

1875
Pseudohypericin

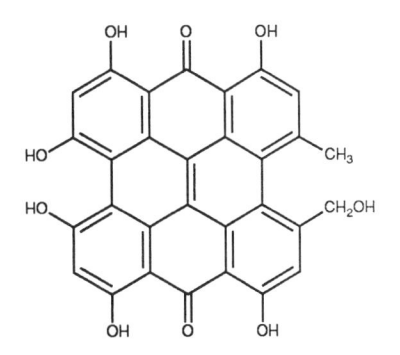

Skeletal type: bianthraquinone

$C_{30}H_{16}O_9$ Mol. wt 520.46

Found in the herbs of *Hypericum triquetrifolium* (Hypericaceae).

Antiretroviral activity both *in vitro* and *in vivo*.

1876
Pseudopurpurin

Skeletal type: anthraquinone

$C_{15}H_8O_7$ Mol. wt 300.23

Found in the roots of several Rubiaceae (mainly as a glucoside): e.g., *Rubia tinctorum*, *Galium* spp., *Asperula* spp., *Sherardia arvensis* and *Relbunium* spp.

Genotoxic in the hamster fibroblast-mutagenicity assay.

1877
Purpurin

Skeletal type: anthraquinone

$C_{14}H_8O_5$ Mol. wt 256.22

Found in the roots of *Rubia tinctorum, R. cordifolia, Galium* spp., *Asperula odorata* and *Relbunium hypocarpum* (all Rubiaceae).

Genotoxic in the hamster fibroblast-mutagenicity assay. This is the second most important pigment in madder plant (*Rubia tinctorum*). Purpurin is also the name of a flavanone from *Tephrosia purpurea*.

1878
Purpurin 1-methyl ether

Skeletal type: anthraquinone

$C_{15}H_{10}O_5$ Mol. wt 270.25

Occurs as a phytoalexin in tissue cultures of *Cinchona* spp. (Rubiaceae).

Antimicrobial activity (e.g., against *Bacillus subtilis* and *Escherichia coli*).

1879
Ramentaceone; 7-Methyljuglone

Skeletal type: naphthoquinone

$C_{11}H_8O_3$ Mol. wt 188.19

Occurs in the root and stem bark of *Diospyros usambarensis*, in *Euclea* spp., and *Maba buxifolia* (all Ebenaceae), and in the aerial parts of *Drosera* spp. (Droseraceae).

Antifungal activity against *Cladosporium cucumerinum*, and also molluscicidal activity.

1880
Rapanone

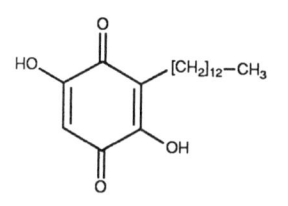

Skeletal type: benzoquinone

$C_{19}H_{30}O_4$ Mol. wt 322.45

Occurs in the bark and wood of *Rapanea maximowiczii* and in *Ardisia* and *Myrsine* spp. (Myrsinaceae), in the stems and twigs of *Aegiceras corniculata* (Aegicerataceae), in the roots of *Connarus monocarpus* (Connaraceae), and in the bulbs of *Oxalis purpurata* (Oxalidaceae).

Anthelmintic activity.

1881
Rhein; Rheic acid; Monorhein; Cassic acid; Rhubarb yellow

Skeletal type: anthraquinone

$C_{15}H_8O_6$ Mol. wt 284.23

Found in the whole plants of *Scrophularia nodosa* (Scrophulariaceae), in the roots of *Rumex* spp., in *Muehlenbeckia hastulata* and *Rheum* spp. (Polygonaceae), in *Cassia senna* (Leguminosae), in the seeds of *Kniphofia aloides* (Liliaceae) and in the leaves and stems of *Haplopappus baylatum* (Compositae).

Moderately antifungal activity against dermatophytes. The diacetyl derivative is used as an antirheumatic.

1882
Rubiadin

Skeletal type: anthraquinone

$C_{15}H_{10}O_4$ Mol. wt 254.25

Occurs in *Rubia tinctorum*, *Morinda citrifolia*, *Galium* spp., *Coprosma* spp., in the wood of *Plocama pendula* (all Rubiaceae), and in the heartwood of *Tectona grandis* (Ebenaceae).

1883
Sennoside A

Skeletal type: bianthrone

$C_{42}H_{38}O_{20}$ Mol. wt 862.77

Found in the leaves of *Cassia senna*, formed during drying, in fruit of *C. angustifolia* (Leguminosae), and in rhizomes of *Rheum palmatum* (Polygonaceae).

Cathartic activity. It is used for the treatment of chronic constipation, and may be found as a component in various herbal remedies.

1884
Tectoquinone; 2-Methylanthraquinone

Skeletal type: anthraquinone

$C_{15}H_{10}O_2$ Mol. wt 222.25

Found in *Acalypha indica* (Euphorbiaceae), in the wood of *Tectona grandis* (Verbenaceae), in the stems of *Morinda lucida* (Rubiaceae), in the bark of *Clausena heptaphylla* (Rutaceae), and also found in various Bignoniaceae.

Repellent against termites and insects in teak and other woods.

1885
Tricrozarin A

Skeletal type: naphthoquinone

$C_{13}H_{10}O_8$ Mol. wt 294.22

Occurs in the fresh bulbs of *Tritonia crocosmaeflora* (Iridaceae).

Antimicrobial against Gram-positive bacteria, fungi and yeasts.

1886
Ubiquinone-10; Coenzyme Q_{10}; Mitoquinone

Skeletal type: benzoquinone

$C_{59}H_{90}O_4$ Mol. wt 863.37

Ubiquinone-10 and related quinones with shorter side chains are universally present throughout the plant kingdom.

Involved in the electron transport in mitochondria. It is used as a cardiovascular agent. Ubiquinones are a part of the respiratory chain-reaction process.

1887
Vitamin K_1; Phylloquinone; 3-Phytomenadione; Phytonadione; Antihaemorrhagic vitamin

Skeletal type: naphthoquinone

$C_{31}H_{46}O_2$ Mol. wt 450.71

Widespread distribution in higher plants, mainly as the *trans* isomer. Important sources are chestnut leaves and alfalfa.

The prothrombogenic vitamin used in the treatment of hypothrombinaemias in human and veterinary medicine.

Chapter 44
Stilbenoids

The stilbenoids are a group of phenolic compounds, biosynthetically interrelated through their common origin from a C_6-C_2-C_6 intermediate. Like the flavonoids (q.v.), they are formed from the condensation of a *p*-hydroxycinnamic acid (C_6-C_3) precursor with three molecules of malonyl coenzyme A, but they differ in that one carbon atom is lost (by decarboxylation) in the process. There are two major groups: the stilbenes, e.g., resveratrol (**1923**), and the phenanthrenes, e.g., batatasin I (**1890**), together with their respective dihydro derivatives, the bibenzyls (e.g., batatasin IV, **1891**) and the dihydrophenanthrenes (e.g., orchinol, **1912**).

Stilbenoids can occur in the free state (as in the heartwoods of trees) or in glycosidic form (e.g., piceid, **1916**). Besides the free hydroxy derivatives, there are compounds with isopentenyl attachment (e.g., **1919**) or with carboxylic acid substitution (e.g., **1908**). An increasing number of oligomeric structures have been recorded in recent years. Thus, ε-viniferin (**1924**), a phytoalexin formed in the grape vine, is a stilbene dimer, while copalliferol B (**1895**), from the bark of a dipterocarp, is a stilbene trimer.

Stilbenoids have a limited distribution in nature. They are common in orchids, both as constitutive constituents of the bulbs and also as phytoalexins formed when the bulbs are infected with fungi. They are also regularly present as heartwood constituents both in conifers (e.g., *Pinus*) and in angiosperm trees (e.g., *Eucalyptus*), and it is usually assumed that the resistance of these woods to fungal attack is due to the presence of these phenolic materials. Other higher plant sources include *Combretum* (Combretaceae), *Dioscorea* (Dioscoreaceae) and *Hydrangea* (Hydrangeaceae). Stilbenoids are regularly present in liverworts, and lunularic acid has a particularly wide distribution in these lower plants.

Stilbenes are particularly readily detected in crude plant extracts because of their intense mauve fluorescence and their characteristic ultraviolet spectral properties. The natural distribution and chemistry of stilbenoids are well documented in reviews by Gorham (1980, 1989 and 1995).

References

Gorham, J. (1980). The stilbenoids. *Progress in Phytochemistry*, **6**, 203–252.
Gorham, J. (1989). Stilbenes and phenanthrenes. In *Methods in Plant Biochemistry*, Volume 1, *Plant Phenolics* (Harborne, J.B., ed.), pp. 159–196. London: Academic Press.
Gorham, J. (1995). *The Biochemistry of the Stilbenoids*. London: Chapman and Hall.

1888
Agrostophyllin

Skeletal type: pyranophenanthrene

$C_{17}H_{14}O_4$ Mol. wt 282.30

Occurs in *Agrostophyllum khasiyanum* (Orchidaceae).

1889
Astringin

Skeletal type: stilbene

$C_{20}H_{22}O_9$ Mol. wt 406.40

Occurs in Canadian *Picea* spp. (Pinaceae), and also in *Angophora cordifolia* and *Eucalyptus* spp. (Myrtaceae).

Protective in heartwood and bark against fungal invasion.

1890
Batatasin I

Skeletal type: phenanthrene

$C_{17}H_{16}O_4$ Mol. wt 284.32

Found in the bulbils of *Dioscorea batatas* and *D. dumetorum* (Dioscoreaceae).

Has a dormancy regulating activity in *Dioscorea*, yam.

1891
Batatasin IV

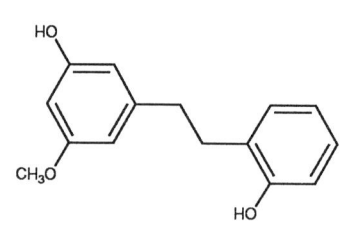

Skeletal type: phenanthrene

$C_{15}H_{16}O_3$ Mol. wt 244.29

Found in *Dioscorea batatas* and in the tubers of *Dioscorea rotundata* (Dioscoreaceae) infected with *Bothyodiplodia theobromae*.

Fungitoxic against *Cladosporium cladosporioides*. Batatasins have dormancy regulating activity in *Dioscorea*, yam.

1892
Blestriarene B

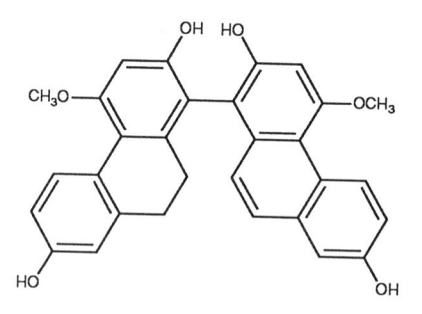

Skeletal type: phenanthrene dimer

$C_{30}H_{24}O_6$ Mol. wt 480.52

Occurs in the tubers of *Bletilla striata* (Orchidaceae), along with the related compounds blestriarene A and C.

Antibacterial activity (against *Staphylococcus aureus* and *S. mutans*).

1893
Chlorophorin

Skeletal type: stilbene

$C_{24}H_{28}O_4$ Mol. wt 380.49

Occurs in the wood of *Chlorophora excelsa* and *C. regia* (Moraceae).

Inhibits drying of varnishes and polyester lacquers.

1894
Coelogin

Skeletal type: phenanthrene

$C_{17}H_{16}O_5$ Mol. wt 300.32

Found in the aerial parts of *Coelogyne ovalis* and the whole plants of *C. cristata* (Orchidaceae).

One of the spasmolytic principles of *Coelogyne ovalis*.

1895
Copalliferol B

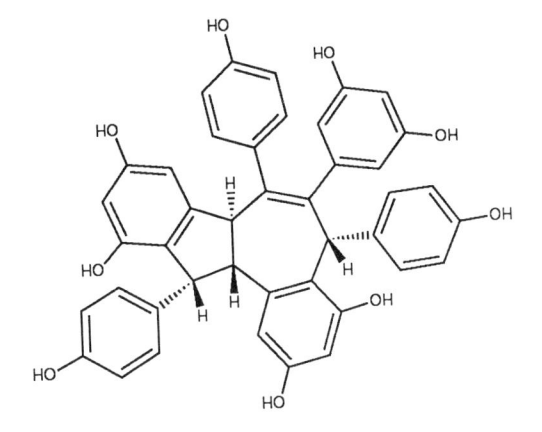

Skeletal type: stilbene trimer

$C_{42}H_{32}O_9$ Mol. wt 680.72

Found in the bark of *Vateria copallifera* (Dipterocarpaceae).

Antibacterial activity against *Staphylococcus* and *Escherichia coli*.

1896
Demethylbatatasin IV

Skeletal type: bibenzyl

$C_{14}H_{14}O_3$ Mol. wt 230.27

Occurs in infected tubers of *Dioscorea rotundifolia, D. bulbifera* and *D. dumetorum* (Dioscoreaceae) as a phytoalexin.

Moderate antifungal activity (against *Aspergillus niger* and *Penicillium sclerotigenum*) and antibacterial activity (against e.g., *Escherichia coli*).

1897
Dihydropinosylvin; 3,5-Dihydroxybibenzyl

Skeletal type: bibenzyl

$C_{14}H_{14}O_2$ Mol. wt 214.27

Found in the heartwood of *Pinus* spp. (Pinaceae), in *Dioscorea batatas*, and in infected tubers of *Dioscorea rotundata* (Dioscoreaceae) as a phytoalexin.

Antifungal and antibacterial activities.

1898
Dihydroresveratrol

Skeletal type: bibenzyl

$C_{14}H_{14}O_3$ Mol. wt 230.27

Found in some *Morus* spp. (Moraceae), and in infected tubers of *Dioscorea bulbifera* and *D. dumetorum* (Dioscoreaceae) as a phytoalexin.

Inhibits germination of fungi (*Cladosporium cladosporioides, Trichophyton mentayophytes*) and also shows antibacterial activity (against *Escherichia coli, Serratia marcescens* and *Staphylococcus aureus*).

1899
4,7-Dihydroxy-2-methoxy-9,10-dihydrophenanthrene

Skeletal type: phenanthrene

$C_{15}H_{14}O_3$ Mol. wt 242.28

Occurs in the tubers of *Bletilla striata* (Orchidaceae).

Antibacterial activity.

1900
Flavidin

Skeletal type: phenanthrene

$C_{15}H_{12}O_3$ Mol. wt 240.26

Found in the aerial parts of *Coelogyne ovalis*, and in the whole plants of *C. flavida, Pholidota articulata* and *Otochilus fusta* (all Orchidaceae).

One of the spasmolytic principles of *Coelogyne ovalis*.

1901
Glepidotin C

Skeletal type: bibenzyl

$C_{19}H_{22}O_3$ Mol. wt 298.39

Occurs in *Glycyrrhiza lepidota* (Leguminosae).

Weak antimicrobial activity (against *smegmatis*).

1902
Gnetin A

Skeletal type: stilbene dimer

$C_{28}H_{22}O_6$ Mol. wt 454.48

Found in the bark of *Gnetum leyboldii* (Gnetaceae).

1903
Hircinol

Skeletal type: phenanthrene

$C_{15}H_{14}O_3$ Mol. wt 242.28

Occurs in the peel extract of healthy tubers of *Dioscorea rotundata* (Dioscoreaceae), and as a phytoalexin from the bulbs of *Loroglossum hircinum* (Orchidaceae).

Inhibits germ tube growth of various fungi such as *Aspergillus niger*. Also, it inhibits enzymatic IAA degradation *in vitro*.

1904
Hydrangenol

Skeletal type: bibenzyl

$C_{15}H_{12}O_4$ Mol. wt 256.26

Occurs in *Hydrangea macrophylla* var. *thunbergii* (Saxifragaceae).

Allergenic principle of *Hydrangea*; it shows antifungal activity.

1905
Isobatatasin I

Skeletal type: phenanthrene

$C_{17}H_{16}O_4$ Mol. wt 284.32

Found in *Dioscorea rotundata* and *Tamus communis* (Dioscoreaceae), and in *Combretum psidioides* (Combretaceae).

Inhibits germ tube growth of several fungi (e.g., *Aspergillus niger* and *Botrytis cinerea*).

1906
Isorhapontin

Skeletal type: stilbene

$C_{21}H_{24}O_9$ Mol. wt 420.42

Occurs in the needles of *Picea abies* and other *Picea* spp. (Pinaceae).

Protective in heartwood and bark against fungal invasion.

1907
Loroglossol

Skeletal type: phenanthrene

$C_{16}H_{16}O_3$ Mol. wt 256.30

Occurs as a phytoalexin in the bulbs of *Loroglossum hircinum* (Orchidaceae).

Inhibits enzymatic IAA degradation *in vitro*, and spore germination of *Phytophthora infestans*. It is weakly active against *Monilinia fruticola* spore germination.

1908
Lunularic acid

Skeletal type: bibenzyl

$C_{15}H_{14}O_4$ Mol. wt 258.28

Found in the roots of *Hydrangea macrophylla* (Saxifragaceae), and in many liverworts, *Lunularia*, *Marchantia* (Hepaticae).

Controls growth and drought resistance in *Lunularia*, and is a dormancy factor. It inhibits elongation of *Avena* coleoptiles and fungal spore germination, and has algicidal activity.

1909
Lunularin

Skeletal type: bibenzyl

$C_{14}H_{14}O_2$ Mol. wt 214.27

Ubiquitous in liverworts (Hepaticae), and also occurs in the heartwood of *Morus laevigata* (Moraceae), and in *Hydrangea macrophylla* (Saxifragaceae).

Inhibits fungal spore germination (e.g., of *Uromyces fabae*).

1910
Marchantin A

Skeletal type: bibenzyl dimer

$C_{28}H_{24}O_5$ Mol. wt 440.50

Occurs in the liverworts *Marchantia polymorpha*, *M. palacea* var. *diptera* and *M. tosana* (Marchantales).

1911
3′-O-Methylbatatasin III

Skeletal type: bibenzyl

$C_{16}H_{18}O_3$ Mol. wt 258.32

Found in the tubers of *Bletilla striata* and in the aerial parts of *Ceologyne ovalis* (both Orchidaceae).

Antibacterial activity against strains of *Streptococcus mutans*.

1912
Orchinol

Skeletal type: phenanthrene

$C_{16}H_{16}O_3$ Mol. wt 256.30

Occurs in *Orchis* spp., *Coeloglossum viride*, *Anacamptis pyramidalis*, *Nigritella nigra*, *Gymnadenia albida* and *Serapis lingua* (Orchidaceae).

Active against soil fungi, and *Candida lipolytica*, *Phytophthora infestans*, *Monilinia fruticola*, and some other fungi. It is a phytoalexin. Orchinol inhibits enzymatic IAA degradation *in vitro*. It restricts invasion of tubers by *Rhizoctonia repens* to a mycorrhizal relationship.

1913
Oxyresveratrol; 2,4,3′,5′-Tetrahydroxystilbene

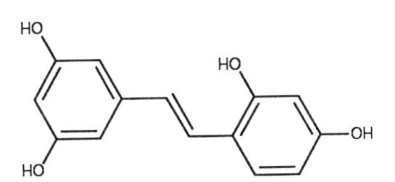

Skeletal type: stilbene

$C_{14}H_{12}O_4$ Mol. wt 244.25

Occurs in *Morus alba*, in the heartwood of *Maclura pomifera*, *Artocarpus lakoocha*, *Chlorophora regia*, and in *Cudrania* spp. (all Moraceae), and in *Veratrum grandiflorum* (Liliaceae).

Antifungal activity, also against human superficial dermatomycosis. It inhibits respiration of rat liver mitochondria at low concentrations. Probably it has a protective role in the wood of *Maclura pomifera*.

1914
Phyllodulcin

Skeletal type: bibenzyl

$C_{16}H_{14}O_5$ Mol. wt 286.29

Found in *Hydrangea macrophylla* var. *thunbergii* (Saxifragaceae).

The sweet principle of *Hydrangea*. It shows antifungal activity.

1915
Piceatannol; 3,3′,4,5′-Tetrahydroxystilbene

Skeletal type: stilbene

$C_{14}H_{12}O_4$ Mol. wt 244.25

Found in the heartwood of *Picea* and *Pinus* spp. (Pinaceae), and in the wood of legume trees, e.g., *Laburnum anagyroides*.

Protects against fungal attack.

1916
Piceid; Polydatin; Resveratrol 3-*O*-β-D-glucoside

Skeletal type: stilbene

$C_{20}H_{22}O_8$ Mol. wt 390.40

Found in the roots of *Polygonum cuspidatum* and *P. multiflorum* (Polygonaceae), in *Picea* spp. (Pinaceae), in *Eucalyptus* spp., and *Angophora cordifolia* (both Myrtaceae).

Antifungal and antibacterial activities. In addition, it inhibits deposition of lipid peroxides and cholesterol in injured liver of rats, similar to resveratrol (q.v.).

1917
Pinosylvin; Stilbene-3,5-diol; 3,5-Dihydroxystilbene

Skeletal type: stilbene

$C_{14}H_{12}O_2$ Mol. wt 212.25

Occurs in the heartwood of over 60 *Pinus* spp. (Pinaceae), in *Dalbergia sisso* (Leguminosae), *Alnus sieboldiana* (Betulaceae) and in *Nothofagus* spp. (Fagaceae). It also occurs after fungal infection, insect attack or desiccation in the sapwood of *Pinus* spp.

Toxic to fungi, bacteria and some animals.

1918
Pinosylvin methyl ether; 5-Methoxy-3-stilbenol

Skeletal type: stilbene

$C_{15}H_{14}O_2$ Mol. wt 226.28

Occurs in the wood of over 60 *Pinus* spp. (Pinaceae), and also in *Alnus sieboldiana* and *A. crispa* (Betulaceae).

Active against wood decaying fungi and *Phytophthora infestans*. It is an antifeedant to the snowshoe hare, *Lepus americanus*.

1919
4-Prenyldihydropinosylvin;
4-Isopentenyldihydropinosylvin

Skeletal type: bibenzyl

$C_{19}H_{22}O_2$ Mol. wt 282.39

Found in *Glycyrrhiza lepidota* (Leguminosae), *Helichrysum umbraculigerum* (Compositae), and *Radula complanata* (Hepaticae).

Antimicrobial activity.

1920
4'-Prenyloxyresveratrol; 4'-Isopentenyloxyresveratrol

Skeletal type: stilbene

$C_{19}H_{20}O_4$ Mol. wt 312.37

Occurs in the shoots of *Morus alba* (Moraceae) infected with *Fusarium solani* f. sp. *mori*.

Antifungal activity.

1921
4-Prenylresveratrol; 4-Isopentenylresveratrol

Skeletal type: stilbene

$C_{19}H_{20}O_3$ Mol. wt 296.37

Occurs as a phytoalexin in infected germinating seeds of peanut, *Arachis hypogaea* (Leguminosae). It is obtained as a mixture of *cis* and *trans* isomers.

Antifungal against *Cladosporium cucumerinum*.

1922
Pterostilbene; Resveratrol 3,5-dimethyl ether

Skeletal type: stilbene

$C_{16}H_{16}O_3$ Mol. wt 256.30

Found in sandalwood, *Pterocarpus santalinus*, and other *Pterocarpus* spp. (Leguminosae), and in *Vitis vinifera* (Vitaceae).

Protects heartwood and bark against fungal invasion.

1923
Resveratrol; 3,5,4'-Trihydroxystilbene

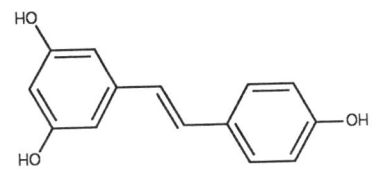

Skeletal type: stilbene

$C_{14}H_{12}O_3$ Mol. wt 228.25

Occurs in *Polygonum cuspidatum* and *P. multiflorum* (Polygonaceae), in *Veratrum grandiflorum* (Liliaceae), in the needles of *Picea glehnii* and *Pinus* spp. (Pinaceae), in the wood of *Eucalyptus wandoo* (Myrtaceae) and of *Nothofagus* spp. (Fagaceae), in *Cassia dentata*, *Intsia bijuga* and *Trifolium dubium* (Leguminosae), in *Artocarpus* and *Morus* spp. and *Cudrania javanensis* (Moraceae), and in *Vitis* spp. (Vitaceae).

Antibacterial and antifungal activities. Inhibits lipid peroxidation induced by ADP, NADPH in rat liver microsomes, and deposition of triglyceride and cholesterol, thus having a protective action and a lipid lowering activity.

1924
ε-Viniferin

Skeletal type: stilbene dimer

$C_{28}H_{22}O_6$ Mol. wt 454.48

Found in the infected leaves of *Vitis vinifera* (Vitaceae) as a phytoalexin.

Spore germination inhibitor of fungi such as *Botrytis cinerea*, *Cladosporium cucumerinum*, and *Plasmopara viticola*.

Chapter 45
Tannins

Tannins occur widely in vascular plants, their occurrence in the angiosperms being particularly associated with woody tissues. By definition, they have the ability to react with protein, forming water-insoluble copolymers. Industrial tannins are substances of plant origin which, because of their ability to crosslink with protein, are capable of transforming raw animal skins into leather. Plant tissues high in tannin (e.g., oak leaves) are largely avoided by most feeders, because of the astringent taste they impart. One of the major functions of tannins in plants is thought to be as a barrier to herbivory, and most tannins that have been purified and studied are biologically active.

Chemically, there are two main types of tannin: condensed tannins, such as procyanidin B4 (**1953**); and hydrolysable tannins, such as agrimoniin (**1926**). The two classes are unevenly distributed in the plant kingdom. The condensed tannins occur almost universally in ferns and gymnosperms, and are widespread among the angiosperms, especially in trees and shrubs. By contrast, hydrolysable tannins are limited to dicotyledonous plants. Both types of tannin can occur together in the same plant, as they do in oak bark and leaf.

Condensed tannins or flavolans are formed biosynthetically by the condensation of catechin (q.v.) units to form dimers and then higher oligomers, with carbon–carbon bonds linking one flavan unit with another by a 4–8 or 6–8 link. The name proanthocyanidin is used alternatively for condensed tannins because, on treatment with hot acid, some of the carbon–carbon linking bonds are broken and anthocyanidin monomers (q.v.) may be formed. Most proanthocyanidins are procyanidins (e.g., **1953**), which yield cyanidin on acid treatment. There are also prodelphinidins (e.g., **1940**) and propelargonidins (e.g., **1925**). Mixed oligomers are also known, such as gambiriin C (**1941**). Most proanthocyanidins isolated so far have up to six flavan units (see **1936**), but there is little doubt that higher oligomers await characterization in plants.

Hydrolysable tannins are mainly of two types: gallotannins, in which a glucose core is surrounded by five or more galloyl ester groups (see, e.g., pentagalloylglucose, **1951**); and ellagitannins such as agrimoniin (**1926**), in which units of hexahydroxydiphenic acid, derived by linkage of two gallic acid groups, are present.

Condensed tannins have been used in medicine to aid the healing of wounds and burns. When applied to the skin, they produce an impervious layer under which the healing process can take place. They are also thought to have some protective value against toxins when taken internally, while in ruminants they minimize the effects of bloat. Hydrolysable tannins are of pharmacological interest because of their antiviral and antitumour properties.

Condensed tannins are readily detected in plant tissues by their formation of a red colour when tissues are heated in solution in the presence of mineral acid. Ellagitannins give a specific carmine-red colour reaction with nitrous acid. The concentration of tannins in plants can be determined by their reaction with a protein (e.g., haemoglobin from blood). Methods of tannin analysis are reviewed by Porter (1989). The chemistry and distribution of plant tannins are comprehensively reviewed by Haslam (1989) and Porter (1994).

References

Haslam, E. (1989). *Plant Polyphenols. Vegetable Tannins Revisited*. Cambridge University Press.
Hemingway, R.W. and Karchesy, J.J. (1989). *Chemistry and Significance of Condensed Tannins*. New York: Plenum Press.
Porter, L.J. (1994). Flavans and proanthocyanidins. In *The Flavonoids – Advances in Research since 1986*. (Harborne, J.B., ed.), pp. 23–55. London: Chapman and Hall.
Porter, L.J. (1989). Tannins. In *Methods in Plant Biochemistry*, Volume 1, *Plant Phenolics*. (Harborne, J.B., ed.), pp. 389–420. London: Academic Press.

1925
Afzelechin-(4α→8)-afzelechin

Skeletal type: propelargonidin dimer

$C_{30}H_{26}O_{10}$ Mol. wt 546.54

Occurs in the bark of *Kandelia candel* (Rhizophoraceae).

Tans proteins.

1926
Agrimoniin

Skeletal type: ellagitannin

$C_{82}H_{54}O_{52}$ Mol. wt 1871.33

Found in the roots of *Agrimonia pilosa* and *A. japonica*, and in *Potentilla kleiniana* (Rosaceae).

Antitumour activity against Sarcoma 180 in mice; it is a moderate inhibitor of induced lipid peroxidation in rat liver mitochondria. *Agrimonia* is used as an antidiarrhoeal, haemostatic and an antiparasitic medicine in Japan and China.

1927
Alnusiin

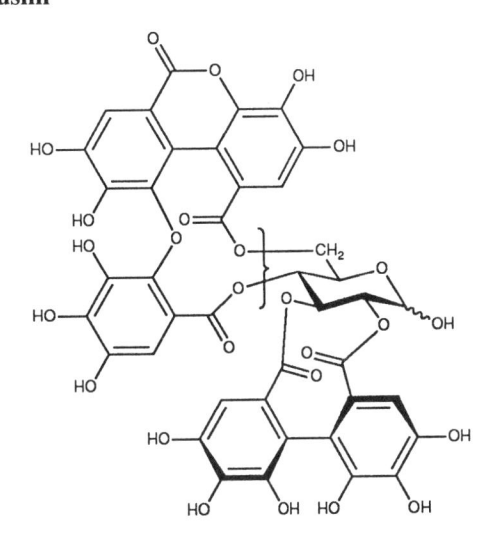

Skeletal type: ellagitannin

$C_{41}H_{26}O_{26}$ Mol. wt 934.66

Found in the fruit of *Alnus sieboldiana* (Betulaceae).

Inhibits induced lipid peroxidation in the mitochondria of fat cells in rats, and the microsomes of rat liver cells, and also shows antitumour activity against Sarcoma 180 in mice.

1928
Casuarictin

Skeletal type: ellagitannin

$C_{41}H_{28}O_{26}$ Mol. wt 936.68

Occurs in *Casuarina stricta* (Casuarinaceae), *Stachyurus praecox* (Stachyuraceae), *Psidium guajava*, *Syzygium aromaticum* and *Eucalyptus viminalis* (Myrtaceae), *Quercus* spp. (Fagaceae), and *Rubus* spp. (Rosaceae).

In vitro antihepatotoxic activity, due to enzyme inhibitory action on glutamine-pyruvic transaminase.

1929
Casuarinin

Skeletal type: ellagitannin

$C_{41}H_{28}O_{26}$ Mol. wt 936.68

Found in *Casuarina stricta* (Casuarinaceae), in *Stachyurus praecox* (Stachyuraceae), in *Psidium guajava*, *Syzygium jambos*, *Feijoa sellowiana* and *Eucalyptus viminalis* (Myrtaceae), in *Osbeckia chinensis* (Melastomataceae), and in the leaves of *Liquidambar formosana* (Hamamelidaceae).

Inhibits lipid peroxidation of rat liver microsomes, and antioxidant efficiency in the rabbit erythrocyte membrane ghost system.

1930
Chebulagic acid

Skeletal type: ellagitannin

$C_{41}H_{30}O_{27}$ Mol. wt 954.66

Occurs in *Terminalia chebula* (Combretaceae).

Inhibits induced lipid peroxidation in the mitochondria of rat liver cells, but enhances adrenocorticotrophic hormone-induced lipolysis in fat cells of rats.

1931
Chebulinic acid

Skeletal type: gallotannin

$C_{41}H_{32}O_{27}$ Mol. wt 956.68

Found in the fruit of *Phyllanthus emblica* (Euphorbiaceae), and in *Terminalia chebula* (Combretaceae).

Inhibits adrenaline-induced lipolysis in fat cells of rats, but enhances adrenocorticotrophic hormone-induced lipolysis and inhibits induced lipid peroxidation in the microsomes of rat liver cells.

1932
Coriariin A

Skeletal type: ellagitannin

$C_{82}H_{58}O_{52}$ Mol. wt 1875.37

Occurs in the leaves of *Coriaria japonica* (Coriariaceae).

Antitumour activity.

1933
Corilagin

Skeletal type: gallotannin

$C_{27}H_{22}O_{18}$ Mol. wt 634.46

Found in *Terminalia chebula* (Combretaceae) and *Acer* spp. (Aceraceae). Further sources are the leaves of *Poupartia fordii* (Anacardiaceae), the leaves of *Sapium japonicum, S. sebiferum, Ricinus communis* and *Aleurites cordata* (all Euphorbiaceae), and *Anogeissus latifolia* (Combretaceae). One of several components (see also pentagalloylglucose) which make up the commercially available "tannic acid" prepared from plant galls of *Rhus semialata* (Anacardiaceae).

In vitro antihepatotoxic activity due to enzyme inhibitory action on glutamine-pyruvic transaminase, and also inhibitory action on induced lipolysis in rat liver microsomes. It inhibits adrenaline-induced lipolysis in fat cells of rats. "Tannic acid" is used for tanning, clarifying wine and beer, as an aid in the manufacture of imitation horn and, in veterinary practice, as an astringent and haemostatic.

1934
3,5-Di-*O*-galloyl-4-*O*-digalloylquinic acid

Skeletal type: gallotannin

$C_{35}H_{28}O_{22}$ Mol. wt 800.59

One of several components which make up the commercially available "tannic acid" prepared from plant galls of *Rhus semialata* (Anacardiaceae).

Inhibits human immunodeficiency virus reverse transcriptase, thus showing anti-AIDS activity. "Tannic acid" also exhibits this activity. Further activities of "tannic acid" are described under corilagin.

1935
Epicatechin-(4β→8)-*ent*-epicatechin

Skeletal type: procyanidin dimer

$C_{30}H_{26}O_{12}$ Mol. wt 578.54

Occurs in the fruit of *Chamaerops humilis* (Palmae).

Tans proteins.

1936
[Epicatechin-(4β→8)]₅-epicatechin

Skeletal type: procyanidin hexamer

$C_{90}H_{74}O_{36}$ Mol. wt 1731.58

Occurs in the bark of *Cinnamomum cassia* (Lauraceae).

Tans proteins

1937
Epigallocatechin-(4β→8)-epicatechin-3-*O*-gallate ester

Skeletal type: prodelphinidin dimer

$C_{37}H_{30}O_{17}$ Mol. wt 746.65

Found in the leaves of the tea plant, *Camellia sinensis* (=*Thea sinensis*) (Theaceae).

Tans proteins.

1938
Eugeniin; Tellimagrandin II

Skeletal type: ellagitannin

$C_{41}H_{28}O_{26}$ Mol. wt 936.68

Found in the buds of *Syzygium aromaticum* (Myrtaceae), in *Tellima grandifolia* (Saxifragaceae), *Rosa* spp. (Rosaceae), *Quercus* spp. (Fagaceae), *Cornus* spp. (Cornaceae), in the leaves of *Coriaria japonica* (Coriariaceae), and in *Fuchsia* spp. (Onagraceae).

Antiviral activity against herpes simplex, and inhibitory action on adrenaline-induced lipolysis of fat cells of rats.

1939
***ent*-Fisetinidol-(4β→8)-catechin-(6→4β)-*ent*-fisetinidol**

Skeletal type: profisetinidin trimer

$C_{45}H_{38}O_{16}$ Mol. wt 834.80

Found in the heartwood of *Schinopsis balansae, S. lorentzii, Rhus leptodictya* and *R. lancea* (Anacardiaceae).

Tans proteins.

1940
Gallocatechin-(4α→8)-epigallocatechin

Skeletal type: prodelphinidin dimer

$C_{30}H_{26}O_{14}$ Mol. wt 610.54

Found in the leaves of *Ribes sanguineum* (Crassulaceae).

Tans proteins.

1941
Gambiriin C; Epiafzelechin-(4β→8)-catechin

Skeletal type: propelargonidin dimer

$C_{30}H_{26}O_{11}$ Mol. wt 562.54

Found in the leaves of *Uncaria gambir* (Rubiaceae).

Tans proteins.

1942
Gemin A

Skeletal type: ellagitannin

$C_{82}H_{56}O_{52}$ Mol. wt 1873.35

Found in *Geum japonicum* (Rosaceae).

Antitumour activity against Sarcoma 180 in mice. It inhibits induced lipolysis in rat liver mitochondria, and adrenaline-induced lipolysis in fat cells of rats.

1943
Geraniin

Skeletal type: ellagitannin

$C_{41}H_{28}O_{27}$ Mol. wt 952.68

Occurs in *Geranium* spp. (Geraniaceae), *Erythroxylum coca* (Erythroxylaceae), *Euphorbia* spp. (Euphorbiaceae), *Acer* spp. (Aceraceae), *Cercidiphyllum* spp. (Cercidiphyllaceae), the leaves of *Coriaria japonica* (Coriariaceae) and *Fuchsia* spp. (Onagraceae), and in the bark and leaves of *Mallotus japonicus* and other Euphorbiaceae.

Inhibits induced lipid peroxidation in rat liver microsomes and the adrenaline-induced lipolysis in fat cells of rats, but enhances adrenocorticotrophic hormone-induced lipolysis in fat cells.

1944
Guibourtinidol-(4α→6)-catechin

Skeletal type: proguibourtinidincyanidin dimer

$C_{30}H_{26}O_{10}$ Mol. wt 546.54

Found in the heartwood of *Acacia luederitzii* (Leguminosae).

Tans proteins.

1945
Isoterchebin; Trapain

Skeletal type: ellagitannin

$C_{41}H_{30}O_{27}$ Mol. wt 954.69

Occurs in the fruit of *Cornus officinalis* (Cornaceae), in *Cytinus hypocistis* (Rafflesiaceae), and in *Trapa japonica* (Trapaceae).

Inhibits induced lipid peroxidation in rat liver mitochondria.

1946
Kandelin A-1; Cinchonain-1a-(4β→8)-catechin

Skeletal type: cinchonain dimer

$C_{39}H_{32}O_{15}$ Mol. wt 740.69

Found in the bark of *Kandelia candel* (Rhizophoraceae).

Tans proteins.

1947
Luteoliflavan-(4β→8)-eriodictyol 5-glucoside

Skeletal type: proluteolinidin

$C_{36}H_{36}O_{16}$ Mol. wt 724.68

Occurs in *Sorghum vulgare* cv. Szegeditoerpe (Gramineae).

Tans proteins.

1948
Mahuannin D;
ent-Apigeniflavan-(2α→7,4α→8)-epiafzelechin

Skeletal type: procyanidin dimer

$C_{30}H_{24}O_9$ Mol. wt 528.52

Occurs in the roots of *Ephedra* spp. (Ephedraceae).

Tans proteins.

1949
Mallotusinic acid

Skeletal type: ellagitannin

$C_{48}H_{32}O_{32}$ Mol. wt 1120.78

Occurs in the bark and leaves of *Mallotus japonicus*, in *Aleurites cordata* and *Alchornea trewioides* and, in trace amounts, in various *Euphorbia* spp. (all Euphorbiaceae).

Inhibits induced lipid peroxidation in rat liver microsomes, but enhances adrenocorticotrophic hormone-induced lipolysis in fat cells of rats.

1950
Pedunculagin

Skeletal type: ellagitannin

$C_{34}H_{24}O_{22}$ Mol. wt 784.57

Found in *Quercus* spp. (Fagaceae), *Rubus* spp. and *Potentilla* sp. (Rosaceae), *Casuarina stricta* (Casuarinaceae), *Stachyurus praecox* (Stachyuraceae), *Camellia japonica* (Theaceae) and *Juglans* spp. (Juglandaceae).

In vitro antihepatotoxic activity due to enzyme inhibitory action on glutamine-pyruvic transaminase. It inhibits induced lipid peroxidation in rat liver mitochondria and adrenaline-induced lipolysis in fat cells of rats.

1951
Pentagalloyl-β-D-glucose

Skeletal type: gallotannin

$C_{41}H_{32}O_{26}$ Mol. wt 940.71

Occurs in the leaves of *Nuphar japonicum* (Nymphaeaceae), in the fruit of *Terminalia chebula* (Combretaceae), in the leaves of *Acer platanoides* (Aceraceae), in the galls of *Quercus infectoria*, in the leaves and galls of *Rhus* spp. and *Cotinus* spp. (both Anacardiaceae), in the leaves of *Fuchsia* and *Epilobium* spp. (both Onagraceae), and in the leaves of *Rosa* spp. (Rosaceae) and *Geranium robertianum* (Geraniaceae).

Antiviral activity against human immunodeficiency virus. It inhibits induced lipid peroxidation in rat liver mitochondria and microsomes.

1952
Proanthocyanidin A2;
Epicatechin-(2β→7, 4β→8)-epicatechin

Skeletal type: procyanidin dimer

$C_{30}H_{24}O_{12}$ Mol. wt 576.52

Occurs in the nuts of *Cola acuminata* (Sterculiaceae), the berries of *Vaccinium vitis-idaea* (Ericaceae), and the fruits of the horse-chestnut, *Aesculus hippocastanum* (Hippocastanaceae) and *Persea gratissima* (Lauraceae).

Tans proteins.

1953
Procyanidin B4; Catechin-(4α→8)-epicatechin

Skeletal type: procyanidin dimer

$C_{30}H_{26}O_{12}$ Mol. wt 578.54

Widespread occurrence, e.g., in the leaves of the raspberry, *Rubus idaeus* (Rosaceae).

An infusion of raspberry leaves containing procyanidin B4 and hydrolysable tannins is used as a gargle, for wounds and ulcers and for children's stomach disorders.

1954
Robinetinidol-(4α→8)-catechin-(6→4α)-robinetinidol

Skeletal type: prorobinetinidin trimer

$C_{45}H_{38}O_{18}$ Mol. wt 866.80

Found in the bark of *Acacia mearnsii* (Leguminosae).

Tans proteins.

1955
Rugosin D

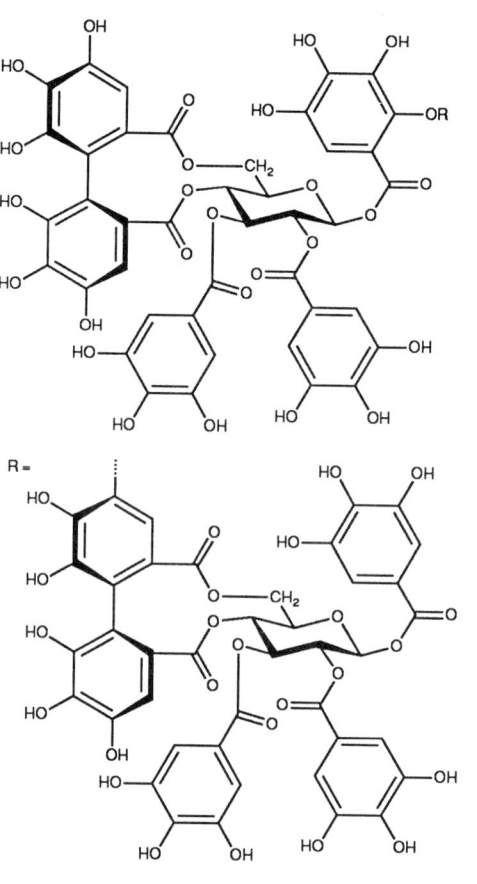

Skeletal type: ellagitannin

$C_{82}H_{58}O_{52}$ Mol. wt 1875.37

Occurs in the flower petals of *Rosa rugosa*, and in *Filipendula ulmaria* (Rosaceae).

Antitumour activity against Sarcoma 180 in mice.

1956
Tellimagrandin I

Skeletal type: ellagitannin

$C_{34}H_{25}O_{22}$　　　　　　　　　　Mol. wt 785.57

Found in *Tellima grandifolia*, *Rosa* spp. and *Geum japonicum* (Rosaceae), in *Quercus* sp. (Fagaceae), *Fuchsia* sp. (Onagraceae), *Casuarina stricta* (Casuarinaceae), *Camellia japonica* (Theaceae), *Stachyurus praecox* (Stachyuraceae), and *Syzygium aromaticum*, *Feijoa sellowiana*, *Psidium guajava* and *Eucalyptus viminalis* (Myrtaceae).

In vitro antihepatotoxic activity due to enzyme inhibitory action on glutamine-pyruvic transaminase. It inhibits adrenaline-induced lipolysis in fat cells of rats.

1957
1,2,3,4-Tetragalloyl-α-D-glucose

Skeletal type: gallotannin

$C_{34}H_{28}O_{22}$　　　　　　　　　　Mol. wt 788.60

Found in the roots of *Nuphar variegatum* (Nympheaceae).

Antibacterial activity against *Staphylococcus aureus* and *Proteus vulgare*.

Chapter 46
Xanthones

The xanthones are a biologically active yet taxonomically restricted group of plant phenols. Biosynthetically, they are related to the flavonoids (q.v.), being formed by the condensation of a phenylpropanoid precursor with two instead of three malonyl coenzyme A units. Benzophenones (see Miscellaneous phenolics) are thought to be biosynthetic intermediates, although the two classes of compound are not usually found together. Most xanthones occur in the free state, either as trihydroxy- (e.g., gentisein, **1968**) or as tetrahydroxy- (e.g., norathyriol, **1982**) derivatives or in partly methylated form (e.g., athyriol, **1958**). Some xanthones have additional isoprenylated substituents (e.g., calophyllin B, **1960**). There are also both *O*-glycosides (e.g., norswertianolin, **1984**) and *C*-glycosides (e.g., lancerin, **1974**) but these are relatively uncommon. More than one system has been used for numbering the substituents in the xanthone nucleus; the IUPAC preferred system is used here (see structure of athyriol, **1958**).

Hydroxyxanthones and their methyl ethers are mainly found in two unrelated plant families, the Gentianaceae and the Guttiferae, where they occur widely in the roots and leaves. There are occasional occurrences in six other families: the Leguminosae, Loganiaceae, Lythraceae, Moraceae, Polygalaceae and Rhamnaceae. By contrast, the *C*-glucoside mangiferin (**1976**) has a much wider distribution in nature. It has been detected in 28 genera from 19 angiosperm families and it also is found in a number of ferns. Hydroxyxanthones, some with chlorine substitution, are also found in lichens and in fungi (Turner and Aldridge, 1983).

Xanthones show considerable biological activity, and it is surprising that none has so far an established use in medicine. Compounds such as bellidifolin (**1959**) inhibit monoamine oxidase activity, while psorospermin (**1985**) exhibits both cytotoxic and antitumour activity. Many other xanthones are recorded as having antimicrobial activity, insecticidal properties, anti-inflammatory effects or tuberculostatic activity. The mammalian metabolism of xanthones has been little studied, except that of mangiferin. When this is fed to rabbits or dogs, it undergoes an unexpected breakdown with the production of euxanthone (1,7-dihydroxyxanthone) and its 7-glucuronide.

Xanthones can be detected in plant extracts by their fluorescent colours on TLC plates when they are examined in ultraviolet light. Mangiferin and its derivatives appear as bright orange colours when examined similarly in the presence of ammonia. They are then further characterized by their ultraviolet and NMR spectra.

References

Natural Distribution

Bennett, G.J. and Lee, H.H. (1989). Xanthones from Guttiferae. *Phytochemistry*, **28**, 967–998.
Hostettmann, K. and Wagner, H. (1972). Xanthone glycosides. *Phytochemistry*, **16**, 821–829.
Turner, W.B. and Aldridge, D.C. (1983). *Fungal Metabolites*, Volume II. London: Academic Press.

Analysis and Biological Properties

Hostettmann, K. and Hostettmann, M. (1989). Xanthones. In *Plant Phenolics*, Volume 1, *Methods in Plant Biochemistry* (Harborne, J.B., ed.), pp. 493–508. London: Academic Press.

1958
Athyriol; 3-Methoxy-1,6,7-trihydroxyxanthone

$C_{14}H_{10}O_6$ Mol. wt 274.23

Occurs in the aerial parts of the fern *Athyrium mesosorum* (Polypodiaceae).

Active as an inhibitor of xanthine oxidase, and hence could be considered for the treatment of gout.

1959
Bellidifolin; 3-Methoxy-1,5,8-trihydroxyxanthone

$C_{14}H_{10}O_6$ Mol. wt 274.23

Found in the aerial parts of *Gentiana lactea* and of *Swertia chirata* (Gentianaceae).

Not only active as a tuberculostatic agent, but also as an inhibitor of the enzyme monoamino oxidase/A. It is antihepatotoxic, and has a strong mutagenic activity against *Salmonella typhimurium*.

1960
Calophyllin B; Guanandin;
6-(3,3-Dimethylallyl)-1,5-dihydroxyxanthone

$C_{18}H_{16}O_4$ Mol. wt 296.33

Occurs in *Calophyllum inophyllum* and *C. bracteatum* (Guttiferae).

Anti-inflammatory activity.

1961
Dehydrocycloguanandin

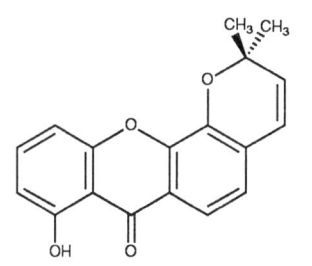

$C_{18}H_{14}O_4$ Mol. wt 294.31

Occurs in *Calophyllum brasiliense*, *C. inophyllum* and *Mesua ferrea* (Guttiferae).

Anti-inflammatory activity in animals.

1962
Demethylbellidifolin; 1,3,5,8-Tetrahydroxyxanthone

$C_{13}H_8O_6$ Mol. wt 260.21

Occurs in the aerial parts of *Gentiana lactea*, and, as the 8-*O*-glucoside, in *Gentiana campestris*, *G. germanica* and *G. ramosa* (Gentianaceae).

Moderate inhibitor of monoamino oxidase/A, and shows weak mutagenic activity against *Salmonella typhimurium*.

1963
6-Deoxyjacareubin

$C_{18}H_{14}O_5$ Mol. wt 310.31

Found in *Calophyllum zeylanicum*, and in the trunk wood of *Kielmeyera speciosa* (Guttiferae).

Antimicrobial, anti-inflammatory and anti-ulcerogenic activities.

1964
3,5-Dimethoxy-1,6-dihydroxyxanthone

$C_{15}H_{12}O_6$ Mol. wt 288.26

Occurs in the roots of *Canscorea decussata* (Gentianaceae).

Tuberculostatic activity *in vitro*.

1965
Euxanthone; Purrenon; 1,7-Dihydroxyxanthone

$C_{13}H_{18}O_4$ Mol. wt 238.29

Found in *Calophyllum, Bonnettia, Garcinia*, and *Haploclathra* spp., in the seeds of *Mammea americana*, and in the heartwood of *Platonia insignis* (all Guttiferae).

Anti-inflammatory activity.

1966
Gambogic acid

$C_{33}H_{44}O_8$ Mol. wt 568.72

Principal acidic component of gamboge, the gum resin of *Garcinia hanburgyi* (Guttiferae). It also occurs in *Garcinia morella*.

No activities for gambogic acid are known, but gamboge has cathartic activity and is used in veterinary medicine as a drastic purgative.

1967
Gentiacaulein; 2,8-Dihydroxy-1,6-dimethoxyxanthone

$C_{15}H_{12}O_6$ Mol. wt 288.26

Occurs in *Gentiana kochiana* and *G. nivalis*; it is the aglycone of gentiaculoside, from *G. acaulis* (Gentianaceae). It also occurs in *Haploclathra paniculata* (Guttiferae).

Inhibits monoamino oxidase/A *in vitro*. Activities have been reported for a compound named gentiacaulein with the structure of 1,3-dihydroxy-7,8-dimethoxyxanthone. A compound of this structure has not yet been isolated from gentianaceous plants; thus the activity probably refers to proper gentiacaulein.

1968
Gentisein; 1,3,7-Trihydroxyxanthone

$C_{13}H_8O_5$ Mol. wt 244.21

Found in *Haploclathra paniculata*, and *Hypericum degenii* (Guttiferae), and in *Gentiana lutea* (Gentianaceae).

Tuberculostatic activity.

1969
Gentisin; 1,7-Dihydroxy-3-methoxyxanthone

$C_{14}H_{10}O_5$ Mol. wt 258.23

Occurs in the roots of *Gentiana lutea* (Gentianaceae), and in *Calophyllum brasiliense*, *Garcinia eugenifolia* and *Mesua ferrea* (Guttiferae).

Moderate mutagenic activity against *Salmonella typhimurium*.

1970
Isoathyriol; 6-Methoxy-1,3,7-trihydroxyxanthone

$C_{14}H_{10}O_6$ Mol. wt 274.23

Occurs in the aerial parts of the fern *Athyrium mesosorum* (Polypodiaceae).

Inhibitory activity on the enzyme xanthine oxidase, and therefore could be considered for the treatment of gout.

1971
Isogentisin; 1,3-Dihydroxy-7-methoxyxanthone

$C_{14}H_{10}O_5$ Mol. wt 258.23

Found in the roots of *Gentiana lutea*, and in *Swertia chirata* (Gentianaceae).

Inhibits monoamino oxidase *in vitro*; additionally, it is active as a mutagen against *Salmonella typhimurium*.

1972
1-Isomangostin

$C_{24}H_{26}O_6$ Mol. wt 410.47

Found in the pericarp of *Garcinia mangostana* (Guttiferae).

Antibacterial and antifungal activities.

1973
Jacareubin

$C_{18}H_{14}O_6$ Mol. wt 326.31

Occurs in the wood of *Calophyllum brasiliense* and of other *Calophyllum* spp., in *Kielmeyera ferruginea* and *Pentadesma butyracea* (all Guttiferae).

Antimicrobial, anti-inflammatory and antiulcerogenic activities.

1974
Lancerin; 4-C-Glucosyl-1,3,7-trihydroxyxanthone

$C_{19}H_{18}O_{10}$ Mol. wt 406.35

Occurs in the whole plants of *Tripterospermum taiwanense*, and in *T. lanceolatum* (Gentianaceae).

Stimulating effect on the central nervous system of rats.

1975
Macluraxanthone

$C_{23}H_{22}O_6$ Mol. wt 394.43

Found in *Garcinia ovalifolia* and in *Rheedia brasiliensis* (Guttiferae), and in the bark of *Maclura pomifera* (Moraceae).

More toxic than rotenone against the larval stages of malarial and yellow-fever mosquitoes, and also shows potent antitermite activity.

1976
Mangiferin; Aphloid; Aphloiol; Chimonin; Euxanthogen

$C_{19}H_{18}O_{11}$ Mol. wt 422.35

Relatively widespread occurrence: e.g., in *Mangifera indica* (Anacardiaceae), *Hiptage madablota* (Malpighiaceae), the genus *Hypericum* (Guttiferae), various Iridaceae, Liliaceae, Gentianaceae, Leguminosae, Flacourtiaceae, Convolvulaceae, Celastraceae and Sapotaceae, and in the fern *Athyrium mesosorum* (Polypodiaceae). It co-occurs in the above plants with isomangiferin (where the C-glucose is at the 4-position) and in O-glycosidic combination.

Anti-inflammatory, antihepatotoxic and antiviral activities. It also stimulates the central nervous system.

1977
Mangostin

$C_{24}H_{26}O_6$ Mol. wt 410.47

Occurs in the fruit peel and the resin of *Garcinia mangostana* (Guttiferae).

Anti-inflammatory and antimicrobial activities. In addition, *in vitro* tests indicate antiulcer activity.

1978
Mesuaxanthone A;
1,5-Dihydroxy-3-methoxyxanthone

$C_{14}H_{10}O_5$ Mol. wt 258.23

Occurs in *Mesua ferrea*, *Kielmeyera speciosa*, *Garcinia xanthochymus*, *Haploclathra* spp. and in the genus *Vismia* (Guttiferae).

Anti-inflammatory activity.

1979
Mesuaxanthone B; 1,5,6-Trihydroxyxanthone

$C_{13}H_8O_5$ Mol. wt 244.21

Occurs in *Mesua ferrea*, *Mammea africana*, *Calophyllum inophyllum* and *C. fragrans*, and *Garcinia* and *Symphonia* spp. (Guttiferae).

Anti-inflammatory activity.

1980
2-O-Methylswertianin;
1,8-Dihydroxy-2,6-dimethoxyxanthone

$C_{15}H_{12}O_6$ Mol. wt 288.26

Found in the aerial parts of *Swertia japonica* (Gentianaceae).

Antihepatotoxic activity in animals.

1981
Morellin

$C_{33}H_{36}O_7$ Mol. wt 544.65

Found in the seeds of *Garcinia morella* (Guttiferae).

Antibiotic activity.

1982
Norathyriol; 1,3,6,7-Tetrahydroxyxanthone

$C_{13}H_8O_6$ Mol. wt 260.21

Found in *Cratoxylum pruniflorum*, *Garcinia mangostana*, *Hypericum androsaemum* and *H. aucheri* (Guttiferae), the genera *Mammea*, *Allanblackia*, *Symphonia* and *Ochrocarpus* (Guttiferae), and several fern genera. Within the Moraceae it is found in *Maclura pomifera*, and in *Clarisa* and *Chlorophora* spp.

Potent inhibitor of the enzyme xanthine oxidase; it shows moderate tuberculostatic activity.

1983
Norswertianin; 1,2,6,8-Tetrahydroxyxanthone

$C_{13}H_8O_6$ Mol. wt 260.21

Found in the aerial parts of *Swertia japonica* and in *S. swertopsis* and *S. perennis* (Gentianaceae).

Moderate mutagenic activity against *Salmonella typhimurium*.

1984
Norswertianolin

$C_{19}H_{18}O_{11}$ Mol. wt 422.35

Occurs in *Swertia purpurascens*, *S. randainensis* and *S. japonica* (Gentianaceae). It was originally described as 1,3,5,8-tetrahydroxyxanthone 1-*O*-glucoside, but recently revised to the structure given above.

Tuberculostatic activity.

Found in *Gentiana lactea* and *Swertia chirata* (Gentianaceae).

Moderate inhibition of the enzyme monoamino oxidase/A *in vitro*, and antihepatotoxic activity in animals.

1985
Psorospermin

$C_{19}H_{16}O_6$ Mol. wt 340.34

Occurs in the roots of *Psorospermum* spp. (Guttiferae).

Antitumour and antileukaemic activities.

1986
Swerchirin; 5-*O*-Methylbellidifolin; 3,5-Dimethoxy-1,8-dihydroxyxanthone

$C_{15}H_{12}O_6$ Mol. wt 288.26

1987
Swertianin; 1,2,8-Trihydroxy-6-methoxyxanthone

$C_{14}H_{10}O_6$ Mol. wt 274.23

Occurs in *Swertia japonica*, *Gentiana bavarica*, *G. brachyphylla* and *G. nivalis* (Gentianaceae).

Moderate mutagenic activity against *Salmonella typhimurium*.

1988
Swertianolin; Bellidifolin-8-*O*-glucoside

$C_{20}H_{20}O_{11}$ Mol. wt 436.38

Found in *Swertia japonica*, *S. tosaensis*, *S. perennis* and *S. purpurascens*, *Gentiana campestris*, *G. germanica* and *G. ramosa* (Gentianaceae).

Tuberculostatic activity.

1989
1,3,5-Trihydroxyxanthone

$C_{13}H_8O_5$ Mol. wt 244.21

Found in *Allanblackia floribunda* (Guttiferae).

Tuberculostatic activity.

1990
Tripteroside; Norathyriol-6-*O*-β-D-glucoside

$C_{19}H_{18}O_{11}$ Mol. wt 422.35

Occurs in the aerial parts of *Tripterospermum taiwanense* (Gentianaceae).

Acts as a depressant on the central nervous system of animals.

Chapter 47
Miscellaneous phenolics

This section includes several uncommon classes of phenolic metabolite. There are the benzophenones, with a basic C_6-C_1-C_6 aromatic skeleton, illustrated here by maclurin (**2015**) and hypercalin B (**2010**). Acrovestone (**1992**) can be regarded as a modified benzophenone which is also isoprenylated. A number of other phenolics in this section have isoprenyl (terpenoid) substitution (q.v.).

The isocoumarins are another minor class of plant compound. Two examples are bergenin (**1995**) a constituent of *Bergenia* in the Saxifragaceae, and 6-methoxymellein (**2016**), a phytoalexin of the carrot. Then there are phthalides, e.g., **1997**, found in the family Umbelliferae, and biphenyls, e.g., aucuparin (**1994**) from the mountain ash. Finally, there are α-pyrones such as kawain (**2012**) and methysticin (**2017**), aromatic lactones, found in plants of the Lauraceae and the Piperaceae.

Most of the miscellaneous phenolics listed here have some significant biological activity. Carcinogenicity is associated with ochratoxin A (**2019**), an isocoumarin mycotoxin of infected cereals, while oestrogenic activity is present in zearalenone (**2034**), a phenolic mycotoxin of infected maize. Other examples of active molecules are robustadial A (**2025**), an antimalarial and cleomiscosin A (**1998**), an anticancer agent. The appropriate reference, which deals with some of the rarer classes of plant phenolic, is the book of Dean (1963).

Reference

Dean, F.M. (1963). *Naturally Occurring Oxygen Ring Compounds*. London: Butterworths.

1991
Acetosyringone

$C_{10}H_{12}O_4$ Mol. wt 196.21

Found in root cultures of *Nicotiana tabacum* and other Solanaceous species; it co-occurs with α-hydroxyacetosyringone.

Exuded from roots and triggers off infection of plants by crown gall disease, *Agrobacterium tumefaciens*.

1992
Acrovestone

$C_{32}H_{42}O_8$ Mol. wt 554.69

Found in the stems and root bark of *Acronychia pedunculata*, and in *A. vestita* (Rutaceae).

Cytotoxic in the human KB tissue culture assay and active against A-549, P-388 and L-1210 cells. *A. pedunculata* is used for treating diarrhoea, asthma, ulcers, and rheumatism.

1993
Arnebinol

$C_{16}H_{20}O_2$ Mol. wt 244.34

Occurs in the roots of *Arnebia euchroma* (Boraginaceae).

Inhibits prostaglandin biosynthesis.

1994
Aucuparin

$C_{14}H_{14}O_3$ Mol. wt 230.27

Found in the heartwood of *Sorbus aucuparia* and *S. decora* (Rosaceae). Identified as a phytoalexin after fungal infection in the loquat, *Eriobotrya japonica* (also Rosaceae).

Antifungal activity.

1995
Bergenin; Ardisic acid B; Corylopsin; Peltaphorin; Vakerin

$C_{14}H_{16}O_9$ Mol. wt 328.28

Occurs in the roots of *Peltophorum pteracarpum* and *P. inerme*, and in *Caesalpinia digyna* (Leguminosae), in *Ardisia hortorum*, *Astilbe macroflora*, and *Bergenia crassifolia* (Saxifragaceae), in *Corylopsis* spp. (Hamamelidaceae), and in *Humiria balsamifera* (Humiriaceae).

Anti-inflammatory activity.

1996
Brazilin

$C_{16}H_{14}O_5$ Mol. wt 286.29

Found in brazil wood, *Caesalpinia echinata*, and in sappan wood, *Caesalpinia sappan* (Leguminosae).

Used as a dye. It is also an acid (yellow)–base (red) indicator.

1997
3-Butylidene-7-hydroxyphthalide; Senkyunolide

$C_{12}H_{12}O_3$ Mol. wt 204.23

Found in the rhizomes of *Ligusticum wallichii* (Umbelliferae).

Increases coronary blood flow in dog heart. The crude drug of the plant rhizomes has haemodynamic and analgesic effects.

1998
Cleomiscosin A; Cleosandrin

$C_{20}H_{18}O_8$ Mol. wt 386.36

Occurs in the stem wood and stem bark of *Soulamea soulameoides*, in the wood of *Simaba multiflora* (both Simaroubaceae), and in the stem wood, bark and twigs of *Matayba arborescens* (Sapindaceae). Under the name of cleosandrin, it was isolated from the seeds of *Cleoma icosandra* (=*C. viscosa*) (Capparidaceae). Two closely related structures, cleomiscosin B and C, are also present in *Cleoma* seeds.

Anticancer and antileukaemic activities. In addition, it shows antihepatotoxic activity *in vitro*.

1999
Daphneticin

$C_{20}H_{18}O_8$ Mol. wt 386.36

Found in the roots and stems of *Daphne tangutica* (=*D. retusa*) (Thymelaeaceae).

Cytotoxic activity against Walker-256 carcino-sarcoma-ascites cells.

2000
5,6-Dehydrokawain; Desmethoxyyangonin

$C_{14}H_{12}O_3$ Mol. wt 228.25

Found in the wood of *Aniba firmula*, in the bark of *Lindera umbellata* (both Lauraceae), in *Piper methysticum* (Piperaceae), and in the rhizomes of *Alpinia speciosa* (Zingiberaceae).

Anticonvulsive and local anaesthetic activities.

2001
Dihydromethysticin

C₁₅H₁₆O₅ — Mol. wt 276.29

Occurs in the rhizomes of *Piper methysticum* (Piperaceae).

Spasmolytic activity.

2002
Euglobal-Ia₁

C₂₃H₃₀O₅ — Mol. wt 386.49

Occurs in the buds of *Eucalyptus globulus* (Myrtaceae).

Granulation inhibitory activity. This is one of a series of related compounds from *Eucalyptus* having the same activity.

2003
Flossonol

C₁₃H₁₆O₃ — Mol. wt 220.27

Found in the roots and stems of *Pararistolochia flosavis* (Aristolochiaceae).

Cytotoxic and antileukaemic activity against PS-cells in culture.

2004
Garcinol

C₃₈H₅₀O₆ — Mol. wt 602.82

Occurs in the stem bark of *Garcinia huillkensis*, in *G. cambogia* and in *G. indica* (Guttiferae).

Antifungal and moderate antibacterial activities.

2005
Goniothalenol; Altholactone

C₁₃H₁₂O₄ — Mol. wt 232.24

Occurs in the stem bark of *Goniothalamus giganteus* and in a *Polyalthia* sp. (Annonaceae).

Cytotoxic in the P388-test and toxic to brine shrimp. The stem bark extract of *G. giganteus* is antileukaemic *in vivo*.

2006
Haematoxylin; Hydroxybrazilin

$C_{16}H_{14}O_6$ Mol. wt 302.29

Occurs in the heartwood of *Haematoxylon campechianum* (Leguminosae).

Reddens on exposure to light, and is used as a dye, as a stain in microscopy, and in the manufacture of ink.

2007
Haemocorin

$C_{32}H_{34}O_{14}$ Mol. wt 642.62

Occurs in *Haemodorum corymbosum*, *Lachnanthes tinctoria* and *Wachendorfia thyrsiflora* (Haemodoraceae). The cellobioside occurs in *H. corymbosum*, while the aglycone of haemocorin is present in the other two spp.

The aglycone shows antitumour, antibacterial and anti-inflammatory activities.

2008
Heritonin

$C_{16}H_{18}O_3$ Mol. wt 258.32

Found in the roots of *Heritonia littoralis* (Sterculiaceae).

Piscicidal activity. The plants are used in the Philippines to kill fish.

2009
(+)-8-Hydroxycalamenene

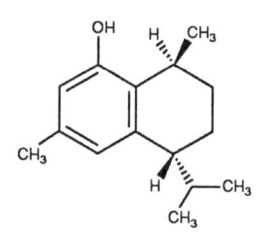

$C_{15}H_{22}O$ Mol. wt 218.34

Occurs in the seeds of *Dysoxylum alliaceum* and *D. acutangulum* (Meliaceae).

Piscicidal and antibacterial activities. Major poisonous principle of *Dysoxylum* plants, which are used in Sumatra as a fish poison.

2010
Hypercalin B

$C_{32}H_{42}O_5$ Mol. wt 506.69

Found in the aerial parts of *Hypericum calycinum* (Guttiferae).

Growth inhibitory activity against Co-115 human carcinoma cell line. It shows molluscicidal activity against the schistosomiasis transmitting snail.

2011
Karwinaphthol B

$C_{17}H_{20}O_4$ Mol. wt 288.35

Occurs in the roots of *Karwinskia humboldtiana* (Rhamnaceae).

Active against *Mycobacterium smegmatis*.

2012
Kawain; Kavain; Gonosan

$C_{14}H_{14}O_3$ Mol. wt 230.27

Occurs in the rhizomes and roots of *Piper methysticum*, also known as kava (Piperaceae).

Spasmolytic, local anaesthetic, anti-inflammatory, antimycotic and antioedemic activities.

2013
Kolanone

$C_{33}H_{42}O_4$ Mol. wt 502.70

Found in the fruit of *Garcinia kola* (Guttiferae).

Broad antimicrobial activity.

2014
Lapachenole

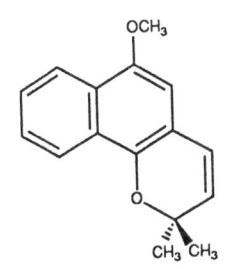

$C_{17}H_{16}O_2$ Mol. wt 252.32

Occurs in the essential oil of *Lippia graveolens* (Verbenaceae), and in the Bignoniaceae.

Carcinogenic. It may be responsible for the antifertility activity of *Lippia graveolens*.

2015
Maclurin; Kino-yellow; Laguncurin; Morintannic acid; Moritannic acid

$C_{13}H_{10}O_6$ Mol. wt 262.22

Occurs in the wood of *Chlorophora tinctoria* (=*Maclura tinctoria*) and in *Morus alba* (Moraceae), in the bark of *Laguncularia racemosa* (Combretaceae), and in the wood of *Acacia* spp. (Leguminosae).

Used for dyeing fabrics.

2016
6-Methoxymellein

$C_{11}H_{12}O_4$ Mol. wt 208.22

Fungal metabolite (e.g., *Aspergillus, Ceratocystis, Penicillium thomii, Sporormia affinis*), but also produced as a phytoalexin by carrots, *Daucus carota* (Umbelliferae) in response to fungal attack.

Antifungal activity.

2017
Methysticin

$C_{15}H_{14}O_5$ Mol. wt 274.28

Found in the rhizomes of *Piper methysticum* (Piperaceae).

Spasmolytic activity.

2018
Nepodin; Musizin

$C_{13}H_{12}O_3$ Mol. wt 216.24

Occurs in the roots of *Rumex obtusifolius*, in *R. nepalensis* (Polygonaceae), and in *Maesopsis eminii* (Rhamnaceae).

Inhibits the growth of *Bacillus subtilis*.

2019
Ochratoxin A

$C_{20}H_{18}NO_6Cl$ Mol. wt 403.82

One of a series of ochratoxins produced by spp. of the fungi *Aspergillus* and *Penicillium* during their infection of stored plant materials, e.g., cereals, peanuts and cottonseed.

Highly toxic and shows carcinogenic activity. It induces nephropathy in animals, and may be the causative agent of nephropathy in pigs and poultry that have consumed contaminated cereals.

2020
Ohioensin-A

$C_{23}H_{16}O_5$ Mol. wt 372.38

Occurs in *Polytrichum ohioense* (Polytrichaceae).

Antitumour and cytotoxic activities.

2021
Paeoniflorin

$C_{23}H_{28}O_{11}$ Mol. wt 480.48

Found in the roots of *Paeonia albiflora* and in *P. lactiflora*, *P. moutan* and *P. officinalis* (Paeoniaceae).

Anti-allergic, anticoagulative and antiplatelet aggregation activities.

2022
Polygonolide

$C_{12}H_{12}O_4$ Mol. wt 220.23

Found in the roots of *Polygonum hydropiper* (Polygonaceae).

Anti-inflammatory activity.

2023
Purpurogallin

$C_{11}H_8O_5$ Mol. wt 220.19

Occurs in glycosidic form in various nutgalls. The diglucoside, dryophantin, is the pigment of the gall produced by *Dryophanta divisa* on the oak *Quercus pedunculata* (Fagaceae).

Red pigment. It is used as an antioxidant in oils, fats, hydrocarbon fuels and lubricants.

2024
Rhododendrin; Betuloside

$C_{16}H_{24}O_7$ Mol. wt 328.36

Found in the leaves of *Rhododendron chrysanthum*, *R. fauriae*, and other *Rhododendron* spp. (Ericaceae), and also in *Betula* spp. (Betulaceae).

Bitter taste. It shows diuretic and diaphoretic activities.

2025
Robustadial A

$C_{23}H_{30}O_5$ Mol. wt 386.49

Occurs together with the stereoisomer, robustadial B, in *Eucalyptus robusta* (Myrtaceae).

Antimalarial activity against *Plasmodium berghei*. The plant is used in China to treat malaria, dysentery and bacterial diseases.

2026
Robustaol A

$C_{25}H_{30}O_9$ Mol. wt 474.52

Found in the leaves of *Eucalyptus berghei* (Myrtaceae).

Antimalarial activity against *Plasmodium berghei*.

2027
Sarothralin

$C_{31}H_{34}O_8$ Mol. wt 534.61

Occurs in the whole plants of *Hypericum japonicum* (Guttiferae).

Antimicrobial activity.

2028
Stypandrol

$C_{26}H_{22}O_6$ Mol. wt 430.46

Occurs in *Stypandra imbricata* (Liliaceae).

Toxic to animals. It is responsible for the toxicity of the plant after ingestion, leading to paralysis and finally death in sheep and goats.

2029
Tremulacin

$C_{27}H_{27}O_{11}$ Mol. wt 527.51

Found in the bark and leaves of *Populus tremula, P. alba* and *P. tremuloides* (Salicaceae).

At naturally occurring concentrations, first-instar development rates of gypsy moth larvae are prolonged, but fourth instar growth rates are reduced. It acts partly as feeding stimulant. In Lepidoptera which are not adapted to *Populus*, it causes degenerative midgut lesions.

2030
Uncinatone

$C_{20}H_{22}O_4$ Mol. wt 326.40

Occurs in the root bark of *Clerodendrum uncinatum* (Verbenaceae).

Antifungal activity against *Cladosporium cucumerinum*.

2031
Vismione D

$C_{25}H_{30}O_5$ Mol. wt 410.52

Found in the root bark of *Psorospermum febrifugum* (Guttiferae).

Antiproliferative activity in a human tumour cell line from adenocarcinoma of the ascending colon.

2032
Xanthochymol

$C_{38}H_{50}O_6$ Mol. wt 602.82

Found in *Garcinia mannii*, *G. ovalifolia* and other *Garcinia* spp., and in *Rheedia madrunno* (all Guttiferae).

Good antibacterial activity, but poor antifungal activity.

2033
Yangonin

$C_{15}H_{14}O_4$ Mol. wt 258.28

Occurs in the rhizomes of *Piper methysticum* (Piperaceae).

Spasmolytic activity.

2034
Zearalenone; Mycotoxin F2; Toxin F2

$C_{18}H_{22}O_5$ Mol. wt 318.37

Mycotoxin of the fungus *Gibberella zeae*, which infects maize.

Induces multiple reproductive deficiencies in animals. Pigs eating plant material infected by this fungus develop oestrogenism and become infertile.

PART V
Terpenoids

Terpenoids

General Introduction

The terpenoids comprise the largest group of natural plant products, and over twenty thousand such structures have been described from plant sources. Camphor, limonene, abscisic acid, aucubin, gossypol, gibberellic acid, digitalin and β-carotene are some typical plant terpenoids. They are all derived biogenetically from the 5-carbon precursor isoprene and, hence, are also known as isoprenoids. In terms of *in vivo* biosynthesis, they are formed from the condensation of two C_5 precursors, dimethylallyl pyrophosphate and isopentenyl pyrophosphate, which gives rise to a C_{10} intermediate, geranyl pyrophosphate. This is the immediate precursor of the monoterpenoids and the related monoterpene lactones, known as iridoids. Geranyl pyrophosphate can condense in turn with another C_5 unit of isopentenyl pyrophosphate to produce the C_{15} intermediate, farnesyl pyrophosphate. This latter compound is the starting point for the synthesis of the C_{15} sesquiterpenoids and, after further modification, of the sesquiterpene lactones. Farnesyl pyrophosphate can undergo further extension by linking with another isopentenyl pyrophosphate residue to produce the C_{20} intermediate, geranylgeranyl pyrophosphate. This is the general precursor of all the plant diterpenoids, with their C_{20} based structures.

Two molecules of the C_{15} intermediate farnesyl pyrophosphate can condense together in a further step in terpenoid biosynthesis, with the formation of squalene, the C_{30} precursor of the largest group of isoprenoids, the triterpenoids. These terpenoids are considered here under eight headings: the triterpenoid and steroid saponins, the phytosterols, the cardenolides and bufadienolides, the cucurbitacins, and then the limonoids and quassinoids, which are bracketed together as nortriterpenoids. Finally, two molecules of geranylgeranyl pyrophosphate (C_{20}) may condense together tail-to-tail to form a C_{40} intermediate, called phytoene, which is the immediate precursor of the yellow carotenoid pigments.

All the main classes of terpenoid are thus included, in sequence of biosynthetic complexity, in this chapter. There are three other minor classes which are not covered in detail here. Polymerization of isopentenyl pyrophosphate can occur in plants, leading to polymers which are commonly secreted in special cells as a milky latex. There are two distinct forms: *cis*-polyisoprene, better known as rubber; and *trans*-polyisoprene, or gutta-percha. Polyisoprene-containing latex is widespread in certain areas of the plant kingdom, and is known in some twelve thousand five hundred species from twenty-two families (Metcalfe, 1983). The sesterterpenes, by contrast, are C_{25}-terpenoids, a small group of mainly fungal metabolites. There are also the hemiterpenes, i.e., C_5 terpenoids, which include isoprene itself, which has been identified as a leaf volatile in, e.g., *Hamamelis japonica* (Hamamelidaceae).

In addition to the terpenoids of this chapter, there are many other secondary metabolites of mixed biosynthetic origin which either contain terpenoid substituents or which are largely of isoprenoid origin. Among the alkaloids, for example, there are nitrogen-containing mono-, sesqui- and diterpenoid derivatives, as well as steroidal alkaloids; these are all covered in different sections of Part III on alkaloids. Phenolic substances may also be partly terpenoid-derived, and isoprene-substituted derivatives may be found in many sections of Part IV. There are, for example, furanocoumarins such as bergapten isoflavonoids such as rotenone, and quinones such as plastoquinone and ubiquinone, both of which have isoprenoid side chains.

The terpenoid literature is very extensive. That relating to individual groups of terpenoid is mentioned in appropriate sections of this chapter; only general references are listed here. There is an encyclopedia of the terpenoids (Glasby, 1982) which covers the literature up to the end of 1979. There are about ten thousand entries but certain important classes, such as the steroids, are not included. A more comprehensive dictionary of terpenoids is scheduled for publication by Chapman & Hall (London) during 1991. The chemistry of terpenoids is the subject of a serial publication *Terpenoids and Steroids*, published by the Royal Society of Chemistry, which appeared in 12 Volumes from 1971 to 1983. This survey of the chemical literature is being continued in the journal *Natural Product Reports*, which began publication in 1984 and has regular updated reviews on each of the main classes of terpenoid. An introductory text on terpenoid chemistry is that of Newman (1972).

The biochemistry and function of plant terpenoids has been reviewed at regular intervals by the Phytochemical Society of Europe in their Symposium Volumes (Pridham, 1967; Goodwin, 1971; Goad *et al.*, 1983; Harborne and Barberan, 1991). Two other volumes on biochemistry and function are those of Runeckles and Mabry (1973) and Nes *et al.* (1984). The methodology of plant terpenoid analysis is reviewed by Charlwood and Banthorpe (1991).

References

Charlwood, B.V. and Banthorpe, D.V. (1991). Volume 7, *The Terpenoids. Methods in Plant Biochemistry* (Dey, P.M. and Harborne, J.B. (eds). London: Academic Press.

Glasby, J.S. (1982). *Encyclopedia of the Terpenoids*, in 2 Volumes. Chichester: Wiley.

Goad, L.J., Rees, H.H. and Threlfall, D.R. (1983). Terpenoid biosynthesis and biochemistry in plants. *Biochemical Society Transactions*, **11,** 497–603.

Goodwin, T.W. (ed.) (1971). *Aspects of Terpenoid Chemistry and Biochemistry*. London: Academic Press.

Harborne, J.B. and Barberan, F.A.T. (eds) (1991). *Ecological Chemistry and Biochemistry of Plant Terpenoids*. Oxford: Clarendon Press.

Hill, R.A. and Connolly, J.D. (eds) (1991). *Dictionary of Terpenoids*. London: Chapman and Hall.

Metcalfe, C.R. (1983). Secretory structures: cells, cavities and canals. In *Anatomy of the Dicotyledons*, Volume 2, 2nd Edition (Metcalfe, C.R. and Chalk, L., eds). Oxford: Clarendon Press.

Nes, W.D., Fuller, G. and Tsai, L.S. (eds) (1984). *Isopentenoids in Plants: Biochemistry and Function*. New York: Marcel Dekker.

Newman, A.A. (ed.) (1972). *Chemistry of Terpenes and Terpenoids*. London: Academic Press.

Pridham, J.B. (ed.) (1967). *Terpenoids in Plants*. London: Academic Press.

Runeckles, V.C. and Mabry, T.J. (eds) (1973). *Terpenoids: Structure, Biogenesis and Distribution*. New York: Plenum Press.

Chapter 48
Monoterpenoids

Monoterpenes, the C_{10} representatives of the terpenoid family, are biosynthetically derived from the union of two isoprene precursors joined head-to-tail. Traditionally, they are components, together with sesquiterpenoids and aromatics, of the plant essential oils. They tend to accumulate in members of certain families, such as the Labiatae, Pinaceae, Rutaceae and Umbelliferae, from which they are commercially produced. However, some of them are ubiquitous and can be found in small amounts in the volatile secretions of most plants. They are particularly associated in the plant with specialized secretory structures, such as oil cells, glandular hairs and resin ducts. Their main functions in plants are for attracting pollinators to flowers, and for protecting green tissues from herbivory and microbial infection.

With about 600 known structures (Glasby, 1982), the monoterpenoids are conveniently divided into four structural categories: acyclic, monocyclic, bicyclic and irregular. The cyclopentanoid monoterpene lactones are biosynthetically related but, in this Dictionary, they are dealt with separately under the heading iridoids (q.v.). The first acyclic monoterpene to be formed is geraniol (**2055**) and other widespread monoterpenes produced from it by simple enzymatic conversion include linalool (**2058**), nerol (**2065**), and citronellol (**2049**).

Many common monocyclic monoterpenes can be formally derived from α-terpineol (**2080**) as a key intermediate, and it is likely that limonene (**2057**), terpinolene (**2081**) and 1,8-cineole (**2045**) are derived from it. The bicyclic monoterpenes are likewise biosynthetically interrelated, especially the most common members of the series such as α-pinene (**2070**), β-pinene (**2071**), borneol (**2036**) and thujone (**2084**). Less is known of the biosynthetic origin of irregular monoterpenes such as the insecticidal pyrethrins (**2075**) and the tropolone γ-thujaplicin (**2083**).

Monoterpenoids can be classified also according to the functional groups present. Thus they can be divided into: unsaturated hydrocarbons such as limonene; alcohols such as linalool; alcohol esters such as linalyl acetate; aldehydes such as citronellal; and ketones such as carvone. Optical isomerism is a common feature and some compounds such as carvone can occur in more than one optically active form. Increasing numbers of acyclic and monocyclic monoterpenes have been detected in plants not only in the free state but also in bound form as glycosides: geranyl glucoside, for example, has been found in rose petals.

Although any given plant rich in essential oils may have only a few major constituents, it may have up to 50 other monoterpenes in lesser amounts. In some plants, there can be considerable chemical variation within the species, i.e., different populations have different mixtures of monoterpenoids (Tetenyi, 1970). In thyme, for example, 13 chemotypes have been recognized among populations growing in France.

Besides occurring very widely in higher plants, monoterpenoids have been detected in some bryophytes and fungi. They also occur occasionally in insects, in defence and pheromonal secretions. The chemical ecology of monoterpenoids is reviewed by Harborne (1988).

Monoterpenoids are commonly strong smelling, colourless oily substances of boiling point 140–180 °C, although a few (e.g., camphor) are crystalline. They are usually separated and identified by combined gas chromatography/mass spectrometry, but thin layer chromatography is a convenient means of preliminary detection.

A major reference to the chemistry of terpenes is Newman (1972) and to the biochemistry is Loomis and Croteau (1980). Comprehensive listings of monoterpenoid structures are those of Devon and Scott (1972) and Glasby (1982). Biological activity is reviewed by Sticher (1972) and by Reynolds (1982).

References

General

Budavari, S. (ed.) (1989). *The Merck Index*, 11th Edition. Rahway: Merck and Co.
Devon, T.K. and Scott, A.I. (1972). *Handbook of Naturally Occurring Compounds*, Volume 2. New York: Academic Press.
Glasby, J.S. (1982). *Encyclopaedia of the Terpenoids*. Chichester: Wiley.
Hegnauer, R. (1964–1973). *Chemotaxonomie der Pflanzen*, Volumes 3–6. Basel: Birkhäuser.
Rauen, H.M. (ed.) (1964). *Biochemisches Taschenbuch I*, 2nd Edition. Berlin: Springer-Verlag.

Chemistry and Biochemistry

Loomis, W.D. and Croteau, R. (1980). In *Biochemistry of Plants*, Volume 4, *Lipids, Structure and Function*, pp. 364–419. New York: Academic Press.
Newman, A.A. (ed.) (1972). *Chemistry of Terpenes and Terpenoids*. New York: Academic Press.

Infraspecific Variation

Tetenyi, P. (1970). *Infraspecific Chemical Taxa of Medicinal Plants*. Budapest: Akademiai Kiado.

Biological Activity

Harborne, J.B. (1988). *Introduction to Ecological Biochemistry*, 3rd Edition. London: Academic Press.
Reynolds, J.E.F. (1989). *Martindale: The Extra Pharmacopoeia*, 29th Edition. London: The Pharmaceutical Press.
Sticher, O. (1977). In *New Natural Products and Plant Drugs with Pharmacological, Biological or Therapeutical Activity* (H. Wagner and P. Wolff, eds). Berlin: Springer-Verlag.

2035
Ascaridole; Ascaridol

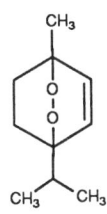

$C_{10}H_{16}O_2$ Mol. wt 168.24

Main constituent of oil of chenopodium, the volatile oil from fresh flowering and fruiting plants of American wormseed, *Chenopodium ambrosioides* var. *anthelminticum* (Chenopodiaceae).

Anthelmintic activity; it is toxic to mammals.

2036
Borneol; *endo*-Borneol; Bornyl alcohol; Camphol

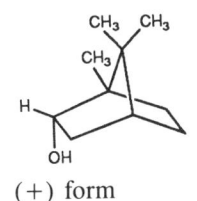

$C_{10}H_{18}O$ (+) form   Mol. wt 154.25

The (+) form is the main constituent of the volatile oil of Borneo camphor, distilled from *Dryobalanops aromatica* (Dipterocarpaceae), and is also present in oil of spike, from the leaves and flowering tops of *Lavandula spica* (Labiatae), oil of rosemary, from fresh flowering tops of *Rosmarinus officinalis* (Labiatae), and the volatile oil of nutmeg, *Myristica fragrans* (Myristicaceae). The (−) form occurs in Ngai camphor, from *Blumea balsamifera* (Compositae).

Used mainly for the manufacture of its esters, but some free borneol is used in perfumery. It is toxic to mammals (affects the central nervous system).

2037
Bornyl acetate; Borneol acetate

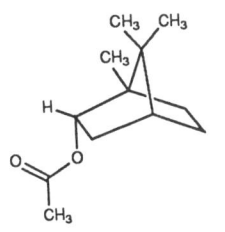

$C_{12}H_{20}O_2$ (+) form   Mol. wt 196.29

Occurs in the volatile oils from the needles of many pine and fir species, e.g., *Pinus montana, P. sylvestris, Abies alba, A. siberica* (Pinaceae), and also in the volatile oils from certain Labiatae, e.g., rosemary, *Rosmarinus officinalis*, and thyme, *Thymus vulgaris*. Bornyl acetate smells of pine needles. Bornyl esters of acetic, formic and isovaleric acids occur in the volatile oil from the rhizome and root of valerian, *Valeriana officinalis* (Valerianaceae).

Expectorant (pine needle oil) and bronchial and nasal inhalant. (+)-Bornyl isovalerate is a sedative.

2038
Camphene

$C_{10}H_{16}$ (−) form   Mol. wt 136.24

Large amounts are present in oil of Siberian fir, distilled from the fresh leaves of *Abies sibirica* (Pinaceae), in Ceylon citronella oil, from the fresh leaves of *Cymbopogon (Andropogon) nardus* (Gramineae), and in the volatile oil of nutmeg, *Myristica fragrans* (Myristicaceae). It also occurs in cypress oil, from the leaves and young branches of *Cupressus sempervirens* (Pinaceae), and in many other essential oils.

Used for the reduction of the cholesterol saturation index in the treatment of gallstones.

2039
Camphor; Bornan-2-one; Camphan-2-one

$C_{10}H_{16}O$ (−) form   Mol. wt 152.24

The (+) form occurs in rectified or Japanese oil of camphor, from the camphor tree, *Cinnamomum camphora* (Lauraceae). The (−) form, Matricaria camphor, occurs in feverfew, *Chrysanthemum (Matricaria) parthenium* (Compositae), and in *Artemisia* spp. (Compositae) and *Lavendula* spp. (Labiatae).

Irritant; it affects the central nervous system and is toxic to humans. It is used commercially as a moth repellent, and as a preservative in pharmaceuticals and cosmetics. Other uses are as a rubefacient and mild analgesic, and as a topical antipruritic.

2040
(−)-Car-3-ene; 3-Carene; Δ³-Carene; Isodiprene

$C_{10}H_{16}$ Mol. wt 136.24

Wide occurrence in essential oils: e.g., in turpentine oils distilled from *Pinus, Picea* and *Abies* spp., and other Pinaceae, especially *Pinus sylvestris* and *P. longifolia*.

Irritant.

2041
Carvacrol; 2-*p*-Cymenol; Isothymol; Isopropyl-*o*-cresol

$C_{10}H_{14}O$ Mol. wt 150.22

Found in several Labiatae oils, e.g., oil of wild marjoram, distilled from the flowering tops of *Origanum vulgare*, and oil of thyme, from the leaves and tops of *Thymus vulgaris*.

Antiseptic (1.5 times the activity of phenol). Also, it shows antifungal and anthelmintic activities. It is used in mouth washes.

2042
Carvone; Carvol

$C_{10}H_{14}O$ (−) form   Mol. wt 150.22

(−)-Carvone is the main constituent in the volatile oil of spearmint, from the flowering tops of *Mentha spicata* (Labiatae), and (+)-carvone is the major constituent of oils of caraway and dill, from the dried ripe fruits of *Carum carvi* and *Anethum graveolens*, respectively (Umbelliferae).

Carminative and antiseptic action. It is used in the manufacture of liqueurs, for flavouring confectionery, and in perfumery and soaps.

2043
Chrysanthemic acid; Chrysanthemumic acid; Chrysanthemum monocarboxylic acid

$C_{10}H_{16}O_2$ Mol. wt 168.24

Esters of chrysanthemic acid with pyrethrolon and cinerolon (pyrethrin I and cinerin I, respectively, q.v.) occur in the flowers of pyrethrum, *Chrysanthemum cinerariifolium* (=*Pyrethrum cinerarifolium*) (Compositae).

Irritates eyes and mucosa. Esters with pyrethrolon and cinerolon have strong insecticidal properties.

2044
Chrysanthenone; 2-Pinen-7-one

$C_{10}H_{14}O$ (+) form   Mol. wt 150.22

Occurs in the essential oil of the garden chrysanthemum, varieties of *Dendranthema indicum* (=*Chrysanthemum indicum*), and related spp. (Compositae).

2045
1,8-Cineole; Eucalyptol; Cajeputol

$C_{10}H_{18}O$ Mol. wt 154.25

The main constituent in oil of eucalyptus, distilled from the fresh leaves of *Eucalyptus globulus* and some other *Eucalyptus* spp. (Myrtaceae), in oil of Levant wormseed, from the flowers of *Artemisia maritima* var. *stechmannia* (=*A. pauciflora*) (Compositae), in oil of cajeput, from the fresh leaves and twigs of *Melaleuca leucadendron* and other *Melaleuca* spp. (Myrtaceae), and in many other essential oils.

Anthelmintic, expectorant and antiseptic activities. Also, it shows cockroach repellent activity. It is used as a flavouring.

2046
Cinerins; (Cinerin I = Chrysanthemum monocarboxylic acid cinerolone ester; Cinerin II = Chrysanthemum dicarboxylic acid monomethyl ester cinerolone ester)

R = CH_3 cinerin I $C_{20}H_{28}O_3$ Mol. wt 316.44
R = $COOCH_3$ cinerin II $C_{21}H_{28}O_5$ Mol. wt 360.46

Found in the flowers of Dalmatian pyrethrum, *Chrysanthemum cinerariifolium* (=*Pyrethrum cinerarifolium*) (Compositae).

Used as an insecticide, but can cause convulsions, diarrhoea, respiratory paralysis, and liver and kidney damage in humans.

2047
Citral; 3,7-Dimethyl-2,6-octadienal (Naturally occurring citral is a mixture of the *trans* isomer, Citral A (=Geranial) and the *cis* isomer, Citral B (=Neral))

trans

cis

$C_{10}H_{16}O$ Mol. wt 152.24

Citral is the main constituent of lemon grass oil, from *Cymbopogon* (or *Andropogon*) *citratus* and *C. flexuosus* (Gramineae). It also occurs in oil of lemon, expressed from the fresh peel of *Citrus limon* (Rutaceae), in oil of sweet orange, from *C. sinensis*, and in oil of Verbena, from *Verbena triphylla* (=*Lippia citriodora*) (Verbenaceae).

Antiseptic activity (5 times stronger than phenol). It has a lemon-like scent, and is used for flavouring foods, in perfumery, and in the synthesis of vitamin A.

2048
Citronellal; 3,7-Dimethyloct-6-enal

$C_{10}H_{18}O$ (+) form   Mol. wt 154.25

Main constituent of Java and Ceylon oils of citronella, distilled from the fresh leaves of *Cymbopogon nardus* (=*Andropogon nardus*) (Gramineae). Also, it occurs in the volatile oils of various *Eucalyptus* spp., e.g., *E. citriodora* (Myrtaceae), in oil of balm, from the leaves and tops of *Melissa officinalis* (Labiatae), in oil of lemon (see citral), and in various other essential oils.

Antiseptic and sedative action (oil of balm). It is used in soap perfumes, and as an insect repellent.

2049
β-Citronellol; Citronellol; Cephrol; 3,7-Dimethyl-6-octen-1-ol

$C_{10}H_{20}O$ (+) form   Mol. wt 156.27

Occurs free and as esters in many essential oils. It is a main constituent in the oil from the leaves of *Boronia citriodora* (Rutaceae), and occurs together with geraniol (q.v.) and their esters in oil of geranium, from the leaves of *Pelargonium odoratissimum* (Geraniaceae), and oil of rose, from the fresh flowers of *Rosa gallica* and *R. damascena* (Rosaceae). The (+) form occurs in Ceylon and Java citronella oils (see citronellal).

Used in perfumery.

2050
Cuminaldehyde; Cuminal; Cumaldehyde; 4-Isopropylbenzaldehyde

$C_{10}H_{12}O$ Mol. wt 148.21

Occurs in the essential oils of eucalyptus, *Eucalyptus globulus* (Myrtaceae), cassia, *Cassia fistula* (Leguminosae), cumin, *Carum carvi* (Umbelliferae), myrrh, *Commiphora abyssinica* (Burseraceae), and other plants.

Used in perfumery, and in the synthesis of cuminaldehyde thiosemicarbazone, which has antiviral activity.

2051
p-Cymene; Cymene; p-Cymol; Dolcymene; Isopropyltoluene

$C_{10}H_{14}$ Mol. wt 134.22

A constituent of oil of American wormseed (see ascaridole), oil of cumin, from the fruit of *Cuminum cyminum* (Umbelliferae), ajowan oil, from the seeds of *Carum copticum* (=*Ligusticum ajowan*= *Trachyspermum ammi*) (Umbelliferae), oil of thyme, from *Thymus vulgaris* (Labiatae), and other volatile oils.

Toxic to mammals (LD_{50} orally in rats 4.75 g kg^{-1}). It is used as a local analgesic in rheumatic conditions.

2052
Diosphenol; Barosma camphor; Buchu camphor; 2-Hydroxypiperitone

$C_{10}H_{16}O_2$ Mol. wt 168.24

Constituent of the volatile oil from buchu leaves, viz., various *Barosma* spp., e.g., *B. betulina*, *B. serratifolia* and *B. crenulata* (Rutaceae).

Diuretic activity.

2053
Dipentene; *dl*-Limonene; Inactive limonene

$C_{10}H_{16}$ Mol. wt 136.24

Occurs in many essential oils, e.g., in oil of bergamot, from the rind of *Citrus aurantium* var. *bergamia* (Rutaceae), in oil of cubeb, from unripe fruit of *Piper cubeba* (Piperaceae), in the volatile oils from the needles of various *Pinus* spp. (Pinaceae), in the oils of citronella, lemon grass and palmarosa, from *Cymbopogon* or *Andropogon* spp. (Gramineae), and in various oils of Umbelliferae fruits.

Skin irritant and sensitizer. It also shows expectorant and sedative activities. Some essential oils containing dipentene are used for flavouring.

2054
Fenchone; 1,3,3-Trimethyl-2-norcamphanone

$C_{10}H_{16}O$ (+) form   Mol. wt 152.24

(+)-Fenchone occurs in oil of fennel, distilled from dried ripe fruit of *Foeniculum vulgare* (Umbelliferae), and in the essential oil of *Lavandula stoechas* (Labiatae). (−)-Fenchone is present in Thuja or white cedar oil, from the leaves of *Thuja occidentalis* (Pinaceae).

Counterirritant. It is used as a food flavour and in perfumes.

2055
Geraniol; Lemonol

$C_{10}H_{18}O$ Mol. wt 154.25

The main constituent of a number of important essential oils, e.g., palmarosa oil, distilled from the leaves of *Andropogon schoenanthus*, and allied spp. (Gramineae), Ceylon oil of citronella (see citronellal), and oil of rose (=otto or attar of rose), from fresh flowers of *Rosa gallica* and *R. damascena* (Rosaceae). It occurs in many other volatile oils, e.g., oil of lemon grass (see citral), and, mainly as esters, in oil of geranium (see geranyl acetate).

Antiseptic activity (7 times stronger than phenol). It has a sweet rose odour, and is used in perfumery and as an insect attractant.

2056
Geranyl acetate

$C_{12}H_{20}O_2$ Mol. wt 196.29

Geranyl acetate and other esters of geraniol (q.v.) occur in oil of geranium, the essential oil from the leaves of *Pelargonium odoratissimum* (Geraniaceae). They also occur in oil of eucalyptus (see cineole), in oil of pettigrain, from the leaves, twigs and unripe fruit of *Citrus vulgaris* (Rutaceae), and in small amounts in oil of lemon (see citral) and oil of rose (see geraniol).

Geranyl esters are used in perfumery, and as an insect attractant.

2057
Limonene; Cajeputene; Cinene; Kautschin

$C_{10}H_{16}$ (+) form   Mol. wt 136.24

(+)-Limonene is the main constituent of the volatile oils expressed from the fresh peel of *Citrus* fruits, e.g., lemon, tangerine, bitter and sweet orange (Rutaceae). It also occurs in oil of orange flowers, neroli oil, and in oils of caraway and dill, see carvone. (−)-Limonene is present in oil of fir, from the needles and young twigs of *Abies alba* (Pinaceae), and in mint oils, from *Mentha* spp. (Labiatae). For (±)-limonene, see dipentene.

Skin irritant, expectorant and sedative activities. Various essential oils containing limonene are used for flavouring food and beverages.

2058
Linalool; Linalol

$C_{10}H_{18}O$ (−) form   Mol. wt 154.25

The main constituent of oil of coriander, from dried ripe fruit of *Coriandrum sativum* (Umbelliferae), and of linaloe oil, distilled from *Bursera delpechiana* and other *Bursera* spp. (Burseraceae). It occurs in lavender oils, from the flowering tops of *Lavandula* spp. (Labiatae), in oil of bergamot, from the fresh peel of *Citrus aurantium* var. *bergamia* (Rutaceae), in oil of orange flowers, *C. sinensis*, and in many other essential oils.

Sedative and fungistatic activities; antiseptic (five times stronger than phenol). It is used in the synthesis of vitamin A, and in perfumery as a substitute for bergamot and French lavender oils.

2059
Linalyl acetate; Bergamol

$C_{12}H_{20}O_2$ (−) form   Mol. wt 196.29

Important constituent of oils of bergamot, see linalool, of pettigrain, from the leaves, twigs and unripe fruit of *Citrus vulgaris* (Rutaceae), and of lavender, from the flower tops of *Lavandula officinalis* (Labiatae). It also occurs in oils of coriander and orange flowers, see linalool.

Used in perfumery.

2060
Menthofuran

C$_{10}$H$_{14}$O   (+) form   Mol. wt 150.22

Occurs in the essential oils of peppermint, *Mentha piperita*, and watermint, *M. aquatica* (Labiatae).

2061
Menthol; Mentol; Peppermint camphor

C$_{10}$H$_{20}$O   (−) form   Mol. wt 156.27

Occurs free and as an ester as the main constituent of peppermint oil, the steam-distilled volatile oil from the fresh flowering plants of *Mentha piperita* and other *Mentha* spp. (Labiatae).

Topical antipruritic, analgesic, and antiseptic activities. It is widely used to relieve symptoms of bronchial and nasal congestion, and internally as a carminative and gastric sedative. It can give rise to hypersensitivity reactions, e.g., contact dermatitis. Uses include flavouring medicines, dentifrices and confectionery.

2062
Menthone; L-Menthone

C$_{10}$H$_{18}$O   (−) form   Mol. wt 154.25

Present in oil of peppermint, see menthol, and in the essential oils of several other *Mentha* spp. (Labiatae).

Mild antiseptic activity (2.25 times stronger than phenol). It is used for flavouring (it has a slight peppermint odour) and in perfumery.

2063
Menthyl acetate

C$_{12}$H$_{22}$O$_2$ (−) form   Mol. wt 198.31

Occurs in oil of peppermint, see menthol.

Used in perfumery because of its floral odour. Menthyl valerate (validol) is used as a sedative, and menthyl salicylate as an ultraviolet light filter in antisunburn creams.

2064
β-Myrcene

C$_{10}$H$_{16}$ Mol. wt 136.24

Constituent of oil of bay, the essential oil distilled from leaves of *Pimenta (Myrcia) acris* (Myrtaceae). It occurs in lesser amounts in many other essential oils.

Used in the manufacture of perfume chemicals.

2065
Nerol

$C_{10}H_{18}O$ Mol. wt 154.25

Usually occurs together with geraniol (q.v.), free and as esters, in many essential oils, e.g., in orange blossom oil, oil of neroli, and in oil of bergamot, see linalool, in oil of rose, see geraniol, and in oil of pettigrain, see linalyl acetate. Neryl acetate is a major constituent in the volatile oil of *Helichrysum angustifolium* (Compositae).

Used in perfumery (odour of sweet rose).

2066
β-Ocimene

$C_{10}H_{16}$ *trans* form Mol. wt 136.24

Widespread occurrence in essential oils, especially from the leaves of basil, *Ocimum basilicum* (Labiatae), and from the fruits of *Evodia rutaecarpa* (Rutaceae). It occurs as both *cis* and *trans* isomers.

2067
Perillaldehyde

$C_{10}H_{14}O$ (−) form Mol. wt 150.22

Best sources are the volatile oils of *Perilla arguta* (Labiatae) and *Sium latifolium*, water parsnip (Umbelliferae). Also, it occurs in mandarin peel oil, *Citrus reticulata* (Rutaceae).

(−)-Perillaldehyde α-syn-oxime ($C_{10}H_{15}NO$) is about 2000 times sweeter than sucrose, and is used as a sweetening agent in Japan ("perilla sugar").

2068
α-Phellandrene; *p*-Mentha-1,5-diene

$C_{10}H_{16}$ (−) form Mol. wt 136.24

The (−) form occurs in many essential oils, e.g., in the oils of several *Eucalyptus* spp. (Myrtaceae), and the (+) form occurs in oil of fennel, see fenchone, and in oil of pepper, from the unripe fruit of black pepper, *Piper nigrum* (Piperaceae).

Can be irritating to, and absorbed through, the skin. Ingestion can cause vomiting and/or diarrhoea. It is used in perfumery.

2069
β-Phellandrene; *p*-Mentha-1(7),2-diene

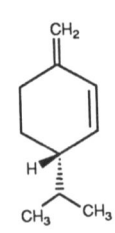

$C_{10}H_{16}$ (+) form Mol. wt 136.24

Common constituent of essential oils. The (−) form occurs in the volatile oils from several *Abies, Picea* and *Pinus* spp. (Pinaceae), whereas the (+) form is the main constituent of the essential oil from the seeds of the water dropwort, *Oenanthe aquatica* (=*Phellandrium aquaticum*) (Umbelliferae).

Expectorant (pine needle oil).

2070
α-Pinene; 2-Pinene; Pinene

$C_{10}H_{16}$ (+) form Mol. wt 136.24

Constituent of dozens of essential oils, especially of oil of turpentine, distilled from the oleoresin of *Pinus palustris* and other *Pinus* spp. (Pinaceae). Also, it occurs in volatile oils from the leaves of many other Pinaceae, of Cupressaceae, Myrtaceae, and Labiatae, and from the peel of *Citrus* spp. (Rutaceae).

Irritant. It can cause skin eruption, delirium, ataxia, and kidney damage. Uses include the manufacture of camphor, insecticides, perfume bases and synthetic pine oil.

2071
β-Pinene; Nopinene

$C_{10}H_{16}$ (+) form Mol. wt 136.24

Occurs in oil of turpentine and in most other essential oils containing α-pinene, but is present in smaller amounts. It is also found in oil of cumin, distilled from the dried fruit of *Cuminum cyminum* (Umbelliferae).

For activity and uses see α-pinene.

2072
Pinocarvone; Carvopinone

$C_{10}H_{14}O$ (−) form Mol. wt 150.22

Occurs in the essential oil of *Eucalyptus globulus* (Myrtaceae).

2073
Piperitone

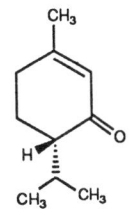

$C_{10}H_{16}O$ (+) form Mol. wt 152.24

The (+) form occurs in the essential oil of *Andropogon iwarancusa* (=*Cymbopogon sennarensis*) (Gramineae), and in the oils of *Mentha* spp. (Labiatae) and *Lippia alba* (Verbenaceae). The (±) form is a constituent of the volatile oil of *Eucalyptus dives* (Myrtaceae).

Used in dentifrices (has an odour reminiscent of camphor and peppermint).

2074
Pulegone

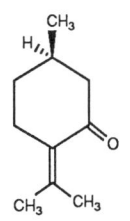

$C_{10}H_{16}O$ (+) form Mol. wt 152.24

The main constituent of oils of pulegium and hedeoma, European and American pennyroyal oils, distilled from the leaves and flowering tops of *Mentha pulegium* and *Hedeoma pulegioides*, respectively (Labiatae). It also occurs in other *Mentha* oils.

For activity and use see piperitone.

2075
Pyrethrins; (Pyrethrin I = Chrysanthemum monocarboxylic acid pyrethrolone ester; Pyrethrin II = Chrysanthemum dicarboxylic acid monomethyl ester pyrethrolone ester)

R = CH$_3$ pyrethrin I C$_{21}$H$_{28}$O$_3$ Mol. wt 328.46
R = COOCH$_3$ pyrethrin II C$_{22}$H$_{28}$O$_5$ Mol. wt 372.47

Found in the flowers of pyrethrum, *Tanacetum cinerariifolium* (Compositae).

Toxic to mammals (LD$_{50}$ orally in rats 1.2 g kg^{-1}). They can cause severe allergic dermatitis. Large quantities can affect the central nervous system. They are used as insecticides.

2076
Rotundifolone; Piperitenone oxide

C$_{10}$H$_{14}$O$_2$ (+) form   Mol. wt 166.22

Occurs in the essential oils of certain mint spp., e.g., *Mentha rotundifolia*, *M. spicata* and *M. longifolia* (Labiatae).

2077
Sabinol

C$_{10}$H$_{16}$O   (+) form   Mol. wt 152.24

Occurs free and as sabinyl acetate in oil of savin, distilled from the fresh tops of *Juniperus sabina* (Cupressaceae). It is also a constituent of essential oils from other *Juniperus* spp.

Toxic. It is used as an emmenagogue and anthelmintic.

2078
α-Terpinene

C$_{10}$H$_{16}$ Mol. wt 136.24

Occurs in oils of marjoram, from the leaves of *Origanum majorana* (Labiatae), of cardamom, from the seeds of *Elettaria cardamomum* (Zingiberaceae), of lemon, see citral, and of *Ocimum* spp. (Labiatae).

Pleasant lemon-like odour.

2079
γ-Terpinene

C$_{10}$H$_{16}$ Mol. wt 136.24

Occurs in oil of ajowan, from seeds of *Carum copticum* (=*Ligusticum ajowan*=*Trachyspermum ammi*) (Umbelliferae).

Found in the turpentine oils of various spp. belonging to the Pinaceae, and in the essential oil of *Ocimum kilimandscharicum* (Labiatae).

Used to mask odours of industrial products.

2080
α-Terpineol; Terpineol; *p*-Menth-1-en-8-ol

C$_{10}$H$_{18}$O (+) form Mol. wt 154.25

Widespread occurrence. The (+) form occurs in oil of pettigrain, see linalyl acetate, and oils of cardamom and marjoram, see α-terpinene. The (−) form occurs in oil of Niaouli, from the leaves of *Melaleuca viridiflora* (Myrtaceae), in the volatile oil of nutmeg, *Myristica fragrans* (Myristicaceae), and in the turpentine oils of various spp. belonging to the Pinaceae. The (±) form is found in cajeput oil, see 1,8-cineole.

Antiseptic activity. It is used in perfumery, and as a flavouring.

2081
Terpinolene

C$_{10}$H$_{16}$ Mol. wt 136.24

2082
β-Thujaplicin

C$_{10}$H$_{12}$O$_2$ Mol. wt 164.21

Found in the heartwood of *Thuja plicata* (Cupressaceae) and as a constituent of the tropolone fraction of the heartwood of many *Cupressus* spp., e.g., *C. sargentii, C. abramsiana* and *C. macrocarpa*.

Antifungal activity.

2083
γ-Thujaplicin

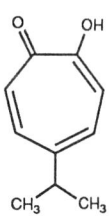

C$_{10}$H$_{12}$O$_2$ Mol. wt 164.21

Found as a constituent of the tropolone fraction of the heartwood of several *Cupressus* spp., e.g., *C. lusitanica*, and of *Thuja plicata* (Cupressaceae).

Antifungal activity.

2084
Thujone; Thujan-3-one

$C_{10}H_{16}O$ (−) form Mol. wt 152.24

Occurs as a mixture of two stereoisomers, α-thujone (shown) and β-thujone, which differ in the stereochemistry of the 4-methyl group. It is found in many essential oils, e.g., oil of white cedar, from the leaves of *Thuja occidentalis* (Cupressaceae), oil of tansy, from the leaves and tops of *Tanacetum vulgare* (Compositae) and, together with thujyl alcohol and acetate, in oil of wormwood, from the leaves and tops of *Artemisia absinthium* (Compositae).

Used as a counterirritant (cedar leaf oil) and anthelmintic (wormwood oil). Ingestion may cause convulsions.

2085
Thymol; *m*-Thymol; Timol; 3-*p*-Cymenol; Thyme camphor; 6-Isopropyl-*m*-cresol

$C_{10}H_{14}O$ Mol. wt 150.22

Best sources are the essential oils of some Labiatae spp., e.g., *Thymus vulgaris* and *Monarda punctata*, and oil of ajowan, see γ-terpinene.

Antiseptic (20 times more active than phenol) and antifungal. It is used for destroying mould, and for preserving botanical and biological specimens, and also in dentistry. Can irritate gastric mucosa.

2086
Thymyl acetate; Thymol acetate; Acetylthymol

$C_{12}H_{16}O_2$ Mol. wt 192.26

Found as a constituent of oil of thyme, distilled from the flowering plant of *Thymus vulgaris* (Labiatae).

Mild irritant. It is used as an antiseptic, rubefacient, carminative and counterirritant.

2087
Umbellulone

$C_{10}H_{14}O$ (−) form Mol. wt 150.22

Found as a constituent of the essential oil of the California bay-tree or myrtle, *Umbellularia californica* (Lauraceae).

2088
Verbenone; Pin-2-en-4-one

$C_{10}H_{14}O$ (+) form Mol. wt 150.22

Occurs in Spanish verbena oil, from the leaves of *Verbena triphylla* (=*Lippia citriodora*) (Verbenaceae).

Toxic to mammals (LD_{50} intraperitoneally in mice 250 mg kg^{-1}).

Chapter 49
Iridoids

The iridoids are a group of bitter tasting monoterpenoid lactones, which are widely present in angiosperms, being found in about 70 families grouped in some 13 orders. They are characterized by or based on a cyclopentanodihydropyran ring system. Iridoids have only recently come into prominence in plants; their chemical instability delayed structural identification, and it was not until the early 1960s that their chemistry was fully established. Iridoids are derived biosynthetically from monoterpene precursors and, like the monoterpenes (q.v.), are formed from geranyl pyrophosphate. The iridoids "proper" are lactone derivatives, often with glucose attachment to the hydroxyl group of the lactone ring. A typical iridoid glucoside is loganin (**2109**), which occurs for example in *Strychnos nux-vomica* fruit to the extent of 4–5% dry weight. The aglycones are usually highly unstable and disintegrate, after hydrolysis with acid, into a blue or black polymer. Herbarium plants containing iridoid glucosides also often turn dark during drying. This dark-blue discoloration is the reason why iridoids were originally called "pseudoindicans". However, some iridoids "proper" occur in plants as stable aglycones, e.g., allamandin (**2089**), fulvoplumierin (**2100**) and plumericin (**2115**).

A second group of iridoids can be distinguished in which the five-membered ring of carboxylic iridoids is opened, giving rise to seco-iridoids, which have, as a result, an additional aldehydo function. The seco-iridoid derived from loganin is secologanin (**2118**), a widespread substance especially common on the Caprifoliaceae. Seco-iridoids have a special role in plants as biosynthetic precursors of terpene alkaloids. For example, condensation of the aldehydo group of secologanin with the amino acid tryptophan produces an alkaloid such as corynantheine, a major base of *Corynanthe johimbe* (Rubiaceae). In general, seco-iridoids tend to co-occur with their related alkaloids.

Within each of the two groups of iridoids represented by loganin and secologanin, respectively, there are many derivatives known. Aucubin (**2093**) is a dihydroxy compound, first isolated from *Aucuba japonica* (Cornaceae) but also well represented in species of the Scrophulariaceae. Asperuloside (**2092**) is a double lactone, named after the source *Asperula odorata* but actually first obtained from madder root, *Rubia tinctorum*. Acylated derivatives are also known, such as catalposide (**2096**), the *p*-hydroxybenzoyl ester of catalpol (**2095**), from *Catalpa* spp. Other ester groupings found attached to hydroxyl groups in iridoids include cinnamic (see harpagoside, **2105**) and acetic (didrovaltratum, **2098**).

A few volatile iridoids without glucose attachment are present in plants, which are called "simple iridoids". A notable example is nepetalactone (**2112**), the active principle of catmint, *Nepeta cataria*, and the substance that causes members of the cat family to have a peculiar fascination for this plant. The actual role of nepetalactone in *Nepeta* spp. appears to be that of an insect feeding deterrent. Structures related to nepetalactone (e.g., dolichodial, **2099**) occur in the defence secretions of ants, stick insects and beetles. Such iridoids are usually volatile, as mentioned before, occurring in plants in the essential oil fraction, and are generally atypical. All surveys of plants for iridoids are based on the detection of the nonvolatile derivatives.

Iridoids are restricted in their occurrence to the dicotyledons, and are mainly found in more advanced families. They are not recorded in the monocotyledons, in gymnosperms or in any lower plant group. Their main purpose seems to be one of providing feeding deterrence, although some are also antimicrobial substances.

Plants containing iridoid and seco-iridoid substances have been used in folk medicine for centuries. Some of these plants were used as bitter tonics, e.g., *Cornus, Menyanthes*, others against inflammations, e.g., eyebright, *Euphrasia*. Nonglucosidic iridoids such as fulvoplumierin and allamandin show antimicrobial and/or antileukaemic activity. Glucosides such as aucubin are active only in the presence of β-glucosidase. The hemiacetal structure of the aglycones appears to play a significant role in the antitumour and antimicrobial activities. Aucubin and various other iridoid glucosides also show laxative and diuretic properties, whereas the valepotriates, the mainly nonglucosidic iridoids occurring in Valerianaceae, are sedative agents.

In large concentrations and in the presence of acids iridoid glycosides are toxic. Bread prepared from wheat contaminated with weed seeds containing iridoids, e.g., *Rhinanthus*, has caused human illness and death. Recently, iridoids such as catalpol have been shown to be ingested from food plants, e.g., *Plantago*, by larvae of the butterfly genus *Euphydryas* and stored in the adults for protection against bird predation.

References

General

Hegnauer, R. (1964–1986). *Chemotaxonomie der Pflanzen*, Volumes 3–7. Basel: Birkhaüser.

Bobbitt, J.M. and Segebarth, K.-P. (1969) The iridoid glycosides and similar substances. In *Cyclopentanoid Terpene Derivatives* (Battersby, A.R. and Taylor, W.J., eds), pp. 1–145. New York: Marcel Dekker.

Plouvier, V. and Favre-Bonvin, J. (1971). Les iridoides et seco-iridoides: repartition, structure, proprietes, biosynthese. *Phytochemistry*, **10**, 1697–1722.

Natural Occurrence

Jensen, S.R., Nielsen, B.J. and Dahlgren, R. (1975). Iridoid compounds, their occurrence and systematic importance in the angiosperms. *Botaniska Notiser*, **128**, 148–180.

El-Neggar, L.J. and Beal, J.L. (1980) Iridoids, a review. *Journal of Natural Products*, **43**, 649–707.

Biological Activity

Sticher, O. (1977) Plant mono-, di- and sesquiterpenoids with pharmacological or therapeutical activity. In *New Natural Products and Plant Drugs with Pharmacological, Biological or Therapeutical Activity* (Wagner, H. and Wolff, P., eds), pp. 137–176. Berlin: Springer-Verlag.

2089
Allamandin

Nonglucosidic iridoid

$C_{15}H_{16}O_7$ Mol. wt 308.29

Found in *Allamanda cathartica* (Apocynaceae).

Allamandin and the chemically related iridoids allamandicin and allamdin from the same plant source have antileukaemic and tumour-inhibiting activities.

2090
Amarogentin

Seco-iridoid glucoside

$C_{29}H_{30}O_{13}$ Mol. wt 586.56

Occurs in the roots of *Gentiana* spp., e.g., *G. lutea*, and in Indian gentian, spp. of *Swertia* (Gentianaceae).

Amarogentin and the closely related seco-iridoids amaroswerin and amaropanin are the bitter principles of gentians, which are used as bitter tonics. They are among the most bitter substances known.

2091
Antirrhinoside

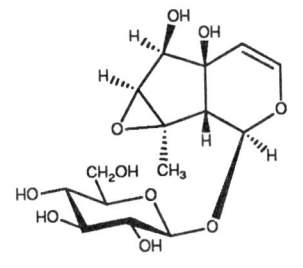

Iridoid glucoside

$C_{15}H_{22}O_{10}$ Mol. wt 362.34

Found in spp. of *Antirrhinum, Linaria* and closely related genera (Scrophulariaceae), e.g., in the leaves of the garden snapdragon, *A. majus*.

2092
Asperuloside; Asperulin; Rubichloric acid

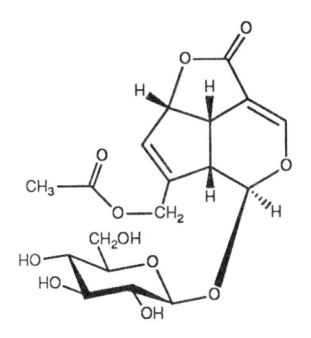

Iridoid glucoside

$C_{18}H_{22}O_{11}$ Mol. wt 414.37

Found in many spp. of the Rubiaceae, e.g., *Asperula odorata* and *Galium aparine*, in the leaves of many *Escallonia* spp. (Saxifragaceae) and in *Daphniphyllum macropodum* (Daphniphyllaceae).

Laxative activity. It shows seed germination and plant growth inhibiting activities.

2093
Aucubin; Aucuboside; Rhinanthin

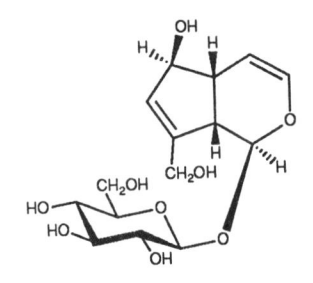

Iridoid glucoside

$C_{15}H_{22}O_9$ Mol. wt 346.34

One of the most common iridoids, occurring in many families of the dicotyledons, e.g., *Aucuba japonica* (Cornaceae) and some *Rhinanthus* spp. (Scrophulariaceae). It is sequestered and stored by the alpine butterfly *Euphydryas cynthia* from the larval food plant *Plantago lanceolata* (Plantaginaceae).

Laxative and diuretic activities. The aglycone, aucubigenin, is toxic to mammals, and has antimicrobial and antitumour activities.

2094
Boschnialactone

Simple iridoid

$C_9H_{14}O_2$ Mol. wt 154.21

Occurs in *Boschniakia rossica* (Orobanchaceae).

Excitatory activity towards cats and other Felidae.

2095
Catalpol

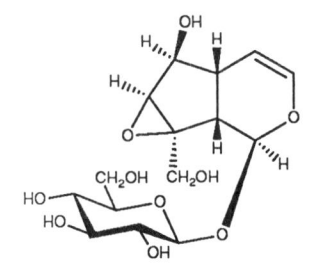

Iridoid glucoside

$C_{15}H_{22}O_{10}$ Mol. wt 362.34

Occurs in spp. of *Catalpa* (Bignoniaceae), *Veronica* (Scrophulariaceae), *Plantago* (Plantaginaceae) and *Buddleja* (Buddlejaceae). Plants containing catalpol often also contain aucubin (q.v.).

Diuretic and laxative activities. Very bitter taste.

2096
Catalposide; Catalpin

Iridoid glucoside

$C_{22}H_{26}O_{12}$ Mol. wt 482.45

Found in *Catalpa speciosa* (Bignoniaceae) and *Veronica persica* (Scrophulariaceae).

Diuretic and laxative properties. Catalposide and some related catalpol esters are feeding attractants to the caterpillars of the Catalpa sphinx moth, *Ceratomia catalpae*, but feeding deterrents to caterpillars of *Lymantria dispar*.

2097
Deoxyloganin

Iridoid glucoside

$C_{17}H_{26}O_9$ Mol. wt 374.40

Found in *Strychnos* spp. (Loganiaceae), *Vinca* spp. (Apocynaceae), and *Menyanthes* spp. (Menyanthaceae).

Laxative.

2098
Didrovaltratum; Didrovaltrate

Nonglucosidic iridoid

$C_{22}H_{32}O_8$ Mol. wt 424.50

Occurs together with many related substances (valepotriates) in the roots of spp. of *Valeriana* and *Centranthus* (Valerianaceae).

Tranquillizing activity, similar to other valepotriates, and improves coordination. It is also used as a weak sedative.

2099
Dolichodial

Simple iridoid

$C_{10}H_{14}O_2$ Mol. wt 166.22

Found in *Teucrium marum* (Labiatae).

Insect repellent and lachrymatory.

2100
Fulvoplumierin

Nonglucosidic iridoid

$C_{14}H_{12}O_4$ Mol. wt 244.25

Found in the bark of some *Plumeria* (=*Plumieria*) spp. (Apocynaceae), e.g., *P. acutifolia* (=*P. acuminata*).

Antibiotic activity, e.g., against strains of *Mycobacterium tuberculosis*.

2101
Gardenoside

Iridoid glucoside

$C_{17}H_{24}O_{11}$ Mol. wt 404.38

Found in *Gardenia jasminoides* f. *grandiflora* and var. *radicans* (Rubiaceae).

Mild laxative.

2102
Genipin

Nonglucosidic iridoid

$C_{11}H_{14}O_5$ Mol. wt 226.23

Found in *Genipa americana* (Rubiaceae).

Increases bile flow.

2103
Geniposide

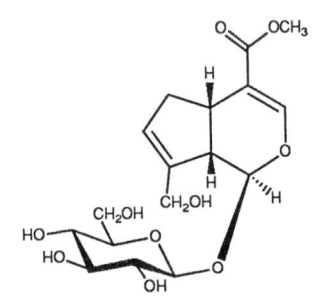

Iridoid glucoside

$C_{17}H_{24}O_{10}$ Mol. wt 388.38

Found in spp. of *Gardenia* (Rubiaceae) and *Cornus* (Cornaceae).

Laxative. The aglycone is genipin (q.v.).

2104
Gentiopicrin; Gentiopicroside

Seco-iridoid glucoside

$C_{16}H_{20}O_9$ Mol. wt 356.34

Found in the roots of *Gentiana lutea* and in other plants of the Gentianaceae.

Antimalarial activity.

2105
Harpagoside

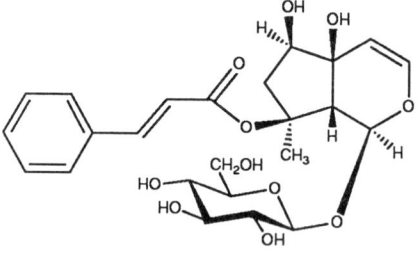

Iridoid glucoside

$C_{24}H_{30}O_{11}$ Mol. wt 494.50

Found in the roots of *Harpagophytum procumbens* (Pedaliaceae), in the rhizomes of *Scrophularia buergeriana* (Scrophulariaceae), and in some *Lamium* spp. (Labiatae).

Analgesic activity. Harpagoside tastes very bitter, whereas harpagide (the same compound but without the cinnamoyl group) tastes slightly sweet.

2106
(+)-Iridodial

Simple iridoid

$C_{10}H_{16}O_2$ Mol. wt 168.24

Found in a *Myoporum* spp. (Myoporaceae).

Insect-repellent activity.

2107
Iridomyrmecin

Simple iridoid

$C_{10}H_{16}O_2$ Mol. wt 168.24

Found in *Actinidia polygama* (Actinidiaceae).

Antibiotic. It shows insecticidal activity.

2108
Isoplumericin

Nonglucosidic iridoid

$C_{15}H_{14}O_6$ Mol. wt 290.28

Found in some *Plumeria* spp. (Apocynaceae).

Inhibits the growth of a number of bacteria and fungi, e.g., strains of *Mycobacterium tuberculosis*.

2109
Loganin; Loganoside

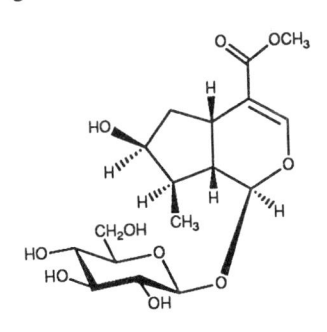

Iridoid glucoside

$C_{17}H_{26}O_{10}$ Mol. wt 390.40

Found in the fruits of *Strychnos nux-vomica* and other *Strychnos* spp. (Loganiaceae), in the rhizomes of *Menyanthes trifoliata* (Menyanthaceae), in the bark of a few *Hydrangea* spp. (Saxifragaceae), and in *Catharanthus roseus* (=*Vinca rosea*) (Apocynaceae).

Laxative. Crude drugs containing loganin are used as a bitter tonic.

2110
Monotropein

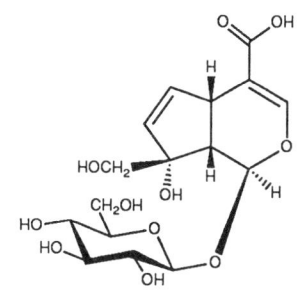

Iridoid glucoside

$C_{16}H_{22}O_{11}$ Mol. wt 390.35

Found in some *Monotropa* spp., e.g., *M. hypopithys* (Monotropaceae), in some *Pyrola* spp. (Pyrolaceae), *Liquidambar* spp. (Hamamelidaceae), *Galium* and *Asperula* spp. (Rubiaceae), and *Globularia* spp. (Globulariaceae).

2111
(+)-Neomatatabiol

Simple iridoid

$C_{10}H_{18}O_2$ Mol. wt 170.25

Found in the leaves of *Actinidia polygama* (Actinidiaceae).

Attracts the male adults of the lacewing, *Chrysopa septempunctata* (Chrysopidae).

2112
Nepetalactone

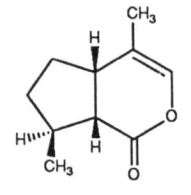

cis-trans   trans-cis

Simple iridoid

$C_{10}H_{14}O_2$ Mol. wt 166.22

A mixture of the *cis-trans* and *trans-cis* isomers occurs in the volatile oil of catnip, *Nepeta cataria* (Labiatae). The *trans-cis* isomer is produced as a male-attracting pheromone by the female vetch aphid, *Megoura viciae*.

Nepetalactone repels insects and has a potent ability to excite cats and other Felidae such as lions and jaguars (but not tigers, however).

2113
Oleuropein

Seco-iridoid glucoside

$C_{25}H_{32}O_{13}$ Mol. wt 540.53

Found in the leaves, bark and fruit of the olive tree, *Olea europaea* and also in some other genera of the Oleaceae.

Hypotensive, coronary dilating, antiarrhythmic and spasmolytic activities. Hydrolysis products of oleuropein, the aglucone and elenolic acid, inhibit the growth of lactic acid bacteria.

2114
Paederoside

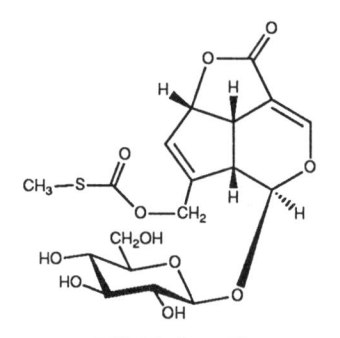

Iridoid glucoside

$C_{18}H_{22}O_{11}S$ Mol. wt 446.44

Occurs in *Paederia scandens* var. *mairei* (Rubiaceae).

Laxative.

2115
Plumericin

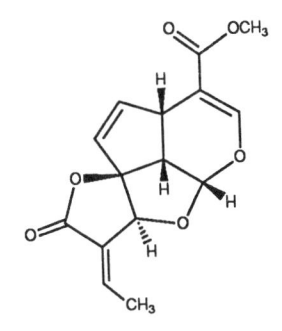

Nonglucosidic iridoid

$C_{15}H_{14}O_6$ Mol. wt 290.28

Found in the roots of *Plumeria multiflora* and *P. rubra* (Apocynaceae).

Inhibits the growth of some fungi and bacteria, e.g., *Mycobacterium tuberculosis*.

2116
Plumieride; Plumieroside

Iridoid glucoside

$C_{21}H_{26}O_{12}$ Mol. wt 470.44

Found in the bark of several *Plumeria* (=*Plumieria*) spp., e.g., *P. acutifolia* (Apocynaceae).

Laxative and weak antifungal activities.

2117
Scandoside methyl ester

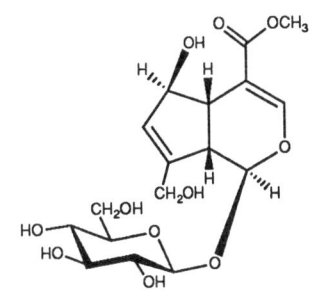

Iridoid glucoside

$C_{17}H_{24}O_{11}$ Mol. wt 404.38

Occurs in *Gardenia* spp. (Rubiaceae).

The aglycone shows potent antitumour activity.

2118
Secologanin; Loniceroside

Seco-iridoid glucoside

$C_{17}H_{24}O_{10}$ Mol. wt 388.38

Found in some *Lonicera* spp. (Caprifoliaceae), and as a minor constituent in *Catharanthus roseus* (=*Vinca rosea*) (Apocynaceae), but probably widespread in plants containing indole alkaloids (q.v.).

Specific precursor of several major types of indole alkaloids and, thus, could be used in the pharmaceutical industry for the production of therapeutically active substances such as the antileukaemic *Catharanthus* alkaloids.

2119
Swertiamarin; Swertiamaroside

Seco-iridoid glucoside

$C_{16}H_{22}O_{10}$ Mol. wt 374.35

Found in *Swertia japonica* (Gentianaceae).

The aglycone, erythrocentaurin (obtained by hydrolysis of swertiamarin with emulsin), is used as a bitter tonic.

2120
Valtratum

Nonglucoside iridoid

$C_{22}H_{30}O_8$ Mol. wt 422.48

Valtratum and many related substances (valepotriates) occur in the roots of some *Valeriana* and *Centranthus* spp. (Valerianaceae).

Tranquillizing effect, similar to other valepotriates, and improves coordination. Also, it acts antagonistically against the hypotensive effect of ethanol. It is used as a weak sedative.

2121
Verbenalin; Verbenaloside; Cornin

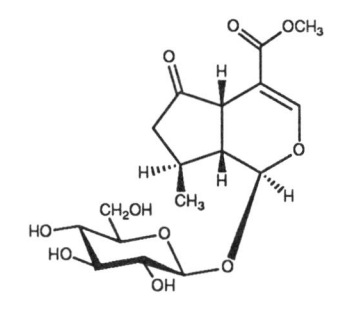

Iridoid glucoside

$C_{17}H_{24}O_{10}$ Mol. wt 388.38

Occurs in some *Verbena* spp., e.g., *V. officinalis* (Verbenaceae), and in *Cornus* spp. (Cornaceae).

Activity on the uterus, resembling that of ergot. It is a weak parasympathomimetic, and shows laxative properties.

Chapter 50
Sesquiterpenoids

The sesquiterpenoids are chemically defined by their formation from a common C_{15} precursor, farnesyl pyrophosphate, and they are thus derived biosynthetically from three isoprene units. They co-occur with the monoterpenoids (q.v.) in plant essential oils, but can be distinguished by their higher boiling points. They can be classified into three main groups according to whether they are acyclic (e.g., farnesol, **2149**), monocyclic (e.g., bisabolol, **2123**) or bicyclic (e.g., β-cadinene, **2126**). Some are simple unsaturated hydrocarbons (e.g., caryophyllene, **2130**), but most have extra functional groups and may be alcohols (e.g., carotol, **2129**), ketones (e.g., hydroxyisopatchoulenone, **2165**), or carboxylic acids (e.g., sclerosporin, **2195**). The derived sesquiterpenoid, abscisic acid (**2122**), which also happens to be a key plant hormone controlling growth and development, actually has all three of these functional groups in its structure.

Sesquiterpenoids can be classified according to their biogenetic origin into some 200 different skeletal types (see, e.g., Nakanishi *et al.*, 1974), and several thousand such compounds are now known in nature. One large group of sesquiterpenoids, those with an additional lactone function, are treated separately in this dictionary, under the heading of sesquiterpene lactones (q.v.).

The main occurrence of sesquiterpenes is in plant essential oils, and some structures — bisabolol, caryophyllene, α-cadinene, β-farnesene, β-selinene — are widely present in the volatiles of plants of such families as the Labiatae, Myrtaceae, Pinaceae and Rutaceae. Other compounds are of more restricted occurrence: carotol (**2129**), for example, is characteristic of the carrot genus *Daucus* (Umbelliferae), while gossypol (**2153**) is confined to the cotton genus *Gossypium* (Malvaceae). Some plant sesquiterpenoids (e.g., rishitin, **2189**) are formed only after the plant has been infected by a micro-organism, and such compounds, called phytoalexins, are by definition antifungal in their activity (Bailey and Mansfield, 1982).

Besides occurring in higher plants, sesquiterpenes are well represented in bryophytes (Asakawa, 1982) and in micro-organisms (Turner and Aldridge, 1983). They are also found in marine animals and are encountered in insects in defence (e.g., germacrene B, **2151**) and in pheromonal (e.g., β-farnesene, **2148**) secretions. One class of insect hormone, the juvenile hormone (JH), is sesquiterpenoid-based and JH mimics (e.g., juvabione, **2171**) and even the hormones themselves (JHIII, **2172**) have been found in plants.

Sesquiterpenoids are commonly colourless oils of high (160–200°C) boiling point; some have pleasant odours, and others are bitter tasting. Optical isomerism is a regular feature of sesquiterpene chemistry. Optical activity is often related to biological activity; thus only the (+)-*cis* isomer of abscisic acid (**2122**) shows hormonal activity. One unexpected aspect of the sesquiterpenoids of bryophytes is the fact that they are usually enantiomeric with the related structures in higher plants. Thus α-chamigrene occurs in the (−) form in higher plants (**2135**) but in the (+) form in bryophytes (Asakawa, 1982).

These lower terpenoids are separated and characterized by similar methods (e.g., gas chromatography/mass spectroscopy) to those used with the monoterpenoids. Many are biologically active and some (e.g., dehydromyodesmone, **2144**) are very toxic to mammals, including humans. Several mycotoxins (e.g., trichothecin, **2205**) are sesquiterpene derivatives.

Comprehensive listings of sesquiterpenoid structure can be found in Glasby (1982). Their biological activity has been reviewed by Herout (1971).

References

Chemistry and Distribution

Asakawa, K. (1992). Chemical constituents of the Hepaticae. *Fortschritte der Chemie organischer Naturstoffe*, **42,** 1–286.

Glasby, J.S. (1982). *Encyclopedia of the Terpenoids*. Chichester: Wiley.

Nakanishi, K., Goto, T., Ito, S., Natori, S. and Nozoe, S. (eds) (1974). *Natural Products Chemistry*, Volume 1. New York: Academic Press.

Turner, W.B. and Aldridge, D.C. (1983). *Fungal Metabolites II*. London: Academic Press.

Biological Activity

Bailey, J.A. and Mansfield, J.W. (eds) (1982). *Phytoalexins*. Glasgow: Blackie.

Herout, V. (1971). Biochemistry of sesquiterpenoids. In *Aspects of Terpenoid Chemistry and Biochemistry* (Goodwin, T.W., ed.), pp. 53–94. London: Academic Press.

2122
Abscisic acid: (*S*)-Abscisic acid; ABA; Abscisin II; Dormin

$C_{15}H_{20}O_4$ Mol. wt 264.33

Found in the young fruit of cotton, *Gossypium hirsutum* (Malvaceae), and in avocado fruit, *Persea gratissima* (Lauraceae) to the extent of 0.76 mg per kg fresh fruit. Present in small amounts, either free or conjugated, in all green plants.

A universal plant hormone which controls stomatal closure and regulates bud dormancy.

2123
α-**Bisabolol**; Bisabolol

$C_{15}H_{26}O$ (−) form Mol. wt 222.37

(−)-α-Bisabolol occurs in chamomile oil, from *Matricaria chamomilla* (Compositae), whereas the (+) form is found in the buds of *Populus balsamifera* (Salicaceae).

(−)-α-Bisabolol has anti-inflammatory activity, but it is much less toxic than guaiazulene (q.v.), which has a similar action.

2124
Botrydial

$C_{17}H_{26}O_5$ Mol. wt 310.40

Metabolite of the fungus *Botrytis cinerea*.

Antibacterial and antifungal activities.

2125
Buddledin A

$C_{17}H_{24}O_3$ Mol. wt 276.38

Occurs together with the related sesquiterpenes buddledin B and buddledin C in the root bark of *Buddleja davidii* (Buddlejaceae).

Buddledins A, B and C exhibit piscicidal activity.

2126
β-**Cadinene**; Cadinene

$C_{15}H_{24}$ (−) form Mol. wt 204.36

β-Cadinene is one of the most common sesquiterpenoids in plants. The cadinenes occur as a mixture of isomers in many essential oils, especially in the leaf or needle oils of conifers, e.g., of *Juniperus communis* (Cupressaceae).

Many essential oils containing cadinenes are used in perfumery.

2127
Canellal

$C_{15}H_{20}O_3$ Mol. wt 248.33

Found in the stem bark of *Canella winterana* (Canellaceae).

Antibiotic activity.

2128
Capsidiol

$C_{15}H_{24}O_2$ Mol. wt 236.36

Phytoalexin produced in the fruit of pepper, *Capsicum frutescens* (Solanaceae), when injected with the fungus *Monilinia fruticola*, and in tobacco leaves, *Nicotiana tabacum* (Solanaceae), infected with tobacco mosaic virus.

Fungitoxic activity.

2129
Carotol

$C_{15}H_{26}O$ (+) form Mol. wt 222.37

Occurs in the oil from the seeds of carrot, *Daucus carota* (Umbelliferae).

2130
β-Caryophyllene; Caryophyllene

$C_{15}H_{24}$ Mol. wt 204.36

β-Caryophyllene occurs as a mixture with α-caryophyllene (=humulene) (q.v.) and isocaryophyllene (q.v.) in many essential oils, e.g., in oil of clove, prepared from the dried flower buds of *Eugenia caryophyllata* (=*Caryophyllus aromaticus*) (Myrtaceae), and oil of copaiba, the oleoresin from spp. of *Copaifera*=*Copaiba* (Leguminosae).

Used in perfumery.

2131
α-Cedrene

$C_{15}H_{24}$ Mol. wt 204.36

Constituent of cedar wood oil, from the wood of the red cedar, *Juniperus virginiana* (Cupressacae), and other *Juniperus* spp.

Cedar wood oil is used in perfumery, and as a clearing agent in microscopy.

2132
α-Cedrol; Cedrol; Cedar camphor; Cypress camphor

$C_{15}H_{26}O$ Mol. wt 222.37

Constituent of cedar wood oil, from the wood of *Juniperus virginiana* (Cupressaceae), and other *Juniperus* spp., and of cypress oil, from *Cupressus sempervirens* (Cupressaceae).

Used in perfumery.

2133
Centdarol

$C_{15}H_{26}O_2$ Mol. wt 238.37

Found in the wood of *Cedrus deodara* (Pinaceae).

Spasmolytic activity.

2134
Chamazulene; Dimethulene; 7-Ethyl-1,4-dimethylazulene

$C_{14}H_{16}$ Mol. wt 184.28

Blue oil produced during steam distillation of chamomile, *Matricaria chamomilla*, wormwood, *Artemisia absinthium* and yarrow, *Achillea millefolium* (Compositae), from sesquiterpene lactones found in these plants, such as matricin, achillin and artabsin (q.v.).

Anti-inflammatory and antipyretic activities.

2135
Chamigrene

α-chamigrene   β-chamigrene

$C_{15}H_{24}$ Mol. wt 204.36

α-Chamigrene and β-chamigrene occur in the fruit oil of *Schisandra chinensis* (Schisandraceae), and β-chamigrene occurs in the essential oil from the leaves of *Chamaecyparis taiwensis* (Cupressaceae).

2136
Cinnamodial; Ugandensidial

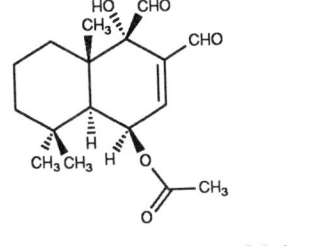

$C_{17}H_{24}O_5$ Mol. wt 308.38

Found in the bark of "taggar", *Cinnamosma fragrans*, and in the heartwood of *Warburgia salutaris* (=*W. ugandensis*) (Canellaceae).

Antifeedant against African armyworms.

2137
α-Copaene; Copaene

$C_{15}H_{24}$ Mol. wt 204.36

Found in African copaiba balsam oil, from *Copaifera* (=*Copaiba*) spp. (Leguminosae), and in various other oils, e.g., from *Sindora* spp. (Leguminosae).

Copaiba is a carminative. The balsam is used in varnishes and in the manufacture of photographic paper.

2138
Cuauhtemone

$C_{15}H_{24}O_3$ Mol. wt 252.36

Found in the aerial parts of the Mexican medicinal shrub "cuauhtematl", *Pluchea odorata* (Compositae).

Plant growth inhibitor.

2139
Cubebene

α-cubebene   β-cubebene

$C_{15}H_{24}$ Mol. wt 204.36

Both α- and β-cubebene are found in cubeb oil, which is prepared from the nearly ripe fruit of *Piper cubeba* (Piperaceae).

Cubeb oil is a flavouring agent.

2140
α-Curcumene

$C_{15}H_{22}$ (−) form   Mol. wt 202.34

Found in the oils of *Curcuma aromatica* and of ginger, *Zingiber officinale* (Zingiberaceae), and many other essential oils.

2141
Cycloeudesmol

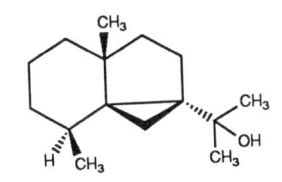

$C_{15}H_{26}O$ Mol. wt 222.37

Occurs in the marine alga *Chondria oppositiclada* (Rhodophyta).

Strong antibiotic activity against *Staphylococcus aureus* and *Candida albicans*.

2142
Daucol

$C_{15}H_{26}O_2$ Mol. wt 238.37

Found in the seed oil of carrots, *Daucus carota* (Umbelliferae).

2143
Dehydrojuvabione

$C_{16}H_{24}O_3$ Mol. wt 264.37

Occurs in the wood and roots of the balsam fir, *Abies balsamea* (Pinaceae).

Insect juvenile hormone activity.

2144
Dehydromyodesmone

Skeletal type: furanoid sesquiterpene ketone

$C_{15}H_{18}O_2$ Mol. wt 230.31

Constituent of the essential oil of *Myoporum deserti* "Theodore" variety (Myoporaceae).

Highly toxic.

2145
Dehydrongaione

Skeletal type: furanoid sesquiterpene ketone

$C_{15}H_{20}O_3$ (−) form   Mol. wt 248.33

Occurs in the stock-poisoning shrub *Myoporum deserti* (Myoporaceae).

Dehydrongaione and several closely related furanoid sesquiterpene lactones from *M. deserti* are highly toxic, and cause liver damage when given intraperitoneally to mice.

2146
Diacetoxyscirpenol; Anguidine

Skeletal type: trichothecane

$C_{19}H_{26}O_7$ Mol. wt 366.42

Metabolite of the fungi *Fusarium diversisporum* and *F. sambusinum*.

2147
Eudesmol; Selineol; Atractylol

α-eudesmol   β-eudesmol

$C_{15}H_{26}O$ Mol. wt 222.37

An isomeric mixture of α- and β-eudesmol occurs in the essential oils of several *Eucalyptus* spp. (Myrtaceae), in the root oil of *Atractylis ovata* (Compositae), and in the oils of many other plants.

2148
Farnesene

α-farnesene

β-farnesene

$C_{15}H_{24}$ Mol. wt 204.36

α-Farnesene occurs in the coating of the peel of apples and pears, *Malus* and *Pyrus* spp. (Rosaceae). β-Farnesene is a constituent of many essential oils, occurring, for example, in the leaf trichomes of *Solanum berthaultii* (Solanaceae).

Oxidation products of farnesene cause scald, a storage disease of apples. Both isomers of farnesene are components of the alarm pheromones of certain aphid species.

2149
***trans,trans*-Farnesol;** Farnesol

C$_{15}$H$_{26}$O Mol. wt 222.37

trans,trans-Farnesol is a constituent of many essential oils, e.g., from the seeds of *Abelmoschus moschatus* (=*Hibiscus abelmoschus*) (Malvaceae), and from many flowers.

Used in many perfumes with floral scent.

2150
Fumagillin; Amebacilin; Fugillin; Fumidil

C$_{26}$H$_{34}$O$_{7}$ Mol. wt 458.56

Metabolite of the fungus *Aspergillus fumigatus*.

Antibiotic, antiparasitic and carcinolytic activities.

2151
(±)-Germacrene B

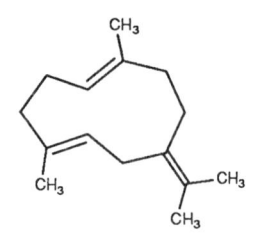

Skeletal type: germacrane

C$_{15}$H$_{24}$ Mol. wt 204.36

Occurs in the oil of the peel of *Citrus junos* (Rutaceae), and in the essential oils of several other plants. Also, it occurs in the volatile secretions of some insects.

2152
Glutinosone

Skeletal type: norsesquiterpene

C$_{14}$H$_{20}$O$_{2}$ Mol. wt 220.31

Phytoalexin produced by the leaves of *Nicotiana glutinosa* (Solanaceae) following infection with tobacco mosaic virus.

Antifungal activity.

2153
Gossypol

Skeletal type: phenolic dimeric sesquiterpene

C$_{30}$H$_{30}$O$_{8}$ Mol. wt 518.57

Occurs in the racemic form in the seeds of cotton, *Gossypium* sp., and in *Montezuma speciosissima* (Malvaceae), and it accumulates in other tissues of the cotton plant when inoculated with the fungus *Verticillium albo-atrum*. The (+) isomer occurs in good yield in *Thespesia populnea* (Malvaceae).

Toxic to nonruminant mammals, to birds, insect larvae, nematodes and other animals. It shows antifungal and antitumour activities. In mammals, it causes loss of body weight, diarrhoea, cardiac irregularity, haemorrhage, oedema, etc. The (±) and (−) forms act as a male contraceptive, since they block sperm formation.

2154
Graphinone; Ovalicin

$C_{16}H_{24}O_5$ Mol. wt 296.37

Graphinone is found in the culture filtrate of the fungus *Graphium* (strain 504–13).

Stimulates the germination of plant seeds. Graphinone is the same compound as ovalicin, which comes from the fungus *Pseudeurotium ovalis* and is claimed to have immunosuppressive, antibiotic and antitumour activities.

2155
Guaiazulene; *S*-Guaiazulene; Azulon; Eucazulen; Kessazulen; Vaumigan

$C_{15}H_{18}$ Mol. wt 198.31

Blue oil produced during the steam distillation of chamomile, *Matricaria chamomilla* (Compositae), and guaiac wood, *Guaiacum officinale* or *G. sanctum* (Zygophyllaceae), from related substances present in these plants.

Anti-inflammatory activity.

2156
Guaiol; Guajol; Guaiac alcohol; Champacol; Champaca camphor

Skeletal type: guaiane

$C_{15}H_{26}O$ (−) form   Mol. wt 222.37

Found in the resin of guaiac wood, *Guaiacum officinale* (Zygophyllaceae), and from sapu wood, *Michelia champaca* (Magnoliaceae). It is also found in various other wood oils.

2157
Helminthosporal

$C_{15}H_{22}O_2$ Mol. wt 234.34

Metabolite of the plant-pathogenic fungus *Helminthosporium sativum* (=*Bipolaris sorokiniana*).

Crop-destroying toxin which produces a seedling blight, foot and root rot, and head blight of cereals.

2158
Helminthosporol

$C_{15}H_{24}O_2$ Mol. wt 236.36

Metabolite of the plant-pathogenic fungus *Helminthosporium sativum* (=*Bipolaris sorokiniana*).

Plant growth regulator.

2159
Hemigossypol

C₁₅H₁₆O₄   Mol. wt 260.29

Phytoalexin produced by stems of cotton, *Gossypium hirsutum* (Malvaceae), when inoculated with the fungus *Verticillium albo-atrum*.

Antifungal activity.

2160
Hernandulcin

C₁₅H₂₄O₂   Mol. wt 236.36

Found in the leaves and flowers of *Lippia dulcis* (Verbenaceae), a plant known to the Aztecs as "sweet herb".

Potentially a natural substitute for sugar, since it is 1000 times sweeter than sucrose.

2161
Himachalol

C₁₅H₂₆O   Mol. wt 222.37

Occurs in the wood of *Cedrus deodara* (Pinaceae).

Spasmolytic activity.

2162
Hirsuitic acid C

C₁₅H₂₀O₄   Mol. wt 264.33

Metabolite produced by the fungus *Stereum hirsutum*.

Antibiotic activity.

2163
Humulene; α-Humulene; α-Caryophyllene; Didymocarpene

Skeletal type: humulane

C₁₅H₂₄   Mol. wt 204.36

Constituent of many essential oils, e.g., of hops, *Humulus lupulus* (Moraceae), and the oils of *Didymocarpus pedicellata* (Gesneriaceae) and *Lindera strychnifolia* leaves (Lauraceae). See also β-caryophyllene.

Used in perfumery.

2164
10β-Hydroxy-6β-isobutyryl furanoeremophilane

C₁₉H₂₈O₄   Mol. wt 320.43

Occurs in *Tetradymia glabrata* (Compositae).

Hepatotoxic (LD_{50} intraperitoneally in mice 400 mg kg^{-1}).

2165
Hydroxyisopatchoulenone;
6β-Hydroxy-1β,10β-guai-4-en-3-one

C₁₅H₂₂O₂ Mol. wt 234.34

Found in *Pleocarphus revolutus* (Compositae).

Cytotoxic.

2166
Illudin M

C₁₅H₂₀O₃ Mol. wt 248.33

Metabolite produced by the poisonous toadstool *Clitocybe illudens* (Tricholomataceae, Basidiomycetes).

Antitumour, antibacterial and antifungal activities. It is a bioluminescent principle.

2167
Illudin S; Lampterol

C₁₅H₂₀O₄ Mol. wt 264.33

Metabolite produced by the poisonous toadstool *Clitocybe illudens* (Tricholomataceae, Basidiomycetes).

Antitumour, antibacterial and antifungal activities. It is a bioluminescent principle.

2168
Ipomeamarone; (+)-Ngaione

Skeletal type: furanosesquiterpene

C₁₅H₂₂O₃ Mol. wt 250.34

Phytoalexin produced by the roots of sweet potato, *Ipomoea batatas* (Convolvulaceae) infected with the fungus *Ceratocystis fimbriata*.

Antifungal activity. It inhibits electron transport and oxidative phosphorylation, and causes respiratory depression in isolated rat mitochondria. Damaged sweet potatoes can accumulate ipomeamarone and its derivatives and, if eaten, may cause toxic effects.

2169
α-Irone

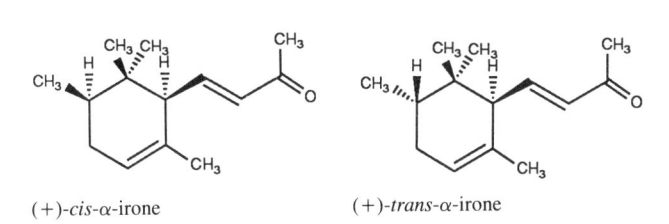

(+)-*cis*-α-irone   (+)-*trans*-α-irone

Skeletal type: norsesquiterpene

C₁₄H₂₂O Mol. wt 206.33

α-Irone occurs together with its isomers β- and γ-irone in the rhizomes of *Iris florentina* (Iridaceae). Irones are responsible for the sweet odour of the flowers of some violets, *Viola* spp. (Violaceae).

Irones are used to make perfumes with the scent of violets.

2170
Isocaryophyllene; γ-Caryophyllene

Skeletal type: caryophyllane

C₁₅H₂₄   Mol. wt 204.36

Occurs in clove oil and many other plant oils, see also β-caryophyllene and humulene.

Used in perfumery.

2171
Juvabione

C₁₆H₂₆O₃   (+) form   Mol. wt 266.38

Occurs in the wood of the balsam fir, *Abies balsamea* (Pinaceae), and of some other *Abies* spp.

Insect juvenile hormone activity.

2172
Juvenile hormone III; JH III

C₁₆H₂₆O₃   Mol. wt 266.38

Found in the grasshopper's cyperus, *Cyperus iria* (Cyperaceae).

Insect hormone activity (interferes with metamorphosis and prevents maturation).

2173
Lacinilene C 7-methyl ether

C₁₆H₂₀O₃   (+) form   Mol. wt 260.34

Found in the frost-killed bracts of cotton, *Gossypium hirsutum* (Malvaceae) and in cotton dust.

In vitro chemotactic activity towards polymorphonuclear leukocytes. It is possibly implicated in the onset of byssinosis, an illness associated with workers inhaling cotton dust.

2174
Lactaroviolin

C₁₅H₁₄O   Mol. wt 210.28

Pigment produced by the fungus *Lactarius deliciosus* (Russulaceae, Basidiomycetes).

Antibiotic activity.

2175
Laserpitin

C$_{25}$H$_{38}$O$_7$ Mol. wt 450.58

Found in the roots of *Laserpitium latifolium* (Umbelliferae).

Used as a flavouring agent in antiquity.

2176
Ledol; Ledum camphor

C$_{15}$H$_{26}$O   Mol. wt 222.37

Constituent of the essential oils from the leaves of *Ledum palustre*, *L. groenlandicum* and *L. columbianum* (Ericaceae).

2177
Longifolene; Junipene; Kuromatsuene

C$_{15}$H$_{24}$ Mol. wt 204.36

Occurs in turpentine oils prepared from *Pinus* spp., e.g., *P. longifolia* (Pinaceae), and in the wood oils of some other species of conifer.

The borane derivative, dilongifolylborane, is used as a chiral hydroborating agent.

2178
Lubimin

C$_{15}$H$_{24}$O$_2$ Mol. wt 236.36

Phytoalexin accumulated by the tubers of potato, *Solanum tuberosum* (Solanaceae), infected with pathogenic fungi. Also, it is produced by fruit of the egg plant, *Solanum melongena*, inoculated with the fungus *Monilinia fruticola*.

Antifungal activity.

2179
Marasmic acid

C$_{15}$H$_{18}$O$_4$ Mol. wt 262.31

Found in the fungus *Marismius conigenus* (Basidiomycetes).

Activity against some bacteria.

2180
Muzigadial

Skeletal type: rearranged drimedane

$C_{15}H_{20}O_3$ Mol. wt 248.33

Found in the bark of the "muziga" tree, *Warburgia salutaris* (=*W. ugandensis*) (Canellaceae).

Antifeedant activity against the armyworms *Spodoptera littoralis* and *S. exempta*. Muzigadial also shows potent antifungal and plant regulatory activities.

2181
Nerolidol; Peruviol

Skeletal type: farnesane

$C_{15}H_{26}O$ (+) form   Mol. wt 222.37

Occurs in oil of neroli, which is oil of orange flowers, *Citrus sinensis* (Rutaceae), in the wood oil of *Myroxylon pereirae* (Leguminosae), in Peruvian balsam, etc.

Used in perfumery (has a flower-like aroma).

2182
Nivalenol

Skeletal type: trichothecane

$C_{15}H_{20}O_7$ Mol. wt 312.33

Metabolite of the fungus *Fusarium nivale*.

Highly toxic to mammals. It is a strong haemorrhagenic agent, and shows antifungal and antibacterial activities.

2183
Patchouli alcohol; Patchouli camphor

$C_{15}H_{26}O$ Mol. wt 222.37

Major component of patchouli oil, which is the essential oil prepared from *Pogostemon cablin* (=*P. patchouly* var. *suavis*) (Labiatae), and from related spp.

Patchouli oil is used to scent cosmetics, shampoos, etc. (gives an oriental fragrance).

2184
Petasin

$C_{20}H_{28}O_3$ Mol. wt 316.44

Found in the leaves and roots of butterbur, *Petasites hybridus* (Compositae).

Spasmolytic agent, 14 times more active than papaverine.

2185
Pinguisone

$C_{15}H_{20}O_2$ Mol. wt 232.33

Occurs in the liverwort *Aneura pinguis* (Hepaticae).

Insect antifeedant.

2186
Polygodial; Tadeonal

$C_{15}H_{22}O_2$ Mol. wt 234.34

Found in water pepper, *Polygonum hydropiper* (Polygonaceae), in *Drymis lanceolata* (=*D. aromatica*) (Winteraceae), and in the bark of *Warburgia stuhlmanii* (Canellaceae).

Polygodial has a sharp peppery taste, and the dried fruit of *D. lanceolata* is used as a pepper substitute. It has potential use as an insect antifeedant.

2187
PR toxin; *Penicillium roqueforti* toxin

$C_{17}H_{20}O_6$ Mol. wt 320.35

Metabolite of the fungus *Penicillium roqueforti* (Eurotiaceae, Ascomycetes).

Highly toxic.

2188
Rhipocephalin

$C_{21}H_{28}O_6$ Mol. wt 376.46

Found in the Caribbean green alga *Rhipocephalus phoenix* (Codiaceae, Chlorophyta).

Toxic to pomacentrid fishes, and it is a feeding deterrent to the herbivorous Caribbean fish *Eupomacentrus leucostictus*.

2189
Rishitin

Skeletal type: norsesquiterpene

$C_{14}H_{22}O_2$ Mol. wt 222.33

Phytoalexin accumulated by the tubers of potato, *Solanum tuberosum* (Solanaceae), infected with the pathogenic fungus *Phytopthora infestans*, and by tobacco, *Nicotiana tabacum* (Solanaceae), infected with *Phytophthora parasitica* var. *nicotiana*.

Antifungal, bactericidal and phytotoxic activities.

2190
Roridin A

Skeletal type: trichothecane

$C_{29}H_{40}O_9$ Mol. wt 532.64

Metabolite of the fungus *Myrothecium roridum*.

Very toxic to mammals. It shows antifungal, antibacterial and cytotoxic activities.

2191
Rugosal

$C_{15}H_{22}O_4$ Mol. wt 266.34

Occurs in the damaged leaves of *Rosa rugosa* (Rosaceae).

Antimicrobial activity.

2192
β-Santalene

$C_{15}H_{24}$ (−) form Mol. wt 204.36

Occurs in oil of santal, sandalwood oil, which is obtained by distillation of the heartwood of *Santalum album* (Santalaceae).

Possesses a cedar-like odour.

2193
α-Santalol

$C_{15}H_{24}O$ (+) form Mol. wt 220.36

One of the main constituents together with β-santalol (q.v.) of sandalwood oil, which is distilled from the heartwood of *Santalum album* (Santalaceae).

Used in perfumery. Sandalwood oil was formerly used as a urinary antiseptic.

2194
β-Santalol

$C_{15}H_{24}O$ (−) form Mol. wt 220.36

One of the main constituents. Together with α-santalol (q.v.) of sandalwood oil, distilled from the heartwood of *Santalum album* (Santalaceae).

Used in perfumery.

2195
Sclerosporin

$C_{15}H_{22}O_2$ Mol. wt 234.34

The major sporogenic substance of *Sclerotinia fruticola*, a pathogenic brown rot fungus on stone fruits.

Induces sporulation on the fungal mycelium (when grown under light) at concentrations as low as 0.001 µg ml^{-1}.

2196
β-Selinene

$C_{15}H_{24}$ Mol. wt 204.36

Found in the oils of the seeds of celery, *Apium graveolens*, and of moon carrot, *Seseli* sp. (Umbelliferae).

2197
Shiromodiol diacetate

$C_{19}H_{30}O_5$ Mol. wt 338.45

Occurs in the leaves of *Lindera triloba* (=*Parabenzoin trilobus*) (Lauraceae).

Insect antifeedant.

2198
Sinensal

α-sinensal

β-sinensal

$C_{15}H_{22}O$ Mol. wt 218.34

Both isomers occur in the peel of oranges, *Citrus sinensis* (Rutaceae).

Flavour constituent (has the typical odour of mandarin peel).

2199
Sirenin

$C_{15}H_{24}O_2$ Mol. wt 236.36

Produced by the female gametes of the water mould *Allomyces*.

Sperm attractant in *Allomyces* (biologically active at concentrations of 10^{-10} M). The name sirenin is also given to a different compound from the algal genus *Etocarpus*.

2200
Solavetivone; Katahdinone

$C_{15}H_{22}O$ Mol. wt 218.34

Phytoalexin accumulated by tubers of potato, *Solanum tuberosum* (Solanaceae), infected with the pathogenic fungus *Phytophthora infestans*, and by tobacco, *Nicotiana tabacum* (Solanaceae), infected with tobacco mosaic virus.

Antifungal activity.

2201
T-2 toxin; Fusariotoxin T-2; Insariotoxin; Mycotoxin T-2

Skeletal type: trichothecane

$C_{24}H_{34}O_9$ Mol. wt 466.54

Metabolite of the fungus *Fusarium tricinctum* (Tuberculariaceae, Deuteromycetes).

Highly toxic, and a caustic skin irritant. It causes nausea, dizziness, diarrhoea, haemorrhaging, etc., and was used for chemical warfare in S.E. Asia.

2202
Tetradymol

$C_{15}H_{22}O_2$ Mol. wt 234.34

Found in the stems and flower buds of *Tetradymia glabrata* (Compositae).

Hepatotoxic: responsible for the death of sheep feeding on *T. glabrata*. LD_{50} orally in mice 250 mg kg^{-1}.

2203
Thujopsene; Widdrene

Skeletal type: thujopsane

$C_{15}H_{24}$ (−) form   Mol. wt 204.36

Occurs in the wood oil of the Japanese hiba tree, *Thujopsis dolabrata* (Cupressaceae).

2204
Trichodermin

Skeletal type: trichothecane

$C_{17}H_{24}O_4$ Mol. wt 292.38

Metabolite of the fungi *Trichoderma virida* and *Myrothecium roridum* (Deuteromycetes).

Antifungal and cytotoxic activities. It is used to treat *Candida albicans* infections.

2205
Trichothecin

Skeletal type: trichothecane

$C_{19}H_{24}O_5$ Mol., wt 332.40

Metabolite of the fungus *Trichothecium roseum* (Moniliaceae, Deuteromycetes).

Mycotoxin with antibiotic activity. LD_{50} intravenously in mice about 300 mg kg^{-1}.

2206
Valerenic acid

$C_{15}H_{22}O_2$ Mol. wt 234.34

Occurs in *Valeriana officinalis* (Valerianaceae).

Spasmolytic activity.

2207
α-Vetivone; Isonootkatone

$C_{15}H_{22}O$ Mol. wt 218.34

Found in oil of vetiver, which is distilled from the roots of vetiver grass, *Vetiveria zizanioides* (=*Anatherum zizanioides*=*Andropogon muricatus*) (Gramineae). See also β-vetivone.

Used in perfumery.

2208
β-Vetivone

$C_{15}H_{22}O$ Mol. wt 218.34

Occurs together with α-vetivone (q.v.) in the essential oil of *Vetiveria zizanioides* (Gramineae).

Could be used in perfumery, but has a weaker odour than α-vetivone.

2209
Verrucarin A

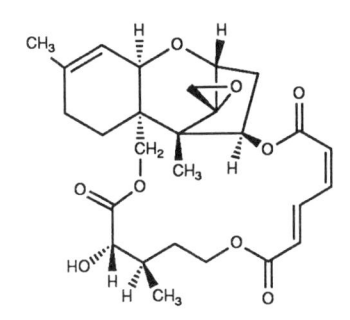

Skeletal type: trichothecane

$C_{27}H_{34}O_9$ Mol., wt 502.57

Metabolite of the fungus *Myrothecium verrucaria*.

Antibiotic and cytotoxic activity. Toxic to mammals (LD_{50} intravenously in mice 1.5 mg kg^{-1}, in rats 0.87 mg kg^{-1}, and in rabbits 0.54 mg kg^{-1}).

2210
Vomitoxin; Deoxynivalenol; Dehydronivalenol

$C_{15}H_{20}O_6$ Mol. wt 296.33

Metabolite of the fungi *Fusarium roseum* and *F. graminearum*.

Toxic (LD_{50} intraperitoneally in mice about 170 mg kg^{-1}). It was used in chemical warfare in S.E. Asia.

2211
Warburganal

$C_{15}H_{22}O_3$ Mol. wt 250.34

Occurs in the bark of *Warburgia salutaris* (=*W. ugandensis*) (Canellaceae).

Warburganal is a potent antifeedant against the armyworms *Spodoptera littoralis* and *S. exempta*. It also shows very potent antifungal, anti-yeast and plant-growth regulatory activities.

2212
α-Ylangene; Ylangene; 8-Isocopaene

$C_{15}H_{24}$ Mol. wt 204.36

Occurs in ylang-ylang oil, the essential oil distilled from fresh flowers of *Cananga odorata* (Annonaceae), in oil of birch buds, *Betula* sp. (Betulaceae), and in oil of *Juniperus oxycedrus* (Cupressaceae).

Ylang-ylang oil is used in perfumery.

2213
Zingiberene

$C_{15}H_{24}$ Mol. wt 204.36

Occurs in the rhizomes of ginger, *Zingiber officinale* (ginger oil), and of *Curcuma* spp. (Zingiberaceae).

Oil of ginger has carminative activity. It is used as a flavouring agent.

Chapter 51
Sesquiterpene lactones

Sesquiterpene lactones are a class of natural sesquiterpenoids (q.v.), which are chemically distinct from other members of the group through the presence of a γ-lactone system. Many have antitumour activity, but their considerable cytotoxicity has so far prevented any useful medicinal application. They are biologically very active; some are highly toxic to mammals (e.g., geigerin) while others (e.g., parthenin) are responsible for the allergic contact dermatitis in humans caused by the plants in which they occur. These lactones are classified biogenetically, according to the carbocyclic skeleton present, into four major groups: germacranolides with a 10-membered ring, e.g., alatolide (**2218**); eudesmanolides with two fused six-membered rings, e.g., alantolactone (**2217**); guaianolides with a five-membered ring fused to a seven and a methyl substituent at C-4*, e.g., archangelolide (**2224**); and pseudoguaianolides, as guaianolides but a methyl substituent at C-5, e.g., ambrosin (**2220**).

Besides these four main skeletal types, there are a variety of other sorts of lactone, formed by further modification of the carbon skeleton during biosynthesis. The xanthanolide dihydrogriesenin (**2263**), for example, is a spiro-compound formed by rearrangement of a guaianolide precursor. Germacranolides are generally recognized as the most primitive class and other skeletal types can be derived biogenetically from them (Fischer *et al.*, 1979). The γ-lactone ring attached, which may contain an additional α-methylene group, can be closed either towards C-6 or C-8; compare alantolactone (**2217**) with C-6 closing and arbusculin A (**2223**) with C-8 closing*. This means that two series of substances can exist within each skeletal type.

Among other structural modifications of sesquiterpene lactones, the incorporations of hydroxyl groups, e.g., anisatin (**2222**), or esterified hydroxyl groups, e.g., arctiopicrin (**2225**) or epoxide rings, e.g., canin (**2240**), are common. A few lactones occur in glycosidic form, e.g., paucin (**2359**), and some contain halogens, e.g., chlorohyssopifolin A (**2244**) or aromatic substituents, e.g., lactucopicrin (**2326**). There is also the possibility of dimerization, as in the case of absinthin (**2214**), which is a guaianolide dimer.

At least 3000 lactones have now been described, the great majority of which have been obtained from a single plant family, the Compositae, where they are characteristic constituents (Heywood *et al.*, 1978; Seaman, 1982). They have also been reported as occasional constituents in 16 other angiosperm families, most notably in members of the Umbelliferae, where they are stereochemically distinct (Holub and Budesinsky, 1986). Otherwise, they have been described from the gymnosperm family, the Cupressaceae, and from a few fungi and liverworts (cf., Picman, 1986). In the Compositae, they are mainly found in the aerial parts of the plant, chiefly in the leaves and flowering heads, in concentrations of about 5% of the dry weight. They are often located specifically in the leaf trichomes or in the surface wax. They are frequently present as mixtures, with up to 15 components per plant species. Infraspecific variation can occur, notably in some *Artemisia* and *Ambrosia* spp. (see Seaman, 1982).

Sesquiterpene lactones are lipid-soluble crystalline substances, readily isolated from dried plant material by extraction into methylene dichloride and separation on silica gel columns. They are conveniently detected after TLC by the use of chromogenic

*For the numbering of the carbon skeleton of guaianolide sesquiterpene lactones, see the formula of achillin (**2215**).

reagents. Some are very bitter to the taste, e.g., absinthin, cnicin and lactupicrin. Proton NMR spectroscopy is the most important means of structural elucidation (Yoshioka *et al.*, 1973), while the determination of stereochemistry frequently relies on the application of circular dichroism (Fischer *et al.*, 1979). Increasing numbers of structures are now being elucidated by means of x-ray diffraction.

The two major references to the chemistry and natural distribution of sesquiterpene lactones are the reviews of Fischer *et al.* (1979) and of Seaman (1982). Both these articles contain extensive listings of structures and sources. The biological and therapeutical activities are discussed in detail by Picman (1986), Rodriguez *et al.* (1976) and Sticher (1977).

References

Chemistry Structures and Occurrence in Plants

Fischer, N.H., Oliver, E.J. and Fischer, H.D. (1979). The biogenesis and chemistry of sesquiterpene lactones (W. Herz, H. Griesebach and G.W. Kirby, eds). *Fortschritte der Chemie organischer Naturstoffe*, **38**, 1.

Seaman, F.C. (1982). Sesquiterpene lactones as taxonomic characters in the Asteraceae. *The Botanical Review*, **48**, 121–595.

Heywood, V.H., Harborne, J.B. and Turner, B.L. (eds) (1978). *The Biology and Chemistry of the Compositae*. London: Academic Press.

Holub, M. and Budesinsky, M. (1986). Sesquiterpene lactones of the Umbelliferae. *Phytochemistry*, **25**, 2015–2026.

Biological Activity and Uses

Picman, A.K. (1986). Biological activities of sesquiterpene lactones. *Biochemical and Systematic Ecology*, **14**, 255–281.

Rodriguez, E., Towers, G.H.N. and Mitchell, J.C. (1976). Biological activities of sesquiterpene lactones. *Phytochemistry*, **15**, 1573–1580.

Sticher, O. (1977). Plant mono-, di- and sesquiterpenoids with pharmacological or therapeutical activity. In *New Natural Products and Plant Drugs with Pharmacological, Biological or Therapeutical Activity* (Wagner, H. and Wolff, P., eds), pp. 137–176. Berlin: Springer-Verlag.

Herout, V. (1971). Biochemistry of sesquiterpenoids. In *Aspects of Terpenoid Chemistry and Biochemistry* (Goodwin, T.W., ed.), pp. 53–94. London: Academic Press.

General

Budavari, S. (ed.) (1989). *The Merck Index*, 11th Edition. Rahway: E. Merck & Co.

Taylor, W.I. and Battersby, A.R. (eds) (1969). *Cyclopentanoid Terpene Derivatives*. New York: Marcel Dekker.

2214
Absinthin: Absinthiin; Absynthin

Skeletal type: dimeric guaianolide

$C_{30}H_{40}O_6$ Mol. wt 496.65

The main bitter principle of wormwood, *Artemisia absinthium* (Herba absinthii) (Compositae).

Wormwood is used as an anthelmintic, a bitter tonic and for flavouring alcoholic beverages (vermouth). Ingestion of absinthin may cause nervousness, convulsions, or even death.

2215
Achillin; Santolin

Skeletal type: guaianolide

$C_{15}H_{18}O_3$ Mol. wt 246.31

Found in yarrow, *Achillea millefolium* and other *Achillea* spp., and in several *Artemisia* spp. (Compositae).

Plant growth inhibitor. During steam distillation, achillin is converted into chamazulene (q.v.), which has anti-inflammatory activity.

2216
Acroptilin; Chlorohyssopifolin C

Skeletal type: guaianolide

$C_{19}H_{23}O_7Cl$ Mol. wt 398.85

Occurs in *Acroptilon repens*, *Centaurea hyssopifolia*, *C. hyrcanica* and *C. linifolia* (Compositae).

Cytotoxic and antitumour activity. It shows strong activity against the pathogenic protozoa *Entamoeba histolytica* and *Trichomonas vaginalis*.

2217
Alantolactone

Skeletal type: eudesmanolide

$C_{15}H_{20}O_2$ Mol. wt 232.33

Occurs in oil of elecampane, *Inula helenium*, and in *I. grandis*, *I. magnifica* and other *Inula* spp. (Compositae).

Helenin (a mixture of alantolactone and isoalantolactone, q.v.) shows antibacterial and antifungal activity. It causes allergic contact dermatitis in humans. It has been used as an expectorant and a cholagogue, as an antiseptic for the urinary tract, as a drug stimulating intestinal secretion, and as a vermifuge against the liver fluke, *Fasciola hepatica*. It shows anti-feedant activity against mammals and insects, and is a potent inhibitor of plant seed germination and seedling growth.

2218
Alatolide

Skeletal type: germacranolide

$C_{19}H_{26}O_6$ Mol. wt 350.42

Found in *Jurinea alata* (Compositae).

Cytotoxic and antitumour activities. It shows antifeedant properties against certain insects.

2219
Amaralin

Skeletal type: pseudoguaianolide

$C_{15}H_{20}O_4$ Mol. wt 264.33

Occurs in the sneezeweed, *Helenium amarum* (Compositae).

Strong analgesic action when used subcutaneously.

2220
Ambrosin

Skeletal type: pseudoguaianolide

$C_{15}H_{18}O_3$ Mol. wt 246.31

Found in Roman wormwood or common ragweed, *Ambrosia artemisiifolia*, in many other *Ambrosia* spp., and in some *Hymenoclea*, *Iva* and *Parthenium* spp. (Compositae).

Cytotoxic and antitumour activities, and active also against the human parasite *Schistosoma haematobium*.

2221
Angustibalin; Helenalin acetate

Skeletal type: pseudoguaianolide

$C_{17}H_{20}O_5$ Mol. wt 304.35

Found in *Balduina angustifolia* (=*Actinospermum angustifolium*) (Compositae).

Cytotoxic and antitumour activities.

2222
Anisatin

$C_{15}H_{20}O_8$ Mol. wt 328.33

Occurs in the seeds of Japanese star anise, *Illicium anisatum* (Illiciaceae).

Toxic to humans.

2223
Arbusculin A

Skeletal type: eudesmanolide

$C_{15}H_{22}O_3$ Mol. wt 250.34

Found in *Artemisia arbuscula* and *A. tridentata* (Compositae).

Plant growth inhibitor.

2224
Archangelolide

Skeletal type: guaianolide

$C_{29}H_{40}O_{10}$ Mol. wt 548.64

Found in the roots and fruit of *Laserpitium archangelica* (Umbelliferae).

Insect antifeedant.

2225
Arctiopicrin

Skeletal type: germacranolide

$C_{19}H_{26}O_6$ Mol. wt 350.42

Found in the leaves of *Arctium minus*, *A. lappa*, *A. nemorosum*, and *A. tomentosum* (Compositae).

2226
Arctolide

Skeletal type: guaianolide

$C_{17}H_{20}O_6$ Mol. wt 320.35

Occurs in the African daisy, *Arctotis grandis* (Compositae).

Cytotoxic and antitumour activities.

2227
Arnicolide A; Dihydrohelenalin acetate

Skeletal type: pseudoguaianolide

$C_{17}H_{22}O_5$ Mol. wt 306.36

Occurs in *Arnica montana* (Compositae).

Cytotoxic and antitumour activities.

2228
Aromaticin

Skeletal type: pseudoguaianolide

$C_{15}H_{18}O_3$ Mol. wt 246.31

Occurs in *Helenium aromaticum* and *H. amarum* (Compositae).

Cytotoxic and antitumour activities.

2229
Artabsin

Skeletal type: guaianolide

$C_{15}H_{20}O_3$ Mol. wt 248.33

Found in wormwood, *Artemisia absinthium*, and sievers wormwood, *A. sieversiana* (Compositae).

During steam distillation of wormwood, artabsin is converted to chamazulene (q.v.) which is anti-inflammatory.

2230
Artecanin; Chrysartemin B

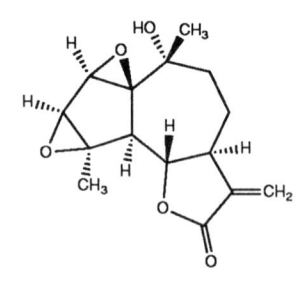

Skeletal type: guaianolide

$C_{15}H_{18}O_5$ Mol. wt 278.31

Occurs in *Artemisia cana* and *Chrysanthemum macrophyllum* (Compositae).

Insect antifeedant.

2231
Arteglasin A

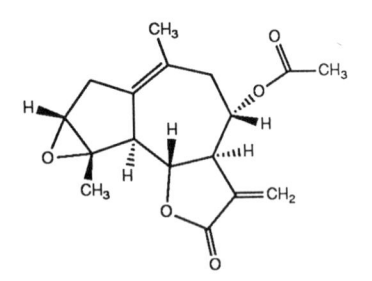

Skeletal type: guaianolide

$C_{17}H_{20}O_5$ Mol. wt 304.35

Occurs in *Artemisia douglasiana* (=*A. ludoviciana*), and in the garden chrysanthemum, *Dendranthema indicum* (=*Chrysanthemum indicum*) (Compositae).

Cytotoxic and antitumour activities. It causes allergic contact dermatitis in humans.

2232
Artemisiifolin

Skeletal type: germacranolide

$C_{15}H_{20}O_4$ Mol. wt 264.33

Found in Roman wormwood or common ragweed, *Ambrosia artemisiifolia*, and in *Centaurea seridis* (Compositae).

2233
Artemisin; 8-Hydroxysantonin

Skeletal type: eudesmanolide

$C_{15}H_{18}O_4$ Mol. wt 262.31

Found in the flower heads of *Artemisia cina*, *A. macrocephala* and *A. maritima* (Compositae).

2234
Artemorin

Skeletal type: germacranolide

$C_{15}H_{20}O_3$ Mol. wt 248.33

Found in *Artemisia verlotorum* and *A. mexicana* (Compositae), and in *Magnolia grandiflora* (Magnoliaceae).

2235
Autumnolide

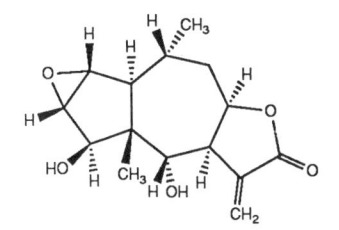

Skeletal type: pseudoguaianolide

$C_{15}H_{20}O_5$ Mol. wt 280.33

Found in sneezeweed, *Helenium autumnale* (Compositae).

Cytotoxic and antitumour activities.

2236
Baileyin

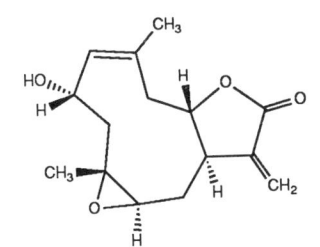

Skeletal type: germacranolide

$C_{15}H_{20}O_4$ Mol. wt 264.33

Found in *Baileya multiradiata* and *B. pleniradiata* (Compositae).

Cytotoxic and antitumour activities.

2237
Bakkenolide A; Fukinanolide

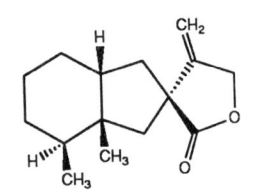

Skeletal type: bakkenolide

$C_{15}H_{22}O_2$ Mol. wt 234.34

Occurs in many butterbur, *Petasites* spp. in *Senecio pyramidatus, Cacalia hastata, Ligularia calthaefolia,* and *Homogyne alpina* (Compositae).

Cytotoxic and antitumour activities. It is an insect antifeedant.

2238
Budlein A

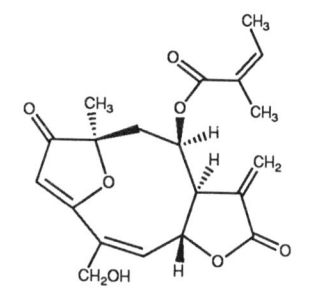

Skeletal type: germacranolide

$C_{20}H_{22}O_7$ Mol. wt 374.40

Found in *Viguiera buddleiaeformis* and *V. angustifolia* (Compositae).

Cytotoxic and antitumour activities.

2239
Calaxin

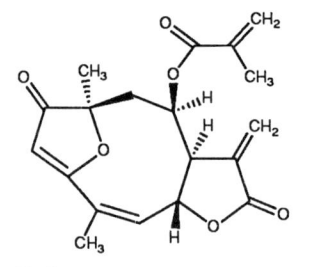

Skeletal type: germacranolide

$C_{19}H_{20}O_6$ Mol. wt 344.37

Occurs in *Helianthus ciliaris* and *Calea axillaris* (Compositae).

Cytotoxic and antitumour activities.

2240
Canin; Chrysartemin A

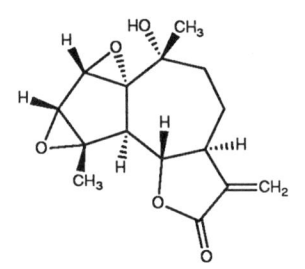

Skeletal type: guaianolide

$C_{15}H_{18}O_5$ Mol. wt 278.31

Occurs in *Artemisia cana* and other *Artemisia* spp., in feverfew, *Tanacetum* (=*Chrysanthemum*) *parthenium*, and in *Handelia trichophylla* (Compositae).

Cytotoxic and antitumour activities. It also shows insect antifeedant and plant growth regulating activities.

2241
Chamissonin

Skeletal type: germacranolide

$C_{15}H_{20}O_4$ Mol. wt 264.33

Occurs in *Ambrosia chamissonis*, *A. acanthicarpa* and *A. dumosa* (Compositae).

2242
Chamissonin diacetate

Skeletal type: germacranolide

$C_{19}H_{24}O_6$ Mol. wt 348.40

Occurs in *Ambrosia acanthicarpa* (Compositae).

Cytotoxic and antitumour activities.

2243
Chlorochrymorin

Skeletal type: guaianolide

$C_{15}H_{19}O_5Cl$ Mol. wt 314.77

Found in *Chrysanthemum morifolium* (Compositae).

Plant growth regulating activity.

2244
Chlorohyssopifolin A; Centaurepensin; Hyrcanin

Skeletal type: guaianolide

$C_{19}H_{24}O_7Cl_2$ Mol. wt 435.31

Occurs in several *Centaurea* spp., e.g., *C. hyssopifolia*, and in *Acroptilon repens* (Compositae).

Cytotoxic and antitumour activities.

2245
Chromolaenide

Skeletal type: germacranolide

$C_{22}H_{28}O_7$ Mol. wt 404.47

Found in *Chromolaena glaberrima* and *Isocarpha oppositifolia* (Compositae).

Antibacterial activity.

2246
Cnicin; Cynisin; Centaurin

Skeletal type: germacranolide

$C_{20}H_{26}O_7$ Mol. wt 378.43

Found in the blessed thistle, *Cnicus benedictus*, and in many *Centaurea* spp. (Compositae).

Antibacterial and antitumour activities. It shows antifeedant activity against certain insects. It has a bitter taste.

2247
Conchosin A

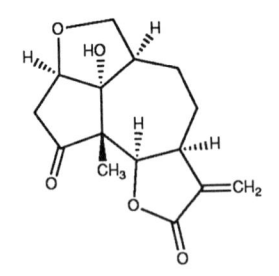

Skeletal type: pseudoguaianolide

$C_{15}H_{18}O_5$ Mol. wt 278.31

Occurs in *Parthenium confertum* (Compositae).

Affects the development of certain insect larvae.

2248
Conchosin B

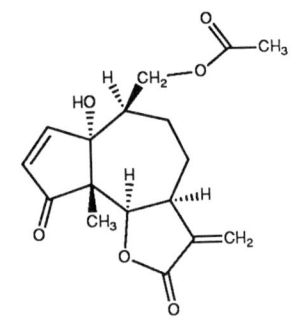

Skeletal type: pseudoguaianolide

$C_{17}H_{20}O_6$ Mol. wt 320.35

Occurs in *Parthenium confertum* (Compositae).

Affects the development of certain insect larvae.

2249
Confertifolin

Skeletal type: drimanolide

$C_{15}H_{22}O_2$ Mol. wt 234.34

Found in *Drimys conifertifolia* and *D. winteri* (Winteraceae), in water-pepper, *Polygonum hydropiper* (Polygonaceae), and in the moss *Cinnamosma fragrans* (Canellaceae).

2250
Confertin; Anhydrocumanin

Skeletal type: pseudoguaianolide

$C_{15}H_{20}O_3$ Mol. wt 248.33

Occurs in *Ambrosia confertiflora, A. dumosa* and *Parthenium schottii* (Compositae).

Affects the development of, and is toxic to, certain insects.

2251
Coriamyrtin

Skeletal type: tutinanolide = picrotoxin

$C_{15}H_{18}O_5$ Mol. wt 278.31

Occurs in the leaves of *Coriaria myrtifolia* and *C. japonica* (Coriariaceae).

Highly toxic to mammals, insects, and other animals. It causes extreme excitation of the central nervous system (stimulation of the respiratory, vasomotor and cardio-inhibitory centres in the brain).

2252
Coronopilin

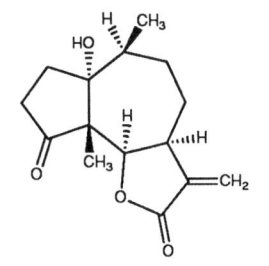

Skeletal type: pseudoguaianolide

$C_{15}H_{20}O_4$ Mol. wt 264.33

Found in many *Ambrosia* spp., e.g., *A. psilostachia* var. *coronopifolia*, in *Hymenoclea salsola*, and in *Iva* spp. and in *Parthenium* spp., e.g., *P. hysterophorus* (Compositae).

Insect antifeedant. It affects the development of certain insects. Probably, this is one of the sesquiterpene lactones in *P. hysterophorus* which causes allergic contact dermatitis in humans and which is partly responsible for the allelopathic properties of this plant.

2253
Costunolide

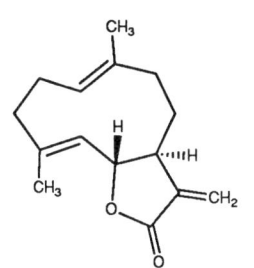

Skeletal type: germacranolide

$C_{15}H_{20}O_2$ Mol. wt 232.33

Occurs in numerous spp. belonging to many genera of the Compositae. The best known source is costus root oil, from *Saussurea lappa*. Also, it occurs in other plant families, e.g., in bay laurel, *Laurus nobilis* (Lauraceae).

Antitumour activity. It causes allergic contact dermatitis. It is active against the parasitic trematode *Schistosoma mansoni*, which causes schistosomiasis in humans.

2254
α-Cyclocostunolide

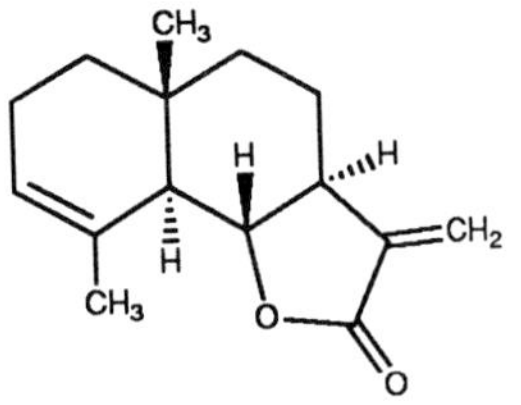

Skeletal type: eudesmanolide

$C_{15}H_{20}O_2$ Mol. wt 232.33

Found in the trunk wood of *Moquinia velutina*, in *Oxylobus oaxacanus* and *Dicoma zeyheri* (Compositae), and in the liverwort *Frullania tamarisci* (Hepaticae).

Schistosomicidal activity (see costunolide).

2255
β-Cyclocostunolide

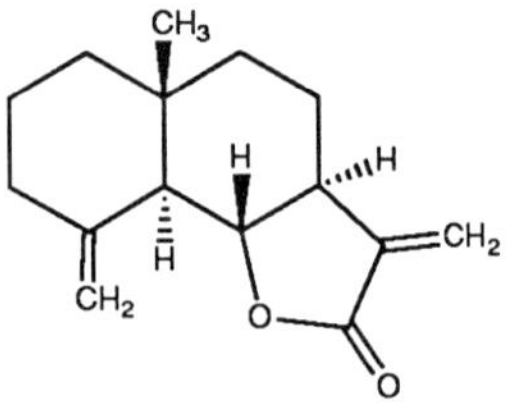

Skeletal type: eudesmanolide

$C_{15}H_{20}O_2$ Mol. wt 232.33

Occurs in *Oxylobus oaxacanus* (Compositae), and in the liverwort, *Frullania tamarisci* (Hepaticeae). (+)-*cis*-β-Cyclocostunolide occurs in *F. dilatata*.

Schistosomicidal activity (see costunolide).

2256
Cynaropicrin

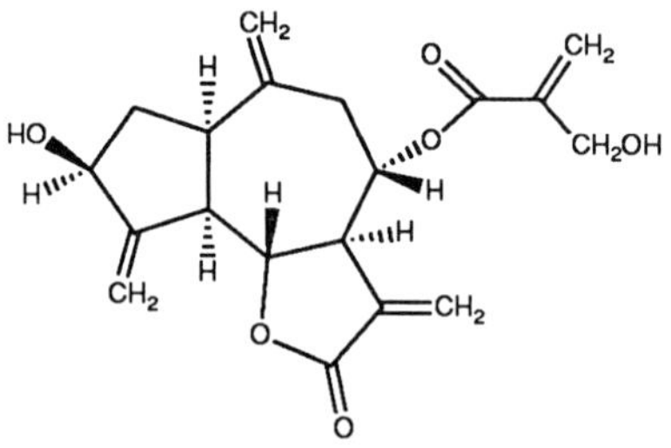

Skeletal type: guaianolide

$C_{19}H_{22}O_6$ Mol. wt 346.39

Found in *Cynara cardunculus*, *C. scolymus*, *Saussurea amara*, *Amberboa muricata*, and several *Centaurea* spp. (Compositae).

Cytotoxic and antitumour activities.

2257
Damsin

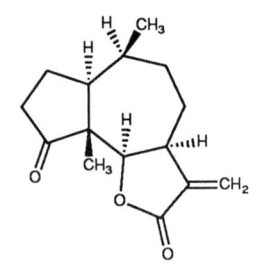

Skeletal type: pseudoguaianolide

$C_{15}H_{20}O_3$ Mol. wt 248.33

Occurs in *Ambrosia maritima* (called "damsissa" in Egypt), and many other *Ambrosia* spp., and in *Parthenium bipinnitifidum* (Compositae).

Cytotoxic, schistosomicidal and some molluscicidal activities.

2258
Dehydrocostus lactone

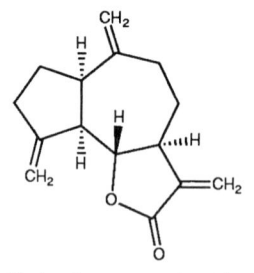

Skeletal type: guaianolide

$C_{15}H_{18}O_2$ Mol. wt 230.31

Found in many spp. of Compositae, e.g., *Stokesia laevis*, *Saussurea lappa*, costus root, *Lychnophora passerina*, *Zaluzania triloba*, and several ironweed, *Vernonia* spp.

Plant growth regulating activity.

2259
Deoxyelephantopin

Skeletal type: germacranolide

$C_{19}H_{20}O_6$ Mol. wt 344.37

Found in *Elephantopus carolinianus* and *E. scaber*.

Cytotoxic and antitumour activities.

2260
8-Deoxylactucin

Skeletal type: guaianolide

$C_{15}H_{16}O_4$ Mol. wt 260.29

Occurs in prickly lettuce, *Lactuca seriola*, and in chicory, *Cichorium intybus* (Compositae).

Cytotoxic and antitumour activity.

2261
Desacetoxymatricarin

Skeletal type: guaianolide

$C_{15}H_{18}O_3$ Mol. wt 246.31

Found in *Matricaria suffruticosa* and several *Achillea* spp. and *Artemisia* spp. (Compositae).

Cytotoxic and antitumour activities. It is a plant growth inhibitor (phytotoxin).

2262
Desacetyleupaserrin

Skeletal type: germacranolide

$C_{20}H_{26}O_6$ Mol. wt 362.43

Found in *Eupatorium semiserratum*, *E. mikanioides*, and several *Helianthus* spp. (Compositae).

Cytotoxic and antitumour activities. It affects the development of certain insect larvae.

2263
Dihydrogriesenin

Skeletal type: secoguaianolide = xanthanolide

$C_{15}H_{18}O_4$ Mol. wt 262.31

Occurs in *Geigera africana* (Compositae).

Toxic to mammals.

2264
Dihydromikanolide

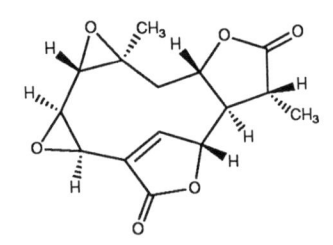

Skeletal type: germacranolide

$C_{15}H_{16}O_6$ Mol. wt 292.29

Occurs in climbing hempweed, *Mikania scandens*, and many other *Mikania* spp. (Compositae).

Antifungal activity, e.g., against the yeast *Candida albicans*.

2265
Drimenin

Skeletal type: drimanolide

$C_{15}H_{22}O_2$ Mol. wt 234.34

Found in the bark of *Drimys winteri* (Winteraceae), and in mosses belong to the genus *Porella* (Porellaceae).

2266
Eleganin

Skeletal type: germacranolide

$C_{22}H_{26}O_9$ Mol. wt 434.45

Found in *Liatris elegans* and *L. scabra* (Compositae).

Cytotoxic and antitumour activities. The name eleganin is also used for a guaianolide sesquiterpenoid.

2267
Elephantin

Skeletal type: germacranolide

$C_{20}H_{22}O_7$ Mol. wt 374.40

Occurs in *Elephantopus elatus* (Compositae).

Cytotoxic and antitumour activities. It also shows plant growth regulating activity.

2268
Elephantopin

Skeletal type: germacranolide

$C_{19}H_{20}O_7$ Mol. wt 360.37

Occurs in *Elephantopus elatus* (Compositae).

Cytotoxic and antitumour activities.

2269
Encelin; Anhydrofarinosin

Skeletal type: eudesmanolide

$C_{15}H_{16}O_3$ Mol. wt 244.29

Found in *Encelia farinosa*, *E. virginensis*, and *Baltimora recta* (Compositae).

Cytotoxic and antitumour activities.

2270
Enhydrin

Skeletal type: germacranolide

$C_{23}H_{28}O_{10}$ Mol. wt 464.48

Found in the leaves of *Enhydra fluctuans*, in *Melampodium longipilum*, *M. perfoliatum*, *Smallanthus fruticosus* and *S. uvedalius* (Compositae).

Antihypertensive activity.

2271
Eremanthin; Vanillosmin

Skeletal type: guaianolide

$C_{15}H_{18}O_2$ Mol. wt 230.31

Found in the heartwood of *Eremanthus elaeagnus* and *E. incanus*, and in several *Lychnophora*, *Vanillosmopsis* and *Vernonia* spp. (Compositae).

Schistosomicidal activity.

2272
Eremantholide A

Skeletal type: germacranolide

$C_{19}H_{24}O_6$ Mol. wt 348.40

Occurs in *Eremanthus bicolor*, *E. elaeagnus*, *E. incanus* and *Centratherum punctatum* (Compositae).

Cytotoxic and antitumour activities.

2273
Eremofrullanolide

Skeletal type: eremophilanolide

$C_{15}H_{20}O_2$ Mol. wt 232.33

Occurs in *Frullania dilatata* (Hepaticae).

2274
Eremophilenolide

Skeletal type: eremophilanolide

$C_{15}H_{22}O_2$ Mol. wt 234.34

Found in the rhizomes of several *Petasites* spp. (Compositae).

Extracts of *Petasites* rhizomes are spasmolytically active.

2275
Erioflorin acetate

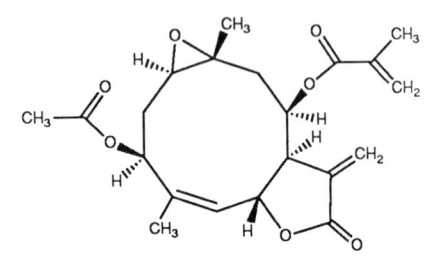

Skeletal type: germacranolide

$C_{21}H_{26}O_7$ Mol. wt 390.44

Occurs in the leaves and stems of *Podanthus ovatifolius* (Compositae).

Cytotoxic and antitumour activities.

2276
Erioflorin methacrylate

Skeletal type: germacranolide

$C_{23}H_{28}O_7$ Mol. wt 416.48

Occurs in *Podanthus ovatifolius* (Compositae).

Cytotoxic and antitumour activities.

2277
Eriolangin

Skeletal type: seco-eudesmanolide

$C_{20}H_{28}O_6$ Mol. wt 364.44

Found in *Eriophyllum lanatum* (Compositae).

Cytotoxic and antitumour activities.

2278
Eupachlorin

Skeletal type: guaianolide

$C_{20}H_{25}O_7Cl$ Mol. wt 412.87

Occurs in the round-leaved thoroughwort, *Eupatorium rotundifolium* (Compositae).

Cytotoxic and antitumour activities.

2279
Eupachlorin acetate

Skeletal type: guaianolide

$C_{22}H_{27}O_8Cl$ Mol. wt 454.91

Occurs in the round-leaved thoroughwort, *Eupatorium rotundifolium* (Compositae).

Cytotoxic and antitumour activities.

2280
Eupachloroxin

Skeletal type: guaianolide

$C_{20}H_{25}O_8Cl$ Mol. wt 428.87

Occurs in the round-leaved thoroughwort, *Eupatorium rotundifolium* (Compositae).

Cytotoxic and antitumour activities.

2281
Eupacunin

Skeletal type: germacranolide

$C_{22}H_{28}O_7$ Mol. wt 404.47

Found in the stems, leaves and flowers of *Eupatorium cuneifolium* and *E. lancifolium* (Compositae).

Cytotoxic and antitumour activities.

2282
Eupacunolin

Skeletal type: germacranolide

$C_{22}H_{28}O_8$ Mol. wt 420.47

Found in the aerial parts of *Eupatorium cuneifolium* and *E. lancifolium* (Compositae).

Cytotoxic and antitumour activities.

2283
Eupacunoxin

Skeletal type: germacranolide

$C_{22}H_{28}O_8$ Mol. wt 420.47

Found in the aerial parts of *Eupatorium cuneifolium* (Compositae).

Cytotoxic and antitumour activities.

2284
Eupaformonin

Skeletal type: germacranolide

$C_{17}H_{22}O_5$ Mol. wt 306.36

Occurs in *Eupatorium formosanum* (Compositae).

Cytotoxic and antitumour activities.

2285
Eupaformosanin

Skeletal type: germacranolide

$C_{22}H_{28}O_8$ Mol. wt 420.47

Occurs in *Eupatorium formosanum* (Compositae).

Cytotoxic and antitumour activities.

2286
Eupahyssopin

Skeletal type: germacranolide

$C_{20}H_{26}O_7$ Mol. wt 378.43

Found in the hyssop-leaved thoroughwort, *Eupatorium hyssopifolium* (Compositae).

Cytotoxic and antitumour activities.

2287
Euparotin

Skeletal type: guaianolide

$C_{20}H_{24}O_7$ Mol. wt 376.41

Occurs in the round-leaved thoroughwort, *Eupatorium rotundifolium* (Compositae).

Cytotoxic and antitumour activities.

2288
Euparotin acetate

Skeletal type: guaianolide

$C_{22}H_{26}O_8$ Mol. wt 418.45

Occurs in the round-leaved thoroughwort, *Eupatorium rotundifolium* (Compositae).

Cytotoxic and antitumour activities.

2289
Eupaserrin

Skeletal type: germacranolide

$C_{22}H_{28}O_7$ Mol. wt 404.47

Found in *Eupatorium semiserratum* and *E. cuneifolium* (Compositae).

Cytotoxic and antitumour activities.

2290
Eupatocunin

Skeletal type: germacranolide

$C_{22}H_{28}O_7$ Mol. wt 404.47

Found in the aerial parts of *Eupatorium cuneifolium* (Compositae).

Cytotoxic and antitumour activities.

2291
Eupatocunoxin

Skeletal type: germacranolide

$C_{22}H_{28}O_8$ Mol. wt 420.47

Occurs in the aerial parts of *Eupatorium cuneifolium* (Compositae).

Cytotoxic and antitumour activities.

2292
Eupatolide

Skeletal type: germacranolide

$C_{15}H_{20}O_3$ Mol. wt 248.33

Found in hemp agrimony, *Eupatorium cannabinum*, and in *E. formosanum* and *Helianthus agrophyllus* (Compositae).

Cytotoxic and antitumour activities.

2293
Eupatoriopicrin

Skeletal type: germacranolide

$C_{20}H_{26}O_6$ Mol. wt 362.43

Found in hemp agrimony, *Eupatorium cannabinum*, in *Chaenactis carphoclinia* and *C. douglassii*, and in *Eriophyllum stachaedifolium* (Compositae).

Cytotoxic and antitumour activities. It also shows antifeedant activity against certain insects.

2294
Eupatoroxin

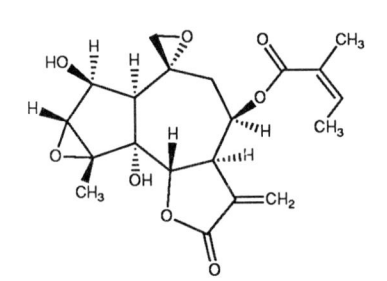

Skeletal type: guaianolide

$C_{20}H_{24}O_8$ Mol. wt 392.41

Found in the round-leaved thoroughwort, *Eupatorium rotundifolium* (Compositae).

Cytotoxic and antitumour activities.

2295
10-*epi*-Eupatoroxin; 10-Epieupatoroxin

Skeletal type: guaianolide

$C_{20}H_{24}O_8$ Mol. wt 392.41

Found in the round-leaved thoroughwort, *Eupatorium rotundifolium* (Compositae).

Cytotoxic and antitumour activities.

2296
Eupatundin

Skeletal type: guaianolide

$C_{20}H_{24}O_7$ Mol. wt 376.41

Found in the round-leaved thoroughwort, *Eupatorium rotundifolium* (Compositae).

Cytotoxic and antitumour activities.

2297
Euponin

Skeletal type: guaianolide

$C_{20}H_{24}O_6$ Mol. wt 360.41

Occurs in *Eupatorium japonicum* (Compositae).

Affects the development of certain insect larvae.

2298
Farinosin

Skeletal type: eudesmanolide

$C_{15}H_{18}O_4$ Mol. wt 262.31

Occurs in *Encelia farinosa* and *E. virginensis* (Compositae).

Cytotoxic and antitumour activities.

2299
Fastigilin B

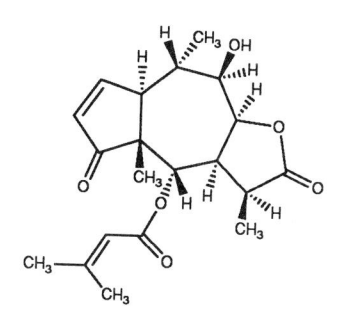

Skeletal type: pseudoguaianolide

$C_{20}H_{26}O_6$ Mol. wt 362.43

Occurs in *Baileya multiradiata* and *Gaillardia fastigiata* (Compositae).

Cytotoxic and antitumour activities.

2300
Fastigilin C

Skeletal type: pseudoguaianolide

$C_{20}H_{24}O_6$ Mol. wt 360.41

Found in *Gaillardia fastigiata*, *Baileya multiradiata* and *Hymenoxys acaulis* (Compositae).

Cytotoxic and antitumour activities.

2301
Florilenalin

Skeletal type: guaianolide

$C_{15}H_{20}O_4$ Mol. wt 264.33

Found in the sneezeweed, *Helenium autumnale* (Compositae).

Cytotoxic and antitumour activities.

2302
Frullanolide

Skeletal type: eudesmanolide

$C_{15}H_{20}O_2$ (−) form   Mol. wt 232.33

The (−) form occurs in the liverwort *Frullania tamarisci*, and the (+) form occurs in *F. dilatata* (Hepaticae).

Causes allergic contact dermatitis in humans.

2303
Gaillardin

Skeletal type: guaianolide

$C_{17}H_{22}O_5$ Mol. wt 306.36

Occurs in *Gaillardia pulchella* and several *Inula* spp. (Compositae).

Cytotoxic and antitumour activities.

2304
Geigerin

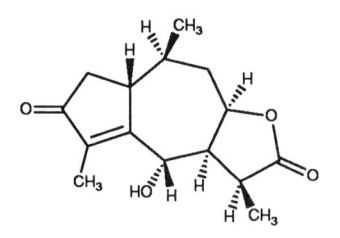

Skeletal type: guaianolide

$C_{15}H_{20}O_4$ Mol. wt 264.33

Occurs in *Geigera africana* (=*G. filifolia*) and *G. aspera* (Compositae).

Very toxic to mammals. Eating *Geigera* spp. has caused mass poisoning among sheep in Africa.

2305
Glaucolide A

Skeletal type: germacranolide

$C_{23}H_{28}O_{10}$ Mol. wt 464.48

Found in many ironweed, *Vernonia*, spp., e.g., *V. glauca* (Compositae).

Feeding deterrent to mammals and insects. It affects the development of certain insect larvae.

2306
Glaucolide B

Skeletal type: germacranolide

$C_{21}H_{26}O_{10}$ Mol. wt 438.44

Found in New York ironweed, *Vernonia glauca* (=*V. noveboracensis*) and many other *Vernonia* spp. (Compositae).

Antifungal activity, e.g., against the yeast *Candida albicans*.

2307
Glechomanolide

Skeletal type: germacranolide

$C_{15}H_{20}O_2$ Mol. wt 232.33

Occurs in ground-ivy, *Glechoma hederacea* (Labiatae).

2308
Goyazensolide

Skeletal type: germacranolide

$C_{19}H_{20}O_7$ Mol. wt 360.37

Occurs in *Eremanthus goyazensis*, *Lychnophora passerina*, *Vanillosmopsis brasiliensis* and *V. pohlii* (Compositae).

Schistosomicidal activity (see costunolide).

2309
Gradolide

Skeletal type: guaianolide

$C_{25}H_{34}O_7$ Mol. wt 446.55

Found in the fruit of *Laserpitium siler* (Umbelliferae).

Insect antifeedant.

2310
Graminiliatrin

Skeletal type: guaianolide

$C_{22}H_{26}O_9$ Mol. wt 434.45

Occurs in the blazing star, *Liatris graminifolia* (Compositae).

Cytotoxic and antitumour activities.

2311
Granilin

Skeletal type: eudesmanolide

$C_{15}H_{20}O_4$ Mol. wt 264.33

Found in *Inula grandis, Ambrosia polystachya, Artemisia ashurbajevii* and *Carpesium abrotanoides* (Compositae).

Antibacterial activity.

2312
Grosshemin; Grossheimin

Skeletal type: guaianolide

$C_{15}H_{18}O_4$ Mol. wt 262.31

Occurs in *Grossheimia macrocephala, Amberboa lipii* (=*Centaurea lippii*), *Cynara scolymus, Chartolepis intermedia*, and *Venidium decurens* (=*Arctotis arctotoides*) (Compositae).

Cytotoxic and antitumour activities. Also, it is an insect antifeedant.

2313
Helenalin

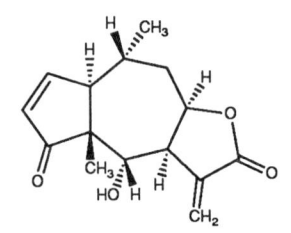

Skeletal type: pseudoguaianolide

$C_{15}H_{18}O_4$ Mol. wt 262.31

Found in many *Helenium* spp., e.g., in sneezeweed, *H. autumnale*, and in many other Compositae, e.g., *Anaphalis, Balduina*, and *Gaillardia* spp.

Intensely poisonous to humans and other mammals, causing paralysis of voluntary and cardiac muscles and fatal gastroenteritis. It is irritant to the nose, eyes and stomach. Helenalin is a potent stimulator of mass cell degranulation. It shows cytotoxic and antileukaemic activities. It is toxic to fish and insects. Also, it shows activity against several fungi pathogenic to humans, as well as molluscicidal and vermifuge activities.

2314
Heliangin; Heliangine

Skeletal type: germacranolide

$C_{20}H_{26}O_6$ Mol. wt 362.43

Occurs in the common sunflower, *Helianthus annuus*, and in the Jerusalem artichoke, *H. tuberosus* (Compositae).

Plant growth regulating activity.

2315
Hiyodorilactone A; Eucannabinolide; Schkuhrin

Skeletal type: germacranolide

$C_{22}H_{28}O_8$ Mol. wt 420.47

Occurs in *Eupatorium sachalinense* (Compositae).

Cytotoxic and antitumour activities.

2316
Hymenoflorin

Skeletal type: pseudoguaianolide

$C_{15}H_{20}O_5$ Mol. wt 280.33

Occurs in *Hymenoxys grandiflora* (Compositae).

Cytotoxic and antitumour activities.

2317
Hymenolin

Skeletal type: pseudoguaianolide

$C_{15}H_{20}O_4$ Mol. wt 264.33

Occurs in *Hymenoclea salsola* (Compositae).

Toxic to mosquito larvae, *Aedes atropalpus*.

2318
Hymenoxon

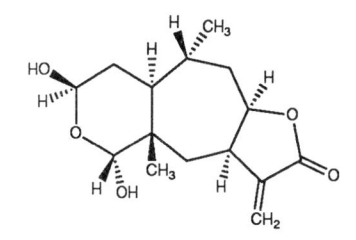

Skeletal type: seco-pseudoguaianolide

$C_{15}H_{22}O_5$ Mol. wt 282.34

Occurs in bitterweed, *Hymenoxys odorata*, in *H. richardsonii* and in *Dugaldia hoopesii* (=*Helenium hoopesii*) (Compositae). Hymenoxon is one of the epimers present in the epimeric mixture, hymenovin, that occurs in these plants.

Highly toxic to mammals (has caused livestock poisoning). It is a potent stimulator of mass cell degranulation. Also, it can produce lethal DNA damage in DNA repair-deficient strains of *Bacillus subtilis*.

2319
Inulicin

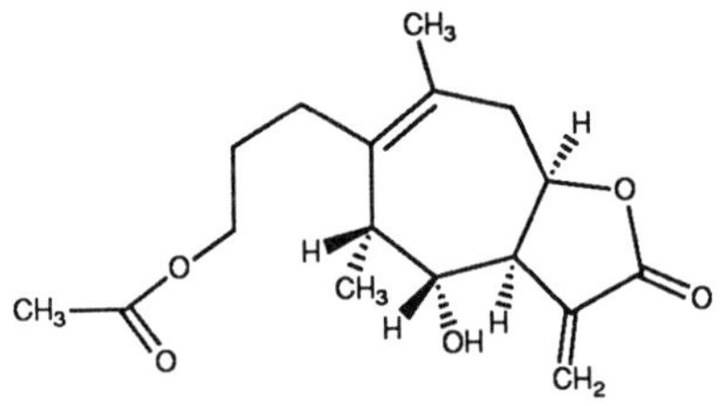

Skeletal type: seco-pseudoguaianolide

$C_{17}H_{24}O_5$ Mol. wt 308.38

Occurs in *Inula japonica* (Compositae).

Antiulcer activity, and capillary-strengthening diuretic properties. It acts as a stimulant of the central nervous system and smooth muscles of the intestine.

2320
Isoalantolactone; Isohelenin

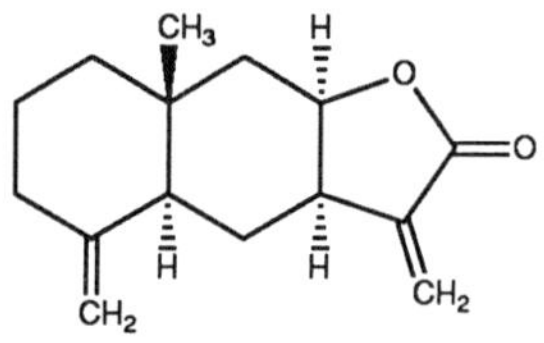

Skeletal type: eudesmanolide

$C_{15}H_{20}O_2$ Mol. wt 232.33

Occurs in many *Inula* spp., e.g., in the essential oil of elecampane, *I. helenium*. Also, it occurs in *Liatris cylindrica* and *Telekia speciosa* (Compositae).

Antibacterial, antifungal and antihelminthic activities. It has been used as a vermifuge. It shows antifeedant activity against, and affects the development of, certain insects. See also alantolactone.

2321
Isohelenol

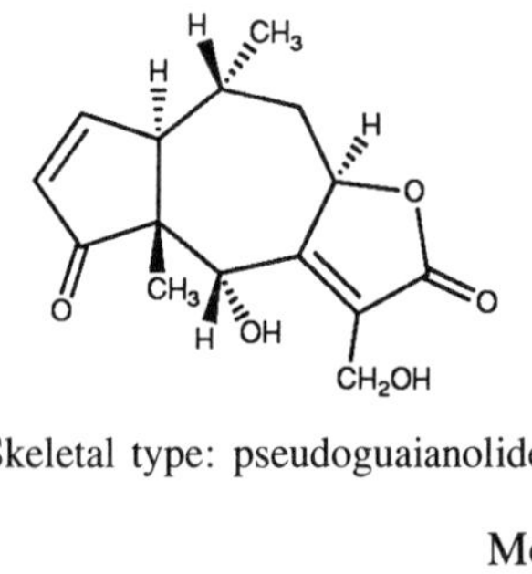

Skeletal type: pseudoguaianolide

$C_{15}H_{18}O_5$ Mol. wt 278.31

Occurs in *Helenium microcephalum* (Compositae).

Cytotoxic and antitumour activities.

2322
Isomontanolide

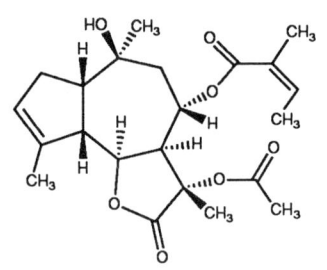

Skeletal type: guaianolide

$C_{22}H_{30}O_7$ Mol. wt 406.48

Found in the fruit of *Laserpitium siler* (Umbelliferae).

Insect antifeedant.

2323
Isotenulin

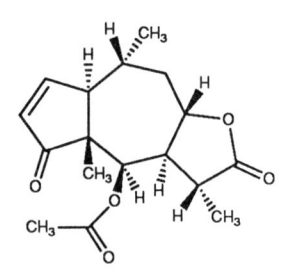

Skeletal type: pseudoguaianolide

$C_{17}H_{22}O_5$ Mol. wt 306.36

Occurs in *Helenium arizonicum* and *H. bigelovii* (Compositae).

Weak analgesic activity when injected subcutaneously.

2324
Ivalin

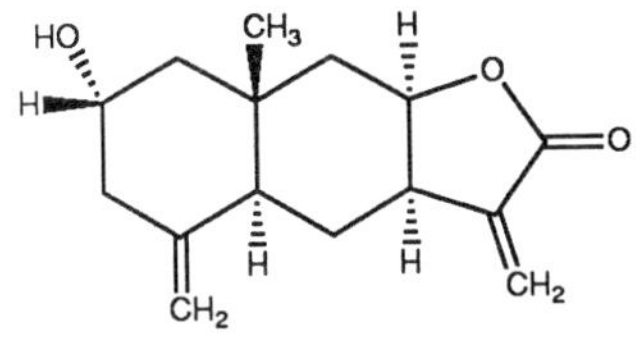

Skeletal type: eudesmanolide

$C_{15}H_{20}O_3$ Mol. wt 248.33

Occurs in *Iva imbricata* and *I. microcephala*, and in *Carpesium*, *Inula*, *Wedelia* and *Zaluzania* spp. (Compositae).

Toxic to mammals.

2325
Lactucin

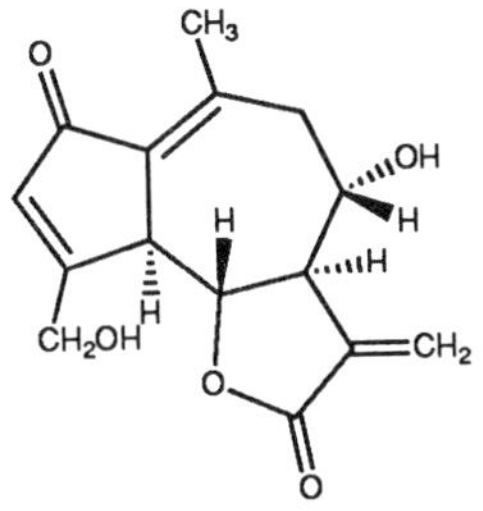

Skeletal type: guaianolide

$C_{15}H_{16}O_5$ Mol. wt 276.29

Occurs in various wild lettuce spp., e.g., *Lactuca canadensis*, *L. serriola* and *L. virosa*, and in chicory, *Cichorium intybus* (Compositae).

Cytotoxic and antitumour activities. It has a sedative action, and is a bitter tonic.

2326
Lactucopicrin; Intibin

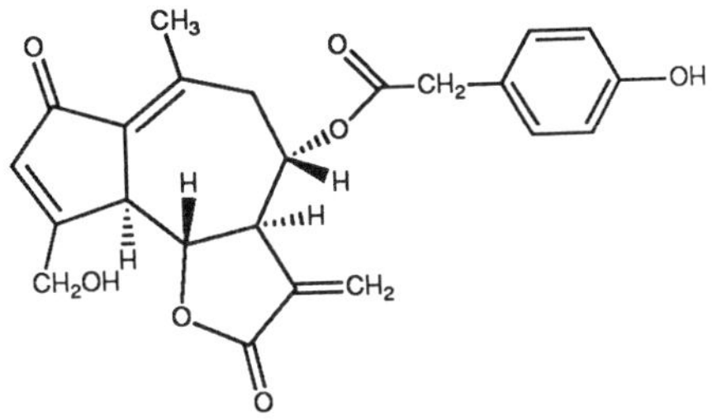

Skeletal type: guaianolide

$C_{23}H_{22}O_7$ Mol. wt 410.43

Occurs in chicory, *Cichorium intybus*, and in various spp. of wild lettuce, e.g., *Lactuca canadensis*, *L. serriola* and *L. virosa* (Compositae).

Hypoglycaemic activity. It is a bitter tonic.

2327
Laserolide

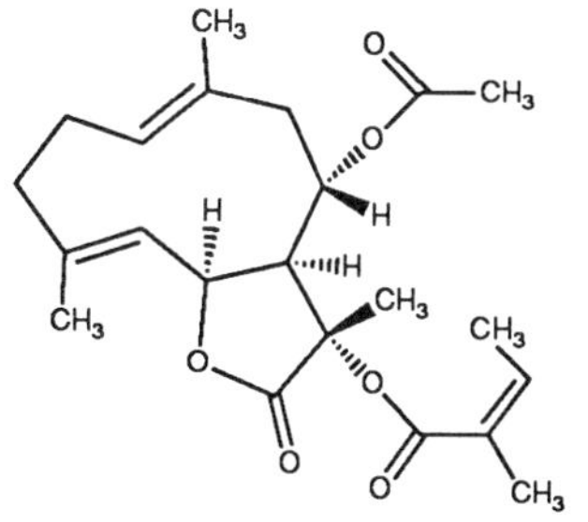

Skeletal type: germacranolide

$C_{22}H_{30}O_6$ Mol. wt 390.48

Found in the roots of *Laser trilobum* (Umbelliferae).

Insect antifeedant.

2328
Laurenobiolide

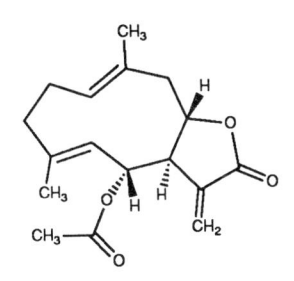

Skeletal type: germacranolide

$C_{17}H_{22}O_4$ Mol. wt 290.36

Found in the roots of bay laurel, *Laurus nobilis* (Lauraceae).

2329
Liatrin

Skeletal type: germacranolide

$C_{22}H_{26}O_8$ Mol. wt 418.45

Occurs in the blazing star, *Liatris chapmanii* (Compositae).

Cytotoxic and antitumour activities.

2330
Ligulatin B; Incanin

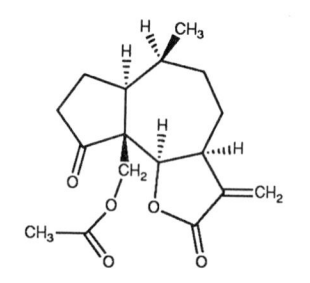

Skeletal type: pseudoguaianolide

$C_{17}H_{22}O_5$ Mol. wt 306.36

Found in *Parthenium ligulatum*, *P. incanum*, *P. schottii* and *P. tomentosum* (Compositae).

Insect antifeedant; it is toxic to certain insects.

2331
Linderane

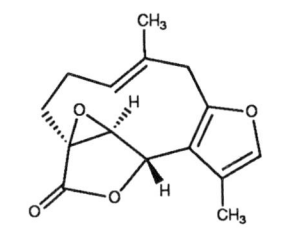

Skeletal type: furogermacranolide

$C_{15}H_{16}O_4$ Mol. wt 260.29

Occurs in the roots of *Lindera strychnifolia* (Lauraceae).

The Chinese drug "Tien tai wu yao" is prepared from the roots of *L. strychnifolia*.

2332
Linifolin A

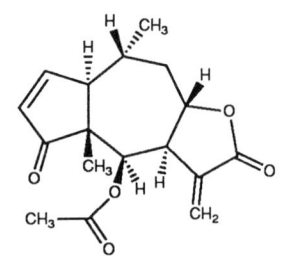

Skeletal type: pseudoguaianolide

$C_{17}H_{20}O_5$ Mol. wt 304.35

Found in a number of sneezeweed spp., e.g., *Helenium linifolium*, *H. alternifolium*, *H. aromaticum*, *H. plantagineum* and *H. scorzoneraefolia*.

Cytotoxic and antitumour activities. It is toxic to certain insect larvae.

2333
Lipiferolide

Skeletal type: germacranolide

$C_{17}H_{22}O_5$ Mol. wt 306.36

Found in the tulip tree, *Liriodendron tulipifera* (Magnoliaceae).

Cytotoxic and antitumour activities. It is an insect antifeedant.

2334
Ludovicin A

Skeletal type: eudesmanolide

$C_{15}H_{20}O_4$ Mol. wt 264.33

Occurs in *Artemisia ludoviciana* subsp. *mexicana* (Compositae).

Cytotoxic and antitumour activities.

2335
Matricin

Skeletal type: guaianolide

$C_{17}H_{22}O_5$ Mol. wt 306.36

Occurs in chamomile, *Matricaria chamomilla*, in several *Achillea* spp., in *Artemisia caruthii*, and in *Jurinea maxima* (Compositae).

Precursor of chamazulene (q.v.), which is anti-inflammatory.

2336
Melampodin A

Skeletal type: germacranolide

$C_{21}H_{24}O_9$ Mol. wt 420.42

Occurs in *Melampodium heterophyllum* and *M. leucanthum* (Compositae).

Insect antifeedant; it also affects insect development and reduces the survival of certain insects.

2337
Melampodinin

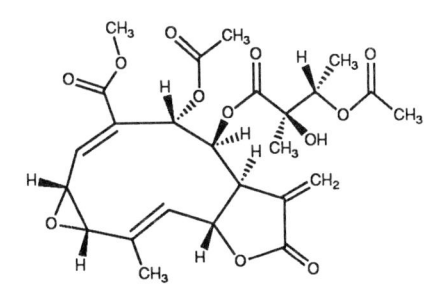

Skeletal type: germacranolide

$C_{25}H_{30}O_{12}$ Mol. wt 522.52

Occurs in *Melampodium americanum*, *M. diffusum* and *M. longipes* (Compositae).

Insect antifeedant; it also affects development of certain insect larvae.

2338
Mellitoxin; 4-Hydroxytutin; Hyenanchin; Ienancin

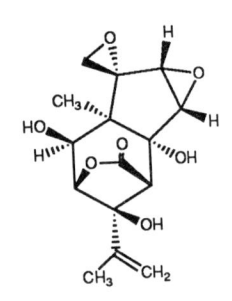

Skeletal type: tutinanolide = picrotoxin

$C_{15}H_{18}O_7$ Mol. wt 310.31

Found in the fruit of *Toxicodendrum globosum* (=*Hyaenanche globosa*) (Euphorbiaceae). First discovered in honey as a metabolic product of the passion vine hopper, *Scolypopa australis*, which feeds on *Coriaria arborea* (Coriariaceae). Here is probably a metabolite of tutin (q.v.).

Toxic principle in honey: it causes extreme excitation of the nervous system.

2339
Mexicanin E

Skeletal type: norpseudoguaianolide

$C_{14}H_{16}O_3$ Mol. wt 232.28

Occurs in several *Helenium* spp., e.g., *H. mexicanum* (Compositae).

Highly toxic to mammals. It shows cytotoxic and antitumour activities.

2340
Mexicanin I

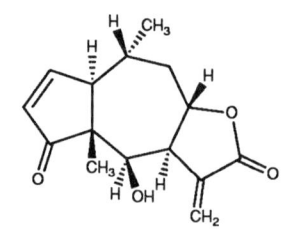

Skeletal type: pseudoguaianolide

$C_{15}H_{18}O_4$ Mol. wt 262.31

Found in sneezeweed, *Helenium autumnale*, and several other *Helenium* spp., in *Gaillardia pinnatifida* and in *Hymenoxys linearis* (Compositae).

Cytotoxic and antitumour activities.

2341
Michelenolide; Costunolide diepoxide

Skeletal type: germacranolide

$C_{15}H_{20}O_4$ Mol. wt 264.33

Found in the root bark of *Michelia compressa* (Magnoliaceae).

Cytotoxic and antitumour activities.

2342
Micheliolide

Skeletal type: guaianolide

$C_{15}H_{20}O_3$ Mol. wt 248.33

Found in *Michelia compressa* (Magnoliaceae).

Cytotoxic and antitumour activities.

2343
Microhelenin A

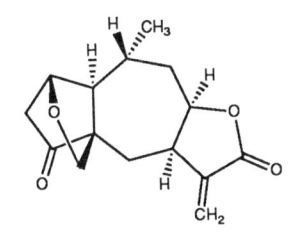

Skeletal type: pseudoguaianolide

$C_{15}H_{18}O_4$ Mol. wt 262.31

Occurs in *Helenium microcephalum* (Compositae).

Cytotoxic and antitumour activities.

2344
Microhelenin C

Skeletal type: pseudoguaianolide

$C_{20}H_{26}O_5$ Mol. wt 346.43

Occurs in *Helenium microcephalum* (Compositae).

Cytotoxic and antitumour activities.

2345
Microlenin

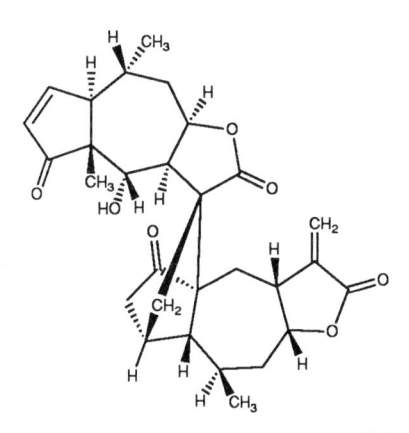

Skeletal type: dimeric pseudoguaianolide

$C_{29}H_{34}O_7$ Mol. wt 494.59

Occurs in *Helenium microcephalum* (Compositae).

Cytotoxic and antitumour activities.

2346
Mikanolide

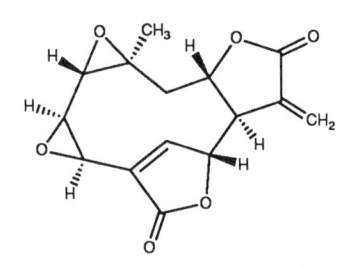

Skeletal type: germacranolide

$C_{15}H_{14}O_6$ Mol. wt 290.28

Found in the climbing hempweed, *Mikania scandens*, and in *M. batatifolia*, *M. cordata*, *M. micrantha*, and *M. monagasensis* (Compositae).

Antibacterial and antifungal activities, e.g., against the yeast *Candida albicans*.

2347
Molephantin

Skeletal type: germacranolide

$C_{19}H_{22}O_6$ Mol. wt 346.39

Occurs in *Elephantopus mollis* (Compositae).
Cytotoxic and antitumour activities.

2348
Molephantinin

Skeletal type: germacranolide

$C_{20}H_{24}O_6$ Mol. wt 360.41

Occurs in *Elephantopus mollis* (Compositae).
Cytotoxic and antitumour activities.

2349
Multigilin

Skeletal type: pseudoguaianolide

$C_{20}H_{24}O_6$ Mol. wt 360.41

Occurs in *Baileya multiradiata* (Compositae).
Cytotoxic and antitumour activities.

2350
Multiradiatin

Skeletal type: pseudoguaianolide

$C_{20}H_{22}O_6$ Mol. wt 358.40

Occurs in *Baileya multiradiata* (Compositae).
Cytotoxic and antitumour activities.

2351
Multistatin

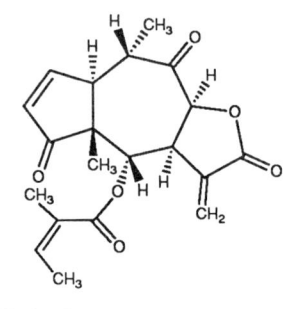

Skeletal type: pseudoguaianolide

$C_{20}H_{22}O_6$ Mol. wt 358.40

Occurs in *Baileya multiradiata* (Compositae).
Cytotoxic and antitumour activities.

2352
Niveusin C; Annuithrin

Skeletal type: germacranolide

$C_{20}H_{26}O_7$ Mol. wt 378.43

Occurs in *Helianthus niveus* subsp. *canescens*, *H. annuus* and *H. maximiliani* (Compositae).

Cytotoxic and antitumour activities.

2353
Nobilin

Skeletal type: germacranolide

$C_{20}H_{26}O_5$ Mol. wt 346.43

Occurs in chamomile, *Chamaemelum nobile* (*Anthemis nobilis*) (Compositae).

Cytotoxic and antitumour activities.

2354
Onopordopicrin

Skeletal type: germacranolide

$C_{19}H_{24}O_6$ Mol. wt 348.40

Found in many *Onopordum* spp., e.g., the Scotch thistle, *O. acanthium*, in *Berkheya speciosa*, and in the lesser burdock, *Arctium minus* (Compositae).

Cytotoxic and antitumour activities. It is an insect antifeedant.

2355
Orizabin

Skeletal type: germacranolide

$C_{19}H_{26}O_7$ Mol. wt 366.42

Occurs in *Tithonia tubaeformis* and *T. tagitiflora* (Compositae).

Cytotoxic and antitumour activities.

2356
Ovatifolin

Skeletal type: germacranolide

$C_{17}H_{22}O_5$ Mol. wt 306.36

Found in the leaves and stems of *Podanthus ovatifolius*, and in *P. mitiqui* (Compositae).

Cytotoxic and antitumour activities.

2357
Parthenin; Parthenicin

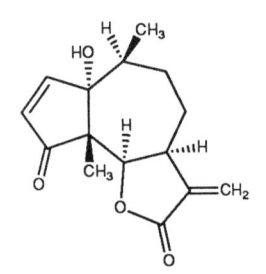

Skeletal type: pseudoguaianolide

$C_{15}H_{18}O_4$ Mol. wt 262.31

Occurs in *Parthenium hysterophorus, Ambrosia psilostachya* and *Iva nevadensis* (Compositae).

Parthenin is the main allergen responsible for the allergic contact dermatitis caused by *P. hysterophorus*. It is a direct cardiac depressant in dogs, is toxic to cattle, and has cytotoxic, antifungal and molluscicidal activity. Parthenin is an insect antifeedant, and is toxic to certain insects. Also, the compound has allelopathic properties.

2358
Parthenolide

Skeletal type: germacranolide

$C_{15}H_{20}O_3$ Mol. wt 248.33

Found in glands on the leaf surface of feverfew, *Tanacetum (Chrysanthemum) parthenium*, in several *Ambrosia* and *Arctotis* spp., and in other Compositae, and also in *Michelia champaca* and *M. lanuginosa* (Magnoliaceae).

Cytotoxic, antitumour, antibacterial and antifungal activities. Active principle of feverfew, which is used in the treatment of migraine.

2359
Paucin

Skeletal type: pseudoguaianolide

$C_{23}H_{32}O_{10}$ Mol. wt 468.51

Occurs in *Baileya pauciradiata* and *B. pleniradiata*, and in several *Hymenoxys* spp. (Compositae).

Cytotoxic and antitumour activities.

2360
1-Peroxyferolide

Skeletal type: germacranolide

$C_{17}H_{22}O_7$ Mol. wt 338.36

Occurs in the tulip tree, *Liriodendron tulipifera* (Magnoliaceae).

Insect antifeedant.

2361
Phantomolin

Skeletal type: germacranolide

$C_{21}H_{26}O_6$ Mol. wt 374.44

Occurs in *Elephantopus mollis* (Compositae).

Cytotoxic and antitumour activities.

2362
Picrotin

Skeletal type: tutinanolide = picrotoxin

$C_{15}H_{18}O_7$ Mol. wt 310.31

Found in "fish berries", the drupes of *Anamirta paniculata* (=*A. cocculus* or *Menispermum occulus*) (Menispermaceae).

Causes extreme excitation of the central nervous system, and has been used to stun fish, and also to treat skin diseases.

2363
Picrotoxinin

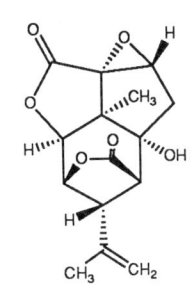

Skeletal type: tutinanolide = picrotoxin

$C_{15}H_{16}O_6$ Mol. wt 292.29

Found in "fish berries", the drupes of *Anamirta paniculata* (=*A. cocculus* or *Menispermum cocculus*) (Menispermaceae).

Picrotoxin, a mixture of picrotin (q.v.) and picrotoxinin, causes extreme excitation of the central nervous system, and has been used to stun fish, and also to treat skin diseases.

2364
Pleniradin

Skeletal type: guaianolide

$C_{15}H_{20}O_4$ Mol. wt 264.33

Occurs in *Baileya pleniradiata* (Compositae).

Cytotoxic and antitumour activities.

2365
Plenolin; Dihydrohelenalin

Skeletal type: pseudoguaianolide

$C_{15}H_{20}O_4$ Mol. wt 264.33

Occurs in *Baileya pleniradiata*, and in sneezeweed, *Helenium autumnale* (Compositae).

Cytotoxic and antitumour activities.

2366
Polhovolide

Skeletal type: guaianolide

$C_{23}H_{32}O_8$ Mol. wt 436.51

Found in the roots of *Laserpitium siler* (Umbelliferae).

Insect antifeedant.

2367
Provincialin

Skeletal type: germacranolide

$C_{27}H_{34}O_{10}$ Mol. wt 518.57

Found in the aerial parts of the blazing star, *Liatris provincialis* (Compositae).

Cytotoxic and antitumour activities.

2368
Pseudoivalin

Skeletal type: guaianolide

$C_{15}H_{20}O_3$ Mol. wt 248.33

Occurs in *Iva microcephala* and *Calocephalus brownii* (Compositae).

Antifungal activity, e.g., against the yeasts *Candida albicans* and *Saccharomyces cerevisiae*.

2369
Pycnolide

Skeletal type: seco-germacranolide

$C_{20}H_{28}O_6$ Mol. wt 364.44

Occurs in *Liatris pycnostachya* (Compositae).

2370
Pyrethrosin; Chrysanthin

Skeletal type: germacranolide

$C_{17}H_{22}O_5$ Mol. wt 306.36

Found in the flowers of pyrethrum, *Chrysanthemum cinerariifolium*, in *C. coccineum* and in *Anthemis cupaniana* (Compositae).

Causes allergic contact dermatitis in humans. It shows molluscicidal and plant growth regulating activities.

2371
Quadrone

$C_{15}H_{20}O_3$ Mol. wt 248.33

Found in the fungus *Aspergillus terreus* (Ascomycetes).

Cytotoxic and antitumour activities.

2372
Quing Hau Sau; Quinghaosu; Artemisinin

$C_{15}H_{22}O_5$ Mol. wt 282.34

Occurs in *Artemisia annua*, quing hao (Compositae).

Antimalarial activity. In clinical use in China.

2373
Radiatin

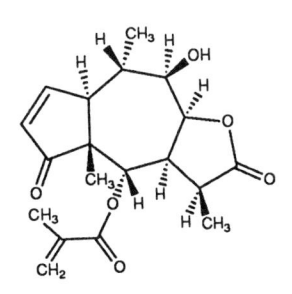

Skeletal type: pseudoguaianolide

$C_{19}H_{24}O_6$ Mol. wt 348.40

Occurs in *Baileya pleniradiata* and *B. multiradiata* (Compositae).

Cytotoxic and antitumour activities.

2374
Ridentin; Ridentin A

Skeletal type: germacranolide

$C_{15}H_{20}O_4$ Mol. wt 264.33

Found in *Artemisia tridentata*, *A. cana* and *A. tripartita* (Compositae).

Cytotoxic and antitumour activities.

2375
Salonitenolide

Skeletal type: germacranolide

$C_{15}H_{20}O_4$ Mol. wt 264.33

Found in the flowers and leaves of *Centaurea salonitana*, in the blessed thistle, *Cnicus benedictus*, in *Berkheya speciosa*, and in *Jurinea maxima* (Compositae).

Cytotoxic and antitumour activities. It is an insect antifeedant.

2376
Santamarin; Santamarine; Balchanin

Skeletal type: eudesmanolide

$C_{15}H_{20}O_3$ Mol. wt 248.33

Found in *Ambrosia confertiflora*, several *Artemisia* spp., in feverfew, *Tanacetum* (*Chrysanthemum*) *parthenium* (Compositae), and in the Magnoliaceae, e.g., *Michelia compressa*.

Cytotoxic and antitumour activities.

2377
α-Santonin; Santonin

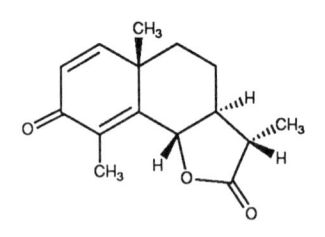

Skeletal type: eudesmanolide

$C_{15}H_{18}O_3$ Mol. wt 246.31

Found widely in *Artemisia* spp. (Compositae).

Cytotoxic and antitumour activities. α-Santonin has been used in the treatment of nervous complaints. It shows anti-helminthic and ascaricidal activities, is an insect antifeedant, and also shows plant growth regulating activity.

2378
β-Santonin

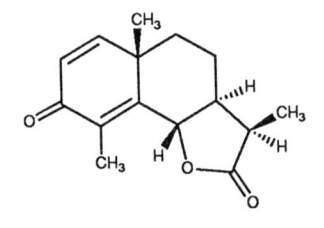

Skeletal type: eudesmanolide

$C_{15}H_{18}O_3$ Mol. wt 246.31

Occurs in many *Artemisia* spp., e.g., *A. caerulescens*, *A. cina*, *A. compacta* and *A. finita* (Compositae).

Formerly used as a vermifuge but, because of its toxicity, it has been excluded from official use in many countries.

2379
Saupirin; Saupirine

Skeletal type: guaianolide

$C_{19}H_{22}O_6$ Mol. wt 346.39

Occurs in the flowers of *Saussurea pulchella* and *S. neopulchella* (Compositae).

Active against protozoa pathogenic to humans, e.g., *Entamoeba histolytica* and *Trichomonas vaginalis*.

2380
Spicatin

Skeletal type: guaianolide

$C_{27}H_{32}O_{10}$ Mol. wt 516.55

Occurs in *Liatris spicata*, *L. pycnostachya* and *L. tenuifolia* (Compositae).

Cytotoxic and antitumour activities. The name spicatin is also used for a limonoid triterpenoid.

2381
Stramonin B

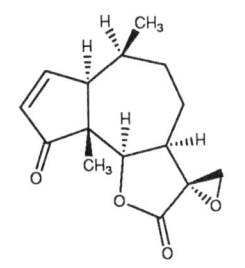

Skeletal type: pseudoguaianolide

$C_{15}H_{18}O_4$ Mol. wt 262.31

Found in *Parthenium tomentosum* var. *stramonium* (Compositae).

Cytotoxic and antitumour activities.

2382
Tagitinin F

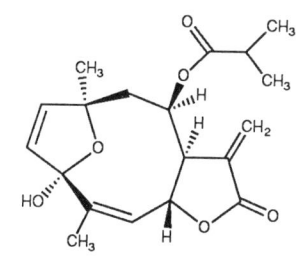

Skeletal type: germacranolide

$C_{19}H_{24}O_6$ Mol. wt 348.40

Occurs in *Tithonia tagitiflora* and *T. diversifolia* (Compositae).

Cytotoxic and antitumour activities.

2383
Tamaulipin A

Skeletal type: germacranolide

$C_{15}H_{20}O_3$ Mol. wt 248.33

Found in *Ambrosia confertiflora* and *A. dumosa* (Compositae).

Cytotoxic and antitumour activities.

2384
Tenulin

Skeletal type: pseudoguaianolide

$C_{17}H_{22}O_5$ Mol. wt 306.36

Found in many *Helenium* spp., e.g., *H. tenuifolium*, *H. amarum*, *H. autumnale*, *H. elegans* and *H. puberulum* (Compositae).

Cytotoxic and antitumour activities, and weak analgesic activity when used subcutaneously. It is toxic to the hamster, mouse and sheep, and barely toxic to cows, but it affects the taste of their milk.

2385
Tetraneurin A

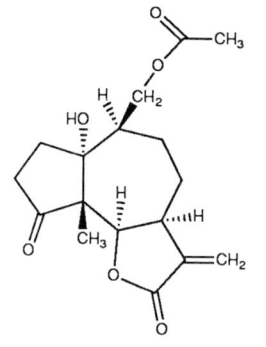

Skeletal type: pseudoguaianolide

$C_{17}H_{22}O_6$ Mol. wt 322.36

Occurs in *Parthenium alpinum*, *P. cineraceum*, *P. confertum*, *P. fruticosum* and *P. hysterophorus* (Compositae).

Tetraneurin A is probably one of the sesquiterpene lactones in *P. hysterophorus* which causes allergic contact dermatitis. It is also an insect antifeedant, and it affects the development of certain insects.

2386
Tetraneurin E

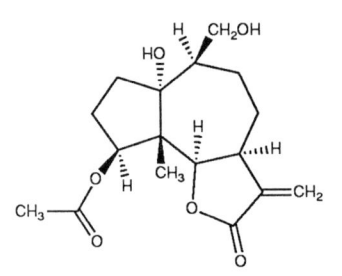

Skeletal type: pseudoguaianolide

$C_{17}H_{24}O_6$ Mol. wt 324.38

Occurs in *Parthenium confertum*, *P. fruticosum*, *P. hispidum*, *P. integrifolium*, and *P. lozanianum* (Compositae).

Affects the development of certain insect larvae.

2387
Thapsigargin

Skeletal type: guaianolide

$C_{34}H_{50}O_{12}$ Mol. wt 650.77

Found in *Thapsia garganica* (Umbelliferae).

Potent activator of cells involved in inflammatory response, e.g., human mast cells, basophils and neutrophils.

2388
Tomentosin

Skeletal type: secoguaianolide = xantholide

$C_{15}H_{20}O_3$ Mol. wt 248.33

Occurs in *Parthenium tomentosum*, and in several *Inula* spp., e.g., elecampane, *I. helenium* (Compositae).

The name tomentosin is also used for a pregnane steroid and, additionally, for a germacranolide sesquiterpenoid.

2389
Trilobolide

Skeletal type: guaianolide

$C_{27}H_{38}O_{10}$ Mol. wt 522.60

Found in the roots of *Laser trilobum* (Umbelliferae).

Cytotoxic and antitumour activities. It is an insect antifeedant.

2390
Tulipinolide

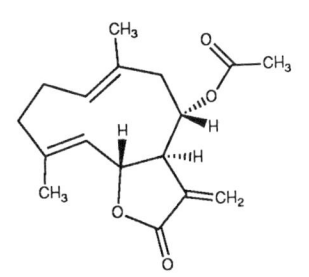

Skeletal type: germacranolide

$C_{17}H_{22}O_4$ Mol. wt 290.36

Occurs in the root bark of the tulip tree, *Liriodendron tulipifera* (Magnoliaceae), and in *Ambrosia camphorata* and *A. dumosa* (Compositae).

Cytotoxic activity.

2391
epi-Tulipinolide

Skeletal type: germacranolide

$C_{17}H_{22}O_4$ Mol. wt 290.36

Found in *Zaluzania pringlei*, *Ambrosia chamissonis* and *A. dumosa* (Compositae), and in the root bark of the tulip tree, *Liriodendron tulipifera* (Magnoliaceae).

Cytotoxic and antitumour activities.

2392
epi-Tulipinolide diepoxide

Skeletal type: germacranolide

$C_{17}H_{22}O_6$ Mol. wt 322.36

Found in the leaves of the tulip tree, *Liriodendron tulipifera* (Magnoliaceae).

Cytotoxic and antitumour activities. It is an insect antifeedant.

2393
Tutin; 2-Hydroxycoriamyrtin

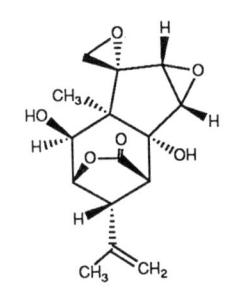

Skeletal type: tutinanolide=picrotoxin

$C_{15}H_{18}O_6$ Mol. wt 294.31

Occurs in many *Coriaria* spp. (called "tutu" in New Zealand), e.g., *C. angustissima* (Coriariaceae), and in *Toxicodendrum capense* (Euphorbiaceae).

Causes extreme excitation of the central nervous system: stimulates the respiratory, vasomotor and cardioinhibitory centres in the brain. It has caused the death of many cattle feeding on *Coriaria* spp. in New Zealand.

2394
Ursiniolide A

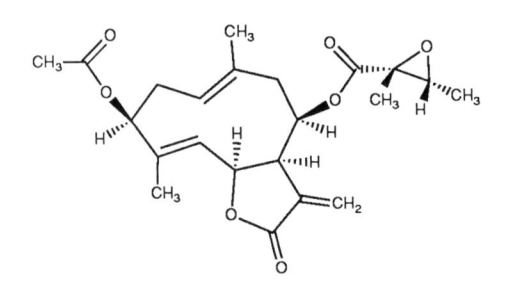

Skeletal type: germacranolide

$C_{22}H_{28}O_7$ Mol. wt 404.47

Found in *Ursinia anthemoides* (Compositae).

Cytotoxic and antitumour activities.

2395
Vermeerin

Skeletal type: seco-pseudoguaianolide

$C_{15}H_{20}O_4$ Mol. wt 264.33

Occurs in *Geigera aspera, G. africana, Hymenoxys anthemoides, H. richardsonii* and *Psilostrophe villosa* (Compositae).

Toxic to mammals. It causes mass poisoning ("vomiting disease") among sheep in Africa feeding on *Geigera* spp.

2396
Vernodalin

Skeletal type: elemanolide

$C_{19}H_{20}O_7$ Mol. wt 360.37

Found in the ironweeds *Vernonia amygdalina* and *V. guineensis* (Compositae).

Cytotoxic and antitumour activities. It is an insect antifeedant.

2397
Vernodalol

Skeletal type: elemanolide

$C_{20}H_{24}O_8$ Mol. wt 392.41

Found in the ironweed *Vernonia anthelmintica* (Compositae).

Insect antifeedant.

2398
Vernoflexin; Vernoflexine; Zaluzanin C senecioate

Skeletal type: guaianolide

$C_{20}H_{24}O_4$ Mol. wt 328.41

Found in several ironweed, *Vernonia*, spp., e.g., *V. flexuosa, V. arkansana*, and *V. chinensis*. (Compositae).

Cytotoxic and antitumour activities.

2399
Vernoflexuoside

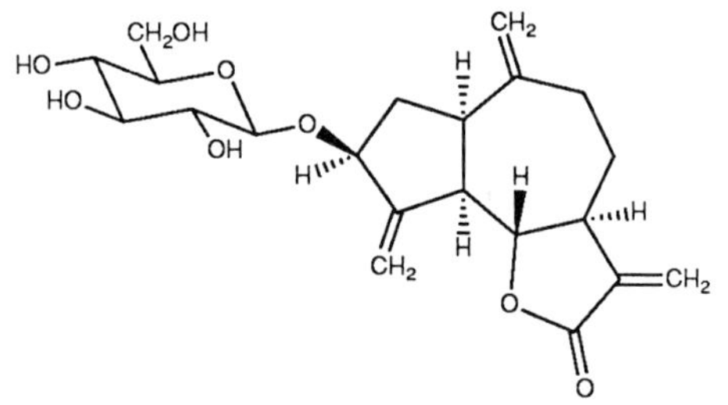

Skeletal type: guaianolide

$C_{21}H_{28}O_8$ Mol. wt 408.46

Occurs in *Vernonia flexuosa* (Compositae).

Cytostatic activity. It produces mitotic disturbances and symptoms of nucleus structure degeneration in onion roots.

2400
Vernolepin

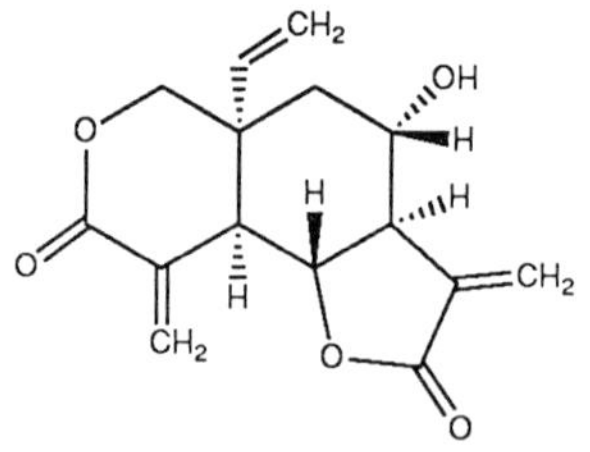

Skeletal type: elemanolide

$C_{15}H_{16}O_5$ Mol. wt 276.29

Found in the ironweeds *Vernonia hymenolepis* and *V. guineensis* (Compositae).

Cytotoxic and antitumour activities. It shows plant growth regulating activity.

2401
Vernolide

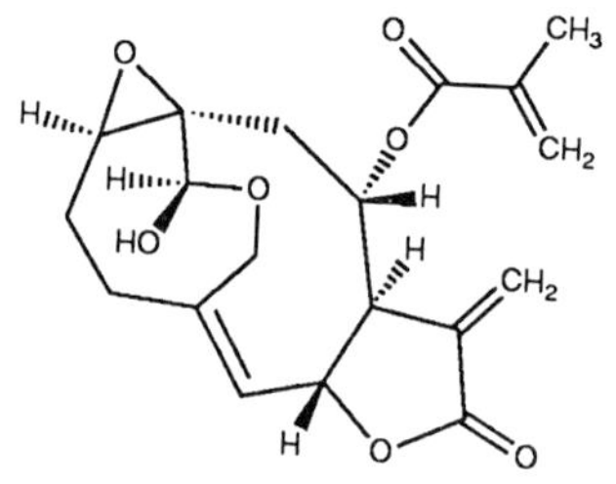

Skeletal type: germacranolide

$C_{19}H_{22}O_7$ Mol. wt 362.39

Found in the ironwoods *Vernonia amygdalina* and *V. colorata* (Compositae).

Cytotoxic and antitumour activities.

2402
Vernomenin

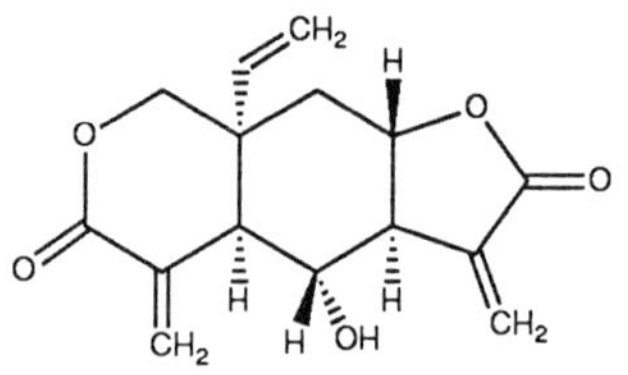

Skeletal type: elemanolide

$C_{15}H_{16}O_5$ Mol. wt 276.29

Occurs in *Vernonia hymenolepis* (Compositae).

Cytotoxic and antitumour activities.

2403
Vernomygdin

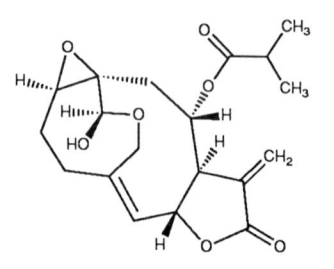

Skeletal type: germacranolide

$C_{19}H_{24}O_7$ Mol. wt 364.40

Occurs in *Vernonia amygdalina* (Compositae).

Cytotoxic and antitumour activities.

2404
Viguiestenin

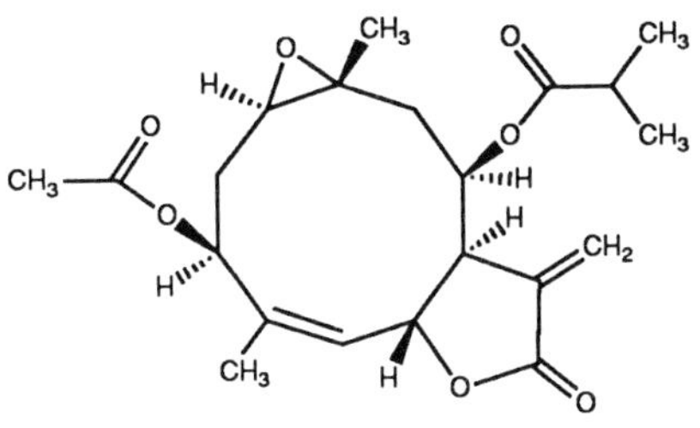

Skeletal type: germacranolide

$C_{21}H_{28}O_7$ Mol. wt 392.46

Occurs in *Viguiera stenoloba* and *V. pinnatilobata* (Compositae).

Cytotoxic and antitumour activities.

2405
Viscidulin B

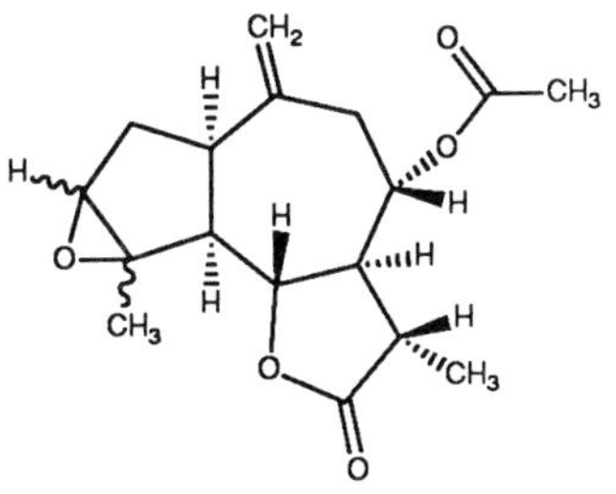

Skeletal type: guaianolide

$C_{17}H_{22}O_5$ Mol. wt 306.36

Found in *Artemisia cana* subsp. *viscidula* (Compositae).

Plant growth regulating activity.

2406
Vulgarin; Judaicin; Tauremisin

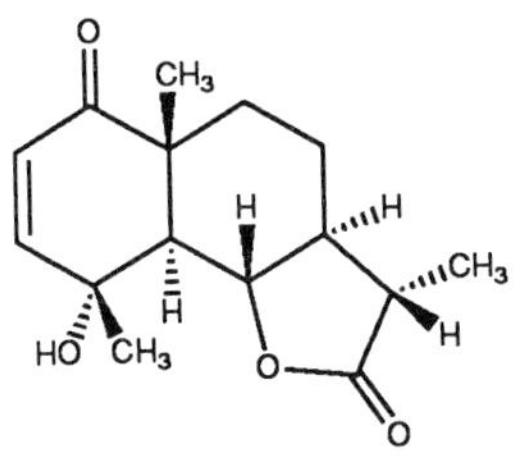

Skeletal type: eudesmanolide

$C_{15}H_{20}O_4$ Mol. wt 264.33

Found in mugwort, *Artemisia vulgaris*, in *A. judaica*, *A. taurica*, and in many other *Artemisia* spp. (Compositae).

Cytotoxic and antitumour activities.

2407
Xanthatin

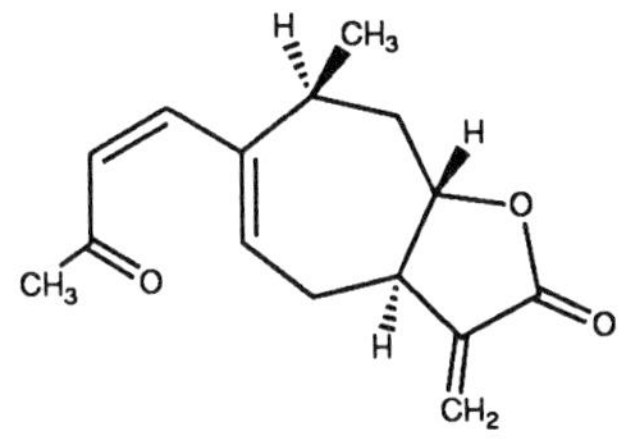

Skeletal type: secoguaianolide = xanthanolide

$C_{15}H_{18}O_3$ Mol. wt 246.31

Occurs in many *Xanthium* spp., e.g., *X. pennsylvanicum*, *X. riparium*, and *X. sibiricum* (Compositae).

Antibacterial and antifungal activities. It affects the development of certain insect larvae.

2408
Xanthinin

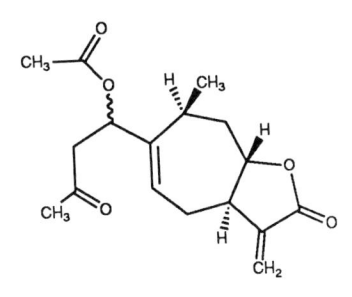

Skeletal type: secoguaianolide = xanthanolide

$C_{17}H_{22}O_5$ Mol. wt 306.36

Found in many *Xanthium* spp., e.g., *X. commune* and *X. orientale*, and in *Agianthus tomentosus* and *Iva ambrosiaefolia* (Compositae).

Activity as a phytohormone. It shows antagonism towards auxin.

2409
Xanthumin

Skeletal type: secoguaianolide = xanthanolide

$C_{17}H_{22}O_5$ Mol. wt 306.36

Found in *Xanthium chasei*, *X. chinense*, *X. occidentale* and *X. strumarium* (Compositae).

Insect antifeedant.

2410
Xerantholide

Skeletal type: guaianolide

$C_{15}H_{18}O_3$ Mol. wt 246.31

Occurs in *Xeranthemum cylindraceum* (Compositae).

Insect antifeedant.

2411
Zaluzanin C

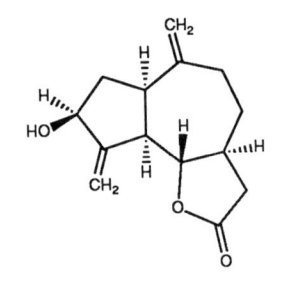

Skeletal type: guaianolide

$C_{15}H_{18}O_3$ Mol. wt 246.31

Found in several *Zaluzania* and *Vernonia* spp., in *Podachaenium eminens* and in *Zinnia acerosa* (Compositae).

Cytotoxic and antitumour activities.

2412
Zexbrevin B

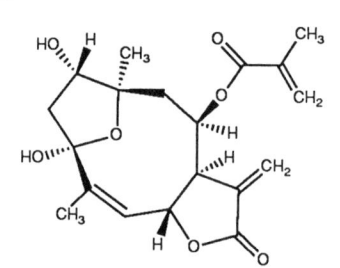

Skeletal type: germacranolide

$C_{19}H_{24}O_7$ Mol. wt 364.40

Occurs in *Zexmenia brevifolia* and *Tithonia tubaeformis* (Compositae).

Cytotoxic and antitumour activities.

Chapter 52
Diterpenoids

The diterpenoids are defined by their biosynthetic origin from the C_{20} precursor geranylgeraniol (**2445**). They are derived from four isoprene units, but modifications occurring during the later stages in synthesis produce a veritable plethora of complex chemical structures. They share many chemical properties with the sesquiterpenoids (q.v.) but are usually crystalline solids rather than liquids. Diterpenoid lactones are known (e.g., ginkgolide A, **2447**) but the lactones do not deserve separate treatment here, as do the sesquiterpene lactones (q.v.).

Diterpenoids can be arbitrarily classified according to the number of ring systems present. Thus, there are bicyclic compounds such as portulal (**2491**), tricyclic compounds such as callicarpone (**2423**) and tetracyclic structures such as podolide (**2489**). In addition, there are acyclic compounds (geranylgeraniol, **2445**, and phytol, **2482**) and macrocyclic substances (casbene, **2427** and neocembrene, **2475**). Many diterpenoids have additional ring systems, in side chains (e.g., ajugarin I,, **2415**), in ester substitution (e.g., candletoxin A, **2424**) or as epoxides (e.g., caryoptin, **2426**). Some have six-membered rings which have undergone aromatization (e.g., carnosol, **2425**), while others have fused five- and seven-membered rings (e.g., gnidicin, **2448**).

Two groups of diterpenoid can be distinguished which have a carboxylic acid function in addition: the resin acids and the gibberellins. The resin acids include abietic (**2413**) and pimaric (**2483**) acids, which occur in both present-day and fossil plant resins. These acids are exuded from the wood of trees, especially conifers, and have a protective function in nature against infection and insect attack. By contrast, the gibberellins are a group of plant hormones, which stimulate growth and which are universal in higher plants. Appreciable amounts can be found in seeds, but otherwise only traces are present in other tissues. Gibberellic acid (**2446**) or GA_3 (the third gibberellin to be discovered) is the most familiar compound, but there are 71 other structures with the same basic skeleton but varying in the substitution pattern. All these gibberellins can occur *in vivo* in conjugated form, and many glucose and alkyl esters and glucosides have been reported. Gibberellic acid is of fungal origin and, although it has been described frequently as occurring in higher plants, the most recent evidence indicates that GA_1, which differs from GA_3 only in lacking the endocyclic 1,2-double bond, is the most common higher plant gibberellin.

As with other terpenoids, a number of diterpenoids are very toxic. Grayanotoxin I (**2452**) is representative of a number of poisonous diterpenes found in *Rhododendron* and other ericaceous plants. Another series of toxins is the phorbol esters of *Croton* oil and *Euphorbia* latex, which are notable for their irritant and co-carcinogenic properties.

Diterpenoids are principally found either in higher plants or in fungi. Phytol is universal in higher plants, where it is attached to the green chromophore of the chlorophylls (q.v.). Practically all other diterpenoids, however, are of very restricted distribution. They are rarely found in glycosidic combination, so that the triglucoside stevioside (**2499**) is exceptional. This compound is also remarkable in being very sweet-tasting, and it is marketed in Japan as a sweetening agent. Another of the rare diterpenoid glycosides is mascaroside (**2469**), which is notable for its very bitter taste.

Diterpenoids are usually isolated from plants as optically active solids, which can exist in both the normal and the antipodal stereochemical configurations. They are separated by thin-layer or column chromatography on silica, using solvents such as petroleum or chloroform. They have to be derivatized (as the trimethyl silyl ethers) before they can be separated by gas chromatography. Much effort has been expended on the separation and identification of the gibberellins, and these are now analysed either by high performance liquid chromatography or, with great sensitivity and precision, by immunoassay.

Reviews of the chemistry of diterpenoids include those of Hanson (1974) and Nakanishi *et al.* (1974). More specific reviews on gibberellins are provided by Macmillan (1983) and Schreiber *et al.* (1989), and on co-carcinogenic diterpenoids by Jury *et al.* (1987).

References

Nakanishi, K., Goto, T., Ito, S., Natori, S. and Nozoe, S. (eds) (1974). *Natural Products Chemistry*, Volume 1. London: Academic Press.

Hanson, J.R. (1972). In *Chemistry of Terpenes and Terpenoids* (Newman, A.A., ed.), pp. 155–206, New York: Academic Press.

Biological Properties

Jury, S.L., Reynolds, T., Cutler, D.F. and Evans, F.J. (eds) (1987). *The Euphorbiales, Chemistry, Taxonomy and Economic Botany*. London: Academic Press.

Macmillan, J. (1983). Gibberellins in higher plants. *Biochemical Society Transactions*, **11**, 528–533.

Schreiber, K., Schutte, H.R. and Sembdner, G. (1989). *Conjugated Plant Hormones: Structure, Metabolism and Function*. Berlin: VEB Deutscher Verlag der Wissenschaften.

2413
Abietic acid

Skeletal type: abietane

$C_{20}H_{30}O_2$ Mol. wt 302.46

Occurs as the primary constituent in many conifer resins.

Assists the growth of lactic and butyric acid bacteria. It is used for the manufacture of varnishes, soaps and plastics.

2414
6-Acetylpicropolin

Skeletal type: clerodane

$C_{24}H_{28}O_9$ Mol. wt 460.49

Found in *Teucrium polium* (= *T. decaisnii*) (Labiatae).

Extremely bitter taste.

2415
Ajugarin I

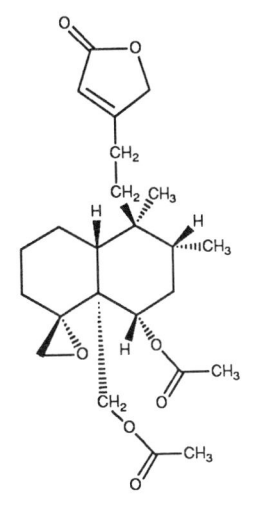

Skeletal type: clerodane

$C_{24}H_{34}O_7$ Mol. wt 434.54

Found in the leaves of *Ajuga remota* (Labiatae).

Ajugarin I, and the chemically related ajugarins II and III from the same plant source, are powerful insect antifeedants, especially against African armyworms.

2416
Allogibberic acid

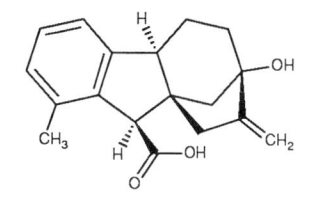

Skeletal type: gibbane

$C_{18}H_{20}O_3$ Mol. wt 284.36

Produced by *Lemna perpusilla* strain 6746 (Lemnaceae) when autoclaved.

Inhibits flowering, increases frond multiplication rate, and decreases frond size in *L. perpusilla*.

2417
Antheridiogen-An

$C_{20}H_{24}O_6$ Mol. wt 360.41

Occurs in the spore cultures of the fern genus *Anemia* (Schizaeaceae), e.g., in *A. phyllitidis* and *A. hirsuta*.

Induces the formation of the antheridium (=male gametophyte) in *Anemia*.

2418
Aphidicolin

$C_{20}H_{34}O_4$ Mol. wt 338.49

Metabolite of the fungus *Cephalosporium aphidicola*.

Antimitotic, antiviral and antibiotic activities. It specifically inhibits DNA α-polymerase.

2419
Asebotoxin II

Skeletal type: grayanotoxin

$C_{23}H_{36}O_6$ Mol. wt 408.54

Occurs together with many related asebotoxins in the flowers of *Pieris japonica* (Ericaceae).

Asebotoxin II and other asebotoxins are highly toxic substances. They are chemically closely related to grayanotoxins (q.v.).

2420
Atractyloside; Atractylin; Potassium atractylate

Skeletal type: kaurane

$C_{30}H_{44}K_2O_{16}S_2$ Mol. wt 803.01

Occurs in *Atractylis gummifera* (Compositae).

Highly toxic to mammals (LD_{50} intramuscularly in rats 431 mg kg^{-1}), with strychnine-like activity. It is used experimentally, because it is a specific inhibitor of ADP transport at the mitochondrial membrane.

2421
Baliospermin

Skeletal type: tigliane

$C_{32}H_{50}O_8$ Mol. wt 562.75

Found in the roots of *Baliospermum montanum* (Euphorbiaceae).

Cytotoxic activity. Extracts of *Baliospermum* plants are drastic purgatives.

2422
Cafestol; Cafesterol

Skeletal type: kaurane

$C_{20}H_{28}O_3$ Mol. wt 316.44

The main constituent of the unsaponifiable portion of coffee bean oil, from *Coffea* spp. (Rubiaceae).

Anti-inflammatory activity.

2423
Callicarpone

Skeletal type: abietane

$C_{20}H_{28}O_4$ Mol. wt 332.44

Occurs in the leaves of *Callicarpa candicans* (Verbenaceae).

Strong piscicidal activity.

2424
Candletoxin A

Skeletal type: tigliane

$C_{35}H_{44}O_9$ Mol. wt 608.74

Occurs in the latex of *Euphorbia poisonii* (Euphorbiaceae).

Toxic.

2425
Carnosol

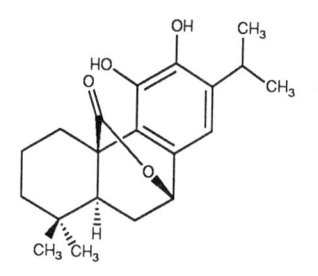

Skeletal type: abietane

$C_{20}H_{26}O_4$ Mol. wt 330.43

Found in various spp. of sage, e.g., *Salvia officinalis*, and in rosemary, *Rosmarinus officinalis* (Labiatae).

Bitter principle. It is used as an aromatic bitter.

2426
Caryoptin

Skeletal type: clerodane

$C_{26}H_{36}O_9$ Mol. wt 492.57

Occurs in *Caryopteris divaricata* (Verbenaceae).

Bitter substance, possessing insect-antifeedant activity.

2427
Casbene

or enantiomer

Skeletal type: casbane

$C_{20}H_{32}$ Mol. wt 272.48

Found in the castor oil plant, *Ricinus communis* (Euphorbiaceae).

Antifungal activity.

2428
Cascarillin

$C_{22}H_{32}O_7$ Mol. wt 408.50

Found in the bark of *Croton eluteria* (Euphorbiaceae).

Used as an aromatic bitter.

2429
Chasmanthin

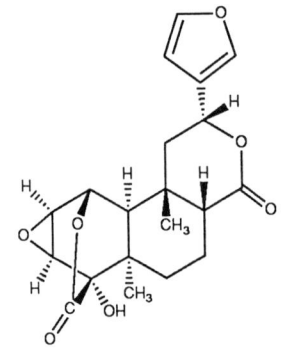

Skeletal type: clerodane

$C_{20}H_{22}O_7$ Mol. wt 374.40

Occurs in the roots of *Jateorhiza palmata* (=*J. columba*) (Menispermaceae).

Bitter principle. The roots of *J. palmata* were formerly used for the preparation of a bitter tonic (Radix Columba).

2430
Clerodendrin A

Skeletal type: clerodane

$C_{31}H_{42}O_{12}$ Mol. wt 606.68

Found in *Clerodendrum trichotomum* (Verbenaceae).

Antifeedant activity against the larvae of *Spodoptera littoralis*.

2431
Clerodin

Skeletal type: clerodane

$C_{24}H_{34}O_7$ Mol. wt 434.54

Found in the leaves and twigs of the Indian bhat tree, *Clerodendrum infortunatum* (Verbenaceae).

C. infortunatum leaf extract is used as a vermifuge.

2432
Coleonol

$C_{22}H_{34}O_7$ Mol. wt 410.51

Occurs in the leaves of *Coleus forskohlii* (=*C. barbatus*) (Labiatae).

Hypotensive activity.

2433
Columbin

Skeletal type: clerodane

$C_{20}H_{22}O_6$ Mol. wt 358.40

Found in the roots of "columba root", *Jateorhiza palmata* (=*J. columba*), and in the seeds of *Dioscoreophyllum cumminsii* (Menispermaceae).

Very bitter substance. The roots of *J. palmata* were formerly used for the preparation of a bitter tonic (Radix Columba).

2434
Cotylenin F

$C_{33}H_{54}O_{11}$ Mol. wt 626.80

Metabolite of a *Cladosporium* sp. (fungi).

Plant growth regulator.

2435
Cyathin A$_3$

$C_{20}H_{30}O_3$ Mol. wt 318.46

Found in the bird's-nest fungus, *Cyathus helenae*.

Antibiotic activity.

2436
Daphnetoxin

Skeletal type: daphnane

$C_{27}H_{30}O_8$ Mol. wt 482.54

Occurs in mezereon, *Daphne mezereum* (Thymelaeaceae).

Highly toxic (LD$_{50}$ orally 0.25 mg kg^{-1}).

2437
7β,12α-Dihydroxykaurenolide

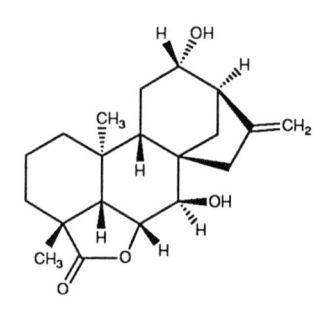

Skeletal type: kaurane

$C_{20}H_{28}O_4$ Mol. wt 332.44

Occurs in the seeds of *Cucurbita pepo* (Cucurbitaceae).

Plant growth regulatory activity.

2438
Diterpenoid EF-D

Skeletal type: tigliane

$C_{27}H_{38}O_7$ Mol. wt 474.60

Occurs in *Euphorbia fortissima* (Euphorbiaceae), together with several other esters of the same diterpenoid.

Skin irritant.

2439
Diterpenoid SP-II

Skeletal type: kaurane

$C_{20}H_{32}O_4$ Mol. wt 336.48

Found in *Sigesbeckia pubescens* (Compositae), together with several other related diterpenoids.

Anti-inflammatory activity. It is a powerful antihypertensive.

2440
Enmein

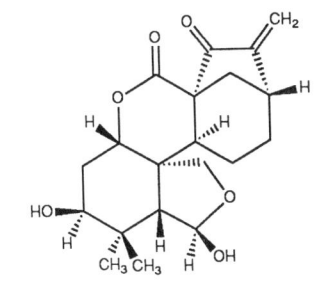

Skeletal type: seco-kaurane

$C_{20}H_{26}O_6$ Mol. wt 362.43

Found in *Plectranthus trichocarpus* (=*Isodon trichocarpus*) (Labiatae).

Bitter taste. *P. trichocarpus* is used for the preparation of a folk remedy against gastro-intestinal disturbances.

2441
Euphorbia factor Ti$_2$

Skeletal type: tigliane

$C_{32}H_{42}O_7$ Mol. wt 538.69

Occurs in *Euphorbia tirucalli* (Euphorbiaceae).

Highly irritant.

2442
Ferruginol

Skeletal type: abietane

$C_{20}H_{30}O$ Mol. wt 286.46

Main constituent of the resin from the New Zealand miro tree, *Podocarpus ferrugineus* (Podocarpaceae).

2443
Forskolin

Skeletal type: labdane

$C_{22}H_{34}O_7$ Mol. wt 410.51

Found in the roots of *Coleus forskohlii* (Labiatae).

Cardiovascular activity.

2444
Fusicoccin H

$C_{26}H_{42}O_8$ Mol. wt 482.62

Minor metabolite of *Fusicoccum amygdali*, a fungus responsible for a wilting disease of peach and almond trees.

Phytotoxic activity.

2445
Geranylgeraniol

acyclic diterpenoid

$C_{20}H_{34}O$ Mol. wt 290.49

Found in oil of linseed, *Linum usitatissimum* (Linaceae), and in the wood oil of the Australian red cedar, *Toona ciliata* (=*Cedrala toona*) (Meliaceae).

The diphosphate of geranylgeraniol is the biogenetic precursor of diterpenes.

2446
Gibberellin A$_3$; Gibberellic acid

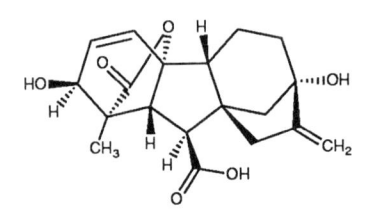

Skeletal type: gibbane

$C_{19}H_{22}O_6$ Mol. wt 346.39

Occurs as a metabolite of the fungus *Gibberella fujikuroi* (=*Fusarium moniliforme*), which is pathogenic to rice. Gibberellin A$_3$ is the best known of the seventy-two gibberellins presently characterized, most of which occur in trace amounts in higher plants, especially in seed tissues.

Gibberellic acid and other gibberellins are plant growth hormones. Gibberellic acid is used commercially to accelerate the first stage of brewing beer, to break seed dormancy, and to produce seedless grapes.

2447
Ginkgolide A

$C_{20}H_{24}O_9$ Mol. wt 408.41

Found in the root bark and leaves of the maidenhair tree, *Ginkgo biloba* (Ginkgoaceae).

Ginkgolide A and related ginkgolides are bitter substances. They are chemically inert, but have bronchodilator and anti-asthmatic activities. They are used medicinally to treat allergic inflammation and asthma. Ginkgolide A is also an insect antifeedant.

2448
Gnidicin

Skeletal type: daphnane

$C_{36}H_{36}O_{10}$ Mol. wt 628.68

Occurs in *Gnidia lamprantha* (Thymelaeaceae).

Antitumour activity.

2449
Gnididin

Skeletal type: daphnane

$C_{37}H_{44}O_{10}$ Mol. wt 648.76

Occurs in *Gnidia lamprantha* (Thymelaeaceae).

Antitumour activity.

2450
Gnidilatin

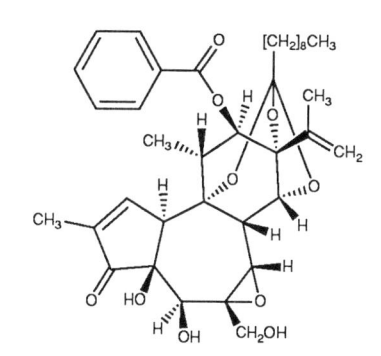

Skeletal type: daphnane

$C_{37}H_{48}O_{10}$ Mol. wt 652.79

Occurs in a number of *Gnidia* spp. (Thymelaeaceae).

Antileukaemic activity.

2451
Gniditrin

Skeletal type: daphnane

$C_{37}H_{42}O_{10}$ Mol. wt 646.74

Found in *Gnidia lamprantha* (Thymelaeaceae).

Antileukaemic activity.

2452
Grayanotoxin I; G-I; Acetylandromedol; Andromedotoxin; Asebotoxin; Rhodotoxin

Skeletal type: grayanotoxin

$C_{22}H_{36}O_7$ Mol. wt 412.53

Some thirty grayanotoxins and related diterpenes have been detected in the leaves of *Kalmia*, *Rhododendron* and *Leucothoe* spp., e.g., *L. grayana*, and other Ericaceae. Also, it is found in the honey of bees which collect nectar from *Rhododendron* spp.

Toxic (LD_{50} intraperitoneally in mice 1.31 mg kg^{-1}). Grayanotoxins exhibit hypotensive activity.

2453
Hallactone A

Skeletal type: modified totarane

$C_{19}H_{22}O_6$ Mol. wt 346.39

Occurs in *Podocarpus hallii* (Podocarpaceae).

Toxic to housefly larvae.

2454
Hallactone B

Skeletal type: modified totarane

$C_{20}H_{24}O_9S$ Mol. wt 440.48

Occurs in *Podocarpus hallii* (Podocarpaceae).

Toxic to housefly larvae.

2455
Huratoxin

Skeletal type: daphnane

$C_{34}H_{48}O_8$ Mol. wt 584.76

Found in the latex of the huru tree, *Hura crepitans* (Euphorbiaceae).

Potent piscicide.

2456
Ineketone

Skeletal type: pimarane

$C_{20}H_{30}O_3$ Mol. wt 318.46

Found in the husks of the rice cultivar *Oryza sativa* cv. "koshihikari" (Gramineae).

Plant seed germination inhibitory activity.

2457
Inflexin

Skeletal type: kaurane

$C_{24}H_{32}O_7$ Mol. wt 432.52

Found in the leaves of *Isodon inflexus* (Labiatae).

Cytotoxic and insect antifeeding activities.

2458
Ingenol

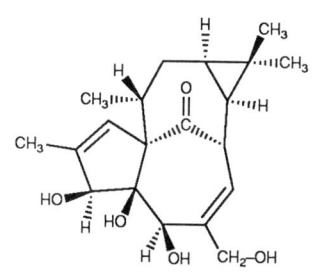

Skeletal type: ingenane

$C_{20}H_{28}O_5$ Mol. wt 348.44

Esters of ingenol occur in some *Euphorbia* spp. (Euphorbiaceae).

Ingenol itself is biologically inactive, but many of its diesters are skin irritants and cocarcinogens, whereas some others show antileukaemic activity.

2459
Ingenol 3,20-dibenzoate

Skeletal type: ingenane

$C_{34}H_{36}O_7$ Mol. wt 556.66

Occur in *Euphorbia esula* (Euphorbiaceae).

Antileukaemic activity *in vivo*.

2460
Inumakilactone A glycoside

$C_{24}H_{30}O_{13}$ Mol. wt 526.50

Found in the seeds of *Podocarpus macrophylla* (Podocarpaceae).

Plant growth inhibitory activity.

2461
Isodomedin

Skeletal type: kaurane

$C_{22}H_{32}O_6$ Mol. wt 392.50

Found in the leaves of *Isodon shikokianus* var. *intermedius* (Labiatae).

Cytotoxic and antibacterial activities. It shows antifeeding activity against the larvae of *Spodoptera exempta*.

2462
Isodonal

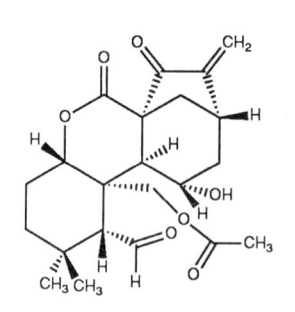

Skeletal type: seco-kaurane

$C_{22}H_{28}O_7$ Mol. wt 404.47

Found in the leaves and stems of *Isodon japonicus* (Labiatae).

Highly specific antibacterial activity against *Bacillus subtilis*.

2463
Jatrophatrione

Skeletal type: jatrophane B

$C_{20}H_{26}O_3$ Mol. wt 314.43

Found in *Jatropha macrorhiza* (Euphorbiaceae).

Antitumour activity.

2464
Jatrophone

Skeletal type: jatrophane A

$C_{20}H_{24}O_3$ Mol. wt 312.41

Occurs in *Jatropha gossypiifolia* (Euphorbiaceae).

Antitumour activity.

2465
Kansuinine B

Skeletal type: jatrophane A

$C_{38}H_{42}O_{14}$ Mol. wt 722.75

Found in *Euphorbia kansui* (Euphorbiaceae).

Highly toxic.

2466
Lathyrol

Skeletal type: lathyrane

$C_{20}H_{30}O_4$ Mol. wt 334.46

Occurs in the seed oil of *Euphorbia lathyrus* (Euphorbiaceae).

Irritant and cocarcinogen.

2467
Mancinellin

Skeletal type: tigliane

$C_{36}H_{52}O_8$ Mol. wt 612.81

Occurs in *Hippomane mancinella* (Euphorbiaceae).

Highly toxic. It is an irritant and a cocarcinogen.

2468
Marrubiin

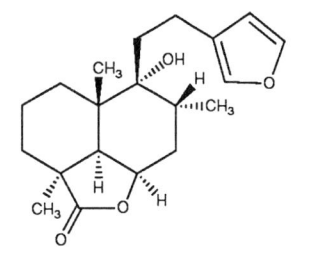

Skeletal type: labdane

$C_{20}H_{28}O_4$ Mol. wt 332.44

Found in hoarhound, *Marrubium vulgare* (Labiatae); it is generated as an artefact from premarrubiin (q.v.) during the extraction process.

Bitter principle. *M. vulgare* is used in folk medicine as an expectorant.

2469
Mascaroside

$C_{26}H_{36}O_{11}$ Mol. wt 524.57

Found in the beans of *Coffea vianneyi* (Rubiaceae).

Extremely bitter taste.

2470
Mezerein

Skeletal type: daphnane

$C_{38}H_{38}O_{10}$ Mol. wt 654.72

Occurs in mezereon, *Daphne mezereum* (Thymelaeaceae).

Antitumour activity.

2471
Montanin

Skeletal type: daphnane

$C_{32}H_{48}O_{8}$ Mol. wt 560.74

Found in the roots of *Baliospermum montanum* (Euphorbiaceae).

Cytotoxic and antileukaemic activities.

2472
Montanin A

$C_{19}H_{20}O_{4}$ Mol. wt 312.37

Found in *Teucrium montanum* (Labiatae).

2473
Montanol

Skeletal type: oxepane

$C_{21}H_{36}O_4$ Mol. wt 352.52

Found in the leaves of *Montanoa tomentosa* (Compositae).

Contragestational activity.

2474
Nagilactone C

Skeletal type: modified totarane

$C_{19}H_{22}O_7$ Mol. wt 362.39

Occurs in a number of *Podocarpus* spp., e.g., *P. nagi* (Podocarpaceae).

Antitumour activity. It also strongly inhibits the expansion and mitosis of plant cells.

2475
Neocembrene; Neocembrene A; Cembrene A

$C_{20}H_{32}$ Mol. wt 272.48

Occurs in several spp. belonging to the Pinaceae, e.g., *Picea obovata*.

Neocembrene is a trail pheromone of the termite *Nasutitermis exitiosus*.

2476
Isopimaric acid

Skeletal type: pimarane

$C_{20}H_{30}O_2$ Mol. wt 302.46

Found in the fruits of *Juniperus rigida* and *J. conferta* (Cupressaceae).

Antibacterial activity.

2477
12-O-2Z,4E-Octadienoyl-4-deoxyphorbol 13-acetate

Skeletal type: tigliane

$C_{30}H_{40}O_7$ Mol. wt 512.65

Occurs in *Euphorbia tirucalli* (Euphorbiaceae).

Strong irritant.

2478
Palmarin

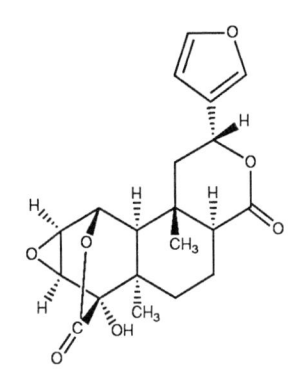

Skeletal type: clerodane

$C_{20}H_{22}O_7$ Mol. wt 374.40

Occurs in the roots of *Jateorhiza palmata* (=*J. columba*) (Menispermaceae).

Bitter principle. The roots of *J. palmata* are the source of Radix Columba, which was formerly used as a bitter tonic.

2479
12-*O*-Palmitoyl-16-hydroxyphorbol 13-acetate

Skeletal type: tigliane

$C_{38}H_{60}O_9$ Mol. wt 660.90

Occurs in the fruit of *Aleurites fordii* (Euphorbiaceae).

Piscicidal activity against kellie fish, *Orizias leptipes*.

2480
Phorbol

Skeletal type: tigliane

$C_{20}H_{28}O_6$ Mol. wt 364.44

Found in the seed oil of *Croton tiglium* (Euphorbiaceae).

Phorbol itself is biologically inactive, but many of its diesters, which also occur in *Croton* oil, are irritants and potent cocarcinogens, whereas some others show antileukaemic activity.

2481
Phorbol 12-tiglate 13-decanoate

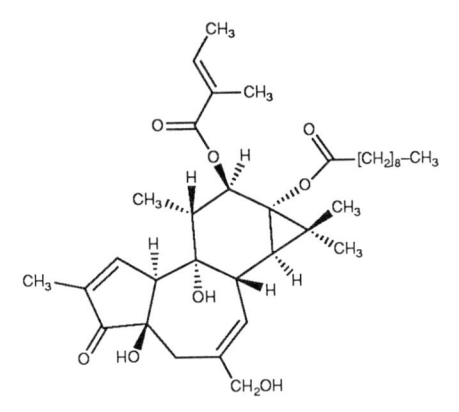

Skeletal type: tigliane

$C_{35}H_{52}O_8$ Mol. wt 600.80

Found in the seed oil of *Croton tiglium* (Euphorbiaceae).

Antileukaemic activity.

2482
Phytol

Acyclic diterpenoid

$C_{20}H_{40}O$ Mol. wt 296.54

Occurs in all chlorophyll-containing plants as an ester of the propionic side chain of chlorophyll.

Used for the preparation of vitamins E and K_1.

2483
(+)-Pimaric acid; Dextropimaric acid; *d*-Pimaric acid; α-Pimaric acid

Skeletal type: pimarane

$C_{20}H_{30}O_2$ Mol. wt 302.46

Occurs in the oleo-resins of *Pinus* spp. (Pinaceae), e.g., in French colophony and galipot, and American rosin.

2484
Pimelea factor P_2; Daphnopsis factor R_1

Skeletal type: daphnane

$C_{37}H_{50}O_9$ Mol. wt 638.81

Found in the leaves and roots of several *Pimelea*, *Daphnopsis* and *Synaptolepis* spp. (Thymelaeaceae).

Strong irritant.

2485
Plaunol B; Diterpenoid CS-B

Skeletal type: clerodane

$C_{20}H_{20}O_6$ Mol. wt 356.38

Found in *Croton sublyratus* (Euphorbiaceae), together with several closely related diterpenoids.

Inhibits ulcer formation in mice.

2486
Plaunol D

$C_{20}H_{22}O_7$ Mol. wt 374.40

Found in several *Croton* spp., e.g., *C. sublyratus* and *C. columnaris* (Euphorbiaceae).

Antipeptic ulcer activity.

2487
Pleuromutilin; Drosophilin B

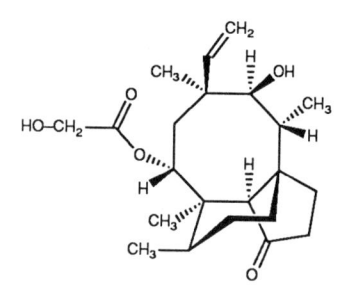

$C_{22}H_{34}O_5$ Mol. wt 378.51

Metabolite of the fungus *Pleurotus mutilis* (Basidiomycetes).

Antibiotic activity against Gram-positive bacteria.

2488
Podolactone B

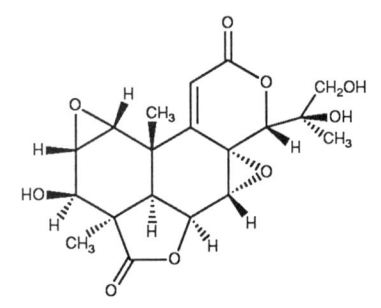

Skeletal type: modified totarane

$C_{19}H_{22}O_9$ Mol. wt 394.39

Found in *Podocarpus neriifolius* (Podocarpaceae).

Strongly inhibits the expansion and mitosis of plant cells.

2489
Podolide

Skeletal type: modified totarane

$C_{19}H_{22}O_5$ Mol. wt 330.39

Found in *Podocarpus gracilior* (Podocarpaceae).

Antitumour activity.

2490
Ponalactone A

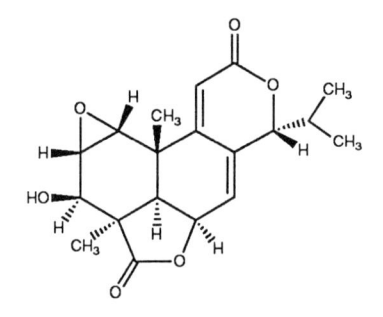

Skeletal type: modified totarane

$C_{19}H_{22}O_6$ Mol. wt 346.39

Occurs in the roots of *Podocarpus nakaii* (Podocarpaceae).

Ponalactone A and its β-glucoside strongly inhibit the expansion and mitosis of plant cells.

2491
Portulal

$C_{20}H_{32}O_4$ Mol. wt 336.48

Occurs in *Portulaca grandiflora* (Portulacaceae).

Plant growth regulatory activity.

2492
Premarrubiin

Skeletal type: labdane

$C_{20}H_{28}O_4$ Mol. wt 332.44

Occurs in hoarhound, *Marrubium vulgare* (Labiatae). It is easily converted into marrubiin (q.v.).

The aerial parts of *M. vulgare* are made into an expectorant in folk medicine.

2493
Resiniferonol; Euphorbia factor RL_{20}

Skeletal type: daphnane

$C_{20}H_{28}O_6$ Mol. wt 364.44

Occurs in *Euphorbia resinifera* (Euphorbiaceae).

Resiniferonol itself is biologically inactive, but many of its diesters are skin irritants, and show cocarcinogenic activity.

2494
Resiniferatoxin; Euphorbia factor RL_9

Skeletal type: daphnane

$C_{37}H_{40}O_9$ Mol. wt 628.73

Occurs in several *Euphorbia* spp., e.g., *E. resinifera* and *E. poisonii* (Euphorbiaceae).

One of the most irritant diterpenoid esters isolated from the Euphorbiaceae or Thymelaeaceae (diester of resiniferonol, q.v.).

2495
Rhodojaponin IV

Skeletal type: grayanotoxin

$C_{24}H_{38}O_8$ Mol. wt 454.57

Found in *Rhododendron japonicum* (Ericaceae).

Highly toxic. Chemically, it is closely related to grayanotoxin I (q.v.).

2496
Shikodonin

$C_{20}H_{26}O_6$ Mol. wt 362.43

Found in *Isodon shikokianus* var. *intermedius* (Labiatae).

Antitumour and insecticidal activities.

2497
Simplexin

Skeletal type: daphnane

$C_{30}H_{44}O_8$ Mol. wt 532.68

Found in *Pimelea simplex* (Thymelaeaceae).

Causes St George disease in cattle: a cardiopulmonary syndrome resulting from chronic intake of the above plant.

2498
Spruceanol

$C_{20}H_{28}O_2$ Mol. wt 300.44

Found in the roots and root bark of *Micrandra spruceana* (=*Cunuria spruceana*) (Euphorbiaceae).

Cytotoxic activity.

2499
Stevioside

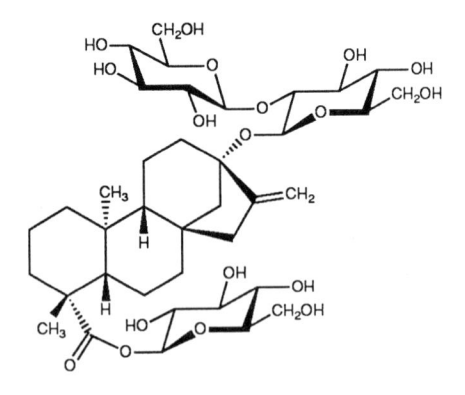

Skeletal type: kaurane

$C_{38}H_{60}O_{18}$ Mol. wt 804.90

Found in the leaves of *Stevia rebaudiana* (Compositae).

This glycoside is about 300 times sweeter than sucrose, and is used as a sweetening agent in some countries. The aglycone, steviol, shows a gibberellin-like activity.

2500
Strigol

$C_{19}H_{22}O_6$ Mol. wt 346.39

Exuded from the roots of cotton, *Gossypium hirsutum* (Malvaceae).

Potent seed germination stimulant of the plant parasite *Striga lutea*.

2501
Synaptolepis factor K₁

Skeletal type: daphnane

$C_{36}H_{54}O_8$ Mol. wt 614.83

Occurs in the leaves, roots and stems of *Synaptolepis* spp. (Thymelaeaceae).

Highly irritant.

2502
Taxodione

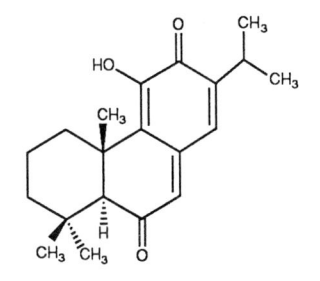

Skeletal type: abietane

$C_{20}H_{26}O_3$ Mol. wt 314.43

Occurs in *Taxodium distichum* (Taxodiaceae).

Antitumour activity

2503
Taxodone

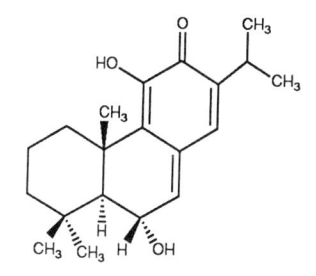

Skeletal type: abietane

$C_{20}H_{28}O_3$ Mol. wt 316.44

Occurs in *Taxodium distichum* (Taxodiaceae).

Antitumour activity.

2504
Terpenoid EA-I

Skeletal type: daphnane

$C_{30}H_{40}O_8$ Mol. wt 528.65

Found in the leaves and latex of *Excoecaria agallocha* (Euphorbiaceae).

Strong piscicidal activity. It is used in New Caledonia for catching fish.

2505
12-Tetradecanoylphorbol 13-acetate;
Croton factor A₁

Skeletal type: tigliane

$C_{36}H_{56}O_8$ Mol. wt 616.84

Occurs in the seed oil of *Croton tiglium* (Euphorbiaceae).

Irritant and cocarcinogen.

2506
Tinyatoxin

Skeletal type: daphnane

$C_{36}H_{38}O_8$ Mol. wt 598.70

Found in the latex of "tinya", *Euphorbia poisonii* (Euphorbiaceae).

Toxin, producing severe skin inflammations.

2507
Tripdiolide

$C_{20}H_{24}O_7$ Mol. wt 376.41

Found in *Tripterygium wilfordii* (Celastraceae).

Antitumour activity.

2508
Triptolide

$C_{20}H_{24}O_6$ Mol. wt 360.41

Found in *Tripterygium wilfordii* (Celastraceae).

Antitumour activity.

2509
Welensalifactor F$_1$

Skeletal type: tigliane

$C_{38}H_{60}O_8$ Mol. wt 644.90

Occurs together with many related diterpenoid diesters in the leaves of welensali, *Croton flavens* (Euphorbiaceae), which is commonly used in Curaçao to prepare a beverage (Welensali tea).

Welensalifactor F$_1$ and related substances have cocarcinogenic activity, and are one of the causes of a high incidence of cancer of the oesophagus in Curaçao.

2510
Zoapatanol

Skeletal type: oxepane

$C_{20}H_{34}O_4$ Mol. wt 338.49

Occurs in the leaves of the Mexican zoapatle tree, *Montanoa tomentosa* (Compositae).

Possesses contragestational activity, and has been used by Mexican women to prepare a tea to induce menses and labour.

Chapter 53
Triterpenoid saponins

The saponins are a group of plant glycosides in which water-soluble sugars are attached to a lipophilic steroid (C_{27}) or triterpenoid (C_{30}) moiety. This hydrophobic/hydrophilic asymmetry means that they have the ability to lower surface tension, and are soap-like. They form foams in aqueous solution and cause haemolysis of blood erythrocytes. They are toxic to cold-blooded animals but not generally to warm-blooded animals. They are widely distributed in the flowering plants, having been identified in more than 100 plant families.

This section contains the structures of triterpenoid (C_{30}) saponins, while the steroid saponins (q.v.) follow in the next section. Triterpenoid aglycones (or sapogenins) are typically pentacyclic (as in aescin, **2512**, a saponin of the horse chestnut) and have a 4,4-dimethyl (or similar) substitution in the left-hand ring. Sugars are commonly attached at the 3-hydroxyl position, as in abrusoside A (**2511**). Disubstitution of the triterpenoid molecule with sugars is a regular feature. Arvenoside A (**2513**), the saponin of *Calendula*, for example, has a disaccharide at the 3-hydroxyl group and a glucose at the 28-carboxylic acid group. The sugar moiety at the 3-hydroxyl may be linear, as in bayogenin 3-cellobioside (**2520**), or branched, as in avenacin A-1 (**2516**) from oats. Saponins with four or five monosaccharide units are common, as in cyclamin (**2526**), while there are occasionally six (phaseoloside D, **2555**) or as many as ten (saponoside D, **2562**). Ester substitution is another structural feature, as in aescin (**2512**), which has acetyl and tiglyl attachments, and very rarely it is possible for a dibasic organic acid to provide a bridge between the sugars substituted at positions 3 and 28 (see tubeimoside, **2569**).

Saponins are frequently present in a given plant as a complex mixture of closely related glycosides based on one or more triterpenoid sapogenins. In lucerne, for example, medicagenic acid 3-triglucoside (**2547**) is only one of at least 11 saponins present. Saponins are usually found in all parts of the plant, with some concentration in root, foliage or seed. They are regularly present in many pasture legume species, and are responsible in part for the condition of bloat in ruminant animals.

Saponins are toxic to insects and molluscs, and some of the most useful natural agents for controlling schistosomiasis snails are saponin in nature (Mott, 1987). Antifungal activity is present in compounds such as α- and β-hederin (**2542**, **2543**) and saponins may aid plants such as ivy and oats to resist microbial infection. From the medicinal viewpoint, the most widely used saponins are the ginsenosides (e.g., **2532**) of ginseng, which are reputed to prolong human life and aid survival to stress (Duke, 1989; Shibata *et al.*, 1985). Triterpenoid saponins generally taste bitter; a few can taste very sweet (glycyrrhizin, **2536**) while gymnemic acid I (**2537**) is a taste-modifier, making sweet foods tasteless.

Saponins are often detected in plant extracts through their haemolytic activity or on the basis of their foam-producing properties. They can be quantified after acid hydrolysis by a colour reaction using either the Liebermann-Burchard (sulfuric acid and acetic anhydride) or the Carr-Price (antimony chloride) reagent. They are separated by high performance liquid chromatography, and then characterized by proton and carbon-13 NMR spectral analysis.

References

Chemistry and Distribution

Hostettman, K. and Marston, A. (1995). *Saponins*. Cambridge University Press.

Mahato, S.B., Sarker, S.K. and Poddar, G. (1988). Triterpenoid saponins. *Phytochemistry*, **27**, 3037–3067.

Mahato, S.B. and Nandy, A.K. (1991). Triterpenoid saponins discovered between 1987 and 1989. *Phytochemistry*, **30,** 1357–1390.

Tschesche, R. and Wulff, G. (1973). Chemie und Biologie der Saponine. In *Progress in the Chemistry of Organic Natural Products*, Volume 30 (Herz, W., Grisebach, H. and Kirby, G.W., eds), pp. 461–666. Vienna: Springer-Verlag.

Biological Activity

Applebaum, S.W. and Birk, Y. (1979). Saponins. In *Herbivores: Their Interaction with Secondary Plant Metabolites* (Rosenthal, G.A. and Janzen, D.H., eds), pp. 539–566. New York: Academic Press.

Duke, J.A. (1989). *Ginseng: A Concise Handbook*. Algonac, Michigan: Reference Publications.

Fenwick, D.E. and Oakenfull, D. (1983). Saponin content of food plants and some prepared foods. *Journal of the Science of Food and Agriculture*, **34**, 186–191.

Hiller, K. (1987). New results on the structure and biological activity of triterpenoid saponins. In *Biologically Active Natural Products* (Hostettmann, K. and Lea, P.J., eds), pp. 167–184. Oxford: Clarendon Press.

Mott, K.E. (ed.) (1987). *Plant Molluscicides*. Chichester: Wiley.

Price, K.R., Johnson, I.T. and Fenwick, G.R. (1987). The chemistry and biological significance of saponins in foods and feedingstuffs. *CRC Critical Reviews in Food Science and Nutrition*, **26**, 27–135.

Shibata, S., Tanaka, O., Shoji, J. and Saito, H. (1985). Chemistry and pharmacology of *Panax*. *Economic and Medicinal Plant Research*, **1**, 218–284.

2511
Abrusoside A

$C_{36}H_{54}O_{10}$ Mol. wt 646.83

Occurs in the leaves of *Abrus fruticulosus* and *A. precatorius* (Leguminosae).

Sweet taste: about 50 times sweeter than sucrose.

2512
Aescin

$C_{54}H_{84}O_{23}$ Mol. wt 1101.24

Found in the horse chestnut, *Aesculus hippocastanum* (Hippocastanaceae). Aescin is also the collective name for the complex mixture of more than 30 different saponins present in the horse chestnut, the main component (about 20%) being the one shown above.

Strong haemolytic activity. It inhibits fungal growth, and shows anti-inflammatory, antiexudative and cancerostatic activities.

2513
Arvenoside A

$C_{48}H_{78}O_{18}$ Mol. wt 943.15

Found in the aerial parts of *Calendula arvensis* (Compositae).

Anti-inflammatory activity.

2514
Astragaloside III

$C_{41}H_{68}O_{14}$ Mol. wt 785.00

Occurs in the roots of *Astragalus membranaceus* (Leguminosae).

Inhibits the formation of lipid peroxidase induced by intraperitoneal administration of adriamycin (15 mg kg^{-1}) in rats.

2515
Astrasieversianin XVI

$C_{47}H_{78}O_{18}$ Mol. wt 931.14

Occurs in *Astragalus sieversianus* (Leguminosae).

Significant antihypertensive activity.

2516
Avenacin A-1

$C_{55}H_{83}NO_{21}$ Mol. wt 1094.28

Found in oats, *Avena sativa* (Gramineae).

Haemolytic activity. It strongly inhibits pathogenic fungi in plants, and probably is responsible for the resistance of oats against "take-all" disease.

2517
Avenacin B-2

$C_{54}H_{80}O_{20}$ Mol. wt 1049.23

Found in oats, *Avena sativa* (Gramineae).

Haemolytic activity. It inhibits pathogenic fungi in plants.

2518
Azukisaponin III

$C_{42}H_{66}O_{15}$ Mol. wt 810.99

Found in azuki beans, *Vigna angularis* (Leguminosae).

2519
Barringtogenol C; Theasapogenol B; Jegosapogenol

$C_{30}H_{50}O_5$ Mol. wt 490.73

Obtained after mild acid hydrolysis of the saponin fractions of *Barringtonia acutangula* (Lecythidaceae), *Aesculus hippocastanum* (Hippocastanaceae), *Styrax japonica* (Styracaceae), and of the leaves and seeds of tea, *Thea sinensis* (Theaceae).

Tea seed saponins are used medicinally.

2520
Bayogenin 3-O-cellobioside

$C_{42}H_{68}O_{15}$ Mol. wt 813.01

Found in *Phytolacca dodecandra* (Phytolaccaceae).

Molluscicidal (LD_{100} for the snail *Biomphalaria glabrata* 0.012 mg ml^{-1}).

2521
Camellidin I

$C_{55}H_{86}O_{25}$ Mol. wt 1147.29

Found in the leaves of *Camellia japonica* (Theaceae).

Antifungal activity.

2522
Camellidin II

$C_{53}H_{84}O_{24}$ Mol. wt 1105.26

Found in the leaves of *Camellia japonica* (Theaceae).

Antifungal activity. It is an antifeedant for larvae of the butterfly *Eurema hecabe mandarina*.

2523
Cimicifugoside

$C_{37}H_{54}O_{11}$ Mol. wt 674.84

Found in some *Cimicifuga* spp., bugbane or bugwort (Ranunculaceae).

Potent and selective inhibitor of nucleoside transport at the plasma membrane site of mammalian cells.

2524
Colubrin

$C_{48}H_{76}O_{18}$ Mol. wt 941.14

Occurs in *Colubrina asiatica* (Rhamnaceae).

Sedative effects in mice.

2525
Colubrinoside

$C_{50}H_{78}O_{19}$ Mol. wt 983.17

Occurs in *Colubrina asiatica* (Rhamnaceae).

Sedative effects in mice.

2526
Cyclamin

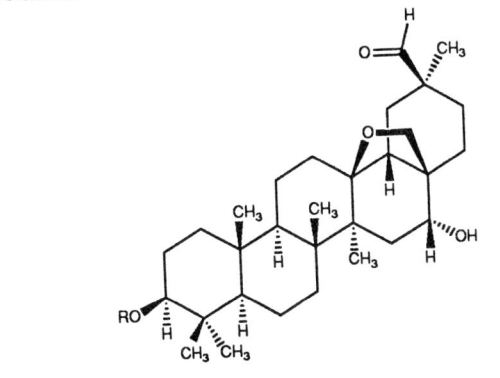

$C_{58}H_{94}O_{27}$ Mol. wt 1223.39

Found in the corms of *Cyclamen europaeum* (Primulaceae).

Very high haemolytic index. It shows cancerostatic and antibiotic activities.

2527
Cyclofoetoside B

$C_{47}H_{80}O_{18}$ Mol. wt 933.14

Occurs in the aerial parts of *Thalictrum foetidum* (Ranunculaceae).

Antitumour activity in rats (about 50 mg kg^{-1} intraperitoneally).

2528
Desglucomusennin; Deglucomusennin; Musennin A

$C_{45}H_{72}O_{16}$ Mol. wt 869.07

Occurs in *Albizia anthelmintica* (Leguminosae).

The saponin fraction of *A. anthelmintica*, of which desglucomusennin is one of the major components, has anthelmintic activity.

2529
Dianoside A

$C_{42}H_{66}O_{15}$ Mol. wt 810.99

Found in *Dianthus superbus* var. *longicalicynus* (Caryophyllaceae).

Analgesic activity.

2530
Echinocystic acid

$C_{30}H_{48}O_4$ Mol. wt 472.71

Obtained by acid hydrolysis of the saponin fractions of *Schefflera capitata* (Araliaceae), *Albizia anthelmintica* (Leguminosae), and many other plants.

The saponin scheffleroside from *S. capitata*, containing echinocystic acid, fucose, galactose and glucuronic acid, has spermicidal activity. The anthelmintic saponins musennin (q.v.) and desglucomusennin (q.v.) also contain echinocystic acid.

2531
Flaccidin B

$C_{41}H_{64}O_{12}$ Mol. wt 748.96

Occurs in *Anemone flaccida* (Ranunculaceae).

Shows inhibitory effects on reverse transcriptase from RNA tumour virus.

2532
Ginsenoside Re

$C_{48}H_{82}O_{18}$ Mol. wt 947.18

Found in the ginseng spp. *Panax pseudoginseng* spp. *himalicus* and *P. ginseng* (Araliaceae).

Extracts of the roots of *Panax* spp., which contain ginsenosides, have been used in China since prehistoric times as stimulant tonics. Ginsenoside Re causes analgesic effects in mice in doses of 10 mg kg^{-1}.

2533
Ginsenoside Rf

$C_{42}H_{72}O_{14}$ Mol. wt 801.04

Found in the roots of ginseng, *Panax ginseng* (Araliaceae).

Tumour-inhibitory activity.

2534
Ginsenoside Rg$_1$

$C_{42}H_{72}O_{14}$ Mol. wt 801.04

Found in the roots of ginseng, *Panax ginseng* (Araliaceae).

Tumour-inhibitory activity; it has been applied successfully in the treatment of gastric cancer.

2535
Glycyrrhetinic acid; Glycyrrhetic acid; α-Glycyrrhetinic acid; Glycyrrhetin

$C_{30}H_{46}O_4$ Mol. wt 470.70

Sapogenin from the dried roots and rhizomes of liquorice, *Glycyrrhiza glabra* (Leguminosae).

Anti-inflammatory and anti-ulcerogenic activities. It also causes increased water retention.

2536
Glycyrrhizin; Glycyrrhizic acid; Glycyrrhinic acid; Glycyrrhizinic acid

$C_{42}H_{62}O_{16}$ Mol. wt 822.96

Saponin from the roots and rhizomes of *Glycyrrhiza glabra* (Leguminosae).

Induces interferons. It has an intensely sweet taste (50 times sweeter than sucrose). It shows antiulcerogenic activity, and is an expectorant. Extracts of *G. glabra* are used in Egypt as a substitute for cortisone, and in general it is used as a pharmaceutical aid.

2537
Gymnemic acid I

$C_{43}H_{66}O_{14}$ Mol. wt 807.00

Occurs in the leaves of *Gymnema sylvestre* (Asclepiadaceae).

Gymnemic acid I and related saponins occurring in *G. sylvestre* are flavour modifiers.

2538
Gypenoside XXV

$C_{47}H_{78}O_{18}$ Mol. wt 931.14

Occurs in *Gynostemma pentaphyllum* (Cucurbitaceae).

Tumour-inhibitory activity.

2539
Gypsogenin; Albasapogenin; Astrantiagenin D; Githagenin; Gypsophilasapogenin

$C_{30}H_{46}O_4$ Mol. wt 470.70

Gypsogenin glycosides occur in many plants, e.g., in *Gypsophylla* spp. (Caryophyllaceae), *Swartzia* spp. (Leguminosae), and in the corn cockle, *Agrostemma githago* (Caryophyllaceae).

Certain gypsogenin glycosides, such as githagin which occurs in the corn cockle, may be toxic to mammals.

2540
Gypsogenin 3-*O*-rhamnosylglucuronide

$C_{42}H_{64}O_{14}$ Mol. wt 792.97

Occurs in *Swartzia madagascariensis* (Leguminosae).

Molluscicidal activity.

2541
Hederagenin 3-*O*-arabinoside

$C_{35}H_{56}O_8$ Mol. wt 604.83

Found in *Hedera helix* (Araliaceae) and *Lonicera nigra* (Caprifoliaceae).

Molluscicidal activity (LD_{100} after 24 h for the snail *Biomphalaria glabrata* 3 mg l^{-1}).

2542
α-Hederin; Helixin; Kalopanaxsaponin A; Kitzuta saponin K_6; Sapindoside A

$C_{41}H_{66}O_{12}$ Mol. wt 750.98

Found in the leaves of ivy, *Hedera helix* (Araliaceae).

Strong haemolytic activity, and also cancerostatic and antifungal activities.

2543
β-Hederin; Eleutheroside K

$C_{41}H_{66}O_{11}$ Mol. wt 734.98

Found in the leaves of ivy, *Hedera helix* (Araliaceae).

Strong haemolytic activity.

2544
Helianthoside A

$C_{53}H_{86}O_{21}$ Mol. wt 1059.27

This glycoside of echinocystic acid (q.v.) occurs in the petals of the sunflower, *Helianthus annuus* (Compositae).

Haemolytic activity. It has a bitter taste.

2545
Ilexolide A

$C_{35}H_{54}O_7$ Mol. wt 586.82

Found in the roots of *Ilex pubescens* (Aquifoliaceae).

Cardiac activity.

2546
Lemmatoxin; Oleanoglycotoxin B

$C_{48}H_{78}O_{18}$ Mol. wt 943.15

Found in the berries of *Phytolacca dodecandra* (Phytolaccaceae), known as "endod" in Ethiopia.

Molluscicidal activity (LD_{90} for the snail *Biomphalaria glabrata* 1.5 mg l^{-1}). It is active against human spermatozoa at concentrations of 50 mg l^{-1}.

2547
Medicagenic acid 3-*O*-triglucoside;
Medicogenic acid 3-*O*-triglucoside

$C_{48}H_{76}O_{21}$ Mol. wt 989.14

At least eleven glycosides of medicagenic acid are present in the roots (and also some in the tops) of alfalfa or lucerne, *Medicago sativa* (Leguminosae), which is used as animal feeding stuff.

The saponins in alfalfa have antinutritional properties. Medicagenic acid glycosides show a wide range of biological activities, e.g., haemolytic activity; they inhibit the growth of *Trichoderma viride*, and retard the development of wheat seedlings.

2548
Musennin

$C_{51}H_{82}O_{21}$ Mol. wt 1031.22

Occurs in *Albizia anthelmintica* (Leguminosae).

The saponin mixture from *Albizia*, of which musennin is one of the main components, has anthelmintic activity.

2549
Notoginsenoside R1

$C_{47}H_{80}O_{18}$ Mol. wt 933.16

Occurs in *Panax notoginseng* (Araliaceae).

The total saponin fraction of *P. notoginseng* has anti-inflammatory activity.

2550
Olaxoside

$C_{48}H_{76}O_{18}$ Mol. wt 941.14

Found in various *Olax* spp. (Olacaceae).

Anti-inflammatory activity.

2551
Oleanoglycotoxin-A

$C_{48}H_{78}O_{18}$ Mol. wt 943.15

Occurs in the berries of *Phytolacca dodecandra* (Phytolaccaceae).

Molluscicidal activity (LD_{100} for the snail *Biomphalaria glabrata* 6 mg l^{-1}). Active against human spermatozoa at a concentration of 50 mg l^{-1}.

2552
Oleanolic acid 3-*O*-glucuronide

$C_{36}H_{56}O_9$ Mol. wt 632.84

Found in *Lonicera nigra* (Caprifoliaceae) and in sugar beet, *Beta vulgaris* (Chenopodiaceae).

Molluscicidal activity (LD_{100} for the snail *Biomphalaria glabrata* 2 mg l^{-1}).

2553
Pfaffic acid

$C_{29}H_{44}O_3$ Mol. wt 440.67

Found in the roots of Brazil ginseng, *Pfaffia paniculata* (Amaranthaceae).

Antitumour activity (inhibits the growth of cultivated tumour cells in concentrations of 4–6 µg ml^{-1}). The roots of *P. paniculata* have been used in folk medicine for antidiabetic purposes.

2554
Pfaffoside A

$C_{40}H_{60}O_{13}$ Mol. wt 748.92

Found in the roots of Brazil ginseng, *Pfaffia paniculata* (Amaranthaceae).

Antitumour activity (inhibits the growth of cultivated tumour cells in concentrations of 30–50 µg ml^{-1}).

2555
Phaseoloside D

$C_{65}H_{104}O_{31}$ Mol. wt 1381.55

Occurs in various species and varieties of beans, *Phaseolus*, e.g., kidney and haricot beans, *Phaseolus vulgaris* (Leguminosae).

2556
Phytolaccoside B

$C_{36}H_{56}O_{11}$ Mol. wt 664.84

Found in pokeweed, *Phytolacca americana* (Phytolaccaceae).

Parasiticidal and molluscicidal activities.

2557
Primulasaponin

$C_{54}H_{99}O_{22}$ Mol. wt 1100.39

Found in the roots and rhizomes of *Primula elatior* (Primulaceae).

Antibiotic and cancerostatic activities.

2558
Propapyriogenin A₂

$C_{30}H_{44}O_5$ Mol. wt 484.68

Obtained by acid hydrolysis of the saponin fractions of *Tetrapanax papyriferum* (Araliaceae) and *Bupleurum rotundifolium* (Umbelliferae).

Anti-inflammatory activity.

2559
Quillaic acid

$C_{30}H_{46}O_5$ Mol. wt 486.70

Obtained from the acid hydrolysis of the saponin fraction from the bark of *Quillaja saponaria* (Rosaceae).

The saponin-rich bark of *Q. saponaria* (9–10%) is used in shampoo liquids, foam producers and in the mineral water industry. Some quillaic acid saponins, such as quillajoside, are strongly haemolytic. Quillajasaponin is used in the film industry.

2560
Saikosaponin A

$C_{42}H_{68}O_{13}$ Mol. wt 781.01

Occurs in the roots of *Bupleurum falcatum* and in *B. chinense* (Umbelliferae).

The saponin fractions of *B. falcatum* and *B. chinense* have anti-inflammatory activity.

2561
Saikosaponin BK1

$C_{48}H_{78}O_{17}$ Mol. wt 927.15

Occurs in *Bupleurum kummingense* (Umbelliferae).

Antileukaemic activity *in vitro*.

2562
Saponoside D

$C_{87}H_{138}O_{49}$ Mol. wt 1968.06

The main saponin of the roots of soapwort, *Saponaria officinalis* (Caryophyllaceae).

Strong surface active properties, forming stable foams when shaken with water.

2563
Senegin II

$C_{70}H_{104}O_{32}$ Mol. wt 1457.60

Found in the roots of *Polygala senega* (Polygalaceae).

A mixture of senegin II and related saponins from *P. senega* is cancerostatic and has expectorant activity.

2564
Soyasaponin A₁

$C_{59}H_{96}O_{29}$ Mol. wt 1269.42

Found in soybeans, the seeds of *Glycine max* (Leguminosae).

The soyasaponin mixture from *G. max* has inhibitory effects on tumour cells, and antithrombic activity.

2565
Soyasaponin I

$C_{48}H_{78}O_{18}$ Mol. wt 943.15

Present in many food plants, especially pulses, e.g., soybeans, *Glycine max*, beans, *Phaseolus* spp., lentils, *Lens culinaris*, and chickpeas, *Cicer arietinum* (all Leguminosae).

Inhibits the formation of lipid peroxidase induced by intraperitoneal administration of adriamycin (15 mg kg^{-1}) in rats.

2566
Spinasaponin A

$C_{42}H_{66}O_{14}$ Mol. wt 794.99

Found in spinach, *Spinacia oleracea* (Chenopodiaceae), together with spinasaponin B, which contains the same sugars but the sapogenol hederagenin instead of oleanolic acid.

2567
Thalicoside A

$C_{42}H_{70}O_{14}$ Mol. wt 799.02

Occurs in *Thalictrum minus* (Ranunculaceae).
Antitumour activity.

2568
Theasaponin

$C_{59}H_{92}O_{27}$ Mol. wt 1233.39

Found in the seeds of the tea plant, *Thea sinensis* (Theaceae).

Strongly haemolytic, and with anti-exudative activity.

2569
Tubeimoside I

$C_{63}H_{98}O_{29}$ Mol. wt 1319.48

Occurs in the bulbs of *Bolbostemma paniculatum* (Cucurbitaceae).

Moderate antitumour activity *in vivo*.

2570
Ursolic acid; Malol; Malolic acid; Micromerol; Prunol; Urson

$C_{30}H_{48}O_3$ Mol. wt 456.71

Found in numerous plants, e.g., *Prunella vulgaris* (Labiatae), many spp. of Ericaceae, e.g., cranberry, *Vaccinium macrocarpon*, and bearberry, *Arctostaphylos uva-ursi*, and in the protective wax-like coating of the skins of apples, *Malus*, pears, *Pyrus*, and fruits of other spp. of Rosaceae.

Cytotoxic and antileukaemic activities.

2571
Virgaureasaponin I; Bellissaponin 2

$C_{59}H_{96}O_{27}$ Mol. wt 1237.42

Occurs in *Solidago virgaurea* and *Bellis perennis* (Compositae).

Inhibits the growth of yeasts, e.g., *Candida albicans*.

2572
Yiamoloside B

$C_{43}H_{68}O_{15}$ Mol. wt 825.02

Found in *Phytolacca octandra* (Phytolaccaceae).

Fungistatic activity.

2573
Ziziphin

$C_{51}H_{80}O_{18}$ Mol. wt 981.20

Found in *Ziziphus jujuba* (Rhamnaceae).

Flavour modifier.

Chapter 54
Steroid saponins

These saponins are similar in structure, biogenesis and biological activity to the triterpenoid saponins (q.v.). They are based on a C_{27} spirostan structure, as in convallagenin A (**2579**) and, unlike the triterpenoid saponins, they lack any substituents at C-4 in the left-hand ring. They are also related to the cardiac glycosides (q.v.). Indeed, the first natural source of steroid saponins was the foxglove, *Digitalis purpurea*, where digitogenin (**2585**) digitonin (**2586**) co-occur with the cardiac glycosides digitalin and digitoxin. Steroid saponins are regularly present in plants of the Agavaceae, Dioscoreaceae, Liliaceae and Scrophulariaceae (especially *Digitalis*).

They resemble the triterpenoid saponins in the number and location of the sugar components. The glycosides can be difficult to purify, so that some aglycones (or sapogenins) such as diosgenin (**2588**) and yamogenin (**2609**) are better known than the corresponding saponins. Diosgenin, isolated from *Dioscorea* roots, is well known as a starting material for the partial synthesis of sex hormones used in contraceptive pills, e.g., progesterone.

Steroid saponins are used as detergents, as foaming agents in fire extinguishers, and as fish poisons. Fish are dazed or killed by the saponin but are not rendered inedible, since the saponins are not toxic to humans, presumably because when taken orally they are not absorbed by the gut. The lack of human toxicity is apparent in the facts that these saponins can occur in a number of food plants (e.g., asparagus), and that extracts of sarsaparilla, containing sarsaparilloside (**2601**), are used as food flavouring.

References

Applebaum, S.W. and Birk, Y. (1979). Saponins. In *Herbivores. Their Interaction with Secondary Plant Metabolites* (Rosenthal, G.A. and Janzen, D.H., eds), pp. 539–566. New York: Academic Press.

Hostettman, K. and Marston, A. (1995). *Saponins*. Cambridge University Press.

Mahato, S.B., Ganguly, A.N. and Sahu, N.P. (1982). Steroid saponins. *Phytochemistry*, **21**, 959–978

Tschesche, R. and Wulff, G. (1973). Chemie und Biologie der Saponine. *Fortschritte in der Chemie organischer Naturstoffe*, **30**, 461–666.

2574
Agavoside A

$C_{33}H_{52}O_9$ Mol. wt 592.78

Occurs in *Agave americana* (Agavaceae).

Agave saponins are active against epidermoid carcinoma of the nasopharynx in tissue culture, so they may have antileukaemic activity. The sapogenin of agavoside A, hecogenin (q.v.), is used in the preparation of steroid hormones.

2575
Asparagoside A

$C_{33}H_{54}O_8$ Mol. wt 578.80

Occurs in *Asparagus officinalis* (Liliaceae).

The sapogenin of asparagoside A, sarsasapogenin (q.v.), is used in the manufacture of compounds of the pregnane series.

2576
Avenacoside A

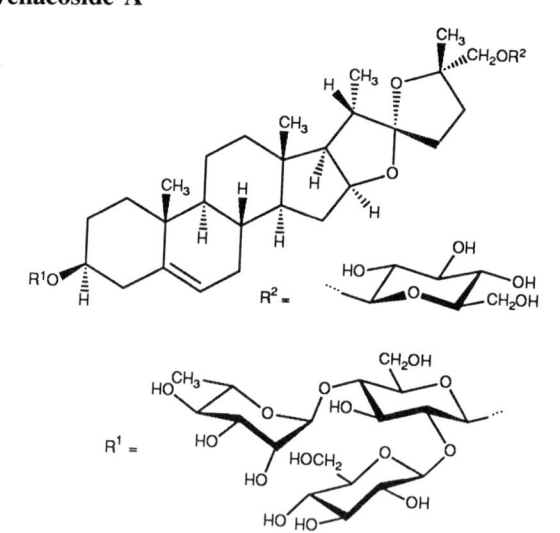

$C_{51}H_{82}O_{23}$ Mol. wt 1063.22

Found in the aerial parts of oats, *Avena sativa* (Gramineae).

During isolation from oats, avenacoside A is transformed to 26-desglucoavenacoside A, because of the presence of a β-glucosidase (avenacosidase) possessing high specificity for the removal of the C-26 glucose moiety. 26-Desglucoavenacoside A has antifungal activity.

2577
Avenacoside B

$C_{57}H_{92}O_{28}$ Mol. wt 1225.36

Found in the aerial parts of oats, *Avena sativa* (Gramineae).

During isolation from oats, avenacoside B is transformed into 26-desglucoavenacoside B, which has antifungal activity (see avenacoside A).

2578
Capsicoside A

$C_{63}H_{106}O_{35}$ Mol. wt 1423.54

Occurs in sweet pepper or paprika, *Capsicum annuum* (Solanaceae).

2579
Convallagenin A

$C_{27}H_{44}O_5$ Mol. wt 448.65

Obtained by acid hydrolysis of convallasaponin A (q.v.), which is present in the flowers of *Convallaria keisuki* (Liliaceae).

2580
Convallamarogenin

$C_{27}H_{42}O_4$ Mol. wt 430.63

Obtained by acid hydrolysis of convallamaroside (q.v.), which is present in the roots of *Convallaria majalis* (Liliaceae).

2581
Convallamaroside; Convallamarin

$C_{57}H_{94}O_{27}$ Mol. wt 1211.38

Occurs in the roots of *Convallaria majalis* (Liliaceae).

Strong haemolytic activity.

2582
Convallasaponin A

$C_{32}H_{52}O_9$ Mol. wt 580.77

Occurs in the flowers of *Convallaria keisuki* (Liliaceae).

2583
Deltonin

$C_{45}H_{72}O_{17}$ Mol. wt 885.07

Found in *Dioscorea deltoidea* (Dioscoreaceae).

Acid hydrolysis of this monodesmoside gives the sapogenin diosgenin (q.v.), which is used in the partial synthesis of hormones having a steroid structure.

2584
Deltoside

$C_{51}H_{84}O_{23}$ Mol. wt 1065.23

Found in *Dioscorea deltoidea* (Dioscoreaceae).

Acid hydrolysis of this bisdesmoside gives diosgenin (q.v.), which is used in the partial synthesis of hormones having a steroid structure.

2585
Digitogenin

$C_{27}H_{44}O_5$ Mol. wt 448.65

Obtained by acid hydrolysis of digitonin (q.v.), which is present in the seeds of the foxglove, *Digitalis purpurea* (Scrophulariaceae).

2586
Digitonin

$C_{56}H_{92}O_{29}$ Mol. wt 1229.35

Found in the seeds of the foxglove, *Digitalis purpurea* (Scrophulariaceae).

Antibiotic and fungistatic activities. It is used in the determination of cholesterol in blood plasma, bile and tissue.

2587
Dioscin

$C_{45}H_{72}O_{16}$ Mol. wt 869.07

Occurs as protodioscin (q.v.) in *Dioscorea* spp. (Dioscoreaceae), *Costus* spp. (Zingiberaceae), *Paris* and *Trillium* spp. (Liliaceae), *Trigonella* spp. (Leguminosae), etc.

The sapogenin, diosgenin (q.v.), is used in the partial synthesis of hormones having a steroid structure.

2588
Diosgenin; Nitogenin

$C_{27}H_{42}O_3$ Mol. wt 414.63

Obtained by acid hydrolysis of many different saponins, e.g., dioscin, deltonin and gracillin (q.v.), from *Dioscorea* spp. (Dioscoreaceae), *Costus* spp. (Zingiberaceae), *Paris* and *Trillium* spp. (Liliaceae) and *Trigonella* spp. (Leguminosae).

Used in the partial synthesis of hormones having a steroid structure, e.g., pregnenolone and progesterone.

2589
Gitogenin

$C_{27}H_{44}O_4$ Mol. wt 432.65

Obtained by acid hydrolysis of gitonin (q.v.) and some other saponins from the foxgloves *Digitalis purpurea* and *D. lanata* (Scrophulariaceae).

2590
Gitonin

$C_{50}H_{82}O_{23}$ Mol. wt 1051.21

Found in the leaves of the foxglove *Digitalis purpurea* (Scrophulariaceae).

2591
Gracillin

$C_{45}H_{72}O_{17}$ Mol. wt 885.07

Occurs as protogracillin (q.v.) in some *Dioscorea* spp. (Dioscoreaceae), and in *Costus speciosus* (Zingiberaceae).

Strong haemolytic activity.

2592
Hecogenin

$C_{27}H_{42}O_4$ Mol. wt 430.63

Obtained by acid hydrolysis of the saponin fractions of *Agave* spp. (Agavaceae).

Used in the preparation of steroid hormones.

2593
Nuatigenin

$C_{27}H_{42}O_4$ Mol. wt 430.63

Obtained by acid hydrolysis of the avenacosides (q.v.) present in oats, *Avena sativa* (Gramineae).

For activity, see avenacosides A and B.

2594
Officinalisnin I

$C_{45}H_{76}O_{19}$ Mol. wt 921.10

Found in the roots of asparagus, *Asparagus officinalis* (Liliaceae).

Bitter taste; it causes the bitterness in asparagus storage roots.

2595
Osladin

$C_{45}H_{74}O_{17}$ Mol. wt 887.09

Occurs in the rhizomes of the fern *Polypodium vulgare* (Polypodiaceae).

Very sweet taste.

2596
Parillin

$C_{51}H_{84}O_{22}$ Mol. wt 1049.23

Found in the roots of sarsaparilla, *Smilax aristolochiaefolia* (Liliaceae).

Strong haemolytic activity. It also shows antibiotic and some cancerostatic activities.

2597
Protodioscin

$C_{51}H_{84}O_{22}$ Mol. wt 1049.23

Found in *Dioscorea gracillima* (Dioscoreaceae).

Acid hydrolysis of this bisdesmoside gives diosgenin (q.v.), which is used in the partial synthesis of steroid hormones.

2598
Protogracillin; Kikubasaponin

$C_{51}H_{84}O_{23}$ Mol. wt 1065.23

Found in the roots of *Dioscorea gracillima* and other *Dioscorea* spp. (Dioscoreaceae).

Acid hydrolysis of this bisdesmoside gives diosgenin (q.v.), which is used in the partial synthesis of steroid hormones.

2599
Ruscogenin; Ruscorectal; Rectolander; Flebopom

$C_{27}H_{42}O_4$ Mol. wt 430.63

Occurs in the rhizomes of *Ruscus aculeatus* (Liliaceae).

Used in the treatment of haemorrhoids. Ruscosides A and B, which are partly identified glycosides of ruscogenin, decrease the cholesterol content of blood, lipid deposition in the aorta, and liver arterial tension. They have antisclerotic and hypotensive activities.

2600
Ruscoside

$C_{50}H_{80}O_{23}$ Mol. wt 1049.19

Found in the rhizomes of *Ruscus aculeatus* (Liliaceae).

The total saponin fraction of *R. aculeatus* has anti-inflammatory activity. Acid hydrolysis of ruscoside yields ruscogenin (q.v.), which is used to treat haemorrhoids.

2601
Sarsaparilloside

$C_{57}H_{96}O_{28}$ Mol. wt 1229.40

Found in the roots and rhizomes of sarsaparilla, *Smilax aristolochiaefolia* (Liliaceae).

Acid hydrolysis gives sarsasapogenin (q.v.), which is used in the manufacture of compounds of the pregnane series. Extracts of sarsaparilla are used for flavouring, e.g., of confectionery.

2602
Sarsasapogenin; Parigenin

$C_{27}H_{44}O_3$ Mol. wt 416.65

Obtained by acid hydrolysis of many saponins from *Smilax* and *Asparagus* spp. (Liliaceae).

Used in the manufacture of compounds of the pregnane series.

2603
Sarsasapogenin 3-*O*-4^G-rhamnosylsophoroside

$C_{45}H_{74}O_{17}$ Mol. wt 887.09

Occurs in *Asparagus curillus* (Liliaceae).

Molluscicidal activity (LD_{100} for the snail *Biomphalaria glabrata* 20 mg l^{-1}).

2604
Smilagenin; Isosarsasapogenin

$C_{27}H_{44}O_3$ Mol. wt 416.65

Obtained from sarsaparilla, *Smilax ornata* (Liliaceae), by acid hydrolysis of the saponin fraction and acid isomerization of sarsasapogenin (q.v.).

Used in the manufacture of compounds of the pregnane series.

2605
Tigogenin

$C_{27}H_{44}O_3$ Mol. wt 416.65

Obtained by acid hydrolysis of saponins present in *Digitalis lanata* (Scrophulariaceae), and the sisal plant, *Agave sisalana* (Amaryllidaceae).

2606
Tigonin

$C_{56}H_{92}O_{27}$ Mol. wt 1197.35

Found in *Digitalis lanata* and *D. purpurea* (Scrophulariaceae), and in *Chlorogalum pomeridianum* (Liliaceae).

2607
Tokoronin

$C_{32}H_{52}O_9$ Mol. wt 580.77

Found in *Dioscorea tokoro* (Dioscoreaceae).

2608
Trigonelloside C; Asparasaponin I

$C_{51}H_{84}O_{22}$ Mol. wt 1049.23

Occurs in fenugreek, *Trigonella foenum-graecum* (Leguminosae).

2609
Yamogenin

$C_{27}H_{42}O_3$ Mol. wt 414.63

Obtained by acid hydrolysis of saponins present in *Trigonella foenum-graecum* (Leguminosae), *Asparagus officinalis* and *Smilax aspera* (Liliaceae).

2610
Yamogenin 3-*O*-neohesperidoside

$C_{39}H_{62}O_{12}$ Mol. wt 722.93

Occurs in *Asparagus plumosus* (Liliaceae).

Molluscicidal activity (LD_{100} for the snail *Biomphalaria glabrata* 25 mg l^{-1}).

Chapter 55
Cardenolides and bufadienolides

The cardenolides and bufadienolides are two related groups of C_{23} and C_{24} steroids, of triterpenoid origin, which are well known for their poisonous properties. The cardenolides are heart poisons which, when used in small doses, are extremely valuable clinically for controlling congestive heart failure. The cardenolides contain a tetracyclic steroidal ring system which is substituted in the 17-position by an α,β-unsaturated γ-lactone (butenolide) ring, as in adonitoxin (**2611**). There are two main series, differentiated by the stereochemistry of the ring junction between the two lower cyclohexane rings. The 5α series (*trans* A/B) occur principally in the Asclepiadaceae, while the 5β series (*cis* A/B) are found by contrast in the Apocynaceae and Scrophulariaceae. A typical 5α-cardenolide is aspecioside (**2616**) from *Asclepias speciosa*, while a typical 5β-cardenolide is the medically important digoxin (**2627**) from the foxglove *Digitalis purpurea*.

The cardenolides usually occur in plants as glycosides, with a sugar attachment commonly at the 3-hydroxyl group or, less frequently, at the 1-hydroxyl, 2-hydroxyl or 11-hydroxyl positions. Over twenty sugars have been isolated from the hydrolysis of cardiac glycosides, and most of these sugars (q.v.) are known naturally only in this association. A distinctly unusual series of cardiac glycosides is found particularly in plants of the milkweed family, where the sugars are attached via adjacent hydroxyl groups at C-2 and C-3 to produce a cyclic bridge system, as is apparent in asclepin (**2615**) and calactin (**2618**). Such bridge glycosides are distinguished from the more usual *O*-glycosides by their marked resistance to acid hydrolysis.

Cardiac glycosides occur principally in the closely related families of the Apocynaceae and Asclepiadaceae. In the milkweed genus *Asclepias*, for example, the compounds are secreted in the latex, and virtually every species that has been examined contains these toxins. Cardiac glycosides are also found in a number of other dicotyledonous families, such as the Cruciferae, Moraceae and Scrophulariaceae (notably in *Digitalis*), and in several monocotyledonous groups, but especially in Liliaceae (*Urginea*).

Bufadienolides can be clearly distinguished from cardenolides by the fact that the lactone ring substituting at C-17 in the steroid nucleus is six-membered rather than five-membered (see, e.g., hellebrin, **2633**). They occur like the cardenolides as *O*-glycosides (e.g., scilliroside, **2643**) but can also be found in the free state (e.g., hellebrigenin 3-acetate, **2632**). Bufadienolides were first characterized as natural products from toad venom, and it is only more recently that they have been recognized as plant poisons as well.

The poisonous nature of cardenolide-containing plants is well documented in the early phytochemical literature, since these plant extracts were used by natives as arrow poisons. In recent times, many livestock deaths have been attributed to the consumption of such plants; as little as one ounce of dried *Asclepias labriformis* leaf is fatal to a sheep. Cardenolides are remarkably nontoxic to Lepidoptera, and some species of Danaid butterfly take in these compounds from dietary sources and store them in the adult body and wings to protect them from birds. The cardenolides have an emetic effect on those birds which predate on these butterflies.

A number of analytical procedures, including reaction with 2,2',4,4'-tetranitrodiphenyl to give blue coloured derivatives, have been developed for detecting and quantifying the cardiac glycosides of plant tissues. They are generally isolated and characterized by techniques which are used with other plant steroids.

The medicinal use of the cardenolide-containing foxglove plant for slowing down the heart muscle dates from Withering in 1785, and certain cardiac glycosides (e.g., digoxin, **2627**) have been in continuous use for treating heart conditions ever since.

References

Detection and Analysis

Seiber, J.N., Lee, S.M. and Benson, J.M. (1983). Cardiac glycosides in species of *Asclepias*. In *Handbook of Natural Toxins*, Volume 1 (Keeler, R.F. and Tu, A.T., eds), pp. 43–83. New York: Marcel Dekker.

Biological and Ecological Properties

Malcolm, S.B. (1991). Cardenolide-mediated interactions between plants and herbivores. In Herbivores, their Interactions with Secondary Plant Metabolites, Volume 1 (Rosenthal, G.A. and Berenbaum, M.R., eds), second edition, pp. 251–296. San Diego: Academic Press.

Seiber, J.N., Lee, S.M. and Benson, J.M. (1984). Chemical characteristics and ecological significance of cardenolides in *Asclepias* species. In *Isopentenoids in Plants* (Nes, W.D., Fuller, G. and Tsai, L.S., eds), pp. 563–588. New York: Marcel Dekker.

2611
Adonitoxin; Adonitoxigenin 3-O-α-L-rhamnoside

Skeletal type: cardenolide

$C_{29}H_{42}O_{10}$ Mol. wt 550.66

Found in *Adonis vernalis* (Ranunculaceae).

Toxic to mammals and other vertebrates.

2612
Adynerin; Adynerigenin 3-O-β-D-diginoside

Skeletal type: cardenolide

$C_{30}H_{44}O_7$ Mol. wt 516.68

Found in the leaves of *Nerium oleander* (Apocynaceae).

Toxic to mammals and other vertebrates.

2613
α-Antiarin; Antiarigenin 3-O-β-D-antiaroside

Skeletal type: cardenolide

$C_{29}H_{42}O_{11}$ Mol. wt 566.66

Occurs in the latex of the upas tree, *Antiaris toxicaria*, and of other *Antiaris* spp. (Moraceae).

Very toxic to vertebrates (LD_{50} intravenously in cats 0.116 mg kg^{-1}). The latex of *A. toxicaria* has been used as an arrow poison.

2614
Antioside; Antiogenin 3-O-α-L-rhamnoside

Skeletal type: cardenolide

$C_{29}H_{44}O_{10}$ Mol. wt 552.67

Occurs in the latex of the upas tree, *Antiaris toxicaria*, and of *A. welwitschii* and *A. decipiens* (Moraceae).

Very toxic to vertebrates. The latex of *A. toxicaria* has been used as an arrow poison.

2615
Asclepin

Skeletal type: cyclic bridged cardiac glycoside

$C_{31}H_{42}O_{10}$ Mol. wt 574.68

Found in the latex of some *Asclepias* spp. (Asclepiadaceae).

Very toxic to vertebrates (LD_{50} intravenously in cats 0.236 mg kg^{-1}).

2616
Aspecioside

Skeletal type: cardenolide

$C_{29}H_{42}O_{10}$ Mol. wt 550.66

Found in the latex of *Asclepias speciosa* and *A. syriaca* (Asclepiadaceae). Also it is found in Monarch butterflies, the larvae of which were reared on plants of these *Asclepias* spp.

Toxic to vertebrates.

2617
Bipindoside; Bipindogenin 3-*O*-β-D-digitaloside

Skeletal type: cardenolide

$C_{30}H_{46}O_{10}$ Mol. wt 566.70

Occurs in *Strophanthus sarmentosus* var. *senegambiae* and *S. thollonii* (Apocynaceae).

Toxic to vertebrates.

2618
Calactin

Skeletal type: cyclic bridged cardiac glycoside

$C_{29}H_{40}O_9$ Mol. wt 532.64

Found in the latex of *Calotropis procera* and *Asclepias curassavica* (Asclepiadaceae). Sequestered when reared on plants containing this cardenolide by the grasshopper *Pekilocerus bufonius*, and by the Monarch butterfly *Danaus plexippus*, as a defence mechanism.

Very toxic to vertebrates (LD_{50} intravenously in cats 0.12 mg kg^{-1}).

2619
Calotropin

Skeletal type: cyclic bridged cardiac glycoside

$C_{29}H_{40}O_9$ Mol. wt 532.64

Found in *Calotropis procera* (Asclepiadaceae). Sequestered by certain insects reared on *Calotropis* as a defence mechanism (see calactin).

Very toxic to vertebrates (LD_{50} intraperitoneally in male Swiss Webster mice 9.8 mg kg^{-1}; LD_{50} intravenously in cats 0.11 mg kg^{-1}).

2620
Cerberoside; Digitoxigenin 3-*O*-gentiobiosylthevetoside; Thevanil; Thevetin B

Skeletal type: cardenolide

$C_{42}H_{66}O_{18}$ Mol. wt 858.99

Found in *Cerbera odollam* and *Thevetia neriifolia* (Apocynaceae).

Toxic to vertebrates. It is used as a cardiotonic.

2621
Convallatoxin; Strophanthidin 3-*O*-α-L-rhamnoside; Convallaton; Corglykon; Korglykon

Skeletal type: cardenolide

$C_{29}H_{42}O_{10}$ Mol. wt 550.66

The major cardiac glycoside from the flowers and leaves of lily of the valley, *Convallaria majalis* (Liliaceae), found also in star of Bethlehem, *Ornithogalum umbellatum* (Liliaceae) and in *Antiaris toxicaria* (Moraceae).

Very toxic to vertebrates (minimum lethal dose intravenously in frogs 0.3 mg kg^{-1}). It is used as a cardiotonic.

2622
Cymarin; Strophanthidin 3-*O*-β-D-cymaroside; *k*-Strophanthin-α

Skeletal type: cardenolide

$C_{30}H_{44}O_9$ Mol. wt 548.68

Found in *Strophanthus kombé* (Apocynaceae), *Adonis vernalis* (Ranunculaceae) and *Castilloa elastica* (Moraceae).

Very toxic to vertebrates (LD_{50} intravenously in cats 0.095 mg kg^{-1}). It is used as a cardiotonic.

2623
Decoside; Decogenin 3-*O*-α-L-oleandroside

Skeletal type: cardenolide

$C_{30}H_{42}O_9$ Mol. wt 546.66

Found in *Strophanthus divaricatus* (Apocynaceae).

Toxic to vertebrates.

2624
Diginatin; Diginatigenin 3-*O*-tridigitoxoside

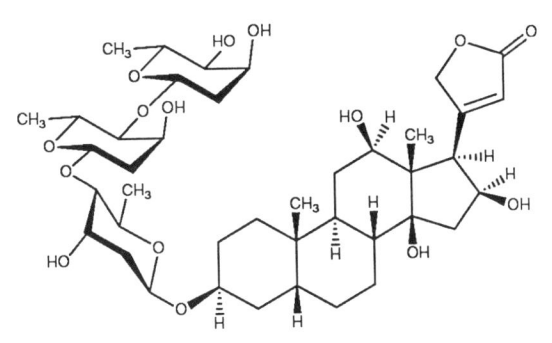

Skeletal type: cardenolide

$C_{41}H_{64}O_{15}$ Mol. wt 796.96

Found in *Digitalis lanata* (Scrophulariaceae).

Toxic to vertebrates. It is used as a cardiotonic.

2625
Digitalin; Gitoxigenin 3-*O*-glucosyldigitaloside; Digitalinum verum; Diginorgin

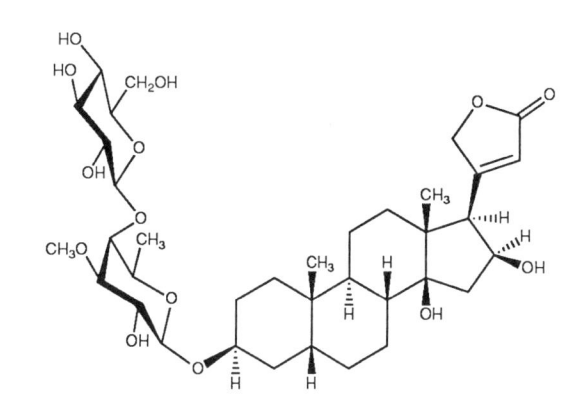

Skeletal type: cardenolide

$C_{36}H_{56}O_{14}$ Mol. wt 712.84

Found in the seeds of *Digitalis purpurea* (Scrophulariaceae) and in the roots of *Adenium honghel* (Apocynaceae).

Toxic to vertebrates. It is used as a cardiotonic.

2626
Digitoxin; Digitoxigenin 3-*O*-tridigitoxoside; Digitophyllin; Cardigin; Carditoxin; Lanatoxin

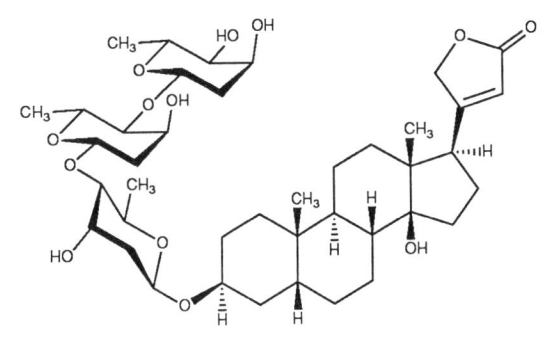

Skeletal type: cardenolide

$C_{41}H_{64}O_{13}$ Mol. wt 764.96

Found in *Digitalis purpurea* (Scrophulariaceae).

Toxic to vertebrates (LD_{50} orally in guinea-pigs 60 mg kg^{-1}; in cats 0.18 mg kg^{-1}; LD_{50} intravenously in cats 0.4 mg kg^{-1}). It is used as a cardiotonic.

2627
Digoxin; Digoxigenin 3-*O*-tridigitoxoside; Cordioxil; Davoxin; Digacin

Skeletal type: cardenolide

$C_{41}H_{64}O_{14}$ Mol. wt 780.96

Found in *Digitalis lanata* and *D. orientalis* (Scrophulariaceae).

Toxic to vertebrates. Derivatives of digoxin are used as a cardiotonic.

2628
Divaricoside; Sarmentogenin 3-*O*-α-L-oleandroside

Skeletal type: cardenolide

$C_{30}H_{46}O_8$ Mol. wt 534.70

Occurs in the seeds and other plant organs of *Strophanthus divaricatus* (Apocynaceae).

Toxic to vertebrates.

2629
Divostroside; Sarmentogenin 3-*O*-α-L-diginoside

Skeletal type: cardenolide

$C_{30}H_{46}O_8$ Mol. wt 534.70

Occurs in the seeds and other plant organs of *Strophanthus divaricatus* (Apocynaceae).

Toxic to vertebrates.

2630
Eriocarpin; Desglucosyrioside

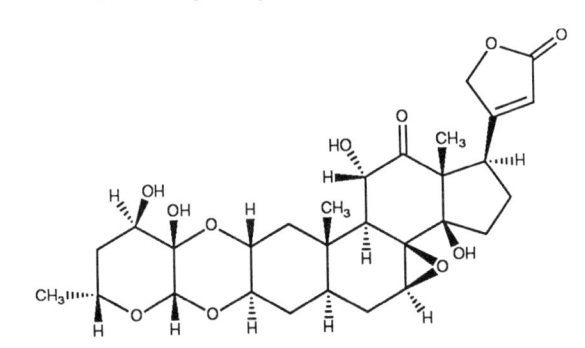

Skeletal type: cyclic bridged cardiac glycoside

$C_{29}H_{38}O_{11}$ Mol. wt 562.62

Found in *Asclepias eriocarpa* and *A. labriformis* (Asclepiadaceae).

Toxic to vertebrates (LD_{50} intraperitoneally in male Swiss Webster mice 6.5 mg kg^{-1}).

2631
Gitoxin; Gitoxigenin 3-O-tridigitoxoside; Anhydrogitalin; Bigitalin; Pseudodigitoxin

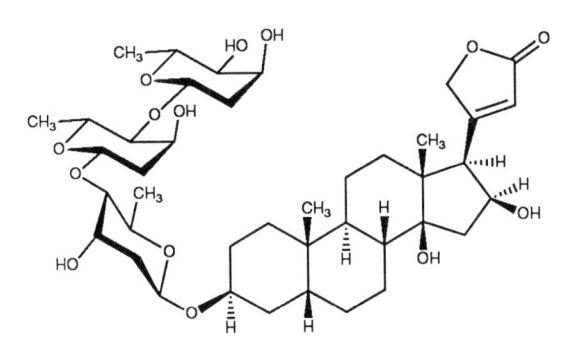

Skeletal type: cardenolide

$C_{41}H_{64}O_{14}$ Mol. wt 780.96

Found in *Digitalis purpurea* and *D. lanata* (Scrophulariaceae).

Toxic to vertebrates. It is used as a cardiotonic.

2632
Hellebrigenin 3-acetate

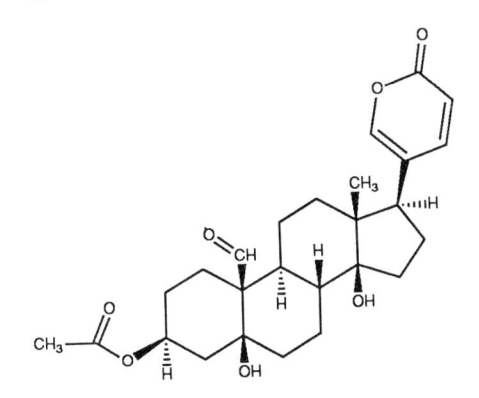

Skeletal type: nonglycosidic bufadienolide

$C_{26}H_{34}O_7$ Mol. wt 458.56

Occurs in *Bersama abyssinica* (Melianthaceae).

Hellebrigenin 3-acetate, and the 3,5-diacetate, exhibit tumour-inhibitory activity.

2633
Hellebrin; Hellebrigenin 3-O-glucosylrhamnoside

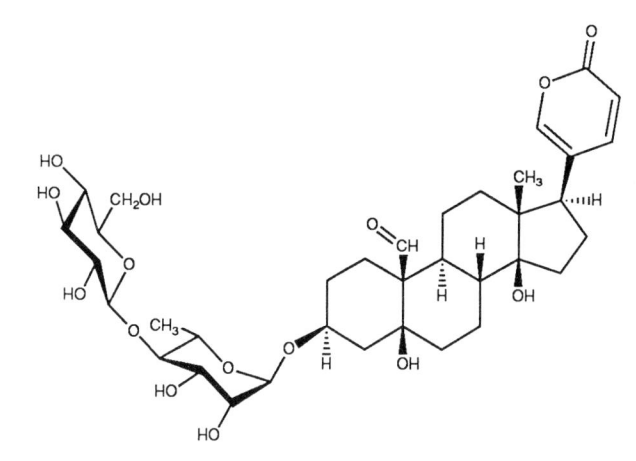

Skeletal type: bufadienolide

$C_{36}H_{52}O_{15}$ Mol. wt 724.81

Occurs in the rhizomes of *Helleborus niger* (Ranunculaceae). Other 3-glycosides of hellebrigenin occur in the bulb of *Urginea depressa* (Liliaceae).

Very toxic to vertebrates (LD_{50} orally in guinea-pigs 0.85 µmol kg^{-1}).

2634
Helveticoside; Strophanthidin 3-O-β-D-digitoxoside; Allioside A; Erysimin; Erysimotoxin

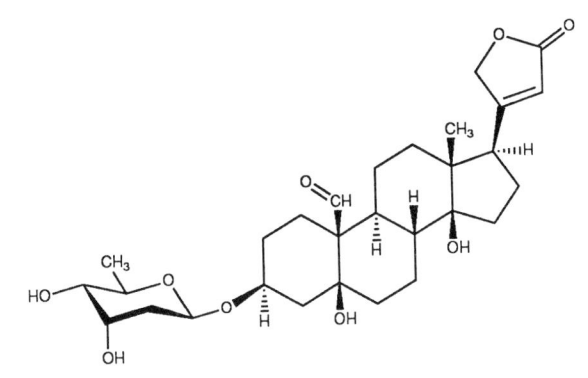

Skeletal type: cardenolide

$C_{29}H_{42}O_9$ Mol. wt 534.66

Occurs in many *Erysimum* spp., e.g., *E. helveticum*, *E. cheiranthoides*, and *E. crepidifolium* (Cruciferae).

Very toxic to mammals and other vertebrates (LD_{100} intravenously in cats 0.104 mg kg^{-1}).

2635
Labriformidin

Skeletal type: cyclic bridged cardiac glycoside

$C_{29}H_{36}O_{11}$ Mol. wt 560.61

Found in *Asclepias labriformis* and *A. eriocarpa* (Asclepiadaceae).

Very toxic to vertebrates (LD_{50} intraperitoneally in male Swiss Webster mice 3.1 mg kg^{-1}).

2636
Labriformin

Skeletal type: cyclic bridged cardiac glycoside

$C_{31}H_{39}NO_{10}S$ Mol. wt 617.73

Occurs in *Asclepias labriformis* and *A. eriocarpa* (Asclepiadaceae).

Very toxic to vertebrates (LD_{50} intraperitoneally in male Swiss Webster mice 9.2 mg kg^{-1}).

2637
Lokundjoside; Bipindogenin 3-*O*-α-L-rhamnoside

Skeletal type: cardenolide

$C_{29}H_{44}O_{10}$ Mol. wt 552.67

Found in some *Strophanthus* spp., e.g., *S. divaricatus* (Apocynaceae).

Toxic to vertebrates.

2638
Musaroside; Sarmutogenin 3-*O*-β-D-digitaloside

Skeletal type: cardenolide

$C_{30}H_{44}O_{10}$ Mol. wt 564.68

Found in *Strophanthus divaricatus* (Apocynaceae).

Toxic to vertebrates.

2639
Ouabain; Ouabagenin 3-O-L-rhamnoside; g-Strophanthin; Acocantherin; Gratibain; Gratus strophanthin; Astrobain; Purostrophan; Strophoperm; Strodival

Skeletal type: cardenolide

$C_{29}H_{44}O_{12}$ Mol. wt 584.67

Found in *Acokanthera ouabaio* and in the seeds of *Strophanthus gratus* (Apocynaceae).

Very toxic to vertebrates (LD$_{50}$ intravenously in rats 14 mg kg^{-1}; intravenously in cats 0.11 mg kg^{-1}; intraperitoneally in male Swiss Webster mice 6.5 mg kg^{-1}). It is used as a cardiotonic.

2640
Rhodexin A; Sarmentogenin 3-O-α-L-rhamnoside

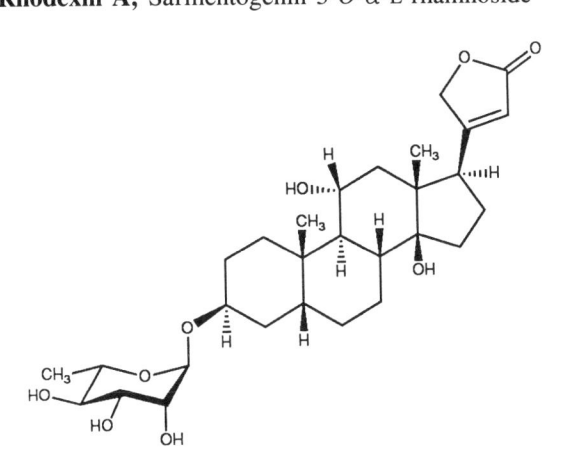

Skeletal type: cardenolide

$C_{29}H_{44}O_9$ Mol. wt 536.67

Occurs in *Rhodea japonica* (Liliaceae).

Toxic to vertebrates.

2641
Sarmentoloside; Sarmentologenin 3-O-(6-deoxy-α-L-taloside)

Skeletal type: cardenolide

$C_{29}H_{44}O_{11}$ Mol. wt 568.67

Found in *Strophanthus sarmentosus* var. *senegambiae* and in *S. divaricatus* (Apocynaceae).

Toxic to vertebrates.

2642
Scillaren A; Scillarenin 3-O-glucosylrhamnoside; Glucoproscillaridin A; Transvaalin

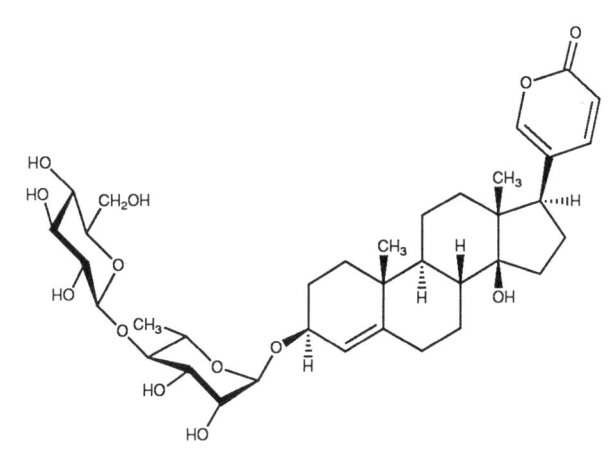

Skeletal type: bufadienolide

$C_{36}H_{52}O_{13}$ Mol. wt 692.81

Occurs in white squill, the white variety of *Urginea maritima* (=*U. scilla* or *Scilla maritima*) (Liliaceae).

Very toxic (LD$_{50}$ intravenously in cats 0.143 mg kg^{-1}). It has a very bitter taste, and is used as a cardiotonic.

2643
Scilliroside; Scillirosidin 3-O-β-D-glucoside

Skeletal type: bufadienolide

$C_{32}H_{44}O_{12}$ Mol. wt 620.70

Occurs in red squill, the red variety of *Urginea maritima* (Liliaceae).

Very toxic (LD_{50} orally in male rats 0.7 mg kg^{-1}), and the compound is an effective raticide.

2644
***k*-Strophanthoside;** Strophanthin; *k*-Strophanthin; Strophanthidin 3-diglucosylcymarose; Combetin; Eustrophinum

Skeletal type: cardenolide

$C_{41}H_{64}O_{19}$ Mol. wt 860.96

Found in *Strophanthus kombé* (Apocynaceae).

Very toxic to vertebrates (LD_{50} intravenously in rats 15 mg kg^{-1}). It is used as a cardiotonic. The seeds of *Strophanthus kombé* have been used in Africa for preparing arrow poisons.

2645
Thevetin A; Cannogenin 3-O-gentiobiosylthevetoside

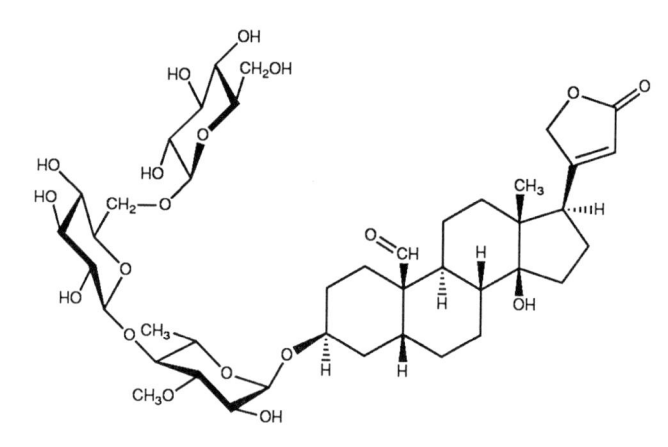

Skeletal type: cardenolide

$C_{42}H_{64}O_{19}$ Mol. wt 872.97

Found in *Thevetia neriifolia* (Apocynaceae).

Very toxic to vertebrates.

2646
Uscharidin

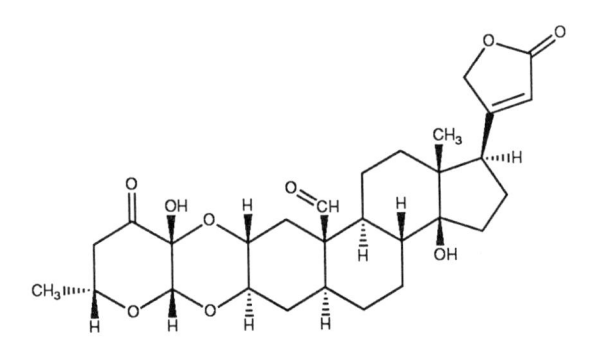

Skeletal type: cyclic bridged cardiac glycoside

$C_{29}H_{38}O_{9}$ Mol. wt 530.62

Found in *Calotropis procera* (Asclepiadaceae).

Toxic to vertebrates (LD_{50} intraperitoneally in male Swiss Webster mice 11.8 mg kg^{-1}; LD_{50} intravenously in cats 1.4 mg kg^{-1}).

2647
Vernadigin; Adonitoxigenin 3-*O*-β-D-diginoside

Skeletal type: cardenolide

$C_{30}H_{44}O_{10}$ Mol. wt 564.68

Occurs in *Adonis vernalis* (Ranunculaceae).

Toxic to vertebrates.

Chapter 56
Phytosterols

Sterols are structurally derived from the perhydrocyclopenta[*a*]phenanthrene ring system, as can be seen in cholesterol (**2655**). At one time, sterols were mainly considered to be animal products, but an increasing number of such compounds have been detected in plant tissues. Indeed, three phytosterols are ubiquitous in higher plants: sitosterol (**2677**), stigmasterol (**2679**) and campesterol (**2654**). They are all essential components of plant membranes. They occur both free and as the simple 3-glucosides. A less common phytosterol is α-spinasterol (**2678**), an isomer of stigmasterol, which is found in spinach and lucerne. Certain sterols are confined to lower plants; one example is ergosterol (**2659**) found in yeast and other fungi. Yet other sterols, such as fucosterol (**2660**), are mainly found in brown algae but may also appear occasionally in higher plants (e.g., in the coconut).

Phytosterols are structurally distinct from animal sterols in having an extra methyl or ethyl substituent in the side chain (compare sitosterol and cholesterol). The discovery of certain animal sterols in plant tissues is therefore very intriguing. The human female sex hormone, oestrone (**2670**), for example, has been found in trace amounts in date palm seed, while the male hormone, testosterone (**2680**), is present in the pollen of Scots pine. A large number of insect moulting hormones have also been uncovered in plants. These are known as phytoecdysones, and some 100 structures have been described (Camps, 1991). Ecdysone (**2658**), the moulting hormone of many insects, itself occurs in plants, but many of the other structures, e.g., podecdysone B (**2671**), are only recorded in the plant kingdom. They appear to represent a novel defence system of plants for protection against insect predation, since they can have a damaging effect on insect feeders.

Phytosterols such as sitosterol are formed in plants from squalene (q.v.), via intermediates such as cycloartenol (**2657**) and 24-methylenecycloartanol (**2666**) which are themselves present in many plants. Sitosterol (**2677**) and several similar derivatives are recognized as essential components of plant cell membranes. They also play an important role in plant cell growth. Phytosterols can be detected in plant extracts using thin layer chromatography followed by spraying with antimony chloride in chloroform. Oestrone can be detected by its characteristic orange colour after spraying with sulfuric acid. These sterols are further characterized and quantified by gas chromatography of their trimethylsilyl ethers.

References

Camps, F. (1991). Plant ecdysteroids and their interaction with insects. In *Ecological Chemistry and Biochemistry of Plant Terpenoids* (Harborne, J.B. and Tomas-Barberan, F.A., eds), pp. 331–376. Oxford: Clarendon Press.

Goad, L.J. (1967). Aspects of phytosterol biosynthesis. In *Terpenoids in Plants* (Pridham, J.B., ed.), pp. 159–190. London: Academic Press.

Goad, L.J. (1991). Inhibition of phytosterol biosynthesis and the consequences for plant growth. In *Ecological Chemistry and Biochemistry of Plant Terpenoids* (Harborne, J.B. and Tomas-Barberan, F.A., eds), pp. 210–229. Oxford: Clarendon Press.

Hill, R.A., Kirk, D.N., Makin, H.L.J. and Murphy, G.M. (1991). *Dictionary of Steroids*. London: Chapman and Hall.
Nes, W.D., Fuller, G. and Tsai, L.-S. (eds) (1984). *Isopentenoids in Plants: Biochemistry and function*. New York: Marcel Dekker.

2648
Ajugalactone

$C_{29}H_{40}O_8$ Mol. wt 516.64

Found in *Ajuga chamaepitys* and *A. reptans* (Labiatae).

Phytoecdysteroid: insect moulting hormone activity. It shows anti-ponasterone A activity.

2649
Ajugasterone C

$C_{27}H_{44}O_7$ Mol. wt 480.65

Found in the root bark of *Vitex madiensis* (Verbenaceae), and in *Rhaponticum carthamoides* (Compositae).

Phytoecdysteroid: insect moulting hormone activity.

2650
Androstenedione; Androtex

$C_{19}H_{26}O_2$ Mol. wt 286.42

Found in the pollen of Scots pine, *Pinus sylvestris* (Pinaceae).

Androgenic activity.

2651
Δ^5-Avenasterol; 29-Isofucosterol

$C_{29}H_{48}O$ Mol. wt 412.70

Occurs in oats, the seeds of *Avena sativa* (Gramineae), and in the marine green alga *Enteromorpha linza* (Chlorophyceae).

2652
Brassicasterol

$C_{28}H_{46}O$ Mol. wt 398.68

Found in the seed oil of rape, *Brassica napus* (Cruciferae).

Structural component of plant cell membranes.

2653
Brassinolide

$C_{28}H_{48}O_6$ Mol. wt 480.69

Found in the pollen of rape, *Brassica napus* (Cruciferae).

Plant growth promoting and regulating activity. This is one of a number of brassinosteroids which promote cell division and cell elongation in plants. It shows anti-ecdysteroid activity: delays imaginal moult when fed to the last instar of the cockroach *Periplaneta americana*.

2654
Campesterol

$C_{28}H_{48}O$ Mol. wt 400.69

Widespread occurrence in higher plants, e.g., in the oil of rape seed, *Brassica napus* (Cruciferae).

Structural component of plant cell membranes.

2655
Cholesterol

$C_{27}H_{46}O$ Mol. wt 386.67

Occurs in the pollen of the date palm, *Phoenix dactylifera* (Palmae), and in many marine red algae (Rhodophyceae). It is a common sterol of animal tissues.

Used as a pharmaceutical aid (emulsifying agent).

2656
Cyasterone

$C_{29}H_{44}O_8$ Mol. wt 520.67

Found in *Ajuga chamaepitys* and *A. reptans* (Labiatae).

Phytoecdysteroid: insect moulting hormone activity. It severely inhibits ovarian development in newly emerged female houseflies. Also, it stimulates protein anabolic activity in mouse liver.

2657
Cycloartenol

$C_{30}H_{50}O$ Mol. wt 426.73

Common in both higher and lower plants.

Sterol precursor in all photosynthetic plants.

2658
Ecdysone; α-Ecdysone

$C_{27}H_{44}O_6$ Mol. wt 464.65

Found in *Lychnis fulgens* (Caryophyllaceae) and in many ferns (Pteridophyta), e.g., *Blechnum minus*, *Polypodium vulgare* and *Pteridium aquilinum*. Ecdysone is a moulting hormone, widely present in insects and crustacea.

Phytoecdysteroid: insect moulting hormone activity. It severely inhibits ovarian development in newly emerged female houseflies.

2659
Ergosterol; Ergosterin

$C_{28}H_{44}O$ Mol. wt 396.66

Occurs in yeasts, e.g., *Saccharomyces cerevisiae* and *Gibberella fujikuroi*, and in many fungi, e.g., wood rotting Basidiomycetes.

When irradiated with UV light, ergosterol forms vitamin D_2 (ergocalciferol).

2660
Fucosterol

$C_{29}H_{48}O$ Mol. wt 412.70

Found in pollen, in coconut, *Cocos nucifera* (Palmae), and in many brown algae, e.g., *Fucus vesiculosus* (Phaeophyceae).

2661
20-Hydroxyecdysone; β-Ecdysone

C₂₇H₄₄O₇   Mol. wt 480.65

Common in higher plants; sometimes present in large quantities, e.g., 2% in flowers of *Serratula inermis* (Compositae) and 3.2% in mature stems of *Diploclisia glaucescens* (Menispermaceae). It is a moulting hormone, widely present in insects and crustacea.

Phytoecdysteroid: insect moulting hormone activity. It inhibits ovarian maturation and egg production in newly emerged female houseflies, and inhibits reproduction of the female confused flour beetle, *Tribolium confusum*. Also, it inhibits the proliferation of sarcoma 180. The acute toxicity of 20-hydroxyecdysone has been reported to be very low.

2662
Inokosterone

C₂₇H₄₄O₇   Mol. wt 480.65

Occurs in the roots of *Achyranthes fauriei* (Amaranthaceae).

Effective phytoecdysteroid. Insect moulting hormone activity. When added to the diet of the silkworm, *Bombyx mori*, it can regulate the length of instars and the extent of silk production. It is an effective growth regulator and abortifacient when fed to females of the tsetse fly, *Glossina morsitans morsitans*. It stimulates protein anabolic activity in mouse liver. Acute toxicity has been reported to be very low.

2663
Lanosterol; Kryptosterol

C₃₀H₅₀O   Mol. wt 426.73

Occurs in fungi, e.g., in *Saccharomyces cerevisiae* and *Gibberella fujikuroi*, and also in the Euphorbiaceae, e.g., the poinsettia, *Euphorbia pulcherrima*.

Direct sterol precursor in animals and nonphotosynthetic plants.

2664
Lophenol

C₂₈H₄₈O   Mol. wt 400.69

Occurs in *Lophocereus schottii* (Cactaceae).

2665
Makisterone B; Callinecdysone B

$C_{28}H_{46}O_7$ Mol. wt 494.66

Found in *Ajuga chamaepitys* (Labiatae) and in the seeds of *Diploclisia glaucescens* (Menispermaceae).

Phytoecdysteroid: acts as a moulting hormone in many species of insects, e.g., in the milkweed bug, *Oncopeltus fasciatus*. It is an effective growth regulator and abortifacient when fed to females of the tsetse fly, *Glossina morsitans morsitans*.

2666
24-Methylenecycloartanol

$C_{31}H_{52}O$ Mol. wt 440.76

Common in plants; often accompanies cycloartenol (q.v.).

Biosynthetic intermediate of other plant phytosterols.

2667
Miroestrol

$C_{20}H_{24}O_6$ Mol. wt 360.41

Occurs in the roots of *Pueraria mirifica* (Leguminosae).

Oestrogenic activity.

2668
Oestradiol-17β; Estradiol; β-Estradiol

$C_{18}H_{24}O_2$ Mol. wt 272.39

Occurs in the seeds of the French bean, *Phaseolus vulgaris* (Leguminosae).

Oestrogenic activity.

2669
Oestriol; Estriol

$C_{18}H_{24}O_3$ Mol. wt 288.39

Found in the flowers of pussy willow, *Salix* sp. (Salicaceae).

Oestrogenic activity.

2670
Oestrone; Estrone; Estrol; Folliculin

$C_{18}H_{22}O_2$ Mol. wt 270.37

Occurs in the pollen and seeds of the date palm, *Phoenix dactylifera* (Palmae), and in seeds of the pomegranate, *Punica granatum* (Punicaceae).

Oestrogenic activity, but considerably less than oestradiol-17β (q.v.).

2671
Podecdysone B

$C_{27}H_{42}O_6$ Mol. wt 462.63

Found in some *Podocarpus* spp. (Podocarpaceae).

Phytoecdysteroid: insect moulting hormone activity.

2672
Pollinastanol

$C_{28}H_{48}O$ Mol. wt 400.69

Occurs in plant pollen, e.g., from *Hypochaeris radicata* and *Taraxacum officinale* (Compositae).

2673
Polypodine B

$C_{27}H_{44}O_8$ Mol. wt 496.65

Occurs in *Polypodium vulgare* (Polypodiaceae), in whole plants of *Lychnis fulgens* (Caryophyllaceae) and *Ajuga reptans* (Labiatae), and in the roots of *Pfaffia iresinoides* (Amaranthaceae) and *Rhaponticum carthamoides* (Compositae).

Phytoecdysteroid; disrupts growth and development of insect larvae, e.g., of the leek moth, *Acrolepiopsis assectella*.

2674
Ponasterone A

$C_{27}H_{44}O_6$ Mol. wt 464.65

Common in plants, e.g., in the leaves of *Podocarpus nakaii* (Podocarpaceae).

Phytoecdysteroid: insect moulting hormone activity. It inhibits ovarian maturation and egg production in newly emerged female houseflies, greatly inhibits larval development in several insects, and regulates the length of instars and the extent of silk production of the silk worm, *Bombyx mori*. Also, it greatly stimulates protein anabolic activity in mouse liver.

2675
Pterosterone

$C_{27}H_{44}O_7$ Mol. wt 480.65

Found in the roots of *Pfaffia iresinoides* (Amaranthaceae), and in the seeds of *Diploclisia glaucescens* (=*Locculus macrocarpus*) (Menispermaceae).

Phytoecdysteroid: insect moulting hormone activity. It greatly stimulates protein anabolic activity in mouse liver.

2676
Schottenol

$C_{29}H_{50}O$ Mol. wt 414.72

The major sterol in the Senita cactus, *Lophocereus schottii* (Cactaceae).

The fruit fly, *Drosophila pachea*, needs this sterol from the Senita cactus in order to make its ecdysone.

2677
Sitosterol; Sitosterin; β-Sitosterol

$C_{29}H_{50}O$ Mol. wt 414.72

Widespread occurrence in higher plants, e.g., in wheat germ, *Triticum*, and in maize or sweetcorn, *Zea mays* (Gramineae).

Antihyperlipoproteinaemic activity. It is an essential component of plant cell membranes.

2678
α-Spinasterol; α-Spinasterin

$C_{29}H_{48}O$ Mol. wt 412.70

Occurs in spinach, *Spinacia oleracea* (Chenopodiaceae), alfalfa, *Medicago sativa* (Leguminosae), senega root, *Polygala senega* (Polygalaceae), and bitter apple, *Citrullus colocynthis* (Cucurbitaceae).

2679
Stigmasterol; Anti-stiffness factor

$C_{29}H_{48}O$ Mol. wt 412.70

Found in many higher plants, e.g., soya beans, *Glycine max*, and Calabar beans, *Physostigma venenosum* (Leguminosae).

Plays a vital structural role in the membranes of plant cells.

2680
Testosterone

$C_{19}H_{28}O_2$ Mol. wt 288.43

Occurs in the pollen of the Scots pine, *Pinus sylvestris* (Pinaceae).

Androgenic activity.

2681
Withaferin A

$C_{28}H_{38}O_6$ Mol. wt 470.61

Found in the leaves of *Acnistus arborescens* and *Withania somnifera*, and in the roots of *W. coagulans* and *W. ashwagandha* (Solanaceae).

Withaferin A and related steroidal lactones or withanolides from solanaceous plants exhibit antibiotic and antitumour activities.

2682
Withanolide D

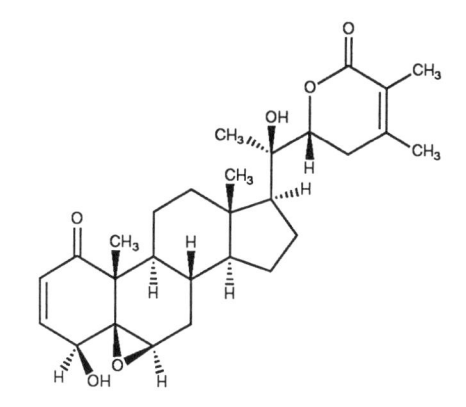

$C_{28}H_{38}O_6$ Mol. wt 470.61

Found in the leaves of *Withania somnifera* (Solanaceae).

Antibiotic and antitumour activities.

Chapter 57
Cucurbitacins

The cucurbitacins are a group of oxygenated tetracyclic triterpenes, which were first characterized as the bitter principles of cucumbers, marrows and squashes, members of the family Cucurbitaceae. Although, subsequently, they have been found elsewhere, plants of this family remain the major source. A typical member is cucurbitacin A (**2687**) which has a methyl (or hydroxymethyl) substituent at the C-9 instead of the more usual C-10 position and is oxygenated in the triterpenoid nucleus at the 2, 3, 9, 11 and 16 positions; all the other cucurbitacins are more or less closely related to it. The cucurbitacins occur both free and in glycosidic combination. Sugars may be attached directly to the steroid nucleus (e.g., cucurbitacin L 2-glucoside, see **2696**), or to the aliphatic side chain (e.g., carnosifloside I, **2684**), or to both ends of the molecule (e.g., carnisofloside III, **2685**).

These bitter agents are widely distributed in the cucumber plant and its relatives, being found in highest concentration in the seeds and roots. Cucurbitacins have been detected occasionally in at least five other plant families, in the Begoniaceae, Cruciferae, Desfontainiaceae, Elaeocarpaceae and Scrophulariaceae. Their intense bitterness means that they are effectively defensive against many herbivores. However, they are not always bitter tasting since two sweet cucurbitacins (e.g., **2686**) have recently been discovered, along with three bitter derivatives, in the rhizomes of *Hemsleya carnisoflora* (Kasai *et al.*, 1987).

Beetles feeding on cucumber plants have become adapted to the presence of cucurbitacins and use them as feeding cues. Occasionally, they absorb them from the diet and store them for defensive purposes (Ferguson *et al.*, 1985). They are generally poisonous to mammals. Their toxicity to humans has so far prevented the therapeutic application of their useful antitumour activity.

References

General

Lavie, D. and Glotter, E. (1971). The cucurbitanes, a group of tetracyclic triterpenes. *Fortschritte der Chemie organischer Naturstoffe*, **29**, 307–362.

Rehm, S. (1960). Die Bitterstoffe der Cucurbitaceen. *Ergebnitzindie Biologie*, **22**, 108–136.

Biological Properties

Ferguson, J.E., Metcalf, R. L. and Fischer, D.C. (1985). Disposition and fate of cucurbitacin B in five species of diabraticites. *Journal of Chemical Ecology*, **11**, 1307–1321.

Kasai, R., Matsumoto, K., Nie, R.L., Morita, T., Awazu, A., Zhou, J. and Tanaka, O. (1987). Sweet and bitter cucurbitane glycosides from *Hemsleya carnosiflora*. *Phytochemistry*, **26**, 1371–1376.

2683
Bryodulcosigenin

$C_{30}H_{50}O_4$ Mol. wt 474.73

Occurs as a glycoside, bryodulcoside, in the roots of *Bryonia dioica* (Cucurbitaceae).

Bryodulcoside has a sweet taste.

2684
Carnosifloside I

$C_{42}H_{68}O_{13}$ Mol. wt 781.01

Found in *Hemsleya carnosiflora* (Cucurbitaceae).

Tasteless, in contrast to other carnosiflosides (q.v.).

2685
Carnosifloside III

$C_{48}H_{78}O_{18}$ Mol. wt 943.15

Found in *Hemsleya carnosiflora* (Cucurbitaceae).

Bitter taste.

2686
Carnosifloside VI

$C_{48}H_{80}O_{18}$ Mol. wt 945.17

Found in *Hemsleya carnosiflora* (Cucurbitaceae).

Sweet taste.

2687
Cucurbitacin A

$C_{32}H_{46}O_9$ Mol. wt 574.72

Found in *Cucumis hookeri*, *C. leptodermis* and *C. myriocarpus* (Cucurbitaceae).

Intensely bitter taste. It is very toxic (LD_{50} intravenously in rabbits 0.7 mg kg^{-1}).

2688
Cucurbitacin B; Amarin; 1,2-Dihydro-α-elaterin

$C_{32}H_{46}O_8$ Mol. wt 558.72

Found in many spp. of Cucurbitaceae, e.g., *Cucumis africanus*, and in some *Iberis* spp. (Cruciferae). It is sequestered from the diet and stored by the spotted cucumber beetle, *Diabrotica undecimpunctata howardii*.

Highly toxic (LD_{50} intravenously in rabbits 0.5 mg kg^{-1}; LD_{10} orally in mice 5 mg kg^{-1}). It shows cytotoxic and antitumour activities. It is a feeding attractant to cucumber beetles, but a feeding deterrent to other insects. It shows antigibberellin activity.

2689
Cucurbitacin C

$C_{32}H_{48}O_8$ Mol. wt 560.74

Found in the bitter cucumber *Cucumis sativus* var. Hanzil (Cucurbitaceae).

Bitter taste.

2690
Cucurbitacin D; Elatericin A

$C_{30}H_{44}O_7$ Mol. wt 516.68

Found in many *Cucumis* spp. and other Cucurbitaceae, in some *Iberis* spp. (Cruciferae), and in *Crinodendron hookerianum* (Elaeocarpaceae).

Cytotoxic and antitumour activities. It is a feeding attractant to cucumber beetles but a deterrent to the honey bee, *Apis mellifera* L., and a species of yellow jacket wasp, *Vespula*.

2691
Cucurbitacin E; α-Elaterin

C$_{32}$H$_{44}$O$_8$ Mol. wt 556.70

Occurs in many spp. of Cucurbitaceae, e.g., *Ecballium elaterium*, and as the 2-*O*-β-D-glucoside in *Citrullus colocynthis*, in many *Iberis* spp. (Cruciferae), and in *Gratiola officinalis* (Scrophulariaceae).

Cytotoxic and antitumour activities. It is toxic (LD$_{50}$ orally in mice 340 mg kg^{-1}). It is a feeding attractant to certain beetles, e.g., the spotted cucumber beetle, *Diabrotica undecimpunctata howardii*, and a feeding deterrent to other insects. Also, it shows antigibberellin activity.

2692
Cucurbitacin F

C$_{30}$H$_{46}$O$_7$ Mol. wt 518.70

Occurs in *Cucumis angolensis* (=*C. dinteri*) (Cucurbitaceae), and in *Crinodendron hookerianum* (Elaeocarpaceae).

Bitter taste.

2693
Cucurbitacin H

C$_{30}$H$_{46}$O$_8$ Mol. wt 534.70

Found in *Acanthosicyos horrida*, *Citrullus naudinianus* and other spp. of Cucurbitaceae, and in *Crinodendron hookerianum* (Elaeocarpaceae).

Bitter taste.

2694
Cucurbitacin I; Ibamarin; Elatericin B

C$_{30}$H$_{42}$O$_7$ Mol. wt 514.67

Occurs in many *Citrullus* spp. (Cucurbitaceae) and *Iberis* spp., e.g., *I. amara* (Cruciferae), and in *Gratiola officinalis* (Scrophulariaceae).

Cytotoxic and antitumour activities. Also, it shows antigibberellin activity.

2695
Cucurbitacin J

$C_{30}H_{44}O_8$ Mol. wt 532.68

Cucurbitacin J and cucurbitacin K, which has the same structure but differs in the configuration of the hydroxyl group at C-24, occur in *Citrullus ecirrhosus*, *C. naudinianus* and *Bryonia dioica* (Cucurbitaceae).

Cucurbitacins J and K show antigibberellin activity.

2696
Cucurbitacin L; 23,24-Dihydrocucurbitacin I

$C_{30}H_{44}O_7$ Mol. wt 516.68

Occurs as the 2-*O*-β-D-glucoside in *Citrullus colocynthis* (Cucurbitaceae).

Bitter taste.

2697
Cucurbitacin O

$C_{30}H_{46}O_7$ Mol. wt 518.70

Occurs as a free aglycone in *Brandegea bigelovii* (Cucurbitaceae).

Cytotoxic activity.

2698
Cucurbitacin P

$C_{30}H_{48}O_7$ Mol. wt 520.71

Occurs as a free aglycone in *Brandegea bigelovii* (Cucurbitaceae).

Cytotoxic activity.

2699
Cucurbitacin Q

$C_{32}H_{48}O_8$ Mol. wt 560.74

Occurs as a free aglycone in *Brandegea bigelovii* (Cucurbitaceae).

Cytotoxic activity.

2700
Cucurbitacin S

$C_{30}H_{42}O_6$ Mol. wt 498.67

Occurs in *Bryonia dioica* (Cucurbitaceae).

Bitter taste.

2701
11-Deoxocucurbitacin I

$C_{30}H_{44}O_6$ Mol. wt 500.68

Occurs in *Desfontainia spinosa* (Desfontainiaceae).

Cytotoxic and antileukaemic activities.

2702
Gratiogenin

$C_{30}H_{48}O_4$ Mol. wt 472.71

Occurs as a glycoside, gratioside, in *Gratiola officinalis* (Scrophulariaceae).

2703
Spinoside A

$C_{39}H_{56}O_{12}$ Mol. wt 716.88

This glycoside of 11-deoxocucurbitacin I (q.v.), and the closely related spinoside B, occur in *Desfontainia spinosa* (Desfontainiaceae).

Both spinosides A and B show cytotoxic activity.

Chapter 58
Nortriterpenoids

Nortriterpenoids are formed in plants from tetracyclic triterpene (q.v.) precursors through oxidation and degradation, so that their basic skeletons contain less than the thirty carbon atoms otherwise present in triterpenoid derivatives. There are two main groups: the limonoids, C_{26} tetranortriterpenes in which the triterpenoid side chain has been degraded by the loss of four carbons; and the quassinoids, C_{20} and C_{19} triterpenoids which have lost all their side chain carbons and others as well.

Limonin (**2724**), a bitter principle of *Citrus* fruits, is a typical limonoid, with a complex ring system which includes two 6-membered lactone rings and two furan rings. Nomilin (**2727**), another bitter compound of *Citrus*, differs from limonin in having a 7-membered lactone ring in its structure. The same ring system is also present in obacunone (**2728**), which is the most widespread limonoid, present in all four families (see below) in which limonoids occur.

Undoubtedly the most chemically complex limonoid is azadirachtin (**2706**), the well known insect antifeedant from the neem tree, *Azadirachta indica*. Its original structure had to be revised twice before the correct one was arrived at, and this was confirmed by x-ray crystallography. Many related triterpenoids also occur in various parts of the neem tree. For example, the seeds contain in addition to azadirachtin A (**2706**), six other azadirachtins (B–G), which are closely related in both structure and antifeedant activity (Rembold, 1989). Azadirachtin is one of the very few natural antifeedants to be developed sufficiently for it to be used in the field to protect crop plants from pest attack (Schmutterer and Ascher, 1987).

The first quassinoid to be examined by chemists was the commercially useful compound quassin (**2731**) from the wood of *Quassia amara*. Its tetracyclic ring system is present, with various modifications, in all the nortriterpenoids of this type. Other quassinoids tend to have side chains attached to the hydroxyl group adjacent to the lactone grouping (e.g., ailanthinone, **2704**) and a few have additional sugar attachments (e.g. bruceoside A, **2711**). Over 75 quassinoids have been described, and these are all from the same plant family, the Simaroubaceae, to which *Quassia* belongs. The limonoids, by contrast, are found in this family and also more abundantly in three related families, the Rutaceae (to which *Citrus* belongs), the Meliaceae and the Cneoraceae.

Limonoids and quassinoids are notable for their very bitter tastes and this applies to most compounds in both series. Limonoids can be detected in plant extracts by TLC and the use of a specific colour test given with the Ehrlich reagent (D.L. Dreyer in Waterman and Grundon, 1983). Radioimmunoassay methods have been developed for the detection of limonin in *Citrus* fruit extracts.

Quassinoids are well known for their antitumour activities, although none has yet been developed as an anticancer drug because of the toxic side-effects. Some quassinoids show antiamoebic properties, and glaucarubin (**2717**) is used as a medicinal preparation in France. Limonoids are chiefly of interest because of the need to counteract the bitter taste which they impart to citrus drinks. More positively, they promise to be of a commercial value as the first members of a new generation of natural insecticides. Azadirachtin is an especially promising candidate, because of its high potency and its ability to disrupt metamorphosis and reproduction as well as feeding in pest insects.

References

Polonski, J. (1973). Quassinoid bitter principles. *Fortschritte der Chemie organischer Naturstoffe*, **30,** 101–150.
Rembold, H. (1989). The azadirachtins – their potential for insect control. *Economic and Medicinal Plant Research*, **3,** 57–72.
Schmutterer, H. and Ascher, K.R.S. (eds) (1987). Natural pesticides from the neem tree and other tropical plants. Eschborn: Deutsche Gesellschaft fur Technische Zusammenarbeit.
Waterman, P.G. and Grundon, M.F. (eds) (1983). *Chemistry and Chemical Taxonomy of the Rutales*. London: Academic Press.

2704
Ailanthinone

Skeletal type: quassinoid

$C_{25}H_{34}O_9$ Mol. wt 478.55

Found in *Pierreodendron kerstingii* and, as a minor constituent, in *Ailanthus altissima* (Simaroubaceae).

Antileukaemic activity *in vivo*, and amoebicidal activity *in vitro*.

2705
Ailanthone

Skeletal type: quassinoid

$C_{20}H_{24}O_7$ Mol. wt 376.41

The main bitter constituent of the seeds and bark of the "Chinese tree of heaven", *Ailanthus altissima* (=*A. glandulosa*) (Simaroubaceae).

Amoebicidal activity.

2706
Azadirachtin

Skeletal type: limonoid

$C_{35}H_{44}O_{16}$ Mol. wt 720.74

Occurs in the neem tree, *Azadirachta indica* (Meliaceae).

Potent insect antifeedant activity, e.g., against some *Spodoptera* and *Locusta* spp.

2707
Bruceantin

Skeletal type: quassinoid

$C_{28}H_{36}O_{11}$ Mol. wt 548.60

Found in the tree *Brucea antidysenterica* (Simaroubaceae).

Antileukaemic activity *in vivo*, and amoebicidal activity *in vitro*.

2708
Bruceantinol

Skeletal type: quassinoid

$C_{30}H_{38}O_{13}$ Mol. wt 606.63

Occurs in *Brucea antidysenterica* (Simaroubaceae).

Antileukaemic activity *in vivo*.

2709
Bruceine B; Brucein B

Skeletal type: quassinoid

$C_{23}H_{28}O_{11}$ Mol. wt 480.48

Found in the seeds of *Brucea amarissima* (Simaroubaceae).

Significant insecticidal activity.

2710
Bruceine D; Brucein D

Skeletal type: quassinoid

$C_{20}H_{26}O_9$ Mol. wt 410.42

Found in *Brucea amarissima* and *B. sumatrana* (Simaroubaceae).

Antileukaemic activity.

2711
Bruceoside A

Skeletal type: quassinoid

$C_{32}H_{42}O_{16}$ Mol. wt 682.69

Found in the seeds of *Brucea javanica* (Simaroubaceae).

Antileukaemic activity.

2712
Castelanone

Skeletal type: quassinoid

$C_{25}H_{34}O_9$ Mol. wt 478.55

Occurs in the root bark of *Castela tweediei* (Simaroubaceae).

In vitro antiviral activity against the oncogenic Rous sarcoma virus.

2713
Chaparrin

Skeletal type: quassinoid

$C_{20}H_{28}O_7$ Mol. wt 380.44

Found in the aerial parts of "chaparro amargoso", *Castela nicholsoni* (Simaroubaceae).

"Chaparro amargoso" is used as a medicine for amoebic dysentery in Mexico.

2714
Chaparrinone

Skeletal type: quassinoid

$C_{20}H_{26}O_7$ Mol. wt 378.43

Found in the seeds of *Quassia undulata* (=*Hannoa klaineana*) and of *Ailanthus altissima* (Simaroubaceae).

In vitro antiviral activity against the oncogenic Rous sarcoma virus. Tigloyloxy-6α-chaparrinone has *in vivo* antileukaemic activity.

2715
Chaparrolide

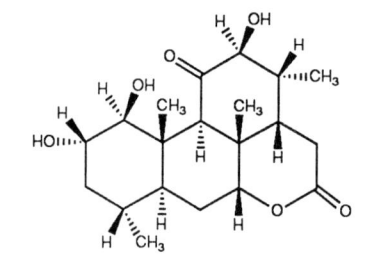

Skeletal type: quassinoid

$C_{20}H_{30}O_6$ Mol. wt 366.46

Found in *Castela nicholsoni* (Simaroubaceae).

Bitter principle.

2716
Eurycomalactone

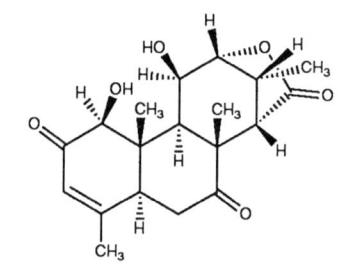

Skeletal type: quassinoid

$C_{19}H_{24}O_6$ Mol. wt 348.40

Occurs in the bark of *Eurycoma longifolia* (Simaroubaceae).

Intensely bitter taste. The bark of *E. longifolia* is used in South-east Asia as an antidysentericum.

2717
Glaucarubin; α-Kirondrin; Glaumeba

Skeletal type: quassinoid

$C_{25}H_{36}O_{10}$ Mol. wt 496.56

The major bitter constituent of the seeds of *Quassia simarouba* (=*Simarouba glauca*), and of the fruits of the "kirondro" tree, *Perriera madagascariensis* (Simaroubaceae).

Amoebicidal activity.

2718
Glaucarubinone

Skeletal type: quassinoid

$C_{25}H_{34}O_{10}$ Mol. wt 494.55

Found in the seeds of *Quassia undulata* (=*Hannoa klaineana*) and of *Quassia simarouba* (=*Simarouba glauca*), and in the bitter fruit of *Perriera madagascariensis* (Simaroubaceae).

Antimalarial (effective at 0.006 µg ml^{-1}), amoebicidal, *in vivo* antileukaemic, and significant insecticidal activities.

2719
Glaucarubolone

Skeletal type: quassinoid

$C_{20}H_{26}O_8$ Mol. wt 394.43

Occurs in the seeds of *Quassia simarouba* (=*Simarouba glauca*), the fruits of *Perriera madagascariensis* and *Quassia undulata* (=*Hannoa klaineana*), and the wood of *Castela nicholsoni* (Simaroubaceae).

Amoebicidal activity, and *in vitro* antiviral activity against the oncogenic Rous sarcoma virus.

2720
Harrisonin

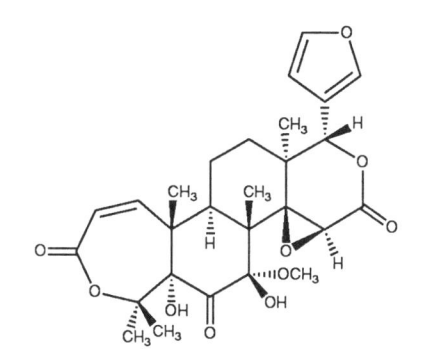

Skeletal type: limonoid

$C_{27}H_{32}O_{10}$ Mol. wt 516.55

Occurs in the roots of *Harrisonia abyssinica* (Simaroubaceae).

Insect antifeedant activity (e.g., against *Spodoptera*). It also has antibacterial activity.

2721
12α-Hydroxyamoorstatin

Skeletal type: limonoid

$C_{28}H_{36}O_{10}$ Mol. wt 532.60

Occurs in the seeds of *Aphanamixis grandifolia* (=*Amoora grandifolia*) (Meliaceae).

Cytotoxic and antileukaemic activities.

2723
Isobruceine A; Isobrucein A

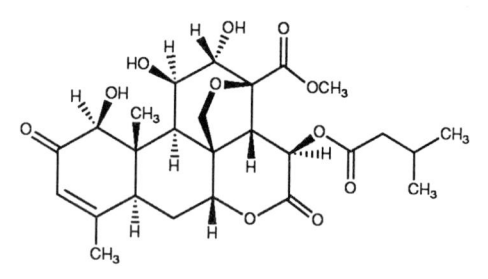

Skeletal type: quassinoid

$C_{26}H_{34}O_{11}$ Mol. wt 522.56

Occurs in *Soulamea tomentosa* (Simaroubaceae).

In vivo antileukaemic activity, and *in vitro* antiviral activity against the oncogenic Rous sarcoma virus. Also it shows insect antifeedant activity, e.g., against the southern armyworm *Spodoptera eridania*.

2722
Ichangin

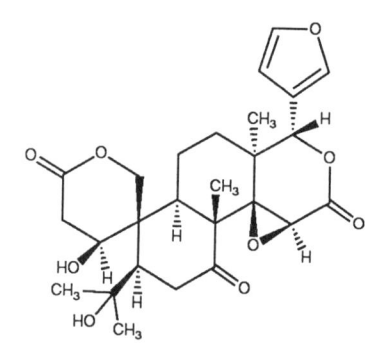

Skeletal type: limonoid

$C_{26}H_{32}O_{9}$ Mol. wt 488.54

Found in *Citrus ichangensis* (Rutaceae).

Bitter taste.

2724
Limonin; Citrolimonin; Dictamnolactone; Evodin; Obaculactone

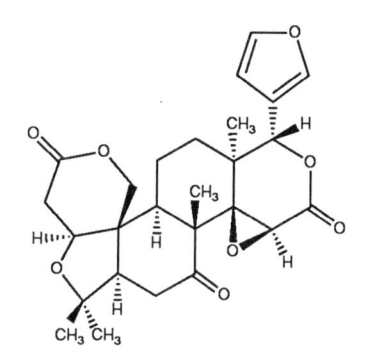

Skeletal type: limonoid

$C_{26}H_{30}O_{8}$ Mol. wt 470.53

Found in *Citrus* spp., e.g., in navel and valencia oranges during ripening, and in lemon, and in other Rutaceae.

Causes delayed bitterness of certain *Citrus* fruit juices.

2725
Neoquassin; Nigakihemiacetal B; Simalikahemiacetal A

Skeletal type: quassinoid

$C_{22}H_{30}O_6$ Mol. wt 390.48

Occurs in the wood of *Quassia amara*, and of various *Picrasma* spp. (Simaroubaceae).

Very bitter taste.

2726
Nigakihemiacetal A

Skeletal type: quassinoid

$C_{22}H_{34}O_7$ Mol. wt 410.51

Occurs in the stem of "nigaki", *Picrasma quassioides* (=*P. ailanthoides*) (Simaroubaceae).

Intensely bitter taste.

2727
Nomilin

Skeletal type: limonoid

$C_{28}H_{34}O_9$ Mol. wt 514.58

Found in the fruits of *Citrus* spp. (Rutaceae).

About twice as bitter as limonin (q.v.); it contributes to the delayed bitterness of *Citrus* fruit juices. It shows modest activity as an insect antifeedant.

2728
Obacunone; Casimirolide

Skeletal type: limonoid

$C_{26}H_{30}O_7$ Mol. wt 454.53

Occurs in many spp. of Rutaceae, e.g., of the genus *Citrus*, in Cneoraceae, e.g., *Cneorum tricoccon*, in Sinaroubaceae, e.g., *Harrisonia abyssinica*, and in Meliaceae, e.g., *Trichilia trifolia*.

Bitter principle.

2729
Picrasin C

Skeletal type: quassinoid

$C_{23}H_{34}O_7$ Mol. wt 422.53

Found in the stem of *Picrasma quassioides* (=*P. ailanthoides*) (Simaroubaceae).

Intensely bitter taste.

2730
Quassimarin C

Skeletal type: quassinoid

$C_{27}H_{36}O_{11}$ Mol. wt 536.59

Occurs in Surinam quassia wood, *Quassia amara* (Simaroubaceae).

Antileukaemic activity.

2731
Quassin; Nigakilactone D; Quassiin

Skeletal type: quassinoid

$C_{22}H_{28}O_6$ Mol. wt 388.47

Occurs in Surinam quassia wood, *Quassia amara*, and in the wood of several *Picrasma* spp. (Simaroubaceae).

Very bitter taste (bitterness threshold 1:60,000). Surinam quassia wood is used as a febrifuge, insecticide, vermicide, and bitter tonic.

2732
Rutaevin

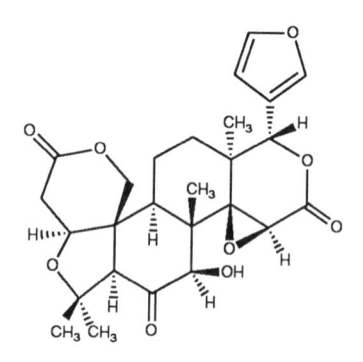

Skeletal type: limonoid

$C_{26}H_{30}O_9$ Mol. wt 486.53

Found in the ripe fruit of *Evodia rutaecarpa* (Rutaceae). Unripe fruit contains only limonin (q.v.).

Nonbitter limonoid, in contrast to the related limonin.

2733
Salannin

Skeletal type: limonoid

$C_{34}H_{44}O_9$ Mol. wt 596.73

Found in the essential oil of *Azadirachta indica* (=*Melia azadirachta*) (Meliaceae).

Insect antifeedant activity.

2734
Samaderin A; Samaderine A

Skeletal type: quassinoid

$C_{18}H_{18}O_6$ Mol. wt 330.34

Occurs in the bark and seeds of *Quassia indica* (=*Samadera indica*) (Simaroubaceae), together with the related substances samaderins B and C.

Samaderins A, B and C show antileukaemic activity. *Quassia indica* is used as a vermifuge and as an insecticide in fly papers.

2735
Simalikilactone D

Skeletal type: quassinoid

$C_{25}H_{34}O_9$ Mol. wt 478.55

Found in *Quassia africana* (Simaroubaceae).

Amoebicidal, antimalarial, antileukaemic and antiviral activities. It shows very potent antifeedant activity against the southern armyworm *Spodoptera eridania* (active at the level of 50 ppm).

2736
Soularubinone

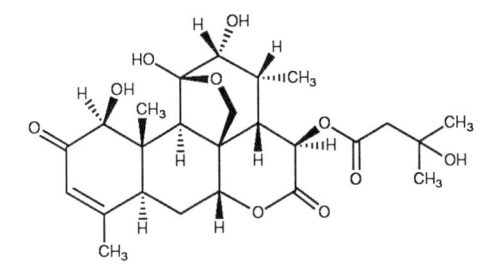

Skeletal type: quassinoid

$C_{25}H_{34}O_{10}$ Mol. wt 494.55

Occurs in the leaves of *Soulamea tomentosa* (Simaroubaceae).

Antimalarial activity, effective at 0.006 µg ml^{-1}, and antileukaemic activity.

2737
Toonacilin

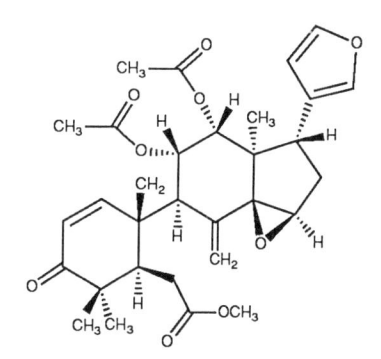

Skeletal type: limonoid

$C_{31}H_{38}O_9$ Mol. wt 554.65

Found in the bark of *Toona ciliata* (=*Cedrela toona*) (Simaroubaceae).

Insect antifeedant activity (against the Mexican bean beetle, *Epilachna varivestis*).

2738
Trichilin A

Skeletal type: limonoid

$C_{35}H_{46}O_{13}$ Mol. wt 674.75

Occurs in *Trichilia roka* (Meliaceae).

Trichilin A and closely related substances from *T. roka* have strong insect antifeedant activity. Like azadirachtin and simalikilactone D (q.v.), trichilins are among the few compounds which are active against the voracious southern armyworm, *Spodoptera eridania*.

2739
Undulatone

Skeletal type: quassinoid

$C_{27}H_{34}O_{11}$ Mol. wt 534.57

Occurs in the root bark of *Hannoa undulata* (Simaroubaceae).

Antileukaemic activity.

2740
Zapoterin; 11β-Hydroxyobacunone

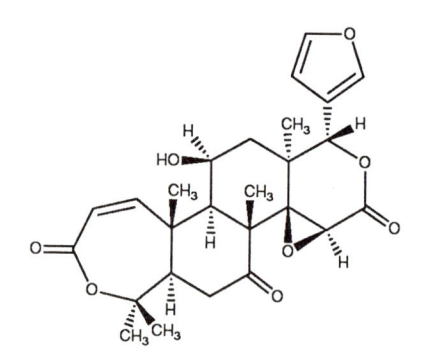

Skeletal type: limonoid

$C_{26}H_{30}O_8$ Mol. wt 470.53

Occurs in the seeds of *Casimiroa edulis* (Rutaceae).

C. edulis is cultivated for its bitter-sweet tasting fruit.

Chapter 59
Miscellaneous triterpenes

This section contains plant triterpenoids which do not fit readily into any other category. They are classified, according to their skeletal type, as dammaranes, euphanes, fernanes, friedelanes, hopanes, lupanes and ursanes (Dev, 1989). Some are widespread in nature, such as α-amyrin (**2741**), β-amyrin (**2742**) and friedelin (**2751**). Squalene (**2757**) is universally present in green plants, since it is the basic precursor of all triterpenoids. Betulin (**2744**) is also a common constituent of the bark of angiosperm trees. The first reference to betulin dates back to 1788 and, hence, it was one of the very first plant products to have been chemically investigated (Hayak *et al.*, 1989). Betulin can be detected in geochemical samples and, like the triterpenoids of the hopane series (e.g., **2752**), is very stable over geological time.

Most of the remaining triterpenoids in this section are more restricted in their natural occurrence. Some, like fernene (**2750**), are confined to the ferns. Others, like dammarenediol-I (**2746**) and dipterocarpol (**2747**) are characteristic components of the large family of dipterocarp trees, which dominate the forests of South-East Asia. Several of the triterpenoids listed have strong antitumour activity (e.g., pristimerin, **2755**), while others (lupeol acetate, **2754**) are antihypoglycaemic.

References

Dev, S. (1989). Terpenoids. In *Natural Products of Woody Plants*, Volume 2 (Rowe, J.W., ed.), pp. 691–807. Berlin: Springer-Verlag.

Hayek, E.W.H., Jordis, U., Moche, W. and Santer, F. (1989). A bicentennial of betulin. *Phytochemistry*, **28**, 2229–2242.

2741
α-Amyrin; α-Amyrenol; Viminalol

Skeletal type: ursane

$C_{30}H_{50}O$ Mol. wt 426.73

Occurs in the latex of many plants, sometimes as acetates, e.g., *Ficus variegata* (Moraceae), and rubber trees. Also, it occurs in *Erythroxylum coca* (Erythroxylaceae) and *Balanophora elongata* (Balanophoraceae). Often, it co-occurs with β-amyrin (q.v.).

2742
β-Amyrin

Skeletal type: oleanane

$C_{30}H_{50}O$ Mol. wt 426.73

Widespread occurrence, including Manila elemi resin, from *Canarium* spp. (Burseraceae). See also α-amyrin.

2743
Asiatic acid

Skeletal type: ursane

$C_{30}H_{48}O_5$ Mol. wt 488.71

Found in *Centella asiatica* (Umbelliferae), in dammar resin, the resinous exudate from *Shorea* spp. (Dipterocarpaceae), in *Centella* also as the triglycoside, asiaticoside.

Asiaticoside promotes wound healing.

2744
Betulin; Betulinol; Betulol; Trochol

Skeletal type: lupane

$C_{30}H_{50}O_2$ Mol. wt 442.73

Occurs in the outer bark of birch, *Betula* spp. (Betulaceae), often in high concentrations; the outer cortical layer of *B. platyphylla* contains, for instance, about 35% betulin. It occurs widely in angiosperms, in trees and shrubs.

Antitumour activity: active against the Walker carcinoma 256 tumour system.

2745
Betulinic acid

Skeletal type: lupane

$C_{30}H_{48}O_3$ Mol. wt 456.71

Widespread occurrence, e.g., in the bark of *Rhododendron arboreum* (Ericaceae).

Antitumour activity: active against the Walker carcinoma 256 tumour system.

2746
Dammarenediol-I

Skeletal type: dammarane

$C_{30}H_{52}O_2$ Mol. wt 444.75

Occurs in commercial dammar resins, and in the resins of *Hopea odorata* and *Shorea vulgaris* (Dipterocarpaceae).

2747
Dipterocarpol; Hydroxydammarane II

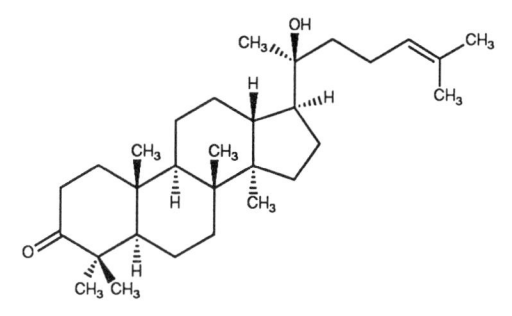

Skeletal type: dammarane

$C_{30}H_{50}O_2$ Mol. wt 442.73

Found in *Dipterocarpus acutangulus* and other *Dipterocarpus* spp. (Dipterocarpaceae), and in the galls of *Pistacia terebinthus* (Anacardiaceae).

Induces growth in the microsymbiont *Frankia* from actinorrhizal root nodules. This has enabled dipterocarps to establish and maintain their mycorrhizal associates so that they dominate lowland forests in SE Asia.

2748
α-Elemolic acid; α-Elemenadienolic acid

Skeletal type: euphane

$C_{30}H_{48}O_3$ Mol. wt 456.71

Occurs in Manila elemi resin, from *Canarium* spp. (Burseraceae).

2749
Euphol

Skeletal type: euphane

$C_{30}H_{50}O$ Mol. wt 426.73

Occurs in the latex of several *Euphorbia* spp. (Euphorbiaceae).

2750
Fernene

Skeletal type: fernane

$C_{30}H_{50}$ Mol. wt 410.73

Occurs in the fern *Dryopteris crassirhizoma* (Aspleniaceae).

2751
Friedelin; Friedelanone

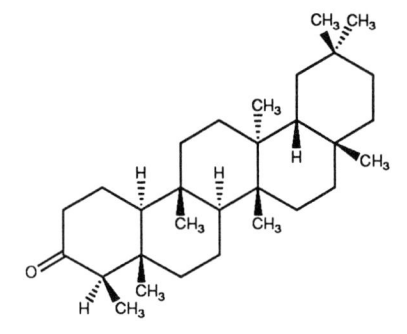

Skeletal type: friedelane

$C_{30}H_{50}O$ Mol. wt 426.73

Occurs in many higher plants, e.g., in *Ceratopetalum apetalum* (Cunoniaceae), and also in a large number of lichens. It is the major triterpene in cork, the outer bark of *Quercus suber* (Fagaceae).

Diuretic activity.

2752
Hopane-29-acetate; Hopan-29-ol acetate; Dryocrassol acetate

Skeletal type: hopane

$C_{32}H_{54}O_2$ Mol. wt 470.78

Found in the fern *Polypodium juglandifolium* (Polypodiaceae).

2753
Lupeol; Fagasterol; Monogynol B; β-Viscol

Skeletal type: lupane

$C_{30}H_{50}O$ Mol. wt 426.73

Occurs in the bark of many plants, especially of species belonging to the Apocynaceae and Leguminosae. The bark of *Phyllanthus emblica* (Euphorbiaceae) contains 2.25% lupeol. It was first isolated from the seed peelings of *Lupinus luteus* (Leguminosae).

Antitumour activity: active against the Walker carcinoma 256 tumour system. It shows antihyperglycaemic and hypotensive activities.

2754
Lupeol acetate

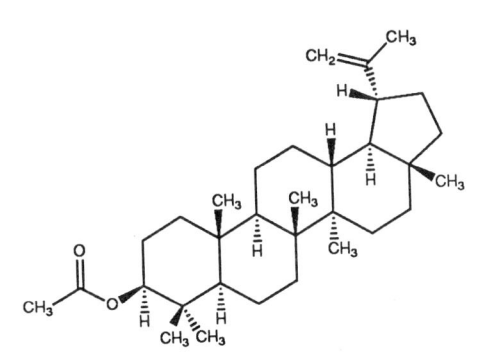

Skeletal type: lupane

$C_{32}H_{52}O_2$ Mol. wt 468.77

Occurs in *Cordia buddleioides* (Boraginaceae), and the date palm, *Phoenix dactylifera* (Palmae).

Antihyperglycaemic and antiulcer activities.

2755
Pristimerin

Skeletal type: friedelane

$C_{30}H_{40}O_4$ Mol. wt 464.65

Found in many species belonging to the Celastraceae, e.g., in the roots of *Pristimera indica*, in *Catha edulis*, *Schaefferia cuneifolia* and *Maytenus* spp.

Potent antitumour activity, antibacterial activity but also toxic. Also it strongly inhibits the germination of lentil seeds.

2756
Sapelin A

Skeletal type: protolimonoid

$C_{30}H_{50}O_4$ Mol. wt 474.73

Found in the wood of *Entandophragma cylindricum* (Meliaceae).

Cytotoxic activity.

2757
Squalene; Spinacene; Supraene

Skeletal type: isoprenoid

$C_{30}H_{50}$ Mol. wt 410.73

This is the immediate precursor of all cyclic triterpenoids. It is found in olive oil (0.1–0.7%), wheat germ oil, yeast, *Baccharis* spp. (Compositae), *Tilia vulgaris* (Tiliaceae), etc.

Bactericidal and antitumour activities; it is also an immunostimulant. Its main use is as an intermediate in the manufacture of pharmaceuticals, aromatics, surface active agents and rubber chemicals.

2758
Taraxasterol; Taraxasterin; Anthesterin; α-Lactucerol

Skeletal type: ursane

$C_{30}H_{50}O$ Mol. wt 426.73

Occurs in the roots of dandelion, *Taraxacum officinale*, in the flowers of *Anthemis nobilis* (Compositae), in *Euphorbia tirucalli* (Euphorbiaceae), and in *Eupatorium cannabinum* (Compositae).

2759
Taraxerol; Alnulin; Skimmiol; Tiliadin

Skeletal type: taraxerane

$C_{30}H_{50}O$ Mol. wt 426.73

Found in the roots of dandelion, *Taraxacum officinale* (Compositae), in *Diospyros* spp. (Ebenaceae), in *Lithocarpus* spp. (Fagaceae) and in *Canarium* spp. (Burseraceae).

Antiulcer and gastric antisecretory activities.

2760
Tingenone

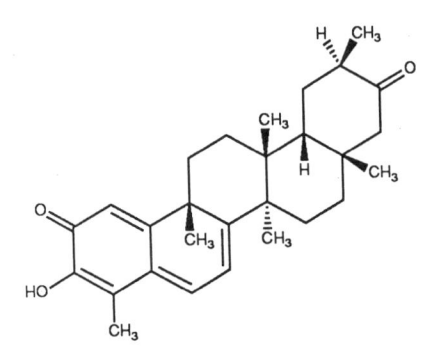

Skeletal type: friedelane

$C_{28}H_{36}O_3$ Mol. wt 420.60

Occurs in many spp. of the Celastraceae, e.g., *Crossopetalum uragoga*, *Schaefferia cuneifolia*, various *Maytenus* spp., and various *Salacia* spp.

Antitrypanosomal activity. It shows a DNA binding effect, and inhibits DNA, RNA and protein synthesis. Also, it strongly inhibits the germination of lentil seeds.

Chapter 60
Carotenoids

Carotenoids, which are C_{40} tetraterpenoids, are an extremely widely distributed group of lipid-soluble pigments, found in all kinds of plants from simple bacteria to yellow-flowered composites. In animals, several carotenoids, especially β-carotene, provide a source, through hydration and splitting of the molecule, of vitamin A, a C_{20} isoprenoid alcohol. Carotenoids, through dietary intake, also provide many brilliant animal colours, as in the flamingo, starfish, lobster and sea urchin. In plants, carotenoids have two principal functions: as accessory pigments in photosynthesis, and as colouring matters in flowers and fruits. In flowers, they mostly appear as yellow colours (daffodil, pansy, marigold) while, in fruits, they may also be orange or red (rose hip, tomato, paprika).

Although there are now 563 known carotenoids (Pfander, 1987), only a few are common in higher plants, and problems of identification can often be resolved, in the first instance, by reference to these common substances. Well known carotenoids are either simple unsaturated hydrocarbons based on lycopene, or their oxygenated derivatives, known as xanthophylls. The chemical structure of lycopene (**2782**) consists of a long chain of eight isoprene units joined head to tail, giving a completely conjugated system of alternate double bonds, which is the chromophore giving it colour. Cyclization of lycopene at one end gives γ-carotene (**2769**) while cyclization at both ends provides the bicyclic hydrocarbon β-carotene (**2768**). β-Carotene isomers (e.g., α- and ϵ-carotene) only differ in the positions of the double bonds in the cyclic end units. The common xanthophylls are either monohydroxycarotenes (e.g., lutein, **2781**; rubixanthin, **2790**), dihydroxy (zeaxanthin, **2793**) or dihydroxyepoxy (violaxanthin, **2792**).

Most of the rarer carotenoids have more complicated structures, and may be more highly hydroxylated (e.g., with keto groups), more highly unsaturated (e.g., with allenic or acetylenic groups), or more extended with additional isoprene residues (giving C_{45} or C_{50} carotenoids). Combined forms of carotenoid occur, especially in flowers and fruits of higher plants, and they are usually xanthophylls esterified with fatty acid residues, e.g., palmitic, oleic, or linoleic acids; see helenien, **2780** and physalein, **2788**. Glycosides are normally rare; in higher plants, the best known is the water-soluble crocin (**2774**), the gentiobiose derivative of crocetin (**2773**), the yellow pigment of meadow saffron. Glycosides of C_{40} carotenoids with rhamnose or glucose as the sugars have been found in various algae.

When isolating a carotenoid from a new higher plant source, the chances are fairly high that it will be β-carotene, since this is by far the most common of all these pigments. In quantitative terms, however, it is not as important as certain xanthophylls. The annual production of carotenoids in plants has been estimated at 10^8 tons year^{-1} and this refers mainly to the synthesis of fucoxanthin (**2779**) (widespread in marine algae) and of lutein, violaxanthin and neoxanthin. These three latter compounds occur, with β-carotene, universally in the leaves of higher plants. Traces of α-carotene, cryptoxanthin or zeaxanthin may also occur in the lipid-soluble fraction of leaf extracts. In flowers too, carotenoid mixtures are the rule rather than the exception. The pigments are often highly oxidized, with epoxides being common and carotenes only being present in traces. The amount of oxidation, however, varies from species to species and, in the corona of certain narcissi, the red colour is due almost entirely to high concentrations (up to 15% of the dry weight) of β-carotene.

Plant carotenoids are of especial interest to nutritionists because of their provitamin A activity (Pitt, 1985). They are widely used as food colourants, and this use is likely to increase (Timberlake and Henry, 1986).

The major reference to the biochemistry of plant carotenoids is Goodwin (1980). There are five chapters on carotenoids in Goodwin (1976) and these provide information on chromatographic and spectral properties. Data on spectroscopic analysis and on the use of carotenoids have been provided successively by Isler (1971), Straub (1976) and Pfander (1987).

References

Names, Synonyms and Chemical Structures
Isler, O. (1971). *Carotenoids*. Basel: Birkhäuser.
Straub, O. (1976). *Key to Carotenoids*, 1st Edition. Basel: Birkhäuser.
Pfander, H. (1987). *Key to Carotenoids*, 2nd Edition. Basel: Birkhäuser.

Natural Occurrence and Identification
Goodwin, T.W. (1980). *The Biochemistry of the Carotenoids*, Volume 1, *Plants*, 2nd Edition. London: Chapman and Hall.
Goodwin, T.W. (ed.) (1976). *Chemistry and Biochemistry of Plant Pigments*, 2nd Edition. London: Academic Press.
Goodwin, T.W. (ed.) (1988). *Plant Pigments*. London: Academic Press.

Provitamin A Activity
Pitt, G.A.J. (1985). In *Fat-soluble Vitamins – their Biochemistry and Applications* (A.T. Diplock) (ed.). London: Heinemann.

Carotenoids as Food Colours
Timberlake, C.F. and Henry, B.S. (1986). Plant pigments as natural food colours. *Endeavour*, **10**, 31–36.
Anon. (1987). *Food Additives: The Numbers Identified*. London: HMSO Publication prepared for the Ministry of Agriculture, Fisheries and Food, UK.

2761
Antheraxanthin; (3*R*,5*R*,6*S*,3′*R*)-Zeaxanthin 5,6-epoxide

C₄₀H₅₆O₃   Mol. wt 584.89

Found in the petals of some flowers, e.g., Californian poppy, *Eschscholzia californica* (Papaveraceae), in the anthers of the tiger lily, *Lilium tigrinum* (Liliaceae), in the pollen of nasturtium, *Tropaeolum majus* (Tropaeolaceae), in many fruits, e.g., sweet red pepper or paprika, *Capsicum annuum* (Solanaceae), papaya, *Carica papaya* (Caricaceae), in the leaves of some higher plants, e.g., of roses, *Rosa* spp. (Rosaceae), and in some algae, e.g., *Ochromonas* spp. (Chrysophyceae).

Yellow food colouring (see lutein).

2762
Azafrin; Escobedin

C₂₇H₃₈O₄   Mol. wt 426.60

Found in the roots of the South American plant "azafranillo", *Escobedia scabrifolia* (Scrophulariaceae) and related spp., in *Alectra parasitica* var. *chitrakutensis* (Scrophulariaceae), and in the rhizomes of *Christisonia bicolor* (Orobanchaceae).

Yellow food colouring.

2763
Bixin

C₂₅H₃₀O₄   Mol. wt 394.52

Found in the seeds and fruit of annatto, *Bixa orellana* (Bixaceae). See also norbixin.

Food colouring in dairy and fat-based foods.

2764
Canthaxanthin; Food orange; β,β-Carotene-4,4′-dione

C₄₀H₅₂O₂   Mol. wt 564.86

Found in the edible mushroom, *Cantharellus cinnabarinus* (Agaricaceae, Hymenomycetes), in blue-green algae (Cyanobacteria), e.g., *Anabaena flos-aquae* (Nostocaceae) and *Tolypothrix tenuis* (Scytonemataceae), and in some Brevibacteriaceae and Micrococci.

Colour additive for food and drugs. It is also used as an oral suntanning agent.

2765
Capsanthin

C₄₀H₅₆O₃   Mol. wt 584.89

Occurs in the ripe fruits of sweet red pepper or paprika, *Capsicum annuum* (Solanaceae), and in some barberries, *Berberis* spp. (Berberidaceae), in the petals of orange lilies, e.g., *Lilium pumilum*, and the anthers of some lilies, e.g., the tiger lily, *Lilium tigrinum* (Liliaceae).

Orange-red food colouring, and flavouring in sauces, salad dressings, sausages, etc.

2766
Capsorubin

$C_{40}H_{56}O_4$ Mol. wt 600.89

Occurs in the ripe fruits of sweet red pepper or paprika, *Capsicum annum* (Solanaceae), and in the petals of orange lilies, e.g., *Lilium pumilum* (Liliaceae).

Uses similar to those of capsanthin (q.v.).

Wide occurrence: in almost all green leaves of higher plants (thus in green vegetables and forage crops), and also in many lower plants, e.g., ferns, mosses, algae, many fungi and some bacteria. It also occurs in many roots, e.g., carrot, *Daucus carota* (Umbelliferae), and sweet potato, *Ipomoea batatas* (Convolvulaceae), in many seeds, e.g., cereals and legumes, in many fruits, e.g., sweet red pepper or paprika, *Capsicum annuum* (Solanaceae), yam, *Dioscorea* spp. (Dioscoreaceae), rose-hips, *Rosa* spp. (Rosaceae), and in the yellow and orange flowers of many plants, e.g., gorse, *Ulex* spp. (Leguminosae).

Provitamin A activity; it is the most important of the carotenoid vitamin A precursors. It is used as a yellow food colouring of fats, e.g., of margarine, and as a sunscreen agent (in erythropoietic protoporphyria).

2767
α-Carotene; (6′R)-β,ε-Carotene

$C_{40}H_{56}$ Mol. wt 536.89

Wide occurrence, but in smaller amounts than β-carotene, e.g., in the roots of carrots *Daucus carota* (Umbelliferae), in the seeds of sweetcorn or maize, *Zea mays* (Gramineae), in palm *Elaeis guineensis* (Palmae), in many fruits, e.g., in tomato, *Lycopersicon esculentum* (Solanaceae), and in persimmon, *Diospyros kaki* (Ebenaceae), in the green leaves of many higher plants, and in the cells of many red algae (Rhodophyta).

Vitamin A precursor, but only half as active as β-carotene (q.v.). It is used as a yellow food colouring.

2768
β-Carotene; Carotene; Solatene; β,β-Carotene

$C_{40}H_{56}$ Mol. wt 536.89

2769
γ-Carotene; β,ψ-Carotene

$C_{40}H_{56}$ Mol. wt 536.89

Found in small amounts in the roots of carrot, *Daucus carota* (Umbelliferae), in the stems of Californian marsh dodders, *Cuscuta* spp. (Convolvulaceae), in the seeds of sweetcorn or maize, *Zea mays* (Gramineae), in the flowers of marigolds, *Calendula* spp. (Compositae), in the fruits of sweet red pepper, *Capsicum annuum* (Solanaceae), of persimmon, *Diospyros kaki* (Ebenaceae), of tomato, *Lycopersicon esculentum* (Solanaceae), and of other plants, in some green algae, e.g., *Chara* spp. (Chlorophyceae), and in fungi, e.g., *Penicillium* spp. (Ascomycetes). Together with β-carotene (q.v.), it is the dominant carotenoid pigment in many fungi belonging to the Phycomycetes, Ascomycetes, Deuteromycetes and Myxomycetes.

Vitamin A precursor, but only half as active as β-carotene (q.v.). It is used as a yellow food colouring.

2770
δ-Carotene; (6R)-ε,ψ-Carotene

$C_{40}H_{56}$ Mol. wt 536.89

Found in the fruits of the sea buckthorn, *Hippophae rhamnoides* (Elaeagnaceae) and of *Gonocaryum pyriforme* (Icacinaceae), in the flowers of the marigold, *Calendula arvensis* (Compositae), in the pollen of nasturtium, *Tropaeolum majus* (Tropaeolaceae), and in traces in the roots of carrots, *Daucus carota* (Umbelliferae).

No provitamin A activity. It is used as a yellow food colouring (see β-carotene).

2771
ζ-Carotene; 7,8,7′,8′-Tetrahydro-ψ,ψ-carotene

$C_{40}H_{60}$ Mol. wt 540.92

Found in the fruits of many plants, e.g., in papaya, *Carica papaya* (Caricaceae), in *Citrus* spp. (Rutaceae), in tomato, *Lycopersicon esculentum* (Solanaceae), in apple, *Malus* spp. (Rosaceae), in American cranberry, *Vaccinium macrocarpon* (Ericaceae), in the flowers of marigolds, *Calendula* spp. (Compositae) and of yellow broom, *Cytisus scoparius* (Leguminosae), in the seeds of maize, *Zea mays* (Gramineae), and in traces in the roots of carrots, *Daucus carota* (Leguminosae). Also, it is found in small amounts in the bacterium *Rhizobium lupini* (Rhizobiaceae).

No provitamin A activity. It is used as a yellow food colouring (see β-carotene).

2772
β-Carotene 5,6-epoxide

$C_{40}H_{56}O$ Mol. wt 552.89

Found in some fruits, e.g., papaya, *Carica papaya* (Caricaceae), orange peel, *Citrus sinensis* (Rutaceae), mango, *Mangifera indica* (Anacardiaceae), apple, *Malus* spp. (Rosaceae), in the green leaves of some plants, and in some yellow flowers, e.g., *Forsythia* spp. (Oleaceae) and coltsfoot, *Tussilago farfara* (Compositae).

Vitamin A precursor, but only half as active as β-carotene (q.v.).

2773
Crocetin

$C_{20}H_{24}O_4$ Mol. wt 328.41

Occurs in saffron (styles of *Crocus sativus*), and in petals of this and other *Crocus* spp. (Iridaceae).

Crocetin esters are used as yellow food colourings (see crocin).

2774
Crocin; α-Crocin; Digentiobiose ester of crocetin

$C_{44}H_{64}O_{24}$ Mol. wt 977.00

Occurs in saffron (styles of *Crocus sativus*), in the petals of *Crocus* spp. (Iridaceae), of the mullein, *Verbascum phlomoides* (Scrophulariaceae), and of *Gardenia* spp. (Rubiaceae).

One of the few truly water-soluble natural carotenoids. It is used as a yellow food colouring.

2775
α-Cryptoxanthin; (3R,6'R)-β,ε-Caroten-3-ol

$C_{40}H_{56}O$ Mol. wt 552.89

Found in the leaves of lucerne, *Medicago sativa* (Leguminosae), in the anthers and petals of nasturtium, *Tropaeolum majus* (Tropaeolaceae), in the seeds of sweetcorn or maize, *Zea mays* (Gramineae), and in many fruits, e.g., in grapefruit, *Citrus paradisi* (Rutaceae), in avocado, *Persea americana* (Lauraceae), and in apricot, *Prunus armeniaca* (Rosaceae).

Yellow food colouring (see lutein).

2776
β-Cryptoxanthin; Cryptoxanthin; Cryptoxanthol; Caricaxanthin; (3R)-β,β-Caroten-3-ol (3R)-3-Hydroxy-β-carotene

$C_{40}H_{56}O$ Mol. wt 552.89

Usually occurs in esterified forms in the petals and fruits of many plants, e.g., in the petals of the yellow lily *Lilium hansonii* (Liliaceae), in the petals and berries of the Chinese lantern plant or bladder cherry, *Physalis alkekengi* (Solanaceae), in the rinds of oranges and other *Citrus* fruits (Rutaceae), and in sweet red peppers, *Capsicum annuum* (Solanaceae). It also occurs as the palmitate, in the fruits of papaya, *Carica papaya* (Caricaceae), in the seeds of maize or corn, *Zea mays* (Gramineae), and in the leaves of some higher plants.

Vitamin A precursor, but only half as active as β-carotene (q.v.). It is used as a yellow food colouring.

2777
Echinenone; Aphanin; β,β-Caroten-4-one; 4-Keto-β-carotene; Myxoxanthin

$C_{40}H_{54}O$ Mol. wt 550.87

Found in fungi belonging to the Hymenomycetes, in blue-green algae (Cyanobacteria), e.g., *Anabaena flos-aquae* and *Oscillatoria rubescens* (Nostocaceae), in some colourless algae, e.g., *Astasia ocellata* (Euglenophyta), and in *Brevibacterium* KY 4313 (Brevibacterium) grown on petroleum.

Provitamin A activity, but only half as active as β-carotene (q.v.).

2778
Flavoxanthin

$C_{40}H_{56}O_3$ Mol. wt 584.89

Occurs in the flowers of the buttercup, *Ranunculus acris* (Ranunculaceae), of the dandelion, *Taraxacum officinale* (Compositae) and of *Chrysanthemum* spp. (Compositae), in the anthers of *Narcissus* spp. (Amaryllidaceae), of *Rosa* spp. (Rosaceae) and of *Tulipa* spp. (Liliaceae), in the pollen of *Acacia dealbata* (Leguminosae), and in the fruits of many plants, e.g., barberry, *Berberis* spp. (Berberidaceae), apricot, *Prunus armeniaca*, and plum, *P. domestica* (Rosaceae), and elderberries, *Sambucus nigra* (Caprifoliaceae).

Yellow food colouring.

2779
Fucoxanthin

$C_{42}H_{58}O_6$ Mol. wt 658.93

Occurs in brown algae (Phaeophyta), e.g., *Fucus vesiculosus*, and in diatoms (Bacillariophyceae) and other Chrysophyta. The yearly production of fucoxanthin by marine algae is some 6.8×10^6 tons, which makes it the most abundant carotenoid in the world.

Important pigment in the photosynthesis of brown algae (Phaeophyta).

2780
Helenien; Adaptinol; Lutein dipalmitate

$R = CH_3[CH_2]_{14}-C(=O)O-$

$C_{72}H_{116}O_4$ Mol. wt 1045.72

Occurs in the flowers of sneezeweed, *Helenium autumnale* (Compositae), and other plants.

Yellow food colouring.

2781
Lutein; $(3R,3'R,6'R)$-β,ϵ-Carotene-3,3'-diol; Vegetable luteol; Xanthophyll

$C_{40}H_{56}O_2$ Mol. wt 568.89

One of the most widely distributed carotenoid alcohols in plants. It occurs in the green leaves of all higher plants, e.g., in cabbages, *Brassica* spp. (Cruciferae), and in the cells of many green algae (Chlorophyta) and red algae (Rhodophyta). It is found mainly as esters in the fruits and yellow flowers of many plants, e.g., pineapple, *Ananas comosus* (Bromeliaceae), in the rind of *Citrus* fruits (Rutaceae), in apricot, peach and plum, *Prunus* spp. (Rosaceae), in apple, *Malus* spp. (Rosaceae), and in American cranberry, *Vaccinium macrocarpon* (Ericaceae), and also in the petals of Aztec marigold, *Tagetes erecta* (Compositae). See also helenein.

Lutein esters are used as food colouring for emulsified products such as salad dressings.

2782
Lycopene; ψ,ψ-Carotene

$C_{40}H_{56}$ Mol. wt 536.89

Occurs in the ripe fruits of tomatoes, *Lycopersicon esculentum* (Solanaceae), of *Citrus* spp. (Rutaceae), of persimmon, *Diospyros kaki* (Ebenaceae), of dog rose, *Rosa canina* (Rosaceae), and in many other fruits and berries. It also occurs in the petals of many orange flowers, e.g., pot marigold, *Calendula officinalis* (Compositae), in very large amounts in the ripe seeds of *Momordica charantia* (Cucurbitaceae), in the roots of swedes, *Brassica rutabaga* (Cruciferae) and in the red varieties of carrot, *Daucus carota* (Umbelliferae). Small amounts occur in some photosynthetic bacteria, e.g., *Rhodopseudomonas* spp. (Rhodospirillaceae).

Orange-red food colouring.

2783
Lycoxanthin; ψ,ψ-Caroten-16-ol; Lycopen-16-ol

$C_{40}H_{56}O$ Mol. wt 552.89

Found in the ripe fruits of tomato, *Lycopersicon esculentum* (Solanaceae), of bittersweet, *Solanum dulcamara* (Solanaceae), of black bryony, *Tamus communis* (Dioscoreaceae), and in other fruits.

Used as a food colouring (see lutein).

2784
Mutatochrome; β-Carotene 5,8-epoxide

$C_{40}H_{56}O$ Mol. wt 552.89

Found in the leaves of higher plants, in many fruits, e.g., in papaya, *Carica papaya* (Caricaceae), in tomato, *Lycopersicon esculentum* (Solanaceae), and in the peel of oranges, *Citrus sinensis* (Rutaceae), in many yellow flowers, e.g., in marigolds, *Calendula* spp. (Compositae), in *Chrysanthemum coronarium* (Compositae), and in *Forsythia* spp. (Oleaceae). Also, it is found in some blue-green algae (Cyanobacteria), e.g., *Oscillatoria agardhii* (Oscillatoriaceae, Nostocales).

Provitamin A activity, but only half as active as β-carotene (q.v.).

2785
Neoxanthin; Foliaxanthin

$C_{40}H_{56}O_4$ Mol. wt 600.89

Universal in the green leaves of higher plants, and also found in some lower plants, e.g., mosses (Bryophyta), green algae (Chlorophyta) and Euglenophyta. It is found mainly as esters in many fruits and yellow flowers, e.g., in sweet red pepper, *Capsicum annuum* (Solanaceae), in peach, *Prunus persica*, plum, *P. domestica* and apple, *Malus* spp. (Rosaceae), in the petals of marsh marigold, *Caltha palustris* (Ranunculaceae), in yellow broom, *Cytisus scoparius* (Leguminosae), in *Forsythia* spp. (Oleaceae), and in *Geum* spp. (Rosaceae).

Yellow food colouring (see lutein)

2786
Neurosporaxanthin; 4′-Apo-β-caroten-4′-oic acid

$C_{35}H_{46}O_2$ Mol. wt 498.75

One of the few xanthophylls present in fungi, e.g., in *Neurospora crassa* (Ascomycetes), *Rhodotorula* spp. (Deuteromycetes) and *Lycogola epidendron* (Myxomycetes).

Provitamin A activity.

2787
Norbixin; Diapocarotene-6,6′-dioic acid

$C_{24}H_{28}O_4$ Mol. wt 380.49

Occurs with bixin (q.v.) in the seed pericarp of annatto, *Bixa orellana* (Bixaceae).

Yellow food colouring.

2788
Physalien; Physalin; Zeaxanthin dipalmitate

$R = CH_3[CH_2]_{14}-C(=O)O-$

$C_{72}H_{116}O_4$ Mol. wt 1045.72

Found in the fruits and petals of the Chinese lantern plant or bladder cherry, *Physalis alkekengi* (Solanaceae).

Yellow food colouring.

2789
Rhodoxanthin

$C_{40}H_{50}O_2$ Mol. wt 562.84

Occurs in the leaves of the pondweed, *Potamogeton natans* (Potamogetonaceae), in the fruit arils of *Taxus baccata* (Taxaceae), in the ferns and allied groups, e.g., *Adiantum* and *Equisetum* spp., in fungi, e.g., *Epicoccum* spp., and in some nonphotosynthetic bacteria, e.g., pink varieties of *Micrococcus tetragenus*.

Colouring material for food, beverages, pharmaceuticals and cosmetics.

2790
Rubixanthin; 3-Hydroxy-γ-carotene; (3R)-β,ψ-Caroten-3-ol

$C_{40}H_{56}O$ Mol. wt 552.89

Occurs in the flowers and fruits of many plants, e.g., in the petals of *Rosa* spp. (Rosaceae), in the sunflower, *Helianthus annuus*, and the French marigold, *Tagetes patula* (both Compositae), in rose-hips, fruits of *Rosa canina*, *R. rubiginosa* and other *Rosa* spp. (Rosaceae), in the rind of tangerines and satsumas, *Citrus reticulata* (Rutaceae), and in apricots, *Prunus armeniaca*, and cloudberry, *Rubus chamaemorus* (both Rosaceae).

Food colouring (see lutein).

2791
Torulene; 3′,4′-Didehydro-β,ψ-carotene

$C_{40}H_{54}$ Mol. wt 534.87

Occurs in many fungi, e.g., *Neurospora crassa* (Ascomycetes), *Puccinia graminis* (Heterobasidiomycetes), and *Rhodotorula* spp. (Deuteromycetes).

Provitamin A activity, but only half as active as β-carotene (q.v.).

2792
Violaxanthin; Zeaxanthin diepoxide

$C_{40}H_{56}O_4$ Mol. wt 600.89

Universal in the green leaves of higher plants and also in certain algae (Chlorophyta, Chrysophyta, Phaeophyta); it occurs mainly as esters in many fruits and flowers, e.g., in sweet red pepper or paprika, *Capsicum annuum* (Solanaceae), in the rind of *Citrus* fruits (Rutaceae), in mango, *Mangifera indica* (Anacardiaceae), in the petals of the Californian poppy, *Eschscholzia californica* (Papaveraceae), and in yellow pansies, *Viola tricolor* (Violaceae).

Yellow food colouring (see lutein).

2793
Zeaxanthin; Zeaxanthol; Anchovyxanthin;
(3R,3'R)-3,3-Dihydroxy-β-carotene;
(3R,3'R)-β,β-Carotene-3,3'-diol

$C_{40}H_{56}O_2$ Mol. wt 568.89

Wide occurrence: although it is a minor component in the leaves of some higher plants, it is a major pigment in green algae (Chlorophyta), and also occurs in some other algae (e.g., Rhodophyta, Euglenophyta, Chrysophyta), and in Cyanobacteria. It is also found in seeds of sweet corn or maize, *Zea mays* (Gramineae), and, usually as esters, in many flowers and fruits, e.g., in the petals of *Crocus sativus* (Iridaceae), in the yellow lily *Lilium hansonii* (Liliaceae), in the fruit of sweet red pepper, *Capsicum annuum* (Solanaceae), and in the peel of bitter oranges, *Citrus aurantium*, and tangerines, *C. reticulata* (Rutaceae). See also physalein.

Yellow food colouring (see lutein).

Index

The index contains (1) the names of the main entries with their entry numbers, (2) the synonyms for these which are followed by '*see*' and the entry number, and (3) the names of significant similar compounds referred to in the texts of some entries which are followed by '*see under*' and the entry number

ABA *see* 2122
Abietic acid 2413
Abietin *see* 1749
Abrin 376
L-Abrine 209
Abrus agglutinin *see under* 376
Abrusoside A 2511
Abscisic acid 2122
(*S*)-Abscisic acid 2122
Abscisin II *see* 2122
Absinthiin *see* 2214
Absinthin 2214
Absynthin *see* 2214
Abyssinone I 1343
Abyssinone V 1344
Abyssinone VI 1201
Acacetin 1417
Acacia *see* 51
Acaciin *see under* 1417
Acacipetalin *see under* 320
Acalyphin 304
Acamelin 1810
Acanthicifoline 830
(−)-Acanthocarpan 1516
Acanthoidine 1144
Acanthoine *see under* 1144
Acanthoside B *see* 1656
Acanthoside D 1592
Acerosin 1418
Acetaldehyde formylmethylhydrazone *see* 280
Acetaldehyde methylformylhydrazone *see* 280
Acetic acid 78
Acetophenone 1710
Acetosyringone 1991
Acetovanillone *see* 1714
1'-Acetoxychavicol acetate 1732
1'-Acetoxyeugenol acetate 1733
1-Acetoxypinoresinol 1593
Acetylandromedol *see* 2452
1-Acetylaspidoalbidine 551
Acetylbenzene *see* 1710
Acetylbenzoylaconine *see* 494
Acetylbrowniine 492
Acetylcaranine 447
O-Acetylcypholophine 1145
7-Acetyl-9-echimidinylretronecine *see* 979
Acetylformic acid *see* 104•
14-Acetylisotalatizidine *see* 508
18-*O*-Acetyllycocotonine *see* 549
3-Acetyl-6-methoxybenzaldehyde 1711
3-Acetylnerbowdine 448
6-Acetylpicropolin 2414
N-Acetyltetrahydroanabasine *see* 883
Acetylthymol *see* 2086
Achilleic acid *see* 79
Achillein *see* 893
Achillin 2215
Acocantherin *see* 2639
Acomonine *see* 513

Aconifine 493
Aconitic acid 79
cis-Aconitic acid *see* 79
Aconitine 494
Acorn sugar *see* 72
Acrifoline 801
Acronidine 981
Acronycidine 982
Acronycine 983
Acrophylline 984
Acroptilin 2216
Acrovestone 1992
Acteoside *see* 1809
Actinidin 377
Actinidine 831
Actinodaphnine 671
Acutumidine 1146
Adaptinol *see* 2780
Adenine 355
Adenocarpine 881
Adenosine 5'-(tetrahydrogen phosphate) *see* 356
Adenosine 5'-triphosphoric acid *see* 356
Adenosine triphosphate 356
Adermine *see* 442
Adiantifoline 672
Adifoline 552
Adinine *see* 552
Adipic acid 80
Adlumidine *see under* 700
(+)-Adlumine 673
Adonite *see* 73
Adonitol *see* 73
Adonitoxigenin 3-*O*-β-D-diginoside *see* 2647
Adonitoxin 2611
Adonitoxinogen 3-*O*-α-L-rhamnoside *see* 2611
D-Adonose *see* 19
Adouétine X 857
Adouétine Y 858
Adouétine Z 859
Adynerigenin 3-*O*-β-D-diginoside *see* 2612
Adynerin 2612
Aegelenine *see* 1009
Aescin 2512
Aesculetin *see* 1305
Aesculin *see* 1306
Aethusin 167
Affinine 553
Affinisine 554
Aflatoxin B_1 1283
Afromosin *see* 1517
Afrormosin 1517
Afzelechin 1345
Afzelechin-(4α→8)-afzelechin 1925
Afzelin *see under* 1469
Agar 45
Agar-agar *see* 45
Agarin *see* 291
Agaritine 415
Agathisflavone 1419

755

Index

Agavoside A 2574
Ageratochromene see under 1275
Agmatine 268
Agrimol C 1712
Agrimol F, G see under 1712
Agrimoniin 1926
Agrimophol 1713
Agroclavine 555
Agrostophyllin 1888
Ailanthinone 2704
Ailanthone 2705
Ajaconine 495
Ajmalicine 556
Ajmaline 557
Ajugalactone 2648
Ajugarin I 2415
Ajugarin II, III see under 2415
Ajugasterone C 2649
Ajugose 23
Aknadicine 674
Aknadinine see under 674
Akuammicine 558
Akuammidine 559
Akuammine 560
Alamarine 675
Alangicine 676
Alangimarckine 678
Alangimarine 677
Alangiside 679
3-Alanine see 211
L-Alanine 210
L-α-Alanine see 210
β-Alanine 211
Alant starch see 55
Alantin see 55
Alantolactone 2217
Alatolide 2218
Albafuran A 1234
Albafuran B see under 1234
Albamycin see 1320
Albanol A 1235
Albasapogenin see 2539
Albine 1048
Albizziin see 212
L-Albizziine 212
Albomaculine 449
Alchorneine 1147
Alchornine 1148
Aldehydoformic acid see 91
3-Aldehydoindole see 432
Alexine 882
Alginic acid 46
Alizarin 1811
Alizarin 2-methyl ether 1812
Alkaloid F30 see 719
Alkanna red see 1813
Alkannin 1813
Alkannin β,β-dimethylacrylate 1814
Allamandicin see under 2089
Allamandin 2089
Allamdin see under 2089
Allicin 142
Alliin 213
Allioside A see 2634
α-Allocryptopine 680
Allogibberic acid 2416
Alloimperatorin 1284
Allomaleic acid see 87
Allosecurinine see under 1192
Allyl disulfide see 143
Allyl sulfide see 144
p-Allylanisole see 1764
Allylcatechol methylene ether see 1797
Allylglucosinolate see 353
Allylguaiacol see 1765

Allyl isothiocyanate see under 353
Alnulin see 2759
Alnusiin 1927
Aloe-emodin 1815
Aloin see 1821
Aloperine 1049
Alpinigenine see under 681
Alpinine 681
Alstonine 561
Alteramine see 1081
Altholactone see 2005
Amabiline 936
α-Amanitin 378
Amaralin 2219
Amaranthin 471
Amarin see 2688
Amarogentin 2090
Amaropanin see under 2090
Amarorine see 624
Amaroswerin see under 2090
Amaryllisine 450
Amataine 562
Ambelline 451
Amber acid see 109
Ambrosin 2220
Amebacilin see 2150
Amentoflavone 1420
Amentoflavone 7,4'-dimethyl ether see 1451
Americine 860
Aminoacetic acid see 235
L-2-Amino-6-amidinohexanoic acid see 245
(S)-2-Amino-4-(aminoxy)butyric acid see 222
p-Aminobenzoic acid 269
(4-Aminobutyl)guanidine see 268
N-(4-Aminobutyl)urea see 272
4-Aminobutyric acid see 214
γ-Aminobutyric acid 214
(S)-α-Amino-β-cyanopropionic acid see 226
2-Amino-2-deoxyglucose see 9
Aminoethanoic acid see 235
4-(2-Aminoethyl)benzene-1,2-diol see 276
4-(2-Aminoethyl)imidazole see 281
3-(2-Aminoethyl)indole see 302
4-(2-Aminoethyl)pyrocatechol see 276
1-Aminoethylbenzene see 294
β-Aminoethylglyoxaline see 281
(+)-α-Amino-L-glutaramic acid see 233
(+)-α-Amino-L-glutaric acid see 232
1-Amino-4-guanidinobutane see 268
L-α-Amino-δ-guanidinovaleric acid see 217
2-Amino-4-(guanidinoxy)butyric acid see 223
(2S,3S)-2-Amino-3-hydroxybutyric acid see 263
(S)-2-Amino-3-hydroxypropionic acid see 262
4-Amino-2-hydroxypyrimidine see 359
2-Aminohypoxanthine see 362
α-Aminoindole-3-propionic acid see 265
(S)-2-Aminoisovaleric acid see 267
(R)-α-Amino-β-mercaptopropionic acid see 227
5-Aminomethyl-3-hydroxyisoxazole see 291
(S)-2-Amino-4-methylvaleric acid see 249
L-α-Amino-γ-oxalylaminobutyric acid 215
L-α-Amino-β-oxalylaminopropionic acid 216
(S)-2-Amino-1-oxo-1-phenylpropane see 274
1-Amino-2-phenylethane see 294
3-Amino-β-proline see 901
(S)-2-Aminopropanoic acid see 210
3-Aminopropanoic acid see 211
L-α-Aminopropionic acid see 210
β-Aminopropionic acid see 211
6-Amino-1H-purine see 355
6-Amino-3H-purine see 355
6-Amino-9H-purine see 355
6-Aminopurine see 355
4-Aminopyrimidin-2-ol see 359
(R)-2-Amino-3-selenomethylpropanoic acid see 254

(S)-α-Aminosuccinamic acid *see* 218
(S)-Aminosuccinic acid *see* 219
L-Aminosuccinic acid amide *see* 218
(S)-α-Amino-δ-ureidovaleric acid *see* 224
Ammoidin *see* 1311
(+)-Ammodendrine 883
Ammoidin *see* 1340
Ammoresinol 1285
Ammothamnine 1050
Ampelopsin 1346
Amphibine A 861
Amphibine B 862
Amphibine C, D *see under* 862
Amurensine 682
Amurensinine *see under* 682
Amurine 683
Amygdalin 305
Amygdalose *see* 26
Amygdaloside *see* 305
Amylum *see* 61
α-Amyrenol *see* 2741
α-Amyrin 2741
β-Amyrin 2742
Amyrolin *see* 1335
(−)-Anabasine 884
Anacardic acid 1661
Anacrotine 937
Anacyclin 168
(−)-Anaferine 885
Anagyine 1051
Anahygrine *see under* 885
Analobine *see* 687
Anantine 1149
Anatabine 886
Anatoxin a 1112
Anchoic acid *see* 82
Anchovyxanthin *see* 2793
Anchusa acid *see* 1813
Anchusin *see* 1813
Ancistrocladine 684
Ancistrocladinine *see under* 684
Ancistrocladisine *see* 684
Androcymbine 1150
Andromedotoxin *see* 2452
Androstenedione 2650
Androtex *see* 2650
Anethole 1734
Aneurine *see* 445
Angelicin 1286
7-Angelyl-9-echimidinylheliotridine *see* 947
7-Angelyl-9-echimidinylretronecine *see* 944
O^7-Angelylheliotridine 938
7-Angelyl-9-lasiocarpylheliotridine *see* 955
7-O-Angelylretronecine *see under* 961
7-Angelyl-9-sarracinylplatynecine *see* 966
7-Angelyl-9-sarracinylretronecine *see* 977
7-Angelyl-9-(−)-viridiflorylretronecine *see* 975
Angoline 685
Angolinine *see* 769
Anguidine *see* 2146
Angularine 939
Angustibalin 2221
Angustifoline 1052
Angustine 563
Anhydrocumanin *see* 2250
Anhydrofarinosin *see* 2269
Anhydrogitalin *see* 2631
Anhydroglycinol 1518
Anibine 887
Aniflorine 985
Animal coniine *see* 271
p-Anisaldehyde 1662
Anisatin 2222
Anise camphor *see* 1734
Anisessine 986

Anisic aldehyde *see* 1662
Anisodamine 1113
Anisotine 987
Ankorine 686
Annofoline 802
Annolobine 687
(−)-Annonaine 688
Annotine 803
Annotinine 804
Annuithrin *see* 2352
Annuloline 1151
Anolobine *see* 687
Anonaine *see* 688
Anopterimine *see under* 496
Anopterimine N-oxide *see under* 496
Anopterine 496
Anthemis glycoside A 306
Anthemis glycoside B 307
Antheraxanthin 2761
Antheridiogen-An 2417
Anthesterin *see* 2758
Anthorine *see* 498
Anthragallol 1816
Anthranolyllcoctonine 497
Anthricin *see* 1605
Anti-stiffness factor *see* 2679
Antiarigenin 3-O-β-D-antiaroside *see* 2613
α-Antiarin 2613
Antiarol 1663
Antibioticum X-465 A *see* 1293
Antierythrite *see* 65
Antihaemorrhagic vitamin *see* 1887
Antiogenin 3-O-α-L-rhamnoside *see* 2614
Antioside 2614
Antirhine 564
Antirrhin *see* 1211
Antirrhinoside 2091
Antiscorbutic vitamin *see* 81
Aphanin *see* 2777
Aphidicolin 2418
Aphloid *see* 1976
Aphloiol *see* 1976
Aphrodine *see* 670
Aphylline 1053
ent-Apigeniflavan-(2α→7,4α→8)-epiafzelechin *see* 1948
Apigenin 1421
Apigenin 7-apiosylglucoside *see* 1424
Apigenin 7,4'-dimethyl ether 1422
Apigenin 6-C-glucoside *see* 1467
Apigenin 7-O-glucoside 1423
Apigenin 8-C-glucoside *see* 1514
Apigenin 4'-methyl ether *see* 1417
Apigenin 7-methyl ether *see* 1449
Apiin 1424
Apiol *see* 1735
Apiole 1735
Apioline *see* 1735
Apiose 1
β-D-Apiose *see* 1
Apioside *see* 1424
Apo-β-caroten-4'-oic acid *see* 2786
Apoatropine 1114
Apocynin 1714
Apodine 565
(−)-Apoglaziovine 689
Apohyoscine 1115
Apoverticine *see* 1110
Apovincamine 566
(−)-Apparicine 567
Arabinitol *see* 62
D-Arabinoketose *see* 19
β-L-Arabinopyranose *see* 2
O-α-L-Arabinopyranosyl-(1→6)-β-D-glucopyranose *see* 44
Arabinose 2
D-Arabinose *see under* 2

L-Arabinose *see* 2
D-Arabinulose *see* 19
Arabite *see* 62
D-Arabitol 62
Arachic acid *see* 113
Arachidic acid 113
Arachidonic acid 114
Aralionine A 863
Aralionine B, C *see under* 863
Arborine 988
Arborinine 989
Arbusculin A 2223
Arbutin 1664
Archangelicin 1287
Archangelolide 2224
Archin *see* 1846
(−)-Arctigenin 1594
Arctiopicrin 2225
Arctolide 2226
Arctuvin *see* 1682
Ardisianone 1817
Ardisic acid B *see* 1995
Arecaidine 888
Arecaine *see* 888
Arecoline 889
Arenaine 832
(−)-Argemonine 690
Argentine 1054
L-Arginine 217
Argyrolobine 1055
Aribine *see* 622
Aricine 568
Aristolindiquinone 1818
Aristolochic acid 416
Aristolochic acid A *see* 416
Aristolochic acid-I *see* 416
Armepavine 691
Arnebinol 1993
Arnebinone 1819
Arnicolide A 2227
Aromadendrin 1347
Aromaticin 2228
(+)-Aromoline 692
Artabotrine *see* 749
Artabsin 2229
Artecanin 2230
Arteglasin A 2231
Artemisiifolin 2232
Artemisin 2233
Artemisinin *see* 2372
Artemorin 2234
Arvenoside A 2513
Asarabacca camphor *see* 1736
Asaricin *see* 1799
Asarin *see* 1736
Asarone *see* 1736
α-Asarone *see under* 1736
β-Asarone 1736
cis-Asarone *see* 1736
trans-Asarone *see under* 1736
Asarum camphor *see* 1736
Ascaridol *see* 2035
Ascaridole 2035
Asclepin 2615
Ascorbic acid 81
L-Ascorbic acid *see* 81
Asebogenin 1348
Asebotin *see under* 1348
Asebotoxin *see* 2452
Asebotoxin II 2419
Asiatic acid 2743
Asiaticoside *see under* 2743
L-Asparagic acid *see* 219
L-Asparagine 218
L-β-Asparagine *see* 218
L-Asparaginic acid *see* 219

Asparagoside A 2575
Asparamide *see* 218
Asparasaponin I *see* 2608
L-Aspartic acid 219
L-Aspartic acid amide *see* 218
Aspecioside 2616
Aspergillic acid 1152
Aspergillin *see* 1174
Asperulin *see* 2092
Asperuloside 2092
Aspidin 1715
Aspidinol 1716
Aspidoalbine 569
Aspidodasycarpine 570
Aspidofractine 571
Aspidospermatine 572
Aspidospermine 573
Asterin *see* 1210
Astilbin *see under* 1412
Astragaloside III 2514
Astrantiagenin D *see* 2539
Astrasieversianin XVI 2515
Astringin 1889
Astrobain *see* 2639
Astrocasine 890
Astrophylline 891
Atalanine 990
trans-Atalantine *see* 990
Atalaphylline 991
Ataline *see under* 990
Athamantin 1288
Atheroline 693
Atherospermidine 694
(+)-Atherospermoline 695
Athyriol 1958
Atisine 498
ATP *see* 356
Atractylin *see* 2420
Atractylol *see* 2147
Atractyloside 2420
Atropamine *see* 1114
Atropine 1116
Atropyltropeine *see* 1114
Aucubigenin *see under* 2093
Aucubin 2093
Aucuboside *see* 2093
Aucuparin 1994
Aurantio-obtusin β-D-glucoside 1820
Aurasperone D 1253
Aureusidin 1203
Aureusin *see under* 1205
Auricularine 574
Auriculine 940
Auriculoside 1349
Aurotensine *see* 746
Australine 892
Austrobailignan 1 1595
Autumnolide 2235
Auxin *see* 286
Avadharidine 499
Avadkharidine *see* 499
Avenacin A-1 2516
Avenacin B-2 2517
Avenacoside A 2576
Avenacoside B 2577
Avenalumin 417
Δ^5-Avenasterol 2651
Avornin *see under* 1848
Awadcharidine *see* 499
Axillarin 1425
Ayapanin *see* 1310
Azadirachtin 2706
Azafrin 2762
Azaleatin 1426
Azalein *see under* 1426

Azelaic acid 82
L-Azetidine 2-carboxylic acid 220
L-2-Azetidine carboxylic acid *see* 220
Azukisaponon III 2518
Azulon *see* 2155

Bacancosin *see* 833
Bacterial vitamin H[1] *see* 269
Baicalein 1427
Baicalein 5,6,7-trimethyl ether 1428
Baicalein 7-O-glucuronide *see* 1429
Baicalin 1429
Baikaine *see* 221
L-Baikiain 221
Baileyin 2236
Bakankoside 833
Bakankosine *see* 833
Bakkenolide A 2237
Balchanin *see* 2376
Balfourodine 992
Balfourodinium 993
Baliospermin 2421
Baloxine 575
Balsoxine 1153
Banisterine *see* 623
Baptifoline 1056
Baptitoxine *see* 1061
Barbaloin 1821
Barosma camphor *see* 2052
Barringtogenol C 2519
Barterin *see* 326
Batatasin I 1890
Batatasin IV 1891
Bayogenin 3-O-cellobioside 2520
BBT *see* 169
(+)-Bebeerine 696
Beet sugar *see* 40
Behenic acid 115
Beiwutine 500
Belladine 452
Belladonnine 1117
α-,β-Belladonnine *see under* 1117
Bellamarine *see* 447
Bellendine 1118
Bellidifolin 1959
Bellidifolin-8-O-glucoside *see* 1988
Bellissaponin 2 *see* 2571
Bengal isinglass *see* 45
1,4-Benzenediol *see* 1682
Benzeneethanamine *see* 294
1,2,3-Benzenetriol *see* 1695
1,3,5-Benzenetriol *see* 1691
Benzoaric acid *see* 1673
Benzoic acid 1665
2H-1-Benzopyran-2-one *see* 1295
1,2-Benzopyrone *see* 1295
2,3-Benzopyrrole *see* 285
Benzoylecgonine 1119
Benzoylmethylecgonine *see* 1123
Benzoylpseudotropeine *see* 1141
Benzoyltropein 1120
Benzyl carbiol *see* 1689
Benzylglucosinolate *see* 349
Benzylisothiocyanate *see under* 349
Benzylthiocyanate *see under* 349
Berbamine 697
Berbenine *see* 697
Berberastine 698
Berberine 699
Bergamol *see* 2059
Bergapten 1289
Bergaptene *see* 1289
Bergenin 1995
Betagarin 1350
Betalamic acid 472

Betalamic acid 472
Betanidin 473
Betanin 474
Betanin sulfate *see* 489
(−)-Betonicine 893
Betulin 2744
Betuloside *see* 2024
Betulinic acid 2745
Betulinol *see* 2744
Betulol *see* 2744
Betuloside *see* 2024
3′,6″-Biapigenin *see* 1499
3′,8″-Biapigenin *see* 1420
6,8″-Biapigenin *see* 1419
8,8″-Biapigenin *see* 1438
Bicolorin *see* 1306
(+)-Bicuculline 700
Biflorin 1254
Biflorine *see* 779
Bigitalin *see* 2631
Bikhaconitine 501
Bilobol 1666
3,3″-Binaringenin *see* 1381
Biochanin A 1519
Biochanin B *see* 1533
Biotexin *see* 1320
Biotin 418
Bipindogenin 3-O-β-D-digitaloside *see* 2617
Bipindogenin 3-O-α-L-rhamnoside *see* 2637
Bipindoside 2617
Bisabolol *see* 2123
α-Bisabolol 2123
Bixin 2763
Bleekerine 576
Blestriarene B 1892
Blestriarenes A, C *see under* 1892
Blood sugar *see* 10
Bocconine 701
Boldine 702
Boldine dimethyl ether *see* 738
Boletic acid *see* 87
Bonafousine 577
Bornan-2-one *see* 2039
Borneol 2036
endo-Borneol *see* 2036
Borneol acetate *see* 2037
(−)-Bornesitol *see* 63
D-Bornesitol *see under* 63
L-Bornesitol 63
Bornyl acetate 2037
Bornyl alcohol *see* 2036
Bornyl formate *see under* 2037
Bornyl isovalerate *see under* 2037
Borrecapine 578
Borreline 579
Borreverine 580
Boschniakine 834
Boschnialactone 2094
Botrydial 2124
Bougainvillein-r-I 475
Boussonin C 1352
Bowman-Birk inhibitor, soybean 379
Bracteatin 1204
Bracteoline 703
Brain sugar *see* 8
Brassicasterol 2652
Brassinolide 2653
Brazilin 1996
Brevicarine *see under* 581
Brevicolline 581
Brevifolin *see* 1731
Bromelain 380
Bromelin *see* 380
Broussin 1351
Broussonin C 1352
Browniine 502

Index

Browniine 14-acetate *see* 492
Bruceantin 2707
Bruceantinol 2708
Brucein B *see* 2709
Brucein D *see* 2710
Bruceine B 2709
Bruceine D 2710
Bruceoside A 2711
Brucine 582
Brugine 1121
Brunsvigine 453
Bryodulcoside *see under* 2683
Bryodulcosigenin 2683
Bucharaine 994
Buchu camphor *see* 2052
Buddledin A 2125
Buddledin B, C *see under* 2125
Budlein A 2238
Bufotenine 270
Bulbocapnine 704
Bullatine G *see* 541
Burseran 1596
Butane-1,4-diamine *see* 297
1,4-Butanedicarboxylic acid *see* 80
Butanedioic acid *see* 109
Butanoic acid *see* 83
Butein 1205
Butein 3,4′-di-*O*-glucoside *see* 1219
But-3-enylglucosinolate *see* 343
But-3-enylisothiocyanate *see under* 343
N-(3-Butenyl)cytisine *see* 1076
5-(3-Buten-1-ynyl)-2,2′-bithienyl 169
Butin 1353
Butin 7,3′-di-*O*-glucoside *see* 1354
Butrin 1354
3-Butylidene-7-hydroxyphthalide 1997
n-Butyric acid *see* 83
Butyric acid 83
Butyrylmallotochromene 1255
Buxamine *see* 1083
Buxamine A, B, C, G *see under* 1083
Buxamine E 1083
Buxomegine *see* 1093
Byakangelicin 1290

(−)-Caaverine 705
Cabucine 583
Cadabine *see* 934
Cadaverine 271
Cadiamine 1057
Cadinene *see* 2126
β-Cadinene 2126
Cafesterol *see* 2422
Cafestol 2422
Caffearine *see* 446
Caffee-tannins *see* 1776
Caffeic acid 1737
Caffeic acid 3-glucoside 1738
Caffeic acid 3-methyl ether *see* 1768
Caffeine 357
1-Caffeoyl-β-D-glucose 1739
N-Caffeoylputrescine 1740
3-Caffeoylquinic acid *see* 1745
5-*O*-Caffeoylshikimic acid 1741
Cajanin 1520
Cajanol 1521
Cajeputene *see* 2057
Cajeputol *see* 2045
Calactin 2618
Calafatimine 706
Calafatine *see under* 706
Calaxin 2239
C-Calebassine *see* 584
Calebassine 584
Callicarpone 2423

Callichiline 585
Calligonine 586
Callinecdysone B *see* 2665
Calmodulin 381
Calophyllin B 1960
Calophyllolide 1291
Calotropin 2619
Calpurnine 1058
Calycanthidine 587
Calycanthine 1154
Calystigine *see* 773
Camellidin I 2521
Camellidin II 2522
Cammaconine 503
Campesenin *see* 1330
Campesterol 2654
Camphan-2-one *see* 2039
Camphene 2038
Camphol *see* 2036
Camphor 2039
Camptothecin 995
Camptothecine *see* 995
(−)-Canadine 707
L-Canaline 222
L-Canavanine 223
Cancentrine 708
Candicine 1155
Candimine 454
Candletoxin A 2424
Cane sugar *see* 40
Canellal 2127
Canescine *see* 597
Canin 2240
Cannabichromene 1256
Cannabidiol 1667
Cannabidiolic acid 1668
Cannabisativine 1156
Cannogenin 3-*O*-gentiobiosylthevetoside *see* 2645
Canthaxanthin 2764
Canthin-6-one 588
Canthinone *see* 588
Canthiumine 864
Cantleyine 835
Capaurimine 10-methyl ether *see* 709
Capaurine 709
Capillarisin 1257
Capillin 170
Capric acid *see* 117
Caproic acid *see* 121
Caprylic acid *see* 131
Caprylic alcohol *see* 159
Capsanthin 2765
Capsicoside A 2578
Capsidiol 2128
Capsorubin 2766
Caracurine V 589
Caranine 455
Carapanaubine 590
N-Carbamylbutane-1,4-diamine *see* 272
N-Carbamylputrescine 272
Carbolic acid *see* 1690
Carbomethoxyibogamine *see* 594
3,3′-Carbonylbiscytosine *see* 1054
(15:1)-Cardanol 1669
Cardelmycin *see* 1320
Cardigin *see* 2626
Cardine *see* 1339
Cardiopetalidine 504
Cardiopetaline *see under* 504
Cardiospermin 308
Carditoxin *see* 2626
(15:1)-Cardol *see* 1666
Carduben *see* 1339
(−)-Car-3-ene 2040
Δ^3-Carene *see* 2040

3-Carene *see* 2040
Caribine 456
Caricaxanthin *see* 2776
Carinatine 457
Carlina oxide 171
Carlinoside 1430
Carmichaeline *see* 528
(−)-Carnegine 710
Carnosifloside I 2684
Carnosifloside III 2685
Carnosifloside VI 2686
Carnosol 2425
Carolinianine 805
Carotatoxin *see* 178
Carotene *see under* 2768
α-Carotene 2767
β-Carotene 2768
β,β-Carotene *see* 2768
(6′R)-β,ε-Carotene *see* 2767
β,ψ-Carotene *see* 2769
γ-Carotene 2769
δ-Carotene 2770
(6R)-ε,ψ-Carotene *see* 2770
ψ,ψ-Carotene *see* 2782
ζ-Carotene 2771
(3R,3′R)-β,β-Carotene-3,3′-diol *see* 2793
(3R,3′R,6′R)-β,ε-Carotene-3,3′-diol *see* 2781
β,β-Carotene-4,4′-dione *see* 2764
β-Carotene 5,6-epoxide 2772
β-Carotene 5,8-epoxide *see* 2784
(3R)-β,β-Caroten-3-ol *see* 2776
(3R)-β,ψ-Caroten-3-ol *see* 2790
(3R,6′R)-β,ε-Caroten-3-ol *see* 2775
ψ,ψ-Caroten-16-ol *see* 2783
β,β-Caroten-4-one *see* 2777
Carotol 2129
Carpacin 1742
Carpaine 894
Carrageen *see* 47
Carrageenan 47
Carrageenin *see* 47
Carthamone 1206
Carubinose *see* 14
Carvacrol 2041
Carvol *see* 2042
Carvone 2042
Carvopinone *see* 2072
Carophyllene *see* 2130
α-Caryophyllene *see* 2163
β-Caryophyllene 2130
γ-Caryophyllene *see* 2170
Carophyllic acid *see* 1765
Caryoptin 2426
Casbene 2427
Cascarillin 2428
Cascarin *see under* 1848
Cascaroside A 1822
Caseadine 711
Caseamine *see under* 711
Caseanadine *see under* 711
Cashmiradelphine 505
Casimiroedine 1157
Casimiroin 996
Casimirolide *see* 2728
Cassaidine 506
Cassaine 507
Cassiamin C 1823
Cassic acid *see* 1881
Cassine 895
Cassinine 836
Cassyfiline 712
(+)-Cassythicine 713
Cassythine *see* 712
Castanin *see* 1517
Castanospermine 896

Castelanone 2712
Casuarictin 1928
Casuarinin 1929
Catalpin *see* 2096
Catalposide 2096
Catapol 2095
(+)-Catechin 1355
Catechin 7-O-β-D-xyloside 1356
Catechin-(4α→8)-epicatechin *see* 1953
Catechinic acid *see* 1355
Catechol 1670
Catechol *see* 1355
Catechuic acid *see* 1355
Catharanthine 591
Catharine 592
Catheduline 2 *see* 837
Catheduline E2 837
D-Cathine 273
D-Cathinone 274
Cathocin *see* 1320
Cathomycin *see* 1320
Caulophylline 1059
Cavinine 458
Ceanothamine A *see* 867
Ceanothamine B *see* 857
Ceanothine B 865
Cedar camphor *see* 2132
α-Cedrene 2131
Cedrol *see* 2132
α-Cedrol 2132
Celabenzene *see* 1158
Celabenzine 1158
Celafurine *see under* 1158
Celapanine 838
Celidoniol *see* 156
Celidonione *see* 157
Cellobiose 24
Cellose *see* 24
Cellulose 48
α-Cellulose *see* 48
Celosianin *see under* 477
Cembrene A *see* 2475
Centaurepensin *see* 2244
Centaurin *see* 2246
Centdarol 2133
(−)-Centrolobine 1743
Cephaeline 714
Cephaeline methyl ether *see* 727
Cephalomannine 1159
Cephalotaxine 1160
Cepharanthine 715
Cephrol *see* 2049
Cerberoside 2620
Cercosporin 1824
Cerebrose *see* 8
Cernuine 806
Cernuoside *see under* 1203
Ceryl alcohol *see* 151
Cetylic acid *see* 133
Cevine *see under* 1108
Ceylon isinglass *see* 45
α-Chaconine 1084
Chaksine 839
Chalconaringenin 1207
Chalepensin 1292
Chalepin *see* 1309
Chalepin acetate *see* 1329
Chamazulene 2134
Chamigrene 2135
Chamissonin 2241
Chamissonin diacetate 2242
Champaca camphor *see* 2156
Champacol *see* 2156
Chanoclavine *see* 593
Chanoclavine-I 593
Chanoclavine-II *see under* 593

Index

Chaparrin 2713
Chaparrinone 2714
Chaparrolide 2715
Chartreusin 1293
Chasmanthin 2429
Chaulmoogric acid 116
Chavicine see under 925
Chavicol see under 1781
Chavidol methyl ether see 1764
Chavicol rutinoside see 1781
Chebulagic acid 1930
Chebulinic acid 1931
Chelidonic acid 195
Chelirubine see 701
Chicoric acid 1744
Chimaphylin 1825
Chanoclavine-II see under 593
Chimonin see 1976
Chinese isinglass see 45
Chinic acid see 105
Chinovose see 16
Chitosamine see 9
Chlorochrymorin 2243
7-Chloroemodin 1826
Chlorogenic acid 1745
Chlorohyssopifolin A 2244
Chlorohyssopifolin C see 2216
Chlorophorin 1893
Chlorophyll a 419
Chlorophyll b 420
Chloroxylonine see 1043
Cholesterol 2655
Cholla gum 49
Chondodendrine see 696
Chromolaenide 2245
Chrysanthemic acid 2043
Chrysanthemin see 1210
Chrysanthemum carboxylic acid see 2043
Chrysanthemumic acid see 2043
Chrysanthenone 2044
Chrysanthin see 2370
Chrysartemin A see 2240
Chrysartemin B see 2230
Chrysatropic acid see 1333
Chrysazin 1827
Chrysin 1431
Chrysin 5,7-dimethyl ether 1432
Chryso-obtusin glucoside 1828
Chrysoeriol 1433
Chrysoeriol 6-C-glucoside see 1465
Chrysophanic acid see 1830
Chrysophanic acid 9-anthrone 1829
Chrysophanol 1830
Chrysophanol 8-O-β-D-glucopyranoside see 1831
Chrysophanol 8-glucoside 1831
Chrysosplenetin 1434
Chrysosplenol C 1435
Chymopapain 382
Cibarian 421
Cichorigenin see 1305
Cichoriin 1294
Cicutoxin 172
Cimicifugoside 2523
Cimifugin 1258
Cinchocatine see 997
Cinchonain-1a-(4β→8)-catechin see 1946
Cinchonan-9-ol see 997
Cinchonidine 997
Cinchonine see under 997
Cinchovatine see 568
Cinegalline 1060
Cinene see 2057
1,8-Cineole 2045
Cinerin I 2046
Cinerin II 2046

Cinnamal see 1746
Cinnamaldehyde 1746
Cinnamic acid 1747
Cinnamic aldehyde see 1746
Cinnamodial 2136
Cinnamoylcocaine 1122
Cinnamoylmethylecgonine see 1122
4′-Cinnamoylmussatioside 1748
Cinnamylcocaine see 1122
Ciratin see 1378
Cirsilineol 1436
Cirsiliol 1437
Citisine see 1061
Citral 2047
Citramalic acid 84
Citric acid 85
Citridic acid see 79
Citrifoliol see 1382
Citrifolioside see 1401
Citrolimonin see 2724
Citronellal 2048
Citronellol see 2049
β-Citronellol 2049
L-Citrulline 224
Cleistanthin A 1597
Cleomiscosin A 1998
Cleomiscosins B, C see under 1998
Cleosandrin see 1998
Clerodendrin A 2430
Clerodin 2431
Clivorine 941
Cneorum chromone A see 1281
Cneorum chromone B see 1260
Cneorum chromone F see 1277
Cnicin 2246
Cocaine 1123
Coclanoline see 782
Codeine 716
β-Codeine see 766
Codonocarpine 1161
Codonopsine 897
Coelogin 1894
Coenzyme Q_{10} see 1886
Coenzyme R see 418
Coffearine see 446
Coffeine see 357
Co-galactoisomerase see 373
Colchicine 1162
Colenol 2432
Coleone A 1832
Colubrin 2524
Colubrinoside 2525
Columbin 2433
Combetin see 2644
Combretin A see under 893
Combretin B see under 893
Concanavalin A 383
Conchinine see under 1038
Conshosin A 2247
Conchosin B 2248
Condelphine 508
Conessine 1085
Confertifolin 2249
Confertin 2250
Confusoside see under 1357
ψ-Conhydrine see 928
γ-Coniceine 898
Coniferaldehyde see 1751
Coniferin 1749
Coniferoside see 1749
Coniferyl alcohol 1750
Coniferyl aldehyde 1751
(+)-Coniine 899
Conquinine see under 1038
Convallagenin A 2579

Convallamarin see 2581
Convallamarogenin 2580
Convallamaroside 2581
Convallasaponin A 2582
Convallaton see 2621
Convallatoxin 2621
Convicine 358
Convolamine 1124
Convoline 1125
Convolvine 1126
Copaene see 2137
α-Copaene 2137
Copalliferol B 1895
Coprine 225
Cordioxil see 2627
Coreopsin see under 1295
Corglykon see 2621
Coriamyrtin 2251
Coriariin A 1932
Corilagin 1933
Corn sugar see 10
Cornin see 2121
Cornudentanone 1833
Coronarian 422
Coronaridine 594
Coronopilin 2252
Corydalis C see 779
Corydinine see 779
Corylopsin see 1995
Corynine see 670
Corytuberine see 760
Cosmosiin see 1423
Costunolide 2253
Costunolide diepoxide see 2341
Cotonefuran 1236
Cotylenin F 2434
o-Coumaric acid 1752
p-Coumaric acid 1753
Coumarin 1295
cis-o-Coumarinic acid lactone see 1295
Coumarone see 1295
p-Coumaroylquinic acid 1754
p-Coumaryl alcohol 1755
Coumermycin A$_1$ 1296
Coumestrol 1522
Couroupitine A see 1044
Co-waldenase see 373
Crataegin see 1306
Crenatine A 866
Crepenynic acid 173
p-Cresol 1671
Crinamine see under 453
Crinasiadine 1163
Crinamine 459
Crinasiatine 1164
Cristacarpin 1523
Crocetin 2773
Crocin 2774
α-Crocin see 2774
Crotalaburnine see 937
Crotaline see 957
Crotanecine 942
Croton factor A$_1$ see 2505
Cryogenine 1165
Cryptolepine 595
Cryptophorine 900
Cryptopleurine 1166
Cryptostrobin see under 1411
Cryptoxanthin see 2776
α-Cryptoxanthin 2775
β-Cryptoxanthin 2776
Cryptoxanthol see 2776
Cuauchichicine 509
Cuauhtemone 2138
Cubebene 2139

Cubebin 1598
Cubebinolide see 1616
Cucoline see 789
Cucurbic acid 196
(1R,2R,3S)-Cucurbic acid see 196
Cucurbitacin A 2687
Cucurbitacin B 2688
Cucurbitacin C 2689
Cucurbitacin D 2690
Cucurbitacin E 2691
Cucurbitacin E 2-O-β-D-glucoside see under 2691
Cucurbitacin F 2692
Cucurbitacin H 2693
Cucurbitacin I 2694
Cucurbitacin J 2695
Cucurbitacin K see under 2695
Cucurbitacin L 2696
Cucurbitacin O 2697
Cucurbitacin P 2698
Cucurbitacin Q 2699
Cucurbitacin S 2700
Cucubitine 901
Cularicine 717
Cularidine 718
Cularimine 719
Cularine 720
Cumaldehyde see 2050
Cuminal see 2050
Cuminaldehyde 2050
Cunaniol see 180
Cupressiflavone 1438
16α-Curan-17-ol see 616
C-Curarine 596
C-Curarine II see 584
α-Curcumene 2140
Curcumin 1756
Curine see 696
Curling factor see 1238
Cuscohygrine 902
Cuskhygrine see 902
Cusparine 998
(+)-Cyanidan-3-ol see 1355
(+)-Cyanidanol see 1355
Cyanidin 1208
Cyanidin 3,5-di-O-glucoside see 1212
Cyanidin 3-O-galactoside 1209
Cyanidin 3-O-glucoside 1210
Cyanidin 3-O-rutinoside 1211
Cyanin 1212
3-Cyano-L-alanine see 226
L-β-Cyanoalanine 226
Cyanocobalamin 423
3-Cyanomethyl-1H-indole see 431
Cyasterone 2656
Cyathin A$_3$ 2435
Cycasin 424
Cyclamin 2526
Cycloartenol 2657
Cyclobuxine see 1086
Cyclobuxine B see under 1086
Cyclobuxine D 1088
α-Cyclocostunolide 2254
β-Cyclocostunolide 2255
Cycloeudesmol 2141
Cyclofoetoside B 2527
Cyclohexanehexol see 67
Cyclohexitol see 67
Cyclokievitone 1524
Cyclopamine 1087
(S)-13-(Cyclopent-2-enyl)tridecanoic acid see 116
Cycloprotobuxine A, D, F see under 1088
Cycloprotobuxine C 1088
Cyclosporin A 384
Cyclovirobuxine C 1089
Cyclovirobuxine D, F see under 1089

Cymarin 2622
β-D-Cymaropyranose see 3
Cymarose 3
D-Cymarose see 3
Cymene see 2051
p-Cymene 2051
2-p-Cymenol see 2041
3-p-Cymenol see 2085
p-Cymol see 2051
Cynarin(e) see under 1757
Cynaropicrin 2256
Cynisin see 2246
Cynoglossophine see 947
Cynometrine 1167
Cyperaquinone 1834
Cypholophine see under 1145
Cypress camphor see 2132
Cypripedin 1835
L-Cysteine 227
L-Cystine 228
Cytisine 1061
Cytochrome c 385
Cytosine 359
Cytovaricin see under 3

DABA see 229
Dactyliferic acid see 1741
Dahlin see 55
Daidzein 1525
Daidzein 7-O-glucoside see 1526
Daidzein 8-C-glucoside see 1574
Daidzein 4'-methyl ether see 1533
Daidzin 1526
Dalbergin 1527
(±)-Dalbergioidin 1528
Dalpanin 1529
Damascenine 1168
Dambose see 67
Dammarenediol-I 2746
Damsin 2257
Dantron see 1827
Daphandrine 721
Daphneticin 1999
Daphnetin 1297
Daphnetoxin 2436
Daphnin see under 1297
Daphnoline 722
Daphnopsis factor R_1 see 2484
Darlingine 1127
Date acid see 1741
Datiscetin 1439
Daturine see 1130
Daucol 2142
Dauricine 723
Davidigenin 1357
Davidioside see under 1357
Davoxin see 2627
De-N-methylicaceine see under 524
Decaline 1169
Decanedioic acid see 106
Decanoic acid 117
Decogenin 3-O-α-L-oleandroside see 2623
Decoic acid see 117
Decoside 2623
Decuroside III 1298
Decuroside IV see under 1298
Decursin 1299
Decursinol 1300
Decylic acid see 117
Deglucomusennin see 2528
Deguelin 1530
Dehydroalbine see 1048
Dehydrocollinusin see 1619
Dehydrocostus lactone 2258

Dehydrocycloguanandin 1961
1″,2″-Dehydrocyclokievitone see 1524
Dehydrodieugenol 1599
(+)-trans-Dehydrodiisoeugenol see 1622
Dehydro-β-erythroidine see under 732
Dehydrofalcarinone 174
(S)-4,5-Dehydrohomoproline see 221
Dehydrojuvabione 2143
5,6-Dehydrokawain 2000
Dehydromatricaria ester 175
Dehydromyodesmone 2144
Dehydrongaione 2145
Dehydronivalenol see 2210
(S)-4,5-Dehydropipecolic acid see 221
Dehydrosafynol 176
Dehydrotremetone 1237
Deidaclin 309
Delartine see 534
Delatine see 522
Delcorine 510
Delcosine 511
Delphinic acid see 93
Delphinidin 1213
Delphinidin 3-(6″-p-coumaroylglucose)-5-(6‴-malonylglucoside) see 1222
Delphinidin 3,3′,5′-tri-O-glucoside 1214
Delphinine 512
Delsemidine see 534
Delsine see 532
Delsoline 513
Deltaline 514
Deltonin 2583
Deltoside 2584
5′-Demethoxydeoxypodophyllotoxin 1600
Demethoxymatteucinol see under 1411
Demethylbatatasin IV 1896
Demethylbellidifolin 1962
Demethylbellidifolin 8-O-glucoside see under 1962
(+)-Demethylcoclaurine 724
N-Demethylconoduramine see 614
O^7-Demethylcularine see 718
4′-Demethyldeoxypodophyllotoxin 1601
7-Demethylencecalin see 1263
Demethylhomopterocarpin see 1555
4′-Demethylpodophyllotoxin 1602
Demethylpterocarpin see 1554
β-Demissidine glycoside see 1090
Demissine 1090
Dendrobine 840
Denudatin B 1603
Denudatine 515
11-Deoxocucurbitacin I 2701
11-Deoxojervine see 1087
15-Deoxyaconitine see 526
Deoxyaconitine see under 500
D-2-Deoxyarabinose see 4
5-Deoxychrysoeriol see 1450
Deoxyelephantopin 2259
6-Deoxy-L-galactose see 7
7-Deoxygermine see 1111
6-Deoxy-D-glucose see 16
Deoxygomisin A 1604
1-Deoxy-chiro-inositol see 76
2-Deoxy-D-chiro-inositol see 72
D-1-Deoxy-muco-inositol see 72
1-Deoxy-myo-inositol see 76
6-Deoxyjacareubin 1963
11-Deoxyjervine see 1087
5-Deoxykaempferol 1440
5-Deoxykaempferol 8-C-glucoside see under 1445
(±)-5-Deoxykievitone 1531
8-Deoxylactucin 2260
Deoxylapachol 1836
Deoxyloganin 2097
Deoxymannojirimycin 903

6-O-(6-Deoxy-α-L-mannopyranosyl)-D-galactopyranose see 36
6-Deoxy-D-mannose see 17
6-Deoxy-L-mannose see 17
3-Deoxymesaconitine see 523
6-Deoxy-3-O-methylgalactose see 5
5-Deoxymyricetin see 1497
Deoxynivalenol see 2210
Deoxynojirimycin see under 903
Deoxynupharidine 841
2-Deoxy-(D)-erythro-pentose see 4
Deoxypeganine 999
D-erythro-2-Deoxypentose see 4
Deoxypodophyllotoxin 1605
5-Deoxyquercetin see 1444
Deoxyribonucleic acid see 360
2-Deoxy-α-D-ribopyranose see 4
2-Deoxy-D-ribose see 4
D-2-Deoxyribose see 4
Deoxyribose 4
Deoxyvasicine see 999
Deoxyvasicinone 1000
Desacetoxymatricarin 2261
Desacetyleupaserrin 2262
Deserpidine 597
26-Desglucoavenacoside A see under 2576
26-Desglucoavenacoside B see under 2577
Desglucomusennin 2528
Desglucosyrioside see 2630
5'-Desmethoxydeoxypodophyllotoxin see 1600
11-Desmethoxyreserpine see 597
Desmethoxyyangonin see 2000
Desmethylnarcotine see 767
Desoxyribonucleic acid see 360
Desoxyribose see 4
Dextropimaric acid see 2483
Dextrose see 10
Dhurrin 310
Diacetoxyscirpenol 2146
Diallyl disulfide 143
Diallyl sulfide 144
(S)-2,4-Diaminobutyric acid see 229
L-α,γ-Diaminobutyric acid 229
(S)-2,6-Diaminohexanoic acid see 250
(S)-2,5-Diaminopentanoic acid see 256
(S)-α,δ-Diaminovaleric acid see 256
Dianoside A 2529
Dianthalexin 425
Dianthramine 426
Diapocarotene-6,6'-dioic acid see 2787
Dicaffeoylquinic acid 1757
Dicaffeoylquinic acids see under 1776
Dicaffeoyltartaric acid see 1744
Dichrin A see 1338
γ-Dichroine see 1010
β-Dichroine see 1010
Dichroine B see 1010
Dicoumarin see 1301
Dicoumarol 1301
Dictamine see 1001
Dictamnine 1001
Dictamnolactone see 2724
Dicumarol see 1301
Dicumol see 1301
Dicysteine see 228
3',4'-Didehydro-β,ψ-carotene see 2791
13,19-Didehydrorosmarinine see 939
2,6-Dideoxy-3-O-methyl-ribo-hexose see 3
2,6-Dideoxy-3-O-methyl-β-L-arabino-hexose see 15
Didrovaltrate see 2098
Didrovaltratum 2098
Didymin see under 1382
Didymocarpene see 2163
Diferulic acid 1758
Diferuloylmethane see 1756
Diffutin 1358

Digacin see 2627
3,5-Di-O-galloyl-4-O-digalloylquinic acid 1934
Digentiobiose ester of crocetin see 2774
Digiferrugineol 1837
Diginatigenin 3-O-tridigitoxoside see 2624
Diginatin 2624
Diginorgin see 2625
Digitalin 2625
Digitalinum verum see 2625
β-D-Digitalopyranose see 5
Digitalose 5
Digitogenin 2585
Digitonin 2586
Digitophyllin see 2626
Digitoxigenin 3-O-gentiobiosylthevetoside see 2620
Digitoxigenin 3-O-tridigitoxoside see 2626
Digitoxin 2626
Digoxigenin 3-O-tridigitoxoside see 2627
Digoxin 2627
Dihydroacacipetalin see 313
Dihydroanhydropodorhizol 1606
Dihydro-L-baikiaine see 258
Dihydrocaffeic acid 1759
Dihydroconiferyl alcohol 1760
Dihydro-o-coumaric acid see 1782
Dihydrocubebin 1607
23,24-Dihydrocucurbitacin I see 2696
1,2-Dihydro-α-elaterin see 2688
Dihydrofisetin see 1372
Dihydrogriesenin 2263
Dihydroharmine see 621
Dihydrohelenalin see 2365
Dihydrohelenalin acetate see 2227
2,3-Dihydro-3α-hydroxytropidine see 1142
L-erythro-Dihydroibotenic acid see 264
1,6-Dihydro-6-iminopurine see 355
3,6-Dihydro-6-iminopurine see 355
Dihydrokaempferol see 1347
Dihydromethysticin 2001
Dihydromikanolide 2264
Dihydromyricetin see 1346
Dihydropinosylvin 1897
1,9-Dihydro-6H-purin-6-one see 363
Dihydroquercetin see 1412
Dihydroresveratrol 1898
Dihydrosamidin 1302
Dihydrosanguinarine 725
3',4'-Dihydroxyacetophenone 1717
1,2-Dihydroxyanthraquinone see 1811
1,8-Dihydroxyanthraquinone see 1827
1,2-Dihydroxybenzene see 1670
2,5-Dihydroxybenzoic acid see 1675
3,5-Dihydroxybibenzyl see 1897
2,3-Dihydroxybutanedioic acid see 110
(3R,3'R)-3,3-Dihydroxy-β-carotene see 2793
5,7-Dihydroxychromone 1259
3,4-Dihydroxycinnamic acid see 1737
6,7-Dihydroxycoumarin see 1305
6,7-Dihydroxycoumarin 7-glucoside see 1294
2,8-Dihydroxy-1,6-dimethoxyxanthone see 1967
1,8-Dihydroxy-2,6-dimethoxyxanthone see 1980
5,7-Dihydroxy-3-ethylchromone see 1269
7,4'-Dihydroxyflavan 1359
5,7-Dihydroxyflavanone see 1400
7,4'-Dihydroxyflavanone see 1386
5,7-Dihydroxyflavone see 1431
5,8-Dihydroxyflavone see 1490
7β,12α-Dihydroxykaurenolide 2437
4,4'-Dihydroxy-2'-methoxychalcone see under 1233
7,8-Dihydroxy-6-methoxycoumarin see 1307
2',6'-Dihydroxy-4'-methoxydihydrochalcone 1360
4,7-Dihydroxy-2-methoxy-9,10-dihydrophenanthrene 1899
7,3'-Dihydroxy-4'-methoxy-8-methylflavan 1361
1,5-Dihydroxy-3-methoxyxanthone see 1978
1,7-Dihydroxy-3-methoxyxanthone see 1969

1,3-Dihydroxy-7-methoxyxanthone see 1971
1,4-Dihydroxy-2-methylanthraquinone 1838
1,8-Dihydroxy-3-methylanthraquinone see 1830
2,5-Dihydroxymethyl-3,4-dihydroxypyrrolidine see 905
7,4′-Dihydroxy-8-methylflavan 1362
2,4-Dihydroxy-5-methylpyrimidine see 371
9,10-Dihydroxyoctadecanoic acid see 118
3,4-Dihydroxyphenethylamine see 276
3,4-Dihydroxy-L-phenylalanine see 231
α-(3,4-Dihydroxyphenyl)-β-aminoethane see 276
2,3-Dihydroxypropanoic acid see 89
α,β-Dihydroxypropionic acid see 89
(9R,10R)-9,10-Dihydroxystearic acid see under 118
(9S,10R)-9,10-Dihydroxystearic acid 118
3,5-Dihydroxystilbene see 1917
6,7-Dihydroxy-3-tiglyloxytropane see 1133
1,7-Dihydroxyxanthone see 1965
Dihydrozeatin 361
Dill apiole see 1761
Dillapiole 1761
Dilongifolylborane see under 2177
DIMBOA glucoside 427
Dimethamine 1062
10,11-Dimethoxyalstonine see 576
2,6-Dimethoxybenzoquinone 1839
Dimethoxycoumarin see 1332
6,7-Dimethoxydictamnine see 1024
7,8-Dimethoxydictamnine see 1043
6,8-Dimethoxydictamnine see 1029
3.5-Dimethoxy-1,6-dihydroxyxanthone 1964
3,5-Dimethoxy-1,8-dihydroxyxanthone see 1986
6,7-Dimethoxy-2,2-dimethylchromene see 1275
2′,6′-Dimethoxy-4′-hydroxyacetophenone 1718
5,2′-Dimethoxy-6,7-methylenedioxyflavanone see 1350
2,6-Dimethoxyphenol 1672
2,6-Dimethoxyquinone see 1839
10,11-Dimethoxystrychnine see 582
Dimethulene see 2134
Dimethyl disulfide 145
trans-2,3-Dimethylacrylic acid see 111
6-(3,3-Dimethylallyl)-1,5-dihydroxyxanthone see 1960
N-(3,3-Dimethylallyl)guanidine see 278
8-(3,3-Dimethylallyl)spatheliachromene 1260
3-(2-(Dimethylamino)ethyl)indole see 275
4-(2-(Dimethylamino)ethyl)phenol see 284
3-(Dimethylaminomethyl)indole see 279
5,6-Dimethylbenzimidazolyl cyanocobamide see 423
N-Dimethylconoduramine see 614
Di-O-methylesculetin see 1332
Dimethylesculetin see 1332
O-Dimethylgalanthine see 457
N,N-Dimethylholafebrine see 1093
3,7-Dimethyl-2,6-octadienal see 2047
3,7-Dimethyl-6-octen-1-ol see 2049
3,7-Dimethyloct-6-enal see 2048
N,N-Dimethylserotonin see 270
N,N-Dimethyltryptamine 275
N,N-Dimethyltyramine see 284
1,3-Dimethylxanthine see 370
3,7-Dimethylxanthine see 369
Dinatin see 1457
1,6-Di(3-nitropropanoyl)-β-D-glucopyranoside see 421
2,6-Di(3-nitropropanoyl)-α-D-glucopyranoside see 422
Dioscin 2587
Dioscorine 904
Diosgenin 2588
Diosmetin 1441
Diosmetin 7-O-rutinoside see 1442
Diosmin 1442
Diosmol see 68
Diosphenol 2052
Diospyrin 1840
2,4-Dioxopyrimidine see 372
Dipentene 2053
Diphenyllobelidione see 915

Diphyllin 1608
Dipropyl disulfide 146
Dipterocarpol 2747
Discarine A see 861
Disenecionyl cis-khellactone 1303
Distichin see under 1464
Distylin see 1412
Ditaine see 600
Diterpenoid CS-B see 2485
Diterpenoid EF-D 2438
Diterpenoid SP-II 2439
β,β′-Dithiodi-L-alanine see 228
Divaricoside 2628
Divicine-β-glucoside see 374
Divostroside 2629
L-Djenikolic acid 230
DMDP 905
DMJ see 903
DMT see 275
DNA 360
DNJ see under 903
Docosanoic acid see 115
cis-13-Docosenoic acid see 120
Dodecanoic acid see 122
Dodecoic acid see 122
Dolcymene see 2051
Dolichodial 2099
Donaxine see 279
Dopa see 231
L-Dopa 231
Dopamine 276
Dopaxanthin 476
Dormin see 2122
Doronine 943
Dracorubin 1363
Dracyclic acid see 1665
Drimenin 2265
Dronabinol see 1701
Droserone 1841
Droserone 5-methyl ether see under 1841
Drosophilin B see 2487
Drummondin A 1261
Drummondins B, C, F see under 1261
Dryocrassol acetate see 2752
Dryophantin see under 2023
Dubamine 1002
Dubinidine 1003
Duboisine see 1130
Dufalone see 1301
Dulcite see 64
Dulcitol 64
Dulcose see 64
Dunnione 1842

Eburnamine 598
Eburnamonine 599
Ecdysone 2658
α-Ecdysone see 2658
β-Ecdysone see 2661
Egconine 1128
Ecgonine cinnamate methyl ester see 1122
Echimidine 944
Echinacoside 1762
Echinenone 2777
Echinocystic acid 2530
Echinone 1843
Echinopsine 1004
Echinorine 1005
Echitamine 600
Eicosanoic acid see 113
cis-5,cis-8,cis-11,cis-14-Eicosatetraenoic acid see 114
Elaeagnine see 586
Elaeocarpidine 1170
(+)-Elaeocarpine 1171
Elaeokanine C 1172

Elatericin A *see* 2690
Elatericin B *see* 2694
α-Elaterin *see* 2691
Elatine 516
Eldeline *see* 514
Eleganin 2266
α-Elemenadienolic acid *see* 2748
Elemicin 1763
α-Elemolic acid 2748
Eleocarpidine *see* 1170
Eleocarpine *see* 1171
Eleokanine C *see* 1172
α-Eleostearic acid 119
Elephantin 2267
Elephantopin 2268
Eleutherin 1844
Eleutheroside K *see* 2543
Ellagic acid 1673
Ellipticine 601
Elymoclavine 602
Embelic acid *see* 1845
Embelin 1845
Emetamine 726
Emetine 727
Emodin 1846
Emodin 8-O-glucopyranoside *see* 1847
Emodin 8-glucoside 1847
Emodin-1-rhamnoside *see* 1848
Encecalin 1262
Encelin 2269
Engeletin *see under* 1347
Enhydrin 2270
Enicoflavine 842
Enmein 2440
L-Ephedrine 277
Epiafzelechin-(4β→8)-catechin *see* 1941
Epiajmaline *see under* 557
(−)-Epicatechin 1364
ent-Epicatechin 1365
Epicatechin-(2β→7,4β→8)-epicatechin *see* 1952
Epicatechin-(4β→8)-*ent*-epicatechin 1935
[Epicatechin-(4β→8)]₅-epicatechin 1936
3-Epicrinamine *see* 462
10-Epieupatoroxin *see* 2295
Epifucose *see* 16
Epigallocatechin 3-gallate 1366
Epigallocatechin-(4β→8)-epicatechin-3-O-gallate ester 1937
Epiheterodendrin *see under* 313
Epilotaustralin *see under* 316
Epilucumin *see under* 317
(+)-Epipinoresinol *see under* 1638
(−)-Epipinoresinol *see under* 1638
Epiproacacipetalin *see under* 320
Epiprogoitrin *see under* 351
D-Epirhamnose *see* 16
Episesartemin *see under* 1650
Epitetraphyllin B *see* 326
Epivoacorine 603
Epivolkenin *see under* 326
Epiyangambin *see under* 1660
6,7β-Epoxy-3α-atropyloxytropane *see* 1115
3α,6α-Epoxy-7β-hydroxytropane *see* 1138
cis-12,13-Epoxy-*cis*-9-octadecenoic acid *see* 141
Epoxyambelline *see* 458
(3′R,4′R)-3′-Epoxyangeloyloxy-4′-acetoxy-3′,4′-dihydroselin 1304
12,13-Epoxyoleic acid *see* 141
6,7-Epoxytropine tropate *see* 1129
Equisetic acid *see* 79
Eremanthin 2271
Eremantholide A 2272
Eremofrullanolide 2273
Eremophilenolide 2274
Ergamine *see* 281
Ergine 604

Erginine *see under* 604
Ergobasine *see* 608
Ergoclavine *see under* 609
Ergocornine 605
Ergocorninine *see under* 605
Ergocristine 606
Ergocristinine *see under* 606
Ergocryptine 607
Ergokryptine *see* 607
Ergometrine 608
Ergometrinine *see under* 608
Ergonovine *see* 608
Ergonovinine *see under* 608
Ergosine 609
Ergosinine *see under* 609
Ergosterin *see* 2659
Ergosterol 2659
Ergostetrine *see* 608
Ergotamine 610
Ergotaminine *see under* 610
Ergotocin *see* 608
Eriocarpin 2630
Eriocitrin 1367
Eriodictin *see under* 1368
Eriodictyol 1368
Eriodyctiol 3′-methyl ether *see* 1379
Eriodyctiol 4′-methyl ether *see* 1377
Eriodyctiol 7-O-neohesperidoside *see* 1394
Eriodyctiol 7-O-rutinoside *see* 1367
Erioflorin acetate 2275
Erioflorin metacrylate 2276
Eriolangin 2277
Erisimotoxin *see* 2634
Erucic acid 120
Erucin *see under* 337
Erysimin *see* 2634
Erysoline *see under* 338
Erysonine 728
Erysotrine 729
Erythrabyssin I *see* 1523
Erythratidine 730
Erythrinin B *see* 1591
Erythrite *see* 65
meso-Erythritol *see* 65
Erythritol 65
Erythrocentaurin *see under* 2119
Erythroglucin *see* 65
α-Erythroidine 731
β-Erythroidine 732
Erythrophleguine 517
Escholine *see* 760
Escobedin *see* 2762
Esculetin 1305
Esculetin dimethyl ether *see* 1332
Esculetol *see* 1305
Esculin 1306
Esculoside *see* 1306
Esdragole *see* 1764
Eseramine 611
Eseridine 612
Eserine *see* 642
Eserine aminoxide *see* 612
Eserine oxide *see* 612
Eskel *see* 1267
Estradiol *see* 2668
β-Estradiol *see* 2668
Estragole 1764
Estriol *see* 2669
Estrol *see* 2670
Estrone *see* 2670
Ethanedioic acid *see* 101
Ethanoic acid *see* 78
Ethyl 2-methylbut-2-enoate 197
Ethyl *p*-methoxycinnamate *see* 1784
7-Ethyl-1,4-dimethylazulene *see* 2134

Ethylacetic acid see 83
Ethylformic acid see 103
Ethylglucosinolate see 341
Ethyl isothiocyanate see under 341
Eucalyptol see 2045
Eucalyptus wax see 166
Eucannabinolide see 2315
Eucazulen see 2155
Euclein see 1840
Eucommin A 1609
(−)-Eudesmin 1610
L-Eudesmin see 1610
Eudesmol 2147
Eudesobovatol A 1611
Eugenic acid see 1765
Eugeniin 1938
Eugenol 1765
Eugenol methyl ether 1766
Euglobal-Ia$_1$ 2002
Euonymit see 64
Eupachlorin 2278
Eupachlorin acetate 2279
Eupachloroxin 2280
Eupacunin 2281
Eupacunolin 2282
Eupacunoxin 2283
Eupaformonin 2284
Eupaformosanin 2285
Eupahyssopin 2286
Euparin methyl ether see 1241
Euparotin 2287
Euparotin acetate 2288
Eupaserrin 2289
Eupatilin 1443
Eupatocunin 2290
Eupatocunoxin 2291
Eupatolide 2292
Eupatoriochromene 1263
Eupatoriopicrin 2293
Eupatoroxin 2294
10-*epi*-Eupatoroxin 2295
Eupatundin 2296
Euphol 2749
Euphorbia factor RL$_9$ see 2494
Euphorbia factor RL$_{20}$ see 2493
Euphorbia factor Ti$_2$ 2441
Euponin 2297
Europine 945
Eurycomalactone 2716
Eustrophinum see 2644
Euxanthogen see 1976
Euxanthone 1965
Evodiamine 613
Evodin see 2724
Evoprenine 1006
Evoxanthidine 1007
Evoxine 1008

Fagaramide 1767
Fagaridine 733
α-Fagarine see 680
β-Fagarine see 1043
γ-Fagarine 1009
Fagarol see under 1648
Fagaronine 734
Fagasterol see 2753
Falaconitine 518
Falcarindiol 177
Falcarinol 178
Falcarinone 179
Fargesone A 1612
Fargesone B see under 1612
Farinosin 2298
Farnesene 2148
Farnesol see 2149

trans,trans-Farnesol 2149
Farrerol 1369
Fastigilin B 2299
Fastigilin C 2300
Favin 386
Fawcettidine 807
Fawcettimine 808
Febrifugine 1010
Fenchone 2054
Fernene 2750
Ferredoxin 387
Ferreirin 1532
Ferruginol 2442
Ferula aldehyde see 1751
Ferulaldehyde see 1751
Ferulic acid 1768
trans-Feruloylgomphrenin-I see 478
Feruloylputrescine see 1806
Fetidine 735
Ficin 388
Ficine 1173
Ficus protease see 388
Ficusin see 1327
Filicic acid see 1719
Filicin see 1719
Filixic acid see 1719
Filixic acid BBB 1719
Finaconitine 519
Fisetin 1444
Fisetin 8-*C*-glucoside 1445
Fisetini-3,4-diol see 1371
Fisetinidol 1370
ent-Fisetinidol-(4β→8)-catechin-(6-4β)-*ent*-fisetinidol 1939
Fisetinidol-4β-ol 1371
Flabellidine 809
Flaccidin B 2531
Flavidin 1900
Flavone 1446
Flavoxanthin 2778
Flebopom see 2599
Flindersiachromone 1264
Flindersiamine 1011
Flindersine 1012
Floribundoside see under 1391
Florilenalin 2301
Floripavine see 787
Flossonol 2003
Fluoroacetic acid see 99
Foetidine see 735
Folic acid 428
Folioxanthin see 2785
Folliculin see 2670
Food orange see 2764
Formic acid 86
Formononetin 1533
Formononetin 7-*O*-glucoside see 1561
Formylformic acid see 91
Formylic acid see 86
3-Formylindole see 432
4-Formylphenol see 1683
Forskolin 2443
Forsythiaside 1769
Forsythoside A see 1769
Fraction I protein see 407
Frangula emodin see 1846
Frangulanine 867
Frangulic acid see 1846
Frangulin A 1848
Frangulin B 1849
Franguloside see under 1848
Fraxetin 1307
Fraxetin-8-glucoside see 1308
Fraxin 1308
Fraxoside see 1308
Friedelanone see 2751

Friedelin 2751
β-D-Fructofuranosyl-*O*-α-D-galactopyranosyl-
 (1→6)-[*O*-α-D-galactopyranosyl-(1→6)]$_2$-α-D-glucopyranoside
 see 43
β-D-Fructofuranosyl-*O*-α-D-galactopyranosyl-(1→2)-α-D-
 glucopyranoside see 42
β-D-Fructofuranosyl-α D-glucopyranoside see 40
β-D-Fructofuranosyl *O*-β-D-glucopyranosyl-(1→6)-α-D-
 glucopyranoside see 25
β-D-Fructopyranose see 6
Fructose see 6
D-Fructose see 6
Fruit sugar see 6
Frullanolide 2302
Frutinone A 1265
Fucoidan 50
Fucoidin see 50
α-D-Fucopyranose see 7
Fucose 7
D-Fucose see under 7
L-Fucose see 7
Fucosterol 2660
Fucoxanthin 2779
Fugapavine see 761
Fugillin see 2150
Fukinanolide see 2237
Fukinotoxin see 959
Fulvine 946
Fulvoplumierin 2100
Fumagillin 2150
Fumaric acid 87
Fumaricine 736
Fumarine see 779
Fumidil see 2150
Funtumine 1091
Furacridone see 1013
Furcatin 1770
N^6-Furfuryladenine see 365
6-Furfurylaminopurine see 365
Furofoline see 1013
Furofoline I 1013
Fusariotoxin T-2 see 2201
Fusicoccin H 2444
Fustin 1372

G-I see 2452
GABA see 214
Gabunamine 614
Gabunine 615
Gaillardin 2303
Galactan see 47
meso-Galactitol see 64
Galactitol see 64
L-Galactomethylose see 7
α-D-Galactopyranose see 8
6-*O*-(β-D-Galactopyranosyl)-D-glucopyranose see 31
O-α-D-Galactopyranosyl-(1→6)-β-D-fructofuranosyl-α-D-
 glucopyranoside see 33
4-*O*-β-D-Galactopyranosyl-D-glucopyranose see 27
6-*O*-β-D-Galactopyranosyl-D-glucose see 31
D-Galactose see 8
L-Galactose see 8
Galactose 8
4-(β-D-Galactosido)-D-glucose see 27
α-D-Galactosyl-(1→6)-β-D-glucosyl-(1→2)-β-D-fructose see 35
α-Galactosyl-(1→6)-α-galactosyl-(1→6)-α-glucosyl-(1→2)-β-
 fructose see 39
α-Galactosyl-(1→6)-α-galactosyl-(1→6)-α-galactosyl-(1→6)-α-
 galactosyl-(1→6)-α-glucosyl-(1→2)-β-fructose see 23
Galangin 1447
Galangin 3,5,7-trimethyl ether 1448
Galantamine see 460
Galanthamine 460
Galanthaminone see 468
Galanthidine see 466

(+)-Galbacin 1613
Galegine 278
Galetin see 1458
Galipine 1014
Gallic acid 1674
Gallocatechin-(4α→8)-epigallocatechin 1940
1-Galloyl-β-D-glucose see 1678
3-Galloylquinic acid see 1702
Gambiriin 1941
Gambogic acid 1966
Gamolenic acid see 127
Garbanzol 1373
Garcinol 2004
Gardenoside 2101
Garryfoline see under 550
Garryine 520
Gaultherin see under 34
Geigerin 2304
Geissoschizoline 616
Geissospermine 617
Gelose see 45
Gelsemicine 618
Gelsemine 619
Gelseminic acid see 1333
Gelucystine see 228
Gemin A 1942
Geneserine see 612
Genipin 2102
Geniposide 2103
Genistein 1534
Genistein 6,8-di-*C*-glucoside see 1564
Genistein 8-*C*-glucoside 1535
Genistein 7-*O*-glucoside see 1536
Genisteol see 1534
Genistin 1536
Genistoside see 1536
Genkwanin 1449
Gentiacaulein 1967
Gentiacauloside see under 1967
Gentianadine 843
Gentianaine 844
Gentianamine 845
Gentianine 846
Gentianose 25
Genticaulin see under 34
Gentiobiose 26
β-Gentiobioside see 305
Gentiocrucine see 844
Gentiodelphin 1215
Gentioflavine 847
Gentiopicrin 2104
Gentiopicroside see 2104
Gentisein 1968
Gentisic acid 1675
Gentisin 1969
Geraldone 1450
Geranial see 2047
Geraniin 1943
Geraniol 2055
Geranyl acetate 2056
Geranylbenzoquinone 1850
Geranylgeraniol 2445
Geranylhydroquinone 1676
Geranyl tiglate see under 111
Germacrene B 2151
Germine 1092
Gerontine see 301
Gerrardine 906
Ghatti gum see 52
Gibberellic acid see 2446
Gibberellin A$_3$ 2446
Gigantine 737
[6]-Gingerdione 1771
[10]-Gingerdione see under 1771
Gingerenone A 1772
[6]-Gingerol 1773

[8]-Gingerol *see under* 1773
[10]-Gingerol *see under* 1773
Ginkgetin 1451
Ginkgoic acid 1677
Ginkgol *see* 1669
Ginkgolic acid *see* 1677
Ginkgolide A 2447
Ginnol *see* 156
Ginnone *see* 157
Ginsenoside Re 2532
Ginsenoside Rf 2533
Ginsenoside Rg$_1$ 2534
Girgensonine 907
Githagenin *see* 2539
Githagin *see under* 2539
Gitogenin 2589
Gitonin 2590
Gitoxigenin 3-*O*-glucosyldigitaloside *see* 2625
Gitoxigenin 3-*O*-tridigitoxoside *see* 2631
Gitoxin 2631
Glabranin 1374
Glabridin 1537
Glaucarubin 2717
Glaucarubinone 2718
Glaucarubolone 2719
Glaucine 738
Glaucolide A 2305
Glaucolide B 2306
Glaumeba *see* 2717
Glaziovine 739
Glechomanolide 2307
Glepidotin A 1452
Glepidotin B 1375
Glepidotin C 1901
Gliotoxin 1174
D-Glucitol *see* 75
Glucoalyssin 328
Glucoberteroin 329
Glucobrassicanapin 330
Glucobrassicin 331
Glucocapparin 332
Glucocheirolin 333
Glucocleomin 334
Glucocochlearin 335
Glucoconringiin 336
Glucoerucin 337
Glucoerysolin 338
Glucogallic acid *see* 1678
β-Glucogallin 1678
Glucoiberin 339
Glucoiberverin 340
Glucojiabutin *see* 335
Glucolepidiin 341
Glucolimnanthin 342
D-Glucomethylose *see* 16
Gluconapin 343
Gluconapoleiferin 344
Gluconasturtiin 345
Glucoproscillaridin A *see* 2642
Glucoputranyivin 346
α-D-Glucopyranose *see* 10
α-D-Glucopyranosido-(1→1)-α-D-glucopyranoside *see* 41
Glucopyranosyl-(1→3)-β-D-fructofuranosyl α-D-glucopyranose *see* 30
2-*O*-β-D-Glucopyranosyl-β-D-glucopyranose *see* 38
3-*O*-β-D-Glucopyranosyl-D-glucopyranose *see* 28
4-*O*-α-Glucopyranosyl-D-glucopyranose *see* 29
4-*O*-β-D-Glucopyranosyl-D-glucopyranose *see* 24
6-*O*-β-D-Glucopyranosyl-D-glucose *see* 26
5-*O*-β-D-Glucopyranosylbetanidin *see* 474
6-*O*-β-D-Glucopyranosylbetanidin *see* 477
5-*O*-β-D-Glucopyranosylisobetanidin *see* 480
2-*O*-β-D-Glucopyranosylnarciclasine *see* 463
β-D-Glucopyranuronic acid *see* 11
Glucoraphanin 347

Glucoraphenin 348
Glucosamine 9
β-D-Glucosamine *see* 9
Glucose 10
D-Glucose *see* 10
Glucose-β-galactoside *see* 31
Glucose-4-β-galactoside *see* 27
1-α-Glucosido-2-β-fructofuranose *see* 40
α-D-Glucosido-(1→1)-α D-glucoside *see* 41
6-(β-D-Glucosido)-D-glucose *see* 26
3-(β-Glucosido)indole *see* 430
Glucosinalbin *see* 352
β-D-Glucosyl caffeate *see* 1739
Glucosylorientin *see under* 1484
β-D-Glucosyloxyazoxymethane *see* 424
p-Glucosyloxybenzaldehyde cyanohydrin *see* 311
p-Glucosyloxymandelonitrile 311
4-*C*-Glucosyl-1,3,7-trihydroxyxanthone *see* 1974
Glucotropaeolin 349
Glucuronic acid 11
β-D-Glucuronic acid *see* 11
L-Glutamic acid 232
Glutamic acid 5-amide *see* 233
L-Glutamine 233
L-Glutaminic acid *see* 232
N-(*N*-L-γ-Glutamyl-L-cysteinyl)glycine *see* 389
L-γ-Glutamyl-L-hypoglycin 234
Glutaric acid 88
L-Glutathione 389
Glutathione-SH *see* 389
Gluten 390
Glutinosone 2152
(−)-Glyceollin I 1538
(−)-Glyceollin II 1539
D-Glyceric acid 89
Glycerin(e) *see* 66
Glycerol 66
Glycine 235
(−)-Glycinol 1540
Glycocol *see* 235
Glycolic acid 90
Glycoperine *see under* 1019
Glycophymine *see* 1016
Glycophymoline 1015
Glycosine *see* 988
Glycosminine 1016
Glycyphyllin 1376
Glycyrrhetic acid *see* 2535
Glycyrrhetin *see* 2535
Glycyrrhetinic acid 2535
α-Glycyrrhetinic acid *see* 2535
Glycyrrhinic acid *see* 2536
Glycyrrhizic acid *see* 2536
Glycyrrhizin 2536
Glycyrrhizinic acid *see* 2536
Glyoxalic acid *see* 91
Glyoxylic acid 91
Gnetin A 1902
Gnidicin 2448
Gnididin 2449
Gnidilatin 2450
Gniditrin 2451
Goitrin *see under* 351
Gomezine *see* 567
Gomisin L$_1$ methyl ether 1614
Gomisin-C *see* 1647
Gomphrenin-I 477
Gomphrenin-II *see under* 477
Gomphrenin-V 478
Goniothalenol 2005
Gonosan *see* 2012
Gossypetin 1453
Gossypetin 8-*O*-glucoside *see* 1454
Gossypin 1454
Gossypitrin *see under* 1453

Gossypol 2153
Gossypose see 35
Goyazensolide 2308
Gracillin 2591
Gradolide 2309
Gramine 279
Graminiliatrin 2310
Granatan-3-one see 1136
Grandidentatin 1774
Grandifoline see 562
Grandisin 1615
Granilin 2311
Grape sugar see 10
Graphinone 2154
Grass wax see 147
Gratibain see 2639
Gratiogenin 2702
Gratioside see under 2702
Gratus strophanthin see 2639
Graveoline 1017
Grayanotoxin I 2452
Greenhartin see 1857
Grevillol 1679
Grifulvin see 1238
Griseofulvin 1238
Grossheimin see 2312
Grosshemin 2312
Guaiac alcohol see 2156
Guaiacol 1680
Guaiazulene 2155
S-Guaiazulene see 2155
Guaiol 2156
Guajol see 2156
Guanandin see 1960
Guanine 362
Guanosine see under 362
Guarine see 357
Guatambuinine see 641
Guibourtinidol-(4α→6)-catechin 1944
L-Gulitol see 75
Gum arabic 51
Gum dragon see 54
Gum ghatti 52
Gum hashab see 51
Gum karaya 53
Gum tragacanth 54
Guvacine 908
Guvacoline see under 908
Gymnemic acid I 2537
Gynesine see 446
Gynocardin 312
Gynokhellan see 1267
Gypenoside XXV 2538
Gypsogenin 2539
Gypsogenin 3-O-rhamnosylglucuronide 2540
Gypsophilasapogenin see 2539
Gyrocarpine 740
Gyromitrin 280

Haemanthaminine 462
Haematoxylin 2006
Haemocorin 2007
Halfordinol 1175
Hallactone A 2453
Hallactone B 2454
β-D-Hamamelopyranose see 12
Hamatine see under 684
Hamamelose 12
Haploperine see 1008
Haplophine see 1009
Haplophyllidine 1018
Haplophytine 620
Haplopine 1019
Harmaline 621
Harman 622

Harmidine see 621
Harmine 623
Harmonyl see 597
Harpagide see under 2105
Harpagoside 2105
Harringtonine 1176
Harrisonin 2720
Hasubanonine 741
Hayatine see under 696
Heavenly blue anthocyanin 1216
Hecogenin 2592
Hederagenin see under 2566
Hederagenin 3-O-arabinoside 2541
Hederin 2542
Hederin 2543
Helenalin 2313
Helenalin acetate see 2221
Helenien 2780
Helenin see under 2217
Heliamine 742
Heliangin 2314
Heliangine see 2314
Helianthoside A 2544
Heliettin 1309
Heliosupine 947
Heliotridine 948
Heliotrine 949
Heliotropin see 1692
9-Heliotrylheliotridine see 949
Helixin see 2542
Hellaridine see 902
Hellebrigenin 3-acetate 2632
Hellebrigenin 3,5-diacetate see under 2632
Hellebrigenin 3-O-glucosylrhamnoside see 2633
Hellebrin 2633
Hellicoside 1775
Helminthosporal 2157
Helminthosporol 2158
Helveticoside 2634
Hemanthidine see 462
Hemigossipol 2159
Hemocorin see under 24
Henna see 1859
Hentriacontane 147
Hentriacontane-14,16-dione 148
Hentriacontanol see 149
Hentriacontan-1-ol 149
Hentriacontanone see 150
Hentriacontan-16-one 150
Hepenolactone see 204
5-(Heptadec-12-enyl)resorcinol 1681
5-Heptadecylene-1-carboxylic acid see 135
1,7-Heptanedicarboxylic acid see 82
Heptanedioic acid see 102
D-glycero-D-galacto-Heptitol see 69
D-glycero-D-manno-Heptitol see 77
D-glycero-D-talo-Heptitol see 77
D-altro-2-Heptulose see 20
Heptylcarbinol see 159
Heraclin see 1289
Herbacetin 1455
Herbacetin 8-methyl ether see 1505
Heritonin 2008
Hernandezine 743
Hernandion see 1605
Hernandulcin 2160
Herniarin 1310
Heroin see under 764
Hesperetin 1377
Hesperetin 7-O-neohesperidoside see 1395
Hesperetin 7-O-rutinoside see 1378
Hesperidin 1378
Hesperitin acid see 1778
Heteratisine 521
Heteroauxin see 286
Heterodendrin 313

Heterophylline see 568
Hetisine 522
Hexacosanol see 151
Hexacosan-1-ol 151
Hexadecanoic acid see 133
cis-9-Hexadecenoic acid see 134
Hexadecylic acid see 133
Hexahydroxycyclohexane see 67
3,5,7,3′,4′,5′-Hexahydroxyflavone see 1478
3,5,6,7,3′,4′-Hexahydroxyflavylium see 1218
3,5,7,3′,4′,5′-Hexahydroxyflavylium see 1213
Hexalupine see under 1080
Hexanedioic acid see 80
Hexanoic acid 121
n-Hexanoic acid see 121
1,3,5-Hexatriyne see 191
trans-Hex-3-enal see 201
cis-Hex-3-en-1-ol see 200
Hexoic acid see 121
arabino-2-Hexulose see 6
Hexylic acid see 121
Hibalactone see 1646
Higenamine see under 724
Hildecarpin 1541
Himachalol 2161
Himbacine 1177
Hinokiflavone 1456
Hinokinin 1616
cis-Hinokiresinol 1617
Hippeastrine 461
Hircinol 1903
Hirsutic acid C 2162
Hirsutidin 1217
Hirsutin see under 1217
Hispaglabridin A 1542
Hispidulin 1457
Histamine 281
L-Histidine 236
Histones 391
Hiyodorilactone A 2315
Holocalin see under 327
Homaline 1178
L-Homoarginine 237
β-Homochelidonine see 680
L-Homocysteine 238
Homoeriodictyol 1379
Homoferreirin 1543
Homoorientin see 1462
Homo-ormosanine see 1064
Homopisatin see 1587
L-Homoproline see 258
L-Homoserine 239
Homovitexin see 1467
Honokiol 1618
Hopan-29-ol acetate see 2752
Hopane-29-acetate 2752
Hordatine A 282
Hordatine B 283
Hordenine 284
5-HT see 298
5-HTP see 243
Humulene 2163
α-Humulene see 2163
Humulon see 1720
Humulone 1720
Huntericine see 599
Hunteriline see 598
Huratoxin 2455
Hydnocarpylacetic acid see 116
Hydrangenol 1904
Hydrangin see 1338
(−)-α-Hydrastine 744
Hydrochinene see 1020
Hydroconchinene see 1020
Hydroconchinine see 1020

o-Hydrocoumaric acid see 1782
Hydroergotocin see 670
Hydroquinidine 1020
Hydroquinol see 1682
Hydroquinone 1682
Hydroquinone-β-D-glucopyranoside see 1664
Hydroxyacetic acid see 90
p-Hydroxyacetophenone 1728
α-Hydroxyacetosyringone see under 1991
10-Hydroxyaconitine see 493
β-Hydroxy-L-alanine see 262
12α-Hydroxyamoorstatin 2721
Hydroxyanagyrine see 1056
6-Hydroxyapigenin see 1504
8-Hydroxyapigenin see 1466
4-Hydroxy-L-arginine see 240
γ-Hydroxy-L-arginine 240
p-Hydroxybenzaldehyde 1683
Hydroxybenzene see 1690
2-Hydroxybenzoic acid see 1698
p-Hydroxybenzoic acid 1684
p-Hydroxybenzylglucosinolate see 352
Hydroxybrazilin see 2006
Hydroxybutanedioic acid see 96
(2R)-2-Hydroxybut-3-enylglucosinolate see 351
(2S)-2-Hydroxybut-3-enylglucosinate see 351
(+)-8-Hydroxycalamenene 2009
11-Hydroxycanthin-6-one 624
3-Hydroxy-γ-carotene see 2790
(3R)-3-Hydroxy-β-carotene see 2776
6α-Hydroxycassamine see 517
2-Hydroxychrysophanol see 1867
p-Hydroxycinnamic acid see 1753
2-Hydroxycoriamyrtin see 2393
7-Hydroxycoumarin see 1338
6-Hydroxycyanidin 1218
Hydroxydammarane II see 2747
4′-Hydroxy-2′,6′-dimethoxyacetophenone see 1718
1-Hydroxy-2,11-dimethoxyaporphine see 750
5-Hydroxy-N,N-dimethyltryptamine see 270
20-Hydroxyecdysone 2661
7-Hydroxyencecalin see 1263
3-Hydroxyepiheterodendrin see under 320
Hydroxyethanoic acid see 90
β-Hydroxyethylbenzene see 1689
7-Hydroxyflavan 1380
6α-Hydroxygermine see 1099
6β-Hydroxy-1β,10β-guai-4-en-3-one see 2165
5-Hydroxy-2-hexenoic acid lactone see 204
(+)-γ-Hydroxy-L-homoarginine 241
4-Hydroxy-L-homoarginine see 241
4-Hydroxyhomopterocarpin 1544
1-Hydroxy-2-hydroxymethylanthraquinone see 1837
6β-Hydroxyhyoscyamine see 1113
3α-Hydroxyhypaconitine see 533
10β-Hydroxy-6β-isobutyryl furanoeremophilane 2164
4-Hydroxyisoeugenol see 1750
Hydroxyisopatchoulenone 2165
6-Hydroxykaempferol 1458
8-Hydroxykaempferol see 1455
6-Hydroxykynurenic acid 429
4β-Hydroxylupanine see 1071
13-Hydroxylupanine 1063
13-Hydroxylupanine 2-pyrrolecarboxylate see 1058
6-Hydroxyluteolin 1459
8-Hydroxyluteolin see 1461
p-Hydroxymandelonitrile see 310
2′-Hydroxy-4′-methoxyacetophenone see 1725
4-Hydroxy-3-methoxyacetophenone see 1714
7-Hydroxy-4′-methoxyflavan see 1351
2-Hydroxymethylanthraquinone 1851
2-Hydroxymethylbenzoic acid 1685
3-C-(Hydroxymethyl)-D-glycerotetrose see 1
2-C-Hydroxymethyl-D-ribose see 12
11β-Hydroxyobacunone see 2740

12-Hydroxy-*cis*-9-octadecenoic acid *see* 137
α-12-Hydroxyoleic acid *see* 137
13-Hydroxy-2-oxosparteine *see* 1063
7-Hydroxypeganine *see* 1045
2-Hydroxypent-4-enylglucosinolate *see* 344
N-(*p*-Hydroxyphenethyl)actinidine 848
4-Hydroxyphenethylamine *see* 303
β-(*p*-Hydroxyphenyl)-L-alanine *see* 266
p-Hydroxyphenylethanolamine *see* 293
2-Hydroxypiperitone *see* 2052
trans-4-Hydroxy-L-proline 242
4-Hydroxyproline betaine *see* 893
2-Hydroxypropanoic acid *see* 95
α-Hydroxypropionic acid *see* 95
5-Hydroxy-2-propylpiperidine *see* 928
2-Hydroxypyrimidine-4(1*H*)-one *see* 372
2-Hydroxypyrimidine-4(3*H*)-one *see* 372
4-Hydroxypyrimidine-4(3*H*)-one *see* 372
α-Hydroxypyrotartaric acid *see* 84
6-Hydroxyquercetin *see* 1491
8-Hydroxyquercetin *see* 1453
12a-Hydroxyrotenone 1545
5-Hydroxysalicylic acid *see* 1675
m-Hydroxysambunigrin *see* 327
8-Hydroxysantonin *see* 2233
12-Hydroxysenecionane-11,16-dione *see* 969
18-Hydroxyseneciphylline *see* 963
12α-Hydroxysparteine *see* 1075
3-Hydroxystachydrine 909
Hydroxysuccinic acid *see* 96
6-Hydroxytremetone 1239
1-Hydroxytriacontane *see* 164
α-Hydroxytricarballylic acid *see* 85
1-Hydroxytropane *see* 1135
5-Hydroxytryptamine *see* 298
5-Hydroxy-L-tryptophan 243
4-Hydroxytutin *see* 2338
3-Hydroxytyramine *see* 276
3-Hydroxy-L-tyrosine *see* 231
4-Hydroxyundecanoic acid lactone *see* 208
Hyenanchin *see* 2338
(−)-Hygrine 910
(−)-Hygroline 911
Hymenocardine 868
Hymenoflorin 2316
Hymenolin 2317
Hymenovin *see under* 2318
Hymenoxon 2318
Hyoscine 1129
Hyoscyamine 1130
Hypaconitine 523
Hypaphorine 625
Hypercalin B 2010
Hypericin 1852
Hypericum red *see* 1852
Hyperin 1460
Hyperoside *see* 1460
Hypnone *see* 1710
L-Hypoglycin 244
Hypoglycin A *see* 244
Hypoglycin B *see* 234
Hypoglycine *see* 244
Hypolaetin 1461
Hypoxanthine 363
Hypoxanthosine *see under* 363
Hyrcanin *see* 2244

IAA *see* 286
Ibamarin *see* 2694
Ibogaine 626
Ibogamine 627
Ibotenic acid 1179
Icaceine 524
Icacine 525
Ichangin 2722

Ichthyothereol *see* 180
Ichthyotherol 180
Icosanoic acid *see* 113
5,8,11,14-Icosatetraenoic acid *see* 114
Idaein *see* 1209
Idein *see* 1209
Ienancin *see* 2338
Ilexolide A 2545
Iliensine *see* 511
Ilienzine *see* 511
Illudin M 2166
Illudin S 2167
1*H*-Imidazole-4-ethanamine *see* 281
2-(Imidazole-4-yl)ethylamine *see* 281
4-Imidazoleethylamine *see* 281
5-Imidazoleethylamine *see* 281
Imperatorin 1311
Inactive limonene *see* 2053
Inamycin *see* 1320
Incanin *see* 2330
Incorporation factor IFP *see* 66
Indaconitine 526
Indian gum *see* 52
Indian tragacanth *see* 53
Indicaine *see* 834
Indican 430
Indicaxanthin 479
Indicine 950
Indigo *see under* 430
Indocybin *see* 296
Indol-3-ylacetic acid *see* 286
Indol-3-ylmethylglucosinolate *see* 331
Indole 285
1*H*-Indole-3-acetic acid *see* 286
Indole-3-acetonitrile 431
Indole-3-aldehyde *see* 432
Indole-3-carboxaldehyde 432
Indoleacetic acid 286
L-Indospicine 245
Indoxyl *see under* 430
Indoxyl-β-D-glucoside *see* 430
Ineketone 2456
Inermin *see* 1554
Inflexin 2457
Ingenol 2458
Ingenol 3,20-dibenzoate 2459
Inokosterone 2662
Inosine *see under* 363
Inosite *see* 67
Inositol *see* 67
i-Inositol *see* 67
meso-Inositol *see* 67
myo-Inositol *see* 67
Insariotoxin *see* 2201
Integerrenine 869
Integerressine 870
Integerrimine 951
Integerrine 871
Intermedine 952
Intibin *see* 2326
Inulicin 2319
Inulin 55
Inuline *see* 497
Inumakilactone A glycoside 2460
Inundatine 810
Ipecoside 745
Ipomeamarone 2168
Irehdiamine I *see* 1095
Irehine 1093
Iridin 1546
(+)-Iridodial 2106
Iridomyrmecin 2107
Irigenin *see under* 1546
Irigenin 7-*O*-glucoside *see* 1546
Irisolidone 1547
Irolone 1548

α-Irone 2169
γ-Irone see under 2169
β-Irone see under 2169
Isatidine 953
Isoajmaline 628
Isoalantolactone 2320
Isoamaranthin see under 477
Isoamidin 1313
Isoamyleneguanidine see 278
8-Isoamylenoxypsoralen see 1311
Isoanagyrine see under 1080
Isoathyriol 1970
Isobatatasin I 1905
Isobebeerine see 747
Isobetanidin see under 475
Isobetanin 480
Isoboldine 746
Isobrucein A see 2723
Isobruceine A 2723
Isobutrin 1219
Isobutyrylmallotochromene 1266
Isocaryophyllene 2170
Isochaksine see under 893
Isochamaejasmin 1381
Isochavicine see under 925
Isochlorogenic acid b 1776
Isochondrodendrine 747
Isocitric acid 92
Isococculidine 748
8-Isocopaene see 2212
Isocoreopsin see under 1353
Isocorydine 749
Isodictamnine 1021
Isodiospyrin 1853
Isodiprene see 2040
Isodomedin 2461
Isodonal 2462
Isodulcite see 17
Isoergine see under 604
Isoeugenol 1777
Isofebrifugine 1022
Isoferulic acid 1778
Isoficine 1180
29-Isofucosterol see 2651
Isofugapavine see 762
Isogentisin 1971
Isohelenin see 2320
Isohelenol 2321
6-Isohexenyl-α-naphthoquinone 1854
Isojuglone see 1859
Isokaempferide see 1470
L-Isoleucine 246
Isoliquiritigenin 1220
Isolobinine 912
Isomagnolol see under 1623
Isomangiferin see under 1976
1-Isomangostin 1972
Isomontanolide 2322
Isonootkatone see 2207
Isoorientin 1462
Isoorientin 2″-O-glucoside see under 1462
Isoorientin 7-methyl ether see 1507
Isopelletierine 923
N^6-(Δ^2-Isopentenyl)adenine 364
Isopentenyldihydropinosylvin see 1919
4′-Isopentenyloxyresveratrol see 1920
4-Isopentenylresveratrol see 1921
Isopicropodophyllone see under 1641
Isopimpinellin 1312
Isopiperine see under 925
Isoplumericin 2108
Isopropyl mercaptan see 163
6-Isopropyl-m-cresol see 2085
Isopropyl-o-cresol see 2041
Isopropylacetic acid see 93

4-Isopropylbenzaldehyde see 2050
Isopropylglucosinolate see 346
Isopropylisothiocyanate see under 346
Isopropyltoluene see 2051
Isopsoralen see 1286
Isoquercitrin 1463
Isorhamnetin 1464
D-Isorhamnose see 16
Isorhapontin 1906
Isorhodeose see 16
Isosafrole 1779
Isosafrole methyl ether see 1742
Isosakuranetin 1382
Isosakuranetin 7-O-neohesperidoside see 1401
Isosakuranin see under 1382
Isosalipurposide see under 1207
Isosamidin 1313
Isosarsasapogenin see 2604
Isoscoparin 1465
Isoscoparin 2″-O-glucoside see under 1465
Isoscutellarein 1466
Isotan B 433
Isotazettine see 470
Isotenulin 2323
Isoterchebin 1945
Isothebaine 750
Isothymol see 2041
Isouvaretin 1383
Isovaleric acid 93
(3S,6S)-3α-Isovaleroxy-6β-hydroxytropane see 1143
Isovitexin 1467
Isowillardiine 247
Ivalin 2324

Jacareubin 1973
Jacobine 954
Jacodine see 970
Jamaicensine see 1052
Jamine 1064
Japan agar see 45
Japan isinglass see 45
Japonine 1023
Jasminidine see under 849
Jasminine 849
Jasmone 198
(−)-Jasmonic acid 199
Jateorrhizine see 751
Jatrophatrione 2463
Jatrophone 2464
Jatrorrhizine 751
Jegosapogenol see 2519
Jervanin-11-one see 1094
Jervine 1094
Jesaconitine 527
JH III see 2172
Jionoside B_1 1780
Judaicin see 2406
Juglanin see under 1469
Juglone 1855
Juliflorine 913
Juliprosopine see 913
Junipene see 2177
Justicidin B 1619
Juvabione 2171
Juvenile hormone III 2172

Kadaya see 53
Kadsurenone 1620
Kadsurin A 1621
Kaempferide 1468
Kaempferol 1469
Kaempferol 3-methyl ether 1470
Kaempferol 4′-methyl ether see 1468
Kalbreclasine 463
Kalopanaxsaponin A see 2542

Kandelin A-1 1946
Kansuinine B 2465
Karacoline see 528
Karakin 434
Karakoline 528
Karaya gum see 53
Karwinaphthol B 2011
Katahdinone see 2200
Katilo see 53
Katine see 273
Kautschin see 2057
Kavain see 2012
Kawain 2012
Kazinol A 1384
Kelamin see 1267
Kelicor see 1267
Kellin see 1267
Keracyanin see 1211
Kessazulen see 2155
4-Keto-β-carotene see 2777
α-Ketoglutaric acid 94
altro-Ketoheptose see 19
D-erythro-Ketopentose see 19
α-Ketopropionic acid see 104
Ketosuccinic acid see 100
Khellactone disenecioate see 1303
Khellin 1267
Khellinin see 1268
Khellol glucoside 1268
Kidney bean lectin see 402
Kievitone 1549
Kigelinone 1856
Kikubasaponin see 2598
Kinetin 365
Kinic acid see 105
Kino-yellow see 2015
α-Kirondrin see 2717
Kitzuta saponin K_6 see 2542
Knightinol 1131
Kokusaginine 1024
Kolaflavanone 1385
Kolanone 2013
Kordofan gum see 51
Korglykon see 2621
α-Kosin 1721
β-Kosin see under 1271
Krukovine see under 695
Kryptosterol see 2663
Kullo see 53
Kunitz inhibitor, soybean 392
Kurchessine 1095
Kuromanin see 1210
Kuromatsuene see 2177
Kusaginin see 1809
Kuteera see 53
Kuwanone G 1471
Kuwanone H 1472

Labriformidin 2635
Labriformin 2636
Lachnophyllum ester 181
Lacinilene C 7-methyl ether 2173
Lactaroviolin 2174
L-Lactic acid 95
Lactoflavin see 443
Lactose 27
α-Lactucerol see 2758
Lactucin 2325
Lactucopicrin 2326
Laevo-dopa see 231
Laevulose see 6
Lagistase see 1673
Laguncurin see 2015
Lambdamycin see 1293
Laminaran 56

Laminaribiose 28
Laminarin see 56
Lamprolobine 1065
Lampterol see 2167
Lanatoxin see 2626
Lancerin 1974
Lanosterol 2663
Lapachenole 2014
Lapachic acid see 1857
Lapachol 1857
β-Lapachone 1858
Lappaconitine 529
Laricin see 1749
Laserolide 2327
Laserpitin 2175
Lasiocarpine 955
9-Lasiocarpylheliotridine see 945
Lasiodine A 872
Lassiocarpine see 955
Lathodoratin 1269
L-Lathyrine 248
Lathyrol 2466
(−)-Laudanidine 752
Laudanine see 752
Laudanine methyl ether see 753
Laudanosine 753
Laurenobiolide 2328
Lauric acid 122
Laurostearic acid see 122
Lawsone 1859
Layor carang see 45
LCR see 664
Leaf alcohol 200
Leaf aldehyde 201
Lecontin see under 1373
Ledol 2176
Ledum camphor see 2176
Leghaemoglobin 393
Leghemoglobin see 393
Legoglobin see 393
Leiocarposide 1686
Lemmatoxin 2546
Lemonol see 2055
Lenthionine 152
Lentil agglutinin see 394
Lentil lectin 394
Leontiformine 1066
Lepargylic acid see 82
Lettuce cotyledon factor see under 1760
Leucaenine see 255
Leucaenol see 255
Leucenine see 255
Leucenol see 255
L-Leucine 249
Leucofisetinidins see under 1371
Leucoharmine see 623
Leurocristine see 664
Leurosidine 629
Leurosine 630
Levodopa see 231
Levoglutamide see 233
Levoglutamine see 233
Levulose see 6
Liatrin 2329
Libanorin 1314
α-Licanic acid 123
Licarin A 1622
Licoisoflavone A 1550
Lignoceric acid 124
Ligulatin B 2330
Lilaline 1181
Limonene 2057
dl-Limonene see 2053
Limonin 2724
Linalol see 2058

Linalool 2058
Linamarin 314
Linarin see under 1417
Linarine see 1036
Linderane 2331
Lindheimerine 530
Linifolin A 2332
Linlyl acetate 2059
Linocaffein see under 1785
Linoleic acid 125
9,12-Linoleic acid see 125
α-Linolenic acid 126
γ-Linolenic acid 127
Linolic acid see 125
Linustatin 315
Lipiferolide 2333
Liquiritigenin 1386
Liquiritin see under 1386
Liriodendrin see under 1592
Liriodenine 754
Lirioresinol B see 1655
Lithospermic acid 1240
Littorine 1132
Lobelanidine 914
Lobelanine 915
Lobelidine see under 916
(−)-Lobeline 916
Lobinine see under 912
Locundjoside 2637
Loganin 2109
Loganoside see 2109
Lonchocarpenin 1551
Longifolene 2177
α-Longilobine see 970
β-Longilobine see 962
Loniceroside see 2118
Lophenol 2664
Lophocerine 755
Lophophorine 756
Loroglossol 1907
Lotaustralin 316
Lotisoflavan 1552
Loturine see 622
Lubimin 2178
Lucenin-2 1473
Luciculin see 535
Lucidin 1860
Lucidin ω-methyl ether 1861
Lucidine B 811
Luciduline 812
Lucumin 317
Lucuminoside see 317
Ludovicin A 2334
(+)-Luguine 757
Lunacridine 1025
Lunacrine 1026
Lunamarine 1027
Lunarine 1182
Lunularic acid 1908
Lunularin 1909
Lupanidine see 1069
Lupanine 1067
Lupeol 2753
Lupeol acetate 2754
Lupinidine see 1078
Lupinine 1068
Lupinus compound LA-1 see 1591
α-Lupulic acid see 1720
β-Lupulic acid see 1722
Lupulone 1722
Lusitanicoside 1781
Luteanine see 749
Lutein 2781
Lutein dipalmitate see 2780
Luteoliflavan-(4β→8)-eriodictyol 5-glucoside 1947

Luteolin 1474
Luteolin 6,8-di-C-glucoside see 1473
Luteolin 6-C-glucoside see 1462
Luteolin 7-O-glucoside 1475
Luteolin 8-C-glucoside see 1484
Luteolin 7-(6″-malonylglucoside) see under 1475
Luteolin 3′-methyl ether see 1433
Luteolin 4′-methyl ether see 1441
Luteolin 7-laminaribioside see under 28
Luteolinidin 1221
Luteone 1553
Luvangetin 1315
Lycaconitine 531
Lycobergine see 816
Lycocernuine 813
Lycoctonine 532
Lycoctonine 18-acetate see 549
Lycodine 814
Lycofawcine 815
Lycoflexine 816
Lyconnotine 817
Lycopen-16-ol see 2783
Lycopene 2782
Lycopericin see 1107
Lycopodine 818
Lycopsamine 956
Lycorenine 464
Lycoricidine 465
Lycoricidinol see 467
Lycorimine see 460
Lycorine 466
Lycoxanthin 2783
Lysergamide see 604
Lysergic acid amide see 604
D-Lysergic acid L-propanolamide see 608
L-Lysine 250
Lythramine 1183
Lyxitol see 62

(−)-Maackiain 1554
Macarpine 758
Macleyine see 779
Macluraxanthone 1975
Maclurin 2015
Macolidine see under 692
Macoline 759
Macrozamin 435
Maculine 1028
Maculosidine 1029
Maculosine 1030
Magellanine 819
Magnoflorine 760
Magnolol 1623
Magnosalicin 1624
Magnoshinin 1625
Mahanimbine 631
Mahuannin D 1948
Majudin see 1289
Makisterone 2665
Malampyrin see 64
Malampyrite see 64
Malic acid 96
Mallotochromene 1270
Mallotophenone 1723
Mallotoxin see 1729
Mallotusinic acid 1949
Malol see 2570
Malolic acid see 2570
Malonic aid 97
Malonylawobanin 1222
Malotoxin see 1155
Malt sugar see 29
Maltobiose see 29
Maltose 29
Malvalic acid 128

Malvidin 1223
Malvidin 3-galactoside *see under* 1223
Malvidin 3-glucoside *see under* 1223
Malvidin 3-rhamnoside *see under* 1223
Malvidin 3,5-diglucoside *see* 1224
Malvin 1224
Mammea-compound A/AA *see* 1316
Mammeisin 1316
Mancinellin 2467
Mandelonitrile β-gentiobioside *see* 305
Mandelonitrile glucoside *see* 322
Mandelonitrile-β-glucuronide *see under* 305
Mandelonitrile vicianoside *see* 325
Mangiferin 1976
Mangostin 1977
Manicol *see* 68
Manihotoxine *see* 314
Manna sugar *see* 68
Mannidex *see* 68
Manniflavanone 1387
Mannite *see* 68
Mannitol *see* 68
D-Mannitol 68
Manno-2-heptulose *see* 13
α-D-Mannoheptulopyranose *see* 13
Mannoheptulose 13
D-Mannoketoheptose *see* 13
L-Mannomethylose *see* 17
α-D-Mannopyranose *see* 14
D-Mannose *see* 14
Mannose 14
Mansonone C 1862
Marasmic acid 2179
Marchantin A 1910
Marein *see under* 1227
Margetine *see* 465
Maritimein *see under* 1225
Maritimetin 1225
Marmelosin *see* 1311
Marrubiin 2468
Mascaroside 2469
Matairesinol 1626
Matricaria camphor *see under* 2039
Matricin 2335
Matrine 1069
Matrine *N*-oxide *see* 1050
Maytansine 1184
Meatsugar *see* 67
Mecambrine 761
(+)-Mecambroline 762
Medicagenic acid *see under* 2547
Medicagenic acid 3-*O*-triglucoside 2547
(−)-Medicarpin 1555
Medicogenic acid 3-*O*-triglucoside *see* 2547
Medioresinol 4′-*O*-β-D-glucopyranoside *see* 1609
Megaphone 1627
Megastachine 820
Melampodin A 2336
Melampodinin 2337
Melampyrum *see* 64
Melannin 1556
Melezitose 30
Melibiose 31
Melicopicine 1031
Melicopine 1032
Melilotic acid 1782
Melilotoside *see under* 1752
Melitose *see* 35
Melitoxin *see* 1301
Melitriose *see* 35
Mellitoxin 2338
Melochinone 1033
Melosatin A 632
Melosatin B 633
p-Menth-1-en-8-ol *see* 2080

p-Mentha-1(7),2-diene *see* 2069
p-Mentha-1,5-diene *see* 2068
Menthofuran 2060
Menthol 2061
Menthone 2062
L-Menthone *see* 2062
Menthyl acetate 2063
Menthyl salicylate *see under* 2063
Menthyl valerate *see under* 2063
Mentol *see* 2061
β-Mercapto-L-alanine *see* 227
1-Mercaptopropane *see* 162
2-Mercaptopropane *see* 163
Mesaconitine 533
Mescaline 287
Mesembranol *see* 636
Mesembranone *see* 635
Mesembrenone 634
Mesembrine 635
Mesembrinol 636
Mesquite gum 57
Mesuaxanthone A 1978
Mesuaxanthone B 1979
Meteloidine 1133
Methanecarboxylic acid *see* 78
Methanedicarboxylic acid *see* 97
Methanethiol *see* 154
Methanoic acid *see* 86
L-Methionine 251
Methoxsalen *see* 1340
10-Methoxyajmalicine *see* 583
6-Methoxyaromadendrin 3-*O*-acetate 1388
4-Methoxybenzaldehyde *see* 1662
m-Methoxybenzylglucosinolate *see* 342
Methoxybrassinin 436
9-Methoxychelerythrine *see* 685
p-Methoxycinnamaldehyde 1783
p-Methoxycinnamic acid ethyl ester 1784
(*R*)-4-Methoxydalbergione 1557
8-Methoxydictamnine *see* 1009
7-Methoxy-2,2-dimethylchromene *see* 1274
5-Methoxy-*N*,*N*-dimethyltryptamine 288
Methoxyhydrastine *see* 766
12-Methoxyibogamine *see* 626
13-Methoxyibogamine *see* 655
N-Methoxy-3-indolylmethylglucosinolate *see* 350
1-Methoxy-3-indolylmethylglucosinolate *see* 350
9-Methoxy-α-lapachone 1863
6-Methoxymellein 2016
5-Methoxy-2-methylfuranonechromone *see* 1282
2-(4-Methoxyphenethyl)chromone 1271
2-Methoxyphenol *see* 1680
5-Methoxypsoralen *see* 1289
8-Methoxypsoralen *see* 1340
5-Methoxy-3-stilbenol *see* 1918
6-Methoxytaxifolin 1389
6-Methoxy-1,3,7-trihydroxyxanthone *see* 1970
3-Methoxy-1,5,8-trihydroxyxanthone *see* 1959
3-Methoxy-1,6,7-trihydroxyxanthone *see* 1958
6-Methoxyumbelliferone *see* 1333
Methyl allyl disulfide 153
Methyl benzoylecgonine *see* 1123
Methyl caffeate 1785
Methyl damasceninate *see* 1168
Methyl disulfide *see* 145
Methyl isopelletierine *see under* 923
Methyl mercaptan 154
Methyl thioalcohol *see* 154
Methylacetic acid *see* 103
N-Methylactinodaphnine *see* 713
N^6-Methyladenine *see* 366
O-Methylaknadinine *see* 741
5-*O*-Methylalloptaeroxylin 1272
Methylallopteroxylin *see* 1272
O-Methylalpinigenine *see* 681
3-Methylamino-L-alanine 252

6-Methylaminopurine 366
Methylanhalonidine see 775
N-Methylanhalonine see 756
2-Methylanthraquinone see 1884
Methylazoxymethanol-β-D-glucoside see 424
Methylazoxymethanol-β-D-primeveroside see 435
3′-O-Methylbatatasin III 1911
5-O-Methylbellidifolin see 1986
Methylbenzoylecgonine 1123
O-Methylbufotenine see 288
trans-Methylbutanedioic acid see 84
3-Methylbutanoic acid see 93
(3-Methylbut-2-enyl)guanidine see 278
8-(3-Methyl-2-butenyl)herniarin see 1321
(2-Methylbutyryl)phloroglucinol see 1724
3-O-Methylcaffeic acid see 1768
1-Methyl-β-carboline see 622
Methylcatechol see 1680
4-Methylcatechol 1687
Methylchavicol see 1764
3-Methylchrysacin see 1830
(+)-N-Methylconiine 917
trans-2-Methylcrotonic acid see 111
Methylcytisine see 1059
5-Methylcytosine 367
Methyl damasceninate see 1168
O-Methyldaphnoline see 721
14-O-Methyldelcosine see 513
3-O-Methyldigitoxose see 3
5-O-Methylembelin 1864
Methylenecycloartanol 2666
α-Methylenecyclopropyl)glycine 253
6,7-Methylenedioxydictamnine see 1028
8,9-Methylene-8-heptadecenoic acid see 128
β-Methylesculetin see 1333
O-Methyleugenol see 1766
6-O-Methyleuparin 1241
N-Methylflindersine 1034
3-O-Methyl-D-fucose see 5
Methylglucosinolate see 332
9-Methyl-3-granatanone see 1136
N-Methylgranatonine see 1136
2-Methyl-2-hydroxybutylglucosinolate see 334
2-Methyl-2-hydroxypropylglucosinolate see 336
3-Methyl-1H-indole see 299
2-O-Methyl-chiro-inositol see 71
3-O-Methyl-D-chiro-inositol see 70
5-O-Methyl-myo-inositol see 74
L-1-O-Methyl-myo-inositol see 63
O-Methylisococculine see 748
Methylisoeugenol 1786
Methylisothiocyanate see under 332
7-Methyljuglone see 1879
N-Methyl-launobine see 704
Methyllycaconitine 534
2-Methylmalic acid see 84
N-Methylmescaline 289
3-O-Methylmorphine see 716
8-Methylnorlobelone see 923
N-Methyloboranine see 781
2′-O-Methylodoratol 1390
O-Methyloxyacanthine see 770
4-Methylphenol see 1671
6′-O-Methylphlebicine see 784
6-Methylpinocembrin see 1411
(2R,6R)-2-Methyl-6-(E-prop-1-enyl)piperidine see 924
(1S)-1-Methylpropylglucosinolate see 335
(EEE)-N-(2-Methylpropyl)hexadeca-2,6,8-trien-10-ynamide 182
Methylprotocatechuicaldehyde see 1709
O-Methylptelefolonium 1035
10-Methylpyranotropane see 1118
Methylresorcinol see 1688
Methylsainfuran see under 1247
Se-Methyl-L-selenocysteine 254
4-(Methylsulfinyl)but-3-enylglucosinolate see 348

4-(Methylsulfinyl)butylglucosinolate see 347
5-Methylsulfinylpentylglucosinolate see 328
3-(Methylsulfinyl)propylglucosinolate see 339
3-(Methylsulfinyl)propylisothiocyanate see under 339
4-(Methylsulfonyl)butylglucosinolate see 338
3-(Methylsulfonyl)propylglucosinolate see 333
3-(Methylsulfonyl)propylisothiocyanate see under 333
2-O-Methylswertianin 1980
(+)-O-Methylthalicberine 763
3-Methyltheobromine see 357
4-(Methylthio)butylglucosinolate see 337
5-(Methylthio)pentylglucosinolate see 329
5-(Methylthio)pentylisothiocyanate see under 329
3-(Methylthio)propylglucosinolate see 340
3-(Methylthio)propylisothiocyanate see under 340
N-Methyl-L-tryptophan see 209
5-Methyluracil see 371
5-O-Methylvisamminol 1273
Methysticin 2017
Mevalonic acid 98
Mevalonolactone see under 98
Mexicanin E 2339
Mexicanin I 2340
Mezcaline see 287
Mezerein 2470
Michelenolide 2341
Micheliolide 2342
Microcrystalline cellulose see 48
Microhelenin A 2343
Microhelenin C 2344
Microlenin 2345
Micromelin 1317
Micromelumin see 1317
Micromerol see 2570
Mikanoidine see 966
Mikanolide 2346
Milk sugar see 27
Millettone 1558
Miltanthin see 778
L-Mimosine 255
Minorine see 663
Miraxanthin-I 481
Miraxanthin-II 482
Miraxanthin-III 483
Miraxanthin-V 484
Miroestrol 2667
Miserotoxin 437
Mistletoe lectins 395
Mitoquinone see 1886
Mitragynine 637
Mitraphylline 638
Mitrinermine see 647
Molephantin 2347
Molephantinin 2348
Monardaein 1226
Monardein see 1226
Monellin 396
Monocrotaline 957
Monofluoroacetic acid 99
Monogynol B see 2753
Monolupine see 1051
Monorhein see 1881
Monotropein 2110
Montanin 2471
Montanin A 2472
Montanol 2473
Montanyl alcohol see 158
Moracin A 1242
Moracins B–Z see under 1242
Morellin 1981
Morelsin see 1600
Morin 1476
Morindone 1865
Morintannic acid see 2015
Moritannic acid see 2015

Morphia *see* 764
Morphine 764
Moruberofuran C *see* 1244
Morusin 1477
Mucara *see* 53
Mucin *see* 1855
Mucronatine *see* 980
Mucronine A 873
Mucronine B 874
Mucronulatol 1559
Mulberrofuran A 1243
Mulberrofuran C 1244
Mulberrofuran G *see* 1235
Multifidol 1724
Multiflorine 1070
Multigilin 2349
Multiradiatin 2350
Multistatin 2351
Murrayanine 639
Murrayin *see* 1334
Musaroside 2638
Musca-aurin-I 485
Musca-aurin-II 486
Muscapurpurin 487
Muscarine 290
L-(+)-Muscarine *see* 290
Muscimol 291
Musculamine *see* 301
Musennin 2548
Musennin A *see* 2528
Mushroom sugar *see* 41
Musizin *see* 2018
Mutatochrome 2784
Muzigadial 2180
Mycose *see* 41
Mycosinol 183
Mycosporin *see* 1185
Mycotoxin F2 *see* 2034
Mycosporine 1185
Mycosporine 1 *see* 1185
Mycotoxin T-2 *see* 2201
Myosmine 918
β-Myrcene 2064
Myricetin 1478
Myricetin 3-O-rhamnoside *see* 1479
Myricitrin 1479
Myricoside 1787
Myricyl alcohol *see* 164
Myristic acid 129
Myristicin 1788
Myristoleic acid 130
Myxoxanthin *see* 2777

Nagarine *see* 493
Nagilactone C 2474
Napelline 535
Napellonine *see* 541
Naphthazarin 1866
Narceine 765
Narciclasine 467
Narcissine *see* 466
Narcosine *see* 766
α-Narcotine 766
(−)-α-Narcotine *see* 766
(−)-β-Narcotine *see under* 766
Narcotoline 767
Naringenin 1391
Naringenin 4′-methyl ether *see* 1382
Naringenin 7-methyl ether *see* 1404
Naringenin 7-O-glucoside *see* 1403
Naringenin 7-O-neohesperidoside *see* 1392
Naringenin 7-O-rutinoside *see* 1393
Naringin 1392
Narirutin 1393
Narwedine 468

Natalensine *see* 462
Natural brown 7 *see* 1855
NDGA *see* 1629
Neocarlinoside 1480
Neocembrene 2475
Neocembrene A *see* 2475
Neochanin *see* 1533
Neoeriocitrin 1394
Neoglucobrassicin 350
Neohesperidin 1395
Neohesperidose 32
Neoisodextropimaric acid 2476
Neoisostegane 1628
Neolinustatin 318
(+)-Neomatatabiol 2111
Neonicotine *see under* 884
Neopine 768
Neoprotoveratrine *see* 1098
Neoquassin 2725
Neoschaftoside 1481
Neoxanthin 2785
Nepetalactone 2112
Nepodin 2018
Neprotine *see* 751
Neral *see* 2047
Neriine *see* 1085
Nerol 2065
Nerolidol 2181
Neryl acetate *see under* 2065
Neurosporaxanthin 2786
Neuridine *see* 301
Nevadensin 1482
(+)-Ngaione *see* 2168
Niacin *see* 439
Niacinamide *see* 438
Nicotinamide 438
Nicotine 919
Nicotinic acid 439
Nicotinic acid amide *see* 438
Nicotinic acid N-methylbetaine *see* 446
Nicotinime *see* 884
Nicotyl amide *see* 438
Nicotyrine 920
β-Nicotyrine *see* 920
Nicouline *see* 1575
Nigakihemiacetal A 2726
Nigakihemiacetal B *see* 2725
Nigakilactone D *see* 2731
Nigelline *see* 1168
Nigrifactin 921
(−)-Nissolin 1560
Nitidine 769
Nitogenin *see* 2588
3-Nitro-1-propyl-β-D-glucopyranoside *see* 437
Nivalenol 2182
Niveusin C 2352
Nobilin 2353
Nodakenetin 1318
Nodakenetin cellobioside *see* 1298
Nodakenetin glucoside *see* 1319
Nodakenin 1319
Nomilin 2727
Nona-2,6-dienal 202
Nonacosan-10-ol 156
Nonacosan-10-one 157
Nonacosane 155
γ-Nonalactone 203
Nonanedioic acid *see* 82
Nopinene *see* 2071
Nor-ψ-ephedrine *see* 273
N-Noracutumine *see* 1146
L-Noradrenaline 292
Norathyriol 1982
Norathyriol-6-O-β-D-glucoside *see* 1990
Norbixin 2787

Index

Nordihydroguaiaretic acid 1629
ψ-Norephedrine see 273
L-Norepinephrine see 292
Norerythrostachaldine 536
Norgine see 46
Norhyoscyamine 1134
(+)-Nornicotine 922
Norobtusifolin 1867
Norpseudoephedrine see 273
Norsantal see 1562
Norswertianin 1983
Norswertianolin 1984
Norsympathol see 293
Norsynephrine see 293
Nortrachelogenin 1630
Norwogonin 1483
Norwogonin 8-methyl ether see 1515
Noscapine see 766
Notoginsenoside R1 2549
Notomycin A$_1$ see 1296
Novobiocin 1320
Nuatigenin 2593
Nucite see 67
Nummularine F 875
α-Nupharidine see 841
Nuttalline 1071
Nyasol see 1617

Obaberine 770
Obaculactone see 2724
Obacunone 2728
α-Obscurine 821
β-Obscurine 822
Obtusifolin see under 1868
Obtusifolin 2-glucoside 1868
Obtustyrene see under 1233
Ochotensine 771
Ochratoxin A 2019
Ochrolifuanine 640
β-Ocimene 2066
Octacosan-1-ol 158
n-Octacosyl alcohol see 158
n-Octadecanoic acid see 138
cis-6,cis-9,cis-12-Octadecatrienoic acid see 127
cis-9,cis-12,cis-15-Octadecatrienoic acid see 126
cis-9,trans-11,cis-13-Octadecatrienoic acid see 136
cis-9,trans-11,trans-13-Octadecatrienoic acid see 119
cis-6-Octadecenoic acid see 135
cis-9-Octadecenoic acid see 132
trans-11-Octadecenoic acid see 140
cis-9,cis-12-Octadecenoic acid see 125
6-Octadecylenic acid see 135
9-Octadecynoic acid see 187
12-O-2Z,4E-Octadienoyl-4-deoxyphorbol 13-acetate 2477
Octalupine see 1063
Octan-1-ol 159
Octanedioic acid see 108
Octanoic acid 131
n-Octanol see 159
Octoic acid see 131
Octoic alcohol see 159
Octopamine 293
n-Octyl alcohol see 159
7-(2-Octylcyclopropenyl)heptanoic acid see 128
8-(2-Octylcyclopropenyl)octanoic acid see 139
Octylic acid see 131
Octylic alcohol see 159
Odoratol 1396
Oestradiol-17β 2668
Oestriol 2669
Oestrone 2670
Officinalisnin I 2594
Ohioensin-A 2020
Oil garlic see 144
Okanin 1227

Olaxoside 2550
β-D-Oleandropyranose see 15
Oleandrose 15
L-Oleandrose see 15
Oleanoglycotoxin B see 2546
Oleanoglycotoxin-A 2551
Oleanolic acid 3-O-glucuronide 2552
Oleic acid 132
Oleuropein 2113
Olivacine 641
Ololiuqui see under 593, 604
Ononin 1561
Onopordopicrin 2354
Ophthalgon see 66
Opianine see 766
Orchinol 1912
Orcin see 1688
Orcinol 1688
Orensine see 881
Oreoselone methyl ether see 1323
Orientin 1484
Orizabin 2355
(−)-Ormosanine 1072
L-Ornithine 256
Orobanchin see 1789
Orobanchoside 1789
Oroboidine see 1058
Orobol 1562
Osajin see under 1569
Oscine see 1138
Osladin 2595
Osmitrol see 68
Osmosal see 68
Osthol 1321
Ostruthin 1322
Otobain 1631
Otobite see 1631
Otonecine 958
Ouabagenin 3-O-α-L-rhamnoside see 2639
Ouabain 2639
Ovalicin see 2154
Ovatifolin 2356
Ovatine 537
Oxalacetic acid 100
Oxalic acid 101
Oxaloacetic acid see 100
Oxoacetic acid see 91
Oxobutanedioic acid see 100
S-Oxodiallyl disulfide see 142
Oxoethanoic acid see 91
2-Oxoglutaric acid see 94
Oxolucidine B 823
19-Oxonorcassaidine see 536
4-Oxo-cis-9,trans-11,trans-13-octadecatrienoic acid see 123
2-Oxopropanoic acid see 104
2-Oxo-11α-sparteine see 1067
10-Oxosparteine see under 1053
Oxosuccinic acid see 100
22-Oxovincaleukoblastine see 664
Oxyacanthine 772
Oxyayanin A 1485
Oxyayanin B 1486
Oxymatrine see 1050
(+)-Oxypeucedanin see 1326
Oxypurine see 363
Oxyresveratrol 1913

PABA see 269
Pachycarpine see under 1078
Pachypodol 1487
Pachyrrhizone 1563
Paederoside 2114
Paeoniflorin 2021
Paeonol 1725
Paeonolide 1726
Paeonoside 1727

Palmarin 2478
Palmatine 773
Palmitic acid 133
Palmitoleic acid 134
Palmitone see 150
12-O-Palmitoyl-16-hydroxyphorbol 13-acetate 2479
Palustrine 1186
Panamine 1073
Panaxynol see 178
Pancratistatin 469
Pandamine 876
Paniculatin 1564
Pantherine see 291
Pantherine see 1185
Pantothenic acid 440
Papain 397
Papaverine 774
Papaya peptidases A and B 398
Papuanine see 668
[6]-Paradol 1790
Paramorphine see 795
Parasorbic acid 204
Paravallarine 1096
Parigenin see 2602
Parillin 2596
Parsley camphor see 1735
Parthenicin see 2357
Parthenin 2357
Parthenolide 2358
Passiflorin see 622
Patchouli alcohol 2183
Patchouli camphor see 2183
Patuletin 1488
Paucin 2359
Paviin see 1308
PEA see 294
Peanut agglutinin see 399
Peanut lectin 399
Pectenine see 710
Pectin 58
Pectin sugar see 2
Pectinose see 2
Pedalitin 1489
Pedunculagin 1950
Peganine 1036
Peimine see 1110
Pelargonidin 1228
Pelargonidin 3-(6″-p-coumaroylglucoside-5-(4‴,6‴-dimalonylglucoside) see 1226
Pelargonidin 3,5-di-O-glucoside see 1229
Pelargonin 1229
(−)-Pelletierine 923
ψ-Pelletierine see 1136
(−)-Pellotine 775
Peltaphorin see 1995
α-Peltatin 1632
β-Pelatin see under 1632
β-Peltatin A methyl ether 1633
Penicillium roqueforti toxin see 2187
Pentadecane 160
n-Pentadecane see 160
Pentagalloyl-β-D-glucose 1951
4,6,3′,4′,5′-Pentahydroxyaurone see 1204
2′,3′,4′,3,4-Pentahydroxychalcone see 1227
3,5,7,3′,4′-Pentahydroxyflavanone see 1412
3,5,7,2′,4′-Pentahydroxyflavone see 1476
3,5,7,3′,4′-Pentahydroxyflavone see 1492
5,6,7,3′,4′-Pentahydroxyflavone see 1459
5,7,3′,4′,5′-Pentahydroxyflavone see 1510
3,5,7,3′,4′-Pentahydroxyflavylium see 1208
3,5,7,3′,4′,5′-Pentahydroxy-3′-methoxyflavylium see 1231
5,6,7,3′,4′-Pentamethoxyflavone see 1506
5,6,7,8,4′-Pentamethoxyflavone see 1508
Pentamethylenediamine see 271
Pentane-1,5-diamine see 271

Pentanedioic acid see 88
1,2,3,4,5-Pentanepentol see 62
Pentanoic acid see 112
Pentaphylline see 1043
Pentathiepane see 152
D-*erythro*-Pentulose see 19
4-Pentenylglucosinolate see 330
11-(3-Pentyloxiranyl)-9-undecenoic acid see 141
Peonidin 1230
Peonidin 3-sophoroside-5-glucoside tri(caffeoylglucose) ester see 1216
Peppermint camphor see 2061
Pereirine see 616
Perforatin A see 1272
Pericalline see 567
Perillaldehyde 2067
(−)-Perillaldehyde-α-syn-oxime see under 2067
1-Peroxyferolide 2360
Perseitol 69
Persitol see 69
Peruviol see 2181
Petasin 2184
Petasitenine 959
Petroselenic acid 135
Petunidin 1231
Petunin see under 1231
Peucedanin 1323
Peucenidin 1324
Peyotline see 775
Pfaffic acid 2553
Pfaffoside A 2554
PHA see 402
Phaeantharine 776
Phaenthine see under 790
Phalloidin 400
Phantomolin 2361
Phaselic acid see 1791
Phaseolic acid 1791
Phaseolin see 1566
(−)-Phaseollidin 1565
(−)-Phaseollin 1566
(−)-Phaseollinisoflavan 1567
Phaseoloside D 2555
Phaseolunatin see 314
Phaseolus substance II see 1549
Phaseoluteone see 1550
Phaseomannite see 67
Phasin see 402
α-Phellandrene 2068
β-Phellandrene 2069
Phenethyl alcohol 1689
Phenethyl caffeate 1792
Phenethylamine 294
Phenol 1690
Phenylacrolein see 1746
β-Phenyl-L-alanine see 257
L-Phenylalanine 257
1-Phenylethanone see 1710
2-(2-Phenylethyl)chromone see 1264
β-Phenylethylamine see 294
2-Phenylethylglucosinolate see 345
2-Phenylethylisothiocyanate see under 345
Phenylformic acid see 1665
1-Phenylhepta-1,3,5-triyne 184
Phenylheptatriyne see 184
1-Phenyl-5-heptene-1,3-diyne 185
(1R,2S)-1-Phenyl-1-hydroxy-2-methylaminopropane see 277
Phenylic acid see 1690
8-Phenyllobelol see 932
2-(3-Phenyl-1-propynyl)furan see 171
Phlegmarine 824
Phloracetophenone 4,6-dimethyl ether see 1731
Phloretin 1397
Phloretin 2′-O-glucoside see 1398
Phloretin 4′-methyl ether see 1348

Index

Phloretin 2′-O-rhamnoside see 1376
Phloridzin 1398
Phlorin see under 1691
Phloroglucin see 1691
Phloroglucinol 1691
Phoratoxin 401
Phorbol 2480
Phorbol 12-tiglate 13-decanoate 2481
Phosphoglyceric acid see under 89
Phrymarolin 1634
PHT see 184
Phycite see 65
Phycitol see 65
Phyllanthin 1635
Phyllanthostatin A 1636
Phyllodulcin 1914
Phylloquinone see 1887
Phillospadine 1187
Physalien 2788
Physalin see 2788
Physcion see under 1869
Physcion 8-gentiobioside 1869
Physcion 8-O-β-D-gentiobioside see 1869
Physcion 1-β-D-glucopyranoside see under 1870
Physcion 8-glucoside 1870
Physcion 8-O-β-D-monoglucoside see 1870
Physcionin see 1870
Physoperuvine 1135
Physosterine see 642
Physostigmine 642
Physostigmine aminoxide see 612
Physostigmine oxide see 612
Physostol see 642
Physovenine 643
Phytochrome 441
Phytoglycogen 59
Phytohaemagglutinin 402
Phytol 2482
Phytolaccanin see 474
Phytolaccoside B 2558
3-Phytomenadione see 1887
Phytonadione see 1887
Piceatannol 1915
Piceid 1916
Picein 1728
Picrasin C 2729
Picropodophyllic acid lactone see 1637
Picropodophyllin 1637
Picrotin 2362
Picrotoxin see under 2363
Picrotoxinin 2363
Pilocarpine 1188
Pilocereine 777
Pilosine 1189
(+)-Pimaric acid 2483
α-Pimaric acid see 2483
d-Pimaric acid see 2483
Pimelea factor P$_2$ 2484
Pimelic acid 102
Pimpinellin 1325
Pin-2-en-4-one see 2088
2-Pinen-7-one see 2044
Pinene see 2070
α-Pinene 2070
β-Pinene 2071
2-Pinene see 2070
Pinguisone 2185
Pinidine 924
Pinitol 70
Pinobanksin 1399
Pinocarvone 2072
Pinocembrin 1400
Pinopalustrin see 1630
Pinoresinol 1638
Pinoresinol dimethyl ether see 1610

Pinosylvin 1917
Pinosylvin methyl ether 1918
L-Pipecolic acid 258
(S)-Pipecolinic acid see 258
Piperidic acid see 214
Piperine 925
1-Piperinoylpiperidine see 925
Piperitenone oxide see 2076
Piperitol glucoside see 1651
Piperitone 2073
Piperlongumine see 926
Piperonal 1692
Piplartine 926
Piptamine see 1072
Piptanthine 1074
Pisatin 1568
Pitasitenine 959
Pitayine see under 1038
Pithecolobine 1190
Plant indican see 430
Plantamajoside 1793
Planteose 33
Plastocyanin 403
Plastoquinone-9 1871
Platyphylline 960
Plaunol B 2485
Plaunol D 2486
Pleiocarpinidine see 598
Pleniradin 2364
Plenolin 2365
Pleuromutilin 2487
Pleurostyline 1191
Plicatic acid 1639
PLMF 6 see 1705
Plumbagin 1872
Plumericin 2115
Plumieride 2116
Plumieroside see 2116
PMCA see 1783
PNA see 399
Podecdysone B 2671
Podolactone B 2488
Podolide 2489
Podophyllinic acid lactone see 1640
Podophyllotoxin 1640
Podophyllotoxone 1641
Podorhizol β-D-glucoside 1642
Polhovolide 2366
Pollinastanol 2672
Polychrom see 1306
Polycladin see 1434
Polydatin see 1916
Polygodial 2186
Polygonolide 2022
Polymannuronic acid see 46
Polypodine B 2673
Polystichin see 1715
Pomiferin 1569
Ponalactone A 2490
Ponasterone A 2674
Poncirin 1401
Ponkanetin see 1508
Populin 1693
Populoside see 1693
Porphyran 60
Portulal 2491
Portulaxanthin 488
Potassium atractylate see 2420
PR toxin 2187
Prangenidine see 1284
Prangolarine 1326
Pratensein 1570
Pratensol see 1519
Pratol see 1533
Prebetanin 489

Precocene 1 1274
Precocene 2 1275
Premarrubiin 2492
Prenyl caffeate 1794
Prenylbenzoquinone 1873
4-Prenyldihydropinosylvin 1919
8-Prenylgalangin see 1452
8-Prenylhomoeriodictyol 1402
5'-Prenylhomoeriodictyol 1402
4'-Prenyloxyresveratrol 1920
4-Prenylresveratrol 1921
Pretazettine 470
Primetin 1490
Primeverose 34
Primin 1874
Primulasaponin 2557
Primulaverin see under 34
Pristimerin 2755
Proacaciberin 319
Proacacipetalin 320
Proacacipetalin 6'-arabinoside see 319
Proanthocyanidin A2 1952
Procyanidin B4 1953
Progoitrin 351
L-Proline 259
(−)-Pronuciferine 778
S-((E)-Prop-1-enyl)-L-cysteine S-oxide 260
S-(Prop-1-enyl)cysteine sulfoxide see 260
Prop-2-enylglucosinolate see 353
1,3-Propanedicarboxylic acid see 88
Propanedioic acid see 97
Propanethial S-oxide 161
Propane-1-thiol 162
Propane-2-thiol 163
1,2,3-Propanetriol see 66
Propanoic acid see 103
Propapyriogenin A_2 2558
trans-S-(1-Propenyl)cysteine S-oxide see 260
p-Propenylanisole see 1734
4-Propenylguaiacol see 1777
Propionic acid 103
Propyl disulfide see 146
Propyl mercaptan see 162
Propylacetic acid see 112
Propylformic acid see 83
(S)-2-Propylpiperidine see 899
Prosophylline see under 927
(+)-Prosopinine 927
Prosopis gum see 57
Prostalidin A 1643
Proteacin 321
Protoalba see 1097
Protoanemonin see under 205
Protocatechuic acid 1694
Protocevine see 1108
Protodioscin 2597
Protogracillin 2598
Protokosin see under 1721
Protopine 779
(+)-Protoquercitol see 72
Protoveratrine A 1097
Protoveratrine B 1098
Protoverine 1099
Provincialin 2367
Provismine see 1339
Prunasin 322
Prunasin xyloside see 317
Prunetin 1571
Prunetol see 1534
Prunin 1403
Prunol see 2570
Pseudaconitine 538
Pseudoacetic acid see 103
Pseudoakuammicine see under 558
Pseudobaptigenin 1572

Pseudochelerythrine see 788
Pseudoconhydrine 928
Pseudocyclobuxine D see under 1086
Pseudodigitoxin see 2631
Pseudoephedrine see under 277
Pseudohyoscyamine see 1134
Pseudohypericin 1875
Pseudoivalin 2368
Pseudonorephedrine see 273
Pseudopelletierine 1136
Pseudopunicine see 1136
Pseudopurpurin 1876
3-Pseudotropanol see 1137
Pseudotropine 1137
Pseudotropine benzoate see 1141
Psilocin 295
Psilocybin 296
Psilocyn see 295
Psilopine see 694
Psoralen 1327
Psoralidin 1573
Psorospermin 1985
Psychotridine 644
Psychotrine 780
Ptaerochromenol 1276
Ptaeroglycol 1277
Pteleatine 1037
Ptelefolonium see 1035
Pterochromenol see 1276
Pteroglycol see 1277
Pterosterone 2675
Pterostilbene 1922
Pteroylglutamic acid see 427
Pteryxin 1328
Puerarin 1574
Pukateine 781
Pulegone 2074
Pulverochromenol 1278
Pungenin see under 1717
Pungenoside see under 1717
Punicic acid 136
Purapuridine see 1101
Purapurine see 1104
Purin-6(1H)-one see 363
1H-Purin-6-amine see 355
Purostrophan see 2639
Purothionins 404
Purpapuridine see 1103
Purpureaside C 1795
Purpurin 1877
Purpurin 1-methyl ether 1878
Purpurogallin 2023
Purrenon see 1965
Putrescine 297
Pycnolide 2369
Pyrethrin I 2075
Pyrethrin II 2075
Pyrethrosin 2370
Pyridine carboxylic acid see 439
Pyridine 3-carboxylic acid see 439
Pyridine carboxylic acid amide see 438
Pyridoxine 442
Pyrimidine-2,4(1H,3H)-dione see 372
2,4-Pyrimidinediol see 372
Pyrocatechin see 1670
Pyrocatechol see 1670
Pyrogallic acid see 1695
Pyrogallol 1695
Pyrogentisic acid see 1682
Pyroibotenic acid see 291
Pyropseudoaconitine see 518
Pyroracemic acid see 104
(S)-2-Pyrrolidine carboxylic acid see 259
3-(2-Pyrrolidinyl)pyridine see 922
α-Pyrufuran 1245

Pyruvic acid 104

Quadrone 2371
Quassiin see 2731
Quassimarin C 2730
Quassin 2731
Quebrachine see 670
(−)-Quebrachitol 71
Quebracho blanco see under 572
Quebracho colarado see under 572
Quercetagetin 1491
Quercetagetin 3,6-dimethyl ether see 1425
Quercetagetin 6-methyl ether see 1488
Quercetagetin 3,6,7,3′-tetramethyl ether see 1434
Quercetagetin 3,7,3′-trimethyl ether see 1435
Quercetin 1492
Quercetin 3-galactoside see 1460
Quercetin 3-O-glucoside see 1463
Quercetin 7-O-glucoside see 1494
Quercetin 3-methyl ether 1493
Quercetin 3′-methyl ether see 1464
Quercetin 5-methyl ether see 1426
Quercetin 7-methyl ether see 1496
Quercetin 3-O-rhamnoside see 1495
Quercetin 3-rutinoside see 1500
Quercetin 3-sophoroside see under 38
Quercetin 3,7,3′-trimethyl ether see 1487
Quercimeritrin 1494
d-Quercitol see 72
L-Quercitol see under 72
vibo-Quercitol see 76
(+)-Quercitol 72
Quercitrin 1495
L-(−)-Quibrachitol see 71
Quillaic acid 2559
Quillajasaponin see under 2559
Quillajoside see under 2559
Quing Hau Sau 2372
Quinghaosu see 2372
Quinic acid 105
Quinidine see under 1038
α-Quinidine see 997
Quinine 1038
β-Quinine see under 1038
Quinotidine see 1020
Quinovatine see 568
α-D-Quinovopyranose see 16
Quinovose 16
Quinquangulin 1279

Radiatin 2373
Raffinose 35
Ragweed pollen allergen Ra5 405
Ramentaceone 1879
Randainol 1644
Ranunculin 205
Rapanone 1880
Raubasine see 556
Raugalline see 557
Raunormine see 597
Raupine see 650
Rauwolfine see 557
Recanescine see 597
Rectolander see 2599
Red oil see 132
Regianin see 1855
Renardine see 972
Rescinnamine 645
Reserpidine see 597
Reserpine 646
Reserpinine see 645
Resiniferonol 2493
Resiniferotoxin 2494
Resorcin see 1696
Resorcinol 1696

Resveratrol 1923
Resveratrol 3,5-dimethyl ether see 1922
Resveratrol 3-O-β-D-glucoside see 1916
Retamine 1075
(+)-Reticuline 782
Retronecine 961
Retrorsine 962
Retrorsine N-oxide see 953
Rhabarberone see 1815
Rhamnetin 1496
Rhamnoliquiritin see under 1386
α-L-Rhamnopyranose see 17
6-O-α-L-Rhamnopyranosyl-D-galactose see 36
2-O-α-L-Rhamnopyranosyl-D-glucopyranose see 32
Rhamnose 17
D-Rhamnose see 17
L-Rhamnose see 17
6-O-α-Rhamnosyl-D-glucose see 37
Rhamnoxanthin see 1848
Rhazine see 559
Rhazinine see 564
Rheadine see 783
Rheic acid see 1881
Rhein 1881
Rheochrysin see 1870
Rhetine see 649
Rheum emodin see 1846
Rhexifoline 850
Rhinanthin see 2093
Rhipocephalin 2188
Rhodeose see 7
Rhodexin A 2640
Rhododendrin 2024
Rhodojaponin IV 2495
ψ-Rhodomyrtoxin 1246
Rhodotoxin see 2452
Rhodoxanthin 2789
Rhoeadine 783
Rhombifoline 1076
Rhombinine see 1051
Rhubarb yellow see 1881
Rhubarberone see 1815
Rhynchophylline 647
Rhyncophylline see 647
Ribalinium 1039
Ribitol 73
meso-Ribitol see 73
D-2-Ribodesose see 4
Riboflavin 443
D-Riboketose see 19
Ribonucleic acid see 368
β-L-Ribopyranose see 18
Ribose 18
D-Ribose see 18
trans-Ribosylzeatin see under 375
Ribulose 19
α-D-Ribulose see 19
Ribulose 1,6-bisphosphate carboxylase-oxygenase see 407
Ricin 406
Ricinine 929
Ricinoleic acid 137
Ridelliine see 963
Riddelline 963
Ridentin 2374
Ridentin A see 2374
Rinderine 964
Rishitin 2189
Rivularine see 938
RNA 368
Robinetin 1497
Robinetinidol-(4α→8)-catechin-(6→4α)-robinetidol 1954
Robinin 1498
Robinobiose 36
Robustadial A 2025
Robustadial B see under 2025

Index

Robustaflavone 1499
Robustaol A 2026
Robustine 1040
Rochessine see 1085
Rodiasine 784
Romanicardic acid see 1677
Roquessine see 1085
Roridin A 2190
Rosmarinic acid 1796
Rosmarinine 965
Rotenone 1575
Rottlerin 1729
Rotundifolone 2076
Roxburghine B 648
Royline see 532
Rubiadin 1882
Rubichloric acid see 2092
Rubisco 407
Rubixanthin 2790
Rubrofusarin 1280
Rugosal 2191
Rugosin D 1955
Ruscogenin 2599
Ruscopine 444
Ruscorectal see 2599
Ruscoside 2600
Ruscoside A see under 2599
Ruscoside B see under 2599
Ruspolinone 930
Rutacridone 1041
Rutacridone epoxide 1042
Rutaecarpine 649
Rutaevin 2732
Rutamarin 1329
Rutamine see 1017
Rutarin 1330
Rutecarpine see 649
Rutin 1500
Rutinose 37
Rutoside see 1500
Ryanodine 539

Sabinol 2077
Sabinyl acetate see under 2077
Saccharabiose see 40
Saccharose see 40
Safrole 1797
Safynol 186
Saikosaponin a 2560
Saikosaponin BK1 2561
Sainfuran 1247
Sainfuran 1576
Sakuranetin 1404
Sakuranin see under 1404
Salannin 2733
Salicin 1967
Salicin benzoate see 1693
Salicoside see 1697
Salicylic acid 1698
Saligenin-β-D-glucopyranoside see 1697
Salipurposide see under 1391
Salonitenolide 2375
(−)-Salsoline 785
(−)-Salsolinol 786
(+)-Salutaridine 787
Salvianolic acid A 1798
Samaderine A 2734
Samaderin B, C see under 2734
Samaderin A see 2734
Sambucin see 1211
Sambunigrin see under 322
Samidin 1331
Sandwicine see under 557
Sanggenon C 1405
Sanggenon D 1406

Sanguinarine 788
β-Santalene 2192
α-Santalol 2193
β-Santalol 2194
Santamarin 2376
Santamarine see 2376
(−)-Santiaguine 931
Santol see 1562
Santolin see 2215
Santonin see 2377
α-Santonin 2377
α-Santonin 2378
Sapelin A 2756
Sapindoside A see 2542
Saponaretin see 1467
Saponoside D 2562
Saraconinine see 1095
Sarisan 1799
Sarmentogenin 3-O-α-L-diginoside see 2629
Sarmentogenin 3-O-α-L-oleandroside see 2628
Sarmentogenin 3-O-α-L-rhamnoside see 2640
Sarmentologenin 3-O-(6-deoxy-α-L-taloside) see 2641
Sarmentoloside 2641
Sarmentosin epoxide 323
Sarmutogenin 3-O-β-D-digitaloside see 2638
Sarothralin 2027
Sarpagine 650
Sarracine 966
Sarsaparilloside 2601
Sarsasapogenin 2062
Sarsasapogenin 3-O-4^G-rhamnosylsophoroside 2603
(−)-Sativan 1577
Sativanine B 877
Sativin see 1577
Saucernetin 1645
Saupirin 2379
Saupirine see 2379
Sauroxine 825
Savinin 1646
SBA see 408
Scandoside methyl ester 2117
Schaftoside 1501
Scheffleroside see under 2530
Schisandrin B see 1604
Schisantherin A 1647
Schkuhrin see 2315
Schottenol 2676
Sciadopitysin 1502
Scillaren A 2642
Scillarenin 3-O-glucosylrhamnoside see 2642
Scilliroside 2643
Scillirosidin 3-O-β-D-glucoside see 2643
Sclerosporin 2195
Scoparone 1332
Scopine tropate see 1129
Scopolamine see 1129
Scopoletin 1333
Scopolin 1334
Scopoline 1138
Scullcapflavone II 1503
Scutellarein 1504
Scutellarein 6-methyl ether see 1457
Scutianine F 878
Sebacic acid 106
Secaclavine see 593
Secologanin 2118
Securinine 1192
(−)-Sedamine 932
α-Sedoheptitol see 77
β-D-Sedoheptulopyranose see 20
Sedoheptulose 20
Selagine 826
L-Selenocystathionine 261
β-Selinene 2196
Selineol see 2147

Seminose *see* 14
Sempervine *see* 651
Sempervirened *see* 651
Sempervirine 651
Senaetnine 967
Senampeline A 968
Senecionine 969
Senecionyl dihydrooreoselol *see* 1314
Seneciphylline 970
Senecivernine 971
Senegin II 2563
Senkirkine 972
Senkyunolide *see* 1997
Sennoside A 1883
Septentriodine *see* 505
Septentrionine 540
Sequoyitol 74
L-Serine 262
Serotonin 298
Serpentine 652
Serratanine A *see* 811
Serratanine B *see* 823
Serratine 828
Serratinidine 827
Sesamin 1648
Sesamol 1699
Sesamolinol 1649
Sesartemin 1650
Seselin 1335
Sexangularetin 1505
Shekanin *see* 1582
Shihunine 1193
Shikimic acid 107
Shikimol *see* 1797
Shikodonin 2496
Shikonin *see under* 1813
Shiromodiol diacetate 2197
Shisonin A *see* 1212
Shogaol 1800
Sigmoidin B 3′-methyl ether *see* 1402
Silandrin 1407
Silicolin *see* 1605
Silybin 1408
Silychristin 1409
Silymarin II *see* 1409
Simalikahemiacetal A *see* 2725
Simalikilactone D 2735
Simplexin 2497
Simplexoside 1651
Sinalbin 352
Sinapaldehyde 1801
Sinapic acid 1802
Sinapic acid choline ester *see* 1803
Sinapine 1803
Sinapyl alcohol 1804
Sinensal 2198
Sinensetin 1506
Sinigrin 353
Sinoacutine *see under* 787
Sinomenine 789
Sirenin 2199
Sitosterin *see* 2677
Sitosterol 2677
β-Sitosterol *see* 2677
Skatole 299
Skimmetin *see* 1338
Skimmianine 1043
Skimmiol *see* 2759
β-Skytanthine 851
Slaframine 933
Smilagenin 2604
Sojagol 1578
Solamargine 1100
Solamidine T *see* 1101
Solancarpidine *see* 1101

Solandrine *see* 1134
Solanidine 1101
α-Solanine 1102
β-,γ-Solanine *see under* 1102
Solanine-S *see* 1104
Solasodamine *see* 1104
Solasodine 1103
Solasodine glycoside *see* 1100
Solasodine-S *see* 1103
Solasonine 1104
Solatene *see* 2768
Solatunine *see* 1102
Solavetivone 2200
Songorine 541
Sonora *see* 57
Sophorabiose *see* 32
Sophoradin *see under* 1410
Sophoraisoflavanone A 1579
Sophoramine 1077
Sophoranone 1410
Sophoricol *see* 1534
Sophorine *see* 1061
Sophorose 38
5-O-β-Sophorosylbetanidin *see* 475
Sorbic oil *see* 204
Sorbifolin *see under* 1504
Sorbin *see* 21
Sorbinose *see* 21
Sorbit *see* 75
Sorbitol *see* 75
D-Sorbitol 75
α-D-Sorbopyranose *see* 21
Sorbose 21
L-Sorbose *see* 21
Soularubinone 2736
Soyasaponin A_1 2564
Soyasaponin I 2565
Soybean agglutinin *see* 408
Soybean lectin 408
(−)-Sparteine 1078
(−)-Sparticarpin 1580
Spathelia bischromene 1281
Spermatheridine *see* 754
Spermidine 300
Spermine 301
Sphaerocarpine *see under* 883
Sphagnum acid 1805
Spheromycin *see* 1320
Spicatin 2380
Spinacene *see* 2757
Spinasaponin A 2566
Spinasaponin B *see under* 2566
α-Spinasterin *see* 2678
α-Spinasterol 2678
Spinoside A 2703
Spinoside B *see under* 2703
Spiradine A 542
Spiramine A 543
Spiramine B, C and D *see under* 543
Spirasine I 544
Spiredine 545
Spirofulvin *see* 1238
Spirolucidine 829
Sporidesmin 1194
Sporidesmin A *see* 1194
Spruceanol 2498
Squalene 2757
Squalidine *see* 951
Stachydrine 934
Stachyose 39
Staphidine 546
Starch 61
Stearic acid 138
Stearolic acid 187
Steganacin 1652

Sterculia see 53
Sterculic acid 139
Stevioside 2499
Stigmasterol 2679
Stilbene-3,5-diol see 1917
Stramonin 2381
Streptonivicin see 1320
Strigol 2500
Strobamine 1139
Strobopinin 1411
Strodival see 2639
Strophanthidin 3-O-β-D-cymaroside see 2622
Strophanthidin 3-O-β-D-digitoxoside see 2634
Strophanthidin 3-diglucosylcymarose see 2644
Strophanthidin 3-O-α-L-rhamnoside see 2621
Strophanthin see 2644
g-Strophanthin see 2639
k-Strophanthin see 2644
k-Strophanthin-α see 2622
k-Strophanthoside 2644
Strophoperm see 2639
Struxine see 669
Strychnicine see 669
Strychnine 653
C-Strychnotoxine see 584
Stypandrol 2028
Styraxin 1653
Subaphyllin 1806
Suberic acid 108
Subsessiline see 562
Succinic acid 109
N-Succinylanthranoyllycoctonine see 531
Sucrose 40
Sudan gum see 51
Sugordomycin $D_{1\alpha}$ see 1296
N-Sulfoindol-3-ylmethylglucosinolate 354
Sulfuretin see 1232
Sulphurein see under 1232
Sulphuretin 1232
Sumatrol 1581
Supinidine 973
Supinine 974
Supraene see 2757
Surinamensin 1654
Suspensaside 1807
Swainsonine 935
Swerchirin 1986
Swertiajaponin 1507
Swertiamarin 2119
Swertiamaroside see 2119
Swertianin 1987
Swertianolin 1988
Symlandine 975
Symphytine 976
Symplocosin see under 1638
Synaptolepis factor K_1 2501
(+)-Syringaresinol 1655
(−)-Syringaresinol di-β-D-glucoside see 1592
(+)-Syringaresinol O-β-D-glucoside 1656
Syringenin see 1804
Syringic acid 1700
Syringin see under 1804

α-T see 188
T-2 toxin 2201
Tabernamine 654
Tabernanthine 655
Tabernoschizine see 567
Tabersonine 656
Tadeonal see 2186
Tagetiin see under 1491
Tagitinin F 2382
Taiwanin B see 1646
Talatisamine see 547
Talatizamine 547

Tamaulipin A 2383
Tangeretin 1508
Tannic acid see under 1933
Taraktophyllin see under 326
Taraxasterin see 2758
Taraxasterol 2758
Taraxerol 2759
L-(+)-Tartaric acid 110
Tauremisin see 2406
Taxifolin 1412
Taxifolin 3-O-acetate 1413
Taxine A 1195
Taxiphyllin see under 310
Taxodione 2502
Taxodone 2503
Taxol 1196
Taxol A see 1196
Taxol B see 1159
Tecomanine see 852
Tecomin see 1857
Tecomine 852
Tecostanine 853
Tecostidine see under 853
Tectoquinone 1884
Tectoridin 1582
Tectorigenin 1583
Tectorigenin 7-O-glucoside see 1582
Teidine see 881
Telepathine see 623
Tellimagrandin I 1956
Tellimagrandin II see 1938
Templetine 1079
Tenulin 2384
Tephrosin 1584
Tephrowatsin A 1414
Terminaline 1105
Ternatins A–F see under 1214
Terpenoid EA-I 2504
Terpineol see 2080
α-Terpineol 2080
α-Terpinene 2078
γ-Terpinene 2079
Terpinolene 2081
α-Terthienyl 188
2,2′:5′,2″-Terthiophene see 188
Testosterone 2680
1,3,4,5-Tetracaffeoylquinic acid 1808
n-Tetracosanoic acid see 124
cis-Tetradec-9-enoic acid see 130
Tetradecanoic acid see 129
12-Tetradecanoylphorbol 13-acetate 2505
Tetradymol 2202
Tetragalloyl-α-D-glucose 1957
Δ^1-Tetrahydrocannabinol 1701
Δ^9-Tetrahydrocannabinol see 1701
7,8,7′,8′-Tetrahydro-ψ,ψ-carotene see 2771
Δ^3-Tetrahydronicotinic acid see 908
2,3,4,5-Tetrahydro-6-propylpyridine see 898
Tetrahydroserpentine see 556
4,6,3′,4′-Tetrahydroxyaurone see 1203
6,7,3′,4′-Tetrahydroxyaurone see 1225
2′,4′,3,4-Tetrahydroxychalcone see 1205
2′,4′,6′,4-Tetrahydroxychalcone see 1207
2′,4,4′,6′-Tetrahydroxydihydrochalcone see 1397
3,5,7,4′-Tetrahydroxy-3′,5′-dimethoxyflavylium see 1223
3,5,7,4′-Tetrahydroxyflavan see 1345
3,7,3′,4′-Tetrahydroxyflavan see 1370
3,5,7,4′-Tetrahydroxyflavanone see 1347
5,7,3′,4′-Tetrahydroxyflavanone see 1368
3,5,7,2′-Tetrahydroxyflavone see 1439
3,5,7,4′-Tetrahydroxyflavone see 1469
5,7,3′,4′-Tetrahydroxyflavone see 1474
3,5,7,4′-Tetrahydroxyflavylium see 1228
5,7,3′,4′-Tetrahydroxyflavylium see 1221
Tetrahydroxyisovaleraldehyde see 1
3,5,7,4′-Tetrahydroxy-3′-methoxyflavylium see 1230

2,4,3′,5′-Tetrahydroxystilbene see 1913
3,3′,4,5′-Tetrahydroxystilbene see 1915
1.2.6.8-Tetrahydroxyxanthone see 1983
1,3,5,8-Tetrahydroxyxanthone see 1962
1,3,6,7-Tetrahydroxyxanthone see 1982
Tetramethylenediamine see 297
(+)-Tetrandrine 790
Tetraneurin A 2385
Tetraneurin E 2386
Tetraphyllin B see under 326
Texasin 1585
Thaliblastine see 791
Thalicarpine 791
Thalicoside A 2567
Thalicrine see 692
Thalicsessine 548
Thalicsimine see 743
Thalictrine see 760
Thalidasine 792
Thalifarazine see under 542
Thalisiline see under 545
Thaliximine see 743
Thalmidine see 763
Thalmine 793
Thalsimine 794
Thapsigargin 2387
Thaumatin 409
Theasapogenol B see 2519
Theasaponin 2568
Theasinensin A 1415
Thebaine 795
Thein see 357
Theine see 357
Theobromine 369
Theocin see 370
Theogallin 1702
Theophylline 370
Thermopsine 1080
Thevanil see 2620
Thevetin A 2645
Thevetin B see 2620
Thiamine 445
Thiarubrine A 189
Thiarubrine B 190
Thioallyl ether see 144
Thiobinupharidine 854
Thionuphlutine A see 854
Thiopropanal S-oxide see 161
L-Threonine 263
Thujan-3-one see 2084
β-Thujaplicin 2082
γ-Thujaplicin 2083
Thujone 2084
Thujopsene 2203
Thujyl acetate see under 2084
Thujyl alcohol see under 2084
Thyme camphor see 2085
Thymine 371
Thyminose see 4
Thymol 2085
m-Thymol see 2085
Thymol acetate see 2086
Thymonin 1509
Thymyl acetate 2086
Tiglic acid 111
Tigloidine 1140
Tigloyloxy-6α-chaparrinone see under 2714
3β-Tigloyloxytropane see 1140
7-Tiglyl-9-(−)-viridiflorylretronecine see 976
Tiglylpseudotropeine see 1140
Tigogenin 2605
Tigonin 2606
Tiliacorine 796
Tiliadin see 2759
Timol see 2085

Tinctorine 1081
Tingenone 2760
Tingitanine see 248
Tinyatoxin 2506
TMP see 1337
Tokoronin 2607
Tomatidine 1106
α-Tomatine see 1107
Tomatine 1107
Tomentosin 2388
Tonka bean camphor see 1295
Tonsillosan see 95
Toonacilin 2737
Toringin see under 1431
Torulene 2791
Toxicarol 1586
α-Toxicarol see 1586
Toxiferine I 657
C-Toxiferine I see 657
C-Toxiferine II see 584
Toxiferine V see 657
Toxiferine XI see 657
Toxin F2 see 2034
Toxol 1248
Toxyl angelate 1249
TR-A see 189
TR-B see 190
9-(+)-Trachelanthylheliotridine see 964
9-(+)-Trachelanthylretronecine see 952
Trachelogenin 1657
Tragacanth see 54
Transvaalin see 2642
Trapain see 1945
Trehalose 41
α,α-Trehalose see 41
Tremetone 1250
Tremulacin 2029
1,2,6-Tri(3-nitropropanoyl)-β-D-glucopyranoside see 434
  see 434
3,3′,4-Tri-O-methylellagic acid 1704
Triacetylene 191
Triangularine 977
Tricetin 1510
Tricetin 3′,5′-dimethyl ether see 1511
Trichilin A 2738
Trichocarpin 1703
Trichodermin 2204
Tricholomic acid 264
Trichosanic acid see 136
Trichosanthin 410
Trichotomine 658
Tricin 1511
Tricontan-1-ol 164
Tricornine 549
Tricothecin 2205
Tricrozarin A 1885
Tricyclodehydroisohumulone 1730
Trideca-1-ene-3,5,7-pentayne 192
Tridecapentaynene see 192
Trifolin see under 1469
Trigenolline see 446
Triglochinin 324
Trigonelline 446
Trigonelloside C 2608
6,3′,4′-Trihydroxyaurone see 1232
1,2,3-Trihydroxybenzene see 1695
1,3,5-Trihydroxybenzene see 1691
3,4,5-Trihydroxybenzoic acid see 1674
2′,4′,4-Trihydroxychalcone see 1220
2′,4′,4-Trihydroxydihydrochalcone see 1357
3,5,7-Trihydroxyflavanone see 1399
3,7,4′-Trihydroxyflavanone see 1373
5,7,4′-Trihydroxyflavanone see 1391
7,3,′,4′-Trihydroxyflavanone see 1353
3,5,7-Trihydroxyflavone see 1447

3,7,4'-Trihydroxyflavone *see* 1440
5,6,7-Trihydroxyflavone *see* 1427
5,7,4'-Trihydroxyflavone *see* 1421
5,7,8-Trihydroxyflavone *see* 1483
5,7,4'-Trihydroxyflavylium *see* 1202
1,2,8-Trihydroxy-6-methoxyxanthone *see* 1987
3,5,4'-Trihydroxystilbene *see* 1923
3,5,4'Trihydroxy-7,3',5'-trimethoxyflavylium *see* 1217
1,3,7-Trihydroxyxanthone *see* 1968
1,5,6-Trihydroxyxanthone *see* 1979
1,3,5-Trihydroxyxanthone 1989
Trilobamine *see* 722
Trilobine 797
Trilobolide 2389
5,6,7-Trimethoxycoumarin 1336
3,3',4-Tri-*O*-methylellagic acid 1704
1,3,3-Trimethyl-2-norcamphanone *see* 2054
4,5',8-Trimethylpsoralen 1337
1,3,7-Trimethylxanthine *see* 357
1,2,6-Tri(3-nitropropanoyl)-β-D-glucopyranoside *see* 434
Trioxalen *see* 1337
Trioxsalen *see* 1337
Tripdiolide 2507
Tripherin *see* 461
Tripteroside 1990
Triptolide 2508
Trisphaerine *see* 461
Tritriacontane 165
Tritriacontane-16,18-dione 166
Trochol *see* 2744
Tropacaine *see* 1141
Tropacocaine 1141
$1\alpha H,5\alpha H$-Tropan-3α-ol *see* 1142
3-(3,6,7-Tropanetriol) tiglate *see* 1133
3β-Tropanol *see* 1137
l-Tropic acid 3α-nortropanyl ester *see* 1134
ψ-Tropine *see* 1137
Tropine 1142
Tropine benzoate *see* 1120
Tropine tropate *see* 1116
Tropylisatropate *see* 1117
Tryptamine 302
Tryptanthrine 1044
L-Tryptophan 265
Tsukushinamine A 1082
Tubeimoside I 1559
(+)-Tubocurarine 798
Tubotoxin *see* 1575
Tubulosine 659
Tulipalin A *see under* 206
Tulipalin B *see under* 207
Tulipinolide 2390
epi-Tulipinolide 2391
epi-Tulipinolide diepoxide 2392
Tuliposide A 206
Tuliposide B 207
Turgorin 1705
Turmeric colour *see* 1756
Turmeric yellow *see* 1756
Turricolol E 1706
Tussilagine 978
Tutin 2393
(−)-Tylocrebine 1197
Tylophorine 1198
Tyramine 303
Tyrosamine *see* 303
L-Tyrosine 266

Ubiquinone-10 1886
UDPG *see* 373
Ugandensidial *see* 2136
Ulexine *see* 1061
Umbellatine *see* 699
Umbelliferone 1338
Umbelliferose 42

Umbellulone 2087
Uncarine A–F *see under* 638
Uncinatone 2030
γ-Undecalactone 208
Undulatone 2739
Uplandicine 979
Uracil 372
δ-Ureido-L-norvaline *see* 224
4-Ureidobutylamine *see* 272
Uridine diphosphate glucose 373
Uridine-5'-diphosphoglucose *see* 373
Ursiniolide A 2394
Ursolic acid 2570
Urson *see* 2570
Urushiol III 1707
Usambarensine 660
Usambarine 661
Usaramine 980
Uscharidin 2646
Usnein *see* 1251
Usniacin *see* 1251
Usnic acid 1251
Usninic acid *see* 1251
Usuramine *see* 980
Uvaretin 1416

Vaccenic acid 140
Vakerin *see* 1995
Valepotriates *see under* 2098
Valerenic acid 2206
Valerianic acid *see* 112
Valerianine 855
Valeric acid 112
n-Valeric acid *see* 112
Valeroidine 1143
Validol *see under* 2063
L-Valine 267
Valtratum 2120
Vanillic acid 1708
Vanillin 1709
Vanilloside *see under* 1709
Vanillosmin *see* 2271
(−)-Variabilin 1587
Vasicine *see* 1036
Vasicinol 1045
Vasicinone 1046
Vaumigan *see* 2155
VCR *see* 664
Veatchine 550
(15β)-Veatchine acetate *see* 537
Vegetable luteol *see* 2781
Vegetable pepsin *see* 397
Veprisinium 1047
Veracevine 1108
(+)-Veraguensin 1658
Veratetrine *see* 1098
Veratramine 1109
Veratridine *see under* 1108
3α-Veratroyl-*N*-hydroxynortropane *see* 1125
O-Veratroylnortropine *see* 1126
O-Veratroyltropine *see* 1124
Verbascose 43
Verbascoside 1809
Verbenalin 2121
Verbenaloside *see* 2121
Verbenone 2088
Vermeerin 2395
Vernadigin 2647
Vernodalin 2396
Vernodalol 2397
Vernoflexin 2398
Vernoflexine *see* 2398
Vernoflexuoside 2399
Vernolepin 2400
Vernolic acid 141

Index

Vernolide 2401
Vernomenin 2402
Vernomygdin 2403
Verrucarin A 2209
Verruculotoxin 1199
Verticine 1110
Vertine see 1165
Very fast death factor see 1112
Vestitol 1588
Vestitone 1589
α-Vetivone 2207
β-Vetivone 2208
Vibeline see 1339
Viburnine see under 599
(−)-Viburnitol 76
Vicenin-2 1512
Vicia cracca lectins 411
Vicianin 325
Vicianose 44
Vicine 374
Vicioside see 374
Vignafuran 1252
Vignatin see 1549
Viguiestenin 2404
Viminalol see 2741
Vinblastine 662
Vincaine see 556
Vincaleukoblastine see 662
4′α-Vincaleukoblastine see 629
Vincamajoridine see 560
Vincamarine see 663
Vincamine 663
Vincamone see under 599
Vinceine see 556
Vincristine 664
Vindoline 665
Vinegar acid see 78
Vinetine see 772
ϵ-Viniferin 1924
Vinine see 590
Vinleurosine see 630
Vinrosidine see 629
(R)-5-Vinyl-2-oxazolidimethione
p-Vinylphenol apiosylglucoside see 1770
Violanthin 1513
Violaxanthin 2792
Virgaureasaponin I 2571
9-Viridiflorylretronecine see 956
Viroallosecurinine see under 1192
Virolin see under 1654
Virosecurinine see under 1192
Visammin see 1267
Viscidulin B 2405
β-Viscol see 2753
Viscotoxin 412
Viscumin 413
Vismione D 2031
Visnacorin see 1282
Visnadin 1339
Visnagen see 1267
Visnagidin see 1282
Visnagin 1282
Visnamine see 1339
Vitamin B_1 see 445
Vitamin B_2 see 443
Vitamin B_6 see 442
Vitamin B_{12} see 423
Vitamin B_x see 269
Vitamin C see 81
Vitamin H see 418
Vitamin K_1 1887
Vitexin 1514
VLB see 662
VLR see 630
Voacamine 666

Voacanginine see 666
Voacorine see under 603
Vobasine 667
Vobtusine 668
Volemitol 77
Volemulose see 20
Volkenin 326
Vomicine 669
Vomitoxin 2210
VRD see 629
Vulcamicina see 1320
Vulgarin 2506
Vulgaxanthin-I 490
Vulgaxanthin-II 491
Vulkamycin see 1320

Warburganal 2211
Warfarin see under 1301
Wedelolactone 1590
Welensalifactor F_1 2509
WGA see 414
Wheat germ agglutinin 414
Widdrene see 2203
Wighteone 1591
Wikstromol see under 1630
Wilfordine 856
Withaferin A 2681
Withanolide D 2682
Withasomnine 1200
Wogonin 1515
Wood sugar see 22
Wrightine see 1085
Wuweizisu B see 1604
Wuweizisu C 1659
Wuweizisu ester see 1657
Wyerone 193
Wyerone acid 194

Xanthatin 2407
Xanthinin 2408
Xanthochymol 2032
Xanthopetaline see 682
Xanthophyll see 2781
Xanthorhamnin see under 1496
Xanthotoxin 1340
Xanthotoxol 1341
Xanthoxylin 1731
Xanthumin 2409
Xanthyletin 1342
Xenognosin A 1233
Xerantholide 2410
Xylopine 799
Xylopinine 800
α-D-Xylopyranose see 22
6-O-D-Xylopyranosyl-D-glucose see 34
Xylose 22
D-Xylose see 22
Xylotenin see 1292

Yageine see 623
Yamogenin 2609
Yamogenin 3-O-neohesperidoside 2160
Yangambin 1660
Yangonin 2033
Yiamoloside B 2572
Ylangene see 2212
α-Ylangene 2212
Yohimbine 670
δ-Yohimbine see 556

Zaluzanin C 2411
Zaluzanin C senecioate see 2398
Zapoterin 2740

Zearalenone 2034
Zeatin 375
trans-Zeatin *see* 375
Zeaxanthin 2793
Zeaxanthin diepoxide *see* 2792
Zeaxanthin dipalmitate *see* 2788
(3*R*,5*R*,6*S*,3′*R*)-Zeaxanthin 5,6-epoxide *see* 2761
Zeaxanthol *see* 2793
Zexbrevin B 2412

Zierin 327
Zingiberene 2213
Ziziphin 2573
Ziziphine A *see* 879
Zizyphine *see* 879
Zizyphine A 879
Zizyphine F 880
Zoapatanol 2510
Zygadenine 1111